MW00837807

Quantum Mechanics

Problems with solutions

Quantum Mechanics

Problems with solutions

Konstantin K Likharev

IOP Publishing, Bristol, UK

ISBN 978-0-7503-1413-8 (ebook)
ISBN 978-0-7503-1414-5 (print)
ISBN 978-0-7503-1415-2 (mobi)

DOI 10.1088/2053-2563/aaf3a6

Version: 20190501

IOP Expanding Physics
ISSN 2053-2563 (online)
ISSN 2054-7315 (print)

British Library Cataloguing-in-Publication Data: A catalogue record for this book is available from the British Library.

Published by IOP Publishing, wholly owned by The Institute of Physics, London

IOP Publishing, Temple Circus, Temple Way, Bristol, BS1 6HG, UK

US Office: IOP Publishing, Inc., 190 North Independence Mall West, Suite 601, Philadelphia, PA 19106, USA

Contents

Appendices

Preface to the EAP Series

Essential Advanced Physics

Essential Advanced Physics (EAP) is a series of lecture notes and problems with solutions, consisting of the following four parts[1]:

- *Part CM*: *Classical Mechanics* (a one-semester course),
- *Part EM*: *Classical Electrodynamics* (two semesters),
- *Part QM*: *Quantum Mechanics* (two semesters), and
- *Part SM*: *Statistical Mechanics* (one semester).

Each part includes two volumes: *Lecture Notes* and *Problems with Solutions*, and an additional file *Test Problems with Solutions.*

Distinguishing features of this series—in brief

- condensed lecture notes (~250 pp per semester)—much shorter than most textbooks
- emphasis on simple explanations of the main notions and phenomena of physics
- a focus on problem solution; extensive sets of problems with detailed model solutions
- additional files with test problems, freely available to qualified university instructors
- extensive cross-referencing between all parts of the series, which share style and notation

Level and prerequisites

The goal of this series is to bring the reader to a general physics knowledge level necessary for professional work in the field, regardless on whether the work is theoretical or experimental, fundamental or applied. From the formal point of view, this level (augmented by a few special topic courses in a particular field of concentration, and of course by an extensive thesis research experience) satisfies the typical PhD degree requirements. Selected parts of the series may be also valuable for graduate students and researchers of other disciplines, including astronomy, chemistry, mechanical engineering, electrical, computer and electronic engineering, and material science.

The entry level is a notch lower than that expected from a physics graduate from an average US college. In addition to physics, the series assumes the reader's familiarity with basic calculus and vector algebra, to such an extent that the meaning of the formulas listed in appendix A, 'Selected mathematical formulas' (reproduced at the end of each volume), is absolutely clear.

[1] Note that the (very ambiguous) term *mechanics* is used in these titles in its broadest sense. The acronym *EM* stems from another popular name for classical electrodynamics courses: *Electricity and Magnetism*.

Origins and motivation

The series is a by-product of the so-called 'core physics courses' I taught at Stony Brook University from 1991 to 2013. My main effort was to assist the development of students' problem-solving skills, rather than their idle memorization of formulas. (With a certain exaggeration, my lectures were not much more than introductions to problem solution.) The focus on this main objective, under the rigid time restrictions imposed by the SBU curriculum, had some negatives. First, the list of covered theoretical methods had to be limited to those necessary for the solution of the problems I had time to discuss. Second, I had no time to cover some core fields of physics—most painfully general relativity[2] and quantum field theory, beyond a few quantum electrodynamics elements at the end of *Part QM*.

The main motivation for putting my lecture notes and problems on paper, and their distribution to students, was my desperation to find textbooks and problem collections I could use, with a clear conscience, for my purposes. The available graduate textbooks, including the famous *Theoretical Physics* series by Landau and Lifshitz, did not match the minimalistic goal of my courses, mostly because they are far too long, and using them would mean hopping from one topic to another, picking up a chapter here and a section there, at a high risk of losing the necessary background material and logical connections between the course components—and the students' interest with them. In addition, many textbooks lack even brief discussions of several traditional and modern topics that I believe are necessary parts of every professional physicist's education[3].

On the problem side, most available collections are not based on particular textbooks, and the problem solutions in them either do not refer to any background material at all, or refer to the included short sets of formulas, which can hardly be used for systematic learning. Also, the solutions are frequently too short to be useful, and lack discussions of the results' physics.

Style

In an effort to comply with the Occam's Razor principle[4], and beat Malek's law[5], I have made every effort to make the discussion of each topic as clear as the time/ space (and my ability :-) permitted, and as simple as the subject allowed. This effort has resulted in rather succinct lecture notes, which may be thoroughly read by a student during the semester. Despite this briefness, the introduction of every new

[2] For an introduction to this subject, I can recommend either a brief review by S Carroll, *Spacetime and Geometry* (2003, New York: Addison-Wesley) or a longer text by A Zee, *Einstein Gravity in a Nutshell* (2013, Princeton University Press).

[3] To list just a few: the statics and dynamics of elastic and fluid continua, the basics of physical kinetics, turbulence and deterministic chaos, the physics of computation, the energy relaxation and dephasing in open quantum systems, the reduced/RWA equations in classical and quantum mechanics, the physics of electrons and holes in semiconductors, optical fiber electrodynamics, macroscopic quantum effects in Bose–Einstein condensates, Bloch oscillations and Landau–Zener tunneling, cavity quantum electrodynamics, and density functional theory (DFT). All these topics are discussed, if only briefly, in my lecture notes.

[4] *Entia non sunt multiplicanda praeter necessitate*—Latin for 'Do not use more entities than necessary'.

[5] 'Any simple idea will be worded in the most complicated way'.

physical notion/effect and of every novel theoretical approach is always accompanied by an application example or two.

The additional exercises/problems listed at the end of each chapter were carefully selected[6], so that their solutions could better illustrate and enhance the lecture material. In formal classes, these problems may be used for homework, while individual learners are strongly encouraged to solve as many of them as practically possible. The few problems that require either longer calculations, or more creative approaches (or both), are marked by asterisks.

In contrast with the lecture notes, the model solutions of the problems (published in a separate volume for each part of the series) are more detailed than in most collections. In some instances they describe several alternative approaches to the problem, and frequently include discussions of the results' physics, thus augmenting the lecture notes. Additional files with sets of shorter problems (also with model solutions) more suitable for tests/exams, are available for qualified university instructors from the publisher, free of charge.

Disclaimer and encouragement

The prospective reader/instructor has to recognize the limited scope of this series (hence the qualifier *Essential* in its title), and in particular the lack of discussion of several techniques used in current theoretical physics research. On the other hand, I believe that the series gives a reasonable introduction to the *hard core* of physics—which many other sciences lack. With this hard core knowledge, today's student will always feel at home in physics, even in the often-unavoidable situations when research topics have to be changed at a career midpoint (when learning from scratch is terribly difficult—believe me :-). In addition, I have made every attempt to reveal the remarkable logic with which the basic notions and ideas of physics subfields merge into a wonderful single construct.

Most students I taught liked using my materials, so I fancy they may be useful to others as well—hence this publication, for which all texts have been carefully reviewed.

[6] Many of the problems are original, but it would be silly to avoid some old good problem ideas, with long-lost authorship, which wander from one textbook/collection to another one without references. The assignments and model solutions of all such problems have been re-worked carefully to fit my lecture material and style.

Preface to *Quantum Mechanics: Problems with solutions*

This volume of the EAP series contains model solutions of the problems formulated in volume 5, *Quantum Mechanics: Lecture Notes*. For the reader's convenience, the problem assignments are reproduced in this volume as well, although the accompanying figures are frequently more detailed and extended to explain the solutions. The appendices A (Selected mathematical formulas) and B (Selected physical constants), common to all parts of the series, are also included in this volume for the reader's convenience.

Since all this series is strongly focused on the development of problem solution skills, the model solutions are rather detailed, and in some cases (especially in the harder problems marked by asterisks) extend and/or enhance the lecture material.

Numbering of formulas within each solution is local, by asterisks; references to formulas in other solutions are clearly indicated. The solutions also have numerous references to formulas in the lecture notes of this (*QM*) part of the EAP series, and occasionally those in other parts of the series. In the latter case, the acronym of the part is included in the reference.

A file with 70 additional problems, which allow shorter solutions and hence are suitable for exams (also with model solutions) is available to university instructors from the publisher by request.

The author tried hard to eliminate all errors in the solutions, but they have not passed a rigorous review by qualified others, and are presented here without warranty.

Acknowledgments

I am extremely grateful to my faculty colleagues and other readers of the preliminary (circa 2013) version of this series, who provided feedback on certain sections; here they are listed in alphabetical order[7]: A Abanov, P Allen, D Averin, S Berkovich, P-T de Boer, M Fernandez-Serra, R F Hernandez, A Korotkov, V Semenov, F Sheldon, and X Wang. (Obviously, these kind people are not responsible for any remaining deficiencies.)

A large part of my scientific background and experience, reflected in these materials, came from my education, and then research, in the Department of Physics of Moscow State University from 1960 to 1990. The Department of Physics and Astronomy of Stony Brook University provided a comfortable and friendly environment for my work during the following 25+ years.

Last but not least, I would like to thank my wife Lioudmila for all her love, care, and patience—without these, this writing project would have been impossible.

I know very well that my materials are still far from perfection. In particular, my choice of covered topics (always very subjective) may certainly be questioned. Also, it is almost certain that despite all my efforts, not all typos have been weeded out. This is why all remarks (however candid) and suggestions from readers will be greatly appreciated. All significant contributions will be gratefully acknowledged in future editions.

Konstantin K Likharev
Stony Brook, NY

[7] I am very sorry for not keeping proper records from the beginning of my lectures at Stony Brook, so I cannot list all the numerous students and TAs who have kindly attracted my attention to typos in earlier versions of these notes. Needless to say, I am very grateful to all of them as well.

Notation

Abbreviations	Fonts	Symbols
c.c. complex conjugate	F, $\mathscr{F}$ scalar variables[8]	$\dot{}$ time differentiation operator (d/dt)
h.c. Hermitian conjugate	$\mathbf{F}$, $\boldsymbol{\mathscr{F}}$ vector variables	∇ spatial differentiation vector (*del*)
	$\hat{F}$, $\hat{\mathscr{F}}$ scalar operators	$\approx$ approximately equal to
	$\hat{\mathbf{F}}$, $\hat{\boldsymbol{\mathscr{F}}}$ vector operators	$\sim$ of the same order as
	F matrix	$\propto$ proportional to
	$F_{jj'}$ matrix element	$\equiv$ equal to by definition (or evidently)
		$\cdot$ scalar ('dot-') product
		$\times$ vector ('cross-') product
		$\overline{}$ time averaging
		$\langle\ \rangle$ statistical averaging
		[,] commutator
		{ , } anticommutator

Prime signs

The prime signs ($'$, $''$, etc) are used to distinguish similar variables or indices (such as j and j' in the matrix element above), rather than to denote derivatives.

Parts of the series

Part CM: Classical Mechanics
Part EM: Classical Electrodynamics
Part QM: Quantum Mechanics
Part SM: Statistical Mechanics

Appendices

Appendix A: Selected mathematical formulas
Appendix B: Selected physical constants

Formulas

The abbreviation Eq. may mean any displayed formula: either the equality, or inequality, or equation, etc.

[8] The same letter, typeset in different fonts, typically denotes different variables.

IOP Publishing

Quantum Mechanics
Problems with solutions
Konstantin K Likharev

Chapter 1

Introduction

***Problem* 1.1.** The actual postulate made by N Bohr in his original 1913 paper was not directly Eq. (1.8) of the lecture notes, but an assumption that at quantum leaps between adjacent large (quasiclassical) orbits with $n \gg 1$, the hydrogen atom either emits or absorbs energy $\Delta E = \hbar\omega$, where ω is its classical radiation frequency—according to classical electrodynamics, equal to the angular velocity of electron's rotation[1]. Prove that this postulate is indeed compatible with Eqs. (1.7) and (1.8).

Solution: Solving the classical Eq. (1.9),

$$m_e \frac{v^2}{r} = \frac{e^2}{4\pi\varepsilon_0 r^2},$$

together with Eq. (1.8),

$$m_e v r = \hbar n,$$

for the electron's velocity v and its orbit's radius r, we get

$$v = \frac{e^2}{4\pi\varepsilon_0}\frac{1}{n\hbar}, \qquad r = \frac{n^2\hbar^2}{m_e} \Big/ \frac{e^2}{4\pi\varepsilon_0}.$$

Hence the angular velocity of electron's rotation is

$$\omega = \frac{v}{r} = \frac{m_e}{\hbar^3}\left(\frac{e^2}{4\pi\varepsilon_0}\right)^2 \frac{1}{n^3}. \qquad (*)$$

On the other hand, for $n \gg 1$, the energy difference between the adjacent energy levels ($\Delta n = \pm 1$) may be calculated from Eqs. (1.12) and (1.13)—which also follow from Eqs. (1.8) and (1.9):

[1] See, e.g. *Part EM* section 8.2.

$$\Delta E \equiv E_{n+1} - E_n \approx \frac{dE_n}{dn} = \frac{E_H}{n^3} = \frac{m_e}{\hbar^2}\left(\frac{e^2}{4\pi\varepsilon_0}\right)^2 \frac{1}{n^3}. \qquad (**)$$

Comparing Eqs. (*) and (**) we see that $\Delta E = \hbar\omega$, i.e. that the actual Bohr's postulate is indeed compatible with Eqs. (1.7) and (1.8).

***Problem* 1.2.** Use Eq. (1.53) of the lecture notes to prove that the linear operators of quantum mechanics are commutative: $\hat{A}_2 + \hat{A}_1 = \hat{A}_1 + \hat{A}_2$, and associative: $(\hat{A}_1 + \hat{A}_2) + \hat{A}_3 = \hat{A}_1 + (\hat{A}_2 + \hat{A}_3)$.

Solution: These relations look obvious, but the reader should remember that in the operators we face a different mathematical entity, and cannot take for granted any properties that have not been postulated. For example, for any two usual functions Ψ_1 and Ψ_2 (which are *c*-numbers at each point of their arguments), we may always write $\Psi_1\Psi_2 = \Psi_2\Psi_1$, but for operators' 'products', such commutation is generally invalid—see, e.g. Eq. (2.14). This is why we should be careful.

First, let us use Eq. (1.53), with the index swap 1↔2, to write

$$(\hat{A}_2 + \hat{A}_1)\,\Psi = \hat{A}_2\Psi + \hat{A}_1\Psi.$$

The operands on the right-hand side of this equation are just functions (not operators!), and obey the rules of the 'usual' algebra. In particular, these terms are commutative, so that the RHS is equal to that of Eq. (1.53). Hence, the left-hand sides of these relations have to be equal as well:

$$(\hat{A}_2 + \hat{A}_1)\,\Psi = (\hat{A}_1 + \hat{A}_2)\,\Psi.$$

Since this relation is valid for an arbitrary function Ψ, it gives the required proof that the operators are commutative as well.

Similarly, we may use Eq. (1.53) twice to write

$$[(\hat{A}_1 + \hat{A}_2) + \hat{A}_3]\,\Psi = (\hat{A}_1 + \hat{A}_2)\,\Psi + \hat{A}_3\Psi = \hat{A}_1\Psi + \hat{A}_2\Psi + \hat{A}_3\Psi.$$

Again, the operands on the right-hand side of this equation are just (complex) functions, and may be regrouped as

$$\hat{A}_1\Psi + \hat{A}_2\Psi + \hat{A}_3\Psi = \hat{A}_1\Psi + (\hat{A}_2\Psi + \hat{A}_3\Psi).$$

Now we may apply Eq. (1.53) twice to the right-hand side of the above relation, to write

$$\hat{A}_1\Psi + (\hat{A}_2\Psi + \hat{A}_3\Psi) = [\hat{A}_1 + (\hat{A}_2 + \hat{A}_3)]\,\Psi.$$

Comparing the initial and final expressions of our calculation, we get

$$[(\hat{A}_1 + \hat{A}_2) + \hat{A}_3]\,\Psi = [\hat{A}_1 + (\hat{A}_2 + \hat{A}_3)]\,\Psi.$$

Since this equality is valid for any Ψ, we may conclude that the linear operators are indeed associative.

***Problem* 1.3.** Prove that for any time-independent Hamiltonian operator $\hat{H}$ and two arbitrary complex functions $f(\mathbf{r})$ and $g(\mathbf{r})$,

$$\int f(\mathbf{r})\hat{H}g(\mathbf{r})d^3r = \int \hat{H}f(\mathbf{r})g(\mathbf{r})d^3r.$$

Solution: Using the fact (discussed in section 1.5 of the lecture notes) that the set of eigenfunctions ψ_n of the given Hamiltonian operator (i.e. the set of stationary states of the system) is full, we may expand the function $g(\mathbf{r})$ and the complex conjugate of the function $f(\mathbf{r})$ in series over the set, just as was done with the function $\Psi(\mathbf{r}, 0)$ in Eq. (1.67):

$$g(\mathbf{r}) = \sum_n g_n\psi_n(r), \quad f^*(\mathbf{r}) = \sum_n f_n\psi_n(r), \qquad \text{so that}$$

$$f(\mathbf{r}) \equiv [f^*(\mathbf{r})]^* = \left[\sum_n f_n\psi_n(r)\right]^* \equiv \sum_n f_n^*\psi_n^*(\mathbf{r}),$$

where f_n and g_n are some (generally, complex) coefficients. Plugging these expressions (with one of the summation indices n denoted as n') into each side of the equality to be proved, and taking the constant coefficients out of the spatial integrals, we may transform them as

$$\int f(\mathbf{r})\hat{H}g(\mathbf{r})d^3r = \sum_{n,n'} f_n^* g_{n'} \int \psi_n^*(\mathbf{r})\hat{H}\psi_{n'}(\mathbf{r})d^3r,$$

$$\int \hat{H}f(\mathbf{r})g(\mathbf{r})d^3r = \sum_{n,n'} f_n^* g_{n'} \int \hat{H}\psi_n^*(\mathbf{r})\psi_{n'}(\mathbf{r})d^3r,$$

Now using Eq. (1.60) with $n \to n'$, $\hat{H}\psi_{n'} = E_{n'}\psi_{n'}$, in the first expression, and its complex conjugate, $\hat{H}\psi_n^* = E_n\psi_n^*$,[2] in the second one, and then employing the orthonormality condition (1.66), we get

$$\int f(\mathbf{r})\hat{H}g(\mathbf{r})d^3r = \sum_{n,n'} f_n^* g_{n'} E_{n'} \int \psi_n^*(\mathbf{r})\psi_{n'}(\mathbf{r})d^3r = \sum_{n,n'} f_n^* g_{n'} E_{n'}\delta_{n,n'} = \sum_n f_n^* g_n E_n,$$

$$\int \hat{H}f(\mathbf{r})g(\mathbf{r})d^3r = \sum_{n,n'} f_n^* g_{n'} E_n \int \psi_n^*(\mathbf{r})\psi_{n'}(\mathbf{r})d^3r = \sum_{n,n'} f_n^* g_{n'} E_n\delta_{n,n'} = \sum_n f_n^* g_n E_n,$$

so that the left-hand and right-hand sides of the relation in question are indeed equal.

[2] The eigenenergies E_n are real numbers, so they do not change at the complex conjugation, and neither are the Hamiltonians of the type (1.41). (In chapter 4 we will see that the statement we are proving is valid for any *Hermitian operator*, in particular *any* Hamiltonian.)

***Problem* 1.4.** Prove that the Schrödinger equation (1.25), with the Hamiltonian operator given by Eq. (1.41), is Galilean form-invariant, provided that the wavefunction is transformed as

$$\Psi'(\mathbf{r}', t') = \Psi(\mathbf{r}, t)\exp\left\{-i\frac{m\mathbf{v}\cdot\mathbf{r}}{\hbar} + i\frac{mv^2t}{2\hbar}\right\},$$

where the prime sign denotes the variables measured in the reference frame 0′ that moves, without rotation, with a constant velocity **v** relatively to the 'lab' frame 0. Give a physical interpretation of this transformation.

Solution: The non-relativistic ('Galilean') transform of the coordinates and time at the transfer between the reference frames is expressed by the following relations[3]:

$$\mathbf{r}' = \mathbf{r} - \mathbf{v}t, \qquad t' = t. \tag{*}$$

The Galilean form-invariance means that the wavefunctions $\Psi'(\mathbf{r}', t')$ and $\Psi(\mathbf{r}, t)$, related as specified in the assignment, satisfy similar Schrödinger equation in the reference frames—respectively, 0 and 0′:

$$\begin{aligned} i\hbar\frac{\partial\Psi'}{\partial t'} &= -\frac{\hbar^2}{2m}\nabla'^2\Psi' + U'(\mathbf{r}', t')\Psi', \quad \text{and} \\ i\hbar\frac{\partial\Psi}{\partial t} &= -\frac{\hbar^2}{2m}\nabla^2\Psi + U(\mathbf{r}, t)\Psi. \end{aligned} \tag{**}$$

For the proof, let us note that the functions $U'(\mathbf{r}', t')$ and $U(\mathbf{r}, t)$ describe the same potential energy of the particle, i.e. must give the same value at the same space–time point:

$$U'(\mathbf{r}', t') \equiv U'(\mathbf{r} - \mathbf{v}t, t) = U(\mathbf{r}, t).$$

(Note also that the wavefunction transform, suggested in the assignment, gives a similar relation for the probability density to find the particle at the same space–time point:

$$w'(\mathbf{r}', t') \equiv |\Psi'(\mathbf{r}', t')|^2 = |\Psi(\mathbf{r}, t)|^2 \equiv w(\mathbf{r}, t),$$

just as it should.)

Next, considering t', at fixed $\mathbf{r}'$, as a function of arguments $\mathbf{r}(t) \equiv \{r_1(t), r_2(t), r_3(t)\}$ and t, we may use the general rule of differentiation of a function of several variables[4], and then Eq. (*) in the form $\mathbf{r} = \mathbf{r}' + \mathbf{v}t'$, to write[5]

$$\frac{\partial}{\partial t'} = \frac{\partial}{\partial t} + \sum_{j=1}^{3}\frac{\partial}{\partial r_j}\frac{\partial r_j}{\partial t'} \equiv \frac{\partial}{\partial t} + \mathbf{v}\cdot\nabla,$$

[3] See, e.g. *Part EM* section 9.1, in particular figure 9.1 and Eq. (9.2).

[4] See, e.g. Eq. (A.23).

[5] This expression is essentially the *convective derivative*, which was discussed several times in this lecture series—see especially *Part CM* section 8.3.

while at fixed t', Eq. (*) yields $\nabla' = \nabla$, so that $\nabla'^2 = \nabla^2$. With these relations, a straightforward differentiation of the left-hand side of the transformation in question, plugged into the first of Eqs. (**), immediately yields the second of these equations, i.e. proves the Galilean form-invariance of the Schrödinger equation.

For the interpretation of the wavefunction's transform, let us apply it to the simplest case of a monochromatic, plane de Broglie wave given by Eqs. (1.29) of the lecture notes, describing a free particle, whose momentum $\mathbf{p} = \hbar\mathbf{k}$ and (kinetic) energy $E = \hbar\omega$ are c-numbers, i.e. have definite values:

$$\Psi(\mathbf{r}, t) = a \exp\left\{i\frac{\mathbf{p}\cdot\mathbf{r}}{\hbar} - i\frac{Et}{\hbar}\right\}.$$

The proved transform shows that in the moving reference frame the wavefunction is a similar plane wave:

$$\begin{aligned}\Psi'(\mathbf{r}', t') &= a \exp\left\{i\frac{\mathbf{p}\cdot\mathbf{r}}{\hbar} - i\frac{Et}{\hbar} - i\frac{m\mathbf{v}\cdot\mathbf{r}}{\hbar} + i\frac{mv^2t}{2\hbar}\right\} \\ &\equiv a \exp\left\{i\frac{\mathbf{p}'\cdot\mathbf{r}'}{\hbar} - i\frac{E't}{\hbar}\right\},\end{aligned}$$

where

$$\mathbf{p}' \equiv \mathbf{p} - m\mathbf{v}, \quad \text{and } E' \equiv E - \mathbf{p}'\cdot\mathbf{v} - \frac{mv^2}{2} = E - \mathbf{p}\cdot\mathbf{v} + \frac{mv^2}{2}.$$

These are exactly the Galilean transform expressions for the momentum and the kinetic energy of the particle, given by the non-relativistic classical mechanics. Indeed, expressing the particle's momentum via its velocity $\mathbf{u}$ (in the lab frame) as $\mathbf{p} = m\mathbf{u}$, so that $E = mu^2/2$, we get

$$\mathbf{p}' = m\mathbf{u} - m\mathbf{v} = m\mathbf{u}', \quad \text{and} \quad E' = \frac{mu^2}{2} - m\mathbf{u}\cdot\mathbf{v} + \frac{mv^2}{2} = \frac{mu'^2}{2},$$

where $\mathbf{u}' = \mathbf{u} - \mathbf{v}$ is the particle's velocity as observed from the moving reference frame.

Problem* 1.5. Prove the so-called *Hellmann–Feynman theorem*[6]:

$$\frac{\partial E_n}{\partial\lambda} = \left\langle\frac{\partial H}{\partial\lambda}\right\rangle_n,$$

[6] Despite the theorem's common name, H Hellmann (in 1937) and R Feynman (in 1939) were not the first ones in the long list of physicists who have (apparently, independently) discovered this fact. Indeed, it may be traced back at least to a 1922 paper by W Pauli, and was carefully proved by P Güttinger in 1931.

where λ is some c-number parameter, on which the time-independent Hamiltonian $\hat{H}$, and hence its eigenenergies E_n, depend.

Solution: Multiplying both parts of the fundamental Eq. (1.60) of the lecture notes by ψ_n^*, and integrating the result over space, we get

$$\int \psi_n^*(\mathbf{r})\hat{H}\psi_n(\mathbf{r})\, d^3r = \int \psi_n^*(\mathbf{r})E_n\psi_n(\mathbf{r})\, d^3r.$$

On the right-hand side of this relation, we may take the constant E_n out of the integral, and then use the orthonormality condition (1.66) to get the following expression for the eigenenergy[7]:

$$E_n = \int \psi_n^*(\mathbf{r})\hat{H}\psi_n(\mathbf{r})d^3r. \tag{*}$$

Let us differentiate both parts of this relation over the parameter λ, taking into account that not only $\hat{H}$ and E_n, but also the eigenfunctions ψ_n may depend on it:

$$\begin{aligned}\frac{\partial E_n}{\partial \lambda} &= \int \frac{\partial}{\partial \lambda}\left[\psi_n^*(\mathbf{r})\hat{H}\psi_n(\mathbf{r})\right]d^3r \\ &\equiv \int\left[\frac{\partial \psi_n^*(\mathbf{r})}{\partial \lambda}\hat{H}\psi_n(\mathbf{r}) + \psi_n^*(\mathbf{r})\frac{\partial \hat{H}}{\partial \lambda}\psi_n(\mathbf{r}) + \psi_n^*(\mathbf{r})\hat{H}\frac{\partial \psi_n(\mathbf{r})}{\partial \lambda}\right]d^3r.\end{aligned} \tag{**}$$

Next, let us spell out the general equality whose proof was the task of problem 1.3, for the particular case when $f(\mathbf{r}) = \psi_n^*(\mathbf{r})$, while $g(\mathbf{r}) = \partial\psi_n(\mathbf{r})/\partial\lambda$:

$$\int \psi_n^*(\mathbf{r})\hat{H}\frac{\partial \psi_n(\mathbf{r})}{\partial \lambda}d^3r = \int \hat{H}\psi_n^*(\mathbf{r})\frac{\partial \psi_n(\mathbf{r})}{\partial \lambda}d^3r.$$

Applying this equality into the last term of Eq. (**), and then using Eq. (1.60), $\hat{H}\psi_n = E_n\psi_n$, in the first term, and its complex conjugate, $\hat{H}\psi_n^* = E_n\psi_n^*$,[8] in the last term, we get

$$\frac{\partial E_n}{\partial \lambda} = \int \psi_n^*(\mathbf{r})\frac{\partial \hat{H}}{\partial \lambda}\psi_n(\mathbf{r})d^3r + E_n\int\left[\frac{\partial \psi_n^*(\mathbf{r})}{\partial \lambda}\psi_n(\mathbf{r}) + \psi_n^*(\mathbf{r})\frac{\partial \psi_n(\mathbf{r})}{\partial \lambda}\right]d^3r. \tag{***}$$

Now let us stop for a minute, and differentiate over λ the wavefunctions' orthonormality condition (1.66), written for the particular case $n' = n$, at the second step using the normalization condition (1.22c):

$$\frac{\partial}{\partial \lambda}\int \psi_n^*(\mathbf{r})\psi_n(\mathbf{r})d^3r \equiv \int\left[\frac{\partial \psi_n^*(\mathbf{r})}{\partial \lambda}\psi_n(\mathbf{r}) + \psi_n^*(\mathbf{r})\frac{\partial \psi_n(\mathbf{r})}{\partial \lambda}\right]d^3r = \frac{\partial}{\partial \lambda}1 \equiv 0.$$

[7] Note that according to Eq. (1.64) of the lecture notes, Eq. (*) means that the Hamiltonian is nothing more than the operator corresponding to a very special observable, the system's energy—the fact which was already mentioned at its introduction in section 1.2.

[8] See the footnote to the model solution of problem 1.3.

But this means that the integral of the last term in Eq. (***) equals zero, and that equality is reduced to

$$\frac{\partial E_n}{\partial \lambda} = \int \psi_n^*(\mathbf{r}) \frac{\partial \hat{H}}{\partial \lambda} \psi_n(\mathbf{r}) d^3r \equiv \left\langle \frac{\partial H}{\partial \lambda} \right\rangle_n,$$

thus proving the Hellmann–Feynman theorem.

Problem* 1.6. Use Eqs. (1.73) and (1.74) of the lecture notes to analyze the effect of phase locking of Josephson oscillations on the dc current flowing through a weak link between two superconductors (frequently called the *Josephson junction*), assuming that an external microwave source applies to the junction a sinusoidal ac voltage with frequency ω and amplitude A.

Solution: Let us assume that the phase locking has happened, so that our dc bias point is already on nth current step (1.76); then for the total voltage across the junction we may write

$$V(t) = n\frac{\hbar\omega}{2e} + A\cos\omega t.$$

Then Eq. (1.73) yields the following differential equation for the Josephson phase evolution,

$$\frac{d\varphi}{d\tau} = n + a\cos\tau, \qquad \text{with } \tau \equiv \omega t \text{ and } a \equiv \frac{2eA}{\hbar\omega},$$

which may be readily integrated:

$$\varphi = a\sin\tau + n\tau + \varphi_0,$$

where φ_0 is a (so far, arbitrary) integration constant.

As a result, the Josephson supercurrent (1.74) is equal to

$$\begin{aligned} I &= I_c \sin(a\sin\tau + n\tau + \varphi_0) \\ &\equiv I_c[\sin(a\sin\tau + n\tau)\cos\varphi_0 + \cos(a\sin\tau + n\tau)\sin\varphi_0]. \end{aligned}$$

Calculating its time-average (i.e. the dc component),

$$\bar{I} = I_c[\overline{\sin(a\sin\tau + n\tau)}\cos\varphi_0 + \overline{\cos(a\sin\tau + n\tau)}\sin\varphi_0],$$

for example as

$$\bar{f} = \frac{1}{2\pi}\int_{-\pi}^{+\pi} f(\tau)d\tau,$$

we see that the first term in the square brackets vanishes due to the asymmetry of the function under the integral, while for the calculation of the second term we may use the well-known integral representation of the Bessel functions of the first kind of an integer order m:[9]

$$J_m(a) = \overline{\cos(a \sin \tau - m\tau)},$$

where m is an arbitrary integer. Taking $m = -n$, we see that the dc current on the nth step is

$$\bar{I} = I_n \sin \varphi_0, \quad \text{where } I_n \equiv I_c J_{-n}(a) \equiv I_c J_{-n}\left(\frac{2eA}{\hbar\omega}\right), \quad \text{so that}$$
$$|I_n| \equiv I_c \left| J_n\left(\frac{2eA}{\hbar\omega}\right)\right|. \qquad (*)$$

Let us assume that the external circuit fixes the dc current through the junction; then the phase shift φ_0 may self-adjust to fit the external current only if it is in the range[10]

$$-|I_n| \leqslant \bar{I} \leqslant +|I_n|.$$

Hence the full size of nth current step is twice the $|I_n|$ given by Eq. (*); a look at the plot of the Bessel functions[11] shows that it oscillates as a function of the ac voltage amplitude A, gradually diminishing at $eA \gg n\hbar\omega$. Exactly this behavior (predicted by B Josephson in his Nobel-prize-winning 1962 paper[12]) was very soon observed experimentally by S Shapiro[13]; as a result, one can frequently meet the term *Shapiro* (or 'Josephson–Shapiro') *steps.*

***Problem* 1.7.** Calculate $\langle x\rangle$, $\langle p_x\rangle$, δx, and δp_x for the eigenstate $\{n_x, n_y, n_z\}$ of a particle placed inside a rectangular, hard-wall box described by Eq. (1.77) of the lecture notes, and compare the product $\delta x \delta p_x$ with the Heisenberg's uncertainty relation.

Solution: Since the spatial factors X, Y and Z of wavefunctions given by Eq. (1.84) of the lecture notes, and by similar relations for Y and Z, are already normalized and real, we may use Eq. (1.23) to write

$$\langle x\rangle = \int_0^{a_x} X^*(x)xX(x)dx = \int_0^{a_x} X^2(x)xdx$$
$$= \frac{2}{a_x}\int_0^{a_x}\left(\sin\frac{\pi n_x x}{a_x}\right)^2 xdx \equiv \frac{1}{a_x}\int_0^{a_x}\left(1 - \cos\frac{2\pi n_x x}{a_x}\right)xdx.$$

[9] See, e.g. Eq. (A.42*a*), taking into account that the imaginary part of the function under the integral on its right-hand side is odd and 2π-periodic.

[10] For a quantitative analysis of stability of such phase locking we would need to take into account other, dissipative component(s) of the current through the junction. However, as the analysis of the simple, standard model of phase-locking (see, e.g. *Part CM* section 5.4) shows, one of two physically different values of the phase difference φ_0, at which Eq. (*) is satisfied (say, $\varphi_0 = \sin^{-1}(I/I_n)$ and $\varphi_0' = \pi - \varphi_0$), is always stable.

[11] See, e.g. *Part EM* figure 2.16.

[12] [1]

[13] [2]

Integrating the second term in the parentheses by parts, we get

$$\langle x\rangle = \frac{1}{a_x}\left(\frac{a_x^2}{2} - \frac{a_x}{2\pi n_x}\int_{x=0}^{x=a_x} x d\sin\frac{2\pi n_x x}{a_x}\right) = \frac{a_x}{2} + \frac{1}{2\pi n_x}\int_0^{a_x}\sin\frac{2\pi n_x x}{a_x}dx$$

$$= \frac{a_x}{2} + \frac{a_x}{(2\pi n_x)^2}\cos\frac{2\pi n_x x}{a_x}\Bigg|_{x=0}^{x=a_x} = \frac{a_x}{2}.$$

This simple result is hardly surprising, because the wavefunctions $X(x)$ are either symmetric or antisymmetric about the central point $a_x/2$—see figure 1.8 of the lecture notes. Acting absolutely similarly, but repeating the integration by parts twice, we get

$$\langle x^2\rangle = \frac{2}{a_x}\int_0^{a_x}\left(\sin\frac{\pi n_x x}{a_x}\right)^2 x^2 dx = a_x^2\left(\frac{1}{3} - \frac{1}{2\pi^2 n_x^2}\right),$$

so that, according to Eqs. (1.33) and (1.34),

$$\delta x = \left(\langle x^2\rangle - \langle x\rangle^2\right)^{1/2} = a_x\left(\frac{1}{12} - \frac{1}{2\pi^2 n_x^2}\right)^{1/2}.$$

Notice that neither $\langle x\rangle$ nor δx depend on other quantum numbers (n_y and n_z), and that the uncertainty of the coordinate is smallest for $n_x = 1$ (in particular for the ground state), with $\delta x_{\min} \approx 0.18\ a_x$, and increases with n_x, approaching the limit $\delta x_{\max} = a_x/\sqrt{12} \approx 0.29\ a_x$ at $n_x \to \infty$.

For the particle's momentum, the calculations are simpler:

$$\langle p_x\rangle = \int_0^{a_x} X^*(x)\hat{p}_x X(x)dx = \frac{2}{a_x}\int_0^{a_x}\sin\frac{\pi n_x x}{a_x}\left(-i\hbar\frac{\partial}{\partial x}\right)\sin\frac{\pi n_x x}{a_x}dx$$

$$= -i\hbar\frac{2\pi n_x}{a_x^2}\int_0^{a_x}\sin\frac{\pi n_x x}{a_x}\cos\frac{\pi n_x x}{a_x}dx$$

$$= -i\hbar\frac{\pi n_x}{a_x^2}\int_0^{a_x}\sin\frac{2\pi n_x x}{a_x}dx = -i\hbar\frac{1}{2a_x}\cos\frac{2\pi n_x x}{a_x}\Bigg|_{x=0}^{x=a_x} = 0.$$

This result could be also predicted in advance, because, as was discussed in section 1.7 of the lecture notes, the standing wave $X(x)$ may be represented as a sum of two traveling waves with equal amplitudes, and equal and opposite momenta $p_x = \pm\hbar k_x = \pm\hbar\pi n_x/a_x$, so that the average momentum vanishes. This reasoning implies that $\langle p_x^2\rangle$ may be calculated from Eq. (1.37), with two possible states having equal probabilities: $W_+ = W_- = ½$:

$$\langle p_x^2\rangle = p_+^2 W_+ + p_-^2 W_- = \frac{1}{2}\left(p_+^2 + p_-^2\right) = \left(\frac{\hbar\pi n_x}{a_x}\right)^2.$$

As a sanity check, this calculation may be confirmed directly from Eqs. (1.33) and (1.34):

$$\langle p_x^2 \rangle = \int_0^{a_x} X^*(x)\hat{p}_x^2 X(x)dx = \frac{2}{a_x}\int_0^{a_x} \sin\frac{\pi n_x x}{a_x}\left(-\hbar^2\frac{\partial^2}{\partial x^2}\right)\sin\frac{\pi n_x x}{a_x}dx$$
$$= \frac{2}{a_x}\hbar^2\left(\frac{\pi n_x}{a_x}\right)^2\int_0^{a_x}\left(\sin\frac{\pi n_x x}{a_x}\right)^2 dx$$
$$= \frac{1}{a_x}\hbar^2\left(\frac{\pi n_x}{a_x}\right)^2\int_0^{a_x}\left(1 - \cos\frac{2\pi n_x x}{a_x}\right)dx$$
$$= \left(\frac{\hbar\pi n_x}{a_x}\right)^2 + \hbar^2\frac{\pi n_x}{2a_x^2}\sin\frac{2\pi n_x x}{a_x}\Bigg|_{x=0}^{x=a_x} = \left(\frac{\hbar\pi n_x}{a_x}\right)^2.$$

Now we can calculate the momentum's uncertainty,

$$\delta p_x = \left(\langle p_x^2 \rangle - \langle p_x \rangle^2\right)^{1/2} = \frac{\hbar\pi n_x}{a_x},$$

and the uncertainty product:

$$\delta x \delta p_x = \hbar\left(\frac{\pi^2 n_x^2}{12} - \frac{1}{2}\right)^{1/2}.$$

This expression shows that for the lowest quantum number, $n_x = 1$, the uncertainty product, $(\delta x \delta p_x)_{\min} \approx 0.568\ \hbar$, is just slightly (about 12%) larger than the Heisenberg's minimum ($0.5\ \hbar$). On the other hand, at $n_x \to \infty$ the uncertainty product grows as $(\pi/\sqrt{12})n_x\hbar \approx 0.907 n_x\hbar$.

***Problem* 1.8.** Looking at the lower (red) line in figure 1.8 of the lecture notes, it seems plausible that the 1D ground-state function (1.84) of the simple potential well (1.77) may be well approximated with an inverted parabola:

$$X_{\text{trial}}(x) = Cx(a_x - x),$$

where C is the normalization constant. Explore how good this approximation is.

Solution: A convenient 'global' measure of the approximation quality is the proximity of the expectation value (1.23) of the system's Hamiltonian, given by the guessed approximation (in the variational method, to be discussed in section 2.9 of the lecture notes, called the *trial function*):

$$\langle H \rangle_{\text{trial}} = \int \psi_{\text{trial}}^*(\mathbf{r})\hat{H}\psi_{\text{trial}}(\mathbf{r})d^3r, \qquad (*)$$

where $\psi_{\text{trial}}(\mathbf{r})$ is properly normalized,

$$\int \psi_{\text{trial}}^*(\mathbf{r})\,\psi_{\text{trial}}(\mathbf{r})\,d^3r = 1,$$

to the genuine ground state energy E_g, which, according to Eq. (1.60), satisfies a similar relation, but with the genuine ground-state wavefunction $\psi_g(\mathbf{r})$:

$$\langle H\rangle_g = \int \psi_g^*(\mathbf{r})\hat{H}\psi_g(\mathbf{r})d^3r = \int \psi_g^*(\mathbf{r})E_g\psi_g(\mathbf{r})d^3r = E_g\int \psi_g^*(\mathbf{r})\psi_g(\mathbf{r})d^3r = E_g.$$

In our 1D case, the normalization condition is

$$\int_0^a X_{\text{trial}}^*(x)X_{\text{trial}}(x)dx \equiv |C|^2\int_0^a x^2(a-x)^2dx = 1,$$

where, for the notation simplicity, $a \equiv a_x$. Working out this simple integral, we get

$$\begin{aligned}|C|^{-2} &= \int_0^a x^2(a-x)^2dx\\ &= \int_0^a (a^2x^2 - 2ax^3 + x^4)\,dx = a^2\frac{a^3}{3} - 2a\frac{a^4}{4} + \frac{a^5}{5} \equiv \frac{a^5}{30}.\end{aligned}$$

Now using the fact that inside our simple quantum-well $U(x) = 0$, so that $\hat{H} = (-\hbar^2/2m)d^2/dx^2$ in all the region where $X_{\text{trial}} \neq 0$, we get

$$\begin{aligned}\langle H\rangle_{\text{trial}} &= \int_0^a X_{\text{trial}}^*(x)\hat{H}X_{\text{trial}}(x)dx = \int_0^a X_{\text{trial}}^*\left(-\frac{\hbar^2}{2m}\right)\left(\frac{d^2}{dx^2}X_{\text{trial}}\right)dx\\ &= |C|^2\int_0^a x(a-x)\left(-\frac{\hbar^2}{2m}\right)(-2)dx\\ &= \frac{30}{a^5}\frac{\hbar^2}{m}\int_0^a x(a-x)dx = \frac{30}{a^5}\frac{\hbar^2}{m}\int_0^a (ax-x^2)dx\\ &= \frac{30}{a^5}\frac{\hbar^2}{m}\left(a\frac{a^2}{2} - \frac{a^3}{3}\right) \equiv 5\frac{\hbar^2}{ma^2}.\end{aligned}$$

Comparing this result with the exact ground state energy given by Eq. (1.85) with $n_x = 1$ and $a_x = a$,

$$E_g = \frac{\pi^2}{2}\frac{\hbar^2}{ma^2} \approx 4.935\frac{\hbar^2}{ma^2},$$

we see that the approximation given by this simple trial function is indeed pretty good, giving a ~1% accuracy, even in the absence of adjustable parameters that are used in the genuine variational method.

***Problem* 1.9.** A particle placed into a hard-wall, rectangular box with sides a_x, a_y, and a_z, is in its ground state. Calculate the average force acting on each face of the box. Can the forces be characterized by certain pressure?

Solution: Directing the coordinates axes along the corresponding sides of the box, we may describe the situation by the boundary problem described by Eq. (1.78b) of the lecture notes, so that the ground state energy E_g of the particle is expressed by Eq. (1.86) with the lowest possible values of the quantum numbers, $n_x = n_y = n_z = 1$:

$$E_g \equiv E_{1,1,1} = \frac{\pi^2\hbar^2}{2m}\left(\frac{1}{a_x^2} + \frac{1}{a_y^2} + \frac{1}{a_z^2}\right).$$

Since this energy (though kinetic by its origin) is a function of the box size only, it may be considered as a contribution to the effective potential energy of the box, and hence the force acting on any of two faces normal to axis x may be calculated as

$$F_x = -\frac{\partial E_g}{\partial a_x} = \frac{\pi^2\hbar^2}{ma_x^3}.$$

Since the area of this face is $A_x = a_y a_z$, the force-to-area ratio is

$$\mathcal{P}_x \equiv \frac{F_x}{A} = \frac{\pi^2\hbar^2}{a_x^3 a_y a_z}.$$

Since the calculations for two other face pairs may be done absolutely similarly, and give similar results (with the proper index replacements), this expression shows that generally

$$\mathcal{P}_x \neq \mathcal{P}_y \neq \mathcal{P}_z,$$

and hence the exerted forces *cannot* be characterized by a unique pressure $\mathcal{P}$, which by definition[14] should be isotropic. Only in the particular case when the box is cubic, with sides $a_x = a_y = a_z \equiv a$ and volume $V = a^3$, then we may speak of pressure:

$$\mathcal{P}_x = \mathcal{P}_y = \mathcal{P}_z \equiv \mathcal{P} = \frac{\pi^2\hbar^2}{ma^5} = \frac{\pi^2\hbar^2}{mV^{5/3}}.$$

Note that the resulting 'equation of state', $\mathcal{P}V^{5/3} = \text{const}$, differs from that of the ideal classical gas ($\mathcal{P}V = \text{const}$). As will be discussed in chapter 8, such 'quantum equations of state' remain the same even if the cubic box is filled by an arbitrary number N of non-interacting particles—either bosons or fermions (though the dependence of the pressure on N is different for these two cases)[15].

***Problem* 1.10.** A 1D quantum particle was initially in the ground state of a very deep, rectangular potential well of width a:

$$U(x) = \begin{cases} 0, & \text{for } -a/2 < x < +a/2, \\ +\infty, & \text{otherwise.} \end{cases}$$

[14] See, e.g. *Part CM* sections 7.2 and 8.1.

[15] As statistical mechanics shows (see, e.g. *Part SM* chapter 3), at sufficiently high temperature the pressure becomes isotropic and classical (with $\mathcal{P}V = \text{const}$), regardless of the shape of the box, the number of the particles, and their quantum properties.

At some instant, the well's width is abruptly increased to a new value $a' > a$, leaving the potential symmetric with respect to the point $x = 0$, and then left constant. Calculate the probability that after the change, the particle is still in the ground state of the system.

Solution: According to Eqs. (1.69) and (1.84) of the lecture notes (with the appropriate shift of the origin), the normalized initial state of the system (before the change of the well's width) is

$$\Psi(x, 0) = \left(\frac{2}{a}\right)^{1/2} \exp\left\{-i\frac{E_g}{\hbar}t\right\} \times \begin{cases} \cos\dfrac{\pi x}{a}, & \text{for } |x| < \dfrac{a}{2}, \\ 0, & \text{otherwise,} \end{cases} \qquad (*)$$

with the ground-state energy E_g given by Eq. (1.85) with $a_x = a$ and $n_x = 1$:

$$E_g = \frac{\pi^2}{2ma^2}.$$

This 'old' state serves as the initial condition for the final state of the system,

$$\Psi(x, t) = \sum_{n=1}^{\infty} c_n \psi_n(x) \exp\left\{-i\frac{E_n}{\hbar}t\right\},$$

where $\psi_n(x)$ are the normalized eigenstates of the new (expanded) well. In particular, according to the same Eq. (1.84), the new ground state is

$$\psi_1(x) = \left(\frac{2}{a'}\right)^{1/2} \cos\frac{\pi x}{a'}.$$

The constant coefficient c_1, which in particular determines the probability $W_1 = |c_1|^2$ of particle's remaining in the ground state, may be found from the 1D version of Eq. (1.68):

$$c_1 = \int_{-\infty}^{+\infty} \psi_1^*(x)\Psi(x, 0)dx,$$

giving

$$\begin{aligned} c_1 &= \frac{2}{(aa')^{1/2}} \int_{-a/2}^{+a/2} \cos\frac{\pi x}{a} \cos\frac{\pi x}{a'} dx \\ &= \frac{2}{(aa')^{1/2}} \int_{0}^{+a/2} \left[\cos \pi x\left(\frac{1}{a} + \frac{1}{a'}\right) + \cos \pi x\left(\frac{1}{a} - \frac{1}{a'}\right)\right] dx \\ &= \frac{2}{\pi(aa')^{1/2}} \left[\left(\frac{1}{a} + \frac{1}{a'}\right)^{-1} \sin\frac{\pi a}{2}\left(\frac{1}{a} + \frac{1}{a'}\right) + \left(\frac{1}{a} - \frac{1}{a'}\right)^{-1} \sin\frac{\pi a}{2}\left(\frac{1}{a} - \frac{1}{a'}\right)\right] \\ &\equiv \frac{4}{\pi} \frac{a^{1/2}a'^{3/2}}{(a'^2 - a^2)} \cos\frac{\pi a}{2a'}, \end{aligned}$$

so that[16]

$$W_1 = |c_1|^2 = \frac{16}{\pi^2} \frac{aa'^3}{(a'^2 - a^2)^2} \cos^2 \frac{\pi a}{2a'}.$$

As a sanity check, if the well is virtually unchanged, $a' = a + \varepsilon \to a$, then $\cos(\pi a/2a') \to \pi\varepsilon/2a$, $(a'^2 - a^2) \to 2a\varepsilon$, so that $c_1 \to 1$, and $W_1 \to 1$, as it should be. On the other hand, if the final well is much wider than the initial one, $a \ll a'$, then $\cos(\pi a/2a') \approx 1$, and $W_1 \approx (16/\pi^2)\, a/a' \ll 1$. This is also reasonable, because the relatively sharp initial distribution gives a contribution to many final eigenfunctions, with a small probability for the particle to be in any particular one of them.

(Additional question for the reader: could a similar problem be rationally formulated for $a' < a$, i.e. a sudden well's *shrinkage* rather than its *extension*?)

***Problem* 1.11.** At $t = 0$, a 1D particle of mass m is placed into a hard-wall, flat-bottom potential well

$$U(x) = \begin{cases} 0, & \text{for } 0 < x < a, \\ +\infty, & \text{otherwise,} \end{cases}$$

in a 50/50 linear superposition of the lowest (ground) state and the first excited state. Calculate:

(i) the normalized wavefunction $\Psi(x, t)$ for an arbitrary time $t \geqslant 0$, and
(ii) the time evolution of the expectation value $\langle x \rangle$ of the particle's coordinate.

Solutions:

(i) Our linear superposition is described with the wavefunction

$$\Psi(x, 0) = C\, [\psi_1(x) + \psi_2(x)],$$

where ψ_1 and ψ_2 are the two lowest-energy eigenfunctions of this problem, which were by-products of the 3D calculation in section 1.7 of the lecture notes—see Eqs. (1.84) and (1.85):

$$\psi_n(x) = \left(\frac{2}{a}\right)^{1/2} \sin\frac{\pi n x}{a}, \qquad E_n = \frac{\pi^2\hbar^2 n^2}{2m}, \qquad \text{with } n = 1,\ 2, \ldots.$$

The coefficient C (or rather its modulus) may be readily calculated from the normalization requirement:

[16] Note that this result would not be affected by adding an arbitrary phase to the wavefunction (*), because it would just shift the phase of the complex coefficient c_1.

$$W \equiv \int_0^a \Psi^*(x, 0)\Psi(x, 0)\, dx$$
$$\equiv |C|^2 \int_0^a [\psi_1(x) + \psi_2(x)]^*[\psi_1(x) + \psi_2(x)]\, dx = 1. \tag{*}$$

Since the wavefunctions $\psi_{1,2}$ are orthonormal,

$$\int_0^a \psi_{1,2}^*(x)\psi_{1,2}(x)dx = 1, \quad \int_0^a \psi_{1,2}^*(x)\psi_{2,1}(x)dx = 0,$$

Eq. (*) yields $|C|^2 = 2$, i.e. $|C| = 1/\sqrt{2}$. Note that the normalization condition leaves the phase of C arbitrary.

According to Eq. (1.69) of the lecture notes, the wavefunction at arbitrary time t is

$$\Psi(x, t) = \frac{1}{a^{1/2}}\left[\sin\frac{\pi x}{a}\exp\{-i\omega_1 t\} + \sin\frac{2\pi x}{a}\exp\{-i\omega_2 t\}\right],$$
$$\text{with} \quad \omega_1 = \frac{E_1}{\hbar}, \quad \omega_2 = \frac{E_2}{\hbar} = 4\frac{E_1}{\hbar}. \tag{**}$$

(ii) Now we may use this wavefunction and the basic Eq. (1.23) to calculate the expectation value of the particle's coordinate:

$$\langle x\rangle = \int_0^a \Psi^*(x, t)\,\hat{x}\,\Psi(x, t)dx$$
$$= \frac{1}{a}\int_0^a \left(\sin\frac{\pi x}{a}\exp\{i\omega_1 t\} + \sin\frac{2\pi x}{a}\exp\{i\omega_2 t\}\right)$$
$$\times x\left(\sin\frac{\pi x}{a}\exp\{-i\omega_1 t\} + \sin\frac{2\pi x}{a}\exp\{-i\omega_1 t\}\right)dx$$
$$= \frac{1}{a}\int_0^a \left(\sin^2\frac{\pi x}{a} + \sin^2\frac{2\pi x}{a} + 2\sin\frac{\pi x}{a}\sin\frac{2\pi x}{a}\cos\omega t\right)xdx,$$
$$\text{with} \quad \omega \equiv \omega_2 - \omega_1.$$

Transforming the product of two sine functions into the difference of cos functions of combinational arguments[17], and working the resulting four integrals by parts[18], we finally get

$$\langle x\rangle = \frac{1}{2}a - \frac{16}{9\pi^2}a\cos\omega t. \tag{***}$$

Evidently, this formula describes sinusoidal oscillations of the particle, with the amplitude $(16/9\pi^2)a \approx 0.18a$, around the middle of the well ($x_0 = a/2$).

[17] See, e.g. Eq. (A.18*c*).

[18] See, e.g. Eq. (A.25).

At least three comments are due here. First, this problem is a good reminder that the quantum-mechanical averaging $\langle ... \rangle$ is by no means equivalent to the averaging over time, and its result may still be a function of time—as in this case.

Second, recall that $\langle x \rangle$ does not oscillate if the system is in either of the involved stationary states, so that the oscillations (***) are the result of the states' interference. The frequency ω of the oscillations is proportional to the difference between the energies of the involved stationary states, in our case

$$\hbar\omega \equiv \hbar(\omega_2 - \omega_1) = E_2 - E_1 = \frac{4\pi^2}{2ma^2} - \frac{\pi^2}{2ma^2} = \frac{3\pi^2}{2ma^2},$$

i.e. to the frequency of the potential radiation at the transition between the corresponding energy levels—see Eq. (1.7) of the lecture notes.

Finally, the above result is exactly valid only if Eq. (**) is taken in its displayed form, which follows from Eq. (1.69) with real coefficients c_n—in our case c_1 and c_2. Generally, these coefficients may have different phases $\varphi_1 \equiv \arg(c_1)$ and $\varphi_2 \equiv \arg(c_2)$. Repeating the above calculations for this case, we may readily find that the time-dependent factor $\cos\omega t$ in Eq. (***) becomes $\cos(\omega t - \varphi)$, where $\varphi \equiv \varphi_2 - \varphi_1$. So, while the common phase of a wavefunction drops out of all expectation values, the mutual phase shift(s) between is its components in a linear superposition do affect the expectation values—in our particular case, of the coordinate.

***Problem* 1.12.** Calculate the potential profiles $U(x)$ for that the following wavefunctions,

(i) $\Psi = c\exp\{-ax^2 - ibt\}$, and
(ii) $\Psi = c\exp\{-a|x| - ibt\}$,

(with real coefficients $a > 0$ and b), satisfy the 1D Schrödinger equation for a particle with mass m. For each case, calculate $\langle x \rangle$, $\langle p_x \rangle$, δx, and δp_x, and compare the product $\delta x \delta p_x$ with the Heisenberg's uncertainty relation.

Solutions: Both these wavefunctions may be represented as the product $\psi_n(x)\exp\{-iE_n t/\hbar\}$, with $E_n = \hbar b$, so that in accordance with the discussion in sections 1.5–1.6 of the lecture notes, we only need to calculate the corresponding functions $U(x)$ from the stationary Schrödinger equation (1.65), which may be rewritten as

$$U(x) = E_n + \frac{1}{\psi_n}\frac{\hbar^2}{2m}\frac{d^2\psi_n}{d^2x}$$

with the given eigenfunctions ψ_n.

(i) In this case, $\psi_n = c\exp\{-ax^2\}$, so that a direct differentiation yields

$$U(x) = E_n + \frac{\hbar^2}{2m}(4a^2x^2 - 2a).$$

Now notice that if we introduce, instead of a, the constant

$$\omega_0 \equiv \frac{2\hbar a}{m}, \qquad (*)$$

the above expression may be rewritten as

$$U(x) = \frac{m\omega_0^2 x^2}{2} + \left(E_n - \frac{\hbar\omega_0}{2}\right),$$

while the corresponding wavefunction becomes

$$\psi_n = c\exp\{-ax^2\} \equiv c\exp\left\{-\frac{m\omega_0 x^2}{2\hbar}\right\}.$$

Since, according to the stationary Schrödinger equation, the origins of E_n and U may be shifted (simultaneously) by an arbitrary constant, we may select this constant so that

$$E_n = \frac{\hbar\omega_0}{2},$$

and $U(x)$ becomes the well-known expression for potential energy of a *harmonic oscillator* of frequency ω_0, and mass m:

$$U(x) = \frac{m\omega_0^2 x^2}{2}.$$

Hence, 'by chance' (actually, not quite :-), we have found one of the eigenstates of this very important 1D system. Later in the course, we will see that this is actually its most important, lowest-energy (*ground*) state, usually marked with quantum number $n = 0$.

Now, after finding the constant c (or rather its modulus) from the normalization condition

$$1 = \int_{-\infty}^{+\infty} \psi_n^* \psi_n dx \equiv |c|^2 \int_{-\infty}^{+\infty} \exp\{-2ax^2\} dx \equiv \left(\frac{\pi}{2a}\right)^{1/2} |c|^2,$$

(at the last step using the well-known, and very important, Gaussian integral[19]), we can use Eq. (1.23) of the lecture notes to calculate the expectation values[20]

$$\langle x \rangle = 0, \quad \langle p_x \rangle = 0, \quad \langle x^2 \rangle = \frac{1}{4a}, \quad \langle p_x^2 \rangle = \hbar^2 a,$$

so that

$$\delta x = \frac{1}{2a^{1/2}}, \quad \delta p_x = \hbar a^{1/2}.$$

[19] See, e.g. Eq. (A.36*b*).

[20] The calculation of two last averages requires one more Gaussian integral, given by Eq. (A.36*c*).

So, the product $\delta x \delta p_x$ equals $\hbar/2$, i.e. has the smallest value allowed by the uncertainty relation (1.35).[21] In the notation (*), customary for the harmonic oscillator description, the above results for the coordinate and momentum variances read

$$\langle x^2 \rangle = \frac{\hbar}{2m\omega_0}, \quad \langle p_x^2 \rangle = \frac{\hbar m \omega_0}{2}.$$

Notice that the averages of the kinetic and potential energies of the oscillator are equal to each other:

$$\left\langle \frac{p_x^2}{2m} \right\rangle = \left\langle \frac{m\omega_0^2 x^2}{2} \right\rangle = \frac{\hbar\omega_0}{4},$$

just as at classical oscillations of the system.

(ii) In this case, $\psi_n = c\exp\{-a|x|\}$, so that a similar calculation of $U(x)$ gives

$$U(x) = \hbar b + e^{a|x|}\frac{\hbar^2}{2m}\frac{d^2}{dx^2}e^{-a|x|}.$$

At $x \neq 0$, this expression gives a constant (equal to $\hbar b + \hbar^2 a^2/2m$), but the point $x = 0$ requires a special calculation, because here the wavefunction has a 'cusp', and is not analytically differentiable. However, using the notions of the sign function sgn(x) and the Dirac's delta-function $\delta(x)$,[22] we can still write formulas valid for all x:

$$\frac{d}{dx}e^{-a|x|} = -a\,\mathrm{sgn}(x)e^{-a|x|}, \quad \frac{d^2}{dx^2}e^{-a|x|} = a[a - 2\delta(x)]\,e^{-a|x|}. \qquad (**)$$

so that, finally, we get a potential $U(x)$ describing (besides an inconsequential constant U_0) an ultimately narrow 1D potential well:

$$U(x) = \hbar b + \frac{\hbar^2}{2m}[a^2 - 2a\delta(x)] \equiv U_0 - \mathscr{W}\delta(x),$$

where

$$U_0 \equiv \hbar b + \frac{\hbar^2 a^2}{2m}, \quad \mathscr{W} \equiv \frac{\hbar^2 a}{m} > 0.$$

In this notation, the eigenfunction and the eigenenergy become

$$\psi = c\exp\{-a\,|x|\} \equiv c\exp\left\{-\frac{m\mathscr{W}}{\hbar^2}\,|x|\right\},$$

$$E = \hbar b = U_0 - \frac{\hbar^2 a^2}{2m} \equiv U_0 - \frac{m\mathscr{W}^2}{2\hbar^2}.$$

[21] Note that this relation also holds for more general Gaussian wave packets, to be discussed in section 2.2.

[22] If you need a reminder, see, e.g. section A.14 of appendix A.

In chapter 2 of the lecture notes, we will see that these results describe the *only* localized eigenstate of such a well; they will be broadly used in this course as the basis for discussion of more complex problems.

Now after the wavefunction's normalization ($cc^* = 1$), Eq. (1.23) of the lecture notes, after a straightforward integration, yields[23]

$$\langle x \rangle = 0, \quad \langle x^2 \rangle = \frac{1}{2a^2}.$$

Calculating the expectation values of p_x and p_x^2, we should be careful not to loose the functions sgn(x) and $\delta(x)$—see Eq. (**):

$$\hat{p}_x \psi = -i\hbar \frac{d}{dx}(ce^{-a|x|}) = -ica\, \mathrm{sgn}(x) e^{-a|x|},$$

$$\hat{p}_x^2 \psi = -\hbar^2 \frac{d^2}{dx^2}(ce^{-a|x|}) = -\hbar^2 ca[a - 2\delta(x)]\, e^{-a|x|}.$$

Now the integration (1.23) yields

$$\langle p_x \rangle = 0, \quad \langle p_x^2 \rangle = \hbar^2 a^2, \quad \text{so that} \quad \delta x \delta p_x = \frac{\hbar}{\sqrt{2}}.$$

We see that for this non-Gaussian eigenfunction, the uncertainty product is substantially (by ~ 40%) larger than the minimum possible value $\hbar/2$.

***Problem* 1.13.** A 1D particle of mass m, moving in the field of a stationary potential $U(x)$, has the following eigenfunction

$$\psi(x) = \frac{C}{\cosh \kappa x},$$

where C is the normalization constant, and κ is a real constant. Calculate the function $U(x)$ and the state's eigenenergy E.

Solution: After calculating the second derivative of the eigenfunction,

$$\frac{d\psi}{dx} = -\frac{C\kappa \sinh \kappa x}{\cosh^2 \kappa x}, \quad \frac{d^2\psi}{dx^2} = \frac{d}{dx}\left(-\frac{C\kappa \sinh \kappa x}{\cosh^2 \kappa x}\right) = \frac{C\kappa^2}{\cosh \kappa x}\left(\frac{2 \sinh^2 \kappa x}{\cosh^2 \kappa x} - 1\right),$$

we may plug it into the 1D version of the stationary Schrödinger equation (1.65):

$$-\frac{\hbar^2}{2m}\frac{d^2\psi}{dx^2} + U(x)\,\psi = E\psi,$$

[23] For the second integration, we may use the table integral given by Eq. (A.34*d*) for $n = 2$.

getting

$$-\frac{\hbar^2\kappa^2}{2m}\frac{C}{\cosh\kappa x}\left(\frac{2\sinh^2\kappa x}{\cosh^2\kappa x}-1\right)+U(x)\frac{C}{\cosh\kappa x}=E\frac{C}{\cosh\kappa x}.$$

Canceling the common multiplier $C/\cosh\kappa x \neq 0$, we may rewrite this relation as follows

$$U(x)-E=\frac{\hbar^2\kappa^2}{2m}\left(\frac{2\sinh^2\kappa x}{\cosh^2\kappa x}-1\right). \qquad (*)$$

The function $U(x)$ and the eigenenergy E are defined to an arbitrary constant (essentially the energy reference level), provided that their difference is definite—as specified by Eq. (*). It is convenient to select this constant so that $U(x)\to 0$ at $x\to\pm\infty$. Since in these limits the expression in the parentheses of Eq. (*) tends to 1, we should associate this constant level with $(-E)$, so that

$$E=-\frac{\hbar^2\kappa^2}{2m}.$$

Now plugging this value back to Eq. (*), we get a result that may be recast into a very simple form:

$$U(x)=\frac{\hbar^2\kappa^2}{2m}\left(\frac{2\sinh^2\kappa x}{\cosh^2\kappa x}-1\right)+E=\frac{\hbar^2\kappa^2}{2m}\left(\frac{2\sinh^2\kappa x}{\cosh^2\kappa x}-1\right)-\frac{\hbar^2\kappa^2}{2m}$$
$$\equiv-\frac{\hbar^2\kappa^2}{2m}\frac{2(\cosh^2\kappa x-\sinh^2\kappa x)}{\cosh^2\kappa x}\equiv-\frac{\hbar^2\kappa^2}{m}\frac{1}{\cosh^2\kappa x}.$$

A plot of this function is shown with the black line in figure below, together with the calculated eigenenergy (dashed horizontal line), both in the units of $\hbar^2\kappa^2/m$, and the eigenfunction $\psi(x)$ (red line). Due to the simple eigenfunction describing the localized state of the particle (which may be shown to be its ground state), this potential is a convenient and popular model for description of 'soft' confinement in one dimension.

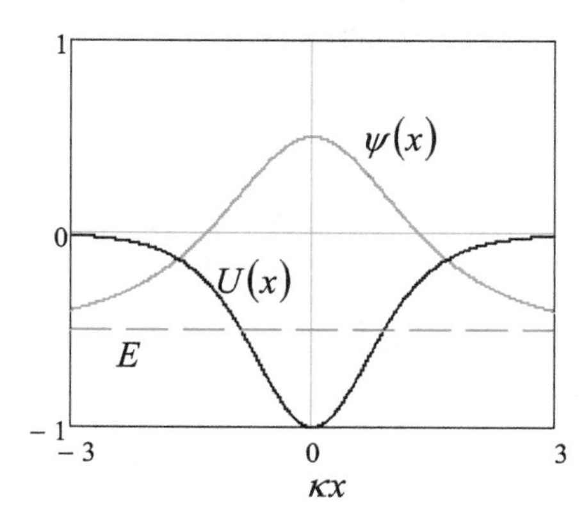

***Problem* 1.14.** Calculate the density dN/dE of the traveling-wave states in large rectangular potential wells of various dimensions: $d = 1$, 2, and 3.

Solution: Let us calculate the number N_3 of 3D states with kinetic energy $\hbar^2k^2/2m$ below a certain fixed value E. For that, we may integrate Eq. (1.90) of the lecture notes over the region of **k**-space satisfying the requirement

$$k^2 < \frac{2mE}{\hbar^2},$$

i.e. over a sphere with the so-called *Fermi radius* $k_F \equiv (2mE)^{1/2}/\hbar$:

$$N_3 = \frac{V}{(2\pi)^3}\int_{k<k_F} d^3k = \frac{V}{(2\pi)^3}\frac{4\pi}{3}k_F^3 \equiv \frac{V}{(2\pi)^3}\frac{4\pi}{3}\frac{(2m)^{3/2}}{\hbar^3}E^{3/2}.$$

From here, the 3D density of states is

$$\frac{dN_3}{dE} = \frac{V}{(2\pi)^3}\frac{4\pi}{3}\frac{(2m)^{3/2}}{\hbar^3}\frac{3}{2}E^{1/2} = V\frac{(2m)^{3/2}}{4\pi^2\hbar^3}E^{1/2}.$$

Note that the density *grows* with energy.

The absolutely similar calculation for the rectangular 2D well, based on Eq. (1.99), yields

$$N_2 = \frac{A}{(2\pi)^2}\int_{k<k_F} d^2k = \frac{A}{(2\pi)^2}\pi k_F^2 \equiv \frac{A}{(2\pi)^2}\pi\frac{2m}{\hbar^2}E,$$

so that the 2D density of states *does not depend* on energy:

$$\frac{dN_2}{dE} = \frac{A}{(2\pi)^2}\pi\frac{2m}{\hbar^2} = A\frac{m}{2\pi\hbar^2}.$$

Finally, for 1D particles, Eq. (1.100) yields

$$N_1 = \frac{l}{2\pi}\int_{|k|<k_F} dk = \frac{l}{2\pi}2k_F \equiv \frac{l}{2\pi}2\frac{(2m)^{1/2}}{\hbar}E^{1/2},$$

so that the 1D density of states,

$$\frac{dN_1}{dE} = \frac{l}{2\pi}2\frac{(2m)^{1/2}}{\hbar}\frac{1}{2}E^{-1/2} = l\frac{(2m)^{1/2}}{2\pi\hbar}E^{-1/2},$$

decreases with energy.

Problem* 1.15. Use the finite difference method with steps $a/2$ and $a/3$ to find as many eigenenergies as possible for a 1D particle in the infinitely deep, hard-wall potential well of width a. Compare the results with each other, and with the exact formula[24].

[24] You may like to start from reading about the finite difference method—see, e.g. *Part CM* section 8.5 or *Part EM* section 2.11.

Solution: The eigenproblem is described by the ordinary differential equation (1.83), which includes the second derivative of the wavefunction $X(x)$. In the finite difference method, we are approximating the derivative with the following finite difference[25]:

$$\frac{d^2X}{dx^2} \approx \frac{X(x-h) + X(x+h) - 2X(x)}{h^2},$$

where h (not to be confused with either $\hbar$ or $\hbar/2\pi$!) is the selected step along axis x.

For $h = a/2$, the only reasonable choice is to select the point x in the middle of the potential well (in the notation of figure 1.8, at $x = a/2$), so that the points $(x - h)$ and $(x + h)$ are on the well's walls, where $X = 0$. Thus Eq. (1.83) turns into a very simple relation

$$\frac{0 + 0 - 2X}{h^2} + k^2X = 0,$$

where $X \equiv X(a/2)$ and $k \equiv k_x$. This homogeneous equation cannot be used to get X, but assuming that $X \neq 0$ (i.e. that the wavefunction is nonvanishing), it gives simple results for the eigenvalue of the standing wave's vector k and hence for the eigenenergy E ($\equiv E_x$) $= (\hbar^2/2m)k^2$:

$$k = \frac{\sqrt{2}}{h} \equiv \frac{2\sqrt{2}}{a} \approx \frac{2.83}{a}, \qquad E = 2\frac{\hbar^2}{2mh^2} \equiv 4\frac{\hbar^2}{ma^2}.$$

These values should be compared with the exact analytical results (1.84)–(1.85) for the lowest (ground) eigenstate ($n_x = 1$):

$$k_1 = \frac{\pi}{a} \approx \frac{3.14}{a}, \qquad E_1 = \pi^2\frac{\hbar^2}{2ma^2} \approx 4.93\frac{\hbar^2}{ma^2}.$$

So, this large step (in the numerical-method lingo, 'coarse mesh') has made the math very simple, but has allowed the calculation of only one, ground eigenstate, and with a relatively poor accuracy: ~10% for k and ~20% for eigenenergy. This could be expected, because such coarse mesh corresponds to the approximation of the genuine sinusoidal solutions (1.84) with a single quadratic parabola.

So it is only natural to explore a slightly finer mesh with $h = a/3$, making a similar approximation for two interleaved segments of the same length $2h = 2a/3$: $x \in [0, 2a/3]$ and $x \in [a/3, a]$. Applying the finite-difference version of Eq. (1.83),

$$\frac{X(x-h) + X(x+h) - 2X(x)}{h^2} + k^2X = 0,$$

to the central points $x_- \equiv h = a/3$ and $x_+ \equiv 2h = 2a/3$ of these two segments, we get two equations for the corresponding wavefunction's values X_- and X_+:

[25] See, e.g. *Part CM* Eq. (8.65) or *Part EM* Eq. (2.220).

$$\frac{0 + X_+ - 2X_-}{h^2} + k^2 X_- = 0, \qquad \frac{X_- + 0 - 2X_+}{h^2} + k^2 X_+ = 0.$$

This system of two linear, homogeneous equation is consistent if its determinant equals zero:

$$\begin{vmatrix} -\dfrac{2}{h^2} + k^2 & \dfrac{1}{h^2} \\ \dfrac{1}{h^2} & -\dfrac{2}{h^2} + k^2 \end{vmatrix} = 0.$$

This quadratic equation for k^2 has two solutions:

$$k^2 = \frac{2}{h^2} \pm \frac{1}{h^2} = (18 \pm 9)\frac{1}{a^2},$$

giving the following two eigenvalue sets:

$$k_- = (18 - 9)^{1/2}\frac{1}{a} = \frac{3}{a}, \qquad E_- = 4.5\frac{\hbar^2}{ma^2};$$
$$k_+ = (18 + 9)^{1/2}\frac{1}{a} \approx \frac{5.20}{a}, \qquad E_+ = 13.5\frac{\hbar^2}{ma^2}.$$

The first of them is just a better approximation for the ground state, with a ~5% accuracy for k and a ~10% accuracy for energy. The second result is a much cruder description of the next (first excited) state, whose exact parameters are given by the same Eqs. (1.84) and (1.85) with $n_x = 2$:

$$k_2 = \frac{2\pi}{a} \approx \frac{6.27}{a}, \qquad E_1 = 4\pi^2\frac{\hbar^2}{2ma^2} \approx 19.7\frac{\hbar^2}{ma^2}.$$

Evidently, an even finer mesh, with a smaller h, would allow a more precise description of more eigenstates, for the price of solving a larger system of homogeneous linear equations[26]. For this particular problem, which has a simple analytical solution, this numerical method makes sense only as a demonstration, but for eigenstates of particles moving in more complex potential profiles $U(x)$, this is one of very few possible approaches. (A different, frequently more efficient numerical approach will be discussed in section 6.1 of the lecture notes—see Eq. (6.7) and its discussion.)

References

[1] Josephson B 1962 *Phys. Lett.* **1** 251
[2] Shapiro S 1963 *Phys. Rev. Lett.* **11** 80

[26] All popular public-domain and commercial software packages, including those listed in section A.16(iv) of appendix A, have efficient standard routines for such solutions.

IOP Publishing

Quantum Mechanics
Problems with solutions
Konstantin K Likharev

Chapter 2

1D wave mechanics

***Problem* 2.1.** The initial wave packet of a free 1D particle is described by Eq. (2.20) of the lecture notes:

$$\Psi(x, 0) = \int a_k e^{ikx} dk.$$

(i) Obtain a compact expression for the expectation value $\langle p \rangle$ of the particle's momentum. Does $\langle p \rangle$ depend on time?
(ii) Calculate $\langle p \rangle$ for the case when the function $|a_k|^2$ is symmetric with respect to some value k_0.

Solutions:

(i) According to the fundamental relation (1.23), and the explicit expression (1.26*b*) for the momentum operator, for we may write

$$\langle p \rangle (t) = \int_{-\infty}^{+\infty} \Psi^*(x, t)\left(-i\hbar\frac{\partial}{\partial x}\right)\Psi(x, t)dx. \qquad (*)$$

According to Eq. (2.27), we may represent the wavefunction of a free particle, with the initial state given in the assignment, as

$$\Psi(x, t) = \int a_k \exp\{i[kx - \omega(k)t]\} dk,$$
$$\text{where } \hbar\omega(k) = E(k) = \frac{\hbar^2 k^2}{2m}, \qquad (**)$$

and the integral is in infinite limits. Plugging this expression and its complex conjugate (in one of the instances, with the replacement $k \to k'$) into Eq. (*), we may transform it as

$$\langle p\rangle(t) = \int_{-\infty}^{+\infty} dx \int dk' \int dk\ a_{k'}^{*}\exp\{-i[k'x - \omega(k')t]\}$$
$$\times\left(-i\hbar\frac{\partial}{\partial x}\right)a_k \exp\{i[kx - \omega(k)t]\}$$
$$= \int_{-\infty}^{+\infty} dx \int dk' \int dk\ a_{k'}^{*}\exp\{-i[k'x - \omega(k')t]\}$$
$$\times (\hbar k)a_k \exp\{i[kx - \omega(k)t]\}$$
$$\equiv \int dk' \int dk\ a_{k'}^{*}a_k(\hbar k) \exp\{i[\omega(k') - \omega(k)]t\}$$
$$\times \int_{-\infty}^{+\infty} dx \exp\{i(k - k')x\}.$$

The last integral is just a delta-function (times 2π),[1] so that we may continue as

$$\begin{aligned}\langle p\rangle(t) &= \int dk' \int dk\ (\hbar k)a_{k'}^{*}a_k \exp\{i[\omega(k') - \omega(k)]t\}\ 2\pi\delta(k - k')\\ &= 2\pi\int (\hbar k)\ a_k^{*}a_k dk = 2\pi\int(\hbar k)\ |a_k|^2\ dk.\end{aligned} \qquad (***)$$

So, the average momentum of a free particle is time-independent (just as in classical mechanics), and is expressed via the amplitude function a_k exactly (besides a numerical normalizing factor) as the average of any function $f(x)$ is expressed via the wavefunction itself:

$$\langle f(x)\rangle(t) = \int_{-\infty}^{+\infty} f(x)\ |\Psi(x, t)|^2\ dx.$$

As will be discussed in section 4.7 of the lecture notes, the reason for this similarity is that the amplitude a_k (or rather a function $\varphi(p) \equiv \varphi(\hbar k)$ proportional to it) plays the role of the wavefunction in the so-called *momentum representation*—an alternative to the *coordinate representation* used in the wave mechanics we are studying now.

(ii) If $|a_k|^2$ is an even function of the difference $(k - k_0)$, we may recast the last form of Eq. (***) as follows:

$$\begin{aligned}\langle p\rangle &= 2\pi\int(\hbar k)\ |a_k|^2\ dk\ \equiv 2\pi\int(\hbar k_0 + \hbar k - \hbar k_0)\ |a_k|^2\ dk\\ &= 2\pi\hbar k_0\int\ |a_k|^2\ dk + 2\pi\hbar\int\ (k - k_0)\ |a_k|^2\ dk.\end{aligned}$$

Since the last integral has infinite limits, we may always represent the integration segment as a limit of $[k_0 - \kappa, k_0 + \kappa]$, at $\kappa \to 0$, i.e. as

$$\lim_{\kappa\to\infty}\int_{-\kappa}^{+\kappa} (k - k_0)\ |a_k|^2\ d(k - k_0) = \lim_{\kappa\to\infty}\int_{-\kappa}^{+\kappa} \tilde{k}\ |a_k|^2\ d\tilde{k},$$
$$\text{where } \tilde{k} \equiv k - k_0.$$

[1] See, e.g. Eq. (A.88).

Since $|a_k|^2$ is an even function of $\tilde{k}$, the whole function under the integral is odd, and the integral vanishes. So our result is reduced to

$$\langle p \rangle = \hbar k_0 2\pi \int |a_k|^2 \, dk. \tag{****}$$

In order to evaluate this integral, let us require the wavefunction to be normalized:

$$\int_{-\infty}^{+\infty} \Psi^*(x, t)\Psi(x, t)\, dx = 1.$$

Plugging in the expansions (**), and transforming the integral exactly as in task (i), we get

$$2\pi \int |a_k|^2 \, dk = 1.$$

So, Eq. (****) is reduced to a very simple and natural form, $\langle p \rangle = \hbar k_0$, which corresponds to the physics discussed in sections 1.1 and 1.7 of the lecture notes—see, e.g. Eqs. (1.14) and the text before Eq. (1.88).

***Problem* 2.2.** Calculate the function a_k, defined by Eq. (2.20) of the lecture notes, for the wave packet with a rectangular spatial envelope:

$$\Psi(x, 0) = \begin{cases} C \exp\{ik_0 x\}, & \text{for } -a/2 \leqslant x \leqslant +a/2, \\ 0, & \text{otherwise.} \end{cases}$$

Analyze the result in the limit $k_0 a \to \infty$.

Solution: Using the Fourier transform reciprocal to Eq. (2.20), we get

$$\begin{aligned} a_k &= \frac{1}{2\pi}\int_{-\infty}^{+\infty} \Psi(x, 0)e^{-ikx}dx = \frac{C}{2\pi}\int_{-a/2}^{+a/2} e^{ik_0 x}e^{-ikx}dx \\ &\equiv \frac{C}{2\pi}\int_{-a/2}^{+a/2} e^{i(k_0-k)x}dx \\ &= \frac{C}{2\pi}\frac{1}{i(k-k_0)}[e^{i(k_0-k)a/2} - e^{-i(k_0-k)a/2}] \\ &\equiv \frac{C}{\pi}\frac{\sin[(k-k_0)a/2]}{k-k_0} \equiv \frac{Ca}{2\pi}\operatorname{sinc}\frac{(k-k_0)a}{2}, \end{aligned}$$

where $\operatorname{sinc}\xi \equiv \sin(\xi)/\xi$ is the well-known function (see figure below), which describes, in particular, the Fraunhofer diffraction on a slit[2].

[2] See, e.g. *Part EM* section 8.4.

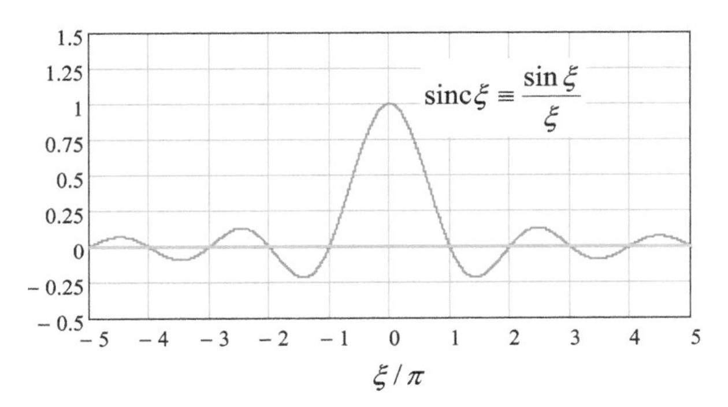

The result shows that, in contrast to the delta-functional amplitude a_k of a sinusoidal ('monochromatic') wavefunction, in our case a_k is a smooth function of k. (Physically this means that the space-restricted monochromatic wavefunction is not monochromatic any more, but is a (coherent) mixture of an infinite number of de Broglie waves with various k.) This a_k has a peak, of a finite height $Ca/2\pi$ at $k = k_0$, and of a width $\Delta k \sim 1/a$. At $k_0 a \to \infty$, this width becomes much smaller than k_0, while the peak's height grows, so that a_k tends to the delta-function of k, which we had for the space-unrestricted sinusoidal wavefunction.

***Problem* 2.3.** Prove Eq. (2.49) of the lecture notes for the 1D propagator of a free quantum particle, starting from Eq. (2.48).

Solution: Following the Gaussian integration routine discussed in section 2.2 of the lecture notes, let us complement the contents of the square brackets in Eq. (2.48) to a full square of (k + constant):

$$k\tilde{x} - \frac{\hbar k^2}{2m}\tilde{t} = -\frac{\hbar \tilde{t}}{2m}\left(k - \frac{m\tilde{x}}{\hbar \tilde{t}}\right)^2 + \frac{m\tilde{x}^2}{2\hbar \tilde{t}} = -\frac{\hbar \tilde{t}}{2m}\tilde{k}^2 + \frac{m\tilde{x}^2}{2\hbar \tilde{t}},$$

where the following natural and convenient notation has been used:

$$\tilde{x} \equiv x - x_0, \quad \tilde{t} \equiv t - t_0, \quad \text{and} \quad \tilde{k} \equiv k - \frac{m\tilde{x}}{\hbar \tilde{t}}. \qquad (*)$$

With this change, Eq. (2.48) may be rewritten as

$$\begin{aligned} G &= \frac{1}{2\pi}\exp\left\{i\frac{m\tilde{x}^2}{2\hbar\tilde{t}}\right\}\int \exp\left\{-i\frac{\hbar\tilde{t}}{2m}\tilde{k}^2\right\}d\tilde{k} \\ &= \frac{1}{2\pi}\exp\left\{i\frac{m\tilde{x}^2}{2\hbar\tilde{t}}\right\}\left(\frac{2m}{\hbar\tilde{t}}\right)^{1/2}\int_{-\infty}^{+\infty}\exp\{-i\xi^2\}d\xi \\ &\equiv \frac{1}{2\pi}\exp\left\{i\frac{m\tilde{x}^2}{2\hbar\tilde{t}}\right\}\left(\frac{2m}{\hbar\tilde{t}}\right)^{1/2}\left[\int_{-\infty}^{+\infty}\cos(\xi^2)d\xi - i\int_{-\infty}^{+\infty}\sin(\xi^2)d\xi\right]. \end{aligned}$$

Each of the *full Fresnel integrals*[3] in the last square brackets is equal to $(\pi/2)^{1/2}$; hence we may write

$$\int_{-\infty}^{+\infty} \cos(\xi^2)d\xi - i\int_{-\infty}^{+\infty} \sin(\xi^2)d\xi = (\pi/2)^{1/2}(1 - i) = \left(\frac{\pi}{i}\right)^{1/2},$$

so that finally

$$G = \frac{1}{2\pi}\exp\left\{i\frac{m\tilde{x}^2}{2\hbar\tilde{t}}\right\}\left(\frac{2m}{\hbar\tilde{t}}\right)^{1/2}\left(\frac{\pi}{i}\right)^{1/2} = \left(\frac{m}{2\pi i\hbar\tilde{t}}\right)^{1/2}\exp\left\{-\frac{m\tilde{x}^2}{2i\hbar\tilde{t}}\right\},$$

which, taking into account the notation (*), is exactly Eq. (2.49).

***Problem* 2.4.** Express the 1D propagator, defined by Eq. (2.44) of the lecture notes, via the eigenfunctions and eigenenergies of a particle moving in an arbitrary stationary potential $U(x)$.

Solution: By definition, the 1D propagator $G(x, t; x_0, t_0)$ is the solution of the time-dependent 1D Schrödinger equation (2.1) with the delta-functional initial condition

$$\Psi(x, t_0) = \delta(x - x_0). \qquad (*)$$

From the discussion in section 1.5, we know that if the potential energy U does not depend on time, the general solution of the equation is given by the 1D version of Eq. (1.69):

$$\Psi(x, t) = \sum_n c_n\psi_n(x)\exp\left\{-i\frac{E_n}{\hbar}(t - t_0)\right\},$$

where $\psi_n(x)$ are the eigenfunctions of the problem, and the coefficients c_n are given by the 1D version of Eq. (1.68):

$$c_n = \int \psi_n^*(x)\Psi(x, t_0)dx.$$

(The initial moment is now denoted as t_0 rather than 0.) Plugging into the last equation the initial condition (*), and integrating, we get $c_n = \psi_n^*(x_0)$, so that, finally,

$$G(x, t; x_0, t_0) = \sum_n \psi_n(x)\psi_n^*(x_0)\exp\left\{-i\frac{E_n}{\hbar}(t - t_0)\right\}.$$

This result shows that in the general case the propagator's dependences on x and x_0 may be different, and only if $U(x) = \text{const}$, it is a function of only the difference $(x - x_0)$—see, for example, Eqs. (2.48) and (2.49), due to the space-translational invariance of the problem.

[3] See, e.g. Eq. (A.37).

***Problem* 2.5.** Calculate the change of the wavefunction of a 1D particle, resulting from a short pulse of an external classical force that may be approximated by the delta-function[4]:

$$F(t) = P\delta(t).$$

Solution: According to the well-known relation $\mathbf{F} = -\nabla U$, a space-independent force $\mathbf{F}(t)$ may be described by the additional potential energy term $U_F(\mathbf{r}, t) = -\mathbf{F}(t) \cdot \mathbf{r}$, in our current 1D case reduced to $U_F(x, t) = -F(t)x$. As a result, the full Hamiltonian of the particle is

$$\hat{H} = \hat{H}_0 + U_F(x, t) = \hat{H}_0 - F(t)x = \hat{H}_0 - Px\delta(t),$$

$$\text{where } \hat{H}_0 \equiv -\frac{\hbar^2}{2m}\frac{\partial^2}{\partial x^2} + U(x, t),$$

so that the Schrödinger equation (2.1) takes the form

$$i\hbar\frac{\partial\Psi}{\partial t} = -\frac{\hbar^2}{2m}\frac{\partial^2\Psi}{\partial x^2} + U(x, t)\Psi - Px\delta(t)\Psi.$$

If the background potential energy $U(x, t)$ is finite at $t = 0$ for all x, and the initial form of the wavefunction is smooth (so that its second derivative over the coordinate is also finite for all x), then during the short interval of the force pulse (which may be symbolically represented as $-0 \leqslant t \leqslant +0$)[5] the first two terms on the right-hand side of the Schrödinger equation are much smaller than the last (diverging) one, and may be neglected:

$$i\hbar\frac{\partial\Psi}{\partial t} = -Px\delta(t)\Psi, \quad \text{i.e. } \frac{\partial\Psi}{\Psi} = i\frac{Px}{\hbar}\delta(t)\partial t, \quad \text{for } -0 \leqslant t \leqslant +0.$$

Integrating both parts of this equation over this infinitesimal time interval, we get

$$\ln\frac{\Psi(x, t = +0)}{\Psi(x, t = -0)} = i\frac{Px}{\hbar}, \quad \text{so that } \Psi(x, t = +0) = \Psi(x, t = -0)\exp\left\{i\frac{Px}{\hbar}\right\}. \quad (*)$$

This is the requested change of the wavefunction. Its physical sense becomes clearer if we represent the initial wavefunction by its Fourier expansion (2.20):

$$\Psi(x, t = -0) = \int a_k e^{ikx} dk;$$

then Eq. (*) yields

$$\Psi(x, t = +0) = \exp\left\{i\frac{Px}{\hbar}\right\}\int a_k e^{ikx} dk \equiv \int a_k e^{ik'x} dk,$$

$$\text{where } k' \equiv k + \frac{P}{\hbar}, \quad \text{i.e. } \hbar k' = \hbar k + P.$$

This result has a simple physical sense: the force pulse changes the effective momentum $p = \hbar k$ of *each monochromatic component* of the particle's wave packet

[4] The constant P is called the force's *impulse*. (In higher dimensionalities, it is a vector—just as the force is.)

[5] If the reader is uneasy with this shorthand notation, he or she may consider a small time interval $-\Delta t/2 \leqslant t \leqslant +\Delta t/2$, and then pursue the limit $\Delta t \to 0$.

by the same constant, equal to the force's impulse P. This result is in full accordance with the correspondence principle, because in classical mechanics the force pulse results in the similar change of particle's momentum, from p to $p' = p + P$. (In higher dimensions, this relation is generalized as $\mathbf{p}' = \mathbf{p} + \mathbf{P}$, both in classical and quantum mechanics.)

Later in the course (in section 5.5 of the lecture notes) we will see that the force-induced multiplier in Eq. (*) is just a particular, coordinate representation of the general momentum shift operator

$$\hat{\mathscr{T}}_P = \exp\left\{i\frac{P\hat{x}}{\hbar}\right\}.$$

(Please do not panic looking at this expression: well before that, in section 4.6, we will discuss what is meant by the exponent of an operator.)

***Problem* 2.6.** Calculate the transparency $\mathscr{T}$ of the rectangular potential barrier

$$U(x) = \begin{cases} 0, & \text{for } x < -d/2, \\ U_0, & \text{for } -d/2 < x < +d/2, \\ 0, & \text{for } d/2 < x, \end{cases}$$

for a particle with energy $E > U_0$. Analyze and interpret the result, taking into account that U_0 may be either positive or negative. (In the latter case, we are speaking about a particle's passage over a rectangular potential well of a finite depth $|U_0|$.)

Solution: Just as has been done for the potential step, we can use the final result of the tunneling problem analysis in section 2.3 of the lecture notes, in particular Eq. (2.71*b*), replacing κ with $(-ik')$, with k' defined by Eq. (2.65):

$$k'^2 \equiv \frac{2m(E - U_0)}{\hbar^2}.$$

The result, valid for both $U_0 < 0 < E$ (a well) and $0 < U_0 < E$ (a barrier), becomes

$$\mathscr{T} = \left|\cos k'd - \frac{i}{2}\left(\frac{k'^2 + k^2}{k'k}\right)\sin k'd\right|^{-2} \equiv \left[1 + \frac{U_0^2}{4E(E - U_0)}\sin^2 k'd\right]^{-1}. \qquad (*)$$

The figure below shows typical results given by this formula. The common feature of these plots is transparency oscillations whose period is clear from the term $\sin^2 k'd$ in the last form of Eq. (*):

$$\Delta(k'd) = \pi.$$

The origin of these oscillations is the (partial) reflection of de Broglie waves at particle passage over the sharp potential cliff, which was been discussed in section 2.3 of the lecture notes—in particular see Eq. (2.71*b*) and figure 2.7a. The reflected wave travels back, is reflected from the opposite cliff, etc, thus forming a standing wave. The constructive interference condition is achieved when the barrier/well width d corresponds to an integer number of standing half-waves, i.e. at $k'd = n\pi$, with $n = 0, 1, 2, \ldots$[6]

[6] An additional task for the reader: explain why in our current problem, in contrast to the resonances inside a potential well described by Eq. (1.77), $n = 0$ is a legitimate value.

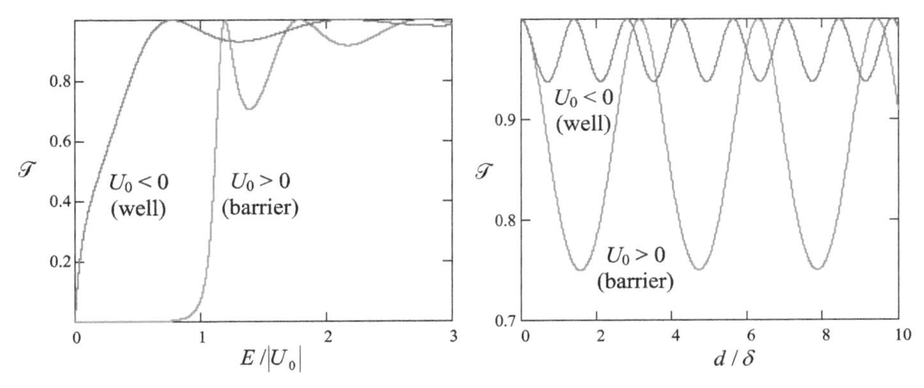

The transparency $\mathscr{T}$ as a function of: (a) the particle's energy (for a fixed ratio $d/\delta = 5$, where δ is defined by Eq. (2.59) of the lecture notes, with $U_0 \to |U_0|$), and (b) the barrier/well width d (for $E = 0.5|U_0|$).

The remarkable fact that for any parameters, $\mathscr{T} = 1$ at all constructive-interference points, was discussed, for a different particular case, in section 2.5.

***Problem* 2.7.** Prove Eq. (2.117) of the lecture notes, for the case $\mathscr{T}_{\text{WKB}} \ll 1$, using the connection formulas (2.104).

Solution: Let us apply the mnemonic rule (i), formulated in section 2.4 of the lecture notes, just after Eq. (2.106), to a relatively thick potential barrier[7], with the transparency $\mathscr{T} \ll 1$. In this case, the partial wave proportional to the coefficient d in Eq. (2.116) is negligibly small at both classical turning points, x_c and x_c' (see figure 2.11), we may rewrite these formulas as

$$\psi_{\text{WKB}} = \begin{cases} \dfrac{a}{k^{1/2}(x)} \exp\left\{i \displaystyle\int_{x_c}^{x} k(x')dx'\right\} \\ + \dfrac{b}{k^{1/2}(x)} \exp\left\{-i \displaystyle\int_{x_c}^{x} k(x')dx'\right\}, & \text{for } x < x_c, \\ \dfrac{c}{\kappa^{1/2}(x)} \exp\left\{- \displaystyle\int_{x_c}^{x} \kappa(x')dx'\right\} \\ \equiv \dfrac{c}{\kappa^{1/2}(x)} \exp\left\{- \displaystyle\int_{x_c}^{x_c'} \kappa(x')dx'\right\} \exp\left\{- \displaystyle\int_{x_c'}^{x} \kappa(x')dx'\right\}, & \text{for } x_c < x < x_c', \\ \dfrac{f}{k^{1/2}(x)} \exp\left\{- \displaystyle\int_{x_c}^{x_c'} \kappa(x')dx'\right\} \\ \times \exp\left\{i\left[\displaystyle\int_{x_c'}^{x} k(x')dx' + \text{const}\right]\right\}, & \text{for } x_c' < x, \end{cases}$$

[7] Of course, it also has to be smooth, i.e. satisfy the WKB approximation conditions (2.96) and (2.107).

where for our current purposes the second term in the top line, and the constant phase shift in the last line are unimportant. According to the mnemonic rule, applied to the classical turning points x_c and x_c',

$$|a| = |b| = |c|, \quad |c| \exp\left\{-\int_{x_c'}^{x} \kappa(x')dx'\right\} = |f|.$$

Now calculating the probability currents (2.95), corresponding to the de Broglie waves propagating to the right, in each of the classically allowed regions, we get

$$I_{x<x_c} = \frac{\hbar}{m}|a|^2, \quad I_{x>x_c'} = \frac{\hbar}{m}|f|^2 = \frac{\hbar}{m}|a|^2 \exp\left\{-2\int_{x_c}^{x_c'} \kappa(x')dx'\right\},$$

so that the barrier transparency is indeed described by Eq. (2.117):

$$\mathscr{T}_{\text{WKB}} \equiv \frac{I_{x>x_c'}}{I_{x<x_c}} = \exp\left\{-2\int_{x_c}^{x_c'} \kappa(x')dx'\right\},$$

***Problem* 2.8.** Spell out the stationary wavefunctions of a harmonic oscillator in the WKB approximation, and use them to calculate $\langle x^2 \rangle$ and $\langle x^4 \rangle$, for the eigenstate number $n \gg 1$.

Solution: In the WKB approximation, the stationary wavefunctions ψ_n are given by Eq. (2.94) of the lecture notes. Taking the lower limit of both WKB integrals at $x = 0$, i.e. at the central point of the oscillator's potential (2.111),

$$U(x) = \frac{m\omega_0^2 x^2}{2},$$

we have to take $a = b$ for symmetric wavefunctions, i.e. for even n, and $a = -b$ for asymmetric wavefunctions, i.e. for odd n—see, e.g. figure 2.35. This gives

$$\psi_n = \frac{C_n}{k_n^{1/2}(x)} \times \begin{cases} \cos \int_0^x k_n(x)dx, & \text{for } n = 0, 2, 4, \ldots, \\ \sin \int_0^x k_n(x)dx, & \text{for } n = 1, 3, 5, \ldots, \end{cases}$$

with $k_n(x)$ given by Eq. (2.82) for the nth stationary state:

$$k_n(x) = \frac{1}{\hbar}\{2m[E_n - U(x)]\}^{1/2} = \frac{1}{\hbar}\left[m\left(2E_n - m\omega_0^2 x^2\right)\right]^{1/2}.$$

According to Eq. (2.262), which coincides with the WKB result (2.114) with the replacement $n' \to n$,

$$2E_n = \hbar\omega_0(2n + 1),$$

so that

$$k_n(x) = \frac{1}{\hbar}\left\{m\left[\hbar\omega_0(2n+1) - m\omega_0^2 x^2\right]\right\}^{1/2} \equiv \frac{m\omega_0}{\hbar}\left(x_n^2 - x^2\right)^{1/2} \equiv \frac{\left(x_n^2 - x^2\right)^{1/2}}{x_0^2},$$

where $x_0 \equiv (\hbar/m\omega_0)^{1/2}$ is the length scale of the harmonic oscillator' wavefunctions (see Eq. (2.276) of the lecture notes), and $\pm x_n$ are the classical turning points in the nth Fock state, defined by the equality $E_n = U(x_n)$; for our quadratic potential,

$$x_n = \left[\frac{\hbar(2n+1)}{m\omega_0}\right]^{1/2} \equiv x_0\,(2n+1)^{1/2}.$$

What remains is to calculate the constant C_n (or rather its modulus) from the normalization condition

$$\int_{-\infty}^{+\infty} |\psi_n(x)|^2\, dx = 1.$$

In the WKB approximation, strictly valid only for $n \gg 1$, the integration limits in this condition should be extended over the classically allowed interval $[-x_n, +x_n]$ only, and the squares of rapidly oscillating sine and cosine functions may be replaced with their average value, ½. As a result, the normalization condition becomes

$$\frac{|C_n|^2}{2}\int_{-x_n}^{+x_n} \frac{dx}{k_n(x)} = 1, \quad \text{i.e.} \quad \frac{|C_n|^2 x_0^2}{2}\int_{-x_n}^{+x_n} \frac{dx}{\left(x_n^2 - x^2\right)^{1/2}} \equiv |C_n|^2 x_0^2\, I = 1,$$

$$\text{with} \quad I \equiv \int_0^1 \frac{d\xi}{(1-\xi^2)^{1/2}},$$

where $\xi \equiv x/x_n$. This integral may be readily worked out, for example, by the substitution $\xi = \sin\varphi$, giving $d\xi = \cos\varphi d\varphi$ and $(1-\xi^2)^{1/2} = \cos\varphi$, so that $I = \pi/2$. Thus, the normalization constant turns out to be independent of the state number:

$$|C_n|^2 = \frac{2}{\pi x_0^2}.$$

By the definition of the expectation values of the observables x^{2m} (where, for our tasks, m equals either 1 or 2), in the nth stationary (Fock) state

$$\langle x^{2m}\rangle = \int_{-\infty}^{+\infty} |\psi_n(x)|^2\, x^{2m} dx.$$

Using the same approximations as have been used to calculate C_n, we get

$$\langle x^{2m}\rangle_{\text{WKB}} = \frac{|C_n|^2}{2}\int_{-x_n}^{+x_n} \frac{x^{2m}dx}{k_n(x)} = \frac{1}{\pi}\int_{-x_n}^{+x_n} \frac{x^{2m}dx}{\left(x_n^2 - x^2\right)^{1/2}}$$

$$= \frac{2x_n^{2m}}{\pi} I_m, \quad \text{where} \quad I_m \equiv \int_0^1 \frac{\xi^{2m}d\xi}{(1-\xi^2)^{1/2}}.$$

These integrals, for $m = 1$ and $m = 2$, may be worked out using the same substitution $\xi = \sin \varphi$, giving

$$I_1 \equiv \int_0^{\pi/2} \sin^2 \varphi d\varphi \equiv \int_0^{\pi/2} \frac{1 - \cos 2\varphi}{2} d\varphi = \frac{\pi}{4},$$

$$I_2 = \int_0^{\pi/2} \sin^4 \varphi d\varphi \equiv \int_0^{\pi/2} \left(\frac{1 - \cos 2\varphi}{2}\right)^2 d\varphi$$

$$\equiv \int_0^{\pi/2} \frac{1 - 2\cos 2\varphi + \cos^2 2\varphi}{4} d\varphi$$

$$\equiv \frac{1}{4}\int_0^{\pi/2} \left(1 + \frac{1}{2} + \frac{1}{2}\cos 4\varphi\right) d\varphi = \frac{3\pi}{16},$$

so that, finally,

$$\langle x^2 \rangle_{\text{WKB}} = \frac{2x_n^2}{\pi} I_1 = \frac{x_0^2(2n+1)}{2} \equiv x_0^2\left(n + \frac{1}{2}\right),$$

$$\langle x^4 \rangle_{\text{WKB}} = \frac{2x_n^4}{\pi} I_2 = \frac{3x_0^4(2n+1)^2}{8} \equiv \frac{3}{2} x_0^2\left(n^2 + n + \frac{1}{4}\right).$$

As will we shown by operator methods later in the course[8], the exact expression for the first of these expectation values is *exactly* the same, while the second one differs by just a constant:

$$\langle x^4 \rangle_{\text{exact}} = \frac{3}{2} x_0^4\left(n^2 + n + \frac{1}{2}\right),$$

so that the WKB approximation indeed gives an asymptotically correct result in the limit $n \to \infty$.

***Problem* 2.9.** Use the WKB approximation to express the expectation value of the kinetic energy of a 1D particle confined in a soft potential well, in its nth stationary state, via the derivative dE_n/dn, for $n \gg 1$.

Solution: We need to calculate

$$\langle T \rangle_n = \left\langle \frac{p^2}{2m} \right\rangle_n = \int_{-\infty}^{+\infty} \psi_n^*(x) \frac{\hat{p}^2}{2m} \psi_n(x) dx \equiv -\frac{\hbar^2}{2m} \int_{-\infty}^{+\infty} \psi_n^*(x) \frac{\partial^2}{\partial x^2} \psi_n(x) dx. \quad (*)$$

As follows from the narrative in section 2.4 of the lecture notes (see, e.g. figure 2.10 and its discussion), for higher stationary states, with $n \gg 1$, the effective depth of the particle's penetration into the classically-forbidden regions is much smaller that the distance, $x_R - x_L$, between these two classical turning points. As a result, we may

[8] A brute-force calculation of these exact values, starting from Eqs. (2.276) and (2.284), and using certain recurrent relations between the Hermite polynomials, is also possible, but more cumbersome.

limit the integration in Eq. (*) to the interval $[x_L, x_R]$. On this interval, we may use Eq. (2.94) of the lecture notes, which may be rewritten as

$$\psi_n(x) = \frac{C_n}{k_n^{1/2}(x)} \sin\left(\int_{x_L}^{x} k_n(x')dx' + \varphi\right), \quad \text{with} \quad k_n^2(x) \equiv \frac{2m[E_n - U(x)]}{\hbar^2}; \quad (**)$$

for our current purposes the value of the constant phase shift φ is not important. Since the WKB approximation is valid only if the sine function in this expression changes much faster than the pre-exponential factor, we may limit the double differentiation in Eq. (*) to this function, getting

$$\langle T \rangle_n = \frac{\hbar^2}{2m} |C_n|^2 \int_{x_L}^{x_R} k_n(x) \sin^2\left(\int_{x_L}^{x} k_n(x')dx' + \varphi\right)dx.$$

At each period of the rapidly oscillating sinusoidal function, its square may be replaced with its average value, equal to ½, so that we get

$$\langle T \rangle_n = \frac{\hbar^2}{4m} |C_n|^2 \int_{x_L}^{x_R} k_n(x)dx.$$

But as we know from Eq. (2.109) of the lecture notes, in the limit $n \gg 1$ this integral equals πn, so that we get a very simple expression:

$$\langle T \rangle_n = \frac{\pi\hbar^2}{4m} |C_n|^2 n.$$

What remains is to calculate $|C_n|^2$ from the normalization condition; with Eq. (**), it gives

$$\int_{-\infty}^{+\infty} \psi_n^*(x)\psi_n(x)dx \equiv |C_n|^2 \int_{x_L}^{x_R} \frac{1}{k_n(x)} \sin^2\left(\int_{x_L}^{x} k_n(x)dx + \varphi\right)dx = 1.$$

With the similar replacement of $\sin^2(\ldots)$ by ½, we get

$$|C_n|^2 = \left[\frac{1}{2}\int_{x_L}^{x_R} \frac{dx}{k_n(x)}\right]^{-1}.$$

In order to calculate this integral, we may spell out Eq. (2.109), in the limit $n \gg 1$, as

$$\int_{x_L}^{x_R} k_n(x)dx \equiv \frac{(2m)^{1/2}}{\hbar}\int_{x_L}^{x_R} [E_n - U(x)]^{1/2}dx = \pi n,$$

and differentiate both parts of the last equality over n—the operation legitimate at $n \gg 1$, when the spectrum is quasi-continuous. In the same limit, the changes of x_L and x_R with n are negligible, and (at the last step, using the expression for $k_n(x)$ again), we get

$$\begin{aligned}\frac{(2m)^{1/2}}{\hbar}\frac{d}{dn}\int_{x_L}^{x_R} [E_n - U(x)]^{1/2}dx &\equiv \frac{(2m)^{1/2}}{\hbar}\frac{dE_n}{dn}\frac{d}{dE_n}\int_{x_L}^{x_R} [E_n - U(x)]^{1/2}dx \\ &\equiv \frac{(2m)^{1/2}}{2\hbar}\frac{dE_n}{dn}\int_{x_L}^{x_R} \frac{dx}{[E_n - U(x)]^{1/2}} \\ &\equiv \frac{m}{\hbar^2}\frac{dE_n}{dn}\int_{x_L}^{x_R} \frac{dx}{k_n(x)} = \pi,\end{aligned}$$

Combining the above relations, we finally get a very simple and elegant result:

$$\langle T \rangle_n = \frac{n}{2}\frac{dE_n}{dn}.$$

For example, for the harmonic oscillator of frequency ω_0, $E_n = \hbar\omega_0(n + ½)$, so that $dE_n/dn = \hbar\omega_0$, and our WKB result yields $\langle T_n \rangle = \hbar\omega_0 n/2$. As will be shown in section 5.4 of the lecture notes, the exact expression is given by Eq. (5.97): $\langle T_n \rangle = E_n/2 = \hbar\omega_0(n + ½)/2$; in the limit $n \gg 1$ these two results are indistinguishable.

***Problem* 2.10.** Use the WKB approximation to calculate the transparency $\mathscr{T}$ of the following triangular potential barrier:

$$U(x) = \begin{cases} 0, & \text{for } x < 0, \\ U_0 - Fx, & \text{for } x > 0, \end{cases}$$

with F, $U_0 > 0$, as a function of the incident particle's energy E.

Hint: Be careful treating the sharp potential step at $x = 0$.

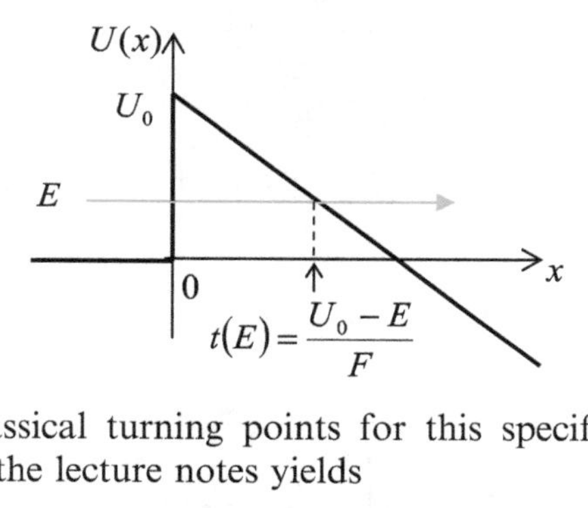

Solution: With the classical turning points for this specific potential (see figure above), Eq. (2.117) of the lecture notes yields

$$\mathscr{T}_{\text{WKB}} = \exp\left\{-\frac{2}{\hbar}\int_0^{t(E)} [2m(U_0 - Fx - E)]^{1/2}dx\right\}$$
$$\equiv \exp\left\{-\frac{2}{\hbar}(2mF)^{1/2}t^{3/2}(E)\int_0^1 \xi^{1/2}d\xi\right\},$$

where $\xi \equiv 1 - x/t(E)$, and $t(E) \equiv (U_0 - E)/F$ is the potential barrier's thickness for a particle of energy E—see figure above. The integral in the last expression is an elementary one, equal to 2/3, so that we get

$$\mathscr{T}_{\text{WKB}} = \exp\left\{-\frac{2}{\hbar}(2mF)^{1/2}t^{3/2}(E)\frac{2}{3}\right\} \equiv \exp\left\{-\frac{4}{3}\frac{(2m)^{1/2}}{\hbar}\frac{(U_0 - E)^{3/2}}{F}\right\}. \quad (*)$$

This is an approximate version (see below) of a formula derived by H Fowler and L Nordheim at the very dawn of quantum mechanics, in 1928. In this form, it is used in solid-state physics and engineering (with $F = -e\mathscr{E}$, where $\mathscr{E}$ is the applied electric field) so often that it even gave its name, the *Fowler–Nordheim tunneling*, to the very effect of electron transfer through a triangular potential barrier[9].

Note, however, that at the sharp (step-like) left border of the barrier, the second condition (2.107) of the WKB approximation validity is not satisfied even in the low-field limit

$$F \ll \frac{1}{\hbar}\left(2mU_0^3\right)^{1/2}, \qquad (**)$$

when $\kappa t(E) \gg 1$, and hence its first condition (2.96) is satisfied for most energies of the interval $0 < E < U_0$. As a result, Eq. (*) is *never* qualitatively correct. In order to correct this deficiency, let us write explicit expressions for the wavefunction of a particle with energy E, in all three relevant regions. If the condition (**) is satisfied, the barrier's transparency, by the order of magnitude given by Eq. (*), is very small, so that inside the barrier (i.e. at $0 \leqslant x \leqslant t(E) \equiv (U_0 - E)/F$) we may not only use the WKB form of the wavefunction, given by the second line of Eq. (2.116) of the lecture notes, but also neglect the second term on its right-hand side (proportional to the coefficient d), because the ratio d/c it is proportional to $\mathscr{T} \ll 1$:

$$\psi(x) = \begin{cases} A\exp\{ikx\} + B\exp\{-ikx\}, & \text{with } k = \dfrac{(2mE)^{1/2}}{\hbar}, & \text{for } x \leqslant 0, \\ \dfrac{c}{\kappa^{1/2}(x)}\exp\left\{-\displaystyle\int_0^x \kappa(x')dx'\right\}, & \text{with } \kappa(x) = \dfrac{\{2m[U(x) - E]\}^{1/2}}{\hbar}, & \text{for } 0 \leqslant x \leqslant t(E), \\ \dfrac{f}{k^{1/2}(x)}\exp\left\{i\displaystyle\int_0^x k(x')dx'\right\}, & \text{with } k(x) = \dfrac{\{2m[E - U(x)]\}^{1/2}}{\hbar}, & \text{for } t(E) \leqslant x. \end{cases}$$

Writing the usual boundary conditions of continuity of the wavefunction and its first derivative at the sharp border $x = 0$, we get a system of two equations for the coefficients A, B and c:[10]

$$A + B = \frac{c}{\kappa^{1/2}(0)}, \qquad ik(A - B) = -c\kappa^{1/2}(0),$$

which yield, in particular,

$$\frac{c}{A} = \frac{2}{\kappa^{-1/2}(0) + i\kappa^{1/2}(0)/k}, \qquad \text{so that } \left|\frac{c}{A}\right|^2 = 4\frac{\kappa(0)}{1 + \kappa^2(0)/k^2} \equiv \frac{4E}{\hbar U_0}[2m(U_0 - E)]^{1/2}.$$

[9] In particular, this is exactly the effect used for writing and erasing bits of information (encoded by the amount of electric charge trapped in a nearly-insulated electrode called the *floating gate*) in the now-ubiquitous flash memories, in particular in 'solid-state drives' (SSD)—more exactly, in their most widespread NAND variety. Another term for the same effect, used in the case of electron tunneling into vacuum is the 'field emission of electrons'.

[10] Note that taking the first derivative of the wavefunction under the barrier, we may skip differentiating the pre-exponential factor, because due to the condition (2.96), the exponential factor changes much faster.

On the other hand, at the border $x = t(E)$, where the WKB condition (2.107) is satisfied, we may use the connection formulas similar to Eqs. (2.106) of the lecture notes, in particular giving

$$\left|\frac{f}{c}\right| = \exp\left\{-\int_0^{t(E)} \kappa(x')dx'\right\} \equiv \mathscr{T}_{\text{WKB}}^{1/2},$$

$$\text{so that} \quad \left|\frac{f}{A}\right|^2 = \left|\frac{c}{A}\right|^2\left|\frac{f}{c}\right|^2 = \frac{4E}{\hbar U_0}[2m(U_0 - E)]^{1/2}\mathscr{T}_{\text{WKB}}.$$

Now, calculating the probability currents corresponding to the incident and passed de Broglie waves, we may find the barrier's transparency[11]

$$\begin{aligned}\mathscr{T} \equiv \frac{I_f}{I_A} = \frac{(\hbar/m)|f|^2}{(\hbar/m)|A|^2\, k} = \frac{4[E(U_0 - E)]^{1/2}}{U_0}\mathscr{T}_{\text{WKB}} \\ \equiv \frac{4[E(U_0 - E)]^{1/2}}{U_0}\exp\left\{-\frac{4}{3}\frac{(2m)^{1/2}}{\hbar}\frac{(U_0 - E)^{3/2}}{F}\right\}.\end{aligned} \qquad (***)$$

For typical energies $E \sim U_0/2$, the pre-exponential factor in this expression is of the order of 2, i.e. is quite noticeable. However, for typical applications, its effect on the result is much smaller than the transparency's uncertainty due to that of parameters U_0 and m. (In solid state situations, m in Eq. (***) should be replaced with the effective mass m_{ef} of the charge carrier in the material of the barrier—see the discussion in section 2.8 of the lecture notes.) This is why using the simpler Eq. (*) for applications is partly justified.

Note also that in some textbooks discussing the Fowler–Nordheim tunneling of electrons from metals or degenerate semiconductors, the above potential profile is modified as

$$U(x) = \begin{cases} 0, & \text{for } x < 0, \\ U_0 - Fx - \dfrac{e^2}{16\pi\kappa\varepsilon_0 x}, & \text{for } x > 0, \end{cases}$$

(where κ is the dielectric constant of the barrier's material), with the corresponding modification (increase) of $\mathscr{T}$, to account for the potential barrier suppression by the image charge effects[12]. However, this modification is qualitatively valid only if the so-called *traversal time* of tunneling τ_t (which will be discussed in section 5.3 of the lecture notes) is much longer than the reciprocal plasma frequency ω_p of the conductor[13], because ω_p^{-1} gives the time scale of the transients (surface plasmon propagation) leading to the image charge field formation[14].

[11] In their original work, H Fowler and L Nordheim derived this formula, in the low-field limit (**), in a different way—using the Airy functions (which were discussed and used in section 2.4 of the lecture notes).

[12] See, e.g. *Part EM* section 2.9, in particular Eq. (2.193), with the replacement $\varepsilon_0 \to \kappa\varepsilon_0$ (see *Part EM* section 3.4).

[13] See, e.g. *Part EM* section 7.2, in particular Eq. (7.37).

[14] See, e.g. the model solution of *Part EM* problem 7.13.

Problem* 2.11. Prove that the element symmetry of the 1D scattering matrix S, describing an arbitrary time-independent scatterer, allows its representation in the form (2.127) of the lecture notes.

Solution: First of all, if the scattering potential does not depend on time, the probability density distribution (for an infinitely wide wave packet) should be also constant in time. In this case, according to Eq. (2.6) of the lecture notes, the values of the probability current I at the points x_1 and x_2 outside the scatterer (see figure 2.12) should be equal, for any combination of the amplitudes A_1, B_2 of the incident waves. Let us consider two particular cases shown in figure below.

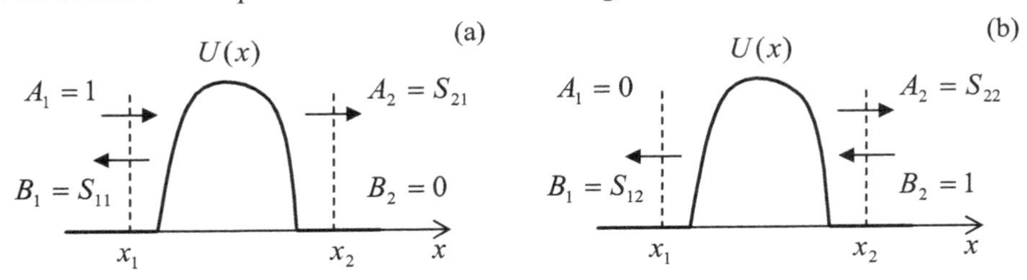

In the situation (a), a unit-amplitude wave is incident from the left ($A_1 = 1$, $B_2 = 0$), while in the case of (b), the situation is opposite ($A_1 = 0$, $B_2 = 1$). According to Eqs. (2.123) and (2.124), we may express the amplitudes of the transmitted and reflected waves in these cases via the scattering matrix elements, as shown by the labels in the figure above. Now using Eq. (2.5) to calculate the probability currents: $I(x_1) = I_{A1} + I_{B1} = (\hbar k/m)(|A_1|^2 - |B_1|^2)$, and $I(x_2) = I_{A2} + I_{B2} = (\hbar k/m)(|A_2|^2 - |B_2|^2)$, and requiring them to be equal in each of the situations (a) and (b), we get two relations:

$$\begin{aligned} &(a): \quad 1 - |S_{11}|^2 = |S_{21}|^2 - 0, \\ &(b): \quad 0 - |S_{12}|^2 = |S_{22}|^2 - 1. \end{aligned} \qquad (*)$$

One more set of relations may be obtained from the fact that all observable results of any Hamiltonian mechanics (including the wave mechanics) of a particle moving in a time-independent potential profile $U(\mathbf{r})$, should be invariant with respect to the time reversal. According to Eqs. (1.23) and (1.69), this invariance requires that at the reversal, the spatial components of 1D wavefunctions change as $\psi(x) \to \psi^*(x)$. At such complex conjugation, a 1D monochromatic traveling wave $C \exp\{ikx\}$ turns into the wave $C^* \exp\{-ikx\}$, propagating in the opposite direction. This means that the two particular cases considered above are now modified, as shown in the figure below.

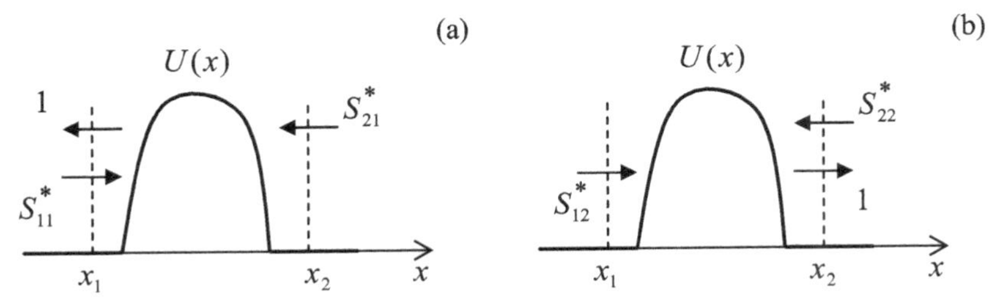

Comparing these cases with the general situation shown in figure 2.12, we see that they require taking:

$$\text{case (a):} \quad A_1 = S_{11}^*, \quad B_1 = 1, \quad A_2 = 0, \quad B_2 = S_{21}^*,$$
$$\text{case (b):} \quad A_1 = S_{12}^*, \quad B_1 = 0, \quad A_2 = 1, \quad B_2 = S_{22}^*,$$

Now applying to these cases the general Eq. (2.123), we get four more relations:

$$\begin{aligned}
&\text{(a):} \quad \begin{cases} 1 = S_{11}S_{11}^* + S_{12}S_{21}^*, \\ 0 = S_{21}S_{11}^* + S_{22}S_{21}^*, \end{cases} \\
&\text{(b):} \quad \begin{cases} 0 = S_{11}S_{12}^* + S_{12}S_{22}^*, \\ 1 = S_{21}S_{12}^* + S_{22}S_{22}^*. \end{cases}
\end{aligned} \qquad (**)$$

Not all of the eight relations of the sets (*) and (**) are independent. Indeed, comparing the first equations of each set, we see that $|S_{21}|^2 \equiv S_{21}S_{21}^*$ has to equal $S_{12}S_{21}^*$, so that

$$S_{12} = S_{21},$$

i.e. the off-diagonal elements of the scattering matrix have to be equal each other. Denoting this single complex number as $t \exp\{i\theta\}$ (with real parameters t and θ), and plugging it into the four inhomogeneous relations of the sets (*) and (**), we see that they give only two independent relations. The first of them,

$$|S_{11}|^2 = |S_{22}|^2,$$

allows the replacement of two complex parameters S_{11} and S_{22} (i.e. of four real parameters) with just three real parameters r, φ_1, and φ_2: $S_{11} = r \exp\{i(\theta + \varphi_1)\}$, $S_{22} = -r \exp\{i(\theta + \varphi_2)\}$. With this notation[15], the second independent relation may be expressed by Eq. (2.127*b*)

$$r^2 + t^2 = 1,$$

evidently expressing the probability current conservation: $\mathscr{R} + \mathscr{T} = 1$.

Now plugging these results into the two homogeneous equations of the set (**), we see that they give just one more new relation:

$$tre^{i\varphi_1} = tre^{-i\varphi_2}.$$

Besides the trivial cases when either $t = 0$ or $r = 0$ (when either the transmitted or reflected wave vanishes, and hence its phase is undetermined), the last relation shows that, apart from a possible but inconsequential shift $2\pi n$, the phases φ_1 and φ_2 are equal and opposite, and may be denoted as $\varphi_1 = \varphi$ and $\varphi_2 = -\varphi$. (This fact may be also expressed as $S_{11} = -S_{22}^*$.) Plugging these results into Eq. (2.124), we get Eq. (2.127*a*) proved.

[15] This notation is motivated by Eqs. (2.121)–(2.122) of the lecture notes, which allow one to interpret the off-diagonal elements of the scattering matrix as *transmission amplitudes*, and their diagonal elements, as *reflection amplitudes*, for two possible directions of the incident wave. (The *amplitudes* t and r should not be confused with the corresponding real *transparency* $\mathscr{T} \equiv t^2$ and *reflectivity* $\mathscr{R} \equiv r^2$.)

***Problem* 2.12.** Prove the universal relations between the elements of the 1D transfer matrix T of a stationary (but otherwise arbitrary) scatterer, mentioned in section 2.5 of the lecture notes.

Solution: First of all, let us use the same argument as in the model solution of the previous problem: the probability current should be the same at the external points x_1 and x_2, for any combination of the amplitudes A_1 and B_1 in the right hand part of the transfer matrix definition—see Eq. (2.125) of the lecture notes. Taking, first, $A_1 = 0,\ B_1 = 1$, we get

$$|T_{22}|^2 - |T_{12}|^2 = 1,$$

while for the second alternative, $A_1 = 1,\ B_1 = 0$,

$$|T_{11}|^2 - |T_{21}|^2 = 1.$$

Two more additional relations may be obtained from the time-inversion arguments spelled out in the solution of the previous problem. They imply, in particular, that Eqs. (2.125) should be valid if we complex-conjugate all wave amplitudes A, B, and simultaneously swap them at each spatial point (to reflect the change of the sign of the wave number k):

$$B_2^* = T_{11}B_1^* + T_{12}A_1^*,$$
$$A_2^* = T_{21}B_1^* + T_{22}A_1^*.$$

Taking the complex conjugate of these equations, and changing the order of lines and columns, we get

$$A_2 = T_{22}^*A_1 + T_{21}^*B_1,$$
$$B_2 = T_{12}^*A_1 + T_{11}^*B_1.$$

Comparing this system with Eqs. (2.125), we see that the matrix elements should satisfy the conditions

$$T_{22}^* = T_{11}, \quad T_{21}^* = T_{12}.$$

An alternative (and simpler) way to obtain all these relations is to plug Eqs. (2.127) of the lecture notes (whose proof was the goal of the previous problem) into Eqs. (2.126). The results may be merged into the following matrix form:

$$\mathrm{T} = \frac{1}{t}\begin{pmatrix} e^{i\theta} & -re^{-i\varphi} \\ -re^{i\varphi} & e^{-i\theta} \end{pmatrix}, \quad \text{with} \quad r^2 + t^2 = 1;$$

one can see that all the above relations between the matrix elements are indeed satisfied.

It is easy to check that the particular results obtained in section 2.5, Eqs. (2.135) and (2.138), satisfy all these relations. Moreover, one may notice that in those examples, the off-diagonal elements of the matrix T are purely imaginary, i.e. $\varphi = \pm\pi/2$. It is easy to prove that this additional property is a corollary of the additional spatial (mirror) symmetry of these particular scattering profiles: $U(x) = -U(x)$.

***Problem* 2.13.** A 1D particle had been localized in a very narrow and deep potential well, with the 'area' $\int U(x)dx$ equal to $-\mathscr{W}$, where $\mathscr{W} > 0$. Then (say, at $t = 0$) the well's bottom is suddenly lifted up, so that the particle becomes free to move. Calculate the probability density, $w(k)$, to find the particle in a state with the wave number k at $t > 0$, and the total final energy of the system.

Solution: As was discussed in the beginning of section 2.6 of the lecture notes, such a well, located at $x = 0$, may be described with a delta-functional potential,

$$U(x) = -\mathscr{W}\delta(x),$$

and the localized state of a particle in the well is described by Eqs. (2.159), (2.161), and (2.162):

$$\Psi(x, t \leqslant 0) = A\exp\left\{-\kappa\,|x| - i\frac{E_{\text{ini}}}{\hbar}t\right\},$$
$$\text{with}\quad \kappa = \frac{m\mathscr{W}}{\hbar^2}\quad \text{and}\quad E_{\text{ini}} = -\frac{\hbar^2\kappa^2}{2m} \equiv -\frac{m\mathscr{W}^2}{2\hbar^2}. \qquad (*)$$

The normalization condition is

$$\int_{-\infty}^{+\infty}|\Psi(x, t \leqslant 0)|^2\,dx \equiv |A|^2\,2\int_0^{\infty}e^{-2\kappa x}dx \equiv \frac{|A|^2}{\kappa} = 1, \quad \text{giving } |A| = \kappa^{1/2}.$$

After the well bottom's lifting, the particle becomes free to move, so that, as was discussed in section 2.2, its wavefunction may be expanded into a sum over either traveling de Broglie waves (as given by Eq. (2.27) of the lecture notes) or, equivalently, standing waves:

$$\Psi(x, t \geqslant 0) = \sum_k(c_kC\cos kx + s_kC\sin kx)\exp\left\{-i\frac{E_k}{\hbar}t\right\}, \quad \text{with } E_k = \frac{\hbar^2k^2}{2m}.$$

For our purposes, the latter (spelled out) form is more convenient. If the coefficient C is selected so that each of the component wavefunctions, $C\cos kx$ and $C\sin kx$, is normalized, the amplitudes c_k and s_k may be calculated from the 1D version of Eq. (1.68):

$$c_k = \int_{-\infty}^{+\infty}C\cos kx\;\Psi(x, 0)\,dx \equiv |A|\,C\int_{-\infty}^{+\infty}\cos kx\;e^{-\kappa|x|}dx,$$
$$s_k = \int_{-\infty}^{+\infty}C\sin kx\;\Psi(x, 0)\,dx \equiv |A|\,C\int_{-\infty}^{+\infty}\sin kx\;e^{-\kappa|x|}dx.$$

The second integral (of an odd function of x, in symmetric limits) equals zero, while the first one may be readily calculated as follows:

$$c_k = |A|C \ 2\,\mathrm{Re} \int_0^{+\infty} e^{ikx} \ e^{-\kappa x} dx = |A|C \ \mathrm{Re}\frac{0-1}{ik-\kappa}$$
$$\equiv 2\,|A|C\frac{\kappa}{k^2+\kappa^2} \equiv 2C\frac{\kappa^{3/2}}{k^2+\kappa^2}. \qquad (**)$$

What remains is to calculate the normalization coefficient C. The most transparent way[16] to do this is to introduce (as was already discussed in chapter 1 of the lecture notes) an artificial, very large segment $-l/2 \leqslant x \leqslant +l/2$, with $\kappa l \gg 1$, requiring the wavefunction to equal zero everywhere outside it, and hence on its boundaries, $x = \pm l/2$. For our eigenfunctions, $C\cos kx$, this gives the following spectrum of possible values, k_n, of the wave number:

$$\cos\frac{k_n l}{2} = 0, \quad \text{i.e. } \frac{k_n l}{2} = \frac{\pi}{2} + \pi n, \quad \text{with } n = 0, 1, 2, \ldots.$$

If l is selected to be large enough, then for all essential wave numbers $k_n \sim \kappa \gg 1/l$, i.e. $n \gg 1$, the first term in the last expression is negligible, so that the spectrum may be well approximated as[17]

$$k_n = \frac{2\pi}{l}n, \qquad (***)$$

and the normalization condition becomes

$$\int_{-l/2}^{+l/2} |C\cos k_n x|^2 \ dx \equiv \frac{|C|^2\, l}{2} = 1, \quad \text{giving } |C| = \left(\frac{2}{l}\right)^{1/2}.$$

With this, Eq. (**) yields the following probability to find the particle in the state with a wave number k:

$$W_k \equiv |c_k|^2 = \frac{8}{l}\frac{\kappa^3}{(k^2+\kappa^2)^2}.$$

Now we may calculate the required probability density $w(k)$ as the ratio of the sum of all probabilities W_k within an elementary interval $dk \ll k$, to the width of this interval. Due to the small distance between the adjacent numbers k_n, the sum may be calculated just as $W_k dn$, where dn is the number of these modes in the interval dk. According to Eq. (***), $dk = (2\pi/l)dn$, so that

$$w(k) \equiv W_k\frac{dn}{dk} = \frac{8}{l}\frac{\kappa^3}{(k^2+\kappa^2)}\frac{l}{2\pi} \equiv \frac{4}{\pi}\frac{\kappa^3}{(k^2+\kappa^2)^2}.$$

Note the cancellation, in this final expression, of the length l of the artificial bounding segment—the necessary condition of the correctness of this normalization

[16] Another way is to recognize that, in a spatially unlimited system, this sum is actually an integral over k, and use the so-called delta-normalization of Ψ. This approach, to be discussed in section 4.7 of the lecture notes, would give identical final results.

[17] This approximation corresponds to the general 1D mode counting rule—see Eq. (1.100) of the lecture notes.

procedure. Another useful sanity check is the calculation of the total probability to find the released particle in the state with *some* $k > 0$:

$$W \equiv \int_0^\infty w(k)dk = \frac{4}{\pi}\int_0^\infty \frac{\kappa^3 dk}{(k^2+\kappa^2)^2} = \frac{4}{\pi}\int_0^\infty \frac{d\xi}{(\xi^2+1)^2},$$

where $\xi \equiv k/\kappa$. This is a table integral[18], equal to $\pi/4$, so that (fortunately: -) $W = 1$, as it should be.

Our result for $w(k)$ shows that the probability density is finite but nonvanishing at $k \to 0$, and rapidly decreases as soon as k is increased beyond the reciprocal spatial extension, κ, of the initial wavefunction. Now we may use it to calculate the total energy of the particle at $t > 0$:

$$\begin{aligned} E_{\text{fin}} &= \int_0^\infty E_k w(k)dk = \int_0^\infty \frac{\hbar^2 k^2}{2m}\frac{4}{\pi}\frac{\kappa^3}{(k^2+\kappa^2)^2}dk \\ &= \frac{4}{\pi}\frac{\hbar^2 k^2}{2m}\int_0^\infty \frac{\xi^2 d\xi}{(\xi^2+1)^2} \\ &\equiv \frac{4}{\pi}\frac{\hbar^2 k^2}{2m}\left[\int_0^\infty \frac{d\xi}{\xi^2+1} - \int_0^\infty \frac{d\xi}{(\xi^2+1)^2}\right], \end{aligned}$$

where $\xi \equiv k/\kappa$ again. The second integral is the same as above (and is equal to $\pi/4$), while the first one is another well-known integral[19], equal to $\pi/2$. As a result, we get an extremely simple formula:

$$E_{\text{fin}} = \frac{\hbar^2\kappa^2}{2m} \equiv |E_{\text{ini}}| \equiv -E_{\text{ini}}.$$

It means, in particular, that the total work done on the system by the force lifting the potential well's bottom, is

$$E_{\text{fin}} - E_{\text{ini}} = 2\,|E_{\text{ini}}|,$$

i.e. twice larger than that (just $|E_{\text{ini}}|$) necessary to do this process very slowly—with a duration Δt much larger than the characteristic time constant $\tau \sim |E_{\text{ini}}|/\hbar$. This two-fold increase is the price for the process speed: at a slow ('adiabatic') well removal, the energy scale of the resulting de Broglie waves is vanishingly low.

***Problem* 2.14.** Calculate the lifetime of the metastable localized state of a 1D particle in the potential

$$U(x) = -\mathscr{w}\delta(x) - Fx, \quad \text{with } \mathscr{w} > 0,$$

using the WKB approximation. Formulate the condition of validity of the result.

[18] See, e.g. Eq. (A.32*c*) with $n = 2$.

[19] See, e.g. Eq. (A.32*a*).

Solution: According to Eqs. (2.159), (2.161), and (2.162) of the lecture notes, and the normalization carried out in the beginning of the previous problem, if $F = 0$, the normalized wavefunction of the (in this case, stable) localized state is

$$\psi_0 = \kappa_0^{1/2} \exp\{-\kappa_0 |x|\}, \quad \text{where } \kappa_0 = \frac{m\mathscr{w}}{\hbar^2},$$

and the corresponding energy is

$$E = -\frac{\hbar^2\kappa_0^2}{2m} = -\frac{m\mathscr{w}^2}{2\hbar^2}. \qquad (*)$$

The application of a nonvanishing force F tilts the potential profile—see figure below.

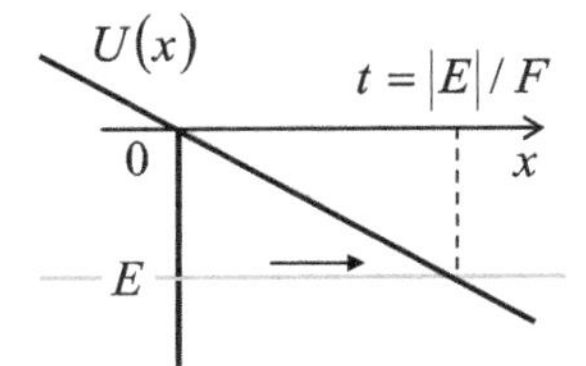

As a result, the localized particle may escape into the classically allowed region $x > t$, where

$$t = \frac{|E|}{F} = \frac{m\mathscr{w}^2}{2\hbar^2 F},$$

by tunneling through the classically forbidden region $0 < x < t$. If the force is sufficiently weak[20],

$$F \ll \frac{m^2\mathscr{w}^3}{\hbar^4}, \qquad (**)$$

the barrier is relatively thick, $\kappa_0 t \gg 1$, the barrier's transparency is low, and we may carry out the lifetime calculation using the WKB approximation.

According to Eq. (2.98) of the lecture notes, in this approximation the wavefunction under the barrier is proportional to

$$\frac{1}{\kappa^{1/2}(x)} \exp\left\{-\int_0^{|x|} \kappa(x')dx'\right\}, \quad \text{with } \frac{\hbar^2\kappa^2(x)}{2m} = U(x) - E \equiv -Fx - E.$$

For the metastable state's energy, given by Eq. (*), this wavefunction virtually coincides with ψ_0 at $|x| \sim 1/\kappa_0 \ll t$, so that the force does not change either the wavefunction's normalization factor, or the energy E substantially. As a result, for the potential barrier region, $0 < x < t$, we may write

$$\begin{aligned}\psi(x) &= \frac{\kappa_0}{\kappa^{1/2}(x)} \exp\left\{-\int_0^{x} \kappa(x')dx'\right\} \\ &\equiv \frac{\kappa_0}{\kappa^{1/2}(x)} \exp\left\{-\int_0^{t} \kappa(x)dx\right\} \exp\left\{-\int_t^{x} \kappa(x')dx'\right\}.\end{aligned}$$

[20] By the way, this is exactly the condition of validity of our results, requested in the assignment.

Now using the first mnemonic rule of the WKB connection formulas, we may write the outgoing de Broglie wave in the classically allowed region ($t < x$) as

$$\psi(x) = \frac{\kappa_0}{k^{1/2}(x)} \exp\left\{-\int_0^t \kappa(x)dx\right\} \exp\left\{i\left[\int_t^x k(x')dx' + \text{const}\right]\right\},$$

$$\text{with} \quad \frac{\hbar^2 k^2(x)}{2m} = Fx + E.$$

The probability current (2.95), corresponding to this wave, is

$$I = \frac{\hbar}{m}\kappa_0^2 \exp\left\{-2\int_0^t \kappa(x)dx\right\} \equiv \frac{m\mathscr{W}^2}{\hbar^3} \exp\left\{-2\int_0^t \kappa(x)dx\right\},$$

so that according to Eq. (2.6) (with the localized wavefunction normalized to 1), the metastable state's lifetime is just $1/I$:[21]

$$\tau = \frac{\hbar^3}{m\mathscr{W}^2} \exp\left\{2\int_0^t \kappa(x)dx\right\} = \frac{\hbar^3}{m\mathscr{W}^2} \exp\left\{\frac{2}{\hbar}\int_{[\ldots]>0} \left[2m\left(\frac{m\mathscr{W}}{\hbar^2} - Fx\right)\right]^{1/2} dx\right\}$$

$$= \frac{\hbar^3}{m\mathscr{W}^2} \exp\left\{\frac{2m^2\mathscr{W}^3}{3\hbar^4 F}\right\}.$$

Note that we could also calculate the limetime simpler, but more crudely, using the WKB formula (2.117) for barrier transparency,

$$\mathscr{T}_{\text{WKB}} = \exp\left\{-2\int_0^t \kappa(x)dx\right\} = \exp\left\{-\frac{2m^2\mathscr{W}^3}{3\hbar^4 F}\right\},$$

and then using Eq. (2.153) with the attempt time is t_a estimated as $2\pi/\omega_a$, with $\hbar\omega_a \equiv |E|$. This approach yields the following result,

$$\tau_{\text{WKB}} = \frac{t_a}{\mathscr{T}_{\text{WKB}}} = \frac{2\pi\hbar}{|E|} \exp\left\{\frac{2m^2\mathscr{W}^3}{3\hbar^4 F}\right\} = \frac{4\pi\hbar^3}{m\mathscr{W}^2} \exp\left\{\frac{2m^2\mathscr{W}^3}{3\hbar^4 F}\right\},$$

with exactly the same tunneling exponent, but a pre-exponential factor wrong by 4π. This is natural, because the left side of the potential barrier (at $x = 0$) is sharp, so that the WKB validity conditions are not satisfied for it. On the other hand, the first approach used above treats this sharpness explicitly, and hence yields the correct pre-exponential factor, though (as was discussed in section 2.4 of the lecture notes) for most practical applications this factor is of minor importance.

***Problem* 2.15.** Calculate the energy levels and the corresponding eigenfunctions of a 1D particle placed into a flat-bottom potential well of width $2a$, with infinitely hard walls, and a transparent, short potential barrier in the middle—see figure below. Discuss the dynamics of the particle in the limit $\mathscr{W} \to \infty$.

---

[21] This is the same integral as in problem 2.10, with the replacement $U_0 - E \to -E \equiv |E|$, and E given by Eq. (*).

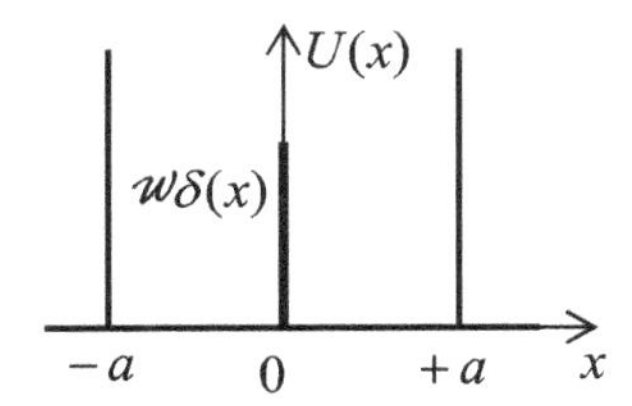

Solution: Selecting the origin of x in the middle of the well, its potential may be described as

$$U(x) = \begin{cases} +\infty, & \text{for } |x| > a, \\ \mathscr{w}\delta(x), & \text{for } |x| < a, \end{cases} \quad \text{with } \mathscr{w} > 0.$$

We already know that the standing-wave eigenfunctions ψ_n of the Schrödinger equation in regions with $U(x) = 0$, in our current case, the segments $-a < x < 0$ and $0 < x < +a$, may be always represented as linear superpositions of the fundamental solutions $\sin kx$ and $\cos kx$. In order to immediately satisfy the boundary conditions $\psi = 0$ at $x = \pm a$, we may take these solutions in the form

$$\psi_n(x) = \begin{cases} C_- \sin k(x + a), & \text{for } -a < x < 0, \\ C_+ \sin k(x - a), & \text{for } 0 < x < +a. \end{cases}$$

What remains is to satisfy the boundary conditions at $x = 0$. Plugging the above solution into Eqs. (2.75) and (2.76) of the lecture notes, we get two equations for the coefficients $C_\pm$:

$$\begin{aligned} k(C_+ - C_-) \cos ka &= \frac{2m\mathscr{w}}{\hbar^2} C_- \sin ka, \\ C_- \sin ka &= -\, C_+ \sin ka. \end{aligned} \qquad (*)$$

The second equation has two types of solutions, corresponding to antisymmetric and symmetric eigenfunctions (with the lowest-energy functions sketched in the figure below):

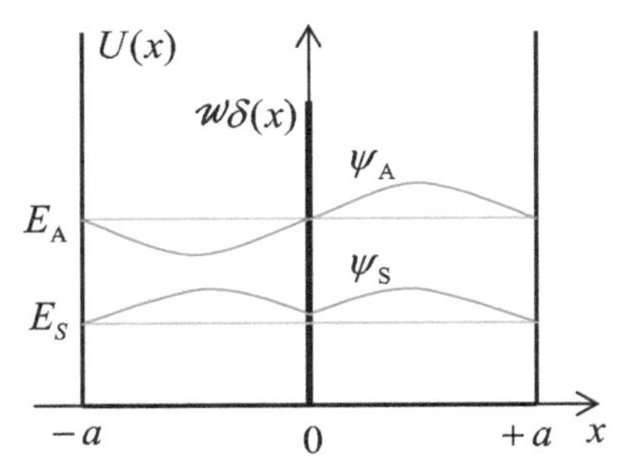

(i) *Antisymmetric solutions* (index A), with

$$(C_+)_\text{A} = (C_-)_\text{A}, \quad \text{i.e. } \psi_\text{A} = C_\text{A} \sin k_\text{A} x,$$

and the eigenvalues independent of $\mathscr{w}$:

$$\sin k_\text{A} a = 0, \quad \text{i.e. } k_\text{A} a = k_n a \equiv \pi n, \quad n = 1, 2, \ldots$$

Note that these values of k, and hence the eigenenergies $E = \hbar^2k^2/2m$ of these antisymmetric states,

$$E_A = E_n \equiv \frac{\hbar^2 k_n^2}{2m} \equiv \frac{\pi^2 n^2}{2ma^2},$$

coincide with those of the single sub-well of width a—see figure 1.8 and its discussion.

(ii) *Symmetric solutions* (index S):

$$(C_+)_S = -(C_-)_S, \quad \text{i.e.} \quad \psi_S = C_S \sin k_S(|x| - a),$$

together with Eq. (*) giving the following characteristic equation for the eigenvalue k_S:

$$\tan k_S a = -\frac{1}{\alpha}, \tag{**}$$

where the parameter α is given by Eq. (2.78) of the lecture notes, with $k = k_S$:

$$\alpha \equiv \frac{m\mathscr{W}}{\hbar^2 k_S}.$$

The figure below shows the graphical solution of Eq. (**) for three representative values of the parameter α, i.e. of the sub-well coupling strength[22]. For each solution, $k_S a$ is confined within the interval

$$\pi n - \frac{\pi}{2} < k_S a < \pi n,$$

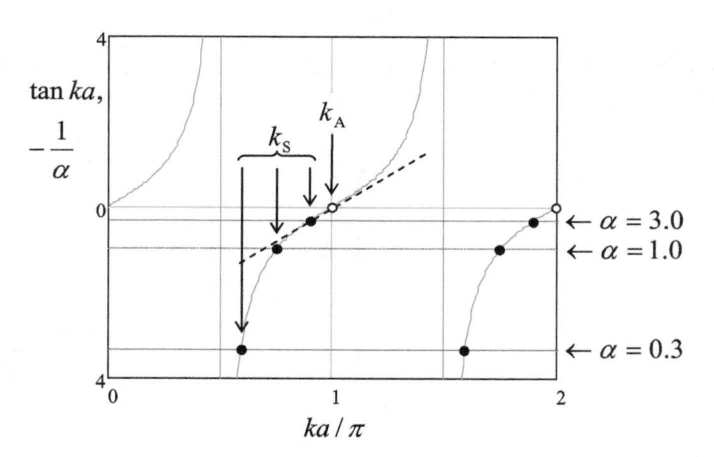

[22] For the eigenvalue classification using this plot, the fact that α depends on k_S is not essential. (For example, one may view the argument ka as a normalized well's width $2a$, which does not change α.)

so that the values of k (and hence of the energy $E = \hbar^2k^2/2m$) for the antisymmetric and symmetric states alternate, with the difference $k_A - k_S$, for each pair of adjacent states, positive but smaller then $\pi/2a$, for any value of α.

In the limit $\alpha \to 0$ (i.e. $\mathscr{W} \to 0$, meaning no partition between the two sub-wells), $k_S \to \pi(2n-1)/2a$, i.e. the symmetric eigenstates approach the shape and the energy of the symmetric states of the full potential well of width $2a$. In the opposite limit of weak sub-well coupling, $\alpha \gg 1$, we have $k_Sa \to \pi n$. In the vicinity of each such point we may approximate tan k_Sa with the difference $(k_Sa - \pi n)$—see the dashed line in the figure above, for $n = 1$. As a result, the characteristic equation (**) in this limit is reduced to

$$k_Sa \approx \pi n - \frac{1}{\alpha},$$

so that the splitting between the wave numbers and eigenenergies of the adjacent symmetric and antisymmetric states is small:

$$k_A - k_S \approx \frac{1}{\alpha a} \ll k_n,$$

$$2\delta_n \equiv E_A - E_S \approx \frac{dE}{dk}(k_A - k_S) = \frac{\pi n\hbar^2}{ma}\frac{1}{\alpha a} \equiv \frac{2E_n}{\pi n\alpha} \ll E_n.$$

The dynamics of the particle placed into such a split well, even in the weak coupling limit $\mathscr{W} \to \infty$, i.e. $\alpha \to \infty$, depends on its initial state. In the simplest case when the initial state corresponds to just one (say, the nth) couple of the adjacent symmetric and antisymmetric eigenstates, with close values of the wave number: $k_A \approx k_S \approx k_n$ and energy: $E_A \approx E_S \approx E_n$, we may analyze the dynamics as follows. At $\alpha \gg 1$, the above expressions for the eigenfunctions may be approximated just as in Eq. (2.169) of the lecture notes (obtained in section 2.6 for a different system—see figure 2.19):

$$\psi_S(x) \approx \frac{1}{\sqrt{2}}[\psi_R(x) + \psi_L(x)], \quad \psi_A(x) = \frac{1}{\sqrt{2}}[\psi_R(x) - \psi_L(x)],$$

where $\psi_{R,L}$ are the normalized ground states of the completely insulated wells:

$$\psi_R(x) = \left(\frac{2}{a}\right)^{1/2} \times \begin{cases} 0, & \text{for } -a < x < 0, \\ \sin k_n |x|, & \text{for } 0 < x < +a, \end{cases}$$

$$\psi_L(x) = \left(\frac{2}{a}\right)^{1/2} \times \begin{cases} \sin k_n |x|, & \text{for } -a < x < 0, \\ 0, & \text{for } 0 < x < +a. \end{cases}$$

As a result, repeating all the arguments of section 2.6, we arrive at the same picture of sinusoidal quantum oscillations of the particle between the two sub-wells (i.e. of the probability of finding on either side of the partition) with the frequency $\omega_n = 2\delta_n/\hbar$. Note that just as in the example analyzed in section 2.6, the time period of these oscillations,

$$\mathcal{T}_n \equiv \frac{2\pi}{\omega_n} = \frac{\pi\hbar}{\delta_n} \approx \frac{2ma^2}{n\hbar}\alpha,$$

is a factor of $\alpha/2\pi \gg 1$ shorter than the lifetime τ (2.152) of the metastable state of the particle in a potential well limited by two delta-functional walls (see figure 2.15) with the similar parameter α.

However, in contrast to the system analyzed in the lecture notes (see figure 2.19), which has just one pair of localized symmetric–antisymmetric states, our current system may have many such pairs. As a result, for an arbitrary initial state of the particle, the quantum oscillations will consist of many simultaneous sinusoidal oscillations, with incommensurate frequencies ω_n.

Problem* 2.16. Consider a symmetric system of two potential wells of the type shown in figure 2.21 of the lecture notes, but now with $U(0) = U(\pm\infty) = 0$—see figure below. What is the sign of the well interaction force due to sharing a quantum particle of mass m, for the cases when the particle is in:

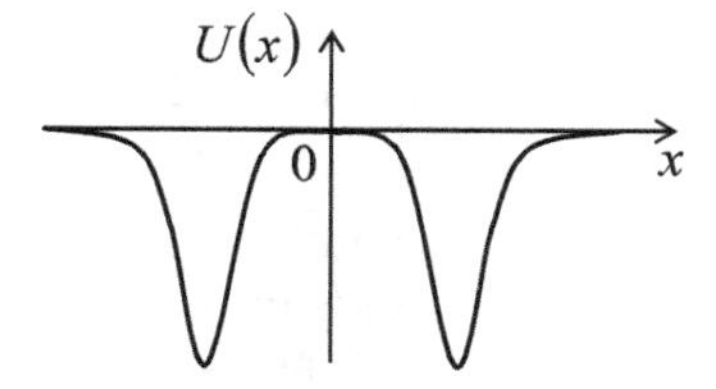

(i) a symmetric localized eigenstate, with $\psi_S(-x) = \psi_S(x)$?
(ii) an asymmetric localized eigenstate, with $\psi_A(-x) = -\psi_A(x)$?

Use an alternative approach to confirm your result for the particular case of delta-functional wells.

Solution: In classical mechanics, a potential field, described by a 1D potential profile $U(x)$, exerts the force $F_p = -dU/dx$ on the particle moving in this field. According to the 3rd Newton law, the force F exerted by the particle on the potential well (i.e. on the source of the field) is equal and opposite:

$$F = -F_p = \frac{dU(x)}{dx}.$$

Due to the correspondence principle, in quantum mechanics the force is described by the corresponding operator, in the coordinate representation equal to dU/dx, whose expectation value $\langle F\rangle$ may be calculated using the general Eq. (2.4). However, if we want to calculate the force exerted just on just one potential well (say, the right one, with $x > 0$—see the figure above), we need to limit the integration by the corresponding semi-axis:

$$\langle F\rangle(t) = \int_0^{\infty} \Psi^*(x, t)\frac{dU}{dx}\Psi(x, t)dx.$$

In the nth stationary state, with its simple time dependence (1.62), this force is time-independent:

$$\langle F\rangle = \int_0^\infty \psi_n^*(x)\frac{dU}{dx}\psi_n(x)dx \equiv \int_0^\infty |\psi_n(x)|^2\,\frac{dU}{dx}dx.$$

Since in a localized stationary state of a 1D system, the probability current (2.5) has to vanish for all x,[23] i.e. the wavefunction's phase φ has to be constant, for the notation simplicity we may always set the phase (which, according to Eq. (2.4), does not affect the expectation value of any physical observable, including F) to zero, and write

$$\langle F\rangle = \int_0^\infty \psi_n^2(x)\frac{dU}{dx}dx \equiv \int_{x=0}^{x=\infty} \psi_n^2(x)dU.$$

Integrating the last expression by parts, we get

$$\begin{aligned}\langle F\rangle &= \psi_n^2(x)U(x)|_{x=0}^{x=\infty} - \int_{x=0}^{x=\infty} U(x)\,d\left(\psi_n^2\right)\\ &\equiv \psi_n^2(x)U(x)|_{x=0}^{x=\infty} - 2\int_0^\infty U(x)\,\psi_n\frac{d\psi_n}{dx}dx.\end{aligned}$$

The first term of the last expression vanishes due to the problem's condition for the function $U(x)$. Plugging into its second term the expression for $U(x)$ following from the stationary Schrödinger equation (2.53) for the nth eigenstate we are considering,

$$U(x) = E_n + \frac{\hbar^2}{2m}\frac{d^2\psi_n}{d^2x}\frac{1}{\psi_n},$$

we get

$$\begin{aligned}\langle F\rangle &= -\int_0^\infty \left[E_n + \frac{\hbar^2}{2m}\frac{d^2\psi_n}{d^2x}\frac{1}{\psi_n}\right] d\left(\psi_n^2\right)\\ &\equiv -E_n\int_{x=0}^{x=\infty} d\left(\psi_n^2\right) - \frac{\hbar^2}{m}\int_0^\infty \frac{d^2\psi_n}{d^2x}\frac{d\psi_n}{dx}dx.\end{aligned}$$

In the second term of the last expression, we may write

$$\frac{d^2\psi_n}{d^2x}\frac{d\psi_n}{dx}dx \equiv \frac{d}{dx}\left(\frac{d\psi_n}{dx}\right)\frac{d\psi_n}{dx}dx \equiv \frac{d\psi_n}{dx}d\left(\frac{d\psi_n}{dx}\right) \equiv \frac{1}{2}d\left[\left(\frac{d\psi_n}{dx}\right)^2\right],$$

[23] Otherwise, according to Eq. (2.6), with say $x = x_1$, and $x_2 \to \infty$, the probability W to find the particle to the right of point x would change with time. Note that (as we already know from section 2.2), this statement is not necessarily true for infinite 1D systems, such as a free particle, because the probability there may 'flow from $-\infty$ to $+\infty$' without accumulation at any finite x. Later in the course, we will also see that even in finite systems of higher dimensions, the probability current density may not vanish in a stationary state, because the probability may 'flow in circles'.

so that the expression for the average force becomes

$$\langle F \rangle = -E_n \int_{x=0}^{x=\infty} d(\psi_n^2) - \frac{\hbar^2}{2m} \int_{x=0}^{x=\infty} d\left[\left(\frac{d\psi_n}{dx}\right)^2\right]$$

$$= -E_n \psi_n^2 \Big|_{x=0}^{x=\infty} - \frac{\hbar^2}{2m}\left(\frac{d\psi_n}{dx}\right)^2 \Bigg|_{x=0}^{x=\infty}.$$

Since we are discussing localized eigenfunctions $\psi_n(x)$, which vanish at $x \to \infty$ together with their derivatives, only the second, lower substitutions (at $x = 0$) in the above expression may be different from zero, and we finally get

$$\langle F \rangle = E_n \psi_n^2\Big|_{x=0} + \frac{\hbar^2}{2m}\left[\left(\frac{d\psi_n}{dx}\right)^2\right]_{x=0}. \qquad (*)$$

This general formula enables us to answer the problem's questions.

(i) For any *antisymmetric* eigenstate, the wavefunction ψ_n itself has to vanish at $x = 0$, so that $\langle F \rangle > 0$, meaning that the wells' interaction in this case is repulsive.

(ii) On the other hand, for any *symmetric* eigenfunction, the derivative $d\psi_n/dx$ vanishes at $x = 0$, so that the second term in Eq. (*) equals zero, while the first one is negative (because for any localized state, $E_n < U(\pm\infty) = 0$). Hence for such an eigenstate, $\langle F \rangle < 0$. Since this is the force exerted on the right well, we may conclude that sharing a particle in a symmetric eigenstate produces an attractive force between the wells.

Note that an alternative way to calculate the well interaction force is to write

$$\langle F \rangle = -\frac{\partial E_n}{\partial a}, \qquad (**)$$

where a is the distance between the wells, with the partial derivative meaning that the shape of each well is assumed to be kept constant at the distance variation. This requirement limits the strict applicability of Eq. (**) to the potential profiles with nonvanishing intervals with $U(x) = \text{const}$ between the wells.

In particular, it may be applied to the system of two delta-functional potential wells considered in section 2.6 of the lecture notes (see figure 2.19):

$$U(x) = -\mathscr{w}\left[\delta\left(x - \frac{a}{2}\right) + \delta\left(x + \frac{a}{2}\right)\right], \quad \text{with } \mathscr{w} > 0.$$

As a reminder, in the limit of distant wells ($\kappa_0 a \gg 1$, where $\kappa_0 \equiv 2m\mathscr{w}\hbar^2$), that analysis gave, for the only pair of localized eigenstates (one asymmetric and the other, symmetric), the following expressions:

$$E_A = E_0 + \delta, \quad E_S = E_0 - \delta, \quad \text{where } \delta \equiv \frac{2m\mathscr{w}^2}{\hbar^2}\exp\{-\kappa_0 a\} > 0,$$

and E_0 does not depend on a. Since δ grows (and hence E_A grows as well, while E_S decreases) as a is reduced, the wells sharing a quantum particle in the asymmetric state repulse each other, while if the particle is in an asymmetric state, the wells attract each other—in an agreement with the conclusions following from Eq. (*).[24]

***Problem* 2.17.** Derive and analyze the characteristic equation for the localized eigenstates of a 1D particle in a rectangular well of a finite depth (see figure below):

$$U(x) = \begin{cases} -U_0, & \text{for } |x| \leqslant a/2, \\ 0, & \text{otherwise.} \end{cases}$$

In particular, calculate the number of the localized states as a function of the well's width a, and explore the limit $U_0 \ll \hbar^2/2ma^2$.

Solution: This problem is conceptually similar to the two problems analyzed in section 2.6, though the quantitative results are different.

(i) The asymmetric eigenfunctions, satisfying the requirement of wavefunction's continuity at $x = \pm a/2$, are

$$\psi_A = C_A \times \begin{cases} \sin kx, & \text{for } |x| \leqslant \dfrac{a}{2}, \\ \operatorname{sgn}(x)\sin\dfrac{ka}{2}\exp\left\{-\kappa\left(|x| - \dfrac{a}{2}\right)\right\}, & \text{for } |x| \geqslant \dfrac{a}{2}, \end{cases}$$

where the parameters k and κ are defined as in, respectively, Eqs. (2.65) and (2.162) of the lecture notes:

$$\frac{\hbar^2 k^2}{2m} = E - U \equiv E + U_0 > 0, \quad \frac{\hbar^2 \kappa^2}{2m} = -E > 0,$$

so that

$$k^2 + \kappa^2 = K^2, \quad \text{where } K^2 \equiv \frac{2mU_0}{\hbar^2} > 0. \qquad (*)$$

[24] As may be readily shown from Eqs. (2.166) and (2.172) (and their graphical solution shown in figure 2.20) of the lecture notes, this qualitative conclusion is valid for any distance between the wells, besides that the asymmetric localized state does not exist at all if a is less that the critical value $a_{\min}$ given by Eq. (2.167).

Using the second boundary condition (of the continuity of the derivative $d\psi/dx$ at $x = \pm a/2$), we get the characteristic equation

$$\tan\frac{ka}{2} = -\frac{k}{\kappa}, \quad \text{i.e.} \tan\frac{ka}{2} = -\frac{ka}{\kappa a} < 0, \qquad (**)$$

whose graphical solution is shown in the left panel of the figure below, for several representative values of the dimensionless parameter κa.

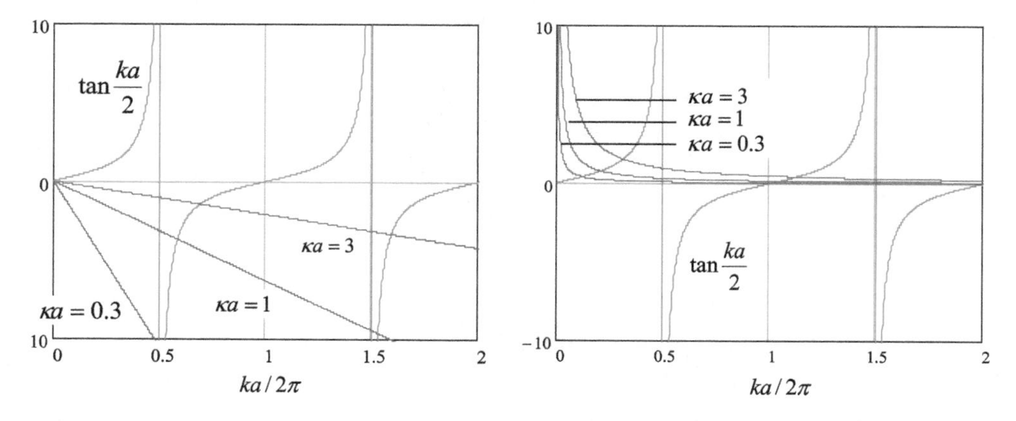

As the plots show, the solutions of Eq. (**), besides the physically unacceptable solution $k = 0$ (which gives a vanishing ψ), may be numbered by integers $n = 1, 2, 3, \ldots$, with

$$n - \frac{1}{2} \leqslant \frac{k_n a}{2\pi} \leqslant n.$$

The lower end of this interval, i.e. $k_n a/2\pi = n - ½$, corresponds to $\kappa \to 0$. This means that the nth solution becomes delocalized, with $k_n = K$, at the following value a_n of the well's width:

$$a = a_n \equiv \frac{2\pi}{K}\left(n - \frac{1}{2}\right),$$

so that the number of the asymmetric states in the well is

$$N_A = \text{floor}\left(\frac{Ka}{2\pi} - \frac{1}{2}\right),$$

where floor(ξ) is the *floor function* (frequently denoted as $\lfloor\xi\rfloor$), defined as the largest integer not greater than the function's argument ξ, whose values may be continuous. In particular, at $a < a_{\min}$, where

$$a_{\min} \equiv a_1 = \frac{\pi}{K} \equiv \frac{\pi\hbar}{(2mU_0)^{1/2}},$$

the well does not have any asymmetric localized states.

(ii) The symmetric eigenfunctions,

$$\psi_S = C_S \times \begin{cases} \cos kx, & \text{for } |x| \leqslant \frac{a}{2}, \\ \cos \frac{ka}{2} \exp\left\{-\kappa\left(|x| - \frac{a}{2}\right)\right\}, & \text{for } |x| \geqslant \frac{a}{2}, \end{cases}$$

lead, in the absolutely similar way, to a different characteristic equation:

$$\tan \frac{ka}{2} = \frac{\kappa}{k} \equiv \frac{\kappa a}{ka} > 0, \qquad (***)$$

whose graphical solution is shown in the right panel of the figure above. As the plots show, Eq. (***) has one solution in each of the complementary intervals:

$$n - 1 \leqslant \frac{k_n a}{2\pi} \leqslant n - \frac{1}{2}.$$

Each of these solutions, besides k_1 (i.e. for $n = 2, 3, \ldots$), gives $\kappa \to 0$ and hence $k_n = K$ (i.e. becomes non-localized) at $k_n a = 2\pi(n - 1)$. Hence, such a solution is impossible at

$$a < a'_n \equiv \frac{2\pi n'}{K}, \qquad \text{where } n' \equiv n - 1 = 1, 2, \ldots,$$

so that the number of remaining symmetric states

$$N_S = \text{floor}\left(\frac{Ka}{2\pi}\right).$$

Notice that if $U_0 > 0$, i.e. $K > 0$, then $N_S > 0$ for any $a > 0$. This means that the lowest localized symmetric eigenfunction (with $k = k_1$) exists in *any* potential well of nonvanishing width and depth.

For a very shallow well, $U_0 \ll \hbar^2/2ma^2$, i.e. $Ka \ll 1$, according to Eq. (*), both ka and κa have to be much less than 1, and for $n = 1$, the characteristic equation (***) is reduced to

$$\frac{k_1 a}{2} = \frac{\kappa}{k_1} \ll 1, \quad \text{i.e.} \quad \kappa = \frac{k_1^2 a}{2} \ll k_0, \qquad (***)$$

giving the following equation for the only remaining energy level:

$$|E_1| \equiv \frac{\hbar^2}{2m}\kappa^2 = \frac{\hbar^2}{2m}\left(\frac{k_1^2 a}{2}\right)^2 \equiv \frac{ma^2}{2\hbar^2}\left(\frac{\hbar^2 k_1^2}{2m}\right)^2 \equiv \frac{ma^2}{2\hbar^2}(U_0 - |E_1|)^2.$$

In this limit $\kappa \ll k \leqslant K$, and hence $|E_1| \ll U_0$, so that we may neglect $|E_1|$ on the right-hand side, thus arriving at the following approximate (but asymptotically correct) result:

$$|E_1| = \frac{m}{2\hbar^2} U_0^2 a^2. \qquad (****)$$

Note that this formula essentially coincides with Eq. (2.162) of the lecture notes, for the energy of the (only) localized state in a delta-functional, i.e. very deep and narrow well:

$$|E| = \frac{m}{2\hbar^2} \mathscr{w}^2.$$

Indeed, if the well has a rectangular shape, the 'area' $\mathscr{w}$ of the delta function is just the product $U_0 a$. The task of the next problem is to generalize this result further—to the case of an arbitrary (but shallow) potential well.

***Problem* 2.18.** Calculate the energy of a 1D particle localized in a potential well of an arbitrary shape $U(x)$, provided that its width a is finite, and the average depth is very small:

$$\left|\bar{U}\right| \ll \frac{\hbar^2}{2ma^2}, \quad \text{where} \quad \bar{U} \equiv \frac{1}{a}\int_{\text{well}} U(x)dx.$$

Solution: Let us select the origin of x in the middle of the well, and integrate both parts of the stationary Schrödinger equation, rewritten as

$$\frac{d^2\psi}{dx^2} = \frac{2m}{\hbar^2}[U(x) - E]\psi,$$

over an interval $[-x_0, +x_0]$, with $x_0 > a/2$. The result is

$$\left.\frac{d\psi}{dx}\right|_{x=+x_0} - \left.\frac{d\psi}{dx}\right|_{x=-x_0} = \frac{2m}{\hbar^2}\int_{-x_0}^{+x_0} [U(x) - E]\,\psi(x)dx. \qquad (*)$$

As we already know from sections 2.3–2.6 of the lecture notes, near the points $x = \pm x_0$, i.e. outside the potential well, the wavefunctions change as $\exp\{-\kappa|x|\}$, where κ is defined by the relation

$$\frac{\hbar^2\kappa^2}{2m} \equiv |E| = -E, \quad \text{i.e.} \quad \kappa = \frac{(2m\,|E\,|)^{1/2}}{\hbar},$$

so that we may rewrite Eq. (*) as

$$\begin{aligned} -\kappa[\psi(+x_0) + \psi(-x_0)] &\equiv -\frac{(2m\,|E\,|)^{1/2}}{\hbar}[\psi(+x_0) + \psi(-x_0)] \\ &= \frac{2m}{\hbar^2}\int_{-x_0}^{+x_0} [U(x) - E]\,\psi(x)dx. \end{aligned} \qquad (**)$$

So far, this is an exact result, valid for any $x_0 > a/2$. Now let us suppose that if $|U|$ satisfies the condition specified in the assignment, $|E|$ is even much smaller. (This assumption, implied by the result of the previous problem for the particular case of constant U, will be confirmed by our final result.) This means that $1/\kappa$ is much larger than a, so that if we select x_0 somewhere within the following wide range,

$$a \ll x_0 \ll 1/\kappa,$$

then within the interval $[-x_0, +x_0]$, the wavefunction is virtually constant. Hence we may cancel it in both parts of Eq. (**), getting simply

$$-\frac{(2m\,|E\,|)^{1/2}}{\hbar}2 = \frac{2m}{\hbar^2}\int_{-x_0}^{+x_0}[U(x) - E]dx$$

$$\equiv -\frac{2m}{\hbar^2}\int_{-a/2}^{+a/2}|U(x)|\,dx + \frac{2m}{\hbar^2}2x_0\,|E\,|.$$

Moreover, since at our choice $x_0 \ll 1/\kappa \equiv \hbar/(2m|E|)^{1/2}$, the last term of the right-part of this relation is negligible in comparison with its left-hand side, and it may be reduced to just

$$-\frac{(2m\,|E\,|)^{1/2}}{\hbar}2 = -\frac{2m}{\hbar^2}\int_{-a/2}^{+a/2}|U(x)|\,dx,$$

giving us the final result[25]

$$|E\,| = \frac{m}{2\hbar^2}\left[\int_{\text{well}}|U(x)|\,dx\right]^2, \quad \text{so that} \quad \frac{|\,E\,|}{|\bar{U}|} = \frac{ma^2}{2\hbar^2}|\bar{U}| \ll 1,$$

confirming that $|E|$ is indeed much smaller than the average value of $|U|$.

For the reader's reference: this scaling of the localized state's energy (as the square of the confinement potential) is specific for 1D problems; as we will see in chapter 3, in a similar 2D problem, $|E|$ is exponentially low, while the 3D localization has a threshold: the confining potential $|U|$ has to reach a certain value before it can house a localized state.

***Problem* 2.19.** A particle of mass m is moving in a field with the following potential:

$$U(x) = U_0(x) + \mathscr{w}\delta(x),$$

where $U_0(x)$ is a smooth, symmetric function with $U_0(0) = 0$, growing monotonically at $x \to \pm\infty$.

(i) Use the WKB approximation to derive the characteristic equation for the particle's energy spectrum, and
(ii) semi-quantitatively describe the spectrum evolution at the increase of $|\mathscr{w}|$, for both signs of this parameter.

[25] In the particular case of a rectangular well, this formula is immediately reduced to Eq. (****) of the previous problem (in the same limit).

Make both results more specific for the quadratic-parabolic potential (2.111): $U_0(x) = m\omega_0^2x^2/2$.

Solutions:

(i) As was demonstrated in section 2.4 of the lecture notes, the 'soft' potential $U_0(x)$ alone may be handled with the WKB approximation very successfully[26], but this approximation is inapplicable to such 'hard' potentials as the delta-functional peak—please have one more look at the condition (2.96). However, we may solve this problem by combining the WKB approach with the delta-functional potential treatment discussed in section 2.3, based on the boundary conditions (2.75)–(2.76). For the delta-functional potential located at $x = 0$, they read

$$\psi_+(0) = \psi_-(0) \equiv \psi(0), \quad \frac{d\psi_+}{dx}(0) - \frac{d\psi_-}{dx}(0) = \frac{2m}{\hbar^2}\mathscr{W}\psi(0), \qquad (*)$$

where $\psi_\pm(x)$ are the wavefunctions at $x \geqslant 0$ and $x \leqslant 0$, respectively.

Due to the symmetry of the potential $U(x)$, the eigenfunctions of our problem should be either symmetric: $\psi_-(-x) = \psi_+(x)$, or antisymmetric: $\psi_-(-x) = -\psi_+(x)$. According to Eq. (*), the *asymmetric* eigenfunctions are not affected by the delta-functional potential peak at all, because for them $\psi(0)$ has to vanish. Hence for these eigenstates, corresponding to even values of the integer n in Eq. (2.109), we still may use the general Bohr–Sommerfeld result (2.110) for an arbitrary smooth potential $U_0(x)$, and the specific result (2.114) for the quadratic potential:

$$E_n = \hbar\omega_0\left(n' + \frac{1}{2}\right), \quad \text{for} \quad n' \equiv n - 1 = 1, 3, 5, \ldots.$$

On the other hand, for *symmetric* eigenfunctions, for which the first of Eqs. (*) is satisfied automatically, and $d\psi_-/dx = -d\psi_+/dx$ at $x = 0$, the second of these boundary conditions may be rewritten as

$$\frac{d\psi_+}{dx}(0) = \frac{m}{\hbar^2}\mathscr{W}\psi_+(0). \qquad (**)$$

For $x > 0$, where $U(x) = U_0(x)$, we may use the connection formulas (2.105), obtained for exactly this situation: the total reflection of a monochromatic de Broglie wave from the classical turning point x_c of a soft potential well—in our current case, of $U_0(x)$. With these formulas, Eq. (2.94) takes the form

$$\psi_+(x) = \frac{a}{k^{1/2}(x)}[e^{-i\varphi(x)} - e^{i\varphi(x)+i\pi/2}], \quad \text{with} \quad \varphi(x) \equiv \int_x^{x_c} k(x')dx' \geqslant 0,$$

$$k^2(x) \equiv \frac{2m}{\hbar^2}[E - U_0(x)].$$

[26] As a reminder, this approximation gives the *exact* result (2.114) for the energy spectrum of the harmonic oscillator.

Now calculating wavefunction's derivative,

$$\frac{d\psi_+}{dx} = \frac{a}{k^{1/2}(x)} i \frac{d\varphi}{dx}[-e^{-i\varphi(x)} - e^{i\varphi(x)+i\pi/2}]$$
$$- \frac{1}{2}\frac{a}{k^{3/2}(x)}\frac{dk}{dx}[e^{-i\varphi(x)} - e^{i\varphi(x)+i\pi/2}],$$

we should note that since $d\varphi/dx = -k$, and $|dk/dx| \sim k/a$, where a is the potential change length scale, the first term on the right-hand side is by the factor $\sim ka$ larger than the second one. But the WKB approximation is only strictly valid at $ka \gg 1$, so that the second term is negligible unless the first one vanishes[27]. As a result, after the cancellation of $a/k^{1/2}(0)$, Eq. (**) yields

$$ik[e^{-i\varphi} + e^{i(\varphi+\pi/2)}] = \frac{m}{\hbar^2}\mathscr{w}[e^{-i\varphi} - e^{i(\varphi+\pi/2)}],$$

where

$$\varphi \equiv \varphi(0) = \int_0^{x_c}\left\{\frac{2m}{\hbar^2}[E - U(x)]\right\}^{1/2} dx \geqslant 0,$$
$$\text{and} \quad k \equiv k(0) = \frac{1}{\hbar}(2mE)^{1/2}, \qquad (***)$$

because $k(0)$ should be understood as a limit of $k(x > 0)$ at $x \to 0$, i.e. calculated taking into account only the 'soft' part $U_0(x)$—which, in our case, vanishes at $x = 0$.

The system of the last three relations defines the eigenenergies E of the symmetric modes. Since

$$i\frac{e^{-i\varphi} + e^{i(\varphi+\pi/2)}}{e^{-i\varphi} - e^{i(\varphi+\pi/2)}} = i\frac{e^{-i\,(\varphi+\pi/4)} + e^{i\,(\varphi+\pi/4)}}{e^{-i\,(\varphi+\pi/4)} - e^{i\,(\varphi+\pi/4)}} = -\text{cotan}\left(\varphi + \frac{\pi}{4}\right),$$

the first of them may be rewritten in a simpler form,

$$\text{cotan}\left(\varphi + \frac{\pi}{4}\right) = -\frac{m\mathscr{w}}{\hbar^2 k}. \qquad (****)$$

This is the characteristic equation whose derivation was our first task.

(ii) Since according to Eqs. (***), both φ and k are functions of E, the characteristic equation (****) does not allow an analytical solution for the arbitrary potential $U_0(x)$. For its semi-quantitative analysis, we may note that since the function cotan $(\varphi + \pi/4)$, with $\varphi \geqslant 0$, turns to 0 at points $(m + 1/4)\pi$, with $m = 0, 1, 2,\ldots$, then if $\mathscr{w} = 0$ (no delta-function at origin), Eq. (****) is satisfied at

$$4\varphi \equiv \frac{1}{\hbar}\oint_C p(x)dx = (4m + 1)\pi = 2\pi\left(n + \frac{1}{2}\right),$$
$$\text{with} \quad n = 2m + 1 = 1, 3, 5, \ldots,$$

[27] Besides that, in our current problem the derivatives dU_0/dx, and hence dk/dx, vanish at the point of our interest, $x = 0$, so that the second contribution to $d\psi_+/dx$ vanishes exactly.

thus returning us to the odd-n subset of the Bohr–Sommerfeld series (2.110). In the particular case of quadratic potential, we may rewrite Eq. (2.113) of the lecture notes as

$$\varphi \equiv \frac{1}{2\hbar}\int_{+x_c}^{+x_c} p(x)dx = \frac{\pi}{2}\frac{E}{\hbar\omega_0},$$

so that for energy we get eigenvalues

$$E = \hbar\omega_0\frac{2}{\pi}\varphi = \hbar\omega_0\frac{2}{\pi}\left(m + \frac{1}{4}\right)\pi = \hbar\omega_0\left(2m + \frac{1}{2}\right),$$

i.e. the subset of the series (2.114) with even $n' \equiv n - 1 = 2m = 0, 2, 4,\ldots$

Now, each continuous branch of the function $\cot an(\varphi + \pi/4)$ is descending with the growth of φ, spanning the values from +1 to $-\infty$ in the interval $0 \leqslant \varphi \leqslant \pi - \pi/4$, and from $+\infty$ to $-\infty$ within intervals $(m - 1/4)\pi \leqslant \varphi \leqslant (m + 3/4)\pi$, with $m = 1, 2, \ldots$—see the black lines in figure below. (The red and blue lines in this figure show the right-hand side of Eq. (****) for the particular case of the quadratic potential, for the positive and negative $\mathscr{w}$, respectively, and for several values of the dimensionless parameter $E_0/\hbar\omega_0$, where $E_0 \equiv m\mathscr{w}^2/2\hbar^2$ is the energy scale imposed by the delta-functional potential—see Eqs. (2.79) and (2.162) of the lecture notes. Semi-quantitatively, these plots are valid for any smooth, symmetric potential $U_0(x)$, monotonically growing at $x \to \infty$.)

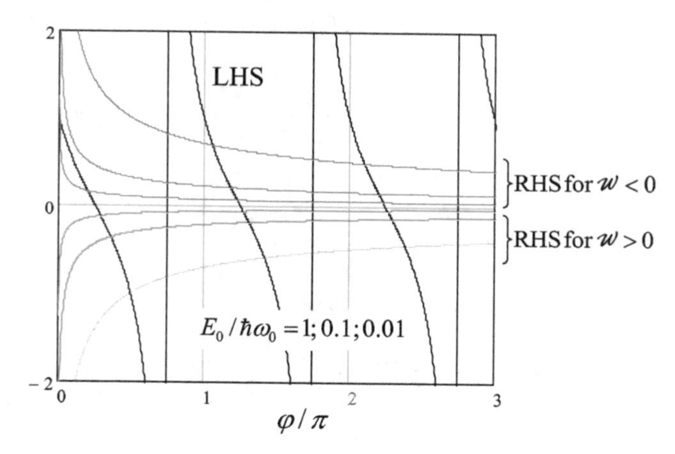

As the plots show, an increase of positive $\mathscr{w}$ (the 'area' of a thin potential barrier) leads to the shift of each eigenvalue toward larger φ and hence larger E, and vice versa, while for $\mathscr{w} < 0$ (corresponding to an additional narrow potential well) each eigenvalue is shifted down. In the former case ($\mathscr{w} > 0$), this trend is unlimited but saturated: at $\mathscr{w} \to +\infty$ we get

$$\varphi \to \left(m + \frac{3}{4}\right)\pi, \qquad \text{for } m = 0, 1, 2, \ldots.$$

For the particular case of the quadratic potential, this relation yields

$$E = \hbar\omega_0 \frac{2}{\pi}\varphi \to \hbar\omega_0 \frac{2}{\pi}\left(m + \frac{3}{4}\right)\pi = \hbar\omega_0\left(2m + \frac{3}{2}\right) = \hbar\omega_0\left(n' + 1 + \frac{1}{2}\right),$$

with even values of $n' = 2m$. This means that due to the barrier, the even-numbered energy levels approach (from below) the odd-numbered levels, with higher n', unaffected by the barrier. (In the figure above, these values correspond to the vertical black lines.)

In the opposite limit of a very deep potential well ($\mathscr{W} \to -\infty$), there is a similar saturation of the phase shifts φ, and hence of the eigenenergies:

$$\varphi \to \left(m - \frac{1}{4}\right)\pi,$$

so that for the quadratic potential

$$E \to \hbar\omega_0 \frac{2}{\pi}\left(m - \frac{1}{4}\right)\pi = \hbar\omega_0\left(2m - \frac{1}{2}\right)$$
$$= \hbar\omega_0\left(n' - 1 + \frac{1}{2}\right), \qquad \text{for } m = 1, 2, 3, \ldots.$$

This expression shows that the even-numbered energy levels approach the odd-numbered levels, with lower n', unaffected by the barrier.

Note, however, that this trend is not valid for the ground state—the symmetric state with $m = 0$. As the figure above shows, the corresponding solution of Eq. (****) exists only for relatively small $|\mathscr{W}|$ (for the quadratic potential, only for the ratio $E_0/\hbar\omega_0$ smaller than ~0.058). However, this particular prediction of the WKB approximation is unreliable, because, as was discussed in section 2.4, its validity condition (2.96) is strictly valid only for $n' \gg 1$. Physically, it is evident that as $(-\mathscr{W})$ grows, the ground state energy should become negative at $E_0/\hbar\omega_0 \sim 1$, and at $E_0/\hbar\omega_0 \gg 1$ (i.e. when the potential $U_0(x)$ has a negligible effect on this state) should approach the value (2.162):

$$E_g \to -E_0 \equiv -\frac{m\mathscr{W}^2}{2\hbar^2}.$$

***Problem* 2.20.** Prove Eq. (2.189) of the lecture notes, starting from Eq. (2.188).

Solution: According to Eqs. (2.94), (2.98), and (2.105) of the lecture notes, within the WKB approximation the localized wavefunction inside one of the wells, say $\psi_L(x)$, may be represented as

$$\psi_L(x) = \begin{cases} \dfrac{2c}{[k(x)]^{1/2}} \sin\left(\displaystyle\int_{x_c}^{x} k(x')dx' + \frac{\pi}{4}\right), & \text{for } x < x_c, \\ \dfrac{c}{[\kappa(x)]^{1/2}} \exp\left\{-\displaystyle\int_{x_c}^{x} \kappa(x')dx'\right\}, & \text{for } x_c < x < 0 < x_c', \end{cases} \qquad (*)$$

where x_c and x_c' are the classical turning points at eigenenergy E_n—see figure 2.21. As we know from the derivation of the WKB formulas, the derivative $d\psi_L/dx$ is dominated, in this approximation, by the exponent—see, e.g. Eq. (2.90). As a result, in the under-barrier region (where the symmetry point $x = 0$ resides), we may write

$$\left.\frac{d\psi_L}{dx}\right|_{x_c<x<0} \approx -\kappa(x)\frac{c}{[\kappa(x)]^{1/2}} \exp\left\{-\int_{x_c}^{x} \kappa(x')dx'\right\}.$$

As a result, the last form of Eq. (2.188) yields

$$\delta = \frac{\hbar^2}{m} |c|^2 \exp\left\{-2\int_{x_c}^{0} \kappa(x')dx'\right\} = \frac{\hbar^2}{m} |c|^2 \exp\left\{-\int_{x_c}^{x_c'} \kappa(x')dx'\right\},$$

where the last step exploits the barrier's symmetry.

The coefficient $|c|^2$ in this relation should be found from the normalization condition,

$$\int_{\substack{\text{inside}\\ \text{left well}}} |\psi_L|^2\, dx = 1,$$

Since the WKB approximation is strictly valid only when the classical path from x_c to x_c' includes many ($n \gg 1$) de Broglie wavelengths λ, at this calculation we can neglect the wavefunction's penetration into the classically forbidden regions (which are of the order of λ) and limit the integration by the classically accessible segment, where the first form of Eq. (*) is applicable:

$$1 = 4\, |c|^2 \int_{k^2(x)>0} \frac{1}{k(x)} \sin^2\left(\int_{x_c}^{x} k(x')dx' + \frac{\pi}{4}\right) dx'.$$

Since $k(x')$ changes little on each de Broglie wavelength, the sine squared in this integral may be replaced with the average of $\sin^2 kx'$ over the wavelength, i.e. with the factor ½. We may also use Eq. (2.33*b*) for the group velocity to write $k(x') = (m/\hbar)v_{gr}(x')$. As a result, the normalization condition becomes

$$1 = \frac{2\,|c|^2\,\hbar}{m} \int_{k^2(x')>0} \frac{dx'}{v_{gr}(x')} \equiv \frac{2\,|c|^2\,\hbar}{m} \int_{k^2(x')>0} \frac{dx'}{(dx_0/dt)_{x_0=x'}}.$$

But the last integral is just the time of the motion of the wave packet's center x_0 (i.e. of the classical position of the particle) from one wall limiting the well to the other one, i.e. half of its oscillation period at energy E_n, i.e. of the tunneling attempt time

t_a. As a result, the normalization condition yields $|c|^2 = m/\hbar t_a$, and for the energy splitting we finally get Eq. (2.189):

$$\delta = \frac{\hbar}{t_a} \exp\left\{-\int_{x_c}^{x_c'} \kappa(x')dx'\right\}.$$

***Problem* 2.21.** For the problem explored in the beginning of section 2.7 of the lecture notes, i.e. the 1D particle's motion in an infinite Dirac comb potential shown in figure 2.24 of the lecture notes,

$$U(x) = \mathscr{W} \sum_{j=-\infty}^{+\infty} \delta(x - ja), \quad \text{with } \mathscr{W} > 0,$$

(where j takes integer values), write explicit expressions for the eigenfunctions at the very bottom, and at the very top of the lowest energy band. Sketch both functions.

Solution: According to Eq. (2.193*b*) of the lecture notes, at the *bottom* of the lowest energy band (i.e. in the ground state of the system), where $e^{iqa} = 1$, the wavefunction is periodic:

$$\psi(x + a) = \psi(x).$$

Moreover, due to the mirror symmetry of the potential profile $U(x)$ with respect to any point $(ja + a/2)$, the wavefunction also should have the same symmetry, in particular

$$\psi(x) = \psi(a - x).$$

Finally, at each segment $ja < x < (j+1)a$, where $U(x) = 0$, the fundamental solutions of the stationary Schrödinger equation are $\sin kx$ and $\cos kx$. Hence we may make an educated guess that at such a segment the eigenfunction has the following simple form:

$$\psi(x) = \psi_j(x) = C \cos k\left(x - ja - \frac{a}{2}\right), \quad \text{at } \left|x - ja - \frac{a}{2}\right| < \frac{a}{2}, \qquad (*)$$

(see figure below), so that in this interval

$$\frac{d\psi_j}{dx} = -Ck \sin k\left(x - ja - \frac{a}{2}\right). \qquad (**)$$

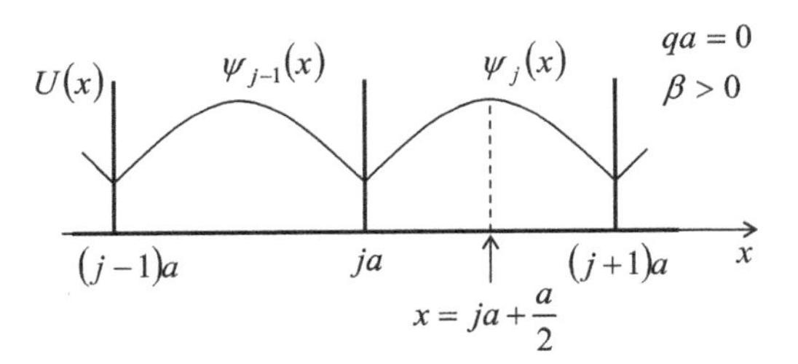

In order to confirm this solution, we may calculate the wave number k following from Eq. (*) and the boundary condition (2.75) at any point $x = ja$.[28] Plugging into this relation the expressions following from Eqs. (*) and (**),

$$\psi(ja) = C\cos\frac{ka}{2}, \quad \frac{d\psi_+}{dx}(ja) \equiv \frac{d\psi_j}{dx}(ja) = Ck\sin\frac{ka}{2},$$

$$\frac{d\psi_-}{dx}(ja) \equiv \frac{d\psi_{j-1}}{dx}(ja) = -Ck\sin\frac{ka}{2},$$

and dividing both parts of the resulting equation by $2Ck$, we get

$$\sin\frac{ka}{2} = \frac{m\mathscr{w}}{\hbar^2 k}\cos\frac{ka}{2}, \quad \text{i.e.} \sin\frac{ka}{2} = \frac{\beta}{ka}\cos\frac{ka}{2}, \qquad (***)$$

where β is the (only) dimensionless parameter of the problem, given by Eq. (2.197) of the lecture notes:

$$\beta \equiv \frac{m\mathscr{w}a}{\hbar^2}.$$

Multiplying both parts of Eq. (***) by $2\sin(ka/2)$, and then using the trigonometric identities[29] $2\sin^2\xi = 1 - \cos 2\xi$ and $2\sin\xi\cos\xi = \sin 2\xi$, we may rewrite it in the form

$$1 - \cos ka = \frac{\beta}{ka}\sin ka.$$

But this is exactly the result given by the general characteristic equation (2.198) of the system:

$$\cos qa = \cos ka + \frac{\beta}{ka}\sin ka, \qquad (****)$$

for our particular value of the quasi-momentum: $qa = 0$.

According to the same Eq. (2.193*b*), the wavefunction corresponding to the *top* of the lowest energy band, i.e. to $e^{iqa} = -1$, changes its sign each lattice period:

$$\psi(x + a) = -\psi(x).$$

Besides that, as the dispersion relation (****) shows (see its plot in figure 2.25 of the lecture notes), at the top points of the lowest band, $\cos qa = -1$ regardless of the parameter β. This is only possible if the wavefunction does not interact with delta-functional potential peaks, i.e. $\psi(ja) = 0$. The only linear combination of $\sin ka$ and

[28] Eq. (*) automatically satisfies the second boundary condition, given by Eq. (2.76).

[29] See, e.g. Eqs. (A.18*d*).

cos ka, satisfying these conditions is the pure sine function, with its nodes at points $x = ja$:

$$\psi(x) = C \sin kx, \quad \text{with } ka = \pi n, \quad \text{for } n = 1, 2, \ldots.$$

(At the lowest energy band, $n = 1$, i.e. $ka = \pi$.) This function, shown in the figure below, with $\psi(ja) = 0$ and

$$\frac{d\psi_+}{dx}(ja) = \frac{d\psi_-}{dx}(ja),$$

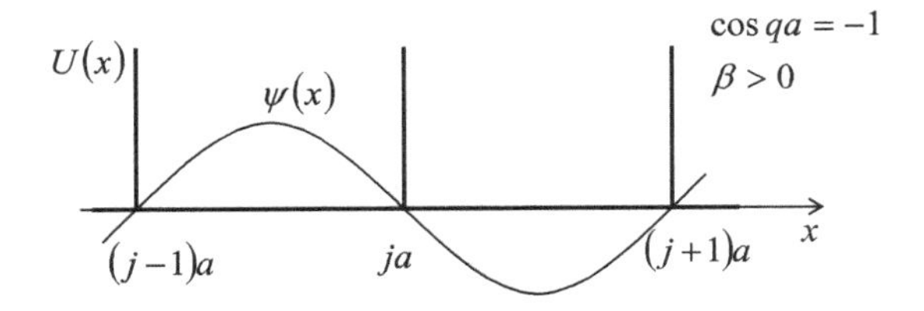

satisfies both boundary conditions (2.75) and (2.76), for any value of the parameter $\mathscr{W}$.

***Problem* 2.22.** A 1D particle of mass m moves in an infinite periodic system of very narrow and deep potential wells that may be described by delta-functions:

$$U(x) = \mathscr{W} \sum_{j=-\infty}^{+\infty} \delta(x - ja), \quad \text{with } \mathscr{W} < 0.$$

(i) Sketch the energy band structure of the system for very small and very large values of the potential well's 'area' $|\mathscr{W}|$, and
(ii) calculate explicitly the ground state energy of the system in these two limits.

Solutions:

(i) This system is similar to the one analyzed in the beginning of section 2.7 of the lecture notes (see figure 2.24 and its discussion), but with the negative sign of $\mathscr{W}$, and hence of the parameter $\beta \equiv ma\mathscr{W}/\hbar^2$—see Eq. (2.197). As a result, its characteristic equation has the same form (2.198),

$$\cos qa = \cos ka + \beta \frac{\sin ka}{ka}, \qquad (*)$$

but now should be analyzed for the case $\beta < 0$. For a comparison of these two cases, the left panel of the figure below shows the plots of the right-hand side of Eq. (*) for two representative values of $|\beta|$, each for two opposite signs of this parameter.

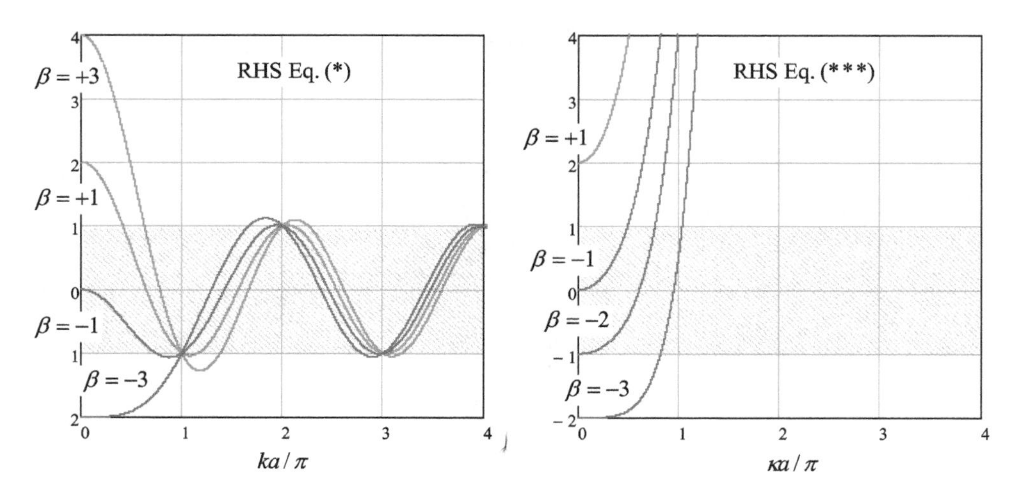

For $\beta > 0$, i.e. for the periodic system of positive delta-functional potential peaks (the 'Dirac comb'), these plots (similar to those shown in figure 2.25 of the lecture notes, but with the parameter β rather than $\alpha \equiv \beta/(ka)$ considered fixed) give the picture of the energy bands (at $-1 < \cos qa < +1$) and gaps, that was discussed in detail in section 2.7, with all energies $E_n(q) > 0$—see the first and the third top panels of the figure below[30]. However, for our current case $\beta < 0$, only the higher energy bands are (qualitatively) similar—see the second and the fourth top panels in the figure below, while the lowest energy band is either completely absent (for $\beta < -2$), or hits the horizontal axis ($E = 0$) at a certain value of the quasi-momentum q.

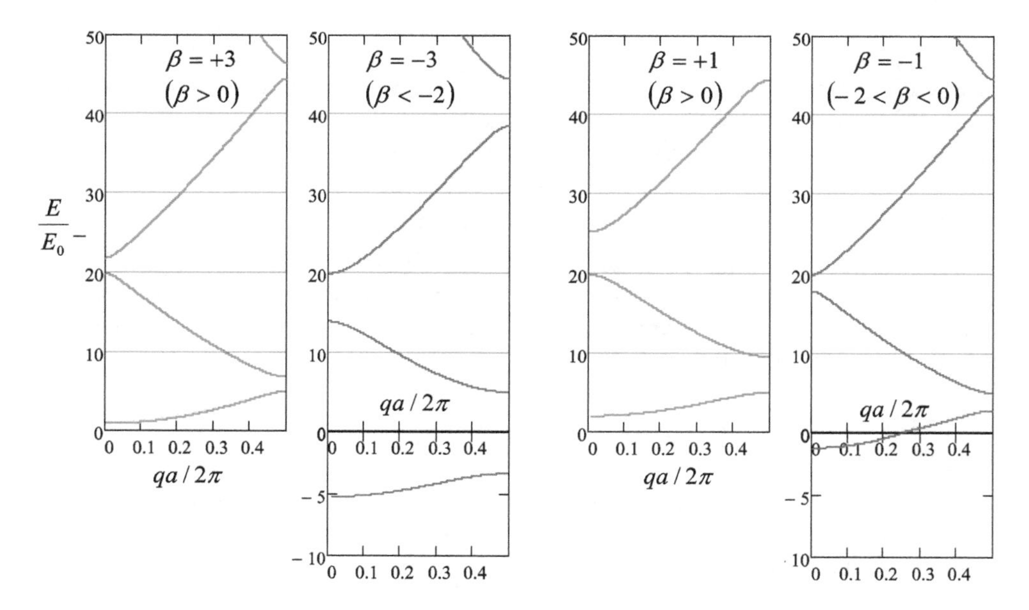

[30] Just to save space, these plots are limited to one half of the first Brillouin zone. In these plots, just as in figure 2.26 of the lecture notes, $E_0 \equiv \hbar^2/2ma^2$ is the natural energy scale of this problem.

The explanation for this behavior is straightforward. In contrast with the case $\mathscr{W} > 0$, when $U(x) \geqslant 0$ at any x, so that the total (potential plus kinetic) energy cannot be negative, in the case $\mathscr{W} < 0$, the potential energy $U(x) \leqslant 0$ at all points, so that $E_n(q)$ can be negative for some β, n, and q. According to the definition of the parameter k (see Eq. (2.54) of the lecture notes),

$$k^2 \equiv \frac{2mE}{\hbar^2},$$

in order to calculate the dispersion curve branches with $E_n(q) < 0$ we have to take $k = \pm i\kappa$, with

$$\kappa^2 \equiv -\frac{2mE}{\hbar^2} \equiv \frac{2m\,|E|}{\hbar^2}. \qquad (**)$$

With this substitution, Eq. (*) takes the form

$$\cos qa = \cosh \kappa a + \beta\frac{\sinh \kappa a}{\kappa a}. \qquad (***)$$

The right-hand side of this equation is plotted, as a function of κa, on the right panel of the figure above, for several values of β. The plots show that for any positive β, this function is always larger than $+1$, so that the equation does not have any real solutions for the quasi-momentum q. Hence, in agreement with the above argument, the dispersion curves cannot spill into the negative energy region. However, for each $\beta < 0$ there is a (single!) range of q where the right-hand side is in the range from -1 to $+1$, giving either (for $\beta < -2$) the whole lowest branch of the dispersion relation, or (for $-2 < \beta < 0$) just its part—see the plots in the lower two subpanels of the figure above, which have been calculated numerically from Eqs. (**) and (***).

(ii) The lowest point $E_1(0)$ of the lowest band, i.e. the ground state energy of the system, may be found from Eqs. (**) and (***) with q equal to zero (actually, to any multiple of 2π), i.e. $\cos qa = 1$:

$$E_g \equiv E_1(0) = -\frac{\hbar^2\kappa^2}{2m}, \quad \text{with} \quad \cosh \kappa a + \beta\frac{\sinh \kappa a}{\kappa a} = 1, \quad \text{for } \beta \leqslant 0. \qquad (****)$$

Since $\sinh \kappa a > \kappa a$, and $\cosh \kappa a > 1$ for any $\kappa a > 0$, the (only) solution of this characteristic equation is real, and hence $E_g < 0$, for any $\beta < 0$. In particular, if $\mathscr{W} \to 0$, i.e. $\beta \to 0$, then $\kappa a \to 0$, and we may use the Taylor expansions $\sinh \kappa a \approx \kappa a$ and $\cosh \kappa a \approx 1 + (\kappa a)^2/2$ to find:

$$\frac{(\kappa a)^2}{2} \approx -\beta, \quad E_g \approx -2\beta E_0 \equiv -\frac{\mathscr{W}}{a}.$$

Note that this is just the average potential energy, $\overline{U} = -\mathscr{W}/a$, of the system. As Eq. (*) with $k = 0$ shows, in this limit the lowest energy band spills to the negative-energy region only at very small values of the quasi-momentum, $qa < (2|\beta|)^{1/2} \ll 1$.

On the other hand, in the opposite limit $\beta \to -\infty$, both hyperbolic functions of κa may be approximated with $\exp\{\kappa a/2\}/2 \gg 1$, and the unity on the right-hand side of the characteristic equation (****) is negligible. This approximation yields

$$\kappa a \approx -\beta, \quad E_g \approx -\beta^2 E_0 \equiv -\frac{\mathscr{w}^2 m}{2\hbar^2}.$$

But this is exactly the energy of the (only) localized eigenstate of a single well—see Eq. (2.162) of the lecture notes. This is natural, because the limit $\beta \to -\infty$ corresponds to a system of deep, virtually uncoupled potential wells. (As Eq. (***) shows, the lowest allowed energy band is exponentially narrow in this limit.)

***Problem* 2.23.** For the system discussed in the previous problem, write explicit expressions for the eigenfunctions of the system, corresponding to:

(i) the bottom of the lowest energy band,
(ii) the top of that band, and
(iii) the bottom of each higher energy band.

Sketch these functions.

Solutions:

(i) As the solution of the previous problem has shown, the wave number k, corresponding to the ground state energy of the system, is imaginary, $k = \pm i\kappa$, for any $\mathscr{w} < 0$, and hence the wavefunction at any interval $ja < x < (j+1)a$ should be a linear combination of sinh κa and cosh κa. Next, according to the Bloch theorem (2.193), at the bottom points of the lowest energy band (i.e. in the ground state of the system), where $e^{iqa} = 1$, the wavefunction has to be periodic:

$$\psi(x + a) = \psi(x).$$

Moreover, due to the mirror symmetry of the potential profile $U(x)$ with respect to any point $(ja + a/2)$, the wavefunction also should have the same symmetry, in particular

$$\psi(x) = \psi(a - x).$$

Hence we may conjecture that the eigenfunction has the following form:

$$\psi(x) = \psi_j(x) = C \cosh \kappa\left(x - ja - \frac{a}{2}\right), \quad \text{at } ja < x < (j+1)a, \qquad (*)$$

(see figure below), so that

$$\frac{d\psi_j}{dx}(x) = C\kappa \sinh \kappa\left(x - ja - \frac{a}{2}\right), \quad \frac{d\psi_j}{dx}(ja) = -\frac{d\psi_{j-1}}{dx}(ja) = -C\kappa \sinh \frac{\kappa a}{2},$$

$$\psi(ja) = C \cosh \frac{\kappa a}{2}.$$

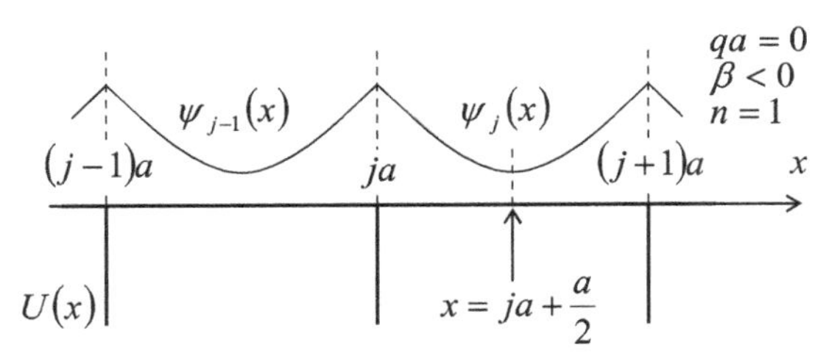

Plugging these expressions into the only essential[31] boundary condition (2.75),

$$\frac{d\psi_j}{dx} - \frac{d\psi_{j-1}}{dx} = \frac{2m}{\hbar^2}\mathscr{w}\psi_j, \quad \text{at } x = ja,$$

with an arbitrary integer j, after the division of both parts by $2C\kappa$, we get the characteristic equation

$$-\sinh\frac{\kappa a}{2} = \frac{m\mathscr{w}}{\hbar^2 k}\cosh\frac{\kappa a}{2}, \quad \text{i.e.} \quad -\sinh\frac{\kappa a}{2} = \frac{\beta}{\kappa a}\cosh\frac{\kappa a}{2}, \qquad (**)$$

where β is the dimensionless parameter of the system, defined by Eq. (2.197) of the lecture notes:

$$\beta \equiv \frac{m\mathscr{w}a}{\hbar^2}.$$

Now multiplying both parts of Eq. (**) by $2\sinh(\kappa a/2)$, and then using the identities[32] $2\sinh^2\xi = \cosh(2\xi) - 1$ and $2\sinh\xi\cosh\xi = \sinh(2\xi)$, we may rewrite this characteristic equation in the form

$$1 = \cosh\kappa a + \frac{\beta}{\kappa a}\sinh\kappa a.$$

But this is exactly the result given by the general characteristic equation of the system, obtained in the solution of the previous problem,

$$\cos qa = \cosh\kappa a + \frac{\beta}{ka}\sinh\kappa a, \quad \text{for } \beta < 0,\ E < 0, \qquad (***)$$

for our particular case $\cos qa = +1$. This agreement confirms our conjecture (*).

(ii) At the top points of the lowest energy band, $\exp\{iqa\} = -1$, so that, according to the Bloch theorem, the eigenfunctions at each period of the system are similar, but with alternating sign:

$$\psi(x + a) = -\psi(x).$$

Also, the eigenfunctions should be, at all points $x \neq ja$, the solutions of the Schrödinger equation with $U(x) = 0$, i.e. have a form of either $\{\sin kx, \cos kx\}$ at

[31] Indeed, our solution (*), by construction, satisfies the boundary condition (2.76): $\psi_{j+1}(x) - \psi_j(x) = 0$ at $x = ja$.
[32] They may be readily proved using either the definition of hyperbolic functions or Eqs. (A.18*d*) and (A.20).

$E > 0$, or $\{\sinh \kappa x, \cosh \kappa x\}$ at $E < 0$. (As was discussed in the previous problem, at the top of the band (i.e. at $\cos qa = -1$), the sign of the eigenenergy E depends on whether the parameter β is larger or smaller than -2.)

Moreover, as the general characteristic equations for $E < 0$ and $E > 0$ show (see the plots of their right-hand sides in the solution of the previous problem), at $\beta < 0$ and at the top points of the lower band, ka is not equal to qa (as it is at for $\beta > 0$), so that the states do 'interact' with the delta-functional potential wells located at $x = ja$, i.e. their wavefunctions cannot be equal to zero at these points. As a result, we may conjecture that the wavefunction has one of the following forms (see their sketches in the figure below):

$$\psi(x) = \psi_j(x) = (-1)^j C \times \begin{cases} \sin k\left(x - ja - \dfrac{a}{2}\right), & \text{for } \beta > -2, \\ \sinh \kappa\left(x - ja - \dfrac{a}{2}\right), & \text{for } \beta < -2, \end{cases} \qquad (****)$$

$$\frac{d\psi_j}{dx}(x) = (-1)^j C \begin{cases} k \cos k\left(x - ja - \dfrac{a}{2}\right), & \text{for } \beta > -2, \\ \kappa \cosh \kappa\left(x - ja - \dfrac{a}{2}\right), & \text{for } \beta < -2. \end{cases} \quad \text{at } ja < x < (j+1)a.$$

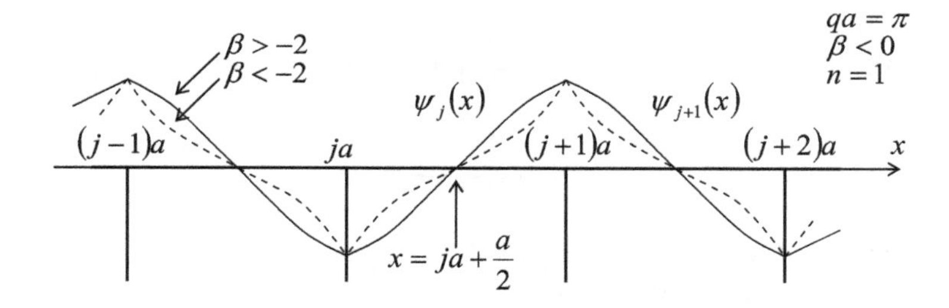

We may readily verify this picture by using the same boundary condition as in part (i),

$$\frac{d\psi_j}{dx} - \frac{d\psi_{j-1}}{dx} = \frac{2m}{\hbar^2} \mathcal{w} \psi_j, \quad \text{at } x = ja,$$

with an arbitrary integer j. For our wavefunctions (****) this condition yields, respectively,

$$(-1)^j Ck \cos\frac{ka}{2} - (-1)^{j-1} Ck \cos\frac{ka}{2} = -\frac{2m}{\hbar^2} \mathcal{w} (-1)^j C \sin\frac{ka}{2}, \quad \text{for } \beta > -2,$$

$$(-1)^j C\kappa \cosh\frac{\kappa a}{2} - (-1)^{j-1} C\kappa \cosh\frac{\kappa a}{2} = -\frac{2m}{\hbar^2} \mathcal{w} (-1)^j C \sinh\frac{\kappa a}{2}, \quad \text{for } \beta < -2.$$

After the division of all terms by $2(-1)^j Ck$, and using the definition of parameter β, these equations are reduced to

$$\cos\frac{ka}{2} = -\frac{\beta}{ka}\sin\frac{ka}{2}, \quad \text{for } \beta > -2,$$
$$\cosh\frac{\kappa a}{2} = -\frac{\beta}{\kappa a}\sinh\frac{\kappa a}{2}, \quad \text{for } \beta < -2.$$

Multiplying these characteristic equations, respectively, by $2\cos(ka/2)$ and $2\cosh(\kappa a/2)$, and using the well-known identities (see the footnote above)

$$2\cos^2\xi = \cos(2\xi) + 1, \quad 2\sin\xi\cos\xi = 2\sin(2\xi);$$
$$2\cosh^2\xi = \cosh(2\xi) + 1, \quad 2\sinh\xi\cosh\xi = 2\sinh(2\xi),$$

we may recast the equations in the form

$$-1 = \cos ka + \beta\frac{\sin ka}{ka}, \quad \text{for } \beta > -2,$$
$$-1 = \cosh \kappa a + \beta\frac{\sin \kappa a}{\kappa a}, \quad \text{for } \beta > -2.$$

But these are exactly the general characteristic equations, derived in the solution of the previous problem, taken for our current case: $\cos qa = -1$. Hence Eqs. (****) indeed give the required eigenfunctions.

(iii) A little bit counter-intuitively, the wavefunctions corresponding to the bottom points of the higher energy bands (with number $n > 1$) differ substantially from those given by both Eqs. (*) and (****), valid for $n = 1$ only. Indeed, as was discussed in the model solution of the previous problem, at those points $E > 0$, so that we have to look at the dispersion relation (2.198), derived in section 2.7 of the lecture notes for this case:

$$\cos qa = \cos ka + \beta\frac{\sin ka}{ka},$$

but with $\beta < 0$. As the blue-line plots of the right-hand side of this characteristic equation in the figure below[33] show, the allowed energy bottoms for $n > 1$ correspond to $\sin ka = 0$ (but $ka \neq 0$) independently of the parameter β. This is only possible if the eigenstate does not interact with delta-functional potential peaks, i.e. if $\psi(ja) = 0$. The only linear combination of $\sin ka$ and $\cos ka$, satisfying this condition is the pure sine function, with its nodes at points $x = ja$:

$$\psi(x) = C\sin kx; \quad \text{with } ka = \pi(n-1), \quad \text{for } n = 2, 3, \ldots$$

[33] These plots were already discussed in the model solution of the previous problem, and are reproduced here just for the reader's convenience.

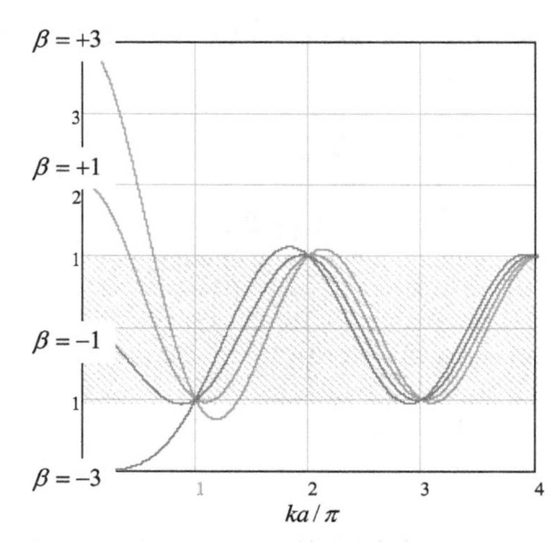

(For example, at the lowest of such energy bands, with $n = 2$, this relation yields $ka = \pi$, giving the wavefunction sketched in the figure below.) Such functions, with $\psi(ja) = 0$ and

$$\frac{d\psi_+}{dx}(ja) = \frac{d\psi_+}{dx}(ja),$$

automatically satisfy both boundary conditions (2.75) and (2.176), for any value of parameter β.[34]

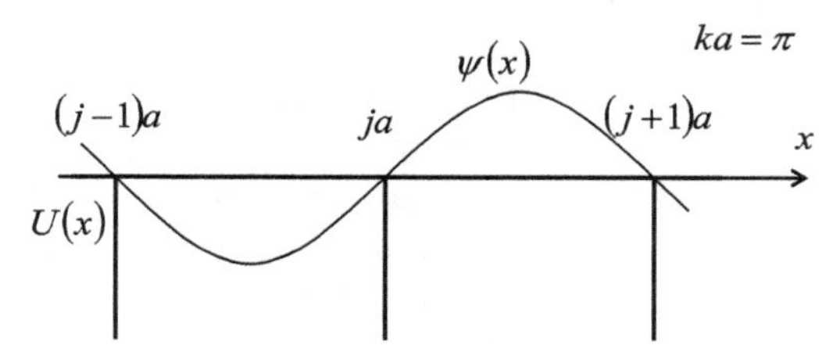

Problem* 2.24. The 1D 'crystal', analyzed in the last two problems, now extends only to $x > 0$, with a sharp potential step to a flat potential plateau at $x < 0$:

$$U(x) = \begin{cases} \mathcal{w}\sum_{j=1}^{+\infty}\delta(x - ja), & \text{with } \mathcal{w} < 0, \quad \text{for } x > 0, \\ U_0 > 0, & \text{for } x < 0. \end{cases}$$

[34] As the red lines in the figure above show, in a similar system, but with $\beta > 0$ (which was studied in section 2.7 of the lecture notes), such simple solutions, with $ka = n\pi$, are implemented at the top, rather than the bottom points of each energy band.

Prove that the system has a set of the so-called *Tamm states*, localized near the 'surface' $x = 0$, and calculate their energies in the limit when U_0 is very large but finite (quantify this condition[35]).

Solution: Let us start with a qualitative picture. A localized wavefunction should be unable to propagate to either $x \to -\infty$ or $x \to +\infty$. This means that the corresponding eigenenergy E should be, first, lower than U_0, and also located inside one of the energy gaps of an infinite 'crystal' with the same parameters as our semi-infinite one—see the figure below[36].

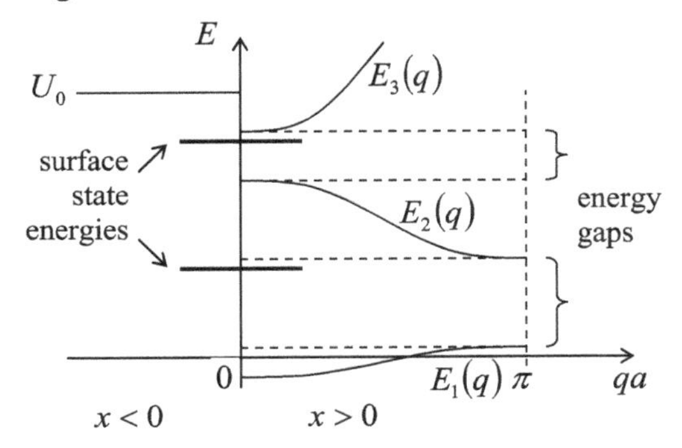

For a quantitative analysis of the Tamm states, let us notice that in the limit $U_0 = \infty$, the simple bottom-of-a-band states discussed in task (iii) of the previous problem (see the dashed line in the figure below, drawn here for the particular case of the second energy band) are not affected by the crystal termination at $x = 0$. Indeed, their wavefunctions vanish at $x = 0$, and hence their parts located at $x \geqslant 0$ are exactly the same as in the similar, but infinite crystal.

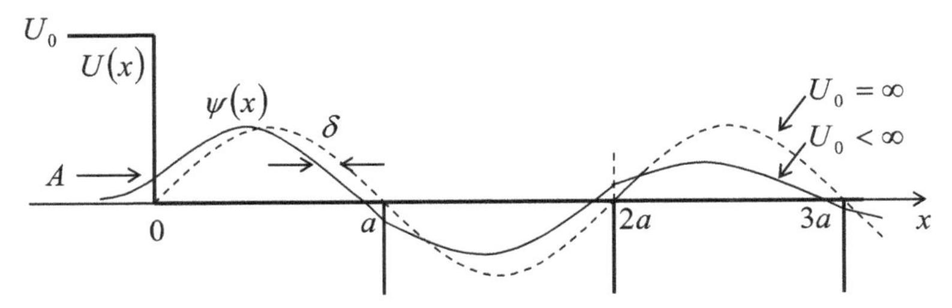

Now, if U_0 is large but finite, as we know from section 2.3 of the lecture notes (see in particular Eq. (2.58) and figure 2.4), the wavefunction penetrates by a small

[35] In applications to electrons in solid-state crystals, the delta-functional potential wells model the attractive potentials of the atomic nuclei, while U_0 represents the workfunction, i.e. the energy necessary for the extraction of an electron from the crystal to the free space—see, e.g. section 1.1(ii) of the lecture notes, and also *Part EM* section 2.6 and *Part SM* section 6.3.

[36] This picture uses a somewhat strange, but very common (and hopefully, self-explanatory) format, displaying the system's energy not only as a function of the quasi-momentum q (for the bulk states), but also (very crudely) as a function of the wavefunction's location in space.

distance into the classically forbidden region $x < 0$—see the solid line in the figure above:

$$\psi(x \leqslant 0) \equiv \psi_-(x) \equiv Ae^{\kappa x}, \quad \frac{d\psi_-}{dx} = A\kappa e^{\kappa x}, \quad \text{with} \quad \frac{\hbar^2\kappa^2}{2m} = U_0 - E,$$

so that

$$\psi_-(0) = A, \quad \frac{d\psi_-}{dx}(0) = A\kappa. \qquad (*)$$

Since the function and its first derivative should stay continuous at $x = 0$, this shift to the left 'pulls' to the left, by a comparable distance $\delta \sim 1/\kappa$, the wavefunction in the allowed regions $x > 0$ as well. This shift leads to some interaction of the wavefunction, now not vanishing at $x = ja$, with the delta-functional potential wells at these points, creating, according to Eq. (2.75) of the lecture notes, 'cusps' (derivative jumps) of the initially smooth wavefunction. These cusps, in turn, result in the decrease of the sinusoidal wavefunction's amplitude, by some factor $0 < \lambda < 1$ (see the solid-line sketch in the figure above), eventually resulting in its full decay at $x \to \infty$, i.e. to the state's localization near the 'surface'.

These handwaving arguments allow us to look for the wavefunction at $x \geqslant 0$ in the following form:

$$\psi(x) = \psi_j(x) = C(-1)^j\lambda^j \sin k(x - ja + \delta), \quad \text{for } ja < x < (j+1)a,$$

so that

$$\psi_j(ja) = C(-1)^j\lambda^j \sin k\delta, \quad \psi_{j-1}(ja) = C(-1)^{j-1}\lambda^{j-1} \sin k(a + \delta),$$

$$\frac{d\psi_j}{dx}(x) = C(-1)^j\lambda^j k \cos k(x - ja + \delta),$$

$$\frac{d\psi_j}{dx}(ja) = C(-1)^j\lambda^j k \cos k\delta, \quad \frac{d\psi_{j-1}}{dx}(ja) = C(-1)^{j-1}\lambda^{j-1} k \cos k(a + \delta),$$

This wavefunction, with the wave number k simply related to the state's energy E,

$$\frac{\hbar^2k^2}{2m} = E,$$

is certainly an exact solution of the Schrödinger equation between the delta-functional wells, so that we need only to satisfy all boundary conditions at these special points $x = ja$ with $j \geqslant 0$. With the account of Eqs. (*), the boundary conditions at $x = 0$,

$$\psi_0(0) - \psi_-(0) = 0, \quad \frac{d\psi_0}{dx}(0) - \frac{d\psi_-}{dx}(0) = 0,$$

yield, after the exclusion of the C/A ratio, one equation for three so-far unknown parameters δ, λ, and k,

$$k \cos k\delta = \kappa \sin k\delta. \qquad (**)$$

Two other equations for these parameters may be found from the boundary conditions (2.75)–(2.76), written for any point $x = ja$, with an integer $j > 0$:

$$\psi_j(ja) - \psi_{j-1}(ja) = 0, \quad \frac{d\psi_j}{dx}(ja) - \frac{d\psi_{j-1}}{dx}(ja) = \frac{2m\mathscr{w}}{\hbar^2}\psi_j(ja).$$

After the substitution of the assumed form of ψ_j, and the cancellation of the common factors $C(-1)^j\lambda^{j-1}$, these two relations become

$$\lambda \sin k\delta + \sin k(a+\delta) = 0, \quad \lambda k \cos k\delta + k \cos k(a+\delta) = \frac{2m\mathscr{w}}{\hbar^2}\lambda \sin k\delta.$$

From these two equations, λ may be readily eliminated, reducing them to just one equation,

$$\sin ka = \frac{2\beta}{ka} \sin k(a+\delta) \sin k\delta, \qquad (***)$$

where β is the dimensionless parameter of the 'crystal', defined by Eq. (2.197) of the lecture notes:

$$\beta \equiv \frac{m\mathscr{w}a}{\hbar^2}.$$

(In our current problem, $\mathscr{w} < 0$, and hence $\beta < 0$, though the Tamm states may also exist at $\mathscr{w} > 0$.) The fact that our wavefunction assumption has led to two j-independent characteristic equations (**) and (***) for two unknown parameters δ and k (assuming that κ is known[37]) proves that this assumption is indeed valid—for arbitrary U_0 and for any energy band number n.

Proceeding to the analysis of these equations, let us notice that Eq. (**), rewritten as

$$\tan k\delta = \frac{k}{\kappa} \equiv \left(\frac{E}{U_0 - E}\right)^{1/2},$$

shows that for any energy in the range of our interest, $0 \leqslant E \leqslant U_0$, the product $k\delta$ is a monotonic function of E, and is confined to the interval $[0, +\pi/2]$. The figure below shows the plots of the left-hand and right-hand sides of Eq. (***) as functions of the product ka, for several values of $k\delta$ from the interval $[0, +\pi/2]$, and a modest negative value of β. (The variation of this parameter does not change topological properties of the equation's solutions.) The plots show that the equation has just one solution for ka somewhat below each value $\pi(n-1)$, corresponding to the bottom of the nth energy band—see the solution of the previous problem. This means that the system

[37] An explicit relation between $k \equiv (2mE)^{1/2}/\hbar$ and $\kappa \equiv [2m(U_0 - E)]^{1/2}/\hbar$ allows finding the state's energy E, and hence all other characteristics of the system.

has just one Tamm state inside each energy gap—see the energy scheme in the beginning of this solution.

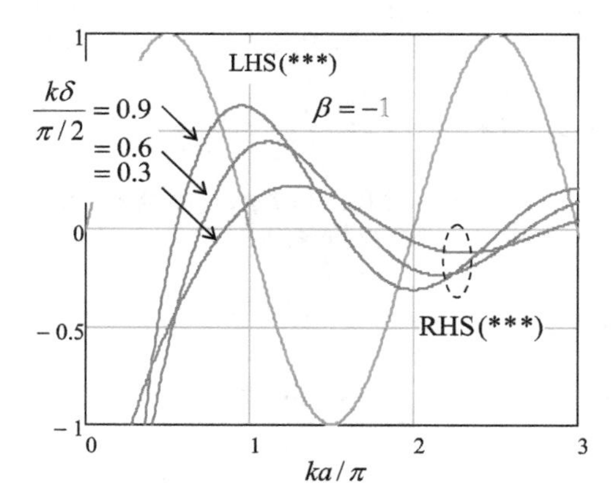

The plots also show the way to solve this equation analytically when $U_0 \gg E$, so that $k\delta \to 0$, and as a result the product ka is only slightly below $\pi(n - 1)$, so that the Tamm state energy is right below the nth energy band's bottom[38],

$$(E_n)_{\min} = \frac{\pi^2\hbar^2}{2ma^2}(n - 1)^2.$$

In this limit, we may take $ka = \pi[(n - 1) - \eta]$, and expand both parts of Eq. (***) in the Taylor series in small parameters $k\delta$ and η, dropping all the terms but the leading ones. Such expansion, after the cancellation of the common multiplier $\cos\pi(n - 1)$, reduces Eq. (***) to

$$\eta \approx -\frac{2\beta}{\pi^2(n - 1)}(k\delta)^2 \approx -\frac{2\beta}{\pi^2(n - 1)}\frac{(E_n)_{\min}}{U_0} \equiv \frac{(n - 1)\,|\mathscr{W}|}{aU_0},$$

so that the distance between the Tamm level and the bottom of the nth energy band is

$$\Delta E \equiv (E_n)_{\min} - E = \frac{\hbar^2}{2ma^2}\{\pi^2(n - 1)^2 - \pi^2[(n - 1) - \eta]^2\}$$
$$\approx \frac{\hbar^2}{ma^2}\eta\pi^2(n - 1) \approx \pi^2(n - 1)^2\frac{\hbar^2}{ma^2}\frac{|\mathscr{W}|}{aU_0}.$$

Note also that in this limit the wavefunction decay parameter λ is very close to 1:

$$\lambda = -\frac{\sin k(a + \delta)}{\sin k\delta} \approx \frac{-\pi\eta + k\delta}{k\delta} = 1 - \frac{\pi\eta}{k\delta}$$
$$\approx 1 - \frac{2\,|\beta|}{\pi(n - 1)}k\delta \approx 1 - \frac{2\,|\beta|}{\pi(n - 1)}\left(\frac{E}{U_0}\right)^{1/2},$$

[38] In the similar limit, but at $\mathscr{W} > 0$, the Tamm state's energy, inside the same bandgap, is close to the *top* of the previous (nth) energy band.

so that the spatial extension $\Delta x \sim a/(1 - \lambda)$ of the Tamm state into the crystal is much larger than δ and a.

These expressions are strictly valid if the dimensionless parameters $k\delta$ and η are much smaller than 1, i.e. when

$$U_0 \gg \frac{(n-1)\,|\mathscr{W}|}{a},\ (E_n)_{\min}.$$

In conclusion, note that these states, named after I Tamm (who was the first to predict their existence in 1932), is just one type of the more general class of *surface states*[39].

***Problem* 2.25.** Calculate the whole transfer matrix of the rectangular potential barrier specified by Eq. (2.68) of the lecture notes, for particle energies both below and above U_0.

Solution: Acting exactly as in section 2.3 of the lecture notes, but either with the account of an additional wave incident from the right, or using the universal relations discussed in section 2.5, we get the following transfer matrices

$$\mathrm{T}_d = \begin{pmatrix} \cosh\kappa d + \dfrac{i}{2}\left(\dfrac{k}{\kappa} - \dfrac{\kappa}{k}\right)\sinh\kappa d & -\dfrac{i}{2}\left(\dfrac{k}{\kappa} + \dfrac{\kappa}{k}\right)\sinh\kappa d \\ \dfrac{i}{2}\left(\dfrac{k}{\kappa} + \dfrac{\kappa}{k}\right)\sinh\kappa d & \cosh\kappa d - \dfrac{i}{2}\left(\dfrac{k}{\kappa} - \dfrac{\kappa}{k}\right)\sinh\kappa d \end{pmatrix}, \quad \text{for } 0 \leqslant E \leqslant U_0,$$

$$\mathrm{T}_d = \begin{pmatrix} \cos k'd + \dfrac{i}{2}\left(\dfrac{k}{k'} + \dfrac{k'}{k}\right)\sin k'd & -\dfrac{i}{2}\left(\dfrac{k}{k'} - \dfrac{k'}{k}\right)\sin k'd \\ \dfrac{i}{2}\left(\dfrac{k}{k'} - \dfrac{k'}{k}\right)\sin k'd & \cos k'd - \dfrac{i}{2}\left(\dfrac{k}{k'} + \dfrac{k'}{k}\right)\sin k'd \end{pmatrix}, \quad \text{for } 0, U_0 \leqslant E.$$

As useful sanity checks, the top left elements (T_{11}) of these expressions are in agreement with Eq. (2.71*b*) of the lecture notes for $\mathscr{T} = |T_{11}|^{-2}$, and the matrices are reduced to Eq. (2.135) for the particular case of a very thin and high barrier ($kd \ll \kappa d \ll 1$). Moreover, it is easy to check that due to the algebraic identity

$$\left[\frac{1}{2}\left(\frac{k}{k'} + \frac{k'}{k}\right)\right]^2 - \left[\frac{1}{2}\left(\frac{k}{k'} - \frac{k'}{k}\right)\right]^2 \equiv 1,$$

the above transfer matrices are in compliance with all general relations for its elements.

[39] For more on surface states, see, e.g. [1]. (Another important member of this class are the *Shockley states*, described by a different theoretical model.)

***Problem* 2.26.** Use the results of the previous problem to calculate the transfer matrix of one period of the periodic Kronig–Penney potential shown in figure 2.31b of the lecture notes (reproduced below).

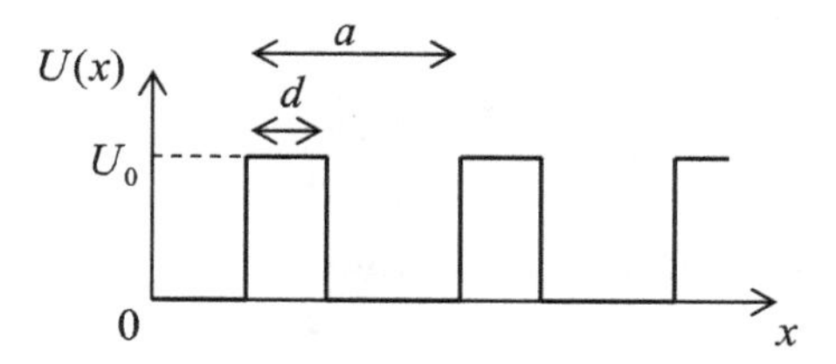

Solution: According to Eq. (2.132) of the lecture notes, in order to calculate the transfer matrix T of one period starting from the potential barrier, it is sufficient to multiply the transfer matrix (2.138) of the free-motion interval of length $(a - d)$,

$$\mathrm{T}_{d-a} = \begin{pmatrix} e^{ik(a-d)} & 0 \\ 0 & e^{-ik(a-d)} \end{pmatrix}.$$

by the matrix T_d calculated in the previous problem. The result, for $E < U_0$, is

$$\mathrm{T} = \begin{pmatrix} \left[\cosh \kappa d + \dfrac{i}{2}\left(\dfrac{k}{\kappa} - \dfrac{\kappa}{k}\right)\sinh \kappa d\right] e^{ik(a-d)} & -\dfrac{i}{2}\left(\dfrac{k}{\kappa} + \dfrac{\kappa}{k}\right)\sinh \kappa d \, e^{-ik(a-d)} \\ \dfrac{i}{2}\left(\dfrac{k}{\kappa} + \dfrac{\kappa}{k}\right)\sinh \kappa d \, e^{ik(a-d)} & \left[\cosh \kappa d - \dfrac{i}{2}\left(\dfrac{k}{\kappa} - \dfrac{\kappa}{k}\right)\sinh \kappa d\right] \\ & \times e^{-ik(a-d)} \end{pmatrix};$$

for $E > U_0$, it is sufficient to make the replacement (2.73): $\kappa \to -ik'$.

At the possible alternative choice of the starting and ending points of the period (the free-motion interval first, and the barrier next), the exponents in T_{12} and T_{21} would be complex-conjugated, with no effect on any observable result.

***Problem* 2.27.** Using the results of the previous problem, derive the characteristic equations for a particle's motion in the periodic Kronig–Penney potential, for both $E < U_0$ and $E > U_0$. Try to bring the equations to a form similar to that obtained in section 2.7 of the lecture notes for the delta-functional barriers—see Eq. (2.198). Use the equations to formulate the conditions of applicability of the tight-binding and weak-potential approximations, in terms of the system's parameters, and the particle's energy E.

Solution: Requiring the difference of matrix T calculated in the previous problem, and the Bloch matrix

$$\begin{pmatrix} e^{iqa} & 0 \\ 0 & e^{iqa} \end{pmatrix},$$

to have zero determinant (cf. Eq. (2.196) of the lecture notes), we get the following characteristic equation

$$\cos qa = \cosh \kappa d \ \cos k(a-d) + \frac{1}{2}\left(\frac{\kappa}{k} - \frac{k}{\kappa}\right)\sinh \kappa d \ \sin k(a-d). \qquad (*)$$

Following the analysis of the periodic system of delta-functional barriers (see figure 2.25 and its discussion), we may consider the right-hand side of this equation as a sinusoidal function of ka, and rewrite Eq. (*) in the following equivalent form:

$$\cos qa = A \cos[k(a-d) + \varphi],$$

where φ is independent of a (and unimportant for our purposes), while

$$A^2 = \cosh^2 \kappa d + \left[\frac{1}{2}\left(\frac{k}{\kappa} - \frac{\kappa}{k}\right)\right]^2 \sinh^2 \kappa d \approx 1 + \left[\frac{1}{2}\left(\frac{k}{\kappa} + \frac{\kappa}{k}\right)\sinh \kappa d\right]^2,$$

so that the characteristic diagram is topologically similar to that shown in figure 2.25 of the lecture notes—see the figure below.

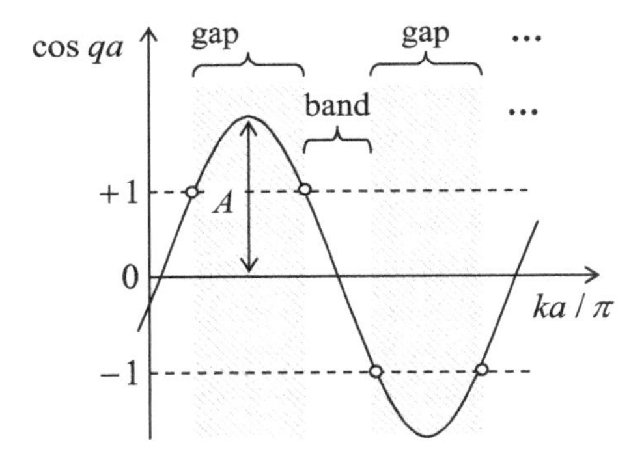

The *tight-binding* approximation is applicable when an allowed energy band is much narrower than the adjacent energy gaps. As the figure above shows, this condition may be represented as $A \gg 1$, giving

$$\frac{1}{2}\left(\frac{k}{\kappa} + \frac{\kappa}{k}\right)\sinh \kappa d \gg 1, \qquad (**)$$

or, in the dimensional units, for $E < U_0$,

$$\frac{U_0}{2[E(U_0 - E)]^{1/2}} \sinh \frac{[2m(U_0 - E)]^{1/2} d}{\hbar} \gg 1,$$

where E should be close to one of the eigenvalues $E^{(n)}$ of isolated potential wells—see Eq. (1.85):

$$E \approx E^{(n)} \equiv \frac{\pi^2 \hbar^2}{2m(a-d)^2} n^2.$$

In the opposite case, $U_0 < E$, Eq. (**) takes the form

$$\frac{1}{2}\left(\frac{k}{k'} - \frac{k'}{k}\right)|\sin k'd| \gg 1, \quad \text{i.e.} \ \frac{U_0}{2[E(E - U_0)]^{1/2}}\left|\sin\frac{[2m(E - U_0)]^{1/2}d}{\hbar}\right| \gg 1,$$

Let us analyze these conditions. If $E < U_0$ (κ is real), virtually the only way to satisfy Eq. (**) is to have sufficiently thick barriers, $\kappa d \gg 1$. (The only other option is to have a very low $E \ll U_0$, which requires an extremely large $a - d \gg d$.) In the opposite case, $U_0 < E$, since $\sin k'd$ is always less than 1, the tight-binding approximation is only possible when E_n is almost exactly equal to U_0.

The *weak-potential* approximation requires, on the opposite, the parameter A to be very close to 1—see the figure above again. This requirement may be rewritten as $(A^2 - 1)^{1/2} \ll 1$, and if $E < U_0$, it reads

$$\frac{1}{2}\left(\frac{k}{\kappa} + \frac{\kappa}{k}\right)\sinh \kappa d \ll 1, \quad \text{i.e.} \ \frac{U_0}{2[E(U_0 - E)]^{1/2}}\sinh\frac{[2m(U_0 - E)]^{1/2}d}{\hbar} \ll 1,$$

where now E is close to the branch anticrossing point—see figure 2.28 and its discussion in section 2.7:

$$E \approx E^{(n)} \equiv \frac{\pi^2\hbar^2}{2ma^2}n^2.$$

This condition may be only satisfied for very thin barriers, $\kappa d \ll 1$.

In the opposite case, $U_0 < E$, the weak-potential condition becomes

$$\frac{1}{2}\left(\frac{k}{k'} - \frac{k'}{k}\right)|\sin k'd| \ll 1, \quad \text{i.e.} \ \frac{U_0}{2[E(E - U_0)]^{1/2}}\sin\left|\frac{[2m(E - U_0)]^{1/2}d}{\hbar}\right| \ll 1,$$

and, if $E \gg U_0$, is satisfied even for thick barriers, because the magnitude of $\sin k'd$ can never be larger than one.

To summarize, if the barrier thickness is appreciable ($d \sim a$), the tight-binding approximation typically works well at $E < U_0$, while at $U_0 \ll E$, the weakly-potential limit is typically applicable. Qualitatively, this is exactly the behavior visible from the characteristic curves of the Mathieu equation—see figure 2.32 in the lecture notes.

***Problem* 2.28.** For the Kronig–Penney potential, use the tight-binding approximation to calculate the widths of the allowed energy bands. Compare the results with those of the previous problem (in the corresponding limit).

Solution: According to Eq. (2.206) of the lecture notes, in the tight-binding limit the allowed energy band width ΔE_n equals 4 $|\delta_n|$, where δ_n is given by Eq. (2.204):

$$\delta_n = \frac{\hbar^2}{m}u_n(x_0)\frac{du_n}{dx}(a - x_0), \qquad (*)$$

where u_n are the localized wavefunctions of an isolated potential well. For the Kronig–Penney potential, the wells are rectangular, and their wavefunctions were calculated in the solution of problem 2.17. In that solution, the well width was denoted as a, and should be replaced with $(a - d)$ in our current notation—see figure 2.31b. With this replacement (but still keeping the origin of x in the well's middle), the solution takes the following form:

(i) Asymmetric wavefunctions, valid for odd $n = 1, 3, \ldots$:

$$u_n = C \times \begin{cases} \sin kx, & \text{for } |x| \leqslant \dfrac{a-d}{2}, \\ \text{sgn}(x)\sin\dfrac{k(a-d)}{2}\exp\left\{-\kappa\left(|x| - \dfrac{a-d}{2}\right)\right\}, & \text{for } \dfrac{a-d}{2} \leqslant |x|, \end{cases}$$

with the following relation between k and κ:

$$\tan\frac{k(a-d)}{2} = -\frac{k}{\kappa},$$

$$\text{giving}\quad \sin\frac{k(a-d)}{2} = -\frac{k}{K}, \qquad \cos\frac{k(a-d)}{2} = \frac{\kappa}{K}.$$

(ii) Symmetric wavefunctions, corresponding to even $n = 0, 2, 4, \ldots$:

$$u_n = C \times \begin{cases} \cos kx, & \text{for } |x| \leqslant \dfrac{a-d}{2}, \\ \cos\dfrac{k(a-d)}{2}\exp\left\{-\kappa\left(|x| - \dfrac{a-d}{2}\right)\right\}, & \text{for } \dfrac{a-d}{2} \leqslant |x|, \end{cases}$$

with a different relation between k and κ,

$$\tan\frac{k(a-d)}{2} = \frac{\kappa}{k},$$

which may be rewritten as

$$\sin\frac{k(a-d)}{2} = \frac{\kappa}{(\kappa^2 + k^2)^{1/2}} = \frac{\kappa}{K}, \qquad \cos\frac{k(a-d)}{2} = \frac{k}{K},$$

where $K^2 \equiv k^2 + \kappa^2 = 2mU_0/\hbar^2$.

Now we may calculate δ_n using Eq. (*) with x_0 anywhere under the barrier, i.e. between $(a - d)/2$ and $[a - (a - d)/2] \equiv (a + d)/2$ from the well center, because the result is independent of the choice[40]:

[40] Note that the sign alternation confirms the results of the general discussion of Eq. (2.204) in section 2.7.

$$\delta_n = (-1)^n \frac{\hbar^2 \kappa}{m} |C|^2 e^{-\kappa d} \times \begin{Bmatrix} \sin^2 \dfrac{k(a-d)}{2}, & \text{for } n = 0, 2, \dots, \\ \cos^2 \dfrac{k(a-d)}{2}, & \text{for } n = 1, 3, \dots, \end{Bmatrix}$$

$$= (-1)^n \frac{\hbar^2 \kappa^3}{mK^2} |C|^2 e^{-\kappa d},$$

so that our task is reduced to the calculation of the normalization coefficients C.

For the symmetric modes, the normalization condition gives

$$1 = \int_{-\infty}^{+\infty} u_n u_n^* dx = 2|C|^2 \left[\int_0^{(a-d)/2} \cos^2 kx dx + \cos^2 \frac{k(a-d)}{2} \int_0^{\infty} \exp\{-2\kappa \tilde{x}\} d\tilde{x} \right]$$

$$= 2|C|^2 \left[\frac{(a-d)}{4} + \frac{\sin k(a-d)}{4k} + \cos^2 \frac{k(a-d)}{2} \frac{1}{2\kappa} \right] = |C|^2 \left[\frac{(a-d)}{2} + \frac{1}{\kappa} \right];$$

a similar calculation for the asymmetric mode gives exactly the same result (though of course one should not forget that the values of k and κ are specific for each n), so that, finally,

$$\Delta E_n = \frac{4\hbar^2 \kappa^2}{mK^2} \left[\frac{\kappa(a-d)}{2} + 1 \right]^{-1} e^{-\kappa d}. \qquad (*)$$

Let us compare this result with that following from the exact characteristic equation derived in the previous problem:

$$\cos qa = A \cos [k(a-d) + \varphi], \quad \text{with} \quad A^2 = 1 + \left[\frac{1}{2} \left(\frac{k}{\kappa} + \frac{\kappa}{k} \right) \sinh \kappa d \right]^2.$$

To solve this transcendental equation in the tight binding limit, $A \gg 1$, we may linearize its right-hand side in the narrow interval of k in which the right-hand side is of the order of 1:

$$\cos qa \approx \frac{d}{dk} \{ A \cos [k(a-d) + \varphi] \}_{\tilde{k}=0} \tilde{k},$$

where $\tilde{k}$ is the deviation of k from the point where the right-hand side equals zero—see the figure below, i.e. where $\cos[k(a-d) + \varphi] = 0$. For the distance Δk between the edges of the same narrow band (where $\cos qa = \pm 1$), this calculation gives

$$\Delta k = 2 \left(\frac{d}{dk} \{ A \cos [k(a-d) + \varphi] \} \right)^{-1}_{\cos[\dots]=0}.$$

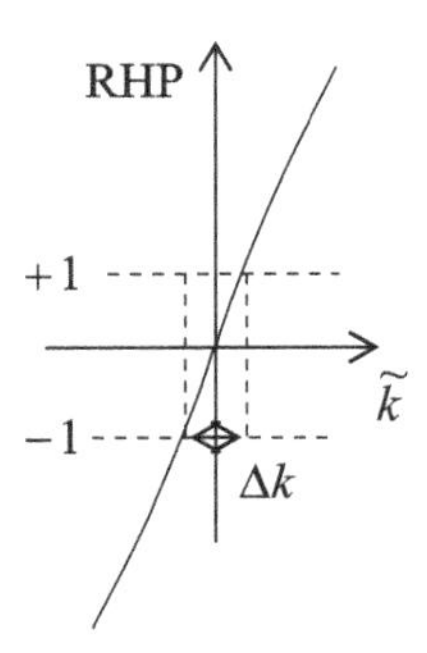

What remains is to recalculate this (small) difference of wave vector values into the difference on the energy scale. We can do that by differentiating relation $E = (\hbar^2 k^2/2m + \text{const})$ over k:

$$\Delta E_n \approx \left|\frac{dE}{dk}\right| \Delta k = \frac{\hbar^2 k}{m}\Delta k = \frac{2\hbar^2 k}{m}\left(\frac{d}{dk}\{A\cos[k(a-d)+\varphi]\}\right)^{-1}_{\cos[\ldots]=0}.$$

Here $A \gg 1$ may be approximated as

$$A = \left\{1 + \left[\frac{1}{2}\left(\frac{k}{\kappa} + \frac{\kappa}{k}\right)\sinh \kappa d\right]^2\right\}^{1/2} \approx \frac{1}{2}\left(\frac{k}{\kappa} + \frac{\kappa}{k}\right)\sinh \kappa d = \frac{K^2}{2\kappa k}\sinh \kappa d \approx \frac{K^2}{4\kappa k}e^{\kappa d},$$

where the last approximate equality is valid at the same condition, $\kappa d \gg 1$, which was used in our calculation of ΔE_n from the tight-binding limit formula. Now, a straightforward (though tedious) differentiation over k yields Eq. (*) obtained from the tight-binding approximation.

***Problem* 2.29.** For the same Kronig–Penney potential, use the weak-potential limit formulas to calculate the energy gap widths. Again, compare the results with those of problem 2.27, in the corresponding limit.

Solution: In this limit, we can use Eq. (2.224) of the lecture notes to write

$$\Delta_n = 2\,|U_n|,$$

where U_n is the nth Fourier coefficient of the function $U(x)$, defined by the Fourier expansion (2.207). The coefficients may be calculated using the reciprocal Fourier transform:

$$U_n = \frac{1}{a}\int_0^a U(x)\exp\left\{i\frac{2\pi x}{a}n\right\}dx.$$

For the Kronig–Penney potential, with the origin of the x-axis aligned with the left edge of the potential barrier[41], this integration gives

[41] The origin's choice affects the phase of the complex coefficient U_n, but not its magnitude, i.e. the energy gap's width.

$$U_n = \frac{U_0}{a}\int_0^d \exp\left\{i\frac{2\pi nx}{a}\right\}dx = \frac{U_0}{2\pi in}\left[\exp\left\{i\frac{2\pi nd}{a}\right\} - 1\right]$$
$$= \frac{U_0}{2\pi in}\left[\left(\cos\left\{\frac{2\pi nd}{a}\right\} - 1\right) + i\sin\left\{\frac{2\pi nd}{a}\right\}\right],$$

so that the nth energy gap equals

$$\Delta_n = 2\left|\frac{U_0}{2\pi in}\left[\left(\cos\left\{\frac{2\pi nd}{a}\right\} - 1\right) + i\sin\left\{\frac{2\pi nd}{a}\right\}\right]\right| \equiv \frac{U_0}{\pi n}\left|\sin\frac{\pi nd}{a}\right|. \qquad (*)$$

Besides the monotonic decrease of the gap with the growth of number n, due to the front factor $1/n$, this expression describes an interesting commensurate effect of the gap suppression at $nd \approx ma$, where m is another integer. At such a parameter relation, the gap location, $\pi n/a$, on the wave vector axis coincides with value $k' \approx k_m = \pi m/d$, corresponding to an over-barrier resonance (see the solution of problem 2.6), which enhances the traveling wave transmission and hence suppresses its interaction with the lattice (which, as was discussed in section 2.7 of the lecture notes, is responsible for the energy gap formation).

In order to compare Eq. (*) with the general characteristic equation derived in problem 2.27, let us take into account that the two edges of the same gap correspond to the two closest roots of the characteristic equation

$$\pm 1 = A\cos\left[k(a - d) + \varphi\right],$$

with the same sign in the left hand part—see figure 2.25 of the lecture notes. At $A \ll 1$, the distance Δk between these points on the k axis is small (see the figure below), and may be approximately calculated by expanding the right hand side of the equation into the Taylor series at the point of its maximum or minimum (where the linear term vanishes), and keeping only the two lowest terms:

$$A \mp \frac{1}{2}\frac{d^2}{dk^2}\{A\cos[k(a - d) + \varphi]\}_{\cos[\dots]=0}\tilde{k}^2 = \pm 1.$$

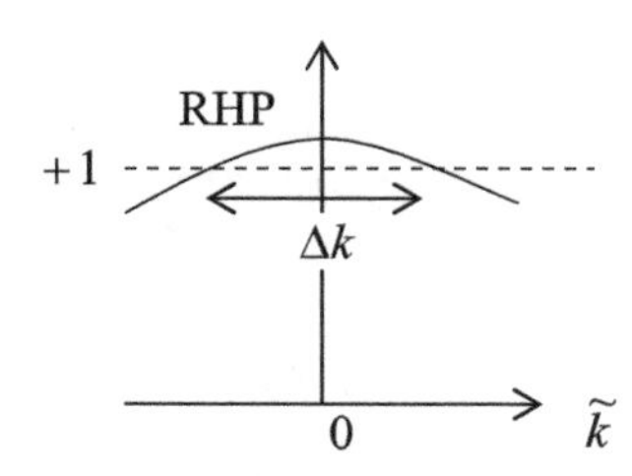

From here, the gap width corresponds to the interval

$$\Delta k = 2(A - 1)^{1/2}\left(\frac{d^2}{dk^2}\{A\cos[k(a - d) + \varphi]\}\right)^{-1}_{\cos[\dots]=0},$$

which may be then recalculated into the energy gap:

$$\Delta_n = \left.\frac{dE_l}{dk}\right|_{k=\pi n/a} \Delta k = \frac{\pi\hbar^2 n}{m}\Delta k.$$

From here, a tedious double differentiation yields Eq. (*).

***Problem* 2.30.** 1D periodic chains of atoms may exhibit what is called the so-called *Peierls instability*[42], leading to the *Peierls transition* to a phase in which the atoms are slightly displaced by $\Delta x_j = (-1)^j \Delta x$, with $\Delta x \ll a$, where j is the atom's number in the chain, and a is its initial period. These displacements lead to the alternation of the coupling amplitudes δ_n (see Eq. (2.204) of the lecture notes) between some values $\delta_n{}^+$ and $\delta_n{}^-$. Use the tight-binding approximation to calculate the resulting change of the nth energy band, and discuss the result.

Solution: In order to describe the band structure, we may use an equation similar to Eq. (2.203) of the lecture notes, but with alternating coupling constants:

$$i\hbar\dot{a}_j = -\delta_n^- a_{j-1} - \delta_n^+ a_{j+1}, \qquad i\hbar\dot{a}_{j+1} = -\delta_n^+ a_j - \delta_n^- a_{j+2}. \qquad (*)$$

The Bloch solution of the type (2.205) now has to accommodate alternating complex amplitudes a:

$$a_j(t) = \begin{Bmatrix} a^+, & \text{for } j \text{ odd,} \\ a^-, & \text{for } j \text{ even,} \end{Bmatrix} \times \exp\left\{ iqx_j - i\frac{\varepsilon_n}{\hbar}t + \text{const} \right\}.$$

(Another way to express the same fact is to say that since the potential profile $U(x)$ is now $2a$-periodic, the Bloch theorem is only valid for this larger period.) Plugging the last expression into Eq. (*), we get a system of two linear equations for two complex amplitudes $a^{\pm}$:

$$\varepsilon_n a^+ = -(\delta_n^- e^{-iqa} + \delta_n^+ e^{iqa})a^-, \qquad \varepsilon_n a^- = -(\delta_n^+ e^{-iqa} + \delta_n^- e^{iqa})a^+.$$

The condition of consistency of this homogeneous system of equations,

$$\begin{vmatrix} \varepsilon_n & \delta_n^- e^{-iqa} + \delta_n^+ e^{iqa} \\ \delta_n^+ e^{-iqa} + \delta_n^- e^{iqa} & \varepsilon_n \end{vmatrix} = 0,$$

solved for the energy deviation ε_n from the uncoupled-limit energy E_n, gives the following dispersion relation:

$$\varepsilon_n = \pm\left[\left(\delta_n^+\right)^2 + (\delta_n^-)^2 + 2\delta_n^+\delta_n^+ \cos 2qa\right]^{1/2}.$$

[42] Named after R Peierls (1907–95), the theorist most famous for the introduction of the notion of holes in semiconductors (and also as one of the main initiators of the Manhattan project).

A more revealing form of the same result may be obtained using the trigonometric identity $\cos 2qa = \cos^2 qa - \sin^2 qa$, and then noticing that the terms under the square root form two full squares:

$$\varepsilon_n = \pm\left[\left(\delta_n^+ + \delta_n^-\right)^2 \cos^2 qa + \left(\delta_n^+ - \delta_n^-\right)^2 \sin^2 qa\right]^{1/2}.$$

It shows that if the coupling alternation is negligible ($\delta_n^+ - \delta_n^- \to 0$), the energy band tends to the sinusoidal form (2.206) with the 'usual' period $\Delta q = 2\pi/a$. However, even a small but nonvanishing asymmetry results in the formation of an additional energy gap (see the numerical plots in the figure below), so that the quasi-momentum period decreases to $(\Delta q)' = \pi/a$. (Again, this is very natural from the point of doubling the spatial period: $a' = 2a$, leading to the quasi-momentum's period $(\Delta q)' = 2\pi/a' = \pi/a$.)[43] At its maxima ($qa = \pi/2 + m\pi$, with m integer), the gap width is

$$\Delta\varepsilon_n = 2\left(\delta_n^+ - \delta_n^-\right). \qquad (*)$$

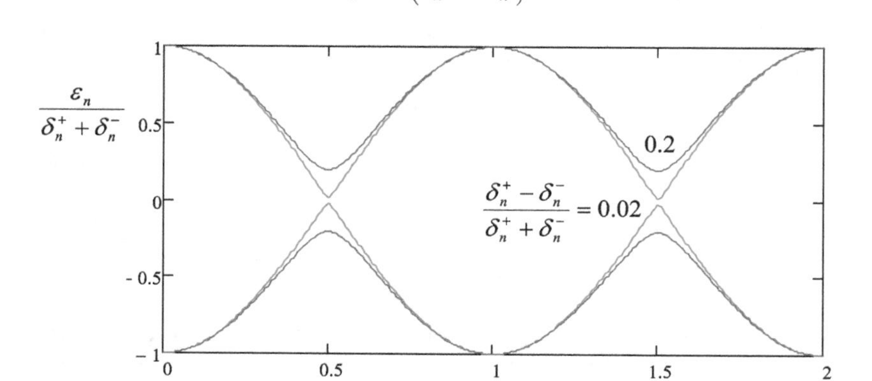

This effect may take place in highly anisotropic (quasi-1D) crystals (such as organic TTF-TCNQ) of atoms with an odd number of electrons in incompleted energy shells (see, e.g. section 3.7 of the lecture notes), and has rather dramatic consequences for their transport properties. Indeed, due to the Fermi statistics of electrons, and their spin ½, their states fill exactly the lower half of the conduction energy band. Such an 'open' Fermi surface enables ready activation of electrons above the surface by even weak applied electric field, and hence their high electric conductivity—see, e.g. *Part SM* section 6.3. However, the Peirce transition separates the lower half-band, completely filled with electrons, from the completely depleted upper half-band with the energy gap (*), suppressing the electron activation, and hence the conductivity. As a result, the conductor turns into what is called the *Peierls dielectric*.

[43] Note that such a gap opening is not a pure quantum phenomenon, but takes place at the propagation of waves of any nature in nearly-periodic systems—see, e.g. *Part CM* problem 6.13.

It is curious that the conductivity electrons are not only affected by the Peierls instability, but also *cause* it. Indeed, as the figure above shows, the Peierls transition leads to the *reduction* of electron energies in the lower (filled) half-band, and hence can make the transition energy-favorable. Note that such self-supporting instabilities are very common in physics—another prominent example is the Cooper pairing of electrons in superconductors.

Problem* 2.31. Use Eqs. (1.73) and (1.74) of the lecture notes to derive Eq. (2.252), and discuss the relation between these Bloch oscillations and the Josephson oscillations of frequency (1.75).

Solution: First, let us combine Eqs. (1.73) and (1.74) to calculate the work of an external voltage source at Josephson phase change between some arbitrary initial (φ_{ini}) and final (φ_{fin}) values, as the integral of its power IV over the time interval Δt of the change:

$$\begin{aligned}\text{Work} &= \int_{\Delta t} IVdt = \int_{\Delta t} (I_c \sin\varphi)\left(\frac{\hbar}{2e}\frac{d\varphi}{dt}\right)dt \\ &= \frac{\hbar I_c}{2e}\int_{\varphi_{\text{ini}}}^{\varphi_{\text{fin}}} \sin\varphi d\varphi = -\frac{\hbar I_c}{2e}(\cos\varphi_{\text{fin}} - \cos\varphi_{\text{ini}}).\end{aligned}$$

We see that the work depends only on the initial and final values of φ (but not on the law phase evolution in time), and hence may be represented as the difference $U(\varphi_{\text{fin}}) - U(\varphi_{\text{ini}})$, where the function

$$U(\varphi) = -E_J \cos\varphi + \text{const}, \quad \text{with } E_J \equiv \frac{\hbar I_C}{2e}, \qquad (*)$$

may be interpreted as the potential energy of the junction (if we consider the Josephson phase as a generalized coordinate).

Besides this energy, the Josephson junction, as a system of two close, nearly isolated (super)conductors, has a certain mutual capacitance C and the associated electrostatic energy $E_C = CV^2/2$. Using Eq. (1.73) again, we may represent it as

$$E_C = \frac{C}{2}V^2 = C\left(\frac{\hbar}{2e}\right)^2\left(\frac{d\varphi}{dt}\right)^2.$$

This expression means that from the point of view at phase φ as the generalized coordinate, E_C should be considered the kinetic energy of the system, whose dependence on the generalized velocity $d\varphi/dt$ is similar to that of a 1D mechanical particle, with an effective mass[44]

$$m_J = C\left(\frac{\hbar}{2e}\right)^2.$$

[44] Since the dimensionality of the generalized coordinate φ is different from [m], that of m_J is different from [kg].

Hence the total energy of the junction, $E_C + U(\varphi)$, is formally similar to that of a 1D non-relativistic particle of the mass m_J, moving along the φ-axis in the sinusoidal potential (*) with the period $a_J = 2\pi$.

However, before using the results of the 1D band theory, discussed in sections 2.6–2.7 of the lecture notes, to this system, we have to resolve one paradox—which, in the mid-1980s, was the subject of a lively scientific discussion. In section 2.6, we implied that the particle's translation by the potential's period a is (in principle) measurable, i.e. the particle positions x and $(x + a)$ are distinguishable—otherwise Eq. (2.193) with $q \neq 0$ would not have much sense. For the Josephson phase φ, the similar assumption is less plausible. Indeed, for example, if we change φ by $a_J = 2\pi$ via changing the phase of one of the superconductors, say φ_1 (see figure 1.7 of the lecture notes) by 2π, then its wavefunction becomes $|\psi| \exp\{i(\varphi_1 + 2\pi)\} = |\psi| \exp\{i\varphi_1\}$, and it is not immediately clear whether these two states may be distinguished. In order to resolve this contradiction, it is sufficient to have a look at Eq. (1.73). It shows that if φ changes in time by 2π (say, by its fast ramp-up), the voltage V across the junction exhibits a pulse with the following 'area':

$$\int V(t)dt = \frac{\hbar}{2e}\int \frac{d\varphi}{dt}dt = \frac{\hbar}{2e}\int d\varphi = \frac{\hbar}{2e}2\pi \equiv \frac{\pi\hbar}{e} \approx 2 \times 10^{-15}\,\text{V}\cdot\text{s}. \qquad (**)$$

Such *single-flux-quantum* (SFQ) pulses[45] not only have been observed experimentally, but even may be used for signaling and ultrafast (sub-THz) computation[46].

Hence, the 2π-shifts of the phase φ are measurable, and in the absence of dissipation the Josephson junction dynamics is indeed similar to that of a 1D particle in a periodic (sinusoidal) potential, and its energy spectrum forms the energy bands and gaps described by the Mathieu equation—see figure 2.31a and Eqs. (2.227)–(2.229) of the lecture notes. Experimentally, the easiest way to verify this picture is to measure the corresponding Bloch oscillations induced by an external current $I_{ex}(t)$. In order to find the frequency of these oscillations, it is sufficient to replace Eq. (2.237), which expresses the 2nd Newton law for the quasi-momentum $\hbar q$, with the charge balance equation

$$\frac{dQ}{dt} = I_{ex}(t), \qquad (***)$$

for the corresponding variable Q, called the *quasi-charge*. This relation tells us that the quasi-charge Q has the simple physical sense of the external electric charge being inserted into the junction by the external current I_{ex}—just like the physical sense of the quasi-momentum $\hbar q$ of a mechanical particle, according to Eq. (2.237), is the contribution to the average particle's momentum by the external force F. (Notice that at such quantum-mechanical averaging of the electric charge, the supercurrent

[45] This term has originated from the fact that the right-hand side of Eq. (**) equals the single quantum unit (Φ_0) of the magnetic flux in superconductors—see section 3.1 below (and/or *Part EM* section 6.4–6.5).

[46] To the best of my knowledge, such computation technology (dubbed RSFQ) still holds the absolute records for the highest speed and smallest energy consumption at an elementary computation—see, e.g. [2] and references therein.

(1.74) drops out from the equation, affecting the phenomena 'only' via its contribution to the energy band structure.)

Since the Josephson-junction analog of the usual wave number $k = (m/\hbar)(dx/dt)$ of a mechanical particle is

$$k_J = \frac{m_J}{\hbar}\frac{d\varphi}{dt} = \frac{m_J}{\hbar}\frac{2e}{\hbar}V = \frac{CV}{2e},$$

and CV is the genuine charge on the capacitor, the analog of q (the quasi-momentum divided by $\hbar$) may be obtained just by the replacement of that product with the quasi-charge Q:

$$q_J = \frac{Q}{2e}.$$

Comparing this expression with Eq. (***), we see that q_J obeys the following equation of motion:

$$\frac{dq_J}{dt} = \frac{I_{ex}(t)}{2e}.$$

so that the role of the mechanical force F is now played by $F_J = \hbar I/2e$. Hence if $I_{ex}(t) = \text{const} = \bar{I}$, we can use Eq. (2.244) with that replacement, and also $a \to a_J = 2\pi$, to get Eq. (2.252) of the lecture notes:

$$f_B \equiv \frac{\omega_B}{2\pi} = \frac{1}{2\pi}\frac{F_J a_J}{\hbar} = \frac{\bar{I}}{2e}.$$

This very simple result has the following physical sense[47]. In the quantum operation mode, the junction is recharged by the external current, following Eq. (***), until its electric charge reaches e (i.e. until the normalized quasi-momentum $q_J a_J = (Q/2e)2\pi$ reaches π—see figure 2.33a of the lecture notes); then one Cooper pair passes through the junction changing its charge to $e - (2e) = -e$, with the same charging energy $Q^2/2C$—the process analogous to crossing the border of the 1st Brillouin zone; then the process repeats again and again[48]. It is paradoxical that Eq. (2.252), describing the frequency of such a *quantum* property of the Josephson phase φ as its Bloch oscillations, does not include the Planck's constant, while Eq. (1.75), describing the *classical* motion of φ, does[49].

In this context, one may wonder which of these two types of oscillations would a dc-biased Josephson junction generate. For the dissipation-free (OK, *virtually* dissipation-free :-) junction, the answer is: the Bloch oscillations (2.252) with the

[47] [3]

[48] Note that the qualitatively similar effect of the single-electron-tunneling (SET) oscillations, with twice higher frequency $f_{SET} = I/e$, takes place, at sufficiently low temperatures, in small 'normal' (non-superconducting) tunnel junctions—see, e.g. *Part EM* section 2.9 and references therein. However, the quantitative descriptions of these effects are rather different, because, in contrast to the Cooper pairs, the electrons in 'normal' conductors do not form a coherent Bose–Einstein condensate.

[49] The phase locking of the Bloch oscillations, as well as that of the SET oscillations, by an external signal of a well characterized frequency may enable fundamental standards of dc current. The experimentally achieved accuracy of such standards is close to 10^{-8}, just a few times worse than that of a less direct way toward such standards—using a Josephson voltage standard and a resistance standard based on the quantum Hall effect.

frequency proportional to the dc *current*. However, any practical junction has some energy losses that may be approximately described by a certain Ohmic conductance G connected in parallel to the junction. Very luckily for Dr Josephson and his Nobel Prize, it turns out to be much easier to fabricate and test junctions with $G \gg 1/R_Q$, where R_Q is the so-called *quantum unit of resistance*

$$R_Q \equiv \frac{\pi\hbar}{2e^2} \approx 6.45 \text{ k}\Omega,$$

the fundamental constant that comes up at analyses of several other effects as well—see, e.g. section 3.2 below. As will be discussed in chapter 7, the dissipation so high provides what is called *dephasing*—the suppression of the quantum coherence between different quantum states of the system—in our current case, between the wavefunctions $u(\varphi - 2\pi j)$ localized at different minima of the periodic potential $U(\varphi)$, and thus make the dynamics of the Josephson phase φ virtually classical, obeying Eqs. (1.73) and (1.74). As was discussed in section 1.6 of the lecture notes, dc biasing of such a junction leads to Josephson oscillations with the frequency (1.75), proportional to the applied dc *voltage*, rather than the current.

Problem* 2.32. A 1D particle of mass m is placed into the following triangular potential well:

$$U(x) = \begin{cases} +\infty, & \text{for } x < 0, \\ Fx, & \text{for } x > 0, \end{cases} \quad \text{with } F > 0.$$

(i) Calculate its energy spectrum using the WKB approximation.
(ii) Estimate the ground state energy using the variational method, with two different trial functions.
(iii) Calculate the three lowest energy levels, and also for the 10th level, with at least 0.1% accuracy, from the exact solution of the problem.
(iv) Compare and discuss the results.

Hint: The values of the first zeros of the Airy function, necessary for task (iii), may be found in many math handbooks, for example, in table 10.13 of the collection by Abramowitz and Stegun—see section A.16(i) of appendix A.

Solutions:

(i) Acting just as in section 2.4 of the lecture notes (see, in particular, figure 2.10 and its discussion), let us calculate the total roundtrip phase shift of a traveling de Broglie wave of energy E_n. The quasiclassical motion from the left classical turning points, $x_L = 0$, to the right point, $x_R = E_n/F$ (see the figure below), yields the shift

$$\Delta\varphi_{\rightarrow} = \int_{x_L}^{x_R} k(x)dx = \frac{1}{\hbar}\int_0^{E_n/F} [2m(E_n - Fx)]^{1/2}dx = \frac{2}{3}(2m)^{1/2}\frac{E_n^{3/2}}{\hbar F}.$$

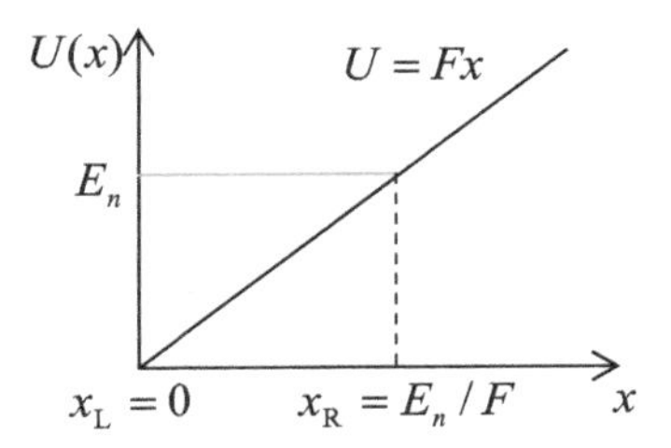

The total phase change of a wave's roundtrip consists of twice that shift (including also the way back from x_L and x_R), plus two shifts due to the wave reflection from the classical turning points. One of these reflections (at $x = x_R$) may be treated quasiclassically, giving the additional shift (in comparison with the 'hard', vertical wall) equal to $\Delta\varphi = \pi/2$. The reflection from the left, vertical potential wall at $x = 0$ does not give such an additional shift. As a result, the total phase change on the roundtrip is

$$\Delta\varphi_{\text{total}} = 2\Delta\varphi_{\rightarrow} + \frac{\pi}{2} = \frac{4}{3}(2m)^{1/2}\frac{E_n^{3/2}}{\hbar F} + \frac{\pi}{2}.$$

Requiring this change to equal $2\pi n$, with $n = 1, 2, \ldots$, we get the WKB spectrum

$$E_n|_{\text{WKB}} = E_0\left[\frac{3\pi}{2}\left(n - \frac{1}{4}\right)\right]^{2/3}, \quad \text{with } E_0 \equiv \left(\frac{\hbar^2 F^2}{2m}\right)^{1/3}.$$

(ii) Looking at the potential profile of the problem (see the figure above), it is clear that the following simple trial function

$$\psi_1(x) = \begin{cases} 0, & \text{for } x < 0, \\ Cxe^{-\lambda x}, & \text{for } x > 0, \end{cases} \quad \text{with } \lambda > 0,$$

may give a reasonable approximation for the ground state of the system. (In particular, it yields the exact, zero value of the wavefunction for $x \leqslant 0$ and for $x = +\infty$, and also ensures the function's continuity at all points.) Its normalization condition is

$$\int_0^\infty |\psi_1|^2\, dx \equiv |C|^2 \int_0^\infty x^2 e^{-2\lambda x} dx = 1.$$

Using the table integral (A.34*d*) with $n = 2$, we readily get

$$|C|^{-2} = \int_0^\infty x^2 e^{-2\lambda x} dx = \frac{1}{(2\lambda)^3}\int_0^\infty \xi^2 e^{-\xi} d\xi = \frac{1}{4\lambda^3}.$$

The expectation value of the Hamiltonian in this trial state is

$$\langle H\rangle_1 \equiv \int_{-\infty}^{+\infty} \psi_1^* \hat{H} \psi_1 dx \equiv \int_{-\infty}^{+\infty} \psi_1^* \left[-\frac{\hbar^2}{2m}\frac{d^2}{dx^2} + U(x) \right] \psi_1 dx$$

$$\equiv \int_0^{\infty} C^* x e^{-\lambda x} \left(-\frac{\hbar^2}{2m}\frac{d^2}{dx^2} + Fx \right) C x e^{-\lambda x} dx$$

$$\equiv |C|^2 \left\{ -\frac{\hbar^2}{2m} \left[\lambda^2 \int_0^{\infty} x^2 e^{-2\lambda x} dx - 2\lambda \int_0^{\infty} x e^{-2\lambda x} dx \right] \right.$$

$$\left. + F \int_0^{\infty} x^3 e^{-2\lambda x} dx \right\}.$$

All these three integrals are of the same type (A.34*d*), with $n = 2$, 1, and 3, respectively. Using the above expression for the normalization constant, we finally get

$$\langle H\rangle_1 = \frac{\hbar^2\lambda^2}{2m} + \frac{3F}{2\lambda}.$$

This expectation value is positive for all $\lambda > 0$, and diverges both at $\lambda \to 0$ and $\lambda \to \infty$, so it certainly has a minimum at some optimum value λ_{opt}, for which

$$\frac{\partial}{\partial\lambda}\langle H\rangle_1 \bigg|_{\lambda=\lambda_{\text{opt}}} = 0.$$

Performing the simple differentiation, we get

$$\lambda_{\text{opt}} = \left(\frac{3mF}{2\hbar^2} \right)^{1/3},$$

$$E_{\text{var1}} \equiv \min_\lambda(\langle H\rangle_{\text{trial}}) \equiv \frac{\hbar^2\lambda_{\text{opt}}^2}{2m} + \frac{3F}{2\lambda_{\text{opt}}} = \left(\frac{3}{2}\right)^{5/3} \left(\frac{\hbar^2 F^2}{m}\right)^{1/3} \approx 2.476\, E_0.$$

Now let us try a somewhat different trial function:

$$\psi_2(x) = \begin{cases} 0, & \text{for } x < 0, \\ Cxe^{-\lambda x^2/2}, & \text{for } x \geqslant 0, \end{cases}$$

also having the proper (zero) boundary values at $x \leqslant 0$ and $x = +\infty$. For this function, the normalization condition is

$$\int_0^{\infty} |\psi_2|^2\, dx \equiv |C|^2 \int_0^{\infty} x^2 e^{-\lambda x^2} dx \equiv |C|^2\, \lambda^{-3/2} \int_0^{\infty} \xi^2 e^{-\xi^2} dx = 1,$$

where $\xi \equiv \lambda^{1/2}x$. The last, dimensionless integral[50] equals $\pi^{1/2}/4$, so that

[50] See, e.g. Eq. (A.36*c*).

$$|C|^2 = \frac{4\lambda^{3/2}}{\pi^{1/2}}.$$

With this normalization, the expectation value of the Hamiltonian in the trial state is

$$\begin{aligned}
\langle H \rangle_2 &\equiv \int_{-\infty}^{+\infty} \psi_2^* \hat{H} \psi_{21} dx \\
&= |C|^2 \int_0^{\infty} x e^{-\lambda x^2/2} \left(-\frac{\hbar^2}{2m} \frac{d^2}{dx^2} + Fx \right) x e^{-\lambda x^2/2} dx \\
&= \frac{4\lambda^{3/2}}{\pi^{1/2}} \left[-\frac{\hbar^2}{2m} \left(-3\lambda \int_0^{\infty} x^2 e^{-\lambda x^2} dx + \lambda^2 \int_0^{\infty} x^4 e^{-\lambda x^2} dx \right) \right. \\
&\quad \left. + F \int_0^{\infty} x^3 e^{-\lambda x^2} dx \right] \\
&\equiv \frac{4\lambda^{3/2}}{\pi^{1/2}} \left[-\frac{\hbar^2}{2m} \left(-3\lambda^{-1/2} \int_0^{\infty} \xi^2 e^{-\xi^2} d\xi + \lambda^{-1/2} \int_0^{\infty} \xi^4 e^{-\xi^2} d\xi \right) \right. \\
&\quad \left. + F\lambda^{-2} \int_0^{\infty} \xi^3 e^{-\xi^2} d\xi \right],
\end{aligned}$$

where the same substitution as above, $\xi \equiv \lambda^{1/2} x$, was used. The first dimensionless integral is the same as above (equal to $\pi^{1/4}/4$), and the remaining two are of the same type[51], equal, respectively, to $(3/8)\pi^{1/2}$ and 1/2, so that

$$\begin{aligned}
\langle H \rangle_2 &= \frac{4\lambda^{3/2}}{\pi^{1/2}} \left[-\frac{\hbar^2}{2m} \left(-3\lambda^{-1/2} \frac{\pi^{1/2}}{4} + \lambda^{-1/2} \frac{3\pi^{1/2}}{8} \right) + F\lambda^{-2} \frac{1}{2} \right] \\
&= \frac{3\hbar^2 \lambda}{4m} + \frac{2F}{\pi^{1/2} \lambda^{1/2}}.
\end{aligned}$$

This expectation value is also positive and diverges both at $\lambda \to 0$ and $\lambda \to \infty$, so it certainly has a minimum at some optimum value λ_{opt}, for which

$$\left. \frac{\partial}{\partial \lambda} \langle H \rangle_2 \right|_{\lambda = \lambda_{\text{opt}}} = 0.$$

Performing a simple differentiation, we get

$$\begin{aligned}
\lambda_{\text{opt}} &= \left(\frac{16 m^2 F^2}{9\pi \hbar^4} \right)^{1/3}, \quad E_{\text{var2}} \equiv \frac{3\hbar^2 \lambda_{\text{opt}}}{4m} + \frac{2F}{\pi^{1/2} \lambda_{\text{opt}}^{1/2}} \\
&= \left(\frac{81}{2\pi} \right)^{1/3} E_0 \approx 2.345 E_0, \quad \text{where } E_0 \equiv \left(\frac{\hbar^2 F^2}{2m} \right)^{1/3}.
\end{aligned}$$

---

[51] See, e.g. Eq. (A.36*a*), with $s = 4$ and $s = 3$, respectively.

The fact that $E_{\text{var2}} < E_{\text{var1}}$ shows that the second trial function provides a better approximation.

(iii) In order to obtain the exact solution to the problem, we can solve the stationary Schrödinger equation

$$-\frac{\hbar^2}{2m}\frac{d^2\psi}{dx^2} + Fx\psi = E_n\psi, \qquad (*)$$

at $x > 0$, with the boundary conditions $\psi(0) = \psi(+\infty) = 0$. Normalizing, just as was done in section 2.4 of the lecture notes, the coordinate x to the constant

$$x_0 \equiv \left(\frac{\hbar^2}{2m dU/dx}\right)^{1/3} \equiv \left(\frac{\hbar^2}{2mF}\right)^{1/3},$$

we may reduce Eq. (*) to the canonical form (2.101) of the Airy equation, with the general solution

$$\psi(\zeta) = C_A \text{Ai}(\zeta) + C_B \text{Bi}(\zeta),$$

where $\zeta \equiv (x - x_R)/x_0$. One of the boundary conditions (at $x \to \infty$) may be satisfied only by taking $C_B = 0$, so that the second one (at $x = 0$, i.e. at $\zeta = -x_R/x_0$) is reduced to the requirement Ai $(-x_R/x_0) = 0$,[52] i.e.

$$-\frac{x_R}{x_0} \equiv -\frac{E_n}{Fx_0} = \zeta_n, \quad \text{i.e. } E_n = -Fx_0\zeta_n = -\left(\frac{\hbar^2 F^2}{2m}\right)^{1/3}\zeta_n = -E_0\zeta_n,$$

where ζ_n is the nth root of the Airy function Ai(ζ). Using the tabulated values of ζ_n, we get:

n	$E_n/E_0\|_{\text{WKB}} = [(3\pi/2)(n - 1/4)]^{2/3}$	$E_{\text{var1}}/E_0 = \min_\lambda \langle H\rangle_1$	$E_{\text{var2}}/E_0 = \min_\lambda \langle H\rangle_2$	$E_n/E_0\|_{\text{exact}} = -\zeta_n$
1	2.320	2.476	2.345	2.338
2	4.081	—	—	4.088
3	5.517	—	—	5.520
10	12.8281	—	—	12.8287

(iv) The table above shows that the variational method results for the ground state depend much on the trial function. Indeed, our second attempt gave an error of just ~0.3%—pretty good for virtually any practical application. Looking at the asymptotic behavior of the Airy functions (see, e.g. the first line of Eq. (2.102) of the lecture notes), we may guess that an even a better trial function could be

$$\psi_{\text{trial}} = Cx \exp\left\{-\frac{2}{3}\zeta^{3/2}\right\}, \quad \text{with } \zeta \equiv \frac{x - x_c}{x_0}, \quad x_0 \equiv \left(\frac{\hbar^2}{2mF}\right)^{1/3},$$

[52] Note that the above WKB result could be also obtained by using this equation with the asymptotic form given by the second of Eqs. (2.106).

and the classical turning point x_c treated as the adjustable parameter. The reader is invited to explore this option (involving less common integrals) as an additional exercise.

On the other hand, even for the ground level ($n = 1$), which is always the hardest task for the WKB approximation, in this case it works surprisingly well, with a relative error close to 1%; the error decreases very fast as we go up the energy level ladder, dropping below 10^{-4} for $n = 10$. Note, however, that while with the variational method, we may be always sure that the genuine ground state energy is below the estimated value, this is not true for the WKB method.

***Problem* 2.33.** Use the variational method to estimate the ground state energy E_g of a 1D particle in the following potential well:

$$U(x) = -U_0 \exp\{-\alpha x^2\}, \qquad \text{with } \alpha > 0, \qquad \text{and } U_0 > 0.$$

Spell out the results in the limits of small and large U_0, and give their interpretation.

Solution: Since any smooth, symmetric potential well $U(x)$ may be Taylor-approximated, near its bottom, with a quadratic parabola, the calculation in the beginning of section 2.9 of the lecture notes shows that a Gaussian function, similar to that given by Eq. (2.270),

$$\psi_{\text{trial}}(x) = C \exp\left\{-\frac{\lambda x^2}{2}\right\}, \qquad \text{with } \lambda > 0,$$

is a reasonable choice for the trial function for our potential. The flow of calculation of the expectation value of the corresponding Hamiltonian

$$\hat{H} = -\frac{\hbar^2}{2m}\frac{d^2}{dx^2} - U_0 \exp\{-\alpha x^2\}$$

is absolutely similar to that used in section 2.9:

$$\begin{aligned}
\langle H\rangle_{\text{trial}} &= \int_{-\infty}^{+\infty} \psi_{\text{trial}}^* \hat{H} \psi_{\text{trial}} dx = |C|^2 \int_{-\infty}^{+\infty} \exp\left\{-\frac{\lambda x^2}{2}\right\} \\
&\times \left(-\frac{\hbar^2}{2m}\frac{d^2}{dx^2} - U_0 \exp\{-\alpha x^2\}\right) \exp\left\{-\frac{\lambda x^2}{2}\right\} dx \\
&= \left(\frac{\lambda}{\pi}\right)^{1/2} \left[\frac{\hbar^2 \lambda}{2m} \int_{-\infty}^{+\infty} \exp\{-\lambda x^2\}\, dx \right. \\
&\left. - U_0 \int_{-\infty}^{+\infty} \exp\{-(\lambda + \alpha)\, x^2\}\, dx - \frac{\hbar^2 \lambda^2}{2m} \int_{-\infty}^{+\infty} x^2 \exp\{-\lambda x^2\}\, dx \right] \\
&= \left(\frac{\lambda}{\pi}\right)^{1/2} \left[\frac{\hbar^2 \lambda}{2m} \frac{\pi^{1/2}}{\lambda^{1/2}} - U_0 \frac{\pi^{1/2}}{(\lambda + \alpha)^{1/2}} - \frac{\hbar^2 \lambda^2}{2m} \frac{\pi^{1/2}}{2\lambda^{3/2}}\right] \\
&\equiv \frac{\hbar^2 \lambda}{4m} - U_0 \left(\frac{\lambda}{\lambda + \alpha}\right)^{1/2}.
\end{aligned}$$

Due to the second term, the last expression is evidently negative, with a negative derivative $\partial\langle H\rangle_{\text{trial}}/\partial\lambda$ at $\lambda = 0$, while due to the first term, it is positive and diverges at $\lambda \to \infty$. Hence, $\langle H\rangle_{\text{trial}}$ as a function of λ has a minimum, corresponding to a localized ('bound') ground state of the system. In the general case, the condition of this minimum,

$$\left.\frac{\partial\langle H\rangle_{\text{trial}}}{\partial\lambda}\right|_{\lambda=\lambda_{\text{opt}}} \equiv \frac{\hbar^2}{4m} - U_0\frac{\alpha}{2\lambda_{\text{opt}}^{1/2}(\lambda_{\text{opt}} + \alpha)^{3/2}} = 0,$$

gives a rather unpleasant 4th-degree-polynomial equation for the dimensionless variable $\xi \equiv \lambda_{\text{opt}}/\alpha$:

$$\xi(\xi + 1)^3 = \left(\frac{U_0}{T_0}\right)^2, \quad \text{where } T_0 \equiv \frac{\hbar^2\alpha}{2m}, \qquad (*)$$

with an extremely bulky general solution[53].

However, we may readily spell out the result in the limits when the depth U_0 of the potential well is much smaller or much larger than the scale T_0 of the kinetic energy of the particle in the well. In the former case, we should have $\xi \ll 1$, so that the left-hand side of Eq. (*) may be approximated with ξ, and this equation yields

$$\xi = \left(\frac{U_0}{T_0}\right)^2, \quad \text{i.e.} \quad \lambda_{\text{opt}} = \alpha\left(\frac{U_0}{T_0}\right)^2 \ll \alpha,$$

$$\text{so that } \langle H\rangle_{\text{min}} = -\frac{U_0^2}{2T_0}, \text{ i.e. } |\langle H\rangle_{\text{min}}| \ll U_0.$$

This result has a simple physical meaning: if the well is shallow, its particle-localization effect is weak, so that the localized wavefunctions are spread far beyond the effective well's width $1/\alpha^{1/2}$. On the scale of this spread, the potential well potential may be well approximated with the delta-function,

$$U(x) \approx -\mathscr{w}\delta(x),$$

whose 'area' $\mathscr{w}$ may be calculated from the delta-function's definition:

$$\int_{-\infty}^{+\infty}\delta(x)dx \equiv -\frac{1}{\mathscr{w}}\int_{-\infty}^{+\infty}U(x)dx \equiv -\frac{1}{\mathscr{w}}\int_{-\infty}^{+\infty}[-U_0\exp\{-\alpha x^2\}]dx$$

$$\equiv \frac{U_0}{\mathscr{w}}\left(\frac{\pi}{\alpha}\right)^{1/2} = 1, \quad \text{thus giving } \mathscr{w} = \left(\frac{\pi}{\alpha}\right)^{1/2}U_0.$$

If we now plug this weight value into Eq. (2.165) of the lecture notes, for the particle's ground state energy in such a delta-functional potential, $E_g = -m\mathscr{w}^2/2\hbar^2$, we get

[53] Mercifully, since in the physically acceptable range $\xi \geqslant 0$, the left-hand side of Eq. (*) is a monotonically growing function of ξ, starting from 0 at $\xi = 0$, this particular equation, for any ratio U_0/E_0, has just one root of our interest.

$$E_g = -\frac{\pi}{2}\frac{mU_0^2}{\hbar^2\alpha} = -\frac{\pi}{4}\frac{U_0^2}{T_0} \approx -0.785\frac{U_0^2}{T_0}.$$

So, in this limit the variational method captures the correct functional dependence of the ground state energy, but is ~60% off the exact result[54].

In the opposite limit of a deep potential well, $U_0 \gg T_0$, the left-hand side of Eq. (*) may be approximated with ξ^4, and this equation yields

$$\xi \equiv \frac{\lambda_{\text{opt}}}{\alpha} = \left(\frac{U_0}{T_0}\right)^{1/2} \gg 1, \quad \text{giving} \quad \langle H\rangle_{\min} = -U_0 + (U_0T_0)^{1/2}.$$

In order to interpret this result, let us use the fact that in this limit the wavefunction's spread is much smaller than the well width, so it 'feels' just the very bottom of the well, where the confining potential may be well approximated with just two leading terms of its Taylor expansion:

$$U(x) = -U_0\exp\{-\alpha x^2\} \approx -U_0 + U_0\alpha x^2.$$

But this is exactly the potential of a harmonic oscillator (offset down by U_0),

$$U(x) = -U_0 + \frac{m\omega_0^2}{2}x^2, \quad \text{with} \quad \frac{m\omega_0^2}{2} = U_0\alpha, \quad \text{i.e. } \omega_0 = \left(\frac{2\alpha U_0}{m}\right)^{1/2},$$

whose exact ground state energy is

$$E_g = -U_0 + \frac{\hbar\omega_0}{2} = -U_0 + \frac{\hbar}{2}\left(\frac{2\alpha U_0}{m}\right)^{1/2} = -U_0 + (U_0T_0)^{1/2}.$$

Thus, as might be expected, in this limit the variational method yields the exact ground-state energy.

***Problem* 2.34.** For a 1D particle of mass m, placed into a potential well with the following profile,

$$U(x) = ax^{2s}, \quad \text{with } a > 0 \text{ and } s > 0,$$

(i) calculate its energy spectrum using the WKB approximation, and
(ii) estimate the ground state energy using the variational method.

Compare the ground state energy results for the parameter s equal to 1, 2, 3, and 100.

[54] The reason of this difference is clear from the comparison of our Gaussian trial function with the exact ground state wavefunction (2.159): $\psi_g \propto \exp\{-\kappa|x|\}$.

Solutions:

(i) In this soft, symmetric confining potential, both classical turning points, x_R and $x_L = -x_R$, may be treated quasiclassically, so that we may use the Bohr–Sommerfeld quantization rule directly in the form of Eq. (2.110) of the lecture notes:

$$\oint p(x)dx \equiv 2\int_{x_L}^{x_R} p(x)dx = 2\pi\hbar\left(n - \frac{1}{2}\right), \quad \text{with } n = 1, 2, \dots, \qquad (*)$$

where in our current case

$$p(x) = \{2m[E_n - U(x)]\}^{1/2} \equiv [2m(E_n - ax^{2s})]^{1/2}$$
$$\equiv (2mE_n)^{1/2}[1 - (x/x_n)^{2s}]^{1/2}.$$

Here $x_n \equiv x_R = -x_L$ is the distance of the classical turning points, for the energy $E = E_n$, from $x = 0$, defined by the following condition:

$$E_n = U(x_n) = ax_n^{2s}, \quad \text{giving } x_n = \left(\frac{E_n}{a}\right)^{1/2s}.$$

Introducing the dimensionless variable $\xi \equiv x/x_n$, so that $dx = x_n d\xi$, and using the potential's symmetry with respect to the origin, we get

$$\int_{x_L}^{x_R} p(x)dx = 2\int_0^{x_n} p(x)dx = 2(2mE_n)^{1/2}x_n\int_0^1 (1 - \xi^{2s})^{1/2}d\xi$$
$$\approx (8mE_n)^{1/2}\left(\frac{E_n}{a}\right)^{1/2s}\int_0^1 (1 - \xi^{2s})^{1/2}d\xi.$$

This is a table integral[55], equal to $(\pi^{1/2}/4s)\Gamma(1/2s)/\Gamma(3/2 + 1/2s)$, so that Eq. (*) yields the following energy spectrum:

$$E_n = E_{\text{WKB}}\,(2n - 1)^{2s/(s+1)}, \quad \text{for } n = 1, 2, \dots, \qquad (**)$$

where $E_{\text{WKB}} \equiv E_1$ is the ground-state energy:

$$E_{\text{WKB}} = \left\{\left(\frac{\hbar^2}{2m}\right)^s a\left[\pi^{1/2}s\frac{\Gamma(3/2 + 1/2s)}{\Gamma(1/2s)}\right]^{2s}\right\}^{1/(s+1)}. \qquad (***)$$

As Eq. (**) shows, for the quadratic-parabolic potential ($s = 1$, i.e. $2s/(s + 1) = 1$), the energy levels are equidistant:

$$E_n = E_1(2n - 1), \quad \text{i.e. } E_{n+1} - E_n = 2E_1 = \text{const},$$

as they should be for a harmonic oscillator—see, e.g. Eq. (2.114). However, as the parameter s grows, i.e. as the particle confinement becomes more rigid, the ratio $2s/(s + 1)$ tends to 2, i.e. the dependence of E_n on n gradually approaches the

[55] See, e.g. Eq. (A.33*b*).

quadratic one, $E_n \propto n^2$, pertinent to the hard-wall well discussed in section 1.7—see Eq. (1.85).

(ii) Since the potential is symmetric with respect to point $x = 0$, and continuous at this (and all other) points, the most natural selection of the ground-state trial function is a Gaussian, for example

$$\psi_{\text{trial}}(x) = C\exp\left\{-\frac{\lambda^2x^2}{4}\right\},$$

with some real λ. The normalization coefficient C may be immediately found either from the standard Gaussian integration of $|\psi_{\text{trial}}|^2$, or just from comparison of this expression with Eq. (2.16) of the lecture notes, in which $\lambda = 1/\delta x$, i.e. $\delta x = 1/\lambda$, giving

$$|C|^2 = \frac{1}{(2\pi)^{1/2}\delta x} = \frac{\lambda}{(2\pi)^{1/2}}.$$

Now the expectation value of the particle's Hamiltonian,

$$\hat{H} = \frac{\hat{p}^2}{2m} + U(x) = -\frac{\hbar^2}{2m}\frac{d^2}{dx^2} + ax^{2s},$$

in the trial state, may be calculated as

$$\begin{aligned}
\langle H\rangle_{\text{trial}} &\equiv \int_{-\infty}^{+\infty}\psi^*_{\text{trial}}\left(-\frac{\hbar^2}{2m}\frac{d^2}{dx^2} + ax^{2s}\right)\psi_{\text{trial}}dx \\
&= \int_{-\infty}^{\infty} C^*\exp\left\{-\frac{\lambda^2x^2}{4}\right\}\left(-\frac{\hbar^2}{2m}\frac{d^2}{dx^2} + ax^{2s}\right)C\exp\left\{-\frac{\lambda^2x^2}{4}\right\}dx \\
&= 2\frac{\lambda}{(2\pi)^{1/2}}\left(\frac{\hbar^2\lambda^2}{4m}\frac{2^{1/2}}{\lambda}\int_0^{\infty}e^{-\xi^2}\,d\xi\right. \\
&\quad \left. - \frac{\hbar^2\lambda^4}{8m}\frac{2^{3/2}}{\lambda^3}\int_0^{\infty}\xi^2e^{-\xi^2}\,d\xi + a\frac{2^{s+1/2}}{\lambda^{2s+1}}\int_0^{\infty}\xi^{2s}e^{-\xi^2}\,d\xi\right).
\end{aligned}$$

All three integrals are of the same well-known type[56], yielding

$$\begin{aligned}
\langle H\rangle_{\text{trial}} &= 2\frac{\lambda}{(2\pi)^{1/2}}\left[\frac{\hbar^2\lambda^2}{4m}\frac{2^{1/2}}{\lambda}\frac{\pi^{1/2}}{2} - \frac{\hbar^2\lambda^4}{8m}\frac{2^{3/2}}{\lambda^3}\frac{\pi^{1/2}}{4} + a\frac{2^{s+1/2}}{\lambda^{2s+1}}\frac{1}{2}\Gamma\left(s + \frac{1}{2}\right)\right]. \\
&\equiv \frac{\hbar^2}{8m}\lambda^2 + a\frac{2^s}{\pi^{1/2}}\Gamma\left(s + \frac{1}{2}\right)(\lambda^2)^{-s}.
\end{aligned}$$

Since for $s > 0$ this expression is positive for any λ^2, and diverges at both $\lambda^2 \to 0$ (due to the second term) and $\lambda^2 \to \infty$ (due to the first term), it always has a minimum at some $\lambda^2 = \lambda^2_{\text{opt}}$, which may be found from the requirement

[56] See, e.g. Eq. (A.36a).

$$\left.\frac{\partial\langle H\rangle_{\text{trial}}}{\partial(\lambda^2)}\right|_{\lambda=\lambda_{\text{opt}}} \equiv \frac{\hbar^2}{8m} - a\frac{2^s}{\pi^{1/2}}\Gamma\left(s+\frac{1}{2}\right)s(\lambda^2)^{-s-1}\Big|_{\lambda=\lambda_{\text{opt}}} = 0,$$

giving

$$\lambda_{\text{opt}}^2 = \left[\frac{8m}{\hbar^2}a\frac{2^s s}{\pi^{1/2}}\Gamma\left(s+\frac{1}{2}\right)\right]^{\frac{1}{s+1}},$$

$$E_{\text{var}} \equiv \langle H\rangle_{\text{trial}}\big|_{\lambda=\lambda_{\text{opt}}} = \left[\left(\frac{\hbar^2}{2m}\right)^s a\frac{s2^{-s}}{\pi^{1/2}}\Gamma\left(s+\frac{1}{2}\right)\right]^{\frac{1}{s+1}}(1+1/s^s).$$

For the quadratic potential, with $s = 1$, and hence $\Gamma(s + 1/2) = \Gamma(3/2) = \pi^{1/2}/2$,[57] both the last expression and Eq. (***) yield the same (and exact!) result

$$E_g = \hbar\left(\frac{a}{2m}\right)^{1/2} \equiv \frac{\hbar\omega_0}{2},$$

where $\omega_0 \equiv (2a/m)^{1/2}$ is the classical frequency of this harmonic oscillator. However, as the table below shows[58], with the growth of the parameter s, the variational method starts to give unreasonably high values of the ground state energy.

	WKB approximation	Variational method with the Gaussian trial function
s	$E_{\text{WKB}}/E_s = [\pi^{1/2}s\,\Gamma(3/2+1/2s)/\Gamma(1/2s)]^{2s/(s+1)}$	$E_{\text{var}}/E_s = (1+1/s^s)[(s\,2^{-s})\,\Gamma(s+½)/\pi^{1/2}]^{1/(s+1)}$
1	1	1
2	0.8671	0.9014
3	0.8008	0.9496
…	…	…
100	0.6236	18.769

This is only natural, because E_{var} always gives an overestimate of the ground state energy, and also because for harder-wall potential wells with higher values of s, the Gaussian, with its relatively slow decay at large x^2, becomes as increasingly inadequate choice for the trial function.

***Problem* 2.35.** Use the variational method to estimate the 1st excited state of the 1D harmonic oscillator.

Solution: As was mentioned in section 2.9 of the lecture notes, this may be done by requiring the new trial function to be orthogonal to the previously calculated ground state's eigenfunction, in our current case given by Eq. (2.275):

[57] See, e.g. Eq. (A.34*e*).

[58] In this table, the energy is normalized by the natural energy unit of this problem, $E_s \equiv [(\hbar^2/2m)^s a]^{1/(s+1)}$.

$$\psi_g = \frac{1}{\pi^{1/4} x_0^{1/2}} \exp\left\{-\frac{x^2}{2x_0^2}\right\}, \quad \text{with} \quad x_0 \equiv \left(\frac{\hbar}{m\omega_0}\right)^{1/2}. \qquad (*)$$

This wavefunction is symmetric, and has no zeros; hence, in the light of the Sturm oscillation theorem mentioned in section 2.9, it is very natural to look for the first excited state's wavefunction in the form

$$\psi_{\text{trial}} = Cx \exp\{-\lambda x^2\}, \quad \text{with} \quad \lambda > 0,$$

because it is asymmetric (and hence automatically orthogonal to ψ_g), and has just one zero. The normalization requirement,

$$\int_{-\infty}^{+\infty} \psi^*_{\text{trial}} \psi_{\text{trial}}\, dx \equiv |C^2| \int_{-\infty}^{+\infty} x^2 \exp\{-2\lambda x^2\}\, dx$$
$$\equiv \frac{|C^2|}{(2\lambda)^{3/2}} 2 \int_0^{+\infty} \xi^2 \exp\{-\xi^2\}\, d\xi = 1,$$

with $\xi \equiv (2\lambda)^{1/2}x$, includes a well-known table integral[59], equal to $\pi^{1/2}/4$, and hence yields

$$|C|^2 = \frac{2(2\lambda)^{3/2}}{\pi^{1/2}}.$$

With this normalization, the Hamiltonian's expectation value is

$$\langle H\rangle_{\text{trial}} \equiv \int_{-\infty}^{+\infty} \psi^*_{\text{trial}} \left(-\frac{\hbar^2}{2m}\frac{d^2}{dx^2} + \frac{m\omega_0^2 x^2}{2}\right) \psi_{\text{trial}} dx$$
$$= |C|^2 \int_{-\infty}^{\infty} x \exp\{-\lambda x^2\} \left(-\frac{\hbar^2}{2m}\frac{d^2}{dx^2} + \frac{m\omega_0^2 x^2}{2}\right) x \exp\{-\lambda x^2\} dx$$
$$\equiv \frac{2(2\lambda)^{3/2}}{\pi^{1/2}} \left[6\lambda \frac{\hbar^2}{2m} \frac{1}{(2\lambda)^{1/2}} 2 \int_0^{\infty} e^{-\xi^2}\, d\xi \right.$$
$$\left. + \left(-4\lambda^2 \frac{\hbar^2}{2m} + \frac{m\omega_0^2}{2}\right) \frac{1}{(2\lambda)^{5/2}} 2 \int_0^{\infty} \xi^4 e^{-\xi^2}\, d\xi \right],$$

with the same notation $\xi \equiv (2\lambda)^{1/2}x$. The former of these two dimensionless integrals is the same as above, and the latter one is of the same type[60], equal to $3\pi^{1/2}/8$. As a result, we get

$$\langle H\rangle_{\text{trial}} = 3\left(\lambda \frac{\hbar^2}{2m} + \frac{1}{4\lambda} \frac{m\omega_0^2}{2}\right).$$

[59] See, e.g. Eq. (A.36*c*).

[60] See, e.g. Eq. (A.36*d*).

The (only) minimum of this function of λ is achieved, not quite surprisingly, at the same value

$$\lambda_{\text{opt}} = \frac{m\omega_0}{2\hbar} \equiv \frac{1}{2x_0^2},$$

as for the ground-state wavefunction (*), so that the resulting 1st excited state's wavefunction is proportional to the same exponent:

$$\psi_1 \equiv (\psi_{\text{trial}})_{\lambda=\lambda_{\text{opt}}} = \frac{2^{1/2}}{\pi^{1/4}x_0^{3/2}}x\exp\left\{-\frac{x^2}{2x_0^2}\right\}.$$

Comparing this expression with Eq. (2.284) of the lecture notes for $n = 1$, and taking into account Eq. (2.282) for H_1, we see that for the harmonic oscillator, the variational method yields the exact expression for $\psi_1(x)$, and hence for the corresponding eigenenergy:

$$E_1 \equiv (\langle H_{\text{trial}}\rangle)_{\lambda=\lambda_{\text{opt}}} = \frac{3\hbar\omega_0}{2}.$$

Note, however, that the further development of this success would require a rapidly increasing volume of calculations. Indeed, as Eqs. (2.282) and (2.284) show, the next eigenfunction, $\psi_2(x)$, is proportional to the Hermite polynomial $H_2(x/x_0) = (4x^2/x_0^2 - 2)$, rather than just to some power of x as $\psi_1(x)$ is, so that looking for it using the variational approach would require at least two adjustable parameters, for example

$$\psi_{\text{trial}} = C(x^2 + \lambda_1)\exp\{-\lambda_2 x^2\}.$$

***Problem* 2.36.** Assuming the quantum effects to be small, calculate the lower part of the energy spectrum of the following system: a small bead of mass m, free to move without friction along a ring of radius R, which is rotated about its vertical diameter with a constant angular velocity ω—see the figure below[61]. Formulate a quantitative condition of validity of your results.

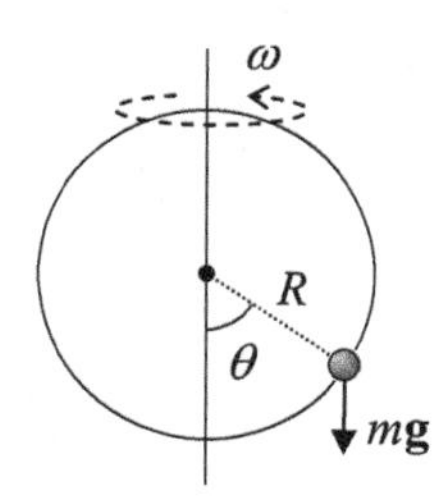

[61] This system was used as the analytical mechanics' 'testbed problem' in *Part CM* of this series, and the reader is welcome to use any relations derived there.

Solution: As was discussed in *Part CM* of this series, the classical Hamiltonian function of the system has the form[62]

$$H = \frac{p_\theta^2}{2mR^2} + U_{\text{ef}}(\theta),$$

with

$$U_{\text{ef}}(\theta) = -mgR\cos\theta - \frac{m}{2}R^2\omega^2\sin^2\theta \equiv -mgR\left(\cos\theta + \frac{\omega^2}{2\Omega^2}\sin^2\theta\right),$$

where p_θ is the generalized momentum corresponding to the generalized coordinate θ (the angle of the bead's deviation from the lowest point of the ring—see the figure above), and $\Omega \equiv (g/R)^{1/2}$ is the frequency of small oscillations of the bead near that point in the case $\omega = 0$ (no ring rotation).

The transition to quantum mechanics may be achieved, as was discussed in chapter 1, by using the corresponding Hamiltonian operator,

$$\hat{H} = \frac{\hat{p}_\theta^2}{2mR^2} + U_{\text{ef}}(\theta), \quad \text{with} \quad \hat{p}_\theta = -i\hbar\frac{\partial}{\partial\theta}. \qquad (*)$$

Since the function $U_{\text{ef}}(\theta)$ is not quite trivial (see the figure below), in the general case the eigenvalues of this Hamiltonian cannot be calculated analytically. However, if the quantum contributions to a system's lowest energies are small, it is sufficient to consider only small vicinities of the minima of this effective potential.

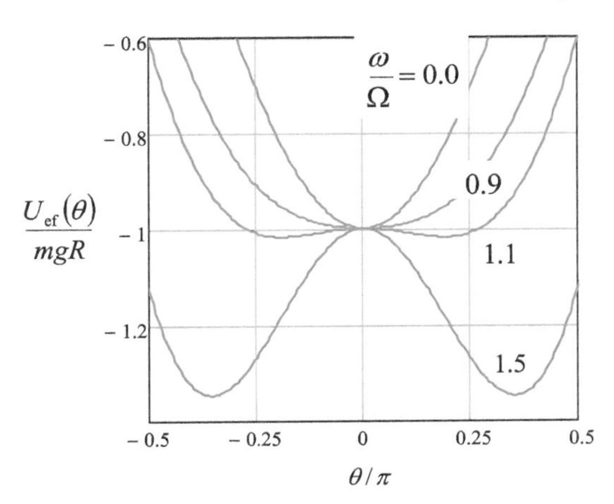

If the ring's rotation is slow, $\omega^2 \ll \Omega^2 \equiv g/R$, the function $U_{\text{ef}}(\theta)$ has only one minimum, at the lower point of the ring: $\theta_0 = 0$. On the other hand, if the rotation velocity ω exceeds the threshold value equal to Ω, there are two similar minima of

[62] This result may be readily obtained using either the Lagrangian formalism in an inertial ('lab') reference frame, or in the rotating (non-inertial) reference frame—see, e.g. *Part CM* sections 2.2 and 4.6, respectively. At the latter approach, the second term in the above expression for U_{ef} is just the additional potential energy of the bead in the field of the centrifugal 'inertial force' $\mathbf{F}_c = -m\mathbf{a}_c \equiv -m\boldsymbol{\omega} \times (\boldsymbol{\omega} \times \mathbf{r})$—see *Part CM* Eq. (4.93) and figure 4.14.

$U_{ef}(\theta)$ at two symmetric points $\theta_1 = \pm\sin^{-1}(\Omega^2/\omega^2) > 0$, corresponding to the bead's rotation at the opposite sides of the ring. Taylor-expanding the effective potential energy near these points, and keeping only two first leading terms of the series, we get

$$U_{ef}(\theta) - U_{min} = \frac{mR^2\tilde{\theta}^2}{2} \times \begin{cases} (\Omega^2 - \omega^2) > 0, & \text{for } \omega^2 < \Omega^2, \quad \text{where } \tilde{\theta} \equiv \theta - \theta_0, \\ (\omega^2 - \Omega^2) > 0, & \text{for } \Omega^2 < \omega^2, \quad \text{where } \tilde{\theta} \equiv \theta - \theta_1. \end{cases} \quad (**)$$

In this approximation, the Hamiltonian (*) is reduced to that of a harmonic oscillator with a frequency equal to either $\Omega_0 \equiv (\Omega^2 - \omega^2)^{1/2}$ (if $\omega^2 < \Omega^2$), or $\Omega_1 \equiv (\omega^2 - \Omega^2)^{1/2}$ (if $\Omega^2 < \omega^2$). Hence the lower part of the energy spectrum is well be described, in both cases, by Eq. (2.262) of the lecture notes:

$$E_n = U_{min} + \left(n + \frac{1}{2}\right) \times \begin{cases} \hbar\Omega_0, & \text{for } \omega^2 < \Omega^2, \\ \hbar\Omega_1, & \text{for } \Omega^2 < \omega^2. \end{cases} \quad (***)$$

These expressions are only correct when the energy E_n is within the range where the expansion (**) is valid, i.e. only if $E_n - U_{min} \ll U_{max}$, giving the condition.

$$n \ll n_{max} \equiv \frac{mR^2}{\hbar} \max[\Omega, \omega].$$

If n_{max} so defined is less then, or even of the order of 1, the quantum effects are strong for all n, and the harmonic-oscillator approximation is not valid at all. Note, however, that in the opposite limit of very strong quantum effects, when $n_{max} \ll 1$, i.e. when $\hbar^2/mR^2 \gg (U_{ef})_{max} - (U_{ef})_{min}$, the system's properties become very simple again. (This *plane rotator* approximation will be discussed in section 3.5 of the lecture notes.)

Finally, let me emphasize that the above result (***) refers to the eigenvalues of the Hamiltonian (*), i.e. the *effective* energy of the bead in the non-inertial reference frame rotating with the ring, but *not* of the 'genuine' mechanical energy in a 'lab' (inertial) reference frame. In this particular system, the latter energy is *not* an integral of motion, and hence does not have stationary values. (The reader to whom this point is not clear is strongly advised to revisit a discussion of this issue in classical mechanics—for example, in *Part CM* sections 2.3 and 4.6 of this series.)

***Problem* 2.37.** A 1D harmonic oscillator, with mass m and frequency ω_0, had been in its ground state; then an additional force F was suddenly applied, and then retained constant in time. Calculate the probability of the oscillator staying in its ground state.

Solution: The ground-state wavefunction of the initial oscillator is given by Eq. (2.275) of the lecture notes, which may be recast as

$$\psi_{\text{ini}}(x) = \frac{1}{\pi^{1/4}x_0^{1/2}} \exp\left\{-\frac{x^2}{2x_0^2}\right\},$$

where $x_0 \equiv (\hbar/m\omega_0)^{1/2}$. Since the wavefunction does not have time to change during the abrupt application of the force, it plays the role of the initial condition, $\Psi(x, 0)$, for the final system, described by the modified Hamiltonian

$$\begin{aligned}\hat{H} &= -\frac{\hbar^2}{2m}\frac{d^2}{dx^2} + \frac{m\omega_0^2}{2}x^2 - Fx \\ &\equiv -\frac{\hbar^2}{2m}\frac{d^2}{dx^2} + \frac{m\omega_0^2}{2}(x - X)^2 + \text{const}, \quad \text{with} \quad X \equiv \frac{F}{m\omega_0^2}.\end{aligned}$$

From the last form of the right-hand side, it is evident that the modified Hamiltonian differs from the initial one only by the shift X of the argument—which of course is just the classically-calculated static extension F/κ of the oscillator's spring, with the elastic constant $\kappa = m\omega_0^2$, by the applied force F. Hence the ground-state wavefunction of the final system differs from the initial one only by this shift:

$$\psi_{\text{fin}}(x) = \frac{1}{\pi^{1/4}x_0^{1/2}} \exp\left\{-\frac{(x - X)^2}{2x_0^2}\right\}.$$

Now we can calculate the requested probability as $W_0 = |c_0|^2$, where the coefficient c_0 is given by the 1D version of Eq. (1.68):

$$\begin{aligned}c_0 &= \int \psi_{\text{fin}}^*(x)\psi_{\text{ini}}(x)dx = \frac{1}{\pi^{1/2}x_0}\int_{-\infty}^{+\infty} \exp\left\{-\frac{x^2 + (x - X)^2}{2x_0^2}\right\}dx \\ &\equiv \frac{1}{\pi^{1/2}}\int_{-\infty}^{+\infty} \exp\left\{-\xi^2 + \xi\xi_X - \frac{1}{2}\xi_X^2\right\} d\xi,\end{aligned}$$

where $\xi \equiv x/x_0$, and $\xi_X \equiv X/x_0 \equiv (F/m\omega_0^2)/(\hbar/m\omega_0)^{1/2}$. This is a Gaussian integral, which may be readily worked out by the same completion to the full square as was repeatedly used in chapter 2:

$$c_0 = \frac{1}{\pi^{1/2}}\int_{-\infty}^{+\infty} \exp\left\{-\left(\xi - \frac{1}{2}\xi_X\right)^2 - \frac{1}{4}\xi_X^2\right\}d\xi = \exp\left\{-\frac{\xi_X^2}{4}\right\}.$$

so that, finally,

$$W_0 = \exp\left\{-\frac{\xi_X^2}{2}\right\} \equiv \exp\left\{-\frac{X^2}{2x_0^2}\right\} \equiv \exp\left\{-\frac{F^2}{2\hbar m\omega_0^3}\right\}.$$

The probability decreases rapidly as soon as the force becomes larger than the so-called *standard quantum limit*

$$F_0 \equiv (\hbar m\omega_0^3)^{1/2};$$

this constant serves as a natural scale for the force effect masking by quantum uncertainty.

***Problem* 2.38.** A 1D particle of mass m had been placed into a quadratic potential well (2.111),

$$U(x) = \frac{m\omega_0^2 x^2}{2},$$

and allowed to relax into the ground state. At $t = 0$, the well was fast accelerated to move with velocity v, without changing its profile, so that at $t \geqslant 0$ the above formula for U is valid with the replacement $x \to x' \equiv x - vt$. Calculate the probability for the system to still be in the ground state at $t > 0$.

Solution: Due to the invariance of the Schrödinger equation with respect to the Galilean transform (whose proof was the task of problem 1.4), in the reference frame moving together with the potential profile, U is the function of the relative coordinate $x' = x - vt$ only, but not of time. As was discussed in section 1.5 of the lecture notes, in such a time-independent profile the stationary state probabilities, in particular that, W_0, of the ground state, cannot change. Hence the system's exit from its ground state can arise only at the moment of the beginning of motion, $t = 0$.

For this moment, the ground state that existed at $t \leqslant 0$, with the wavefunction given by Eq. (2.275) of the lecture notes,

$$\Psi(x, 0) = \psi_0(x) = \frac{1}{\pi^{1/4} x_0^{1/2}} \exp\left\{-\frac{x^2}{2x_0^2}\right\}, \quad \text{where} \quad x_0 = \left(\frac{\hbar}{m\omega_0}\right)^{1/2}, \qquad (*)$$

serves as the initial condition, so that to calculate the requested probability W_0, we may apply Eq. (1.68) written in the moving reference frame:

$$W_0 = |c_0|^2, \quad \text{with } c_0 = \int_{-\infty}^{+\infty} \psi_0^*(x')\, \Psi'(x', 0) dx'.$$

Here $\psi_0(x')$ is given by the same Eq. (*), with the replacement $x \to x'$, because in the moving reference frame the potential $U(x')$, and hence the ground state wavefunction, are exactly the same as they are in the lab frame at $t \leqslant 0$. However, the initial wavefunction $\Psi'(x', 0)$ has to be recalculated from $\Psi(x, 0)$ using the wavefunction transform whose proof was the subject of the same problem 1.4; for the 1D case

$$\Psi'(x', t') = \Psi(x, t) \exp\left\{-i\frac{mvx}{\hbar} + i\frac{mv^2 t}{2\hbar}\right\}.$$

For $t = 0$, when $x' = x$, this transform is reduced to

$$\Psi'(x', 0) = \Psi(x, 0) \exp\left\{-i\frac{mvx}{\hbar}\right\} = \psi_0(x') \exp\left\{-i\frac{mvx'}{\hbar}\right\}$$

so that

$$c_0 = \int_{-\infty}^{+\infty} \psi_0^*(x')\psi_0(x') \exp\left\{-i\frac{mvx'}{\hbar}\right\} dx'$$
$$= \frac{1}{\pi^{1/2}x_0}\int_{-\infty}^{+\infty} \exp\left\{-\frac{x'^2}{x_0^2} - i\frac{mvx'}{\hbar}\right\}dx'.$$

This is a standard Gaussian integral, with the structure absolutely similar, for example, to that in Eq. (2.21) of the lecture notes, which was worked out in detail in section 2.2. An absolutely similar calculation yields

$$c_0 = \exp\left\{-\frac{v^2}{4v_0^2}\right\}, \qquad \text{so that } W_0 = |c_0|^2 = \exp\left\{-\frac{v^2}{2v_0^2}\right\},$$
$$\text{where} \quad v_0 \equiv \omega_0 x_0 = \left(\frac{\hbar\omega_0}{m}\right)^{1/2}.$$

This result shows that if the motion velocity v is much lower than the natural quantum-mechanical scale v_0 of the particle's 'motion' in its ground state[63], then $W_0 \to 1$, i.e. the oscillator remains in its ground state with an almost 100% probability. If, conversely, $v \gg v_0$, then $W_0 \to 0$, meaning that the abrupt start of the potential well's motion almost certainly (with the probability $1 - W_0 \to 1$) 'shakes up' the oscillator into a linear superposition of its excited states.

***Problem* 2.39.** A 1D harmonic oscillator had initially been in its ground state. At a certain moment of time, its spring constant κ is abruptly increased, so that its frequency $\omega_0 = (\kappa/m)^{1/2}$ is increased by a factor of α, and then is kept constant at the new value. Calculate the probability that after the change, the oscillator is still in its ground state.

Solution: According to Eq. (2.275) of the lecture notes, the ground state of the initial system is

$$\psi_{\text{ini}}(x) = \left(\frac{m\omega_0}{\pi\hbar}\right)^{1/4} \exp\left\{-\frac{m\omega_0 x^2}{2\hbar}\right\}.$$

Since this wavefunction does not have time to change during the abrupt parameter change, it plays the role of the initial condition, $\Psi(x,0)$, for the new system (the oscillator with the new spring constant). Hence we can use the 1D version of Eq. (1.68) to calculate the overlap integral c_0 of this function with the similar ground state eigenfunction of the finite system (in which we should make replacement $\omega_0 \to \alpha\omega_0$):

[63] For example, it is easy (and hence left for the reader:-) to use Eq. (*) to prove that the expectation value of the observable $(p/m)^2$, i.e. of the square of particle's velocity, in the ground state equals $v_0^2/2$.

$$
\begin{aligned}
c_0 &= \int_{-\infty}^{+\infty} \psi_{\text{fin}}^*(x)\psi_{\text{ini}}(x)dx \\
&= \int_{-\infty}^{+\infty} \left(\frac{m\omega_0}{\pi\hbar}\right)^{1/4} \exp\left\{-\frac{m\omega_0 x^2}{2\hbar}\right\}\left(\frac{m\alpha\omega_0}{\pi\hbar}\right)^{1/4} \exp\left\{-\frac{m\alpha\omega_0 x^2}{2\hbar}\right\}dx \\
&= \alpha^{1/4}\left(\frac{m\omega_0}{\pi\hbar}\right)^{1/2} \int_{-\infty}^{+\infty} \exp\left\{-(1+\alpha)\frac{m\omega_0 x^2}{2\hbar}\right\}dx \\
&\equiv \frac{\alpha^{1/4}}{\pi^{1/2}(1+\alpha)^{1/2}} \int_{-\infty}^{+\infty} \exp\left\{-\frac{\xi^2}{2}\right\}d\xi = \frac{2^{1/2}\alpha^{1/4}}{(1+\alpha)^{1/2}}.
\end{aligned}
$$

From this result, the probability of the oscillator staying in the ground state is:

$$W_0 = |c_0|^2 = \frac{2\alpha^{1/2}}{1+\alpha}.$$

This function is plotted in the figure below. As a sanity check, at $\alpha = 1$ (i.e. no change at all), $W_0 = 1$—as it should be. If the spring constant *has* been changed, $W_0 < 1$ both at $\alpha > 1$ (as in the problem's assignment), and at $\alpha < 1$, i.e. at the spring constant reduction.

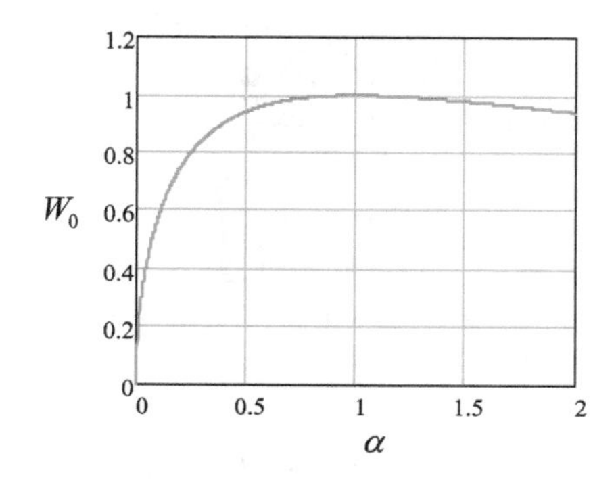

***Problem* 2.40.** A 1D particle is placed into the following potential well:

$$U(x) = \begin{cases} +\infty, & \text{for } x < 0, \\ m\omega_0^2 x^2/2, & \text{for } x \geqslant 0. \end{cases}$$

(i) Find its eigenstates and eigenenergies.
(ii) This system had been allowed to relax into its ground state, and then the infinite potential wall at $x < 0$ was rapidly removed, so that the system was instantly turned into the usual harmonic oscillator (with the same m and ω_0). Find the probability for the oscillator to be in its ground state.

Solutions:

(i) The stationary Schrödinger equation of the initial system at $x > 0$ coincides with that of the usual harmonic oscillator, and is hence satisfied by any of its eigenfunctions—see Eq. (2.284) of the lecture notes. However, the infinite potential at $x < 0$ imposes the boundary condition $\psi_n(0) = 0$, which is satisfied only by asymmetric eigenfunctions with odd quantum numbers $n = 2m + 1$ ($m = 0, 1, 2, \ldots$). Taking into account that the wavefunctions should be now normalized on the segment $0 < x < +\infty$, rather than $-\infty < x < +\infty$, we finally get

$$\psi_m(x) = \begin{cases} 0, & \text{for } x < 0, \\ \dfrac{1}{[2^{2m}(2m+1)!]^{1/2}\pi^{1/4}x_0^{1/2}} \exp\left\{-\dfrac{x^2}{2x_0^2}\right\} H_{2m+1}\left(\dfrac{x}{x_0}\right), & \text{for } x > 0, \end{cases}$$

where x_0 is given by Eq. (2.76), and the Hermite polynomials $H_n(\xi)$ may be defined by Eq. (2.281).

(ii) Taking into account that, according to Eq. (2.282), $H_1(\xi) = 2\xi$, for the ground state of the system (with $m = 0$), the above result is reduced to

$$\psi_0(x) = \begin{cases} 0, & \text{for } x < 0, \\ \dfrac{1}{\pi^{1/4}x_0^{1/2}} \dfrac{2x}{x_0} \exp\left\{-\dfrac{x^2}{2x_0^2}\right\}, & \text{for } x > 0. \end{cases}$$

After the fast removal of the wall, this function plays the role of the initial condition $\Psi(x, 0)$ for the resulting harmonic oscillator, so that we may calculate the requested probability as $W_g = |c_g|^2$, with the coefficient c_g calculated according to the 1D version of Eq. (1.68):

$$c_g = \int \psi_g^*(x)\psi_0(x)dx,$$

where $\psi_g(0)$ is the ground-state wavefunction of the usual harmonic oscillator, given by Eq. (2.275):

$$\psi_g(x) = \frac{1}{\pi^{1/4}x_0^{1/2}} \exp\left\{-\frac{x^2}{2x_0^2}\right\}.$$

As a result, we get

$$c_g = \frac{1}{\pi^{1/2}x_0}\int_0^\infty \frac{2x}{x_0} \exp\left\{-\frac{x^2}{x_0^2}\right\}dx \equiv \frac{1}{\pi^{1/2}}\int_0^\infty e^{-\xi}d\xi = \frac{1}{\pi^{1/2}},$$

so that, finally, $W_g = 1/\pi \approx 0.318$.

***Problem* 2.41.** Prove the following formula for the propagator of the 1D harmonic oscillator:

$$G(x,\ t;\ x_0,\ t_0) = \left(\frac{m\omega_0}{2\pi i\hbar\ \sin[\omega_0(t - t_0)]}\right)^{1/2}$$
$$\times \exp\left\{\frac{im\omega_0}{2\hbar\ \sin[\omega_0(t - t_0)]}\left[\left(x^2 + x_0^2\right)\cos[\omega_0(t - t_0)] - 2xx_0\right]\right\}.$$

Discuss the relation between this formula and the propagator of a free 1D particle.

Solution: According to its definition, given by Eq. (2.44) of the lecture notes (see also Eqs. (2.45) and (2.46) and their discussion), the propagator $G(x,\ t;\ x_0,\ t_0)$ of a 1D quantum system has to satisfy two conditions:

(i) if considered as a function of x and t only, it should obey the Schrödinger equation of the system, and
(ii) it has to approach $\delta(x - x_0)$ at $t \to t_0$.

For our case, condition (i) may be checked by a direct differentiation of G over x (twice) and t, and plugging the results into the Schrödinger equation (2.261):

$$i\hbar\frac{\partial G}{\partial t} = \hat{H}G \equiv -\frac{\hbar^2}{2m}\frac{\partial^2 G}{\partial x^2} + \frac{m\omega_0^2 x^2}{2}G.$$

In order to check condition (ii) we may notice that in the limit $(t - t_0) \ll 1/\omega_0$, the propagator coincides with that of the free particle, given by Eq. (2.49) of the lecture notes, for which condition (ii) is satisfied by its construction—see section 2.2.

***Problem* 2.42.** In the context of the Sturm oscillation theorem, mentioned in section 2.9 of the lecture notes, prove that the number of eigenfunction zeros of a particle confined in an arbitrary but finite potential well, always increases with the corresponding eigenenergy.

Hint: You may like to use the suitably modified Eq. (2.186).

Solution: Repeating the calculation that has led to Eq. (2.186), but now for two stationary states, with numbers n and n', and $E_{n'} > E_n$, for the x-segment limited by two adjacent zeros, x_m and x_{m+1}, of the stationary wavefunction corresponding to the lower energy, $\psi_n(x)$, we get

$$(E_{n'} - E_n)\int_{x_m}^{x_{m+1}} \psi_n\psi_{n'}dx = \frac{\hbar^2}{2m}\left[\frac{d\psi_n}{dx}\psi_{n'}\right]_{x_m}^{x_{m+1}}, \quad \text{where} \quad \psi_n(x_m) = \psi_n(x_{m+1}) = 0. \quad (*)$$

Since the zero points x_m and x_{m+1} are adjacent, the function $\psi_n(x)$ does not change its sign between them. Since the wavefunctions are defined to an arbitrary complex multiplier $\exp\{i\varphi\}$, with real φ, let it select this constant[64] so that $\psi_n(x)$ is real and positive on the interval $x_m < x < x_{m+1}$. Then $d\psi_n/dx$ has to be positive (or equal zero) at $x = x_m$ and negative (or equal zero) at $x = x_{m+1}$—see the figure below.

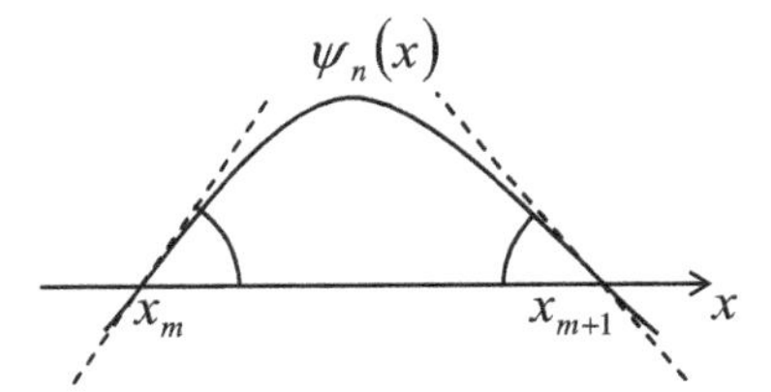

Let us assume for a minute that the function $\psi_{n'}(x)$, corresponding to the larger energy $E_{n'} > E_n$, also does not have a zero at this interval; in this case we may also make this function real and positive on the whole interval $[x_m, x_{m+1}]$ by the appropriate choice of its phase factor. Then the left-hand side of Eq. (*) is positive, while its right-hand side is either negative or equal to zero. Hence our assumption has been wrong, i.e. the function $\psi_{n'}(x)$ has at least one zero at the interval $x_m < x < x_{m+1}$. (It may be useful for the reader to revisit figures 1.8 and 2.35 of the lecture notes to see how spectacularly this general result works for the particular cases of hard and soft confinement.)

Now let us apply this result to each inter-zero interval of the function $\psi_n(x)$, noticing that it is also valid for infinite intervals, with $x_m \to -\infty$ and/or $x_{m+1} \to +\infty$. (In these cases, the product $(d\psi_n/dx)\psi_{n'}$ in Eq. (*) equals zero at the corresponding end of the interval; note that the zero(s) of the function $\psi_{n'}$ at such an interval still have to be finite.) If the function $\psi_n(x)$ has M finite zeros x_m, there are $(M + 1)$ of such intervals, and hence the function $\psi_{n'}(x)$ has at least $(M + 1)$ finite zeros. So the statement made in the assignment is indeed correct[65].

Problem* 2.43. Use the WKB approximation to calculate the lifetime of the metastable ground state of a 1D particle of mass m in the 'pocket' of the potential profile

$$U(x) = \frac{m\omega_0^2}{2}x^2 - \alpha x^3.$$

Contemplate the significance of this problem.

[64] According to Eq. (2.5) of the lecture notes, the phase φ of a stationary wavefunction of a confined state, with the probability current $I = 0$, cannot depend on x.

[65] Other facts necessary for the full proof of the Sturm oscillation theorem, namely that M grows exactly by 1 at each step of the energy spectrum ladder, and equals zero for the ground state, require more refined arguments. However, they are virtually evident from the WKB-based Bohr–Sommerfeld quantization rule (2.110). Indeed, each new half-wave of the wavefunction corresponds to the increase of $\Delta\varphi_{\rightarrow}$, defined by Eq. (2.108), by π, of the $\Delta\varphi_{\leftarrow} = \Delta\varphi_{\rightarrow}$ also by π, i.e. the increase of the total wave change (2.109) by 2π, i.e. to the increase of the quantum number n by 1.

Solution: This potential profile, sketched in the figure below for the case $\alpha > 0$,[66] forms a soft potential well at $x \sim 0$, from which the particle may tunnel into the unrestricted half-space $x > x_0$. As a result, even the ground state of the particle in the well is metastable.

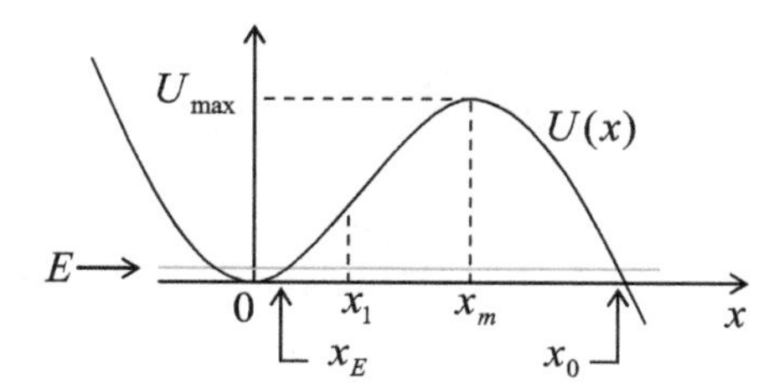

As was discussed in section 2.5 of the lecture notes, the very notion of lifetime τ of such a state is valid only if the potential barrier's transparency $\mathscr{T}$, calculated at the state's energy E, is much less than 1. For a smooth potential like ours, we may crudely estimate the transparency, using the WKB-approximation-based Eq. (2.117), as

$$-\ln \mathscr{T} \sim \frac{2}{\hbar}\{2m[U_{\max} - E]\}^{1/2} x_0.$$

Calculating x_0 and $U_{\max}$ for our potential from conditions $U(x) = 0$ at $x = x_0$, and $(dU/dx) = 0$ at $x = x_m$,

$$x_0 = \frac{m\omega_0^2}{2\alpha}, \qquad U_{\max} \equiv U(x_m) = \frac{\left(m\omega_0^2\right)^3}{54\alpha^2},$$

$$\text{so that} \quad x_m = \frac{m\omega_0^2}{3\alpha} \equiv \frac{2}{3}x_0,$$

and plugging these results to the above estimate, we see that the condition $\mathscr{T} \ll 1$ requires that

$$E \ll U_{\max}.$$

(In this limit we may use Eq. (2.274) to write

$$E = \frac{\hbar\omega_0}{2},$$

because for the small distances from the potential well's bottom, at which the ground-state wavefunction is localized, the second, cubic term of the potential is negligible.) Due to this condition, the WKB expression,

$$-\ln \mathscr{T} = \frac{2(2m)^{1/2}}{\hbar}\int_{()\geqslant 0}\left(\frac{m\omega_0^2}{2}x^2 - \alpha x^3 - \frac{\hbar\omega_0}{2}\right)^{1/2} dx, \qquad (*)$$

which follows from the general Eq. (2.117) for our potential profile, may be simplified.

[66] This choice of the sign makes all the notation simpler. (All the results for negative α, i.e. for the potential $U(x) = m\omega_0^2x^2/2 + \alpha' x^3$, with $\alpha' = -\alpha > 0$, are evidently similar, with the coordinate inversion $x \to -x$.)

In the crudest approximation, in which the ground-state energy $E = \hbar\omega_0/2$ is neglected completely in comparison with $U(x)$, the integral is simple:

$$-\ln \mathscr{T}_0 \equiv \frac{2(2m)^{1/2}}{\hbar}\int_0^{x_0}\left(\frac{m\omega_0^2}{2}x^2 - \alpha x^3\right)^{1/2} dx \equiv \frac{27U_{\max}}{\hbar\omega_0}\int_0^1 \xi(1-\xi)^{1/2}d\xi,$$

where $\xi \equiv x/x_0$. The last integral may be readily worked out (for example, using the new substitution $\alpha \equiv 1 - \xi$) and is equal to 4/15, so that we get

$$-\ln \mathscr{T}_0 = \frac{36}{5}\frac{U_{\max}}{\hbar\omega_0},$$

which shows that, indeed, if $\hbar\omega_0 \ll U_{\max}$, then $-\ln \mathscr{T}_0 \gg 1$, so that $\mathscr{T}_0 \ll 1$.

Now, using Eq. (2.153) of the lecture notes, we can estimate the metastable lifetime τ as $t_a/\mathscr{T}_0$, where t_a is the period between particle's 'attempts' to pass through the barrier. In our case, these attempts are not clearly separated in time, so we may only crudely estimate t_a as the period $2\pi/\omega_0$ of the classical oscillations at the bottom of the potential well, so that

$$\tau \sim \tau_0 \equiv \frac{t_a}{\mathscr{T}_0} = \frac{2\pi}{\omega_0}\exp\left\{\frac{36}{5}\frac{U_{\max}}{\hbar\omega_0}\right\}. \qquad (**)$$

As was discussed in section 2.3 in the context of a rectangular potential barrier, this expression, with the correct exponent[67], is satisfactory for most practical applications. However, due to the approximate nature of this estimate, we cannot expect its pre-exponential factor to be exact, and even be sure that it is independent of other parameters of the problem. In order to make a more exact calculation, we may need to take into account the small ground-state energy $E = \hbar\omega_0/2 \ll U_{\max}$, in the first nonvanishing approximation. Looking at the figure above, it is clear the effect of non-zero E on the WKB integral (*) is strongest at $x \sim 0$, where the function $U(x)$ grows slowly. In this region, $x_E \leqslant x \leqslant x_1$, where x_E is the left classical turning point, defined by condition

$$E = U(x_E), \quad \text{i.e.} \ \frac{\hbar\omega_0}{2} = \frac{m\omega_0^2 x_E^2}{2}, \quad \text{so that } x_E = \left(\frac{\hbar}{m\omega_0}\right)^{1/2},$$

and x_1 is any middle point (see figure above) satisfying two strong conditions:

$$x_E \ll x_1 \ll x_0, \quad \text{so that } \frac{\hbar\omega_0}{2} \ll \frac{m\omega_0^2 x_1^2}{2} \ll \alpha x_1^3, \qquad (***)$$

we may ignore the potential's anharmonicity αx^3. On the other hand, in the complementary region $x_1 \leqslant x \leqslant x_0$, the anharmonic term has to be treated exactly,

[67] Note the proximity of the numerical coefficient under the exponent, $36/5 = 7.20$, for this cubic-parabolic barrier, to that for the quadratic-parabolic barrier, $2\pi \approx 6.28$—see Eq. (2.119) of the lecture notes, which is correctly described by the WKB approximation at $(U_{\max} - E) \gg \hbar\omega_0$.

but the effects of non-zero energy $E = \hbar\omega_0/2$ may be described in the linear approximation. As a result, the correction to our crude result may be calculated as

$$\Delta(-\ln \mathscr{T}) \equiv (-\ln \mathscr{T}) - (-\ln \mathscr{T})_0$$
$$\approx \frac{2(2m)^{1/2}}{\hbar}\left[\int_{x_E}^{x_1}\left(\frac{m\omega_0^2}{2}x^2 - \frac{\hbar\omega_0}{2}\right)^{1/2} dx - \int_{x_E}^{x_1}\left(\frac{m\omega_0^2}{2}x^2\right)^{1/2} dx\right.$$
$$\left. + \int_{x_1}^{x_0} \frac{\hbar\omega_0}{2}\frac{\partial}{\partial E}\left(\frac{m\omega_0^2}{2}x^2 - \alpha x^3 - E\right)^{1/2}_{E=0} dx\right].$$

Performing the differentiation inside the last integral, and using the notation introduced above to bring the integrals to a dimensionless form, we get

$$-\ln \mathscr{T} \approx 2\int_1^{x_1/x_E \gg 1} (\zeta^2 - 1)^{1/2} d\zeta - 2\int_1^{x_1/x_E \gg 1} \zeta d\zeta - \int_{x_1/x_0 \ll 1}^{1} \frac{d\xi}{\xi(1 - \xi)^{1/2}},$$

where $\zeta \equiv x/x_E$, and $\xi \equiv x/x_0$. The second of these integrals is elementary, while the other two may be also readily worked out: the first one, using the new substitution $\zeta \equiv \cosh \alpha$, and the last one, using the new substitution $\beta \equiv (1 - \xi)^{1/2}$. The final result,

$$\Delta(-\ln \mathscr{T}) \approx -\ln \frac{2x_1}{x_E} - \ln \frac{4x_0}{x_1} \equiv -\ln \frac{8x_0}{x_E} \equiv -\ln \left(\frac{864\, U_{\max}}{\hbar\omega_0}\right)^{1/2},$$

does not depend on the exact choice of the auxiliary parameter x_1 (as it should be for the correctness of our procedure), and we get the corrected WKB expression

$$\mathscr{T} = \left(\frac{864\, U_{\max}}{\hbar\omega_0}\right)^{1/2} \exp\left\{-\frac{36}{5}\frac{U_{\max}}{\hbar\omega_0}\right\}.$$

The second issue we have to attend to is the attempt time, which was just guessed in our crude estimate (**). The most prudent approach here is to determine the lifetime τ directly from the probability decay equation (2.151), just as was done in section 2.3 (for a quantitatively different problem). For that, let us note that within the range (***), the particle's wavefunction may be described from two different points of view. On one hand, it has to correctly describe the ground-state (2.275) of the system:

$$\psi(x) = W^{1/2}\left(\frac{m\omega_0}{\pi\hbar}\right)^{1/4} \exp\left\{-\frac{m\omega_0 x^2}{2\hbar}\right\}, \qquad \text{for } x \ll x_0,$$

where W is the probability to find the particle in this metastable state. On the other hand, under our soft potential barrier, the wavefunction should be well described by the WKB approximation:

$$\psi(x) = \frac{c}{\kappa^{1/2}} \exp\left\{-\int_{x_E}^{x} \kappa(x) dx\right\},$$

$$\text{where} \quad \frac{\hbar^2\kappa^2(m)}{2m} \equiv U(x) - E \approx \frac{m\omega_0^2 x^2}{2} - \frac{\hbar\omega_0}{2}, \qquad \text{for} \quad x_E \ll x \ll x_0.$$

Calculating this simple integral, similar to the first integral in the expression for $\Delta(-\ln \mathscr{T})$, and requiring the two expressions for $\psi(x)$ to coincide at $x_E \ll x$, we get

$$c = \left(\frac{m\omega_0}{2\pi\hbar}\right)^{1/2} W^{1/2}.$$

But due to the WKB connection formulas, discussed in section 2.4 of the lecture notes, applied to the classical turning point x_0, the coefficient c has to be simply related to the WKB amplitude f of the traveling wave behind the barrier, defined by the last line of Eq. (2.116): $|f| = |c|\mathscr{T}^{1/2}$. Hence the probability current (2.95) behind the barrier is

$$I = \frac{\hbar}{m}|f|^2 = \frac{\hbar}{m}|c|^2\mathscr{T} = \frac{\hbar}{m}\left(\frac{m\omega_0}{2\pi\hbar}\right)W\mathscr{T} \equiv \frac{\omega_0}{2\pi}W\mathscr{T},$$

and the total probability conservation law (2.6), in our case $dW/dt = -I$, yields the decay law

$$\frac{dW}{dt} = -\frac{W}{\tau},$$

$$\text{with } \tau = \frac{2\pi}{\omega_0}\mathscr{T}^{-1} = \frac{2\pi}{\omega_0}\left(\frac{\hbar\omega_0}{864U_{\max}}\right)^{1/2}\exp\left\{\frac{36}{5}\frac{U_{\max}}{\hbar\omega_0}\right\}, \qquad \text{for } \omega_0\tau \gg 1.$$

Comparing the first expression for τ with Eq. (2.153) of the lecture notes, we see that our initial guess of the attempt time, $t_a = 2\pi/\omega_0$, was correct, despite the highly non-classical nature of the ground state of the harmonic oscillator. (This is not too surprising in view of the fact that the WKB approximation yields the correct result (2.114) for all its energy levels, including the ground-state one.)

Finally, let us discuss why this problem is very important. Let a particle be confined at a minimum of an *arbitrary* (but smooth) potential $U_0(x)$. Let us gradually deform this potential, for example by application of an additional force F, which 'tilts' it as

$$U(x) = U_0(x) - Fx,$$

so that at some critical value F_c of the force, the minimum finally disappears. At F below, but very close to this critical value, the 'pocket' of energies $U_{\min} < E < U_{\max}$ is very shallow, and the spatial extension of the pocket is very small, so that the potential $U(x)$ in its vicinity may be expanded into the Taylor series at its minimum, with only a few leading terms being essential. The linear term of the expansion, by definition, disappears at the minimum of the potential energy (say, $x = 0$), so that the leading term is quadratic, and may be always represented as $m\omega_0^2x^2/2$, as in the potential of the solved problem.

However, this term cannot describe the potential barrier—and hence the finite lifetime of the metastable state. For the minimal description of this effect we need to keep the next, cubic term in the Taylor series, thus arriving at the model analyzed above. Hence, our result for τ is valid for the metastable ground state in virtually any sufficiently smooth potential $U(x)$, near the critical point of the potential well's disappearance. (The exception would be a very special function $U(x)$, whose third derivative vanishes exactly at the point where the first one does.)

References

[1] Davison S and Stęślicka M 1992 *Basic Theory of Surface States* (Oxford: Clarendon)
[2] Bunyk P *et al* 2001 *Int. J. on High Speed Electronics and Systems* **11** 257
[3] Averin D *et al* 1985 *Sov. Phys.—JETP* **61** 407

IOP Publishing

Quantum Mechanics
Problems with solutions
Konstantin K Likharev

Chapter 3

Higher dimensionality effects

***Problem* 3.1** A particle of energy E is incident (in the figure below, within the plane of drawing) on a sharp potential step:

$$U(\mathbf{r}) = \begin{cases} 0, & \text{for } x < 0, \\ U_0, & \text{for } 0 < x. \end{cases}$$

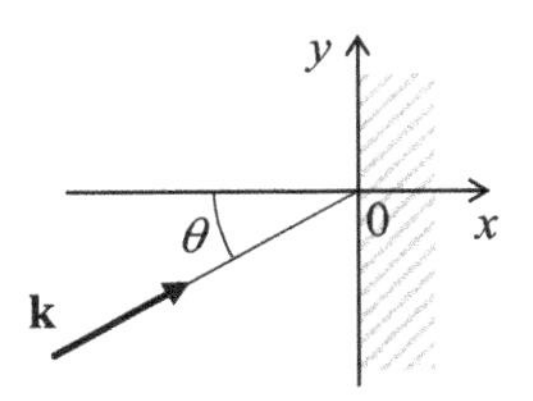

Calculate the particle's reflection probability $\mathscr{R}$ as a function of the incidence angle θ; sketch and discuss this function for various magnitudes and signs of U_0.

Solution: In order to satisfy the boundary conditions at all points along the step's plane $x = 0$, the y-dependence of the wavefunctions at $x \leqslant 0$ and $0 \leqslant x$ should be the same. Thus the appropriate solutions of the Schrödinger equation in these two regions are[1]

$$\psi(\mathbf{r}) = \begin{cases} (A \exp\{ik_x x\} + B \exp\{-ik_x x\}) \exp\{ik_y y\}, & \text{for } x < 0, \\ C \exp\{ik_x' x\} \exp\{ik_y y\}, & \text{for } 0 < x. \end{cases}$$

where the wave vector components may be found from the natural generalization of Eqs. (2.54) and (2.57) of the lecture notes:

[1] In classical mechanics, the fact that k_y is the same at $x < 0$ and $x > 0$, corresponds to the conservation of the y-component of the particle's momentum, due to the absence of a force in this direction: $F_y = -\partial U(\mathbf{r})/\partial y = 0$.

doi:10.1088/2053-2563/aaf3a6ch3

$$\frac{\hbar^2(k_x^2 + k_y^2)}{2m} = E, \quad \frac{\hbar^2(k_x'^2 + k_y^2)}{2m} = E - U_0. \tag{*}$$

Note that these expressions are valid even if E is so low that $k'^2_x < 0$ (for $U_0 > E$, this is the case for any angle θ); in this case the we may take $k'_x = i\kappa$, with real $\kappa > 0$, so that at $0 < x$ the wavefunction decays as $\exp\{-\kappa x\}$.

Thus the problem is reduced to the similar 1D problem that was solved in section 2.3 of the lecture notes (see figure 2.4 and its discussion), and we can use the first of Eqs. (2.63), which yields

$$\mathscr{R} \equiv \left|\frac{B}{A}\right|^2 = \left|\frac{k_x - k_x'}{k_x + k_x'}\right|^2. \tag{**}$$

However, due to Eqs. (*), and the evident geometric relation (see the figure above)

$$\frac{k_y}{k_x} = \tan\theta, \tag{***}$$

we may see that if $k_y \neq 0$ (i.e. $\theta \neq 0$), the energy dependence of x-components of the wave number is now different from the 1D case. Indeed, from Eqs. (*) and (***) we can readily get

$$\hbar k_x = (2mE)^{1/2}\cos\theta, \quad \hbar k_x' = [2m(E\cos^2\theta - U_0)]^{1/2},$$

so that Eq. (**) yields

$$\mathscr{R} = \left|\frac{\cos\theta - (\cos^2\theta - U_0/E)^{1/2}}{\cos\theta + (\cos^2\theta - U_0/E)^{1/2}}\right|^2, \quad \text{for } \frac{U_0}{E} < \cos^2\theta. \tag{****}$$

The plots in the figure below show this reflection probability as a function of the incidence angle θ for several values of the ratio U_0/E. If U_0 is negative, Eq. (****) is valid for any angle, and describes a gradual increase of the reflection from the potential 'step-down' (see the discussion in section 2.3 of the lecture notes) with the growth of θ. (Note that the reflection always becomes total, $\mathscr{R} \to 1$, at $\theta \to \pi/2$, i.e. at the 'grazing-angle' incidence.) Another visible trend is that the reflection is generally lower for smaller steps, and vanishes at $U_0 \to 0$.

Both these trends are also valid if U_0 is positive (i.e. for the potential 'step-up'), but is still less than the particle's energy E. Here Eq. (****) also describes a growth of $\mathscr{R}$ with the incidence angle, but now the growth is faster, and the reflection becomes a total one at a final 'critical' value $\theta_c = \arccos(U_0/E)^{1/2}$. At larger angles (and at any angle for $U_0 > E$), k_x' is purely imaginary, and Eq. (**) yields

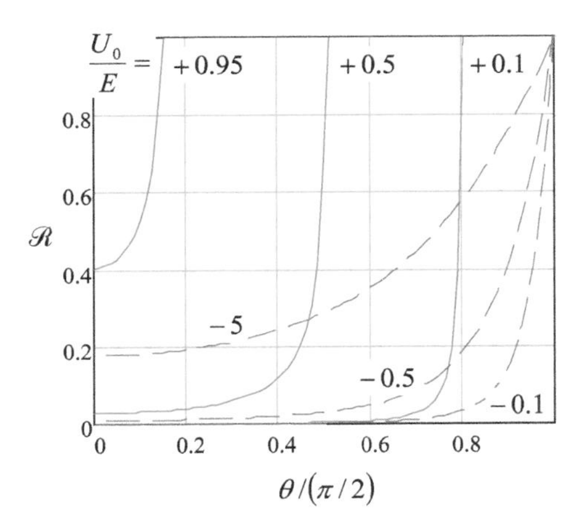

$$\mathscr{R} = 1, \quad \text{for } \cos^2\theta < \frac{U_0}{E},$$

describing the total internal reflection, completely similar to that of the electromagnetic waves[2].

Moreover, Eq. (****) is an analog of the Fresnel formulas valid for electromagnetic waves[3]. However, due to the scalar nature of the de Broglie waves, there is only one such formula in quantum mechanics, rather than two in electrodynamics—for two possible electromagnetic wave polarizations.

Problem* 3.2. Analyze how are the Landau levels (3.50) modified by an additional uniform electric field $\mathscr{E}$, directed along the plane of the particle's motion. Contemplate the physical meaning of your result, and its implications for the quantum Hall effect in a gate-defined Hall bar. (The area $l \times w$ area of such a bar (see figure 3.6 of the lecture notes) is defined by metallic gate electrodes parallel to the 2D electron gas plane—see the figure below. The negative voltage V_g, applied to the gates, squeezes the 2D gas from the area under the gates into the complementary, Hall-bar part of the plane.)

gate $V_g < 0$ $\mathscr{B}$ $V_g < 0$ gate

w

2D electron gas plane →

semiconductor

[2] See, e.g. *Part EM* section 7.4.

[3] See, e.g. *Part EM* Eqs. (7.91) and (7.95).

Solution: The constant electric field directed along a certain coordinate axis (say, x) creates an additional potential

$$\Delta U = -q\mathscr{E}x.$$

Repeating the calculations carried out in the beginning of section 3.2 of the lecture notes, with the account of this additional potential, we see that Eq. (3.47) is now modified as follows:

$$-\frac{\hbar^2}{2m}\frac{d^2}{dx^2}X_k + \frac{\hbar^2}{2m}\left[\frac{q}{\hbar}\mathscr{B}(x - x_0')\right]^2 X_k - q\mathscr{E}xX_k = EX_k,$$

where, as in section 3.2,

$$x_0' \equiv x_0 + \frac{\hbar k}{q\mathscr{B}}.$$

This equation may be rewritten in a form similar to the initial one,

$$-\frac{\hbar^2}{2m}\frac{d^2}{dx^2}X_k + \frac{\hbar^2}{2m}\left(\frac{q}{\hbar}\mathscr{B}\tilde{x}\right)^2 X_k = \tilde{E}X_k,$$

but with an additional shift of the reference point:

$$\tilde{x} \equiv x - x_0'', \quad \text{where } x_0'' \equiv x_0' + \frac{m}{q\mathscr{B}^2}\mathscr{E},$$

and, more importantly, a different constant on the right-hand side:

$$\tilde{E} \equiv E + q\mathscr{E}x_0'' - \frac{m\mathscr{E}^2}{2\mathscr{B}^2}.$$

This means that Eq. (3.50) of the lecture notes is now valid for the parameter $\tilde{E}$, rather than for eigenenergy E, and the genuine energy spectrum now depends on x_0'', i.e. on the position of wavefunction's center:

$$E_n = \hbar\omega_c\left(n + \frac{1}{2}\right) - q\mathscr{E}x_0'' + \text{const}.$$

The physical interpretation of this result is straightforward: the whole set of Landau levels moves up or down together with the electrostatic potential energy the particle *would have* if it was classically-localized at the center $\{x_0'', y_0\}$ of its wavefunction:

$$E_n = \hbar\omega_c\left(n + \frac{1}{2}\right) + U\left(x_0'', y_0\right), \qquad (*)$$

where in our particular case $U(x, y) = -q\mathscr{E}x + \text{const}$. It is virtually evident that Eq. (*) is valid for any external potential $U(x, y)$, if its potential changes in space smoothly enough. Indeed, an analysis shows that Eq. (*) is asymptotically correct if the potential's curvature is sufficiently small; for example, for $U = U(x)$ and relatively low Landau levels, $n \sim 1$, the potential should satisfy the following condition[4]:

$$\left|\frac{\partial^2 U}{\partial x^2}\right| \ll \frac{1}{r_L}\left|\frac{\partial U}{\partial x}\right|, \qquad (**)$$

where r_L is the Landau radius (3.51): $r_L \equiv (\hbar/|q\mathscr{B}|)^{1/2}$. For the usual quantum Hall experiments, with $|q| = e \approx 1.6 \times 10^{-19}$ C and $\mathscr{B}$ of a few tesla, the Landau radius is of the order of 10 nm, while the walls of the potential well $U(x)$ created by the negative gate voltage V_g are smeared by a distance of the order of the gate electrode distance d from the 2D electron gas plane—typically of the order of a few hundred nm. Hence condition (**) is reasonably fulfilled for the potential well forming the Hall bar. As a result, one may analyze the quantum Hall effect in such a bar using the picture of *space-dependent Landau levels* $E_n(x, y)$, repeating the potential well's profile—see the figure below.

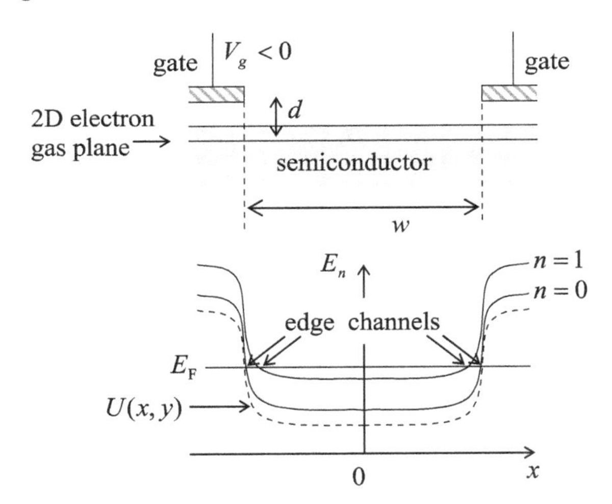

As was discussed in section 3.2 of the lecture notes, at sufficiently low temperatures, the electron states corresponding to the parts of these levels submerged below the Fermi energy E_F are fully occupied, while those above it are empty. As I hope the reader knows from undergraduate physics (and as will be discussed in *Part SM* chapter 6), the electric-field-driven electron transport may take place only at the Fermi surface, because it requires repeated pick-ups of small portions of energy from the driving field and their discarding to scattering centers. Hence, at the quantum Hall effect the transport is only possible in quasi-1D *edge channels* (of a small width $\sim r_L$) at the crossing of each Landau-level surface $E_n(x, y)$ with the Fermi energy plane $E = E_F = \text{const}$.

[4] If this condition is not valid, the electric field may also affect the distance between the Landau levels—see, e.g. the next problem.

Detailed analyses (for whose description this series does not have enough time/space) show that electrons traveling along these channels cannot be back-scattered by (unavoidable) small inhomogeneities of the sample. This fact is exactly the origin of the unprecedented accuracy of the Hall resistance R_H (3.56), which is so unusual for solid state physics.

***Problem* 3.3.** Analyze how the Landau levels (3.50) are modified if a 2D particle is confined in an additional 1D potential well $U(x) = m\omega_0{}^2x^2/2$.

Solution: With the additional potential, the Schrödinger equation (3.41) takes the form

$$-\frac{\hbar^2}{2m}\left(\mathbf{n}_x\frac{\partial}{\partial x} + \mathbf{n}_y\frac{\partial}{\partial y} - i\frac{q}{\hbar}\mathbf{A}\right)^2 \psi + \frac{m}{2}\omega_0^2 x^2 \psi = E\psi.$$

With the same choice of the vector-potential as in Eq. (3.44),[5] and the Fourier expansion (3.45), instead of Eq. (3.47) we now get

$$-\frac{\hbar^2}{2m}\frac{d^2}{d\tilde{x}^2}X_k + \left[\frac{\hbar^2}{2m}\left(\frac{q}{\hbar}\mathscr{B}\tilde{x}\right)^2 + \frac{m}{2}\omega_0^2 x^2\right]X_k = EX_k,$$

$$\text{with} \quad \tilde{x} \equiv x - x_0', \quad x_0' \equiv x_0 + \frac{\hbar k}{q\mathscr{B}}.$$

Now the two terms inside the square brackets, both quadratic-parabolic functions of x, may be merged:

$$\frac{\hbar^2}{2m}\left(\frac{q}{\hbar}\mathscr{B}\tilde{x}\right)^2 + \frac{m}{2}\omega_0^2 x^2 = \frac{m}{2}\left(\omega_c^2 \tilde{x}^2 + \omega_0^2 x^2\right) \equiv \frac{m}{2}\omega_{ef}^2 \tilde{\tilde{x}}^2 + \text{const},$$

where ω_{ef} is the effective frequency, defined by the following relation:

$$\omega_{ef}^2 \equiv \omega_c^2 + \omega_0^2, \qquad \text{where } \omega_c \equiv \frac{q\mathscr{B}}{m}, \tag{*}$$

and $\tilde{\tilde{x}} \equiv x - x_k$ is the coordinate x referred to a certain origin x_k, which depends on our arbitrary choice of x_0, and hence is itself arbitrary. As a result, besides an arbitrary (and inconsequential) choice of the energy and coordinate offsets, the Schrödinger equation is again reduced to that of a harmonic oscillator, and hence has a similar energy spectrum,

[5] If the additional potential U changes along a different axis, it would be natural to change the magnetic field gauge so that the vector-potential would change in the same direction.

$$E_n = \hbar\omega_{\text{ef}}\left(n + \frac{1}{2}\right),$$

but now with the modified (increased) frequency defined by Eq. (*). Hence the 'soft' confinement increases the distance between the Landau levels.

***Problem* 3.4.** Find the eigenfunctions of a spinless, charged 3D particle moving in 'crossed' (mutually perpendicular), uniform electric and magnetic fields, with $\mathscr{E} \ll c\mathscr{B}$. For each eigenfunction, calculate the expectation value of the particle's velocity in the direction normal to both fields, and compare the result with the solution of the corresponding classical problem.

Hint: Generalize the Landau's solution for 2D-confined particles, discussed in section 3.2 of the lecture notes.

Solution: Just as was done in section 3.2, let us direct axis z along the applied magnetic field; then we may use the same choice (3.44) of the vector-potential:

$$A_x = 0, \quad A_y = \mathscr{B}(x - x_0), \quad A_z = 0.$$

Let us direct the x-axis along the electric field; then we may select its electrostatic potential in the form

$$\phi(\mathbf{r}) = -\mathscr{E}x, \qquad \text{so that } U(\mathbf{r}) = q\phi(\mathbf{r}) = -q\mathscr{E}x.$$

With these choices, the Schrödinger equation (3.27) takes the form

$$-\frac{\hbar^2}{2m}\left\{\frac{\partial^2}{\partial x^2} + \left[\frac{\partial}{\partial y} - \frac{iq}{\hbar}\mathscr{B}(x - x_0)\right]^2 + \frac{\partial^2}{\partial z^2}\right\}\psi - q\mathscr{E}x\psi = E\psi.$$

This equation is evidently satisfied by the following eigenfunction (which is a natural generalization of the function used in Eq. (3.45) of the lecture notes)[6]:

$$\psi_k = X_k(x)\exp\{ik_y(y - y_0) + ik_z(z - z_0)\}, \tag{*}$$

where the function $X_k(x)$ satisfies the 1D equation

$$-\frac{\hbar^2}{2m}\left\{\frac{d^2}{dx^2} - \left[k_y - \frac{q}{\hbar}\mathscr{B}(x - x_0)\right]^2 - k_z^2\right\}X_k - q\mathscr{E}xX_k = E_kX_k.$$

This equation may be rewritten in the form of Eq. (3.47):

[6] Here the index k symbolizes the set of c-number parameters k_y, k_z, x_0, y_0, and z_0.

$$-\frac{\hbar^2}{2m}\frac{d^2}{d\tilde{x}^2}X_{\mathbf{k}} + \frac{\hbar^2}{2m}\left(\frac{q\mathscr{B}}{\hbar}\tilde{x}\right)^2 X_{\mathbf{k}} = \tilde{E}X_{\mathbf{k}},$$

where $\tilde{x} \equiv x - x_0''$ is the coordinate x, offset by some value (now depending on both applied fields):

$$x_0'' \equiv x_0' + \frac{m\mathscr{E}}{q\mathscr{B}^2} \equiv x_0 + \frac{\hbar k_y}{q\mathscr{B}} + \frac{m\mathscr{E}}{q\mathscr{B}^2}. \qquad (**)$$

and $\tilde{E}_k$ is the eigenenergy E_k, offset by a constant. As was discussed in section 3.2, Eq. (**) is satisfied by eigenfunctions of the 1D harmonic oscillator with the frequency ω_c equal to the cyclotron frequency of particle's motion in the applied magnetic field—see Eq. (3.48) of the lecture notes.

Now we may combine Eqs. (3.20) and (3.25) of the lecture notes to calculate the operator of the particle's velocity along axis y, normal to both applied fields:

$$\hat{v}_y = \frac{1}{m}\left(-i\hbar\frac{\partial}{\partial y} - qA_y\right) = \frac{1}{m}\left[-i\hbar\frac{\partial}{\partial y} - q\mathscr{B}(x - x_0)\right].$$

Using Eq. (**), we may rewrite this result as

$$\hat{v}_y = \frac{1}{m}\left[-i\hbar\frac{\partial}{\partial y} - q\mathscr{B}\left(\tilde{x} + \frac{\hbar k_y}{q\mathscr{B}} + \frac{m\mathscr{E}}{q\mathscr{B}^2}\right)\right] \equiv \frac{1}{m}\left(-i\hbar\frac{\partial}{\partial y} - \hbar k_y\right) - \frac{q\mathscr{B}}{m}\tilde{x} - \frac{\mathscr{E}}{\mathscr{B}}.$$

For the properly normalized eigenfunction (*), the expectation value of the operator $\partial/\partial y$ is ik_y, so that the expectation value of the expression in the last parentheses vanishes. Also, due to the symmetry of the confining potential of a harmonic oscillator and the resulting symmetry of its eigenfunctions[7],

$$X_k(-\tilde{x}) = \pm X_k(\tilde{x}), \quad \text{i.e.} \ | X_k(-\tilde{x}) | = | X_k(\tilde{x}) |,$$

the expectation value of its coordinate equals zero for any eigenstate:

$$\langle\tilde{x}\rangle \equiv \int_{-\infty}^{+\infty} X_k^*(\tilde{x})\tilde{x}X_k(\tilde{x})d\tilde{x} = \int_{-\infty}^{+\infty} | X_k(\tilde{x}) |^2 \tilde{x}d\tilde{x} = 0,$$

so that we finally get

$$\langle v_y\rangle = -\frac{\mathscr{E}}{\mathscr{B}}, \quad \text{with} \ \left|\langle v_y\rangle\right| \ll c. \qquad (***)$$

(The last strong equality explains the condition $\mathscr{E} \ll c\mathscr{B}$ in the assignment; if it is not fulfilled, the analysis of this problem requires relativistic quantum mechanics.)

Very counter-intuitively, this simple result is valid for *any* eigenfunction (*) of the system, i.e. *any* set of parameters k_y, k_z, x_0, y_0, and z_0![8] This fact becomes (slightly :-) less surprising if we recall the classical solution of this problem[9]: it shows that the

[7] See Eqs. (2.281) and (2.284) and/or figure 2.35 of the lecture notes.

[8] In particular, it gives the average velocity of the particle's motion along the edge channels that were discussed in the solution of problem 3.2.

[9] See, e.g. *Part EM* section 9.6 (iii), and in particular Eq. (9.168) and figure 9.12.

trochoid-like trajectory of the particle 'drifts', in the direction normal to both vectors $\mathscr{E}$ and $\mathscr{B}$, exactly with the velocity expressed by Eq. (***), independently of initial conditions[10]. Of course, the instant velocity v_y of the classical particle, besides the average drift component (***), generally has another component, oscillating with the cyclotron frequency, whose amplitude and phase do depend on the initial conditions. But the same may be true for the expectation value $\langle v_y \rangle$ in quantum mechanics, if the initial state of the particle is a superposition of two or more eigenstates (*), rather than just one of them, as was implied in the calculation of Eq. (***).

***Problem* 3.5.** Use the Born approximation to calculate the angular dependence and the full cross-section of scattering of an incident plane wave, propagating along the x-axis, by the following pair of similar point inhomogeneities:

$$U(\mathbf{r}) = \mathscr{W}\left[\delta\left(\mathbf{r} - \mathbf{n}_z\frac{a}{2}\right) + \delta\left(\mathbf{r} + \mathbf{n}_z\frac{a}{2}\right)\right].$$

Analyze the results in detail. Derive the condition of the Born approximation's validity for such delta-functional scatterers.

Solution: Plugging this $U(\mathbf{r})$ into the general Born integral, given by Eq. (3.86) of the lecture notes, we get the following scattering function:

$$f(\mathbf{k}, \mathbf{k}_i) = -\frac{m\mathscr{W}}{2\pi\hbar^2}\left(\exp\left\{-i\mathbf{q}\cdot\mathbf{n}_z\frac{a}{2}\right\} + \exp\left\{i\mathbf{q}\cdot\mathbf{n}_z\frac{a}{2}\right\}\right).$$

Since in this problem (see the figure below)

$$\mathbf{q}\cdot\mathbf{n}_z \equiv (\mathbf{k} - \mathbf{k}_i)\cdot\mathbf{n}_z = \mathbf{k}\cdot\mathbf{n}_z = k\cos\Theta,$$

where Θ is the angle between the direction of the vector $\mathbf{k}$ (and hence the observation direction) and the axis z,[11] our result may be rewritten as

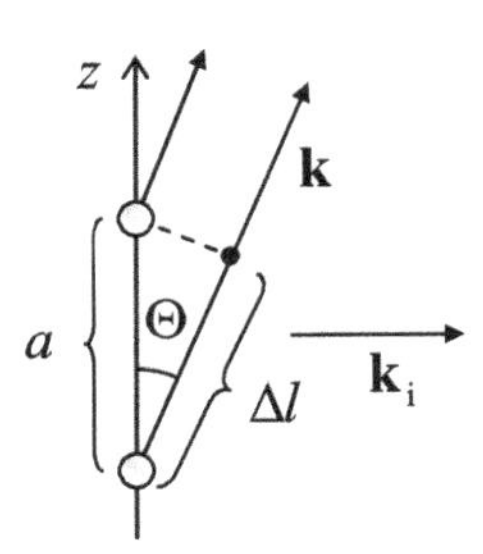

[10] Even the reader unfamiliar with this general classical result should readily recognize its following particular, simple case: a linear, uniform motion of the particle along axis y is possible only with such velocity v_y that the electric and magnetic components of the Lorentz force cancel each other, so that the total force vanishes: $\mathbf{F} = q(\mathscr{E} + \mathbf{v}\times\mathscr{B}) \equiv q(\mathscr{E}\mathbf{n}_x + v_y\mathbf{n}_y \times \mathscr{B}\mathbf{n}_z) \equiv \mathbf{n}_x q(\mathscr{E} + v_y\mathscr{B}) = 0$, giving the result identical to Eq. (***).

[11] Note that Θ is different from what is usually called the scattering angle θ (between the vectors $\mathbf{k}$ and $\mathbf{k}_i$).

$$f(\mathbf{k}, \mathbf{k}_i) = -\frac{m\mathscr{w}}{\pi\hbar^2}\cos\left(\frac{ka}{2}\cos\Theta\right),$$

so that, according to Eq. (3.84), the differential cross-section of scattering is

$$\frac{d\sigma}{d\Omega} \equiv |f(\mathbf{k}, \mathbf{k}_i)|^2 = \left(\frac{m\mathscr{w}}{\pi\hbar^2}\right)^2\cos^2\left(\frac{ka}{2}\cos\Theta\right) = \frac{1}{2}\left(\frac{m\mathscr{w}}{\pi\hbar^2}\right)^2[1 + \cos(ka\cos\Theta)]. \quad (*)$$

Now we can calculate the total cross-section, using spherical coordinates with the axis z taken for the polar axis:

$$\sigma = 2\pi\int_0^\pi \frac{d\sigma}{d\Omega}\sin\Theta d\Theta = \sigma_1\int_{-1}^{+1}[1 + \cos(ka\xi)]d\xi = 2\sigma_1\left(1 + \frac{\sin ka}{ka}\right),$$

where σ_1 is the energy-independent total cross-section of each point scatterer:

$$\sigma_1 \equiv \frac{1}{\pi}\left(\frac{m\mathscr{w}}{\hbar^2}\right)^2.$$

This situation is of course just a variety of the Young-type experiment (cf. figure 3.1 of the lecture notes), and Eq. (*) is a particular case of Eq. (3.11) with $|a_1| = |a_2|$, and the alternative path lengths difference $l_2 - l_1 = \Delta l = a\cos\Theta$—see the figure above. For this particular geometry, the scattered wave is symmetric about axis z. This is natural, because in the Born approximation the role of the incident wave, propagating along axis x, is reduced to the excitation of spherical secondary waves ψ_s from all (in our case, just two) partial scatterers. As a result of interference of these two spherical waves, the scattered wave's intensity oscillates with the angle Θ, reaching its maxima at

$$\cos\Theta_n = \frac{2\pi}{ka}n \equiv \frac{\lambda}{a}n, \quad \text{with } n = 0, 1, 2, \dots,$$

i.e. at the angles at that the path difference Δl between the two waves is a multiple of the de Broglie wavelength $\lambda = 2\pi/k$.

However, at low particle energies ($ka \ll 1$, i.e. $a \ll \lambda$) this 'constructive interference' condition may be satisfied only for $n = 0$, and the scattering is spherically symmetric, and energy-independent: $\sigma = 4\sigma_1$, the factor of 4 arising from the *coherent* addition of two waves in all scattering directions. On the other hand, at high energies ($ka \gg 1$) the intensity of the scattered wave oscillates very rapidly with the angle, so that the total cross-sections of the scatterers add up as if they were incoherent: $\sigma = 2\sigma_1$.

In order to estimate the Born approximation's validity condition, let us replace the delta-functional scatterer with one of a finite (though small) size $R \ll a, k^{-1}$, and a potential of the magnitude $\sim U_0$, so that $\mathscr{w} \sim U_0R^3$. According to Eq. (3.77) of the lecture notes, in order to have $|\psi_s| \ll |\psi_i|$ *inside* the scatterer, we should have

$$U_0 \ll \frac{\hbar^2}{mR^2}, \qquad \text{i.e. } \sigma_1^{1/2} \equiv \frac{m\mathscr{w}}{\sqrt{\pi}\hbar^2} \ll R.$$

For a fixed $\mathscr{w}$ (and hence σ_1), and $R \to 0$, this condition is *never fulfilled*. This means that we cannot take the above expression for σ_1 too literally (unless it is indeed much less than R^2, where R is the physical size of the 'point' scatterer).

However, the interference pattern as such, i.e. the functional dependence of the intensity on the angle Θ, has a much broader validity. Indeed, in order the Born approximation to be correct on this issue, it is sufficient for the wave scattered by one point not to interfere with the incident wave scattering by another point. For that, in the integral (3.72) we can approximately replace Ud^3r by $\mathscr{w}$, ψ with ψ_i, and the denominator by a. Then the generic requirement of the Born approximation, $|\psi_s| \ll |\psi_i|$ gives a much milder condition,

$$\sigma_1^{1/2} \equiv \frac{m\mathscr{w}}{\sqrt{\pi}\hbar^2} \ll a,$$

which does not involve the scatterer's size R.

***Problem* 3.6.** Complete the analysis of the Born scattering by a uniform spherical potential (3.97), started in section 3.3 of the lecture notes, by calculation of its total cross-section. Analyze the result in the limits $kR \ll 1$ and $kR \gg 1$.

Solution: The scattering intensity has the axial symmetry about the axis of incident wave's propagation, so that the total cross-section may be calculated as

$$\sigma = 2\pi \int_0^{\pi} \frac{d\sigma}{d\Omega} \sin\theta d\theta, \quad \text{with} \quad q = 2k \sin\frac{\theta}{2}, \qquad (*)$$

with the differential cross-section given by the last of Eqs. (3.98) of the lecture notes:

$$\frac{d\sigma}{d\Omega} = R^4 u_0^2 \left(\frac{qR \cos qR - \sin qR}{q^3} \right)^2,$$

where

$$u_0 \equiv \frac{U_0}{\hbar^2/2mR^2}$$

is the dimensionless parameter characterizing the scattering potential U_0, which was already used in figure 3.10 of the lecture notes.

The easiest way to calculate the integral in Eq. (*) is to notice that since

$$dq = k \cos\frac{\theta}{2} d\theta, \quad \text{and} \quad qdq = 2k^2 \sin\frac{\theta}{2} \cos\frac{\theta}{2} d\theta \equiv k^2 \sin\theta d\theta,$$

the product $\sin\theta d\theta$ may be replaced with qdq/k^2, so that

$$\sigma = \frac{2\pi}{k^2} R^4 u_0^2 \int_0^{2k} \frac{(\sin qR - qR \cos qR)^2}{q^6} q dq \equiv \frac{2}{(kR)^2} \sigma_g u_0^2 \int_0^{2kR} \frac{(\sin \xi - \xi \cos \xi)^2}{\xi^5} d\xi,$$

where $\xi \equiv qR$, and $\sigma_g \equiv \pi R^2$ is the largest geometric cross-section of the sphere, i.e. its cross-section as 'seen' by the incident particles. This is a table integral[12], finally giving

$$\frac{\sigma}{u_0^2 \sigma_g} = \frac{2}{(kR)^2} \left[\frac{2(2kR)kR \sin(4kR) + \cos(4kR) - 2(2kR)^2 - 1}{8(2kR)^4} + \frac{1}{4} \right]. \quad (**)$$

The figure below shows the cross-section given by Eq. (**) as the function of the dimensionless product kR, proportional to $E^{1/2}$. In the low-energy limit ($kR \to 0$), it tends to the energy-independent value 4/9, which could be readily obtained without the general integration, just by using the fact (discussed in section 3.3 of the lecture notes) that in this limit the scattering is isotropic, with $d\sigma/d\Omega = \sigma_0 u_0^2/9\pi$:

$$\sigma|_{kR \ll 1} = 4\pi \frac{d\sigma}{d\Omega}\bigg|_{kR \ll 1} = \frac{4}{9} u_0^2 \sigma_g.$$

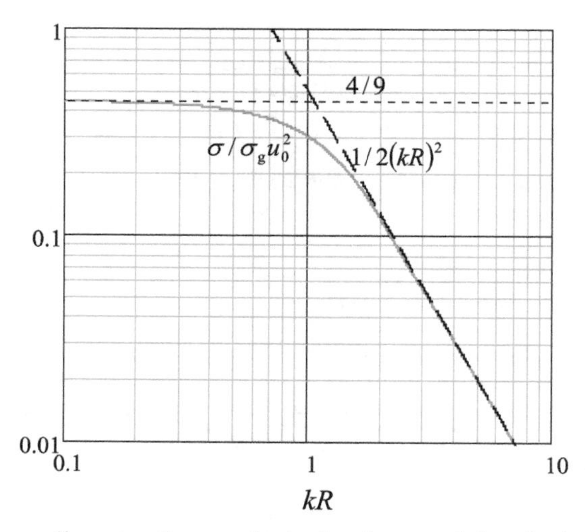

Note also that according to the analysis in the model solution of the previous problem, the Born approximation is only valid if the parameter u_0 is much smaller than 1, so that the calculated full cross-section is much smaller than σ_g.[13]

In the opposite, high-energy limit $kR \gg 1$, our general result (**) is reduced to

$$\sigma|_{kR \gg 1} = \frac{u_0^2 \sigma_g}{2(kR)^2},$$

[12] See, e.g. Eq. (A.30*b*).

[13] As will be discussed in section 3.8 of the lecture notes, the exact (i.e. beyond-Born) theory of scattering by an opaque sphere (which may be described by our current model with $u_0 \gg 1$) gives, in this limit, the total cross-section $\sigma = 4\sigma_0$.

i.e. the cross-section decreases as $1/E$. This happens because, as figure 3.10 of the lecture notes shows, a substantial diffraction takes place only at $qR \sim 1$, i.e. at small scattering angles $\theta \approx q/k \sim 1/kR$, and hence within a small solid angle $\sim 1/(kR)^2 \ll 1$.

***Problem* 3.7.** Use the Born approximation to calculate the differential cross-section of particle scattering by a very thin spherical shell, whose potential may be approximated as

$$U(r) = \mathscr{w}\,\delta(r - R).$$

Analyze the results in the limits $kR \ll 1$ and $kR \gg 1$, and compare them with those for a uniform sphere considered in section 3.3 of the lecture notes.

Solution: Plugging the given (spherically-symmetric) potential into Eq. (3.90). we get

$$f(\mathbf{k}, \mathbf{k}_i) = -\frac{2mR^2\mathscr{w}}{\hbar^2}\frac{\sin qR}{qR} \equiv -\frac{2mR^2\mathscr{w}}{\hbar^2}\,\text{sinc}\,qR,$$

so that the differential cross-section (3.84) of the shell is

$$\frac{d\sigma}{d\Omega} \equiv |\,f(\mathbf{k}, \mathbf{k}_i)\,|^2 = \left(\frac{2mR^2\mathscr{w}}{\hbar^2}\right)^2 \text{sinc}^2 qR. \qquad (*)$$

Here the function sinc is defined as

$$\text{sinc}\,\xi \equiv \frac{\sin \xi}{\xi},$$

its square describing, in particular, the famous *Fraunhofer diffraction pattern*—see the solid red line in the figure below, the dashed blue line showing its envelope $1/\xi^2$.[14]

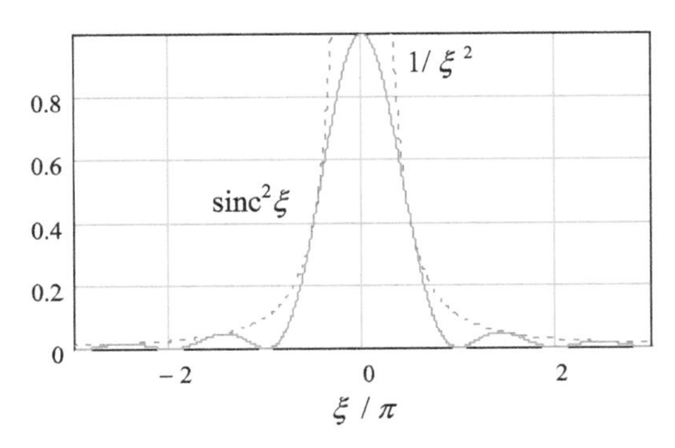

In a qualitative (but not quantitative!) similarly to Eq. (3.98) of the lecture notes, which gives the differential cross-section of a uniform sphere in the same Born approximation, Eq. (*) also describes an infinite set of zero-scattering points $q_n = \pi n/R$, with $n = \pm 1, \pm 2, \ldots$, now exactly periodic in q, besides the forward-scattering point $q \equiv 2k \sin(\theta/2) = 0$, i.e. $\theta = 0$.

[14] This function was discussed in *Part EM* sections 8.4–8.8.

However, just as in the case of the uniform sphere, these diffraction minima (which may be observed as rings of the constant values of the diffraction angle θ) are implemented only at sufficiently large values of the product kR: in order to have N rings, kR has to be larger than πN. In the opposite case $kR \ll 1$, the product qR is much less than 1 for any angle θ, so that $\text{sinc}(qR) \approx 1$, and the scattering is virtually isotropic, with the total cross-section

$$\sigma = 4\pi \frac{d\sigma}{d\Omega} = 4\pi \left(\frac{2mR^2 \mathscr{w}}{\hbar^2} \right)^2 \equiv 4\sigma_g \beta^2, \qquad (**)$$

where $\sigma_g \equiv \pi R^2$ is the geometric cross-section of the sphere, and β is the dimensionless parameter characterizing $\mathscr{w}$, defined similarly to the one used in chapter 2 of the lecture notes for the discussion of 1D problems—see Eq. (2.197):

$$\beta \equiv \frac{\mathscr{w}}{\hbar^2/2mR}.$$

With the proper replacements $a \to R$, $U_0 \to \mathscr{w}R$, Eq. (3.75) of the lecture notes shows that Eq. (**) is valid only if $\beta \ll 1$, i.e. if the total cross-section of scattering is much smaller than its geometric cross-section. Note also that $d\sigma/d\Omega$ and σ do not depend on the sign of $\mathscr{w}$; as will be shown in the solution of problem 3.38, this independence only holds in the Born approximation.

***Problem* 3.8.** Use the Born approximation to calculate the differential and full cross-sections of electron scattering by a screened Coulomb field of a point charge Ze, with the electrostatic potential

$$\phi(\mathbf{r}) = \frac{Ze}{4\pi\varepsilon_0 r} e^{-\lambda r},$$

neglecting spin interaction effects, and analyze the result's dependence on the screening parameter λ. Compare the results with those given by the classical ('Rutherford') formula[15] for the unscreened Coulomb potential ($\lambda \to 0$), and formulate the condition of the Born approximation's validity in this limit.

Solution: Applying Eq. (3.90) of the lecture notes to the spherically-symmetric scattering potential

$$U(r) = -e\phi(\mathbf{r}) = -\frac{C}{r} e^{-\lambda r}, \quad \text{with } C \equiv \frac{Ze^2}{4\pi\varepsilon_0},$$

we get

[15] See, e.g. *Part CM* section 3.5, in particular Eq. (3.73).

$$f(\mathbf{k}, \mathbf{k}_i) = -\frac{2m}{\hbar^2 q}\int_0^\infty U(r)\sin(qr)\,rdr = \frac{2mC}{\hbar^2 q}I, \qquad \text{with } I \equiv \int_0^\infty e^{-\lambda r}\sin(qr)dr.$$

This integral may be easily worked out by representing $\sin(qr)$ as $\text{Im}[\exp\{iqr\}]$:

$$I = \text{Im}\int_0^\infty e^{(-\lambda+iq)r}dr = \text{Im}\frac{e^{(-\lambda+iq)r}}{-\lambda+iq}\Bigg|_0^\infty = \frac{q}{\lambda^2+q^2},$$

so that Eq. (3.84) yields the following differential cross-section of scattering:

$$\frac{d\sigma}{d\Omega} = \left(\frac{2mC}{\hbar^2 q}\frac{q}{\lambda^2+q^2}\right)^2 \equiv \frac{\beta^2}{(\lambda^2+q^2)^2}, \qquad \text{where } \beta \equiv \frac{C}{\hbar^2/2m}. \qquad (*)$$

In order to calculate the full cross-section σ from Eq. (3.85), it is instrumental to use the geometric relation $q = 2k\sin(\theta/2)$, which was repeatedly discussed in section 3.3 of the lecture notes (in particular, see figure 3.9b), and also the trigonometric identity $\sin^2(\theta/2) = (1 - \cos\theta)/2$, to get

$$\begin{aligned}\sigma &\equiv \oint\frac{d\sigma}{d\Omega}d\Omega = 2\pi\int_0^\pi\frac{d\sigma}{d\Omega}\sin\theta d\theta = -2\pi\beta^2\int_{\theta=0}^{\theta=\pi}\frac{d(\cos\theta)}{[\lambda^2+2k^2(1-\cos\theta)]^2}\\ &= 2\pi\beta^2\int_{-1}^{1}\frac{d\xi}{[\lambda^2+2k^2(1-\xi)]^2} \qquad (**)\\ &= \frac{2\pi\beta^2}{2k^2}\frac{1}{\lambda^2+2k^2(1-\xi)}\Bigg|_{\xi=-1}^{\xi=+1} \equiv \frac{4\pi\beta^2}{\lambda^2(\lambda^2+4k^2)}.\end{aligned}$$

In the limit of strong screening, $\lambda \to \infty$, there is a growing range of values k^2 (and hence of the electron energies E), in which we may neglect q^2 in comparison with λ^2 in the denominator of Eq. (*), so that with the growth of λ, the scattering becomes virtually energy- and angle-independent. In this limit, Eq. (**) yields

$$\sigma \approx \frac{4\pi\beta^2}{\lambda^4} = \frac{4\pi}{\lambda^4}\left(\frac{2mC}{\hbar^2}\right)^2,$$

i.e. the full cross-section is energy-independent and scales as $a^4 \sim A^2 \propto 1/\lambda^4 \to 0$, where $a \sim 1/\lambda$ is the effective screening radius, i.e. $A \propto a^2$ is the potential's 'physical' cross-section.

In the opposite limit of negligible screening ($\lambda \to 0$), i.e. of the unscreened Coulomb potential, Eq. (*) yields

$$\frac{d\sigma}{d\Omega} = \left(\frac{2mC}{\hbar^2 q^2}\right)^2 = \left[\frac{mC}{2\hbar^2 k^2\sin^2(\theta/2)}\right]^2.$$

Noticing that $(\hbar k)^2/2m$ is just the energy E of scattered particles, we see that this result *exactly* coincides with the Rutherford formula

$$\frac{d\sigma}{d\Omega} = \left(\frac{C}{4E}\right)^2 \frac{1}{\sin^4(\theta/2)}.$$

This coincidence is quite remarkable, taking into account completely different conceptual structure of the classical and quantum calculations. Note that in the limit $\lambda \to 0$, the full cross-section (**) diverges as $A \propto a^2 = 1/\lambda^2$, just as it does in the classical theory.

The divergence of the effective screening radius $a \sim 1/\lambda$ also affects the application, in this limit, of Eq. (3.77) of the lecture notes for the Born approximation validity. Indeed, since our potential diverges at $r \to 0$, i.e. does not have an immediately apparent magnitude scale U_0, we may estimate this scale by identifying the magnitude of the integral $\int U(\mathbf{r})d^3r$ with U_0a^3. For our particular potential,

$$\int U(\mathbf{r})d^3r = -4\pi C \int_0^\infty \frac{1}{r}e^{-\lambda r}r^2 dr = -\frac{4\pi C}{\lambda^2} \sim Ca^2,$$

so that $U_0 \sim C/a$. With this estimate, Eq. (3.77) yields the following limitation on the constant C:

$$C \ll \frac{\hbar^2}{ma} \max[ka, 1].$$

But in the limit $\lambda \to 0$, i.e. $a \to \infty$, ka becomes much larger than 1 for virtually any values of k, so that for the unscreened Coulomb potential we may replace max[ka, 1] with just ka, and the validity condition takes the form

$$C \ll \frac{\hbar^2}{ma}ka \equiv \frac{\hbar^2 k}{m}, \quad \text{i.e.} \quad \frac{Ze^2}{4\pi\varepsilon_0} \ll \frac{\hbar^2 k}{m} \equiv \frac{\hbar p}{m} \equiv \hbar v,$$

where v is the velocity of the scattered electron. (With our estimate's accuracy, the difference between the phase and group velocities, given by Eq. (2.33b), is insignificant.) Using the definition of the *fine structure constant* (which will be repeatedly discussed later in this course):

$$\alpha \equiv \frac{e^2}{4\pi\varepsilon_0\hbar c} \approx \frac{1}{137},$$

this condition may be rewritten as

$$Z\alpha \ll \frac{v}{c}. \qquad (***)$$

Since all our results so far are valid only for non-relativistic particles, with $v/c \ll 1$,[16] the Born approximation may be valid only for relatively light atoms[17]; for the most important case of a hydrogen atom ($Z = 1$), Eq. (***) takes the form

[16] We will return to this important problem, generalizing it for arbitrary velocities, in chapter 9.

[17] Note that for heavy atoms the potential model explored in this problem is very approximate anyway; a much better approximation is given by the so-called *Fermi–Thomas model*—see chapter 8.

$$\frac{v}{c} \gg \alpha \approx \frac{1}{137}.$$

***Problem* 3.9.** A quantum particle with electric charge Q is scattered by a localized distributed charge with a spherically-symmetric density $\rho(r)$, and zero total charge. Use the Born approximation to calculate the differential cross-section of the *forward scattering* (with the scattering angle $\theta = 0$), and evaluate it for the scattering of electrons by a hydrogen atom in its ground state.

Solution: According to the electrostatics, the potential energy of a point charge Q in an external electric field is $U(\mathbf{r}) = Q\phi(\mathbf{r})$, where $\phi(\mathbf{r})$ is the electrostatic potential of the field. If the field is created by a static charge distribution $\rho(\mathbf{r})$, its potential satisfies the Poisson equation[18]

$$\nabla^2\phi = -\frac{\rho}{\varepsilon_0}. \qquad (*)$$

In our case the charge distribution is spherically-symmetric, and so is the potential distribution: $\phi(\mathbf{r}) = \phi(r)$. In this case the Laplace operator is (relatively :-) simple,

$$\nabla^2 = \frac{1}{r^2}\frac{d}{dr}\left(r^2\frac{d}{dr}\right),$$

enabling us to write an explicit expression for $\phi(r)$ as a double integral over radius:

$$\phi(r) = -\frac{1}{\varepsilon_0}\int_0^r \frac{dr'}{r'^2}\int_0^{r'} \rho(r'')r''^2 dr'' + \text{const},$$

where the constant, for convenience, may be selected from the usual condition $\phi(\infty) = 0$. The resulting expression for $U(r) = Q\phi(r)$ may be plugged into Eq. (3.90) of the lecture notes, written for the forward-scattering limit $\mathbf{k} \to \mathbf{k}_i$ (i.e. at $\mathbf{q} \equiv \mathbf{k} - \mathbf{k}_i \to 0$):

$$f(\mathbf{k}_i, \mathbf{k}_i) = -\frac{2m}{\hbar^2}\int_0^\infty U(r)r^2 dr,$$

and the resulting triple integral may be reduced to a single one via a double integration by parts. (This exercise is highly recommended to the reader.)

However, the same final result may be obtained more simply from Eq. (3.86) by recalling that (as was already noted in that section of the lecture notes) the integral

$$U_\mathbf{q} \equiv \int U(\mathbf{r})e^{-i\mathbf{q}\cdot\mathbf{r}}d^3r$$

in this expression is just the (complex) amplitude in the 3D Fourier expansion of the function $U(\mathbf{r})$ over monochromatic plane wavefunctions $e^{-i\mathbf{q}\cdot\mathbf{r}}$. The Laplace operator's action on such a function is equivalent to its multiplication by the factor $(i\mathbf{q}) \cdot (i\mathbf{q}) \equiv -q^2$, so that Eq. (*) immediately yields the following simple relation

[18] See, e.g. *Part EM* Eq. (1.41).

between $U_{\mathbf{q}}$ and the similarly defined Fourier amplitude $\rho_{\mathbf{q}}$ of the charge distribution:

$$U_{\mathbf{q}} = Q\phi_{\mathbf{q}} = \frac{Q}{\varepsilon_0 q^2}\rho_{\mathbf{q}},$$

and the Born integral (3.86) takes the form

$$f(\mathbf{k}, \mathbf{k}_i) = -\frac{mQ}{2\pi\hbar^2\varepsilon_0}\frac{1}{q^2}\rho_{\mathbf{q}} = -\frac{mQ}{2\pi\hbar^2\varepsilon_0}\frac{1}{q^2}\int \rho(\mathbf{r})e^{-i\mathbf{q}\cdot\mathbf{r}}d^3r.$$

This expression is valid for an arbitrary localized charge distribution. If it is spherically-symmetric, the formula may be reduced to a 1D integral over radius r, exactly as was done for an arbitrary potential profile in Eq. (3.90) of the lecture notes:

$$f(\mathbf{k}, \mathbf{k}_i) = -\frac{2mQ}{\hbar^2\varepsilon_0}\frac{1}{q^3}\int_0^\infty \rho(r)\sin(qr)\,rdr.$$

At the nearly-forward scattering, $q \to 0$, the sine function under the integral may be well approximated by just two leading terms of its Taylor series at $q \approx 0$:

$$\sin qr \approx qr - \frac{1}{3!}(qr)^3.$$

The integral of the first of them, $\int\rho(r)r^2dr$, is proportional to the net charge of the scattering center and, by the problem's condition, equals zero. This leaves us with the integral of the second term:

$$f(\mathbf{k}, \mathbf{k}_i) = -\frac{2mQ}{\hbar^2\varepsilon_0}\frac{1}{q^3}\int_0^\infty \rho(r)\left(-\frac{q^3r^3}{6}\right)rdr \equiv \frac{mQ}{3\hbar^2\varepsilon_0}\int_0^\infty \rho(r)\,r^4dr,$$

which does not depend on q, and hence is valid for the purely-forward scattering ($\mathbf{k} = \mathbf{k}_i$) as well. Now the corresponding differential cross-section may be found from Eq. (3.84):

$$\left.\frac{d\sigma}{d\Omega}\right|_{\theta=0} \equiv |f(\mathbf{k}_i, \mathbf{k}_i)|^2 = \left[\frac{mQ}{3\hbar^2\varepsilon_0}\int_0^\infty \rho(r)\,r^4dr\right]^2. \qquad (**)$$

For the hydrogen atom, the charge distribution $\rho(\mathbf{r})$ consists of a proton's contribution, which is well approximated by $e\delta(\mathbf{r})$, and hence does not contribute to our integral, and that of the electron: $\rho(\mathbf{r}) = -e|\psi(\mathbf{r})|^2$. According to Eq. (3.174), (3.179), and (3.208) of the lecture notes, in the ground state, with the quantum numbers $n = 1$, $l = 0$, and $m = 0$,

$$\rho(r) = -e\,|Y_0^0(\theta, \varphi)|^2|\mathcal{R}_{1,0}(r)|^2 = -e\frac{1}{4\pi}\frac{4}{r_0^3}e^{-2r/r_0},$$

with $r_0 = r_B$, so that the required integration is easy[19]:

[19] See, e.g. Eq. (A.34e) with $n = 4$.

$$\int_0^\infty \rho(r) r^4 dr = -\frac{e}{\pi r_B^3} \int_0^\infty e^{-2r/r_B} r^4 dr$$

$$= -\frac{e}{\pi r_B^3}\left(\frac{r_B}{2}\right)^5 \int_0^\infty \xi^4 e^{-\xi} d\xi = \frac{e}{\pi r_B^3}\left(\frac{r_B}{2}\right)^5 4! \equiv -\frac{3}{4\pi} e r_B^2.$$

Now taking into account Eq. (1.10) for the Bohr radius,

$$r_B = \frac{4\pi\varepsilon_0 \hbar^2}{e^2 m_e},$$

Eq. (**), with $m = m_e$ and $Q = -e$, yields a remarkably simple result:

$$\left.\frac{d\sigma}{d\Omega}\right|_{\theta=0} = r_B^2.$$

***Problem* 3.10.** Reformulate the Born approximation for the 1D case. Use the result to find the scattering and transfer matrices of a 'rectangular' (flat-top) scatterer with

$$U(x) = \begin{cases} U_0, & \text{for } |x| < d/2, \\ 0, & \text{otherwise.} \end{cases}$$

Compare the results with the those of the exact calculations carried out in chapter 2, and analyze how their relation changes in the eikonal approximation.

Solution: Just as in the 3D case, discussed in section 3.3 of the lecture notes, the solution of the 1D version of the Born equation (3.66),

$$\left(\frac{d^2}{dx^2} + k^2\right)\psi_s = \frac{2m}{\hbar^2} U(x)\psi_0,$$

may be expressed as

$$\psi_s(x) = \frac{2m}{\hbar^2}\int U(x')\psi_0(x')G(x, x')dx'$$

via a 1D Green's function that satisfies the following equation:

$$\left(\frac{\partial^2}{\partial x^2} + k^2\right)G(x, x') = \delta(x - x').$$

At $x \neq x'$, the equation is evidently satisfied by the usual monochromatic de Broglie waves propagating from the source, i.e. *from* the point x':

$$G(x, x') = \begin{cases} C_+ \exp\{ik(x - x')\}, & \text{for } x > x', \\ C_- \exp\{-ik(x - x')\}, & \text{for } x < x'. \end{cases}$$

(As in the 3D case, the waves converging *upon* the source point x', are irrelevant.) The coefficients $C_\pm$ in this relation may be found from the boundary conditions at the source point $x = x'$. First, the Green's function (as any wavefunction) should be continuous, giving $C_+ = C_- \equiv C$, so that we may write

$$G(x, x') = C \exp\{ik\,|x - x'|\}.$$

Second, integrating the equation for G over a small interval of x around point x', just as was done at the derivation of Eq. (2.75), we get another boundary condition

$$\left.\frac{\partial G}{\partial x}\right|_{x=x'+0} - \left.\frac{\partial G}{\partial x}\right|_{x=x'-0} = 1,$$

giving $C = 1/2ik$, so that finally the 1D Born approximation is

$$\psi_s(x) = \frac{m}{ik\hbar^2}\int U(x')\psi_0(x')\exp\{ik\,|x - x'|\}\,dx'.$$

For the 'rectangular' scatterer specified in the assignment, and an incident wave of a unit amplitude incident from the left, $\psi_0(x) = \exp\{ikx\}$, this approximation yields

$$\psi_s(x) = \frac{mU_0}{ik\hbar^2}\int_{-d/2}^{+d/2}\begin{Bmatrix}\exp\{ikx'\}\exp\{ik(x - x')\}, & \text{for } x \geqslant +d/2\\ \exp\{ikx'\}\exp\{ik(x' - x)\}, & \text{for } x \leqslant -d/2\end{Bmatrix}dx'.$$

(For our current purposes, we do not care too much about the wave inside the scattering region.) Carrying out these elementary integrations, and adding the incident wave to the scattered one, for the total wavefunction outside the scatterer we get

$$\psi(x) = e^{ikx} - i\alpha \times \begin{cases} e^{ikx}, & \text{for } x \geqslant +d/2,\\ \dfrac{\sin kd}{kd}e^{-ikx}, & \text{for } x \leqslant -d/2.\end{cases} \qquad (*)$$

where I have used a convenient dimensionless parameter

$$\alpha \equiv \frac{mU_0 d}{k\hbar^2} \equiv \frac{U_0}{E}\frac{kd}{2},$$

which reduces to our old friend (2.78), $\alpha = m\mathscr{w}/k\hbar^2$, for a short scatterer, with $d \ll 1/k$ and $\mathscr{w} \equiv U_0 d$. (Please note that according to Eq. (*), the Born approximation is only valid if $\alpha \ll \max\,[1, kd]$, i.e. only if $U_0/E \ll \max[1/kd, 1]$.)

Comparing Eq. (*) with the definition of the complex amplitudes used in the discussion of the 1D transfer and scattering matrices (see Eq. (2.120) of the lecture notes), with the reference points $x_1 = -d/2$ and $x_2 = +d/2$,

$$\psi_{1,2}(x) = A_{1,2}\exp\left\{ik\left(x \pm \frac{d}{2}\right)\right\} + B_{1,2}\exp\left\{-ik\left(x \pm \frac{d}{2}\right)\right\},$$

we get

$$A_1 = \exp\left\{-ik\frac{d}{2}\right\}, \quad B_1 = -i\alpha\frac{\sin ka}{ka}\exp\left\{ik\frac{d}{2}\right\},$$
$$A_2 = (1 - i\alpha)\exp\left\{ik\frac{d}{2}\right\}, \quad B_2 = 0.$$

Now using the definition (2.123) of the scattering matrix components (for the case $B_2 = 0$), we get

$$S_{11} = \frac{B_1}{A_1} = -i\alpha\frac{\sin kd}{kd}e^{ikd}, \quad S_{21} = \frac{A_2}{A_1} = (1 - i\alpha)\, e^{ikd}.$$

As a sanity check, for a short scatterer, and $\alpha \ll 1$, the above results for S_{11} and S_{21} coincide with the results obtained in section 2.5 of the lecture notes—see Eqs. (2.133). The two remaining elements of the scattering matrix,

$$S_{22} = -i\alpha\frac{\sin kd}{kd}e^{ikd}, \quad S_{12} = (1 - i\alpha)\, e^{ikd},$$

may be found either by repeating our calculations for a wave incident from the right, or by using the general relations discussed in section 3.5.

Now we can use Eqs. (2.126) to calculate the transfer matrix. Keeping only the main terms, proportional to $\alpha^0 \equiv 1$ and $\alpha^1 \equiv \alpha$, we get

$$\mathrm{T}_{\mathrm{Born}} \approx \begin{pmatrix} (1 - i\alpha)\, e^{ikd} & -i\alpha\dfrac{\sin kd}{kd} \\ i\alpha\dfrac{\sin kd}{kd} & (1 + i\alpha)\, e^{-ikd} \end{pmatrix}. \qquad (**)$$

(Note that for this transfer matrix, the first two general relations derived in the solution of problem 2.12 are *violated* in the second order in α, because of this approximate treatment. This situation is similar to that with the optical theorem discussed in section 3.3 of the lecture notes, which is also violated in the Born approximation.)

The exact transfer matrix for this scatterer, which was calculated in the solution of problem 2.25, reads (for the relevant case $E > U_0$):

$$\mathrm{T} = \begin{pmatrix} \cos k'd + \dfrac{i}{2}\left(\dfrac{k}{k'} + \dfrac{k'}{k}\right)\sin k'd & -\dfrac{i}{2}\left(\dfrac{k}{k'} - \dfrac{k'}{k}\right)\sin k'd \\ \dfrac{i}{2}\left(\dfrac{k}{k'} - \dfrac{k'}{k}\right)\sin k'd & \cos k'd - \dfrac{i}{2}\left(\dfrac{k}{k'} + \dfrac{k'}{k}\right)\sin k'd \end{pmatrix}.$$

In the Born approximation limit, i.e. at $U_0 \to 0$,

$$\frac{k}{k'} \equiv \left(\frac{E}{E - U_0}\right)^{1/2} \approx 1 + \frac{U_0}{2E} \equiv 1 + \frac{\alpha}{kd}, \text{ so that } \frac{k}{k'} - \frac{k'}{k} \approx \frac{2\alpha}{kd}, \quad \frac{k}{k'} + \frac{k'}{k} \approx 2,$$

and the matrix takes the form

$$\mathrm{T} = \begin{pmatrix} \cos k'd + i \sin k'd & -i\alpha \dfrac{\sin k'd}{kd} \\ i\alpha \dfrac{\sin k'd}{kd} & \cos k'd - i \sin k'd \end{pmatrix} \equiv \begin{pmatrix} e^{ik'd} & -i\alpha \dfrac{\sin k'd}{kd} \\ i\alpha \dfrac{\sin k'd}{kd} & e^{-ik'd} \end{pmatrix}.$$

Comparing the last form of this relation with Eq. (**), we can see that the main deficiency of the Born approximation[20] is that, by construction, it ignores the changes of the propagation speed of the incident and scattered waves, due to the scatterer's potential. (This deficiency disappears only if the difference between kd and $k'd$ is much smaller than 1, which is, for large d, a very tough call.)

This drawback of the Born approximation is corrected in the eikonal approximation—see Eq. (3.102) of the lecture notes. Indeed, applying that formula to our current problem, we see that it yields a result similar to Eq. (**), but with the replacement

$$kd \to k''d, \qquad \text{where } k'' \equiv k - \frac{mU_0}{\hbar^2 k} \equiv k\left(1 - \frac{U_0}{2E}\right).$$

At $U_0/E \to 0$, the wave number participating in the exact matrix T,

$$k' \equiv \frac{[2m(E - U_0)]^{1/2}}{\hbar} \equiv k\left(1 - \frac{U_0}{E}\right)^{1/2} \approx k\left(1 - \frac{U_0}{2E} - \frac{U_0^2}{8E^2} - \ldots\right),$$

tends to the so-corrected value k'' much faster than to the uncorrected (free-space) value k.

***Problem* 3.11.** In the tight-binding approximation, calculate the lowest eigenenergies and the corresponding eigenstates of a particle placed into a system of three similar, symmetric, weakly-coupled potential wells located in the vertices of an equilateral triangle.

Solution: Since in this system, just like in the 1D chain analyzed in section 2.6 of the lecture notes, each potential well is coupled (equally) with just two neighbors, we

[20] In addition, the Born approximation gives, in the diagonal elements of the transfer matrix, a term proportional to α, which is not apparent in its exact form. It is easy to check, however, that such a term appears in the first order of the Taylor expansion of $\exp\{ik'd\}$ in this small parameter.

may repeat all arguments that had led us to Eq. (2.206) of the lecture notes, and rewrite that relation as

$$E = E_1 - 2\delta_1 \cos \alpha,$$

where $\alpha \equiv qa$, and δ_1 is given by Eq. (2.204) for the ground state (with $n = 1$). However, in contrast with the infinitely long chain the phase shift α, which participates in Eq. (2.205),

$$a_j(t) = a \exp\left\{ i\alpha j - i\frac{\varepsilon_1}{\hbar}t + \text{const} \right\},$$

could take any value, in our current ring-like, periodic system, a shift by three positions ($\Delta j = 3$) should lead to a physically indistinguishable probability amplitude a_j, so that 3α should be a multiple of $2\pi m$, where m is any integer. This gives us only three physically distinguishable values of α, and only two different values of energy:

$$\alpha = 0, \qquad \text{with } E = E_g \equiv E_1 - 2\delta_1,$$

$$\alpha = \pm\frac{2\pi}{3}, \qquad \text{with } E = E_e \equiv E_1 + \delta_1.$$

Hence, due to well coupling, the initial ground-state energy level is split into two sublevels, with the lower (genuine ground-state) one being non-degenerate, and the higher sublevel doubly-degenerate.

Please note a deep analogy between these quantum states and the states (with $n = 0$ and ± 1) of the particle on a ring, which was analyzed in section 3.5—see, e.g. the eigenenergy diagram shown in figure 3.18, for $\Phi = 0$. Similarly to that system, a magnetic field with a nonvanishing component normal to the plane of the site's triangle would lift the higher level's double degeneracy, if the particle is electrically charged.

***Problem* 3.12.** Figure below shows a fragment of a periodic 2D lattice, with the red and blue points showing the positions of different local potentials—say, different atoms.

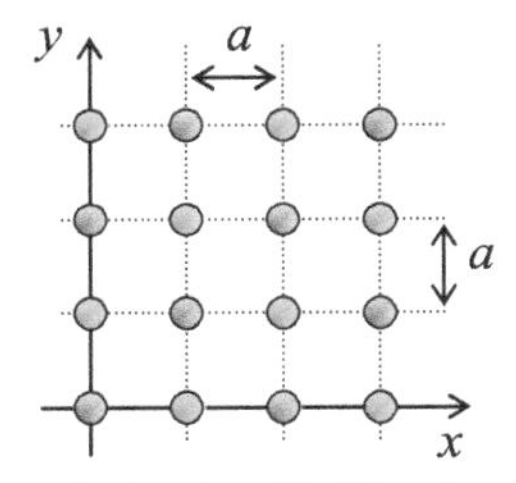

(i) Find the reciprocal lattice and the 1st Brillouin zone.
(ii) Calculate the wave number k of a monochromatic de Broglie wave incident along axis x, at that the lattice creates the first-order diffraction peak within the $[x, y]$ plane, and the direction toward this peak.

(iii) Semi-qualitatively, describe the evolution of the intensity of the peak when the local potentials, represented by the different points, become similar.

Solution: The primitive cell of the lattice evidently has to consist of at least two points—one 'red' and one 'blue'. A natural (though by no means not unique) selection of the cell and the primitive vectors is shown in the figure below:

$$\mathbf{a}_1 = \{2a, \ \ 0\}; \qquad \mathbf{a}_2 = \{a, \ \ a\}.$$

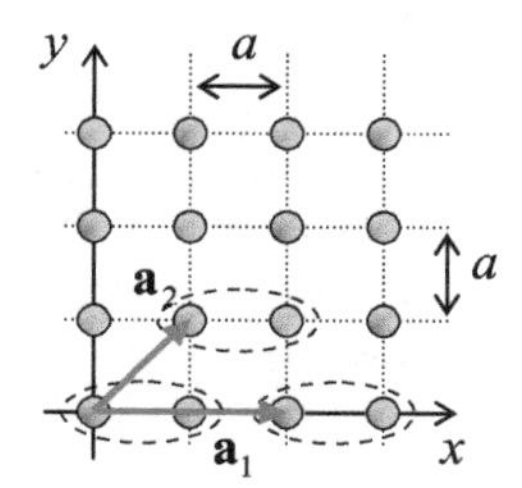

For this 2D lattice, we may define the third primitive vector $\mathbf{a}_3$ formally, for example as a normalized vector product of $\mathbf{a}_1$ and $\mathbf{a}_2$, getting[21]

$$\mathbf{a}_3 \equiv \frac{\mathbf{a}_1 \times \mathbf{a}_2}{a} = \frac{1}{a}\begin{vmatrix} \mathbf{n}_x & \mathbf{n}_y & \mathbf{n}_z \\ 2a & 0 & 0 \\ a & a & 0 \end{vmatrix} = \{0, \ \ 0, \ \ 2a\},$$

so that the products participating in the key relations (3.111) are

$$\mathbf{a}_2 \times \mathbf{a}_3 = \begin{vmatrix} \mathbf{n}_x & \mathbf{n}_y & \mathbf{n}_z \\ a & a & 0 \\ 0 & 0 & 2a \end{vmatrix} = \left\{2a^2, -2a^2, 0\right\}; \quad \mathbf{a}_1 \cdot (\mathbf{a}_2 \times \mathbf{a}_3) = 4a^3;$$

$$\mathbf{a}_3 \times \mathbf{a}_1 = \begin{vmatrix} \mathbf{n}_x & \mathbf{n}_y & \mathbf{n}_z \\ 0 & 0 & 2a \\ 2a & 0 & 0 \end{vmatrix} = \left\{0, 4a^2, 0\right\}.$$

With these expressions, the first two of Eqs. (3.111) yield the following primitive vectors of the reciprocal lattice:

$$\mathbf{b}_1 = \frac{\pi}{a}\{+1, -1\}, \quad \mathbf{b}_2 = \frac{\pi}{a}\{0, +2\}.$$

This result shows that the reciprocal lattice is also of square type, but turned by $\pi/4$ relative to the axes $\mathbf{n}_x$ and $\mathbf{n}_y$—see the figure below. Naturally, the 1st Brillouin zone of this is also a square—see the dashed lines in that figure below.

[21] There is no need to be overly concerned by this choice, because for 2D lattices the third primitive vector $\mathbf{a}_3$ is a formal (mathematical) construct, which enables us to use the 3D relations (3.111) directly, and it is only important to keep it linearly independent of $\mathbf{a}_1$ and $\mathbf{a}_2$.

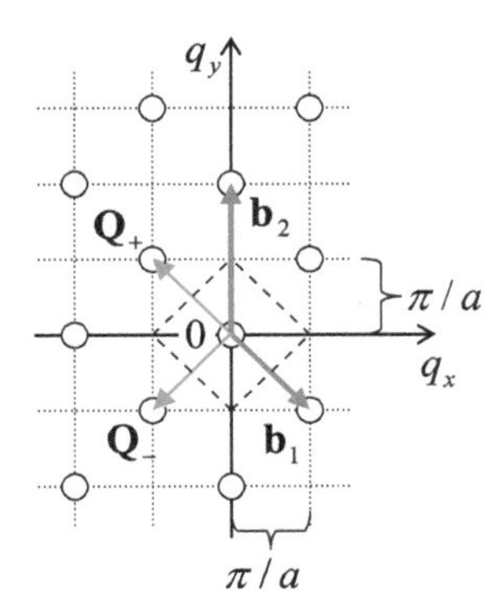

(ii) Of all the reciprocal lattice vectors

$$\mathbf{Q} \equiv l_1\mathbf{b}_1 + l_2\mathbf{b}_2 \equiv \frac{\pi}{a}\{l_1, 2l_2 - l_1\},$$

four vectors

$$\mathbf{Q} = \frac{\pi}{a}\{\pm 1, \pm 1\},$$

corresponding to the combinations $\{l_1 = 0, l_2 = \pm 1\}$ and $\{l_1 = \pm 1, l_2 = \pm 1\}$, have the smallest nonvanishing magnitude $Q = \sqrt{2}\pi/a$.[22] The red arrows in the figure above show the two of them,

$$\mathbf{Q}_+ = -\mathbf{b}_1 = \frac{\pi}{a}\{-1, +1\}, \quad \text{and} \quad \mathbf{Q}_- = -\mathbf{b}_1 - 2\mathbf{b}_2 = \frac{\pi}{a}\{-1, -1\},$$

that are responsible for the lowest order of diffraction for particles incident from the left, with $\mathbf{k}_i = \{k, 0\}$. The figure below shows that the lowest-order elastic scattering (with $|\mathbf{k}| = |\mathbf{k}_i| \equiv k$) is by angles $\pm\pi/2$, and that the resonant value of the incident wave vector is

$$k = \frac{Q_\pm}{\sqrt{2}} \equiv \frac{\pi}{a}.$$

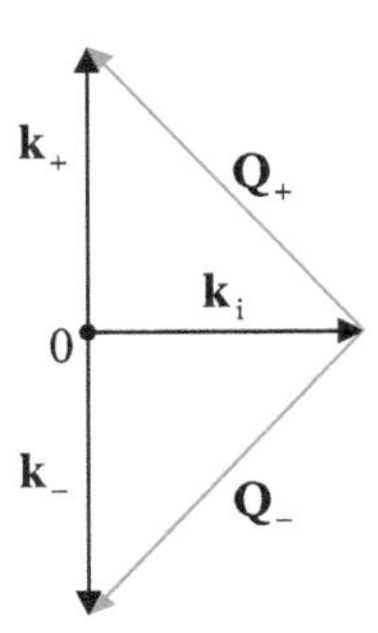

[22] As figure 3.9 of the lecture notes shows, vector $\mathbf{q} = \mathbf{Q} = 0$ corresponds to $\mathbf{k} = \mathbf{k}_0$, i.e. to direct propagation of incidents particles, without scattering.

(iii) According to Eq. (3.84), the diffraction intensity (as characterized by the differential cross-section of scattering) in a particular direction scales as modulus square of the Born integral (3.86). For our particular case ($\mathbf{q} = \mathbf{Q}_\pm$), the function under this integral is

$$U(\mathbf{r})\exp\{-i\mathbf{Q}_\pm \cdot \mathbf{r}\} = U(\mathbf{r})\exp\left\{-i\frac{\pi}{a}(-x \pm y)\right\} \equiv U(\mathbf{r})\exp\left\{i\frac{\pi}{a}x\right\} \exp\left\{\pm i\frac{\pi}{a}y\right\}.$$

The positions of the local potentials, symbolized by the red and blue points in the top figure in the assignment, in each primitive cell differ by $\Delta x = a$, corresponding to equal and opposite values of the factor $\exp\{i\pi x/a\}$. Hence if these potentials become equal, the Born integral along axis x vanishes for any fixed y, so that the whole integral is equal to zero, i.e. this particular diffraction maximum disappears.

This result is natural, because in this case the system may be represented with a simple square lattice (the 2D version of the one shown in figure 3.11a of the lecture notes), having a single-point primitive cell and a different (simpler) set of primitive vectors. The reader is challenged to work out this simpler case, and in particular to calculate the lowest nonvanishing diffraction peak's position and the necessary value of k.

***Problem* 3.13.** For the 2D hexagonal lattice (see figure 3.12b of the lecture notes):

(i) find the reciprocal lattice $\mathbf{Q}$ and the 1st Brillouin zone;
(ii) use the tight-binding approximation to calculate the dispersion relation $E(\mathbf{q})$ for a 2D particle moving in a potential with such periodicity, with an energy close to the eigenenergy of the axially-symmetric states quasi-localized at the potential minima;
(iii) analyze and sketch (or plot) the resulting dispersion relation $E(\mathbf{q})$ inside the 1st Brillouin zone.

Solutions:

(i) Let us choose the primitive vectors of the direct Bravais lattice $\mathbf{R}$ as shown by the blue arrows in the figure below:

$$\mathbf{a}_1 = \frac{a}{2}(-\mathbf{n}_x + \sqrt{3}\,\mathbf{n}_y), \quad \mathbf{a}_2 = \frac{a}{2}(\mathbf{n}_x + \sqrt{3}\,\mathbf{n}_y),$$

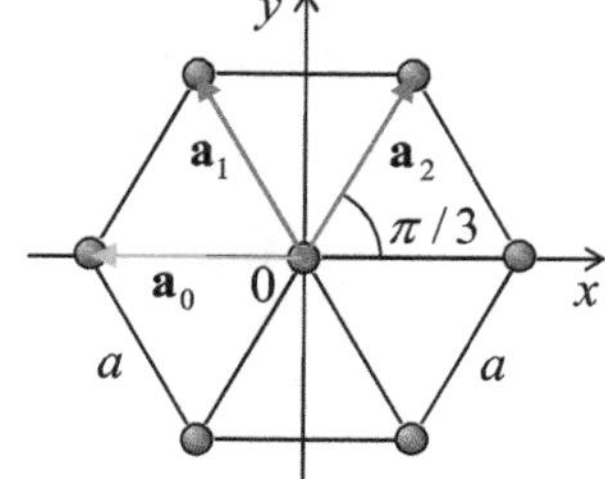

where a is the so-called *lattice constant*—in this case, the distance between the closest points. Making also a natural choice $\mathbf{a}_3 = a\mathbf{n}_z$ (so that this artificial primitive vector is linearly independent of the substantial two), we get

$$\mathbf{a}_2 \times \mathbf{a}_3 = \frac{a^2}{2}(\sqrt{3}\mathbf{n}_x - \mathbf{n}_y), \quad \mathbf{a}_3 \times \mathbf{a}_1 = \frac{a^2}{2}(-\mathbf{n}_x - \sqrt{3}\mathbf{n}_y), \quad \mathbf{a}_1 \cdot (\mathbf{a}_2 \times \mathbf{a}_3) = -\frac{\sqrt{3}a^3}{2},$$

so that for the primitive vectors of the reciprocal lattice $\mathbf{Q}$, Eqs. (3.111) yield

$$\mathbf{b}_1 = 2\pi\frac{1}{\sqrt{3}a}\left(-\sqrt{3}\mathbf{n}_x + \mathbf{n}_y\right), \quad \mathbf{b}_2 = 2\pi\frac{1}{\sqrt{3}a}\left(\sqrt{3}\mathbf{n}_x + \mathbf{n}_y\right).$$

The vectors $\mathbf{b}_{1,2}$ have equal lengths, $b = 2\pi/(\sqrt{3}a/2)$, and the angle between them is a multiple of $\pi/3$; hence the reciprocal lattice $\mathbf{Q}$ is also hexagonal, with the lattice constant equal to b, but turned with respect to the direct lattice—see the figure below. Now we can form the 1st Brillouin zone by connecting the central point of the lattice ($q_x = q_y = 0$) with its six nearest neighbors, and drawing the perpendicular to the middle of each segment—see the dashed lines in the figure below. We see that the zone (shaded in this figure) is a hexagon whose side's closest approach to the origin is $q_{\min} = b/2 = (2/\sqrt{3})(\pi/a)$ and the farthest is $q_{\max} = (2\sqrt{3})q_{\min} = (4/3)(\pi/a)$.

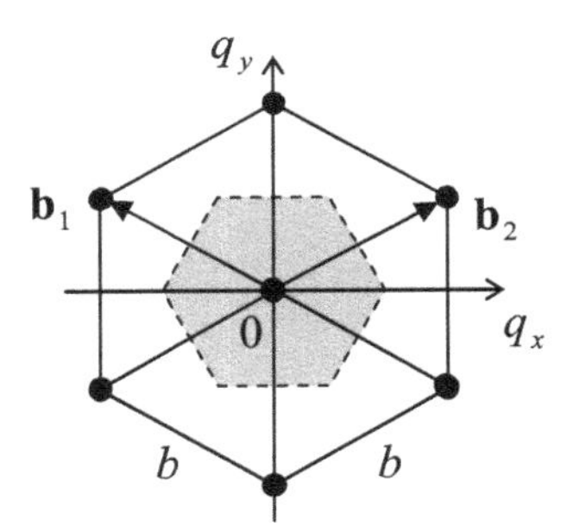

(ii) In the direct hexagonal lattice $\mathbf{R}$, each quasi-localized state has six closest similar neighbors, at the same distance a, so that the tight-binding approximation should account for six corresponding state couplings. For axially-symmetric localized states, all coupling coefficients δ_n are equal, so that instead of Eq. (3.118) of the lecture notes (derived for a square lattice), we have to write

$$E = E_n - \delta_n \sum_{\substack{j=\text{nearest}\\ \text{neighbors}}} e^{i\mathbf{q}\cdot\mathbf{r}_j} = E_n - 2\delta_n(\cos \mathbf{q}\cdot\mathbf{a}_0 + \cos \mathbf{q}\cdot\mathbf{a}_1 + \cos \mathbf{q}\cdot\mathbf{a}_2),$$

where $\mathbf{a}_0 \equiv -a\mathbf{n}_x$ (see the green arrow in the first figure above). Plugging into this relation the Cartesian representations of the vectors $\mathbf{a}_{1,2}$ (spelled out above) and $\mathbf{q} = q_x\mathbf{n}_x + q_y\mathbf{n}_y$, we finally get

$$E - E_n = -2\delta_n\left(\cos q_x a + 2\cos\frac{q_x a}{2}\cos\frac{\sqrt{3}q_y a}{2}\right).$$

(iii) The energy given by this formula is shown in the figure below as a color-coded contour plot, with the dashed line showing the boundary of the hexagonal 1st Brillouin zone calculated in task (i)—cf. the previous figure. We see that the function $E(\mathbf{q})$ has the periodicity of the honeycomb lattice, with the 'summits' and 'passes' along that boundary. Note that finding this zone from the plot of $E(\mathbf{q})$ would be even simpler than that via the reciprocal lattice construction! (Practically, it is prudent to use both ways and compare the results, as a sanity check.)

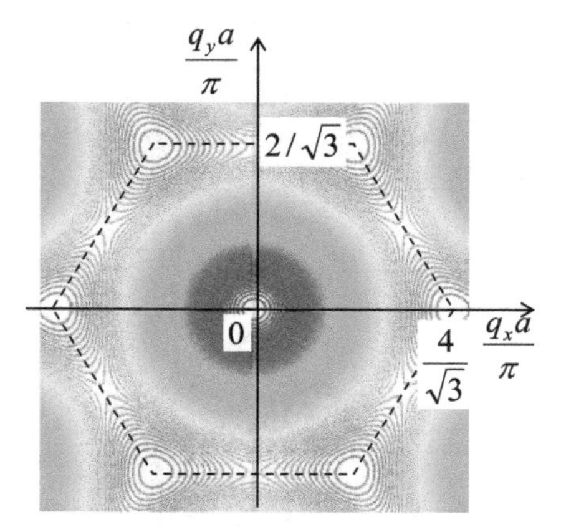

***Problem* 3.14.** Complete the tight-binding calculation of the band structure of the honeycomb lattice, started in the end of section 3.4 of the lecture notes. Analyze the results; in particular prove that the Dirac points $\mathbf{q}_D$ are located in the corners of the 1st Brillouin zone, and express the velocity v_n, participating in Eq. (3.122), in terms of the coupling energy δ_n. Show that the final results do not change if the quasi-localized wavefunctions are not axially-symmetric, but are proportional to $\exp\{im\varphi\}$, where φ is the angle in the lattice plane—as they are, with $m = \pm 1$, for the $2p_z$ electrons of carbon atoms in graphene, which are responsible for its transport properties.

Solution: Let us select the two-point primitive cell of the lattice as shown in figure 3.12a of the lecture notes, which is partly reproduced in the left panel of the figure below, with the letters α and β marking the probability amplitudes of the quasi-localized wavefunctions participating in Eqs. (3.119). The right two panels show the vectors $\mathbf{r}_j$ and $\mathbf{r}'_{j'}$ participating in Eqs. (3.120)–(3.121), conveniently numbered to have $\mathbf{r}'_j = -\mathbf{r}_j$. Due to that symmetry, the double sum in Eq. (3.121) may be simplified as

$$\Sigma = \sum_{j,j'=1}^{3} e^{i\mathbf{q}\cdot(\mathbf{r}_j-\mathbf{r}_{j'})} = 3 + 2\sum_{\substack{j,j'=1\\(j'<j)}}^{3} \cos\left[\mathbf{q}\cdot(\mathbf{r}_j - \mathbf{r}_{j'})\right]$$

$$= 3 + 2\cos\left[\mathbf{q}\cdot(\mathbf{r}_2 - \mathbf{r}_1)\right] + 2\cos\left[\mathbf{q}\cdot(\mathbf{r}_3 - \mathbf{r}_1)\right] + 2\cos\left[\mathbf{q}\cdot(\mathbf{r}_3 - \mathbf{r}_2)\right].$$

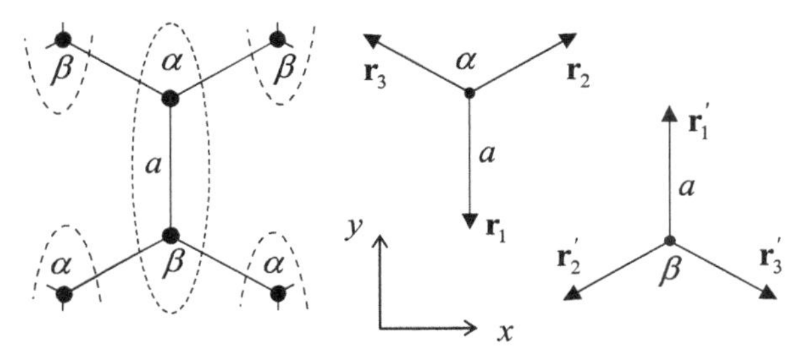

Selecting the coordinate axes as shown at the bottom of the same figure above, and taking into account that all the angles between the vectors $\mathbf{r}_j$ are multiples of $2\pi/3$, the vector differences participating in the last formula for Σ may be expressed in the Cartesian coordinates as

$$\mathbf{r}_2 - \mathbf{r}_1 = \frac{\sqrt{3}}{2}a\left(\mathbf{n}_x + \sqrt{3}\,\mathbf{n}_y\right), \quad \mathbf{r}_3 - \mathbf{r}_1 = \frac{\sqrt{3}}{2}a\left(-\mathbf{n}_x + \sqrt{3}\,\mathbf{n}_y\right), \quad \mathbf{r}_3 - \mathbf{r}_2 = -\sqrt{3}\,a\mathbf{n}_x,$$

where a is the distance between closest points of the lattice (see the figure above), so that Eq. (3.121) yields[23]:

[23] This result was first obtained (as a part of a theoretical analysis of the usual graphite) by P Wallace in 1947. Note that the shape of each energy sheet is similar to that for the hexagonal lattice, calculated in the previous problem (besides a proportional re-scaling of q_x and q_y); however, the existence of the top sheet is specific for the honeycomb direct lattice, with its two-point primitive cell.

$$\frac{E_\pm - E_n}{\delta_n} = \pm\, \Sigma^{1/2} = \pm\left\{3 + 2\cos\left[\frac{\sqrt{3}}{2}\left(q_x a + \sqrt{3}\, q_y a\right)\right]\right.$$
$$\left. + 2\cos\left[\frac{\sqrt{3}}{2}\left(-q_x a + \sqrt{3}\, q_y a\right)\right] + 2\cos\left(-\sqrt{3}\, q_x a\right)\right\}^{1/2}$$
$$\equiv \pm\left[3 + 4\cos\left(\frac{\sqrt{3}}{2} q_x a\right)\cos\left(\frac{3}{2} q_y a\right) + 2\cos\left(\sqrt{3}\, q_x a\right).\right]^{1/2}$$

The left panel of the figure below shows the 'global' 2D contour plot of the lower energy sheet, proportional to $-\Sigma^{1/2}$, on the **q**-plane, and its right panel, a local 3D plot of both sheets, i.e. $\pm\Sigma^{1/2}$, near one of the special 'Dirac' points (where $\Sigma = 0$, i.e. the sheets touch), namely the one with

$$q_x|_{\mathrm{D}} = \frac{4\pi}{3\sqrt{3}\, a}, \quad q_y|_{\mathrm{D}} = 0. \tag{*}$$

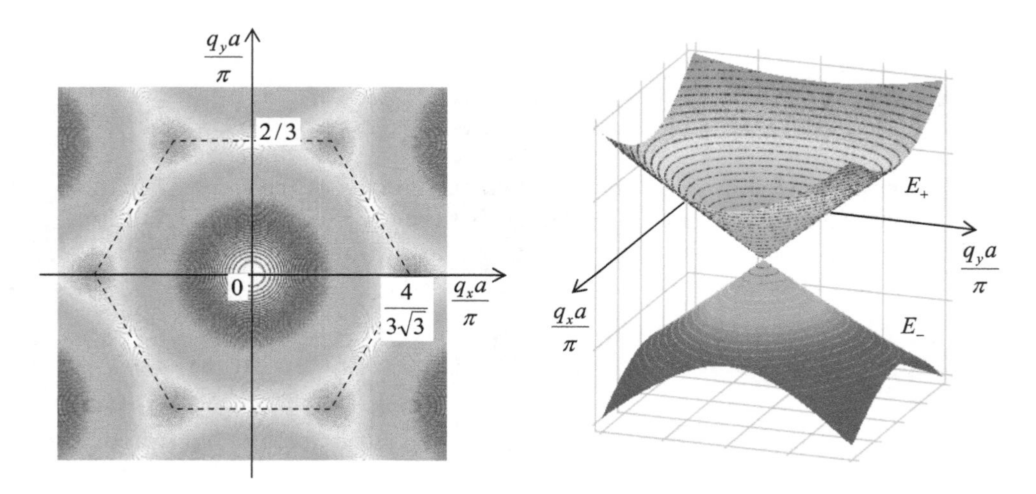

As the global plot shows, there are six such Dirac points at the same distance from the origin, separated by angles $\pi/3$, i.e. located in the corners of a honeycomb cell. The physics of these points is as follows. If the Bloch wave's quasi-momentum is directed along any of the vectors $\mathbf{q}_\mathrm{D}$ (see, for example, the red arrows in the figure below), the wave propagates from a particular point of a primitive cell (say, that with the localized wavefunction of amplitude β) straight to similar points of other primitive cells. As we know from 1D band theory, at the 1st Brillouin zone boundary, the phase shift between the probability amplitudes of the adjacent points equals π (plus an inconsequential multiple of 2π). Hence both possible phase shifts (either 0 or π) between the probability amplitudes α and β, describing the two energy sheets, lead to the phase patterns which differ only by a common rotation by $\pm\pi/3$ (see the figure below) and hence have the same energy: $E_+(\mathbf{q}_\mathrm{D}) = E_-(\mathbf{q}_\mathrm{D})$.

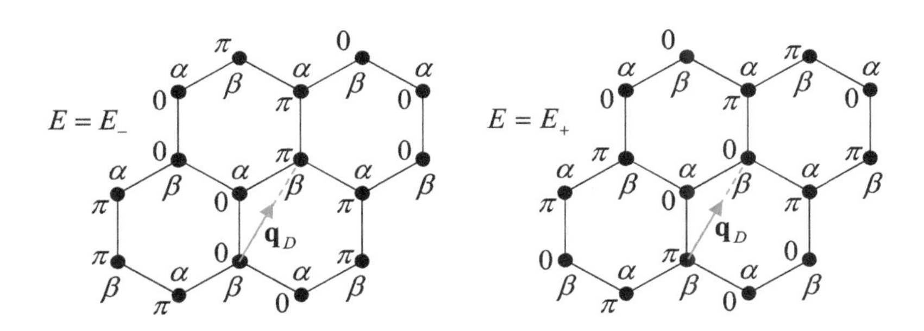

However, any deviation of the quasi-momentum $\mathbf{q}$ from $\mathbf{q}_D$ (in either magnitude or direction) immediately breaks this balance. Indeed, expanding the function $\Sigma(q_x, q_y)$ in the Taylor series near any of the Dirac points, for example the point described by Eq. (*), and neglecting all the terms beyond the quadratic ones, we see that at the Dirac point the constant and linear terms cancel, giving

$$\Sigma_{\mathbf{q}\approx\mathbf{q}_D} \approx \frac{9}{4}\left(\tilde{q}_x^2 + \tilde{q}_y^2\right) a^2, \quad \text{where } \tilde{q}_x \equiv q_x - q_x|_D, \quad \tilde{q}_y \equiv q_y - q_y|_D.$$

This result shows that the dispersion relation $E_\pm(\mathbf{q})$ near that point is indeed linear and isotropic[24], as Eq. (3.122) of the lecture notes has promised:

$$E_\pm|_{\mathbf{q}\approx\mathbf{q}_D} \approx E_n \pm \hbar v_n \,|\, \tilde{\mathbf{q}} \,|,$$

with the dispersion-free velocity

$$v_n = \frac{3}{2}\delta_n \frac{a}{\hbar}.$$

For the $2p_z$ electrons in graphene ($a \approx 0.142$ nm), $\delta_n \approx 2.8$ eV, and this result yields $v_n \approx 0.9 \times 10^6 \text{ m s}^{-1}$.

The simple theory described above is applicable to such electron states despite the fact that their quasi-localized functions $u_n(\mathbf{r})$ are not axially-symmetric, but are proportional to $\exp\{im\varphi\}$, with $m = \pm 1$. Indeed, as Eq. (2.204) implies, in order to take account of such proportionality, each of the coefficients δ_n participating in Eqs. (3.119) has to be multiplied by the factor $\exp\{im(\varphi - \varphi')\}$, where φ is the angle of the vector connecting two adjacent lattice points number j and j', and φ' is the angle of the opposite vector. Evidently, $\varphi - \varphi' = \pi$, so that the additional factor in δ_n is just $\exp\{im\pi\}$, i.e. either +1 or −1, depending on whether m is even (as for axially-symmetric states) or odd (as for $m = \pm 1$). Since the dispersion relation (3.121) has the sign $\pm$ before δ_n anyway, the sign of m does not affect it.

[24] An additional exercise for the reader: describe the motion of electrons with $E \approx E_n$ in an additional uniform magnetic field $\mathscr{B} = \mathscr{B}\mathbf{n}_z$, assuming that it is not too strong ($\mathscr{B}a^2 \ll \Phi'_0$), and using the quasi-classical approximation analogous to Eq. (2.237).

***Problem* 3.15.** Examine basic properties of the so-called *Wannier functions*, defined as

$$\phi_{\mathbf{R}}(\mathbf{r}) \equiv \text{const} \times \int_{\text{BZ}} \psi_{\mathbf{q}}(\mathbf{r}) e^{-i\mathbf{q}\cdot\mathbf{R}} d^3q,$$

where $\psi_{\mathbf{q}}(\mathbf{r})$ is the Bloch wavefunction (3.108), $\mathbf{R}$ is any vector of the Bravais lattice, and the integration over the quasi-momentum $\mathbf{q}$ is extended over any (e.g. the first) Brillouin zone.

Solution: According to the Wannier function's definition, it is just the 3D Fourier image of the Bloch function (i.e. of any extended eigenfunction of the stationary Schrödinger equation for a particle in a periodic potential), considered as a function of its quasi-momentum $\mathbf{q}$. (Notice the index attached to $\psi(\mathbf{r})$ to emphasize its dependence on $\mathbf{q}$; this index was just implied in sections 2.7 and 3.4 of the lecture notes.) Let us plug Eq. (3.108) of the lecture notes (with the index $\mathbf{q}$ attached to the periodic functions $u(\mathbf{r})$ as well) into the Wannier function's definition:

$$\phi_{\mathbf{R}}(\mathbf{r}) = \text{const} \times \int_{\text{BZ}} u_{\mathbf{q}}(\mathbf{r}) e^{i\mathbf{q}\cdot\mathbf{r}} e^{-i\mathbf{q}\cdot\mathbf{R}} d^3q \equiv \text{const} \times \int_{\text{BZ}} u_{\mathbf{q}}(\mathbf{r}) e^{i\mathbf{q}\cdot(\mathbf{r}-\mathbf{R})} d^3q.$$

Since the functions $u_{\mathbf{q}}(\mathbf{r})$ are, by definition, invariant with respect to the translation by any Bravais lattice vector $\mathbf{R}'$,

$$u_{\mathbf{q}}(\mathbf{r} + \mathbf{R}') = u_{\mathbf{q}}(\mathbf{r}),$$

we may calculate

$$\begin{aligned}\phi_{\mathbf{R}+\mathbf{R}'}(\mathbf{r} + \mathbf{R}') &= \text{const} \times \int_{\text{BZ}} u_{\mathbf{q}}(\mathbf{r} + \mathbf{R}') e^{i\mathbf{q}\cdot(\mathbf{r}+\mathbf{R}'-\mathbf{R}-\mathbf{R}')} d^3q \\ &\equiv \text{const} \times \int_{\text{BZ}} u_{\mathbf{q}}(\mathbf{r}) e^{i\mathbf{q}\cdot(\mathbf{r}-\mathbf{R})} d^3q \equiv \phi_{\mathbf{R}}(\mathbf{r}).\end{aligned}$$

This result means that the Wannier function depends only on the distance between $\mathbf{r}$ and $\mathbf{R}$, i.e. is repeated near each point of the Bravais lattice.

Next, the Wannier functions centered at different Bravais lattice points $\mathbf{R}$, and based on (mutually orthogonal) eigenfunctions $\psi_{\mathbf{q},n}(\mathbf{r})$ of particular energy bands n, form a full orthonormal basis:

$$\begin{aligned}\int_{\text{PC}} \phi^*_{\mathbf{R},n}(\mathbf{r}) \phi_{\mathbf{R}',n'}(\mathbf{r}) d^3r &\propto \int_{\text{PC}} d^3r \int_{\text{BZ}} d^3q \int_{\text{BZ}'} d^3q' \psi^*_{\mathbf{q},n}(\mathbf{r}) e^{i\mathbf{q}\cdot\mathbf{R}} \psi_{\mathbf{q}',n'}(\mathbf{r}) e^{-i\mathbf{q}'\cdot\mathbf{R}'} \\ &\propto \int_{\text{BZ}} d^3q \int_{\text{BZ}'} d^3q'\, e^{i\mathbf{q}\cdot\mathbf{R}} e^{-i\mathbf{q}'\cdot\mathbf{R}'} \int_{\text{PC}} \psi^*_{\mathbf{q},n}(\mathbf{r}) \psi_{\mathbf{q}',n'}(\mathbf{r}) d^3r \\ &\propto \int_{\text{BZ}} d^3q \int_{\text{BZ}'} d^3q'\, e^{i\mathbf{q}\cdot\mathbf{R}} e^{-i\mathbf{q}'\cdot\mathbf{R}'} \delta(\mathbf{q} - \mathbf{q}') \delta_{n,n'} \propto \delta_{\mathbf{R},\mathbf{R}'}\, \delta_{n,n'},\end{aligned}$$

where the $\mathbf{r}$-integration is over the primitive cell (PC) of the Bravais lattice. (The set may be readily made orthonormal, with the appropriate choice of the constant in the Wannier function definition.)

In order to appreciate the practical value of the Wannier functions' basis, notice that the Bloch functions (as all eigenfunctions of any time-independent Hamiltonian) are only defined to an arbitrary phase coefficient $\exp\{i\varphi\}$, where the choice of the phase φ, for any $\mathbf{q}$, is arbitrary. Hence the shape of Wannier functions depends on the choice of the function $\varphi(\mathbf{q})$; in particular, this function may be hand- (or rather computer-[25]) crafted to make $\phi_{\mathbf{q},n}$ maximally localized near the point $\mathbf{r} = \mathbf{R}$. This property of the Wannier functions, together with their evident independence of the quasi-momentum $\mathbf{q}$, and hence of the eigenenergy $E_n(\mathbf{q})$, makes their basis convenient for several applications, for example, for numerical simulations of the electron structure in condensed matter and of the electron transport in nanostructures[26].

***Problem* 3.16.** Evaluate the long-range electrostatic interaction (the so-called *London dispersion force*) between two similar, electrically-neutral atoms or molecules, modeling each of them as an isotropic 3D harmonic oscillator with the electric dipole moment $\mathbf{d} = q\mathbf{s}$, where $\mathbf{s}$ is the oscillator's displacement from its equilibrium position.

Hint: Represent the total Hamiltonian of the system as a sum of Hamiltonians of independent 1D harmonic oscillators, and calculate their total ground-state energy as a function of distance between the dipoles[27].

Solution: According to classical electrostatics, the potential energy of interaction between two electric dipoles is[28]

$$U_{\text{int}} = \frac{1}{4\pi\varepsilon_0 r^3}(d_{1x}d_{2x} + d_{1y}d_{2y} - 2d_{1z}d_{2z}),$$

where the z-axis is directed along the vector $\mathbf{r}$ connecting the dipoles with moments $\mathbf{d}_1$ and $\mathbf{d}_2$. In the single-particle model of an electrically-neutral molecule, $\mathbf{d}_{1,2} = q\mathbf{s}_{1,2}$, where $\mathbf{s}_1 = \{x_1, y_1, z_1\}$ and $\mathbf{s}_2 = \{x_2, y_2, z_2\}$ are the displacements of the effective particles, with electric charges q, from the oppositely charged, immobile centers. In the isotropic 3D oscillator model, the interaction energy U_{int} should be added to the sum of the potential energies (3.123) of the oscillators, so that the total potential energy is

$$U = \frac{m\omega_0^2}{2}\left(x_1^2 + y_1^2 + z_1^2 + x_2^2 + y_2^2 + z_2^2\right) + \frac{q^2}{4\pi\varepsilon_0 r^3}(x_1x_2 + y_1y_2 - 2z_1z_2).$$

[25] See, e.g. [1].

[26] See, e.g. http://www.wannier-transport.org/wiki/index.php/.

[27] This explanation of the interaction between electrically-neutral atoms was put forward in 1930 by F London, on the background of a prior (1928) work by C Wang. Note that in some texts this interaction is (rather inappropriately) referred to as the 'van der Waals force', though it is only one, long-range component of the van der Waals model—see, e.g. *Part SM* section 4.1.

[28] See, e.g. *Part EM* Eq. (3.16), which uses a different notation ($\mathbf{p}$) for the dipole moments.

Since the dipole approximation is only valid at large distances $r \gg a \sim x_0 \equiv (\hbar/m\omega_0)^{1/2}$, when the second term is relatively small, U has a stable minimum (at the point $\mathbf{s}_1 = \mathbf{s}_2 = 0$), which does not depend on r, so that the classical electrodynamics cannot describe the London dispersion force. However, such force appears in quantum mechanics (and also in statistical physics at temperature $T \neq 0$), in which Heisenberg's uncertainty relation forbids the oscillators from fully static positions at these potential minima. In order to quantify this effect, let us notice that in new coordinates, defined as

$$\mathbf{s}_\pm \equiv \frac{\mathbf{s}_1 \pm \mathbf{s}_2}{\sqrt{2}}, \quad \hat{\mathbf{p}}_\pm \equiv \frac{\hat{\mathbf{p}}_1 \pm \hat{\mathbf{p}}_2}{\sqrt{2}}, \quad \text{so that } \mathbf{s}_{1,2} = \frac{\mathbf{s}_+ \pm \mathbf{s}_-}{\sqrt{2}}, \quad \hat{\mathbf{p}}_{1,2} = \frac{\hat{\mathbf{p}}_+ \pm \hat{\mathbf{p}}_-}{\sqrt{2}},$$

the potential energy U, and the system's Hamiltonian as a whole fall apart into sums of coordinate and momentum components squared:

$$\begin{aligned}\hat{H} &= \frac{1}{2m}\left(\hat{p}_{1x}^2 + \hat{p}_{1y}^2 + \hat{p}_{1z}^2 + \hat{p}_{2x}^2 + \hat{p}_{2y}^2 + \hat{p}_{2z}^2\right) + U \\ &= \frac{1}{2m}\left(\hat{p}_{+x}^2 + \hat{p}_{+y}^2 + \hat{p}_{+z}^2 + \hat{p}_{-x}^2 + \hat{p}_{-y}^2 + \hat{p}_{-z}^2\right) \\ &+ \frac{m\omega_0^2}{2}\left(x_+^2 + y_+^2 + z_+^2 + x_-^2 + y_-^2 + z_-^2\right) \\ &+ \frac{q^2}{8\pi\varepsilon_0 r^3}\left(x_+^2 - x_-^2 + y_+^2 - y_-^2 - 2z_+^2 + 2z_-^2\right).\end{aligned}$$

This is just the sum of Hamiltonians of six independent 1D harmonic oscillators with the following eigenfrequencies:

$$\begin{aligned}&\text{for } x_+ \text{ and } y_+: && \omega_{t+} = \omega_0(1 + \mu)^{1/2} \approx \omega_0(1 + \mu/2 - \mu^2/8 + \ldots), \\ &\text{for } x_- \text{ and } y_-: && \omega_{t-} = \omega_0(1 - \mu)^{1/2} \approx \omega_0(1 - \mu/2 - \mu^2/8 + \ldots), \\ &\text{for } z_+: && \omega_{l+} = \omega_0(1 - 2\mu)^{1/2} \approx \omega_0(1 - \mu - \mu^2/2 + \ldots), \\ &\text{for } z_-: && \omega_{l-} = \omega_0(1 + 2\mu)^{1/2} \approx \omega_0(1 + \mu - \mu^2/2 + \ldots), \\ & && \text{where } \mu \equiv \frac{q^2}{4\pi\varepsilon_0 r^3 m\omega_0^2} \ll 1.\end{aligned}$$

Here the Taylor expansions in small parameter μ are extended to the third (quadratic) terms, because the sum of the first-order terms vanishes, and the first nonvanishing correction to the ground-state energy of the oscillator system is proportional to μ^2:

$$\begin{aligned}E_g &= \sum_\omega \frac{\hbar\omega}{2} = \frac{\hbar}{2}(2\omega_{t+} + 2\omega_{t-} + \omega_{l+} + \omega_{l-}) \\ &\approx \frac{\hbar\omega_0}{2}\left[6 - \mu^2\left(4\frac{1}{8} + 2\frac{1}{2}\right)\right] \equiv 3\hbar\omega_0 - \frac{3}{4}\hbar\omega_0\mu^2.\end{aligned}$$

The first term of the last expression is evidently the ground-state energy of two far-separated 3D oscillators, while the second term may be interpreted as the effective potential of their interaction:

$$U_{\text{ef}} = -\frac{3}{4}\hbar\omega_0\mu^2 \equiv -\frac{3}{4}\frac{\hbar q^4}{(4\pi\varepsilon_0)^2 m^2\omega_0^3}\frac{1}{r^6} \equiv -\frac{3}{4}Z^4\left(\frac{m_e}{m}\right)^2\left(\frac{x_0}{r}\right)^6 E_H,$$

where E_H is the Hartree energy (1.13), and $Z \equiv q/e$. (The factor $Z^4(m_e/m)^2$ may be much larger than 1 in large atoms/molecules, where the polarization is due to a simultaneous displacement of $N \gg 1$ electrons, so that $Z^4 \propto N^4$, while $(m/m_e) \propto N^2$.)

Note the following features of the result for U_{ef}:

- the interaction is always attractive—for any sign of the charge q;
- the interaction potential is proportional to $1/r^6$, i.e. drops with the distance much faster that that ($U \propto r^{-3}$) for atoms with permanent (field-independent) dipole moments.
- U_{ef} is proportional to $\hbar$, emphasizing again the quantum nature of the long-range attraction—at least when the thermal excitation of the oscillators is negligible.

Later, in the course (in chapters 5 and 6), we will explore a completely different way to derive this formula and its generalization to an arbitrary single-particle model of the atom. The latter result will show that the listed general features of the London dispersion force do not depend on the model details.

***Problem* 3.17.** Derive expressions for the eigenfunctions and the corresponding eigenenergies of a 2D particle of mass m, free to move inside a thin round disk of radius R. What is the degeneracy of each energy level? Calculate five lowest energy levels with an accuracy better than 1%.

Solution: We may start the solution of this problem (which is just the 2D version of the problem solved at the end of section 3.6 of the lecture notes) from Eq. (3.147) of the lecture notes, with $U(\rho) = 0$,

$$-\frac{\hbar^2}{2m}\left[\frac{1}{\rho\mathcal{R}}\frac{d}{d\rho}\left(\rho\frac{d\mathcal{R}}{d\rho}\right) - \frac{m^2}{\rho^2}\right] = E. \qquad (*)$$

For our problem, this equation should be solved with the boundary condition

$$\mathcal{R}(R) = 0,$$

due to the continuity of the wavefunction $\psi = \mathcal{R}(\rho)\mathcal{F}(\varphi) \propto \mathcal{R}(\rho)e^{im\varphi}$ at the whole border $\rho = R$, for all values of the angle φ. After the introduction of the dimensionless argument $\xi \equiv k\rho$, and with the usual definition of the free-particle's wave number k (as $\hbar^2k^2/2m \equiv E$), Eq. (*) is reduced to the canonical form of the Bessel equation[29]:

$$\frac{d^2\mathcal{R}}{d\xi^2} + \frac{1}{\xi}\frac{d\mathcal{R}}{d\xi} + \left(1 - \frac{m^2}{\xi^2}\right)\mathcal{R} = 0.$$

Its general solution is a linear combination of the Bessel functions of the first and second kind, of the same integer order m:

$$\mathcal{R}(\xi) = C_1J_m(\xi) + C_2Y_m(\xi).$$

However, the functions $Y_m(\xi)$ diverge at $\xi \to 0$,[30] while the wavefunction has to stay finite in the center of the disk. Hence, the coefficient C_2 should equal zero, and the boundary condition is reduced to the following requirement:

$$J_m(kR) = 0.$$

This condition, which plays the role of a characteristic equation, is satisfied if the product kR is equal to any nontrivial root of the Bessel function, $\xi_{m,l} \neq 0$, where the index $l = 1, 2, 3,\ldots$ numbers the roots. As a result, the eigenfunctions are

$$\psi_{m,l} = C_lJ_m\left(\frac{\rho}{R}\xi_{m,l}\right)e^{im\varphi},$$

and the energy spectrum is

$$E_{m,l} = E_0\xi_{m,l}^2, \quad \text{where } E_0 \equiv \frac{\hbar^2}{2mR^2}.$$

Each energy level with $m \neq 0$ is doubly degenerate, due to two possible signs of m. Indeed, the Bessel function $J_{-m}(\xi)$ equals $(-1)^m J_m(\xi)$, and hence has the same roots $\xi_{m,l}$ as $J_m(\xi)$. So, the eigenenergies of the states with the opposite signs of m are exactly equal, while their eigenfunctions are different, due to different azimuthal factors $\mathcal{F}(\varphi) = \exp\{im\varphi\}$.

Evidently, the lowest energy levels correspond to the smallest roots $\xi_{m,l}$. Using a table of the roots[31], we get the following approximate values for five lowest levels:

m	l	$\xi_{n,l}$	$E_{m,l}/E_0 = (\xi_{n,l})^2$
0	1	2.405	5.78
1	1	3.832	14.68
2	1	5.136	26.38
0	2	5.520	30.47
3	1	6.380	40.70

[29] See, e.g. *Part EM* section 2.7.

[30] See, e.g. *Part EM* Eq. (2.152) and/or figure 2.19.

[31] See, e.g. *Part EM* table 2.1—or virtually any math handbook listed in section A.16(ii) of appendix A.

Note the sudden 'intrusion' of the second root of $J_0(\xi)$ ($m = 0$, $l = 2$) into the initially orderly sequence of the first roots of $J_m(\xi)$, very similar to that of the similar 3D problem, solved at the end of section 3.6 of the lecture notes.

The comparison of these two problems also shows that the lowest eigenenergies of the 2D system (a particle inside the disk of radius R) are lower than those of the 3D system—a particle of the same mass inside a sphere of the same radius. Let me challenge the reader: could you predict this fact before doing calculations?

***Problem* 3.18.** Calculate the ground-state energy of a 2D particle of mass m, localized in a very shallow flat-bottom potential well

$$U(\rho) = \begin{cases} -U_0, & \text{for } \rho < R, \\ 0, & \text{for } \rho > R, \end{cases} \quad \text{with } 0 < U_0 \ll \frac{\hbar^2}{mR^2}.$$

Solution: Starting from the case of an arbitrary (but positive) U_0, and repeating the arguments made in the model solution of the previous problem (with E replaced with the difference $E - (-U_0) \equiv E + U_0$, i.e. with the wave number k now defined by the relation

$$\frac{\hbar^2 k^2}{2m} = E + U_0 \geqslant 0,$$

due the different position of the well's 'floor'—see the figure below), we get the following radial wavefunctions inside the well:

$$\mathcal{R}(\rho) = C_- J_m(k\rho), \quad \text{for } \rho \leqslant R. \qquad (*)$$

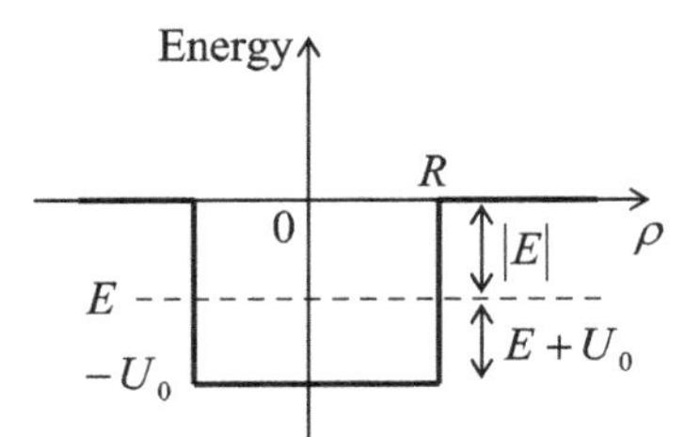

Outside of the well (at $\rho > R$), where $U(\rho) = 0$, Eq. (3.147) of the lecture notes may be similarly reduced to the modified Bessel equation[32],

$$\frac{d^2\mathcal{R}}{d\xi^2} + \frac{1}{\xi}\frac{d\mathcal{R}}{d\xi} - \left(1 + \frac{m^2}{\xi^2}\right)\mathcal{R} = 0, \quad \text{with } \xi \equiv \kappa\rho, \quad \text{where } \frac{\hbar^2\kappa^2}{2m} = |\, E \,| \geqslant 0,$$

whose general solution is a linear superposition of the modified Bessel functions $I_m(\xi)$ and $K_m(\xi)$. Since the former functions diverge at $\xi \to \infty$ (i.e. at $\rho \to \infty$), $\mathcal{R}(\rho)$ should be proportional to the latter function alone:

[32] See, e.g. *Part EM* Eq. (2.155) and its discussion.

$$\mathcal{R}(\rho) = C_+ K_m(\kappa\rho), \qquad \text{for } \rho \geqslant R. \qquad (**)$$

At the well's wall (at $\rho = R$), the wavefunction and its radial derivative have to be continuous. Plugging Eqs. (*) and (**) into these boundary conditions, we get two equations:

$$C_- J_m(kR) = C_+ K_m(\kappa R), \qquad C_- kJ'_m(kR) = C_+ \kappa K'_m(\kappa R),$$

where the prime denotes the function's derivative over its total argument. The condition of compatibility of these two linear homogeneous linear equations for two constants $C_\pm$ gives the following characteristic equation for energy E:

$$\begin{vmatrix} J_m(kR) & -K_m(\kappa R) \\ kJ'_m(kR) & -\kappa K'_m(\kappa R) \end{vmatrix} = 0, \quad \text{i.e. } \kappa J_m(kR) K'_m(\kappa R) = k K_m(\kappa R) J'_m(kR). \quad (***)$$

Now let us consider the ground state, with the lowest, zero value of the magnetic quantum number m, for the limit of very low U_0, and hence of very low $|E|$ and $U_0 + E$, which are both contained between 0 and U_0 (see the figure above):

$$E + U_0,\ |E| \ll E_0, \qquad \text{where } E_0 \equiv \frac{\hbar^2}{mR^2}.$$

According to the above definition of the parameters k and κ, this means that both arguments, kR and κR, of the Bessel functions are much smaller than 1. In this case, the functions may be approximated as[33]

$$J_0(kR) \approx 1 - \left(\frac{kR}{2}\right)^2 \approx 1, \quad K_0(\kappa R) \approx -\left[\ln\left(\frac{\kappa R}{2}\right) + \gamma\right] \equiv \ln \frac{2e^{-\gamma}}{\kappa R},$$

(where $\gamma \approx 0.5771$ is the Euler constant), so that

$$J'_0(kR) \approx -\frac{kR}{2}, \quad K'_0(\kappa R) \approx -\frac{1}{\kappa R},$$

and Eq. (***) is reduced to

$$\kappa \frac{1}{\kappa R} = k \ln\left(\frac{2e^{-\gamma}}{\kappa R}\right)\frac{kR}{2}, \quad \text{i.e.} \quad \frac{(E + U_0)}{E_0} \ln\left(\frac{2E_0}{|E|\, e^{2\gamma}}\right)^{1/2} = 1.$$

Since in the limit we are analyzing, the factor before the last logarithm is much smaller than 1, the ln function should be large, so that its argument has to be extremely large. (Recall that this function grows very slowly at large values of its argument.) Since the sum of $|E|$ and $(E + U_0)$ is equal to U_0, i.e. is fixed (see the figure above), this is only possible if $|E|$ is much smaller than not only E_0, but than $(E + U_0)$ as well, so that we may neglect E in the numerator of the pre-logarithm fraction, getting

[33] See, e.g. *Part EM* Eqs. (2.146) and (2.157).

$$E = -|E| = -\frac{2E_0}{e^{2\gamma}}\exp\left\{-\frac{2E_0}{U_0}\right\}$$
$$\approx -0.631\, E_0 \exp\left\{-\frac{2E_0}{U_0}\right\}, \quad \text{for } U_0 \ll E_0. \qquad (****)$$

This formula shows that a bound (localized) state, though with an exponentially small $|E|$, exists for any U_0, however small. Qualitatively, this is also true for a similar 1D system (see the solutions of problems 2.17 and 2.18), but there the ground state's energy level is much deeper than the exponentially shallow level (****): $|E| \sim U_0^2/E_0 \equiv E_0(U_0/E_0)^2$.

***Problem* 3.19.** Estimate the energy E of the localized ground state of a particle of mass m, in an axially-symmetric 2D potential well of a finite radius R, with an arbitrary but very small potential $U(\rho)$. (Quantify this condition.)

Solution: Just as in the previous problem, we may argue that the ground state has the 'magnetic' quantum number m equal to 0, so that Eq. (3.147) of the lecture notes for its radial wavefunction $\mathcal{R}(\rho)$ takes the form

$$-\frac{\hbar^2}{2m}\frac{1}{\rho}\frac{d}{d\rho}\left(\rho\frac{d\mathcal{R}}{d\rho}\right) + U(\rho)\mathcal{R} = E\mathcal{R}, \qquad (*)$$

with a negative E. Just as in the previous problem, at distances $\rho > R$, where $U(\rho) = 0$, this equation may be reduced to the modified Bessel equation,

$$\frac{d^2\mathcal{R}}{d\xi^2} + \frac{1}{\xi}\frac{d\mathcal{R}}{d\xi} - \mathcal{R} = 0, \quad \text{with } \xi \equiv \kappa\rho, \quad \text{where } \frac{\hbar^2\kappa^2}{2m} = -E \geqslant 0.$$

as has the same solution $\mathcal{R} = CK_0(\kappa\rho)$, so that at intermediate distances ρ, with $R < \rho \ll 1/\kappa$ we may use the same approximation:

$$\mathcal{R} \approx C\ln\left(\frac{2e^{-\gamma}}{\kappa\rho}\right) \equiv -C\ln\rho + \text{const}, \quad \frac{d\mathcal{R}}{d\rho} \approx -\frac{C}{\rho}, \quad \frac{1}{\mathcal{R}}\frac{d\mathcal{R}}{d\rho} \approx -\frac{1}{\rho}\ln\frac{2e^{-\gamma}}{\kappa\rho}. \qquad (**)$$

On the other hand, from the solution of the previous problem we may expect that if $|U(\rho)|$ is sufficiently small, the magnitude of the ground state energy $-E \equiv |E|$ is even smaller, so that at $\rho < R$ we may neglect, in Eq. (*), the full energy in comparison with the potential energy:

$$-\frac{\hbar^2}{2m}\frac{1}{\rho}\frac{d}{d\rho}\left(\rho\frac{d\mathcal{R}}{d\rho}\right) + U(\rho)\mathcal{R} = 0.$$

Let us quantify the smallness of the potential U by assuming that the scale of its magnitude is much smaller than the kinetic energy scale $E_0 \equiv \hbar^2/mR^2$—again just as in the previous problem. Then we may integrate the last equation from $\rho = 0$ (where

$d\mathcal{R}/d\rho = 0$ because of the azimuthal symmetry of the wavefunction) to the same intermediate value of the 2D radius, $R < \rho \ll 1/\kappa$, neglecting the small change of $\mathcal{R}$ in the second term, i.e. taking it to be equal to the value $\mathcal{R}(0)$ at the center of the well. The elementary integration yields

$$\frac{d\mathcal{R}}{d\rho} = \mathcal{R}(0)\frac{2m}{\hbar^2\rho}\int_0^R U(\rho')\rho' d\rho', \quad \text{so that } \frac{1}{\mathcal{R}}\frac{d\mathcal{R}}{d\rho} = \frac{2m}{\hbar^2\rho}\int_0^R U(\rho')\rho' d\rho'.$$

The last expression still differs from the last expression in Eq. (**) by the logarithm, but since it is a very slow function of its argument, we may require these two results to be close to each other at $\rho \sim R$, getting the following approximate estimate[34]

$$\frac{2m}{\hbar^2}\int_0^R U(\rho)\rho d\rho \sim -\ln\frac{1}{\kappa R},$$

finally giving

$$\kappa \sim \frac{1}{R}\exp\left\{\frac{\hbar^2}{2m}\bigg/\int_0^R U(\rho)\rho d\rho\right\},$$
$$\text{i.e.} \quad -E = \frac{\hbar^2\kappa^2}{2m} \sim \frac{\hbar^2}{2mR^2}\exp\left\{\frac{\hbar^2}{m}\bigg/\int_0^R U(\rho)\rho d\rho\right\}. \tag{***}$$

In order for this estimate to be valid, $-E$ should be much smaller than the scale of U, and hence than $E_0 \equiv \hbar^2/mR^2$, so that the integral in the exponent of Eq. (***) has to be negative by sign and small by magnitude:

$$0 < -\int_0^R U(\rho')\rho' d\rho' \ll \frac{\hbar^2}{m} \equiv E_0 R^2,$$

essentially repeating the assumption already made above.

Note that Eq. (***) yields the correct exponent (but of course only a rough estimate of the pre-exponential coefficient) for the particular system considered in the previous problem, $U(\rho) = U_0 = \text{const}$, because in this case

$$\int_0^R U(\rho)\rho d\rho = -U_0\int_0^R \rho d\rho \equiv -U_0\frac{R^2}{2},$$
$$\text{so that} \quad \frac{\hbar^2}{m}\bigg/\int_0^R U(\rho)\rho d\rho = -\frac{\hbar^2}{m}\bigg/U_0\frac{R^2}{2} \equiv -\frac{2E_0}{U_0}.$$

***Problem* 3.20.** Spell out the explicit form of the spherical harmonics $Y_4^0(\theta, \varphi)$ and $Y_4^4(\theta, \varphi)$.

Solution: According to Eqs. (3.165)–(3.171) of the lecture notes, any $Y_l^l(\theta, \varphi) = C_l^l \sin^l\theta\, e^{il\varphi}$, so that $Y_4^4(\theta, \varphi) = C_4^4 \sin^4\theta\, e^{4i\varphi}$. The normalization, using Eq. (3.173), is straightforward:

[34] Taking into account the difference between $2e^{-\gamma} = 1.23\ldots$ and 1 under the logarithm would be beyond the accuracy of this estimate.

$$|C_4^4|^{-2} = \int_0^{2\pi} d\varphi \int_0^{\pi} \sin^8\theta \sin\theta d\theta = 4\pi \int_0^{+1} (1 - \xi^2)^4 d\xi = 4\pi \frac{128}{315},$$

so that, finally,

$$Y_4^4(\theta, \varphi) = \frac{3}{16}\left(\frac{35}{2\pi}\right)^{1/2} \sin^4\theta \, e^{4i\varphi}.$$

Now proceeding to $Y_4^0(\theta, \varphi)$: since $Y_l^m(\theta, \varphi) \propto e^{im\varphi}\Theta(\theta)$, the azimuthal-angle factor of the spherical harmonic is independent of l, and for $m = 0$ is just a constant. Hence, we need to calculate only the polar factor $\Theta(\theta) \propto P_4^0(\cos\theta) = P_4(\cos\theta)$. From the Rodrigues formula (3.165), leaving the numerical coefficient aside for a while, we have

$$P_4(\xi) \propto \frac{d^4}{d\xi^4}(\xi^2 - 1)^4.$$

The differentiation is not as hard as one could imagine, because

$$(\xi^2 - 1)^4 \equiv \xi^8 - 4\xi^6 + 6\xi^4 - 4\xi^2 + 1,$$

and the two lowest-power terms are not important, because they gradually disappear at the sequential differentiation:

$$\begin{aligned}
\frac{d}{d\xi}(\xi^2 - 1)^4 &= 8\xi^7 - 4 \cdot 6\xi^5 + 6 \cdot 4\xi^3 - 4 \cdot 2\xi, \\
\frac{d^2}{d\xi^2}(\xi^2 - 1)^4 &= 8 \cdot 7\xi^6 - 4 \cdot 6 \cdot 5\xi^4 + 6 \cdot 4 \cdot 3\xi^2 - 4 \cdot 2, \\
\frac{d^3}{d\xi^3}(\xi^2 - 1)^4 &= 8 \cdot 7 \cdot 6\xi^5 - 4 \cdot 6 \cdot 5 \cdot 4\xi^3 + 6 \cdot 4 \cdot 3 \cdot 2\xi, \\
\frac{d^4}{d\xi^4}(\xi^2 - 1)^4 &= 8 \cdot 7 \cdot 6 \cdot 5\xi^4 - 4 \cdot 6 \cdot 5 \cdot 4 \cdot 3\xi^2 + 6 \cdot 4 \cdot 3 \cdot 2 \\
&\equiv 2^4 3 \cdot (35\xi^4 - 30\xi^2 + 3),
\end{aligned}$$

so that

$$Y_4^0(\theta, \varphi) = C_4^0(35\cos^4\theta - 30\cos^2\theta + 3).$$

The normalization is a little bit more tedious but still very much doable:

$$\begin{aligned}
|C_4^0|^{-2} &= \int_0^{2\pi} d\varphi \int_0^{\pi} (35\cos^4\theta - 30\cos^2\theta + 3)^2 \sin\theta d\theta \\
&= 4\pi \int_0^{+1} \left(35\xi^4 - 30\xi^2 + 3\right)^2 d\xi = 4\pi \frac{64}{9},
\end{aligned}$$

so that, finally,

$$Y_4^0(\theta, \varphi) = \frac{3}{16\pi^{1/2}}(35\cos^4\theta - 30\cos^2\theta + 3).$$

***Problem* 3.21.** Calculate $\langle x \rangle$ and $\langle x^2 \rangle$ in the ground state of the planar and spherical rotators of radius R. What can you say about the averages $\langle p_x \rangle$ and $\langle p_x^2 \rangle$?

Solution: Since for the planar rotator, the 2D radius $\rho \equiv (x^2 + y^2)^{1/2}$ is fixed at value R, its square ρ^2 is definitely R^2, so that

$$\langle \rho^2 \rangle = R^2.$$

Due to the axial symmetry of the ground state's wavefunction (corresponding to the azimuthal quantum number equal to zero), we may write $\langle x \rangle = 0$ and $\langle x^2 \rangle = \langle y^2 \rangle$, so that $\langle \rho^2 \rangle \equiv \langle x^2 \rangle + \langle y^2 \rangle = 2\langle x^2 \rangle$, and

$$\langle x^2 \rangle = \frac{1}{2}\langle \rho^2 \rangle = \frac{R^2}{2}.$$

For the spherical rotator, with fixed $r^2 \equiv x^2 + y^2 + z^2 = R^2$,

$$\langle r^2 \rangle = R^2.$$

Due to the spherical symmetry of the ground state, $\langle x \rangle = 0$ and $\langle x^2 \rangle = \langle y^2 \rangle = \langle z^2 \rangle$, so that $\langle r^2 \rangle \equiv \langle x^2 \rangle + \langle y^2 \rangle + \langle z^2 \rangle = 3\langle x^2 \rangle$, and

$$\langle x^2 \rangle = \frac{1}{3}\langle r^2 \rangle = \frac{R^2}{3}.$$

Using the same symmetry arguments, we may write $\langle p_x \rangle = 0$ (for both systems), $\langle p_x^2 \rangle = \langle p^2 \rangle/2$ for the planar rotator, and $\langle p_x^2 \rangle = \langle p^2 \rangle/3$ for the spherical rotator. However, calculating $\langle p^2 \rangle$ exclusively from the angular motion of the particle would be wrong. (In the ground, s-state of both systems, in which ψ does not have any angular dependence, this would give us $\langle p^2 \rangle = 0$ and $\langle p_x^2 \rangle = 0$, in a clear contradiction with Heisenberg's uncertainty principle.) Actually, the momentum's uncertainty in such systems is dominated by the lateral confinement of the motion (please revisit section 1.8 of the lecture notes if needed) and cannot be calculated exactly unless the confinement potential, which forces the particle to be a rotator, is specified quantitatively. We can, however, use the uncertainty relation for the following estimate: $\delta p_x \sim \delta p \sim \hbar/\delta R$, where δR is the radial width of the potential well providing the radial confinement. Since for the validity of rotators' models we need $\delta R \ll R$,[35] we may write $\delta x \delta p_x \sim R\hbar/\delta R \gg \hbar$.

***Problem* 3.22.** A spherical rotator, with $r \equiv (x^2 + y^2 + z^2)^{1/2} = R = \text{const}$, of mass m is in a state with the following wavefunction: $\psi = \text{const} \times (1/3 + \sin^2\theta)$. Calculate its energy.

---

[35] This condition ensures, in particular, that the above calculations of $\langle x^2 \rangle$ are correct.

Solution: Rewriting the wavefunction as

$$\psi = \text{const} \times \left(\frac{1}{3} + 1 - \cos^2\theta\right) \equiv \text{const} \times \left(\frac{4}{3} - \cos^2\theta\right)$$
$$\equiv \text{const} \times \left[1 - \frac{1}{3}(3\cos^2 - 1)\right],$$

and comparing the terms of this linear superposition with Eq. (3.174) and the third line of Eq. (3.176) of the lecture notes, we may write

$$\psi = \text{const} \times \left[(4\pi)^{1/2} Y_0^0 - \frac{1}{3}\left(\frac{16\pi}{3}\right)^{1/2} Y_2^0\right] \equiv \text{const} \times (4\pi)^{1/2}\left[Y_0^0 - \frac{2}{3\sqrt{3}} Y_2^0\right].$$

This expression shows that our state is a linear combination of two eigenfunctions of the rotator, both with the magnetic quantum number m equal to zero, but with different orbital numbers: one with $l = 0$, and another with $l = 2$. The ratio of the probabilities W_0 and W_2 of these eigenstates equals that of the squares of the probability amplitude magnitudes:

$$\frac{W_2}{W_0} = \left(\frac{2}{3\sqrt{3}}\right)^2 \equiv \frac{4}{27}.$$

Requiring the sum of the probabilities to equal 1, we get

$$W_0 = \frac{1}{1 + 4/27} = \frac{27}{31}, \qquad W_2 = \frac{4/27}{1 + 4/27} = \frac{4}{31}.$$

From here, and using Eq. (3.163) of the lecture notes for the eigenenergies of the rotator, we get

$$E = W_0 E_0 + W_2 E_2 = \frac{27}{31} \cdot 0 + \frac{4}{31} \cdot \frac{\hbar^2 \cdot 2(2+1)}{2mR^2} \equiv \frac{12}{31}\frac{\hbar^2}{mR^2}.$$

***Problem* 3.23.** According to the discussion in the beginning of section 3.5 of the lecture notes, the eigenfunctions of a 3D harmonic oscillator may be calculated as products of three 1D 'Cartesian oscillators'—see, in particular Eq. (3.125), with $d = 3$. However, according to the discussion in section 3.6, the wavefunctions of the type (3.200), proportional to the spherical harmonics Y_l^m, are also eigenstates of this spherically-symmetric system. Represent the wavefunctions (3.200) of:

(i) the ground state of the oscillator, and
(ii) each of its lowest excited states,

as linear combinations of products of 1D oscillator's wavefunctions. Also, calculate the degeneracy of the nth energy level of the oscillator.

Solutions:

(i) The ground state of a system is always degenerate, and according to (3.125) of the lecture notes (with $d = 3$) is merely the product,

$$\psi_0(\mathbf{r}) = \psi_0(x)\,\psi_0(y)\,\psi_0(z),$$

of the ground-state wavefunctions ψ_0 of the 1D Cartesian oscillators—see Eq. (2.275):

$$\psi_0(x) = C_0 \exp\left\{-\frac{x^2}{2x_0^2}\right\}, \quad \text{with } C_0 = \frac{1}{\pi^{1/4}x_0^{1/2}}, \quad x_0 \equiv \left(\frac{\hbar}{m\omega_0}\right)^{1/2}, \qquad (*)$$

and similarly for other two coordinates. We may rewrite $\psi_0(\mathbf{r})$ in the form (3.200) with $l = m = 0$:[36]

$$\psi_0(\mathbf{r}) = \mathcal{R}_{1,0}(r)\,Y_0^0(\theta, \varphi), \quad \text{with } Y_0^0 = \text{const}, \quad \text{and } \mathcal{R}_{1,0}(r) \propto \exp\left\{-\frac{m\omega_0 r^2}{2\hbar}\right\},$$

and verify, by the direct differentiation, that this radial function indeed satisfies Eq. (3.181) with $l = 0$, $U(r) = m\omega_0{}^2r^2/2$, and $E = \hbar\omega_0/2$.

(ii) In the Cartesian oscillator language, three lowest excited states of the 3D oscillator correspond to three possible combinations of the three indices n_j (see Eq. (3.124) of the lecture notes with $d = 3$) with the lowest nonvanishing sum $n \equiv n_x + n_y + n_z = 1$:

$$\psi_{1x}(\mathbf{r}) \equiv \psi_1(x)\psi_0(y)\psi_0(z), \quad \psi_{1y}(\mathbf{r}) \equiv \psi_0(x)\psi_1(y)\psi_0(z), \quad \text{and} \quad \psi_{1z}(\mathbf{r}) \equiv \psi_0(x)\psi_0(y)\psi_1(z). \qquad (**)$$

Here the wavefunction ψ_1 of the first excited state of a 1D oscillator may be calculated, for example, from Eq. (2.284) with $n = 1$, and the second of Eqs. (2.282):

$$\psi_1(x) = C_1 x \exp\left\{-\frac{x^2}{2x_0^2}\right\}, \quad \text{with } C_1 = \frac{2^{1/2}}{\pi^{1/4}x_0^{3/2}},$$

and similarly for two other coordinates. Using Eq. (*) and expressing the Cartesian coordinates in the pre-exponential factors via the spherical coordinates, we may write

[36] The apparently unnatural increase of n by 1 in the first index of the radial functions reflects the difference in the conventions accepted for 1D harmonic oscillators (where the ground state is traditionally denoted with quantum number $n = 0$), and for the spherical-wavefunctions representation (3.200), where it is convenient to start counting the principal quantum number n from 1.

$$\psi_{1x}(\mathbf{r}) \pm i\psi_{1y}(\mathbf{r}) = C_1C_0^2(x \pm iy)\exp\left\{-\frac{m\omega_0 r^2}{2\hbar}\right\}$$
$$= \text{const} \times \mathcal{R}_{2,1}(r)\sin\theta\ e^{\pm i\varphi},$$
$$\psi_{1z}(\mathbf{r}) = C_1C_0^2 z\exp\left\{-\frac{m\omega_0 r^2}{2\hbar}\right\}$$
$$= \text{const} \times \mathcal{R}_{2,1}(r)\cos\theta, \quad \text{with } \mathcal{R}_{2,1}(r) \propto r\exp\left\{-\frac{m\omega_0 r^2}{2\hbar}\right\}.$$

Comparing these expressions with Eqs. (3.175), we see that their angular parts are proportional, respectively, to the spherical harmonics $Y_1^{\pm 1}(\theta, \varphi)$ and $Y_1^0(\theta, \varphi)$. Finally, it is straightforward to verify, by the direct differentiation, that the radial function $\mathcal{R}_{2,1}(r)$, defined above, satisfies Eq. (3.181), again with $U(r) = m\omega_0^2 r^2/2$, but now with $l = 1$ and $E = 5\hbar\omega_0/2$. Thus, the stationary wavefunctions (3.200) with $n = 2$, $l = 1$, and all three possible values $m = \{-1, 0, +1\}$ may be indeed expressed as linear combinations of the Cartesian wavefunction products (**).

Finally, the degeneracy g of nth energy level of the system, following from Eq. (3.124) of the lecture notes, with $d = 3$,

$$E_n = \hbar\omega_0\left(\frac{3}{2} + n\right),$$

equals the number of different sets $\{n_x, n_y, n_z\}$ of these non-negative quantum numbers, with the fixed sum $n = n_x + n_y + n_z$. In the usual combinatorics lingo, this is just the number of different ways to place n *indistinguishable* balls into three *distinct* boxes. According to Eq. (A.7), this number is

$$g = M_n^{(3)} = {}^{n+2}C_2 \equiv \frac{(n+2)!}{2!n!} \equiv \frac{(n+1)(n+2)}{2}.$$

In particular, for the ground state with $n_x = n_y = n_z = 0$ and hence $n = 0$, this formula yields $g = 1$, while for the first excited state with $n = 1$, it gives $g = 3$, reflecting the wavefunctions spelled out above.

Note that this result for g, even after the necessary increase of n by 1 to compensate for the notation difference[37], differs from Eq. (3.204) of the lecture notes, which would give, in particular, $g = 4$ for $n = 1$. There is no contradiction here: the latter formula was derived, and is valid only for the Coulomb potential (3.190), while the radial functions of the 3D harmonic oscillator, with $n \equiv n_x + n_y + n_z \geqslant 1$, have very different properties. For example, in our particular case $n = 1$, the radial equation (3.181) with $U(r) = m\omega_0^2 r^2/2$, does *not* have a solution corresponding to a 2*s*-state with $l = 0$ and $E = 5\hbar\omega_0/2$.

[37] As a reminder, the traditional ground state numbers (also used in this course) are $n = 0$ for the harmonic oscillator, but $n = 1$ for the Bohr atom problem.

***Problem* 3.24.** Calculate the smallest depth U_0 of a spherical, flat-bottom potential well

$$U(\mathbf{r}) = \begin{cases} -U_0, & \text{for } r < R, \\ 0, & \text{for } R < r, \end{cases}$$

at which it has a bound (localized) eigenstate. Does such a state exist for a very narrow and deep well $U(\mathbf{r}) = -\mathscr{w}\,\delta(\mathbf{r})$, with a positive and finite $\mathscr{w}$?

Solution: Just as in the similar 2D situation (see problems 3.18 and 3.19), the lowest eigenenergy (and hence the smallest possible value of U_0) corresponds to the s-state, with $l = m = 0$. The functional form of its wavefunction inside the well may be calculated just as in the particle-inside-a-sphere problem discussed at the end of section 3.6 of the lecture notes, besides an energy offset:

$$\psi = A\frac{\sin kr}{r}, \quad \text{with } \frac{\hbar^2 k^2}{2m} \equiv E - U(r) \equiv E + U_0, \quad \text{for } r \leqslant R. \qquad (*)$$

The corresponding solution outside the well is mathematically the same (just with $U(r) = 0$), but since a localized state's energy E cannot be positive (to avoid outward traveling waves), it is more adequately represented as

$$\psi = C\frac{\exp\{-\kappa r\}}{r}, \quad \text{with } \frac{\hbar^2 \kappa^2}{2m} \equiv -E > 0, \quad \text{for } r \geqslant R. \qquad (**)$$

The ratio C/A and the energy E may be, as usual, calculated from two continuity conditions (of the wavefunction and its radial derivative) at $r = R$.

However, carrying out the full calculation (which would lead to a transcendental characteristic equation) is unnecessary here, because the assignment only asked us about the *minimum* value of U_0. As U_0 tends to this value, the function becomes delocalized, i.e. $\kappa \to 0$. In this limit, the radial derivative of the product $f(r) \equiv r\psi(r) = Ce^{-\kappa r}$ at $r = R + 0$ has to vanish. Due to the derivative's continuity, the same should be also true for the same product inside the sphere (here equal to $A \sin kr$), at $r = R - 0$, so that k should satisfy the requirement

$$\left(\frac{d}{dr}\sin kr\right)_{r=R} = 0, \quad \text{i.e. } \cos kR = 0.$$

The lowest value of $k > 0$ that satisfies this equation is $k_{\min} = \pi/2R$. According to Eq. (**), at $U_0 \to (U_0)_{\min}$, i.e. at $\kappa \to 0$, the eigenenergy E tends to 0, so that the second of Eqs. (*) yields[38]

$$(U_0)_{\min} = \frac{\hbar^2 k_{\min}^2}{2m} = \frac{\pi^2 \hbar^2}{8mR^2}.$$

[38] A (simple) additional exercise for the reader: use the similar argumentation to prove that in order to have n spherically-symmetric (s-) states, the well's depth U_0 has to be larger than $U_n = (U_0)_{\min}(2n - 1)^2$.

It is instructive to compare this result for the localization in a 3D case with the solutions of similar problems in 1D (problem 2.17) and 2D (problem 3.18): in contrast with the 3D case, at lower dimensionalities there is *no* any lower bound on U_0 for the localization[39].

In the delta-functional approximation $U(\mathbf{r}) = -\mathscr{w}\,\delta(\mathbf{r})$, we should take

$$\mathscr{w} = -\int_{r<R} U(r)d^3r = \frac{4\pi}{3}U_0R^3,$$

so that our result for $(U_0)_{\min}$ may be rewritten as

$$\mathscr{w}_{\min} = \frac{\pi^3\hbar^2}{6m}R. \qquad (***)$$

We see that if $R \to 0$, then $\mathscr{w}_{\min} \to 0$, i.e. a sufficiently small-scale well has a localized state for any finite $\mathscr{w}$ (but *not* any finite U_0!).

A thoughtful reader may be surprised by the fact that neither at this point, nor anywhere else in this course, are the localization properties of this 3D delta-functional potential, $U(\mathbf{r}) = -\mathscr{w}\delta(\mathbf{r})$, discussed in more detail. The reason is that most of these properties are not universal: due to the (integrable) divergence of the wavefunctions of the type (**) at $r \to 0$, its interaction with a short-range potential $U(\mathbf{r})$ depends on the 'internal design' of the potential, not only on its 3D integral, $-\mathscr{w}$.[40] This is why properties of short-range interactions are frequently described by less natural, but also less ambiguous models, for example

$$U(r) = 0, \quad \text{for } r > 0; \qquad \left.\frac{df}{dr}\right|_{r\to 0} \to -\kappa_0 f,$$

where, as above, $f(r) \equiv r\psi(r)$, while κ_0 is a given parameter. The ground (s-)state of this model is described by Eqs. (**) with $\kappa = \kappa_0$, and hence with $E = -\hbar^2\kappa_0^2/2m$.

***Problem* 3.25.** A 3D particle of mass m is placed into a spherically-symmetric potential well with $-\infty < U(r) \leqslant U(\infty) = 0$. Relate its ground-state energy to that of a 1D particle of the same mass, moving in the following potential well:

$$U'(x) = \begin{cases} U(x), & \text{for } x \geqslant 0, \\ +\infty, & \text{for } x \leqslant 0. \end{cases}$$

[39] Historically, this difference had interesting implications for the development of the theory of superconductivity, where the weak attraction between electrons (fermions) leads to their binding into Cooper pairs (effective bosons capable of the Bose–Einstein condensation) only because the Fermi–Dirac statistics confines them to a quasi-2D momentum space at the Fermi surface—see, e.g. chapter 3 in [2].

[40] A clear illustration of this fact is given by Eq. (***): it shows that the state-localization ability of the flat-bottom potential well, even in the limit $R \to 0$, depends not only on $\mathscr{w}$, but also on another parameter, R.

In the light of the found relation, discuss the origin of the difference between the solutions of the previous problem and problem 2.17.

Solution: As was discussed in section 3.6 of the lecture notes, the ground state of a spherically-symmetric system is always an s-state, with the wavefunction

$$\psi(\mathbf{r}) = Y_0^0(\theta, \varphi)\mathcal{R}(r) \equiv \frac{1}{(4\pi)^{1/2}}\mathcal{R}(r),$$

corresponding, in particular, to the orbital quantum number $l = 0$. In this case, Eq. (3.181) for the radial function is reduced to

$$-\frac{\hbar^2}{2mr^2}\frac{d}{dr}\left(r^2\frac{d\mathcal{R}}{dr}\right) + U(r)\mathcal{R} = E\mathcal{R}. \qquad (*)$$

In the beginning of section 3.1, it was shown that in the particular case of a free particle, i.e. $U(r) = 0$, a similar equation (3.3) for the radially-symmetric wavefunction $\psi(r)$ yields a 1D Schrödinger equation (also with $U(r) = 0$) for the function $f(r) \equiv r\psi(r)$. Inspired by this fact, let us look for the solution of Eq. (*) in the similar form:

$$\mathcal{R}(r) = \frac{f(r)}{r}. \qquad (**)$$

Indeed, this substitution, followed by the cancellation of the common factor $1/r$, yields:

$$-\frac{\hbar^2}{2m}\frac{d^2f}{dr^2} + U(r)f = Ef, \quad \text{for } r \geqslant 0. \qquad (***)$$

Due to the condition $U(r) \leqslant U(\infty) \equiv 0$, the ground-state energy E corresponding to this equation has to be negative < 0, so that at large distances from the center, where $U(r) \to 0$, the ground-state wavefunction decays exponentially (see, e.g. the solution of the previous problem) and hence we have to require $f(r) \equiv r\psi(r) \to 0$ at $r \to \infty$. On the other hand, in order to keep the wavefunction (**) finite at $r \to 0$, $f(0)$ has to equal zero.

But Eq. (***) and the above boundary conditions are exactly those satisfied by the ground-state wavefunction $\psi(x)$ and energy E of the 1D system mentioned in the assignment; hence the values of E of these 3D and 1D ground states are also equal. This result allows one to re-use the solutions of some key 1D problems for the 3D spherically-symmetric problems[41].

Note, however, that this 3D $\leftrightarrow$ 1D mapping is valid only if the 1D potential at $x < 0$ is positively-infinite, thus enforcing the boundary condition $f(0) = 0$ on the 1D wavefunction. For example, the solutions of the apparently similar problems 2.17 (1D) and 3.24 (3D), on the particle motion in a flat-bottom potential well of depth U_0, are radically different. As a reminder, in the 1D case a localized ground

[41] Actually, the analogy may be extended to all asymmetric eigenfunctions $\psi(x) = -\psi(-x)$ and their energies.

state exists for any, arbitrarily small U_0,[42] while in the 3D case there is a minimal value,

$$(U_0)_{\min} = \frac{\pi^2\hbar^2}{8mR^2}$$

(where R is well's radius) necessary for such localization. The reason for this difference is that (as the solution of problem 2.17 shows) the 1D well's ground state wavefunction is symmetric, with $f(0) \neq 0$, and hence does not satisfy the 3D $\leftrightarrow$ 1D mapping condition $\psi(0) = \lim_{r\to 0} f(r)/r < \infty$.

***Problem* 3.26.** Calculate the smallest value of the parameter U_0, for which the following spherically-symmetric potential well,

$$U(r) = -U_0 e^{-r/R}, \qquad \text{with } U_0,\ R > 0,$$

has a bound (localized) eigenstate.

Hint: You may like to introduce the following new variables: $f \equiv r\mathcal{R}$ and $\xi \equiv Ce^{-r/2R}$, with an appropriate choice of the constant C.

Solution: For this potential, Eq. (3.181) of the lecture notes, with $l = 0$ (corresponding to the ground state of the system, which is always an s-state) and $E \to 0$ (corresponding to the particle localization threshold), takes the form

$$-\frac{\hbar^2}{2mr^2}\frac{d}{dr}\left(r^2\frac{d\mathcal{R}}{dr}\right) + U(r)\mathcal{R} = 0.$$

Introducing the first replacement suggested in the Hint, $\mathcal{R} \equiv f/r$,[43] so that $r^2 d\mathcal{R}/dr = r^2 d(f/r)/dr = rdf/dr - f$, and $d(r^2 d\mathrm{R}/dr)/dr = rd^2f/dr^2$, we get

$$-\frac{\hbar^2}{2m}\frac{d^2f}{dr^2} + U(r)f = 0.$$

Now using the second of the suggested replacements, $\xi \equiv Ce^{-r/2R}$, so that $d\xi = -(Ce^{-r/2R}/2R)dr = -\xi dr/2R$, i.e. $dr = -2Rd\xi/\xi$, and our particular function $U(r) = -U_0 e^{-r/R} \equiv -U_0\xi^2$, we see that if we choose C as

$$C \equiv \left(\frac{8mR^2U_0}{\hbar^2}\right)^{1/2}, \qquad (**)$$

our differential equation is reduced to

$$\frac{d^2f}{d\xi^2} + \frac{1}{\xi}\frac{df}{d\xi} + f = 0.$$

[42] See also the solution of problem 2.18 for a more general potential.

[43] As a reminder, this substitution was used in section 3.1 of the lecture notes to derive Eq. (3.4), and also in the solutions of two previous problems.

But this is just the canonical form of the Bessel equation of order $\nu = 0$,[44] and its solution, finite at $\xi = 0$ (i.e. at $r \to \infty$), is the Bessel function $J_0(\xi)$, so that our radial wavefunction (up to a constant multiplier) is

$$\mathcal{R} \equiv \frac{f}{r} = \frac{1}{r} J_0(Ce^{-r/2R}).$$

In order for this function to be finite at $r \to 0$, i.e. at $Ce^{-r/2R} \to C$, this constant has to coincide with one of the roots of the function $J_0(\xi)$. According to Eq. (**), the smallest possible depth U_0 of the potential well corresponds to its smallest, first root $\xi_{01} \approx 2.405$,[45] so that we finally get

$$C_{\min} \equiv \left[\frac{8mR^2(U_0)_{\min}}{\hbar^2}\right]^{1/2} = \xi_{01}, \quad \text{i.e. } (U_0)_{\min} = \frac{\xi_{01}^2 \hbar^2}{8mR^2}.$$

Comparing this result with the solutions of the two previous problems, we see that they differ only by the numerical constant: $\pi^2 \approx 9.87 \leftrightarrow \xi_{01}{}^2 \approx 5.78$. Hence in our current case of the exponentially decaying confining potential, its minimum depth, necessary for particle localization, is approximately twice smaller than that in the case of a flat-bottom potential well with the same radius R. Of course, in the former case (of the exponential potential) the definition of the well's radius is to a certain extent conditional.

***Problem* 3.27.** A particle moving in a certain central potential $U(r)$, with $U(r) \to 0$ at $r \to \infty$, has a stationary state with the following wavefunction:

$$\psi = Ar^{\alpha}e^{-\beta r} \cos\theta,$$

where A, α, and $\beta > 0$ are constants. Calculate:

(i) the probabilities of all possible values of the quantum numbers m and l,
(ii) the confining potential, and
(iii) the state's energy.

Solutions:

(i) Comparing the angular part, $\cos\theta$, of the given wavefunction with the second line of Eq. (3.175) of the lecture notes, we may see that it coincides (to a constant multiplier) with the spherical harmonic $Y_1{}^0$, indicating that this is the state with $l = 1$ and $m = 0$, so that the probabilities of all other numbers l and m vanish.

(ii)–(iii) Plugging in the expression for the radial factor of the wavefunction[46],

[44] See, e.g. *Part EM* section 2.7, in particular, Eq. (2.130) with $\nu = 0$.
[45] See, e.g. the top left cell of *Part EM* table 2.1.
[46] Due to the linearity of the Schrödinger equation, the constant multiplier may be dropped for this calculation.

$$\mathcal{R}(r) = r^{\alpha} e^{-\beta r},$$

into Eq. (3.181) of the lecture notes, with the above-found value $l = 1$, and performing a straightforward differentiation, we get

$$U(r) = E + \frac{\hbar^2}{2m}\left[\frac{\alpha^2 + \alpha - 2}{r^2} - \frac{2\beta(\alpha + 1)}{r} + \beta^2\right].$$

Due to the imposed condition $U(\infty) = 0$, we have to assign the constant term, proportional to β^2, to the state's energy E, so that, finally,

$$E = -\frac{\hbar^2\beta^2}{2m}, \quad U(r) = \frac{\hbar^2}{2m}\left[\frac{\alpha^2 + \alpha - 2}{r^2} - \frac{2\beta(\alpha + 1)}{r}\right].$$

As a sanity check, the radial function $\mathcal{R}_{2,1}(r)$, given by the second of Eqs. (3.209) of the lecture notes, also corresponding to $l = 1$ (and to $n = 2$), is proportional to our current $\mathcal{R}(r)$ for the particular case $\alpha = 1$ (so that $\alpha^2 + \alpha - 2 = 0$), so that in this case we recover the Coulomb potential (3.190), with $C = 2\hbar^2\beta/m$. With this value of C, the first of Eqs. (3.192) yields $E_0 = 4\hbar^2\beta^2/m$, so that the above eigenenergy E is equal to $-E_0/8$, as it should be for $n = 2$, according to Eq. (3.201).

***Problem* 3.28.** Use the variational method to estimate the ground-state energy of a particle of mass m, moving in the following spherically-symmetric potential:

$$U(\mathbf{r}) = ar^4.$$

Solution: As was discussed in section 3.6 of the lecture notes, for the motion in any spherically-symmetric potential, the ground state is always an s-state, with a spherically-symmetric wavefunction $\psi(r)$. Since, in addition, our confining potential is continuous everywhere, including the origin $r = 0$, the most natural simple trial function is a 3D Gaussian:

$$\psi_{\text{trial}} = C \exp\left\{-\frac{\lambda^2}{4}r^2\right\},$$

with some real λ. Calculating the constant C (or rather its modulus) from the normalization requirement,

$$\int \psi^*_{\text{trial}}(r)\, \psi_{\text{trial}}(r)\, d^3r = 1,$$

we get

$$|C|^{-2} = \int \exp\left\{-\frac{\lambda^2}{2}r^2\right\}d^3r \equiv \int \exp\left\{-\frac{\lambda^2}{2}(x^2+y^2+z^2)\right\}dxdydz$$

$$= \left(\int_{-\infty}^{+\infty} \exp\left\{-\frac{\lambda^2}{2}x^2\right\}dx\right)^3$$

$$= \left(\frac{2^{1/2}}{\lambda}\int_{-\infty}^{+\infty} \exp\{-\xi^2\}d\xi\right)^3 = \left(\frac{2^{1/2}}{\lambda}\pi^{1/2}\right)^3 \equiv \frac{(2\pi)^{3/2}}{\lambda^3}.$$

Now we can calculate the expectation value of the system's Hamiltonian

$$\hat{H} = \frac{\hat{p}^2}{2m} + U(\mathbf{r}) = -\frac{\hbar^2}{2m}\frac{1}{r^2}\frac{d}{dr}\left(r^2\frac{d}{dr}\right) + ar^4,$$

(in the second form of this expression the angular part of the general Laplace operator[47] has been dropped due to the spherical symmetry of the trial state), corresponding to our trial function:

$$\langle H\rangle_{\text{trial}} = \int \psi^*_{\text{trial}}(r)\hat{H}\psi_{\text{trial}}(r)d^3r$$

$$= 4\pi|C|^2\int_0^{\infty} \exp\left\{-\frac{\lambda^2}{4}r^2\right\}\left[-\frac{\hbar^2}{2m}\frac{1}{r^2}\frac{d}{dr}\left(r^2\frac{d}{dr}\right) + ar^4\right]\exp\left\{-\frac{\lambda^2}{4}r^2\right\}r^2dr$$

$$= 4\pi\frac{\lambda^3}{(2\pi)^{3/2}}\int_0^{\infty}\left[-\frac{\hbar^2}{2m}\frac{\lambda^2}{2}\left(\frac{\lambda^2r^2}{2} - 3\right) + ar^4\right]\exp\left\{-\frac{\lambda^2}{2}r^2\right\}r^2dr$$

$$= 4\pi\frac{\lambda^3}{(2\pi)^{3/2}} \times \begin{bmatrix} \dfrac{3\hbar^2\lambda^2}{4m}\left(\dfrac{2^{1/2}}{\lambda}\right)^3\displaystyle\int_0^{\infty}\xi^2e^{-\xi^2}d\xi - \dfrac{\hbar^2\lambda^4}{8m}\left(\dfrac{2^{1/2}}{\lambda}\right)^5\displaystyle\int_0^{\infty}\xi^4e^{-\xi^2}d\xi \\ +a\left(\dfrac{2^{1/2}}{\lambda}\right)^7\displaystyle\int_0^{\infty}\xi^6e^{-\xi^2}d\xi \end{bmatrix}.$$

All these three integrals belong to the same well-known group[48] and are expressed via the values of the gamma-function in half-integer points, equal to $\pi^{1/2}$ multiplied by certain rational factors, so that the result is rather simple[49]:

[47] See, e.g. Eq. (A.67).

[48] See, e.g. Eqs. (A.36*a*).

[49] Actually, the first term of the final expression, describing the kinetic energy, might be also calculated just as $|C|^2$, by the integration in the Cartesian coordinates. This is why it is not surprising that it is exactly three times the corresponding term calculated for the similar but 1D Gaussian function in problem 2.34.

$$\langle H \rangle_{\text{trial}} = 4\pi \frac{\lambda^3}{(2\pi)^{3/2}} \left[\frac{3\hbar^2\lambda^2}{4m} \left(\frac{2^{1/2}}{\lambda} \right)^3 \frac{\pi^{1/2}}{4} - \frac{\hbar^2\lambda^4}{8m} \left(\frac{2^{1/2}}{\lambda} \right)^5 \frac{3\pi^{1/2}}{8} + a \left(\frac{2^{1/2}}{\lambda} \right)^7 \frac{15\pi^{1/2}}{16} \right]$$
$$\equiv \frac{3\hbar^2}{8m}\lambda^2 + \frac{15a}{\lambda^4}.$$

Now minimizing the expectation value by the adjustment of the parameter λ to the optimal value calculated from requirement

$$\left. \frac{\partial \langle H \rangle_{\text{trial}}}{\partial(\lambda^2)} \right|_{\lambda=\lambda_{\text{opt}}} \equiv \left. \left(\frac{3\hbar^2}{8m} - \frac{30a}{\lambda^6} \right) \right|_{\lambda=\lambda_{\text{opt}}} = 0,$$

we get

$$\lambda_{\text{opt}} = \left(\frac{80ma}{\hbar^2} \right)^{\frac{1}{6}}, \quad E_{\text{var}} \equiv \langle H \rangle_{\text{trial}}|_{\lambda=\lambda_{\text{opt}}} = \frac{9}{8} \left(\frac{10\hbar^4 a}{m^2} \right)^{1/3}.$$

***Problem* 3.29.** Use the variational method, with the trial wavefunction $\psi_{\text{trial}} = \text{const}/(r+a)^b$, where both $a > 0$ and $b > 1$ are fitting parameters, to estimate the ground-state energy of the hydrogen-like atom/ion with the nuclear charge $+Ze$. Compare the solution with the exact result.

Solution: Due to the spherical symmetry of the trial function, in the system's Hamiltonian,

$$\hat{H} = \frac{\hat{p}^2}{2m} - \frac{C}{r} \equiv -\frac{\hbar^2}{2m}\nabla^2 - \frac{C}{r}, \quad \text{where } C \equiv \frac{Ze^2}{4\pi\varepsilon_0},$$

we may keep only the radial part of the Laplace operator:

$$\hat{H} = -\frac{\hbar^2}{2m} \frac{1}{r^2} \frac{d}{dr} \left(r^2 \frac{d}{dr} \right) - \frac{C}{r},$$

so that the expectation value of energy in the trial state is

$$\langle H\rangle_{\text{trial}} \equiv \int \psi^*_{\text{trial}}(\mathbf{r})\,\hat{H}\psi_{\text{trial}}(\mathbf{r})d^3r$$
$$= |A|^2\int_0^\infty 4\pi r^2 dr\,(r+a)^{-b}\left[-\frac{\hbar^2}{2m}\frac{1}{r^2}\frac{d}{dr}\left(r^2\frac{d}{dr}\right)-\frac{C}{r}\right](r+a)^{-b}$$
$$= 4\pi\,|A|^2\int_0^\infty\left\{\frac{\hbar^2}{2m}b[2r(r+a)^{-2b-1}-(b+1)r^2(r+a)^{-2b-2}]-C(r+a)^{-2b}r\right\}dr$$
$$\equiv 4\pi\,|A|^2$$
$$\times\left\{\begin{array}{l}\frac{\hbar^2 b}{2m}\left[2\left(\int_a^\infty r'^{-2b}dr' - a\int_a^\infty r'^{-2b-1}dr'\right)\right.\\ \left.\quad-(b+1)\left(\int_a^\infty r'^{-2b}dr' - 2a\int_a^\infty r'^{-2b-1}dr' + a^2\int_a^\infty r'^{-2b-2}dr'\right)\right]\\ -C\left(\int_a^\infty r'^{-2b+1}dr' - a\int_a^\infty r'^{-2b}dr'\right)\end{array}\right\},$$

where $r' \equiv r + a$, so that $r = r' - a$. Due to the given conditions $a > 0$ and $b > 1$, all these integrals converge at both limits, with zero contributions from the upper of them, giving

$$\langle H\rangle_{\text{trial}} = 4\pi\,|A|^2$$
$$\times\left\{\begin{array}{l}\frac{\hbar^2 b}{2ma^{2b-1}}\left[2\left(\frac{1}{2b-1}-\frac{1}{2b}\right)-(b+1)\left(\frac{1}{2b-1}-\frac{1}{b}+\frac{1}{2b+1}\right)\right]\\ -\frac{C}{a^{2b-2}}\left(\frac{1}{2b-2}-\frac{1}{2b-1}\right)\end{array}\right\} \qquad (*)$$
$$\equiv 4\pi\,|A|^2\left[\frac{\hbar^2 b}{2ma^{2b-1}(2b-1)(2b+1)}-\frac{C}{2a^{2b-2}(b-1)(2b-1)}\right].$$

The calculation of the normalization coefficient A is similar and less bulky:

$$1 = \int \psi^*_{\text{trial}}(\mathbf{r})\,\psi_{\text{trial}}(\mathbf{r})d^3r$$
$$= 4\pi\,|A|^2\int_0^\infty (r+a)^{-2b}\,r^2 dr = 4\pi\,|A|^2\int_a^\infty r'^{-2b}\,(r'-a)^2 dr'$$
$$\equiv 4\pi\,|A|^2\left(\int_a^\infty r'^{-2b+2}\,dr' - 2a\int_a^\infty r'^{-2b+1}\,dr' + a^2\int_a^\infty r'^{-2b}\,dr'\right)$$
$$= \frac{4\pi\,|A|^2}{a^{2b-3}(b-1)(2b-1)(2b-3)}.$$

Plugging the resulting expression for $|A|^2$ into Eq. (*), we get

$$\langle H\rangle_{\text{trial}} = \frac{\hbar^2 b(b-1)(2b-3)}{2ma^2(2b+1)} - \frac{C(2b-3)}{2a}.$$

Now, let us optimize this expression, starting from the parameter a. Since for $b > 1$ all parentheses in it are positive, $\langle H\rangle_{\text{trial}}$ is a smooth function of a, going from $+\infty$ at $a \to 0$ to -0 at $a \to \infty$, and thus having just one (negative) minimum, which may be calculated from the condition

$$\left.\frac{\partial\langle H\rangle_{\text{trial}}}{\partial a}\right|_{a=a_{\text{opt}}} = 0, \quad \text{i.e.} \ -\frac{\hbar^2 b(b-1)(2b-3)}{m a_{\text{opt}}^3(2b+1)} + \frac{C(2b-3)}{2a_{\text{opt}}^2} = 0.$$

This equation yields

$$a_{\text{opt}} = \frac{2b(b-1)}{2b+1} r_0, \quad \min_a\langle H\rangle_{\text{trial}} = -\frac{(2b-3)(2b+1)}{8b(b-1)} E_0,$$

where $r_0 \equiv \hbar^2/mC$ and $E_0 \equiv \hbar^2/mr_0^2 \equiv m(C/\hbar)^2$—see Eq. (3.192) of the lecture notes.

Proceeding to the optimization over b, we may notice that in the allowed range of this parameter, $1 < b < \infty$, the fraction in the last displayed expression always grows with b (see the figure below), approaching 1/2 asymptotically at $b \to \infty$. In this limit, the expectation value approaches the exact ground state energy:

$$\min_{a,b}\langle H\rangle_{\text{trial}} = E_g = -\frac{1}{2}E_0$$

—see Eq. (3.201) with $n = 1$. (Note that in this limit, $a_{\text{opt}}/r_0 \to b \to \infty$.)

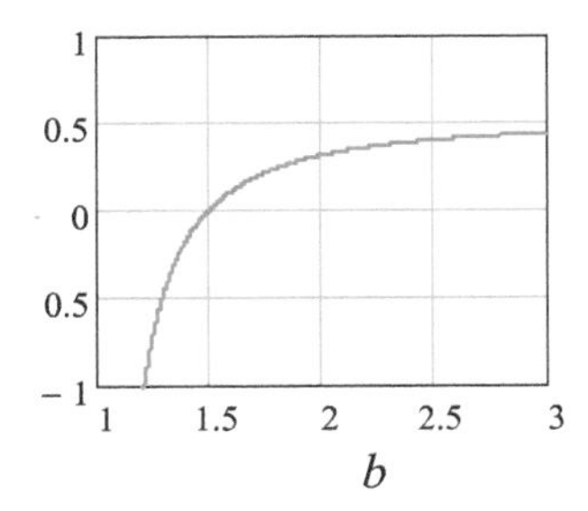

The fact that a good fitting requires the power in the expression for ψ_{trial} to tend to infinity should not be surprising, because only in this limit does the trial function resemble the genuine, exponential ground-state wavefunction—see Eq. (3.208) of the lecture notes. What is indeed counter-intuitive is that the fitting by such a different function enables finding the *exact* value of the ground-state energy[50]. This fact demonstrates the power of the variational method with more than one fitting parameter—at the cost of longer calculations.

***Problem* 3.30.** Calculate the energy spectrum of a particle moving in a monotonic, but otherwise arbitrary attractive spherically-symmetric potential $U(r) < 0$, in the approximation of very large orbital quantum numbers l. Formulate the quantitative

[50] A good additional exercise: check whether the optimized trial function yields the exact expectation values of the observables r, $1/r$, $1/r^2$, and $1/r^3$—see Eqs. (3.210)–(3.211), with $n = 1$ and $l = 0$.

condition(s) of validity of your theory. Check that for the Coulomb potential $U(r) = -C/r$, your result agrees with Eq. (3.201) of the lecture notes.

Hint: Try to solve Eq. (3.181) of the lecture notes approximately, introducing the same new function, $f(r) \equiv r\mathcal{R}(r)$, as was already used in section 3.1 and in the solutions of a few earlier problems.

Solution: Plugging in the suggested substitution,

$$\mathcal{R}(r) \equiv \frac{f(r)}{r},$$

into Eq. (3.181), we reduce it to the form similar to the Schrödinger equation of a 1D particle,

$$-\frac{\hbar^2}{2m}\frac{d^2 f}{dr^2} + U_{\text{ef}}(r)f = Ef, \qquad (*)$$

with the following effective potential energy:

$$U_{\text{ef}}(r) \equiv U(r) + \frac{\hbar^2 l(l+1)}{2mr^2}.$$

Note that this is the same potential which participates in the classical theory of the orbital ('planetary') motion[51], besides that the square of the angular momentum is replaced for its eigenvalue (3.178):

$$L^2 = \hbar^2 l(l+1).$$

The classical theory gives a hint how our current problem may be solved. Indeed, if the magnitude of the attractive potential $U(r) < 0$ does not grow too fast (faster than $1/r^2$) at $r \to 0$, the effective potential has a minimum at a certain radius r_0, determined by the following condition:

$$\left.\frac{dU_{\text{ef}}(r)}{dr}\right|_{r=r_0} \equiv \left.\frac{dU(r)}{dr}\right|_{r=r_0} - \frac{\hbar^2 l(l+1)}{mr_0^3} = 0. \qquad (**)$$

In the classical case, r_0 is the radius of a circular orbit of the particle, at which the attractive force $F = -dU/dr$ provides exactly the necessary centripetal acceleration $\omega^2 r = (L/mr^2)^2 r \equiv (L^2/mr^3)/m$, so that r does not change in time, i.e. the radial component of its momentum, $p_r \equiv mv_r = mdr/dt$, equals zero. On the other hand, in quantum mechanics, according to the Heisenberg principle, r and p_r cannot be exactly fixed simultaneously, so that the radial motion has to be quantized.

Let us assume that this quantization does not bring the particle's energy E much higher than the effective potential energy's minimum,

[51] See, e.g. *Part CM* sections 3.4–3.5, in particular Eq. (3.44).

$$U_{\text{ef}}(r_0) = U(r_0) + \frac{\hbar^2 l(l+1)}{2mr_0^2}.$$

Then the motion takes place at $r \approx r_0$, and we may replace the genuine effective potential with its Taylor expansion near the minimum:

$$U_{\text{ef}}(r) \approx U_{\text{ef}}(r_0) + \frac{\kappa}{2}(r - r_0^2), \quad \text{with } \kappa \equiv \left.\frac{d^2 U_{\text{ef}}(r)}{dr^2}\right|_{r=r_0}$$
$$= \left.\frac{d^2 U(r)}{dr^2}\right|_{r=r_0} + \frac{3\hbar^2 l(l+1)}{mr_0^4}. \qquad (***)$$

In this approximation, Eq. (*) describes the usual 1D harmonic oscillator with frequency

$$\omega_0 = \left(\frac{\kappa}{m}\right)^{1/2} = \left[\frac{1}{m}\left.\frac{d^2 U(r)}{dr^2}\right|_{r=r_0} + \frac{3\hbar^2 l(l+1)}{m^2 r_0^4}\right]^{1/2},$$

whose energy spectrum is given by Eq. (2.262), $E_r = \hbar\omega_0(n_r + 1/2)$, with $n_r = 0, 1, 2, \ldots$. As a result, the total energy of the system becomes

$$E = U_{\text{ef}}(r_0) + \hbar\omega_0\left(n_r + \frac{1}{2}\right) \equiv U(r_0) + \frac{\hbar^2 l(l+1)}{2mr_0^2} + \hbar\omega_0\left(n_r + \frac{1}{2}\right).$$

The validity of this result is determined by that of the approximation (***). If the potential $U(r)$ is sufficiently smooth, the approximation is valid if the radial spread δr of the wavefunction is much smaller than r_0. As we know from section 2.9 of the lecture notes, the spread may be estimated as

$$\delta r \sim \left[\frac{\hbar}{m\omega_0}(2n_r + 1)\right]^{1/2}.$$

Since, according to Eq. (**), r_0 grows with l faster than $\delta r \propto \omega_0^{-1/2}$, generally this theory works well for large values of l, and not very high values of n_r, though the exact condition depends of the particular function $U(r)$.

For the Coulomb potential, $U(r) = -C/r$, the above general relations yield

$$U_{\text{ef}}(r) = -\frac{C}{r} + \frac{\hbar^2 l(l+1)}{2mr^2}, \quad \frac{C}{r_0^2} - \frac{\hbar^2 l(l+1)}{mr_0^3} = 0,$$
$$\text{so that } r_0 = \frac{\hbar^2 l(l+1)}{mC}, \quad U(r_0) = -\frac{C^2 m}{\hbar^2 l(l+1)},$$
$$\frac{\hbar^2 l(l+1)}{2mr_0^2} = \frac{C^2 m}{2\hbar^2 l(l+1)}, \quad \hbar\omega_0 = \hbar\left[-\frac{2C}{mr_0^3} + \frac{3\hbar^2 l(l+1)}{m^2 r_0^4}\right]^{1/2} = \frac{C^2 m}{\hbar^2 [l(l+1)]^{3/2}},$$

and the calculated energy spectrum is

$$E = -\frac{1}{2}m\left(\frac{C}{\hbar}\right)^2\left[\frac{1}{l(l+1)} - \frac{2n_r + 1}{[l(l+1)]^{3/2}}\right].$$

For $l \gg 1$, n_r the expression in the last square brackets may be Taylor-expanded in small parameters n_r/l and $1/l$, giving

$$\left[\frac{1}{l(l+1)} - \frac{2n_r + 1}{[l(l+1)]^{3/2}}\right] \equiv \frac{1}{l^2}\frac{(1 + 1/l)^{1/2} - 2n_r/l - 1/l}{(1 + 1/l)^{3/2}} \approx \frac{1}{l^2}\frac{(1 + 1/2l) - 2n_r/l - 1/l}{1 + 3/2l}$$
$$\approx \frac{1}{l^2}\left(1 - \frac{2n_r}{l} - \frac{2}{l}\right).$$

But these leading terms of the expansion coincide with the similar expansion of the exact formula for this bracket, $1/n^2$ (see Eq. (3.201) of the lecture notes), provided that we take

$$n = l + n_r + 1, \quad \text{with } n_r = 0, 1, 2, \ldots . \qquad (****)$$

Indeed,

$$\frac{1}{n^2} \equiv \frac{1}{(l + n_r + 1)^2} \equiv \frac{1}{l^2}\frac{1}{(1 + n_r/l + 1/l)^2} \approx \frac{1}{l^2}\frac{1}{1 + 2n_r/l + 2/l}$$
$$\approx \frac{1}{l^2}\left(1 - \frac{2n_r}{l} - \frac{2}{l}\right).$$

So, for the Coulomb potential the approximate theory gives the result coinciding with the exact one at $l \gg n_r$, 1. Note also that the above calculation gave us, as a byproduct, a very interesting formula (****) which sheds new light on the famous l-degeneracy of the hydrogen atom' energy spectrum. Namely, a decrease of the background energy $U_{\text{ef}}(r_0)$ due to a decrease of the orbital number l by one, is exactly equal and opposite to the addition $\hbar\omega_0$ to the energy of the radial motion, due to an increase of the radial quantum number n_r by one, so that the total energy E does not change[52]. The above general expression for E shows clearly that such exact compensation is a 'mathematical accident', and is violated for even small deviation of the attractive potential $U(r)$ from the Coulomb law, thus lifting the l-degeneracy. As was discussed in section 3.7 of the lecture notes, this is exactly what happens in the atoms of heavier elements, due to radius-dependent shielding of the positive potential of their nuclei by the negative electric charge of other electrons.

[52] Graphically, we may consider this compensation as a trade-off of the number of 'wiggles' (read zeros) of the radial and angular wavefunctions, at the same n, and hence the same total energy of the system. Rather amazingly, in this form this exact compensation takes place even at lower values of l, where the radial functions differ rather substantially from those of the harmonic oscillator—please have one more look at figure 3.22 of the lecture notes.

***Problem* 3.31.** An electron had been in the ground state of a hydrogen-like atom/ion with nuclear charge Ze, when the charge suddenly changed to $(Z + 1)e$.[53] Calculate the probabilities for the electron of the changed system to be:

(i) in the ground state, and
(ii) in the lowest excited state.

Solutions: According to Eqs. (3.174), (3.200), and (3.208) of the lecture notes, the electron's wavefunction before the nuclear change was

$$\psi_g(\mathbf{r}) = \left(\frac{1}{4\pi}\right)^{1/2} \frac{2}{r_0^{3/2}} e^{-r/r_0} = \left(\frac{1}{4\pi}\right)^{1/2} \frac{2Z^{3/2}}{r_B^{3/2}} e^{-Zr/r_B}, \qquad (*)$$

where r_0 is given by the second of Eqs. (3.192) with $m = m_e$ and the Coulomb interaction constant $C = Ze^2/4\pi\varepsilon_0$:

$$r_0 = \frac{\hbar^2}{m_e C} = \frac{4\pi\varepsilon_0\hbar^2}{m_e Z e^2} \equiv \frac{r_B}{Z},$$

r_B being the Bohr radius—see Eq. (1.10).

(i) The ground state wavefunction $\psi_g{}'(\mathbf{r})$ after the change of Z is given by the same formula, but with the replacement

$$r_0 \to r_0{}' \equiv \frac{r_B}{Z+1}. \qquad (**)$$

According to Eq. (1.68), the probability W_g for the electron to be in the ground state of the new ion is determined by the wavefunction overlap integral

$$c_g = \int \psi_g{}'^*(\mathbf{r})\psi_g(\mathbf{r})d^3r = \frac{4Z^{3/2}(Z+1)^{3/2}}{r_B^3}\int_0^\infty \exp\left\{-r\left(\frac{Z}{r_B} + \frac{Z+1}{r_B}\right)\right\} r^2 dr$$

$$\equiv \frac{4Z(Z+1)}{(2Z+1)^3}\int_0^\infty \xi^2 e^{-\xi} d\xi,$$

where $\xi \equiv (2Z + 1)r/r_B$. This is a well-known integral[54], equal to $\Gamma(3) = 2! = 2$, so that, finally,

$$c_g = \frac{8Z^{3/2}(Z+1)^{3/2}}{(2Z+1)^3}, \quad \text{and } W_g \equiv |c_g|^2 = \frac{2^6 Z^3(Z+1)^3}{(2Z+1)^6}.$$

[53] Such a fast change happens, for example, at the beta-decay, when one of nucleus' neurons suddenly turns into a proton, emitting a high-energy electron and a neutrino, which leave the system very fast (instantly on the atomic time scale), and do not affect directly the atom transition's dynamics.
[54] See, e.g. Eq. (A.34*a*) with $s = 3$.

(ii) Due to the spherical symmetry of the initial wavefunction (*), of the four degenerate lowest excited states (all with $n = 2$, but with either $l = 0$ and $m = 0$, or with $l = 1$ and $m = 0, \pm1$), it has a nonvanishing overlap integral only with the s-state (with $l = 0$). Its wavefunction is given by Eqs. (3.174), (3.200), and the first of Eqs. (3.209) of the lecture notes, again with the replacement (**):

$$\psi_e(\mathbf{r}) = \left(\frac{1}{4\pi}\right)^{1/2} \frac{1}{(2r_0')^{3/2}}\left(2 - \frac{r}{r_0'}\right) e^{-r/2r_0'}$$
$$= \left(\frac{1}{4\pi}\right)^{1/2} \frac{(Z+1)^{3/2}}{(2r_B)^{3/2}}\left[2 - \frac{(Z+1)r}{r_B}\right] e^{-(Z+1)\,r/2r_B},$$

so that the overlap integral is

$$c_e = \int \psi_g'^*(\mathbf{r})\psi_e(\mathbf{r})d^3r$$
$$= \frac{2Z^{3/2}(Z+1)^{3/2}}{2^{3/2}r_B^3}\int_0^\infty \left[2 - \frac{(Z+1)r}{r_B}\right]\exp\left\{-r\left(\frac{Z}{r_B} + \frac{Z+1}{2r_B}\right)\right\}r^2dr$$
$$\equiv 2^{5/2}\frac{Z^{3/2}(Z+1)^{3/2}}{(3Z+1)^3}\left[2\int_0^\infty \xi^2 e^{-\xi}d\xi - \frac{2(Z+1)}{3Z+1}\int_0^\infty \xi^3 e^{-\xi}d\xi\right],$$

where now $\xi \equiv (3Z+1)r/2r_B$. The first integral is the same as the one in the first task, and equals 2, while the second one is of the same type, but with $s = 4$, and equals $\Gamma(4) = 3! = 6$, so that

$$c_e = 2^{5/2}\frac{Z^{3/2}(Z+1)^{3/2}}{(3Z+1)^3}\left(2\cdot 2 - \frac{2(Z+1)}{3Z+1}\cdot 6\right) \equiv -\,2^{11/2}\frac{Z^{3/2}(Z+1)^{3/2}}{(3Z+1)^4},$$
$$W_e \equiv |\,c_e\,|^2 = 2^{11}\frac{Z^3(Z+1)^3}{(3Z+1)^8}.$$

The above results show that if Z is low, the probabilities W_g and W_e are comparable; for example for $Z = 1$, $W_g = 2^9/3^6 \approx 0.702$, while $W_e = 1/4 = 0.25$.[55] However, in the limit $Z \gg 1$, the probability W_g of staying in the ground state is close to 1:

$$W_g = \frac{(1+1/Z)^3}{(1+1/2Z)^6} \approx \frac{1 + 3/Z + 3/Z^2}{1 + 6/2Z + 15/2Z^2} \approx 1 - \frac{9}{2Z^2} \equiv 1 - \frac{4.5}{Z^2} \to 1,$$

while the probability of the atom's excitation is small; in particular

$$W_e = 2^{11}\frac{Z^3(Z+1)^3}{(3Z+1)^8} \approx \frac{2^{11}}{3^8Z^2} \approx \frac{0.312}{Z^2} \to 0.$$

This is very natural, because at $Z \gg 1$, the *relative* change of Z is small.

[55] The small difference between $W_g + W_e$ and 1 gives the total probability of excitation of higher (s-)states.

***Problem* 3.32.** Due to a very short pulse of an external force, the nucleus of a hydrogen-like atom, initially at rest in its ground state, starts moving with velocity $\mathbf{v}$. Calculate the probability W_g that the atom remains in its ground state. Evaluate the energy to be given, by the pulse, to a hydrogen atom in order to reduce W_g to 50%.

Solution: Repeating the argumentation used in the model solution of problem 2.38, we may use the Galilean transform

$$\Psi'(\mathbf{r}', t') = \Psi(\mathbf{r}, t) \exp\left\{-i\frac{m\mathbf{v}\cdot\mathbf{r}}{\hbar} + i\frac{mv^2 t}{2\hbar}\right\},$$

whose proof was the subject of problem 1.4, to conclude that immediately after the application of the force pulse (say, at $t = +0$) its wavefunction, in the reference frame moving with the atom at $t > 0$, is

$$\psi_{+0}(\mathbf{r}) = \psi_{-0}(\mathbf{r}) \exp\left\{-i\frac{m\mathbf{v}\cdot\mathbf{r}}{\hbar}\right\},$$

where ψ_{-0} is the wavefunction immediately before the pulse, i.e. that of the ground state of the atom. Hence the overlap integral (1.68) for the ground state is

$$c_g = \int \psi_g^*(\mathbf{r})\psi_{+0}(\mathbf{r})d^3r = \int \left|\psi_g(\mathbf{r})\right|^2 \exp\left\{-i\frac{m\mathbf{v}\cdot\mathbf{r}}{\hbar}\right\} d^3r.$$

Now using Eqs. (3.174), (3.200), and (3.208) of the lecture notes, and directing the z-axis along the vector $\mathbf{v}$, so that $\mathbf{v}\cdot\mathbf{r} = vz = vr\cos\theta$, in the corresponding spherical coordinates we get

$$\begin{aligned}
c_g &= \int \frac{1}{4\pi}\frac{4}{r_0^3} \exp\left\{-\frac{2r}{r_0} - i\frac{mvr\cos\theta}{\hbar}\right\} d^3r \\
&= \frac{2}{r_0^3}\int_0^\infty \exp\left\{-\frac{2r}{r_0}\right\} r^2 dr \int_{-1}^{+1} \exp\left\{-i\frac{mvr\cos\theta}{\hbar}\right\} d(\cos\theta) \\
&= -\frac{2}{r_0^3}\int_0^\infty \exp\left\{-\frac{2r}{r_0}\right\} r^2 dr \left(\frac{\hbar}{mvr} 2\sin\frac{mvr}{\hbar}\right) \\
&\equiv -\frac{4\hbar}{mvr_0^3}\operatorname{Im}\int_0^\infty \exp\left\{-\frac{2r}{r_0} + i\frac{mvr}{\hbar}\right\} r dr \\
&= -\frac{4\hbar}{mvr_0^3}\operatorname{Im}\left[\left(-\frac{2}{r_0} + i\frac{mv}{\hbar}\right)^{-2}\right] \equiv \left[1 + \left(\frac{mvr_0}{2\hbar}\right)^2\right]^{-2},
\end{aligned}$$

so that the probability of staying in the ground state is

$$W_g = \left|c_g\right|^2 = \left[1 + \left(\frac{mvr_0}{2\hbar}\right)^2\right]^{-4} \equiv \frac{1}{\left(1 + v^2/4v_0^2\right)^4}, \qquad (*)$$

where $v_0 \equiv \hbar/m r_0$ is the natural scale of the velocity of the electron's 'motion' in the ground state. (Indeed, as the model solution of the next problem shows, $v_0 = \langle p^2\rangle^{1/2}/m$.)

For the hydrogen atom (with $m = m_e \approx 0.91 \times 10^{-30}$ kg and $r_0 = r_B \approx 0.53 \times 10^{-10}$ m), the velocity is close to 2.2×10^6 m s^{-1}. According to Eq. (*), in order to get $W_g = 1/2$, the velocity v should be equal to $2 \cdot (2^{1/4} - 1)^{1/2} v_0 \approx 1.90 \times 10^6$ m s^{-1}, giving the atom (whose mass is dominated by that of its single-proton nucleus, $m_p \approx 1.67 \times 10^{-30}$ kg) the kinetic energy $T \approx m_p v^2/2 \approx 3.02 \times 10^{-15}$ J ≈ 18.9 keV. (Since this energy is much larger than the change of the electron's energy, which is of the order of the Hartree energy unit $E_H \approx 27$ eV, the electron cannot affect this calculation significantly.)

***Problem* 3.33.** Calculate $\langle x^2\rangle$ and $\langle p_x^2\rangle$ in the ground state of a hydrogen-like atom/ion. Compare the results with the Heisenberg's uncertainty relation. What do these results tell about the electron's velocity in the system?

Solution: The simplest way to solve this problem[56] is to notice that due to the spherical symmetry of the ground state's wavefunction (corresponding to quantum numbers $n = 1$, $l = 0$, and $m = 0$), we can write

$$\langle r^2\rangle \equiv \langle x^2\rangle + \langle y^2\rangle + \langle z^2\rangle = 3\langle x^2\rangle.$$

Hence we can calculate the first average as

$$\langle x^2\rangle = \frac{1}{3}\langle r^2\rangle = \frac{1}{3}\int \psi^* r^2 \psi d^3r = \frac{1}{3}\int |Y_0^0|^2 d\Omega \int_0^\infty |\mathcal{R}_{1,0}(r)|^2 r^4 dr.$$

Since the spherical functions are normalized, the integral over the solid angle equals 1, while for the radial integral we may use Eq. (3.208) and a table integral[57] to get a surprisingly simple result:

$$\langle x^2\rangle = \frac{1}{3}\int_0^\infty |\mathcal{R}_{1,0}(r)|^2 r^4 dr = \frac{4}{3r_0^3}\int_0^\infty \exp\left\{-\frac{2r}{r_0}\right\} r^4 dr = \frac{r_0^2}{24}\int_0^\infty e^{-\xi}\xi^4 d\xi = r_0^2.$$

The second assignment may be addressed similarly, though requires a bit more caution, because it involves differentiation:

$$\langle p_x^2\rangle = \frac{1}{3}\langle p^2\rangle = \frac{1}{3}\int \psi^* \hat{p}^2 \psi d^3r = \frac{1}{3}\int |Y_0^0|^2 d\Omega \int_0^\infty r^2 dr\ \mathcal{R}_{1,0}^*(-i\hbar\nabla)^2 \mathcal{R}_{1,0}.$$

At this point, it is important to remember that even though $\mathcal{R}_{1,0}$ (as a radial function) depends on r only, the operator ∇^2 is still different from d^2/dr^2—see Eq. (A.67) with $\partial/\partial\theta = \partial/\partial\varphi = 0$. As a result, we get

[56] A more straightforward solution, using integration in spherical coordinates (with x^2 replaced with $r^2\sin^2\theta\cos^2\varphi$, etc), is also doable, but a bit more bulky.

[57] See, e.g. Eq. (A.34*d*) with $n = 4$.

$$\langle p_x^2 \rangle = -\frac{4\hbar^2}{3r_0^3}\int_0^\infty r^2 dr\, e^{-r/r_0}\left\{\frac{1}{r^2}\left[\frac{d}{dr}\left(r^2\frac{d}{dr}e^{-r/r_0}\right)\right]\right\}$$
$$= \frac{4\hbar^2}{3r_0^3}\int_0^\infty \exp\left\{-\frac{2r}{r_0}\right\}\left(\frac{r^2}{r_0^2} - 2\frac{r}{r_0}\right)dr.$$

The last integral falls into a sum of two integrals, similar to the one already worked out, with $n = 2$ and $n = 1$, and yields

$$\langle p_x^2 \rangle = \frac{4\hbar^2}{3r_0^2}\left[\int_0^\infty \xi^2 e^{-2\xi} d\xi - 2\int_0^\infty \xi e^{-2\xi} d\xi\right] = \frac{\hbar^2}{3r_0^2}.$$

Thus the rms fluctuation product,

$$\delta x \delta p_x = \frac{\hbar}{\sqrt{3}} \approx 0.577\hbar,$$

is slightly (~15%) higher than the minimum $\hbar/2$ allowed by the Heisenberg's uncertainty relation.

Note that due to the spherical symmetry of the system, the rms value of the total momentum is expressed by a very simple formula,

$$\delta p \equiv \langle p^2 \rangle^{1/2} = \left[3\langle p_x^2 \rangle\right]^{1/2} = \frac{\hbar}{r_0},$$

and that due to the classical relation $\mathbf{v} = \mathbf{p}/m$, the result of division of this result by the particle's mass m,

$$v_0 \equiv \frac{\delta p}{m} = \frac{\hbar}{m r_0},$$

may be interpreted as the rms velocity of the electron in a hydrogen-like atom. For the hydrogen atom (with $m = m_e \approx 0.91 \times 10^{-30}$ kg and $r_0 = r_B \approx 0.53 \times 10^{-10}$ m), the velocity is close to 2.2×10^6 m s^{-1}. The fact that this velocity is much lower than c justifies, once again, the non-relativistic analysis that was discussed in section 3.6 of the lecture notes.

***Problem* 3.34.** Use the Hellmann–Feynman theorem (see problem 1.5) to prove:

(i) the first of Eqs. (3.211) of the lecture notes, and
(ii) the fact that for a spinless particle in an arbitrary spherically-symmetric attractive potential $U(r)$, the ground state is always an s-state (with the orbital quantum number $l = 0$).

Solutions:

(i) Let us note that Eq. (3.181) for the radial part $\mathcal{R}_{n,l}$ of the eigenfunction may be considered as a 1D Schrödinger equation for the following effective Hamiltonian:

$$\hat{H} = -\frac{\hbar^2}{2mr^2}\frac{d}{dr}\left(r^2\frac{d}{dr}\right) + \frac{\hbar^2 l(l+1)}{2mr^2} + U(r). \qquad (*)$$

Temporarily taking the potential in the form $U(r) = -(C - \lambda)/r$ (so that it coincides with the genuine Coulomb potential (3.190) at $\lambda = \lambda_0 \equiv 0$), we get

$$\frac{\partial \hat{H}}{\partial \lambda} = \frac{1}{r},$$

so that the Hellmann–Feynman theorem, with the proper index generalization, $n \to \{n, l\}$, yields

$$\frac{\partial E_{n,l}}{\partial \lambda} = \left\langle \frac{\partial \hat{H}}{\partial \lambda} \right\rangle_{n,l} = \left\langle \frac{1}{r} \right\rangle_{n,l}. \qquad (**)$$

On the other hand, with our temporary replacement $C \to C - \lambda$, Eq. (3.201) reads

$$E_{n,l} = -\frac{m(C-\lambda)^2}{2n^2\hbar^2}, \quad \text{so that} \quad \left.\frac{\partial E_{n,l}}{\partial \lambda}\right|_{\lambda=0} = \frac{mC}{n^2\hbar^2}. \qquad (***)$$

Comparing Eqs. (**) and (***), we get

$$\left\langle \frac{1}{r} \right\rangle_{n,l} = \frac{1}{n^2}\frac{Cm}{\hbar^2}.$$

But according to the second of Eqs. (3.192), the last fraction is just $1/r_0$, thus giving us the first of the results (3.211).

(ii) Now let us temporarily consider, in the same Eq. (3.181), the quantum number l as a continuous parameter. Now the Hellmann–Feynman theorem yields

$$\frac{\partial E_{n,l}}{\partial l} = \left\langle \frac{\partial \hat{H}}{\partial l} \right\rangle_{n,l} = \frac{\hbar^2}{2m}\left\langle \frac{1}{r^2}\frac{\partial[l(l+1)]}{\partial l} \right\rangle_{n,l} = \frac{\hbar^2}{2m}\left\langle \frac{2l+1}{r^2} \right\rangle_{n,l}.$$

For all allowed values $l \geqslant 0$, the operator in the last bracket is a positively defined form, so that its expectation value cannot be negative for any quantum state $\{n, l\}$, and hence

$$\frac{\partial E_{n,l}}{\partial l} \geqslant 0, \quad \text{for any } l \geqslant 0,$$

in particular showing that the ground state always corresponds to $l = 0$.

***Problem* 3.35.** For the ground state of a hydrogen atom, calculate the expectation values of $\mathscr{E}$ and $\mathscr{E}^2$, where $\mathscr{E}$ is the electric field created by the atom, at distances $r \gg r_0$ from its nucleus. Interpret the resulting relation between $\langle \mathscr{E} \rangle^2$ and $\langle \mathscr{E}^2 \rangle$, at the same observation point.

Solution: The net electric field $\mathscr{E}$ of the atom is the sum of the field $\mathscr{E}_n$ of its nucleus, with the electric charge $q = +e$, and that ($\mathscr{E}_e$) of the electron, with the equal and opposite charge, $q' = -e$. At distances r much larger than the size of the nucleus, $\mathscr{E}_n$ may be calculated as the radial field of a point charge q:

$$\mathscr{E}_n(\mathbf{r}) = \mathbf{n}_r\mathscr{E}_n, \quad \text{with } \mathscr{E}_n = \frac{e}{4\pi\varepsilon_0 r^2}.$$

Since, according to Eqs. (3.174), (3.200), and (3.208), the ground-state wavefunction $\psi = \psi_{1,0,0}$ of the electron is spherically symmetric:

$$\psi(\mathbf{r}) = Y_0^0(\theta, \varphi)\mathcal{R}_{1,0}(r) = \frac{1}{(4\pi)^{1/2}}\frac{2}{r_0^{3/2}}e^{-r/r_0}, \qquad (*)$$

so is the expectation value of its electric charge density:

$$\langle\rho\rangle(\mathbf{r}) = -e\,|\psi(r)|^2 = -\frac{e}{4\pi}\frac{4}{r_0^3}e^{-2r/r_0}.$$

Since the relation between the charge density and the electric field it induces in the free space is linear, the expectation value (i.e. the ensemble average) of the electric field created by the electron may be calculated from the static inhomogeneous Maxwell equation[58], with this averaged charge density as the source:

$$\nabla\cdot\langle\mathscr{E}_e\rangle = \frac{\langle\rho\rangle}{\varepsilon_0} \equiv -\frac{e}{4\pi\varepsilon_0}\frac{4}{r_0^3}e^{-2r/r_0}.$$

According to the vector algebra[59], this average field is also spherically-symmetric and radial, $\langle\mathscr{E}_e(\mathbf{r})\rangle = \mathbf{n}_r\langle\mathscr{E}_e(\mathbf{r})\rangle$, with its magnitude obeying the ordinary, first-order differential equation

$$\frac{1}{r^2}\frac{d}{dr}(r^2\langle\mathscr{E}_e\rangle) = -\frac{e}{4\pi\varepsilon_0}\frac{4}{r_0^3}e^{-2r/r_0}.$$

This equation may be readily integrated (with the boundary condition $\langle\mathscr{E}_e(0)\rangle = 0$, imposed by the spherical symmetry of the field), giving

$$\langle\mathscr{E}_e\rangle(\mathbf{r}) = -\frac{e}{4\pi\varepsilon_0 r^2}\frac{4}{r_0^3}\int_0^r r'^2 e^{-2r'/r_0}dr' = \frac{e}{4\pi\varepsilon_0 r^2}\left[e^{-2r'/r_0}\left(\frac{2r'^2}{r_0^2} + \frac{2r'}{r_0} + 1\right)\right]_0^r$$

$$\equiv \frac{e}{4\pi\varepsilon_0 r^2}\left[e^{-2r/r_0}\left(\frac{2r^2}{r_0^2} + \frac{2r}{r_0} + 1\right) - 1\right],$$

so that the net average field of the atom.

$$\langle\mathscr{E}\rangle(\mathbf{r}) = \mathscr{E}_n(\mathbf{r}) + \langle\mathscr{E}_e\rangle(\mathbf{r}) = \frac{e}{4\pi\varepsilon_0}\left(\frac{2}{r_0^2} + \frac{2}{rr_0} + \frac{1}{r^2}\right)e^{-2r/r_0},$$

[58] See, e.g. *Part EM* Eq. (1.27).

[59] See, e.g. Eq. (A.68), with $\partial/\partial\theta = \partial/\partial\varphi = 0$.

and hence its square,

$$\langle\mathscr{E}\rangle^2(\mathbf{r}) = \left(\frac{e}{4\pi\varepsilon_0}\right)^2\left(\frac{2}{r_0^2} + \frac{2}{rr_0} + \frac{1}{r^2}\right)^2 e^{-4r/r_0} \to 4\left(\frac{e}{4\pi\varepsilon_0 r_0^2}\right)^2 e^{-4r/r_0}, \quad \text{at } r \gg r_0, \quad (**)$$

exponentially drop at large distances.

The calculation of $\langle\mathscr{E}^2\rangle$ at arbitrary distances $r \sim r_0$ is more cumbersome, so let us immediately use the condition $r \gg r_0$. At such distances, in the classical electrodynamics[60], the net electric field of an electrically-neutral system consisting of a point charge $q = +e$, located at the origin, and an electron charge $q' = -e$, located at point $\mathbf{r}'$, with $r' \sim r_0 \ll r$, is well approximated by the field,

$$\mathscr{E}(\mathbf{r};\, \mathbf{r}') = \frac{1}{4\pi\varepsilon_0}\frac{3\mathbf{r}(\mathbf{r}\cdot\mathbf{p}) - \mathbf{p}r^2}{r^5}, \qquad (***)$$

of an electric dipole $\mathbf{p} = q{\cdot}0 + q'{\cdot}\mathbf{r}' = -e\mathbf{r}'$, so that

$$\begin{aligned}\mathscr{E}^2(\mathbf{r};\, \mathbf{r}') &= \left(\frac{e}{4\pi\varepsilon_0}\right)^2\left(\frac{3\mathbf{r}(\mathbf{r}\cdot\mathbf{r}') - \mathbf{r}'r^2}{r^5}\right)^2 \\ &= \left(\frac{e}{4\pi\varepsilon_0}\right)^2\frac{3r^2(\mathbf{r}\cdot\mathbf{r}')^2 + r'^2r^4}{r^{10}} = \left(\frac{e}{4\pi\varepsilon_0}\right)^2\frac{r'^2}{r^6}(3\cos^2\theta + 1),\end{aligned}$$

where $\cos\theta \equiv (\mathbf{r}'{\cdot}\mathbf{r})/r'r$, meaning that θ may be interpreted by the angle between the vectors $\mathbf{r}'$ and $\mathbf{r}$.

Due to the correspondence principle, at the transfer to quantum mechanics, we still may use the above expression for $\mathscr{E}^2$, but have to understand it as a linear operator acting on the electron's wavefunction $\psi(\mathbf{r}')$, so that the expectation value of the corresponding observable at point $\mathbf{r}$, in the ground state of the atom, has to be calculated as

$$\langle\mathscr{E}^2\rangle(\mathbf{r}) = \int\psi^*(\mathbf{r}')\,\mathscr{E}^2(\mathbf{r};\, \mathbf{r}')\,\psi(\mathbf{r}')d^3r' = \int\mathscr{E}^2(\mathbf{r};\, \mathbf{r}')\ |\,\psi(\mathbf{r}')\,|^2\, d^3r'.$$

Taking the direction of vector $\mathbf{r}$ for the polar axis, and then using Eq. (*), we get

$$\begin{aligned}\langle\mathscr{E}^2\rangle(\mathbf{r}) &= \int_0^{2\pi}d\varphi\int_0^{\pi}\sin\theta\, d\theta\int_0^{\infty}r'^2dr'|\,Y_0^0(\theta,\varphi)\,|^2|\,\mathcal{R}_{1,0}(r')\,|^2\,\mathscr{E}^2(\mathbf{r};\, \mathbf{r}') \\ &= 2\left(\frac{e}{4\pi\varepsilon_0}\right)^2\frac{1}{r^6r_0^3}\int_0^{\pi}(3\cos^2\theta + 1)\sin\theta d\theta\int_0^{\infty}r'^4e^{-2r'/r_0}dr' \\ &\equiv \frac{1}{16}\left(\frac{e}{4\pi\varepsilon_0}\right)^2\frac{r_0^2}{r^6}\int_{-1}^{+1}(3\xi^2 + 1)d\xi\int_0^{\infty}\zeta^4e^{-\zeta}d\zeta,\end{aligned}$$

[60] See, e.g. *Part EM* section 3.1, in particular Eq. (3.13).

where $\xi \equiv \cos\theta$ and $\zeta \equiv r/(r_0/2)$. The first integral is elementary and equal to 4, while the second one is a table integral[61] equal to $4! \equiv 24$, so that, finally,

$$\langle \mathscr{E}^2 \rangle(\mathbf{r}) = 6\left(\frac{e}{4\pi\varepsilon_0 r_0^2}\right)^2 \left(\frac{r_0}{r}\right)^6, \quad \text{for } r \gg r_0. \qquad (****)$$

Comparison of this result with Eq. (**) shows a rather dramatic difference between the square of the average electric field of the atom, and the average of its square: while the former expectation value drops exponentially with distance r from the atom, the latter one decreases much more slowly, as $1/r^6$. The interpretation of this difference is offered by the (frequently, very useful) notion of *quantum fluctuations* of the field: since $|\psi(\mathbf{r}')|^2$ may be interpreted as the density of the probability of finding the electron at point $\mathbf{r}'$, we may say that its 'random motion' in the $\mathbf{r}'$ space creates random fluctuations of the dipole field $\mathscr{E}$, which, according to Eq. (***), decays with distance only as $1/r^3$, so that its square drops as $1/r^6$, as described by Eq. (****). However, due to the randomness of the spatial orientations of the vector $\mathbf{r}'$, and hence of the field vector $\mathscr{E}$, the bulk of these fluctuations is averaged out from $\langle \mathscr{E} \rangle$, leaving behind only the exponentially small 'tail' (**). This behavior, very typical for quantum mechanics, has already been encountered[62], and will be met several more times later in the course.

Note also that Eq. (****) is intimately related with the attractive London dispersion force with the effective potential $U_{\text{ef}} \propto 1/r^6$ between two neutral atoms/molecules at large distances—see problems 3.16, 5.15, and 6.18.

***Problem* 3.36.** Calculate the condition at which a particle of mass m, moving in the field of a very thin spherically-symmetric shell, with

$$U(\mathbf{r}) = \mathscr{W}\delta(r - R), \quad \text{with } \mathscr{W} < 0,$$

has at least one localized ('bound') stationary state.

Solution: Repeating the initial arguments of the model solution of problem 3.24, with the only difference that now the product $r\psi(r)$ is given by a linear combination of two exponential functions similar to Eq. (**) of that solution even at $r < R$ (this linear combination has to vanish at $r = 0$ to avoid the divergence of ψ, i.e. to be proportional to $\sinh\kappa r$), we may look for the ground-state eigenfunction in the form

$$\psi(r) = \frac{1}{r} \times \begin{cases} C_- \sinh \kappa r, & \text{for } r \leqslant R, \\ C_+ e^{-\kappa r}, & \text{for } r \geqslant R, \end{cases} \quad \text{with } \frac{\hbar^2\kappa^2}{2m} \equiv -E > 0.$$

In the view of the discussion in the solution of problem 3.25, the relations between the coefficients $C_\pm$ may be found from the boundary conditions at $r = R$, using

[61] See, e.g. Eq. (A.34*d*) with $n = 4$.

[62] For example, the Gaussian wave packet (2.16) of a free 1D particle, as well as the ground state (2.275) of a 1D harmonic oscillator, have $\langle x \rangle^2 = 0$, but $\langle x^2 \rangle = \delta x^2 > 0$, so if the particle in these situations is charged, its electric field has similar properties.

Eqs. (2.75) and (2.76) of the lecture notes, with the proper replacement $\psi \to r\psi$. These conditions yield, correspondingly:

$$\kappa\left(-C_+e^{-\kappa R} - C_- \cosh \kappa R\right) = \frac{2m}{\hbar^2}\mathscr{w}C_+e^{-\kappa R}, \qquad C_+e^{-\kappa R} - C_- \sinh \kappa R = 0.$$

The condition of consistency of this system of two linear, homogeneous equations,

$$\begin{vmatrix} \left(-\kappa - \dfrac{2m\mathscr{w}}{\hbar^2}\right)e^{-\kappa R} & -\kappa \cosh \kappa R \\ e^{-\kappa R} & -\sinh \kappa R \end{vmatrix} = 0,$$

gives us the following characteristic equation for κ:

$$\kappa R\,(\coth \kappa R + 1) = -\frac{2m\mathscr{w}\,R}{\hbar^2}. \qquad (*)$$

The product $\kappa R \coth \kappa R$ equals 1 at $\kappa R = 0$, and only grows with the argument κR (which should be positive because of its definition), and the second term in the parentheses of Eq. (*) only increases this trend. Hence this equation may have a solution only if the magnitude of its right-hand side is larger than 1, i.e. if[63]

$$|\mathscr{w}| > \frac{\hbar^2}{2mR}, \qquad (**)$$

i.e. if the dimensionless parameter $\beta \equiv \mathscr{w}/(\hbar^2/2mR)$, already used in the model solution of problem 3.7 (and is negative in our current problem), is below -1.

This is the condition we were seeking for. Note that in the limit $R \to 0$, we may associate this potential with a 3D delta-function, $U(\mathbf{r}) = \mathscr{w}_{3D}\delta(\mathbf{r})$,[64] with the parameter $\mathscr{w}_{3D}$ defined by the following condition:

$$\mathscr{w}_{3D} = \int_0^\infty U(\mathbf{r})d^3r = \int_0^\infty 4\pi r^2 U(r)dr = 4\pi R^2 \mathscr{w}.$$

Plugging this relation into Eq. (**), we get the following condition,

$$|\mathscr{w}_{3D}| > \mathscr{w}_{\min}, \quad \text{with } \mathscr{w}_{\min} = \frac{2\pi\hbar^2 R}{m},$$

which is *always* satisfied at $R \to 0$ (of course only if $\mathscr{w}$, and hence $\mathscr{w}_{3D}$ are negative). At small but nonvanishing R, the last expression for $\mathscr{w}_{\min}$ is qualitatively similar, but quantitatively different from Eq. (***) in the model solution of problem 3.24. This is one more illustration of the statement made at the end of that solution: the particle localization properties of a potential well of a very small size cannot be fully characterized by just the 'area' of the 3D delta-function, even if the potential is axially symmetric—as it is in both these problems.

[63] At the border of this range, at $2m\mathscr{w}R = \hbar^2$, the characteristic equation yields $\lambda = 0$, i.e. $\kappa R = 0$, so that the solution ceases to be localized: $\mathscr{R}(\infty) \neq 0$.

[64] Of course, $\mathscr{w}_{3D}$ should not be confused with $\mathscr{w}$; these parameters even have different dimensionalities—respectively, J m^{-3} and J m^{-1}.

***Problem* 3.37.** Calculate the lifetime of the lowest metastable state in the same spherical-shell potential as in the previous problem, but now with $\mathscr{W} > 0$, for sufficiently large $\mathscr{W}$. (Quantify this condition.)

Solution: We may follow the approach used in section 2.5 of the lecture notes to solve a similar 1D problem—see figure 2.17 and its discussion. If $\mathscr{W}$ is large enough, the Schrödinger equation inside the shell may be approximately satisfied with a spherically-symmetric standing wave, vanishing at $r = R$. As was discussed at the end of section 3.6 of the lecture notes, the lowest eigenfunction is a product of the spherical harmonic with $l = 0$ and $m = 0$ (which is just a constant) by a radial function, proportional to the lowest spherical Bessel function of the first kind, $j_0(kr) = \sin kr/kr$—see the first of Eqs. (3.186). The boundary condition yields $k = \pi/R$, and after an easy normalization, we get

$$\psi|_{r\leqslant R} = A\frac{\sin kr}{r} \equiv \frac{A}{2r}(e^{ikr} - e^{-ikr}), \quad \text{with } A = \left(\frac{W}{2\pi R}\right)^{1/2},$$

where W (not to be confused with the 1D delta-function's 'area' $\mathscr{W}$) is the total probability to find the particle inside the sphere.

Outside the sphere, we may take only the first term of the s-wave solution (3.6),

$$\psi|_{r\geqslant R} = C\frac{\exp\{ikr\}}{r},$$

which describes the outward traveling wave, due to 'leakage' from the sphere. The coefficient C may be found exactly as in the 1D case; since for a spherically-symmetric function, $\nabla = \mathbf{n}_r\partial/\partial r$, the calculation is literally the same, and gives the same result:

$$C = \frac{A}{2}\frac{1}{1 + i\alpha} \approx \frac{A}{2i\alpha}, \tag{*}$$

where α is the dimensionless coefficient defined by Eq. (2.78):

$$\alpha \equiv \frac{m\mathscr{W}}{k\hbar^2}.$$

(The second step in Eq. (*) is legitimate, because our calculation is valid only if $|C| \gg A$, i.e. when $\alpha \ll 1$; this is the requested condition of validity of our analysis.)

Now everything it ready to calculate the total probability current outside the sphere:

$$I = 4\pi r^2 j, \quad \text{where } j = \frac{\hbar}{m}\text{Im}\left(\psi^*\frac{d}{dr}\psi\right) = \frac{\hbar}{m}\frac{|C|^2}{r^2}k,$$

which does not depend on the exact value of the radius $r > R$ we are calculating it at:

$$I = 4\pi\frac{\hbar k}{m}\frac{1}{4\alpha^2}\frac{W}{2\pi R} \equiv \frac{1}{\tau}W, \quad \text{where } \tau \equiv \alpha^2\frac{2mR^2}{\hbar k}.$$

(With the last form of the expression for I, the continuity equation (1.48) takes the form

$$\frac{dW}{dt} = -I = -\frac{W}{\tau},$$

so that the τ so defined is indeed the required lifetime of the metastable state.) The above result for τ may be represented in a more transparent form

$$\frac{\tau}{\hbar} = \frac{\beta^2}{4\pi}\frac{1}{E} \gg \frac{1}{E},$$

where $\beta \equiv 2mR\mathcal{W}/\hbar^2 \gg 1$ is the dimensionless parameter already used in the model solutions of problem 3.7 and two previous problems for the similar potential, and $E = \pi^2\hbar^2/2mR^2$ is the eigenenergy of this (lowest) metastable state—see Eq. (3.188) with $l = 0$ and $n = 1$, i.e. $\xi_{l,n} = \pi$, so that $k = k_{l,n} = \xi_{l,n}/R = \pi/R$.

***Problem* 3.38.** A particle of mass m and energy E is incident on a very thin spherical shell, whose localized states were the subject of two previous problems.

(i) Derive the general expressions for the differential and total cross-sections of scattering.
(ii) Spell out the contribution σ_0 to the full cross-section σ, due to the spherically-symmetric component of the scattered de Broglie wave.
(iii) Analyze the result for σ_0 in the limits of very small and very large magnitudes of $\mathcal{W}$, for both signs of this parameter. In particular, in the limit $\mathcal{W} \to +\infty$, relate the result to the metastable state's lifetime τ calculated in the previous problem.

Solutions:

(i) According to Eqs. (3.222) and (3.224) of the lecture notes, for this axially-symmetric problem, both $d\sigma/d\Omega$ and σ are fully defined by the set of the complex amplitudes A_l of the so-called 'partial waves', i.e. the spherical-harmonic components of the scattered de Broglie wave

$$\psi_s = a_i \sum_{l=1}^{\infty} \mathcal{R}_l(r) P_l(\cos\theta), \quad \text{with } \mathcal{R}_l(r) \to (-i)^{l+1} A_l \frac{e^{ikr}}{kr} \quad \text{for } r \to \infty, \qquad (*)$$

where a_i is the amplitude of the incident wave, which may be also represented as a sum over the spherical harmonics (reduced to the Legendre polynomials P_l), using the expansion (3.225):

$$\psi_i = a_i \sum_{l=0}^{\infty} i^l (2l+1) j_l(kr) P_l(\cos\theta).$$

(Here $j_l(\xi)$ are the spherical Bessel functions of the first kind—see Eqs. (3.185)–(3.186) of the lecture notes, and $k = (2mE)^{1/2}/\hbar$ is the wave number of the incident and elastically-scattered waves.)

For our particular scatterer, described by the 1D delta-functional potential

$$U(\mathbf{r}) = \mathscr{W}\delta(r - R),$$

which vanishes at all points with $r \neq R$, the radial functions obey the simple Eq. (3.183) both at $r \leqslant R$ and $r \geqslant R$. As was discussed in sections 3.6 and 3.8 of the lecture notes, the appropriate solutions in these regions are

$$\mathcal{R}_l(r) = \begin{cases} C_l j_l(kr), & \text{for } r \leqslant R, \\ A_l h_l^{(1)}(kr), & \text{for } R \leqslant r, \end{cases}$$

where the coefficient A_l is the same as in the asymptote (*), and $h_l^{(1)}(\xi) \equiv j_l(\xi) + iy_l(\xi)$ is the spherical Hankel function of the first kind—see the first of Eqs. (3.215). By its construction, the lth component of the total wavefunction (including the contributions by ψ_i and ψ_s) is proportional to

$$a_i\left[\, i^l(2l + 1)j_l(kr) + \mathcal{R}_l(kr)\right],$$

so that it already satisfies the boundary conditions at $r = 0$ and $r \to \infty$, and we only need to impose on it the conditions at $r = R$, due to the delta-functional potential of the spherical shell. We may derive these boundary conditions, for example, by using the fact discussed in the model solution of problem 3.25: the product $f_l(r)$ of the lth radial function by r satisfies a 1D Schrödinger equation with the effective potential

$$U_{\text{ef}}(r) = U(r) + \frac{\hbar^2 l(l + 1)}{2mr^2}.$$

The second term on the right-hand side is continuous at $r = R$, so that it does not affect the boundary conditions at that point, which are therefore the same as in the 1D case—see Eqs. (2.75)–(2.76):[65]

$$f_l(R + 0) - f_l(R - 0) = 0, \quad \frac{d}{dr}f_l(r)|_{r=R+0} - \frac{d}{dr}f_l(r)|_{r=R-0} = \frac{2m\mathscr{W}}{\hbar^2}f_l(R).$$

Plugging in the above expressions for the total wavefunction into these boundary conditions, we get, respectively,

$$A_l h_l^{(1)}(kR) - C_l j_l(kR) = 0,$$

$$A_l\frac{d}{dr}\left[rh_l^{(1)}(kr)\right]_{r=R} - C_l\frac{d}{dr}\left[rj_l(kr)\right]_{r=R} = \frac{2m\mathscr{W}R}{\hbar^2}\left[\, i^l(2l + 1)j_l(kR) + A_l h_l^{(1)}(kR)\right].$$

Solving this simple system of two linear equations for the coefficients A_l and C_l, we get, in particular,

$$A_l = \frac{\beta\, i^l(2l + 1)j_l(kR)}{d\left[\xi\, h_l^{(1)}(\xi)\right]/d\xi|_{\xi=kR} - \left[h_l^{(1)}(kR)/j_l(kR)\right] d\left[\xi\, j_l(\xi)\right]/d\xi|_{\xi=kR} - \beta h_l^{(1)}(kR)}, \quad (**)$$

[65] Note that the model solution of the previous problem used an alternative way to derive the corresponding boundary conditions for $\psi_l = f_l/r$.

where the dimensionless, real coefficient β, characterizing the relative strength of the delta-functional potential, was defined in the model solution of problem 3.7 (see also the solutions of the two previous problems):

$$\beta \equiv \frac{\mathscr{w}}{\hbar^2/2mR}.$$

Eq. (**), together with Eqs. (3.222) amd (3.224) of the lecture notes gives the complete (though not immediately transparent) solution of the scattering problem.

(ii) The spherically-symmetric component of the scattered wavefunction (with $l = 0$) is proportional to the coefficient A_0. Using the first column of Eq. (3.186), we get

$$\xi j_0(\xi) = \sin \xi, \quad \frac{d}{d\xi}\left[\xi j_0(\xi)\right] = \cos \xi;$$

$$\xi h_0^{(1)}(\xi) \equiv \xi\left[j_0(\xi) + i y_0(\xi)\right] = -ie^{i\xi}, \quad \frac{d}{d\xi}\left[\xi h_0^{(1)}(\xi)\right] = e^{i\xi},$$

and Eq. (**) with $l = 0$ is reduced to

$$A_0 = \frac{\beta \operatorname{sinc} kR}{1 + i(\cot kR + \beta/kR)} e^{-ikR},$$

(where $\operatorname{sinc}\xi \equiv \sin\xi/\xi$, as in the solution of problem 3.7), giving the following contribution to the full cross-section (3.224):

$$\sigma_0 = \frac{4\pi}{k^2}|A_0|^2 = \frac{4\sigma_g}{(kR)^2}\frac{\beta^2 \operatorname{sinc}^2 kR}{1 + (\cot kR + \beta/kR)^2}, \qquad (***)$$

where $\sigma_g \equiv \pi R^2$ is the 'geometric' (classical) cross-section of the shell.

(iii) The two panels of the figure below show, in the appropriate semi-log scale, the ratio σ_0/σ_g as a function of the dimensionless product $kR \propto E^{1/2}$, for several representative values and two opposite signs of the parameter β (i.e. of $\mathscr{w}$).

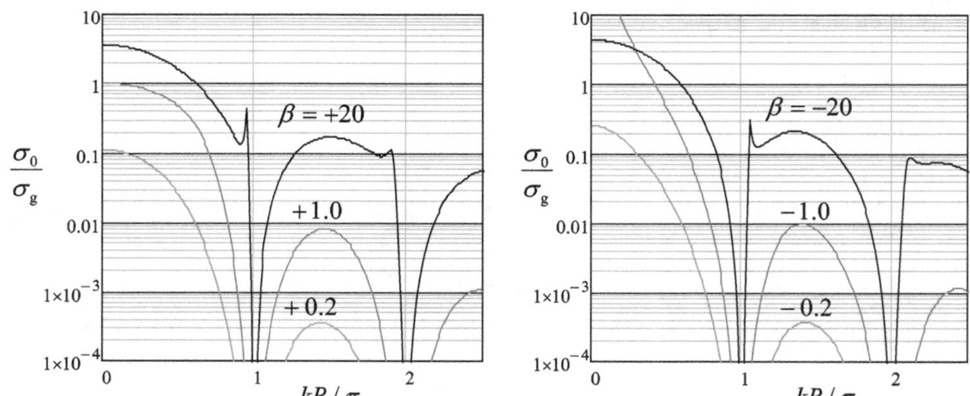

In the weak-potential limit $\beta \to 0$, the cross-section does not depend on the sign of this parameter:

$$\sigma_0 \approx \sigma_g(2\beta)^2 \text{sinc}^4 kR, \qquad \text{at } |\beta| \ll 1,$$

typically for the Born approximation. Indeed, at $kR \to 0$, when the scattering by any object is spherically-symmetric, and hence σ_0 dominates the total cross-section σ, our result tends to the value $\sigma/\sigma_g = 4\beta^2$, which was calculated, in that approximation, in problem 3.7.

However, as $|\beta|$ becomes comparable to, or larger than 1, i.e. beyond the Born approximation limit, σ_0 does depend on the sign of β, and its dependence on the parameter kR (i.e. on the particle's energy) shows several non-trivial effects. Most interestingly, as the left panel of the figure above shows, at $\beta \gg 1$ the dependences exhibit sharp resonance peaks at values of kR slightly below each πn, with $n = 1, 2, \ldots$.[66] The physics of this effect[67] should be clear from the model solution of the previous problem (or rather its straightforward extension to an arbitrary metastable s-state with $m = 0$): at $\beta = \infty$, when the spherical shell is impenetrable, it has localized states with $\psi_n \propto \sin k_n r/r$, and $k_n = \pi n/R$—see also Eq. (3.188) of the lecture notes, for $l = 0$. If β is large but finite, such state is metastable, i.e. the amplitude of this standing de Broglie wave cannot persist on its own; however, it may be sustained via its weak coupling with the incident wave. As at the classical resonance[68], and also at the resonant tunneling in 1D quantum systems discussed in chapter 2 of the lecture notes of this course (see, e.g. figure 2.16 and its discussion), if the frequencies $\omega = E/\hbar$ of these de Broglie waves (and hence their energies E and the wave numbers $k = (2mE)^{1/2}/\hbar$) are close, the amplitude of the induced standing wave strongly increases, and so does the scattering intensity.

The figure below shows a zoom-in on the vicinity of the lowest resonance (with $n = 1$); at $\beta \gg 1$, its features may be readily derived analytically from Eq. (***). Indeed, the exact resonance is reached at the point $(kR)_{\text{res}}$ that makes the parentheses in the denominator of this expression vanish:

$$\cot (kR)_{\text{res}} + \frac{\beta}{(kR)_{\text{res}}} = 0;$$

for $\beta \gg 1$ and $(kR)_{\text{res}} \approx k_1 R = \pi$, when $\cot(kR) \approx -1/(\pi - kR)$, this equation yields

$$(kR)_{\text{res}} \approx \pi\left(1 - \frac{1}{\beta}\right).$$

At that point, $\text{sinc}(kR)_{\text{res}} \equiv \text{sinc}(kR)_{\text{res}} \approx (\pi/\beta)/\pi \equiv 1/\beta$, so that the height of the resonance maximum is independent of β:

[66] As was already mentioned in section 2.5 of the lecture notes, such resonance functions of the incident particle's energy are sometimes called the *Breit–Wigner distributions* (or 'cross-sections', or 'functions').

[67] It belongs to the same group of resonant effects as the *Ramsauer–Townsend* effect, discovered (apparently, independently) by C Ramsauer and J S Townsend in the mid-1920s at scattering of low-energy electrons by noble-gas atoms, though due to the specificity of our current model, its results are somewhat different from their experimental observations.

[68] See, e.g. *Part CM* section 3.1.

$$\sigma_{\text{res}} \approx \frac{4\sigma_g}{(kR)^2_{\text{res}}} \frac{\beta^2 \text{sinc}^2(kR)_{\text{res}}}{1 + 0^2} \approx \frac{4\sigma_g}{\pi^2}\beta^2\left(\frac{1}{\beta}\right)^2 = \frac{4}{\pi^2}\sigma_g.$$

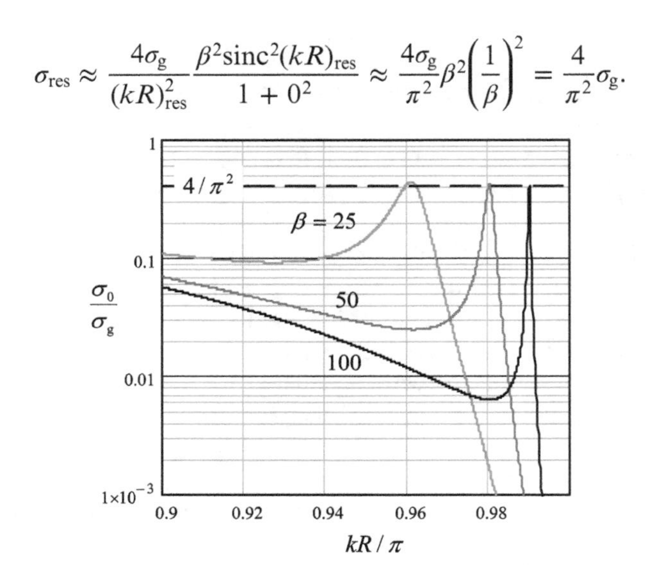

As the above figures show, these asymptotic analytical expressions for $(kR)_{\text{res}}$ and σ_{res} are in nice correspondence with the numerical results for $\beta_{\text{res}} \gg 1$. Moreover, it is easy to get an expression for the resonance width $\Delta(kR)$, just as was done in section 2.5 for the 1D resonant tunneling—see the derivation of Eq. (2.142). Indeed, Eq. (***) shows that σ_0 decreases two-fold from its resonance value σ_{res} when the deviation of the parentheses in its denominator from 0 becomes equal to ± 1. Since at $\beta \gg 1$ this deviation is dominated by the first term, $\cot(kR)$, we should require $\cot[(kR)_{\text{res}} \pm \Delta(kR)/2]$ to be equal to ± 1.[69] With the asymptotic expression $\cot(kR) \approx -1/(\pi - kR)$ already used above, and the anticipated condition $|\Delta(kR)| \ll \pi - (kR)_{\text{res}} \approx \pi/\beta \ll 1$, this requirement readily yields

$$\Delta(kR) \approx \frac{2\pi^2}{\beta^2},$$

confirming the above assumption, and again in good agreement with the numerical plots. The resulting energy width of the resonance,

$$\Delta E \approx \frac{dE}{dk}\Delta k = \frac{d}{dk}\left(\frac{\hbar^2 k^2}{2m}\right)\Delta k = \frac{\hbar^2 k}{m}\Delta k \equiv 2E\frac{\Delta(kR)}{kR} \approx 2E\frac{2\pi^2/\beta^2}{\pi} \equiv \frac{4\pi}{\beta^2}E \ll E,$$

is in the same relation with the metastable state's lifetime $\tau = \hbar\beta^2/4\pi E$, calculated in the previous problem, as for the similar 1D problem:

[69] Here the division of $\Delta(kR)$ by two stems from the usual definition of this width as the FWHM—see section 2.5.

$$\Delta E \cdot \tau = \hbar,$$

again emphasizing the generality of this relation (with the reservations discussed in section 2.5 of the lecture notes)[70].

As the right panel of the first figure above shows, at $\beta < 0$ (i.e. at $\mathscr{w} < 0$), the resonances at $k \approx n\pi/R$ with $n = 1, 2,\ldots$ are virtually similar to those at $\beta < 0$, though located on the other side of the asymptotic values $n\pi/R$. However, the low-energy scattering may be significantly stronger. Indeed, in the limit $kR \to 0$, Eq. (***) is reduced to a simple expression,

$$\frac{\sigma_0}{\sigma_g} = \frac{4\beta^2}{(1+\beta)^2}, \qquad \text{for } kR \ll 1, \mid \beta + 1 \mid,$$

plotted in the figure below. According to this expression (valid also for the full cross-section σ, due to the dominance of the spherically-symmetric scattering in this limit), at $|\beta| \to \infty$, the ratio σ_0/σ_g tends to the value 4 (describing the low-energy scattering by an impenetrable sphere - see section 3.8 of the lecture notes, in particular figure 3.25b), independently of the sign of β. However, at finite negative β the scattering is always stronger, with a peak near $\beta = -1$. The origin of this peak becomes clear if we revisit the solution of problem 3.36: as β tends to this value from below, the energy of the localized eigenstate, and hence the corresponding eigenvalue of k tend to zero, enabling effective interaction of the wide-spread wavefunction of the state with the low-k incident wave.

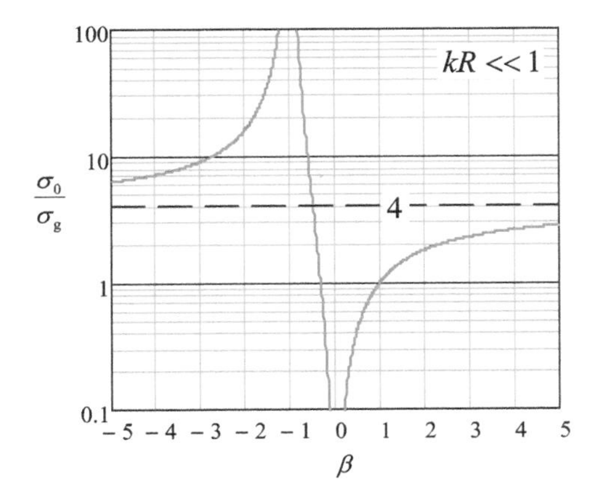

***Problem* 3.39.** Calculate the spherically-symmetric contribution σ_0 to the total cross-section of particle scattering by a uniform sphere of radius R, described by the potential

---

[70] Note that all these results for parameters of the resonant scattering at $kR \approx \pi n$ are also valid for the full cross-section σ, because the contributions of the higher components σ_l, with $l > 0$, become comparable with σ_g only near the corresponding non-spherical resonances, i.e. at $k \approx \xi_{l,n}/R$—see, e.g. the table following Eq. (3.188) of the lecture notes.

$$U(r) = \begin{cases} U_0, & \text{for } r < R, \\ 0, & \text{otherwise.} \end{cases}$$

Analyze the result, and give an interpretation of its most remarkable features.

Solution: Let us first assume that U_0 is lower than the particle's energy E. (Note that even in this case, U_0 may be either positive or negative.) Then, according to the discussion in section 3.8 of the lecture notes, we may look for the solution of the scattering problem in the form

$$\psi = a_\mathrm{i} \times \begin{cases} \sum_{l=0}^{\infty} \left[i^l (2l+1) j_l(kr) + A_l h_l^{(1)}(kr) \right] P_l(\cos\theta), & \text{for } R \leqslant r, \\ \sum_{l=0}^{\infty} B_l j_l(k'r) P_l(\cos\theta), & \text{for } r \leqslant R, \end{cases} \qquad (*)$$

where k and k' are the de Broglie wave numbers, respectively, outside of the sphere and inside it, defined as usual by:

$$\frac{\hbar^2 k^2}{2m} \equiv E, \quad \frac{\hbar^2 k'^2}{2m} \equiv E - U_0.$$

Note that in contrast with the solution of the previous problem, the incident wave (represented as a sum over spherical harmonics, using Eq. (3.225) of the lecture notes) is taken into account explicitly only in the upper line of Eq. (*), i.e. outside the sphere, while inside it, it is included in the single sum with the scattered wave. (The motivation for this approach is that the spherical Bessel functions $y_l(kr)$, and hence $h_l^{(1)}(kr)$, do not have finite values at $r \to 0$, while the functions $j_l(kr)$ do.)

The boundary conditions on the sphere's surface (the continuity of the wavefunction and its radial derivative at $r = R$) do not mix different spherical harmonics of the solution, and since we are only interested in the spherically-symmetric contribution σ_0 to the total cross-section, proportional to $|A_0|^2$, we may limit our analysis to the corresponding component of the wavefunction (*):

$$\psi_0 = a_\mathrm{i} \times \begin{cases} \left[j_0(kr) + A_0 h_0^{(1)}(kr) \right] \equiv \left(\sin kr - iA_0 e^{ikr} \right)/kr, & \text{for } R \leqslant r, \\ B_0 j_0(k'r) \equiv B_0 \sin k'r/k'r, & \text{for } r \leqslant R. \end{cases}$$

For this component, the boundary conditions yield two linear equations for the coefficients A_0 and B_0,

$$\left(\sin kR - iA_0 e^{ikR} \right)/k = B_0 \sin k'R/k', \quad \cos kR + A_0 e^{ikR} = B_0 \cos k'R,$$

whose solution yields, in particular:

$$A_0 = \frac{k' \sin kR \cos k'R - k \cos kR \sin k'R}{k \sin k'R + ik' \cos k'R} e^{-ikR},$$

so that the second of Eqs. (3.231), with $l = 0$, yields

$$\sigma_0 = \frac{4\pi}{k^2}|A_0|^2 \equiv \frac{4\sigma_g}{(kR)^2}|A_0|^2$$
$$= \frac{4\sigma_g}{(kR)^2}\left|\frac{k' \sin kR \cos k'R - k \cos kR \sin k'R}{k \sin k'R + ik' \cos k'R}\right|^2. \qquad (**)$$

Now reviewing the above calculation, we see that it remains valid in the case $U_0 > E$, where k' is imaginary ($k' = i\kappa$ with real κ, so that $\sin k'r = i\sin \kappa r$, $e^{ikr} = e^{-\kappa r}$, etc.), and we may use Eq. (**) for any value of U_0. This result is plotted on two panels of the figure below, as a function of the dimensionless product $kR \propto E^{1/2}$, for several positive and negative values of the dimensionless parameter $u_0 \equiv U_0/(\hbar^2/2mR^2)$.

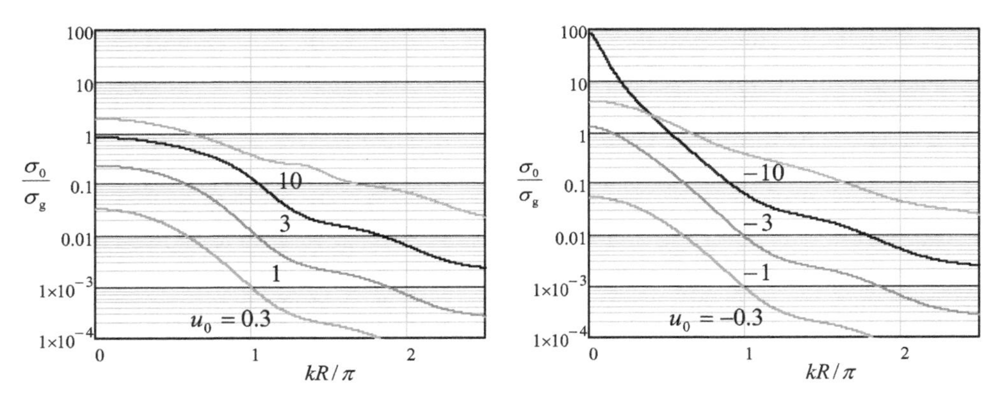

As the plots show, in contrast to the previous problem, the energy dependences of σ_0 are rather uneventful for any U_0, showing a fast decrease of this cross-section's component as kR becomes larger than ~1. (It does not make much sense to analyze this dependence in this high-energy region in detail, because here the actual total cross-section σ may be significantly contributed by other spherical-harmonic components σ_l with $l > 0$—see, e.g. figure 3.25b of the lecture notes.) However, as the plots show, the low-energy scattering (which is dominated by the calculated σ_0) has a rather non-trivial dependence on U_0, especially in the region $U_0 < 0$, where $\sigma \approx \sigma_0$ may be much larger than the 'visible' cross-section $\sigma_g \equiv \pi R^2$ of the sphere. Indeed, in the limit $kR \to 0$ we may readily simplify Eq. (**) to get

$$\frac{\sigma_0}{\sigma_g} \to 4 \cdot \left|1 - \frac{\tan k'R}{k'R}\right|^2.$$

The figure below shows this result as a function of U_0, which in this limit equals $-\hbar^2 k'^2/2m$.

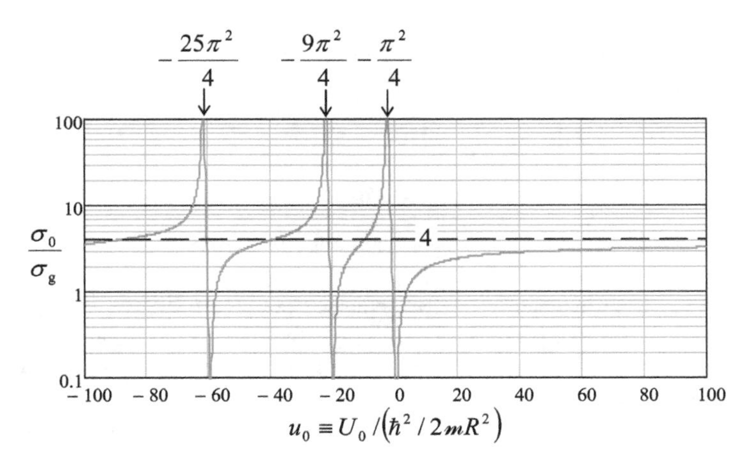

The plot shows that at $U_0 > 0$ (i.e. when the spherical scatterer is a flat-top potential 'bump'), the cross-section monotonically grows from the Born-approximation value $(4/9)u_0^2\sigma_g \ll \sigma_g$ (which was calculated in the solution of problem 3.6), to the asymptotic value $4\sigma_g$, which was calculated and discussed in section 3.8 of the lecture notes—see, e.g. figure 3.25b. However, at $U_0 < 0$, i.e. when the scatterer is a potential well, the cross-section, on the way to the same asymptotic value $4\sigma_g$, exhibits a series of sharp (formally, infinite) resonances at the values $U_0 = -U_n$, where

$$U_n \equiv \frac{\pi^2\hbar^2}{8mR^2}(2n-1)^2, \quad \text{with } n = 1, 2, 3, \ldots,$$
$$\text{i.e. at } k'R = \frac{\pi}{2}, \frac{3\pi}{2}, \frac{5\pi}{2}, \ldots \qquad (***)$$

The physics of these resonances is that (as was discussed in the model solution of problem 3.24) a sufficiently deep potential well, with $|U_0| < U_n$, has n localized s-states. As U_0 approaches $-U_n$, the eigenenergy E_n of the highest (nth) state tends to zero, and the effective radius R_{ef} of its eigenfunction tends to infinity. Naturally, such a nearly-extended state strongly interacts with the long de Broglie wave of the incident, low-energy particle, with the scattering's cross-section $\sigma \sim R_{ef}^2 \to \infty$.

***Problem* 3.40.** Use the finite difference method with the step $h = a/2$ to calculate as many eigenenergies as possible, for a particle confined to the interior of:

(i) a square with side a, and
(ii) a cube with side a,

with hard walls. For the square, repeat the calculations, using a finer step: $h = a/3$. Compare the results for different values of h with each other and with the exact formulas.

Hint: It is advisable to first solve (or review the solution of) the similar 1D problem in chapter 1, or start from reading about the finite difference method[71]. Also: try to exploit the problem's symmetry.

Solutions:

(i) The 2D finite-difference approximation of $\nabla^2\psi(x, y)$ is[72]

$$\frac{\psi(x-h, y)+\psi(x+h, y)+\psi(x, y-h)+\psi(x, y+h)-4\psi(x, y)}{h^2},$$

where h is the spatial step (the 'mesh size'). For $h = a/2$, the only natural choice of the five involved points is shown in the figure below. Taking into account that due to the boundary conditions, the values of ψ at all these points except the central point A, i.e. on the walls of the confining square, vanish, the finite-difference version of the 2D stationary Schrödinger equation reads

$$-\frac{\hbar^2}{2m}\frac{0+0+0+0-4\psi_A}{h^2} = E\psi_A.$$

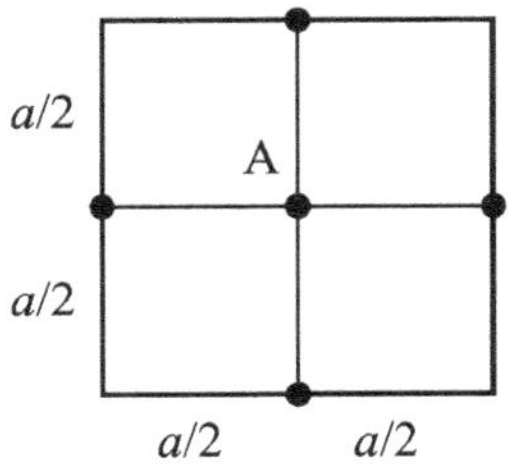

Canceling $\psi_A \neq 0$ and plugging in $h = a/2$, we get

$$E = 8\frac{\hbar^2}{ma^2},$$

the value to be compared with the exact result for the ground state, which follows from the 2D version of Eq. (1.86) with $n_x = n_y = 1$:

$$E_{1,1} = \pi^2\frac{\hbar^2}{2ma^2}(1^2 + 1^2) \equiv \pi^2\frac{\hbar^2}{ma^2}.$$

We see that the relative error of the numerical method is about 20% ($8 \leftrightarrow \pi^2 \approx 9.87$).

For the finer step $h = a/3$, due to the obvious symmetry of the square problem (see the figure below), we may distinguish four significantly different (linearly-independent) modes:

$$(1, 1): \psi_A = \psi_B = \psi_C = \psi_D,$$
$$(1, 2): \psi_A = \psi_B = -\psi_C = -\psi_D,$$

[71] See, e.g. *Part CM* section 8.5 or *Part EM* section 2.11.

[72] See, e.g. *Part CM* Eq. (8.66) or *Part EM* Eq. (2.221).

(2, 1): $\psi_A = -\psi_B = \psi_C = -\psi_D$,
(2, 2): $\psi_A = -\psi_B = -\psi_C = \psi_D$.

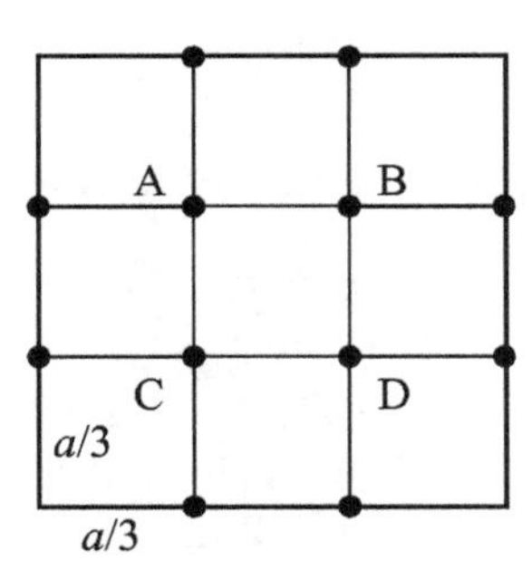

Here the numbers in parentheses correspond to the values of quantum numbers n_x and $n_{y,}$ in the analytical solution of the problem (see section 1.7 of the lecture notes),

$$\psi(x, y) = \frac{2}{a^2} \sin \frac{\pi n_x x}{a} \sin \frac{\pi n_y y}{a},$$

that provide the similar wavefunction's symmetry.

This symmetry enables us to solve the problem, for each mode, by writing a finite difference equation for just one internal point, e.g. the point A. For example, for the lowest (1, 1) mode, the stationary Schrödinger equation becomes

$$-\frac{\hbar^2}{2m}\left(\frac{0 + \psi_B + \psi_C + 0 - 4\psi_A}{h^2}\right) \to -\frac{\hbar^2}{2m}\left(\frac{0 + \psi_A + \psi_A + 0 - 4\psi_A}{h^2}\right) = E_{1,1}\psi_A.$$

From this equation (with $h = a/3$) we get:

$$E_{1,1} = 9\frac{\hbar^2}{ma^2}.$$

This is a substantially better approximation of the exact result than what we had with the initial step $h = a/2$, with an error below 10%.

In the same way, for the (1, 2) mode we get

$$-\frac{\hbar^2}{2m}\left(\frac{0 + \psi_B + \psi_C + 0 - 4\psi_A}{h^2}\right) \to -\frac{\hbar^2}{2m}\left(\frac{0 + \psi_A - \psi_A + 0 - 4\psi_A}{h^2}\right) = E_{1,2}\psi_A,$$

giving the following result:

$$E_{1,2} = 18\frac{\hbar^2}{ma^2}.$$

For the mode (2, 1), we evidently get the same result: $E_{2,1} = E_{1,2}$, indicating that this energy level is doubly degenerate—just as it is in the exact theory. However, the above numerical value of the eigenenergy for these excited states is much farther from the exact result,

$$E_{1,2} = \frac{\pi^2\hbar^2}{2ma^2}(1^2 + 2^2) \approx 24.6\frac{\hbar^2}{ma^2},$$

than for the ground state. (The relative error here is ~35% instead of ~10%.)

This trend (at fixed step size) continues as we go to higher energy levels. Indeed, for the highest mode (2, 2) that we can describe with the step so coarse, we get

$$-\frac{\hbar^2}{2m}\left(\frac{0 + \psi_B + \psi_C + 0 - 4\psi_A}{h^2}\right) \to -\frac{\hbar^2}{2m}\left(\frac{0 - \psi_A - \psi_A + 0 - 4\psi_A}{h^2}\right) = E_{2,2}\psi_A,$$

resulting in

$$E_{2,2} = 27\frac{\hbar^2}{ma^2}$$

—a rather mediocre approximation of the exact result

$$E_{2,2} = \frac{\pi^2\hbar^2}{2ma^2}(2^2 + 2^2) \approx 39.5\frac{\hbar^2}{ma^2}.$$

So, we clearly see the general trend: a finite-difference scheme with n internal points allows us to find n eigenstates, with a better accuracy achieved for states with lower-energy states, with smoother wavefunctions.

(ii) In 3D, the calculation is similar, besides that now the Laplace operator is approximated as[73]

$$\nabla^2\psi(x, y, z) \approx \frac{1}{h^2}\times\left\{\begin{array}{l}\psi(x-h, y, z) + \psi(x+h, y, z) + \psi(x, y-h, z) \\ + \psi(x, y+h, z) + \psi(x, y, z-h) + \psi(x, y, z+h) - 6\psi(x, y, z)\end{array}\right\}.$$

The result for the crude step $h = a/2$,

$$E = 12\frac{\hbar^2}{ma^2},$$

is ~ 20% lower than the exact value given by Eq. (1.86) with $n_x = n_y = n_z = 1$:

$$E_{1,1,1} = \pi^2\frac{\hbar^2}{2ma^2}(1^2 + 1^2 + 1^2) \approx 14.8\frac{\hbar^2}{ma^2}.$$

References

[1] Mostofi A *et al* 2008 *Comput. Phys. Commun.* **178** 685

[2] Tinkham M 1996 *Introduction to Superconductivity* 2nd ed (McGraw-Hill)

[73] See, e.g. *Part EM* Eq. (2.222).

IOP Publishing

Quantum Mechanics
Problems with solutions
Konstantin K Likharev

Chapter 4

Bra–ket formalism

***Problem* 4.1.** Prove that if $\hat{A}$ and $\hat{B}$ are linear operators, and C is a c-number, then:

(i) $(\hat{A}^\dagger)^\dagger = \hat{A}$;
(ii) $(C\hat{A})^\dagger = C^*\hat{A}^\dagger$;
(iii) $(\hat{A}\hat{B})^\dagger = \hat{B}^\dagger\hat{A}^\dagger$;
(iv) the operators $\hat{A}\hat{A}^\dagger$ and $\hat{A}^\dagger\hat{A}$ are Hermitian.

Solutions: In order to prove that two operators are equivalent, it is sufficient to prove that all their matrix elements in some orthonormal, full basis u_j are equal. For task (i), this is very simple to do, using Eqs. (4.25) of the lecture notes twice (back and forth), and the fact that for any c-number, in particular any bra–ket, $(c^*)^* = c$:

$$\langle u_j|(\hat{A}^\dagger)^\dagger|u_{j'}\rangle = \langle u_{j'}|\hat{A}^\dagger|u_j\rangle^* = \left(\langle u_j|\hat{A}|u_{j'}\rangle^*\right)^* = \langle u_j|\hat{A}|u_{j'}\rangle,$$

thus proving the corresponding operator relation.

For task (ii), the calculation may be similar, taking into account that according to Eq. (4.19), any c-number multipliers may be moved into/out of any bra–ket combination at will, and that the complex conjugate of a product of c-numbers is equal to the product of their complex conjugates:

$$\begin{aligned}\langle u_j|(C\hat{A})^\dagger|u_{j'}\rangle &= \langle u_{j'}|(C\hat{A})|u_j\rangle^* = (C)^*\langle u_{j'}|\hat{A}|u_j\rangle^* = C^*\langle u_{j'}|\hat{A}|u_j\rangle^* \\ &= C^*\langle u_j|\hat{A}^\dagger|u_{j'}\rangle = \langle u_j|(C^*\hat{A}^\dagger)\,|u_{j'}\rangle.\end{aligned}$$

Note that the proved operator equality may be formulated verbally: for c-numbers, the Hermitian conjugation is equivalent to the simple complex conjugation.

For task (iii), the proof is also similar, but a bit longer, involving the use of the closure relation (4.44) twice (back and forth):

$$\langle u_j|(\hat{A}\hat{B})^\dagger|u_{j'}\rangle = \langle u_{j'}|\hat{A}\hat{B}\,|u_j\rangle^* = \sum_{j''}\left(\langle u_{j'}|\hat{A}|u_{j''}\rangle\langle u_{j''}|\hat{B}|u_j\rangle\right)^* = \sum_{j''}\langle u_{j'}|\hat{A}\,|u_{j''}\rangle^*\,\langle u_{j''}|\hat{B}\,|u_j\rangle^*$$
$$= \sum_{j''}\langle u_{j''}|\hat{B}\,|u_j\rangle^*\langle u_{j'}|\hat{A}\,|u_{j''}\rangle^* = \sum_{j''}\langle u_j|\hat{B}^\dagger|u_{j''}\rangle\langle u_{j''}|\hat{A}^\dagger|u_{j'}\rangle = \langle u_j|\hat{B}^\dagger\hat{A}^\dagger|u_{j'}\rangle.$$

Finally, for task (iv) we may simply use the relations proved in tasks (ii) and (iii) to show that both operator products in question do satisfy the definition (4.22) of the Hermitian operator:

$$(\hat{A}\hat{A}^\dagger)^\dagger = (\hat{A}^\dagger)^\dagger\hat{A}^\dagger = \hat{A}\hat{A}^\dagger, \quad (\hat{A}^\dagger\hat{A})^\dagger = \hat{A}^\dagger(\hat{A}^\dagger)^\dagger = \hat{A}^\dagger\hat{A}.$$

***Problem* 4.2.** Prove that for any linear operators $\hat{A}$, $\hat{B}$, $\hat{C}$, and $\hat{D}$,

$$[\hat{A}\hat{B},\, \hat{C}\hat{D}] = \hat{A}\{\hat{B},\, \hat{C}\}\,\hat{D} - \hat{A}\hat{C}\{\hat{B},\, \hat{D}\} + \{\hat{A},\, \hat{C}\}\hat{D}\hat{B} - \hat{C}\{\hat{A},\, \hat{D}\}\hat{B}.$$

Solution: Using the associative law of multiplication, i.e. the ability to remove parentheses just as in the usual scalar products, we may represent the left-hand side of the given equality as

$$(\hat{A}\hat{B})(\hat{C}\hat{D}) - (\hat{C}\hat{D})(\hat{A}\hat{B}) \equiv \hat{A}\hat{B}\hat{C}\hat{D} - \hat{C}\hat{D}\hat{A}\hat{B},$$

and its right-hand side as

$$\hat{A}(\hat{B}\hat{C} + \hat{C}\hat{B})\hat{D} - \hat{A}\hat{C}(\hat{B}\hat{D} + \hat{D}\hat{B}) + (\hat{A}\hat{C} + \hat{C}\hat{A})\hat{D}\hat{B} - \hat{C}(\hat{A}\hat{D} + \hat{D}\hat{A})\hat{B}$$
$$\equiv \hat{A}\hat{B}\hat{C}\hat{D} + \hat{A}\hat{C}\hat{B}\hat{D} - \hat{A}\hat{C}\hat{B}\hat{D} - \hat{A}\hat{C}\hat{D}\hat{B} + \hat{A}\hat{C}\hat{D}\hat{B} + \hat{C}\hat{A}\hat{D}\hat{B} - \hat{C}\hat{A}\hat{D}\hat{B} - \hat{C}\hat{D}\hat{A}\hat{B}.$$

All adjacent terms of the last form, besides the first one and the last one, mutually cancel, and we arrive at the same expression as for the left-hand side.

***Problem* 4.3.** Calculate all possible binary products $\sigma_j\sigma_{j'}$ ($j, j' = x, y, z$) of the Pauli matrices, defined by Eqs. (4.105) of the lecture notes:

$$\sigma_x \equiv \begin{pmatrix} 0 & 1 \\ 1 & 0 \end{pmatrix}, \quad \sigma_y \equiv \begin{pmatrix} 0 & -i \\ i & 0 \end{pmatrix}, \quad \sigma_z \equiv \begin{pmatrix} 1 & 0 \\ 0 & -1 \end{pmatrix},$$

and their commutators and anticommutators (defined similarly to those of the corresponding operators). Summarize the results, using the Kronecker delta and Levi-Civita permutation symbols[1].

Solution: A straightforward multiplication of the matrices, using the basic rule (4.52), yields

$$\sigma_x\sigma_y = \begin{pmatrix} 0 & 1 \\ 1 & 0 \end{pmatrix}\begin{pmatrix} 0 & -i \\ i & 0 \end{pmatrix} = \begin{pmatrix} i & 0 \\ 0 & -i \end{pmatrix} \equiv i\sigma_z, \quad \sigma_y\sigma_x = \begin{pmatrix} 0 & -i \\ i & 0 \end{pmatrix}\begin{pmatrix} 0 & 1 \\ 1 & 0 \end{pmatrix} = \begin{pmatrix} -i & 0 \\ 0 & i \end{pmatrix} \equiv -i\sigma_z,$$

[1] See, e.g. Eqs. (A.82) and (A.83).

$$\sigma_x^2 \equiv \sigma_x\sigma_x = \begin{pmatrix} 0 & 1 \\ 1 & 0 \end{pmatrix}\begin{pmatrix} 0 & 1 \\ 1 & 0 \end{pmatrix} = \begin{pmatrix} 1 & 0 \\ 0 & 1 \end{pmatrix} \equiv \mathrm{I}.$$

Acting absolutely similarly, for other pairs of indices we get

$$\sigma_y\sigma_z = i\sigma_x, \quad \sigma_z\sigma_y = -i\sigma_x, \quad \sigma_z\sigma_x = i\sigma_y, \quad \sigma_x\sigma_z = -i\sigma_y, \quad \sigma_y^2 = \sigma_z^2 = \mathrm{I}.$$

From here, the commutator of σ_x and σ_y is

$$[\sigma_x, \sigma_y] \equiv \sigma_x\sigma_y - \sigma_y\sigma_x = 2i\sigma_z;$$

absolutely similarly, we get

$$[\sigma_y, \sigma_z] = 2i\sigma_x, \quad [\sigma_z, \sigma_x] = 2i\sigma_y.$$

The index permutation on the left-hand side of any of these relations changes the sign of its right-hand side; also, the self-commutators $[\sigma_j, \sigma_j]$ are equal to zero by definition. All these results may be conveniently summarized using the Levi-Civita and the Kronecker delta symbols:

$$\sigma_j\sigma_{j'} = i\sigma_{j''}\varepsilon_{jj'j''} + \mathrm{I}\delta_{jj'}, \text{ i.e. } [\sigma_j, \sigma_{j'}] = 2i\sigma_{j''}\varepsilon_{jj'j''}, \qquad (*)$$

where the indices j, j', and j'' may take any values of the set $\{1, 2, 3\}$.

The same matrix products may be readily used to calculate their anticommutators:

$$\{\sigma_j, \sigma_{j'}\} \equiv \sigma_j\sigma_{j'} + \sigma_{j'}\sigma_j = 2\mathrm{I}\delta_{jj'}.$$

***Problem* 4.4.** Calculate the following expressions,

(i) $(\mathbf{c} \cdot \boldsymbol{\sigma})^n$, and then
(ii) $(b\mathrm{I} + \mathbf{c} \cdot \boldsymbol{\sigma})^n$,

for the scalar product $\mathbf{c} \cdot \boldsymbol{\sigma}$ of the Pauli matrix vector $\boldsymbol{\sigma} \equiv \mathbf{n}_x\sigma_x + \mathbf{n}_y\sigma_y + \mathbf{n}_z\sigma_z$ by an arbitrary c-number vector $\mathbf{c}$, where $n \geqslant 0$ is an integer, and b is an arbitrary scalar c-number.

Hint: For task (ii), you may like to use the binomial theorem[2], and then transform the result in a way enabling you to use the same theorem backwards.

Solutions:

(i) First, let us calculate

$$\begin{aligned}(\mathbf{c} \cdot \boldsymbol{\sigma})^2 &\equiv (c_x\sigma_x + c_y\sigma_y + c_z\sigma_z)(c_x\sigma_x + c_y\sigma_y + c_z\sigma_z) \\ &= c_x^2\sigma_x^2 + c_y^2\sigma_y^2 + c_z^2\sigma_z^2 + c_xc_y\{\sigma_x, \sigma_y\} + c_yc_z\{\sigma_y, \sigma_z\} + c_zc_x\{\sigma_z, \sigma_x\}.\end{aligned}$$

[2] See, e.g. Eq. (A.12).

From the results of the previous problem, this expression is just

$$c_x^2\mathrm{I} + c_y^2\mathrm{I} + c_z^2\mathrm{I} \equiv c^2\mathrm{I},$$

where c is the modulus of the vector $\mathbf{c}$. Now we can use this result to calculate the expression (i):

$$(\mathbf{c}\cdot\boldsymbol{\sigma})^n = \begin{cases} [(\mathbf{c}\cdot\boldsymbol{\sigma})^2]^{n/2} = (c^2\mathrm{I})^{\,n/2} \equiv c^n I, & \text{for even } n, \\ (\mathbf{c}\cdot\boldsymbol{\sigma})^{n-1}\mathbf{c}\cdot\boldsymbol{\sigma} = (c^{n-1}\mathrm{I})\,\mathbf{c}\cdot\boldsymbol{\sigma} \equiv c^n\dfrac{\mathbf{c}\cdot\boldsymbol{\sigma}}{c}, & \text{for odd } n. \end{cases}$$

(ii) Here we can first use the binomial theorem and then the above result for the expression (i):

$$(b\mathrm{I} + \mathbf{c}\cdot\boldsymbol{\sigma})^n = \sum_{k=0}^{n} {}^nC_k (b\mathrm{I})^{n-k}(\mathbf{c}\cdot\boldsymbol{\sigma})^k = \sum_{k=0}^{n} {}^nC_k b^{n-k} \times \begin{cases} c^k\mathrm{I}, & \text{for even } k, \\ c^k\dfrac{\mathbf{c}\cdot\boldsymbol{\sigma}}{c}, & \text{for odd } k. \end{cases}$$

Now we may use the sign-alternating property of the factor $(-1)^k$ to represent this sum in the following, more regular form:

$$\sum_{k=0}^{n} {}^nC_k b^{n-k} c^k \left[\frac{\mathrm{I} + \mathbf{c}\cdot\boldsymbol{\sigma}/c}{2} + (-1)^k \frac{\mathrm{I} - \mathbf{c}\cdot\boldsymbol{\sigma}/c}{2}\right]$$

$$= \frac{\mathrm{I} + \mathbf{c}\cdot\boldsymbol{\sigma}/c}{2}\sum_{k=0}^{n} {}^nC_k b^{n-k} c^k + \frac{\mathrm{I} - \mathbf{c}\cdot\boldsymbol{\sigma}/c}{2}\sum_{k=0}^{n} {}^nC_k b^{n-k}(-c)^k.$$

This trick enables us to apply the binomial theorem again, now backwards, to each of the two sums on the right-hand side of the last equality, and get the final result in a more compact form:

$$(b\mathrm{I} + \mathbf{c}\cdot\boldsymbol{\sigma})^n = \frac{\mathrm{I} + \mathbf{c}\cdot\boldsymbol{\sigma}/c}{2}(b+c)^n + \frac{\mathrm{I} - \mathbf{c}\cdot\boldsymbol{\sigma}/c}{2}(b-c)^n$$

$$= \frac{(b+c)^n + (b-c)^n}{2}\mathrm{I} + \frac{(b+c)^n - (b-c)^n}{2}\frac{\mathbf{c}\cdot\boldsymbol{\sigma}}{c}.$$

Note that this result is very general (and hence very important), because *any* 2×2 matrix may be represented in the form $(b\mathrm{I} + \mathbf{c}\cdot\boldsymbol{\sigma})$—see Eq. (4.106) of the lecture notes.

***Problem* 4.5.** Use the solution of the previous problem to derive Eqs. (2.191) of the lecture notes for the transparency $\mathscr{T}$ of a system of N similar, equidistant, delta-functional potential barriers.

Solution: As was discussed in section 2.7 of the lecture notes, the transparency may be found as $|T_{11}|^{-2}$, where T is the transfer matrix (2.190) of the system:

$$\mathrm{T} = \underbrace{\mathrm{T}_\alpha \mathrm{T}_a \mathrm{T}_\alpha \ldots \mathrm{T}_\alpha}_{N+(N-1)\text{ operands}}, \quad \text{with } \mathrm{T}_\alpha = \begin{pmatrix} 1 - i\alpha & -i\alpha \\ i\alpha & 1 + i\alpha \end{pmatrix}, \quad \text{and} \quad \mathrm{T}_a = \begin{pmatrix} e^{ika} & 0 \\ 0 & e^{-ika} \end{pmatrix}.$$

First of all, let us note that an addition of one more operand T_a to the front of this product, turning it into a pure power of the product $\mathrm{T}_a\mathrm{T}_\alpha$, adds only a phase multiplier to the matrix element we need:

$$T_{11}^{+} \equiv \{(\mathrm{T}_a\mathrm{T}_\alpha)^N\}_{11} = (\mathrm{T}_a\mathrm{T})_{11} = \left\{\begin{pmatrix} e^{ika} & 0 \\ 0 & e^{-ika} \end{pmatrix}\begin{pmatrix} T_{11} & T_{12} \\ T_{21} & T_{22} \end{pmatrix}\right\}_{11}$$

$$= \begin{pmatrix} T_{11}e^{ika} & T_{12}e^{ika} \\ T_{21}e^{-ika} & T_{22}e^{-ika} \end{pmatrix}_{11} = T_{11}e^{ika},$$

not changing the modulus of this element, and hence the resulting expression for transparency, so that it is sufficient for us to calculate the matrix $\mathrm{T}^+ \equiv (\mathrm{T}_a\mathrm{T}_\alpha)^N$.

Now let us spell out the elementary product:

$$\mathrm{T}_a\mathrm{T}_\alpha = \begin{pmatrix} e^{ika} & 0 \\ 0 & e^{-ika} \end{pmatrix}\begin{pmatrix} 1 - i\alpha & -i\alpha \\ i\alpha & 1 + i\alpha \end{pmatrix} \equiv \begin{pmatrix} (1 - i\alpha)e^{ika} & -i\alpha e^{ika} \\ i\alpha e^{-ika} & (1 + i\alpha)e^{-ika} \end{pmatrix}.$$

As any 2×2 matrix, this one may be represented as a linear combination of the three Pauli matrices and the identity matrix, with certain c-number coefficients:

$$\mathrm{T}_a\mathrm{T}_\alpha = b\mathrm{I} + \mathbf{c}\cdot\boldsymbol{\sigma} = \begin{pmatrix} b + c_z & c_x - ic_y \\ c_x + ic_y & b - c_z \end{pmatrix}.$$

In our case, a comparison of two last displayed equalities yields

$$b = \cos ka + \alpha \sin ka, \qquad c_x = \alpha \sin ka, \qquad c_y = \alpha \cos ka,$$
$$c_z = i(\sin ka - \alpha \cos ka),$$

so that

$$b^2 - 1 = (\cos ka + \alpha \sin ka)^2 - 1 \equiv 2\alpha \sin ka \cos ka + (\alpha^2 - 1)\sin^2 ka.$$

Now calculating the modulus square of the vector $\mathbf{c}$,

$$c^2 \equiv c_x^2 + c_y^2 + c_z^2 = (\alpha \sin ka)^2 + (\alpha \cos ka)^2 - (\sin ka - \alpha \cos ka)^2$$
$$\equiv 2\alpha \sin ka \cos ka + (\alpha^2 - 1)\sin^2 ka,$$

we see that $c^2 = b^2 - 1$. If we consider Eq. (2.191*b*) of the lecture notes,

$$\cos qa = \cos ka + \alpha \sin ka, \qquad (*)$$

as the definition of q, then $b = \cos qa$, and

$$c^2 = b^2 - 1 = \cos^2 qa - 1 = (\pm i \sin qa)^2.$$

Hence we may take $c = \pm i \sin qa$ with either sign, so that

$$b + c = e^{\pm iqa}, \quad b - c = e^{\mp iqa}.$$

Now we may use the result of task (ii) of the previous problem, giving (with the notation replacement $n \to N$):

$$\begin{aligned}\mathrm{T}^+ = (b\mathrm{I} + \mathbf{c}\cdot\boldsymbol{\sigma})^N &= \frac{(b+c)^N + (b-c)^N}{2}\mathrm{I} + \frac{(b+c)^N - (b-c)^N}{2}\frac{\mathbf{c}\cdot\boldsymbol{\sigma}}{c}\\ &= \mathrm{I}\cos Nqa \pm i\frac{\mathbf{c}\cdot\boldsymbol{\sigma}}{c}\sin Nqa.\end{aligned}$$

From here, for the only (top left) matrix element we are interested in, we get:

$$\begin{aligned}T_{11}^+ &= \cos Nqa \pm i\frac{(\mathbf{c}\cdot\boldsymbol{\sigma})_{11}}{c}\sin Nqa = \cos Nqa \pm i\frac{c_z}{c}\sin Nqa\\ &= \cos Nqa + i\frac{\sin ka - \alpha \sin ka}{\sin qa}\sin Nqa.\end{aligned}$$

Since, according to Eq. (*), q is always either purely real or purely imaginary, the first term of the last sum is always purely real, and the second one is purely imaginary. (As soon as $\sin Nqa$ becomes imaginary, so does $\sin qa$.) Thus we can write

$$\mathscr{T} = |T_{11}|^{-2} = |T_{11}^+|^{-2} = \left[(\cos Nqa)^2 + \left(\frac{\sin ka - \alpha \sin ka}{\sin qa}\sin Nqa\right)^2\right]^{-1},$$

thus proving Eq. (2.191*a*). As we have seen in section 2.7, this expression is very convenient for exploring the basic features of the resonant tunneling and 1D band theory.

***Problem* 4.6.** Use the solution of problem 4.4(i) to spell out the following matrix: $\exp\{i\theta\mathbf{n}\cdot\boldsymbol{\sigma}\}$, where $\boldsymbol{\sigma}$ is the Pauli matrix vector, $\mathbf{n}$ is a c-number vector of unit length, and θ is a c-number scalar.

Solution: As was discussed in section 4.6 of the lecture notes, operator functions and hence matrix functions (such as the exponent in this problem) are defined by their Taylor expansions. In our current case,

$$\begin{aligned}\exp\{i\theta\,\mathbf{n}\cdot\boldsymbol{\sigma}\} &\equiv \sum_{k=0}^{\infty}\frac{1}{k!}(i\theta\,\mathbf{n}\cdot\boldsymbol{\sigma})^k\\ &= \sum_{\substack{k=0\\(k\text{ even})}}^{\infty}\frac{1}{k!}(i\theta)^k(\mathbf{n}\cdot\boldsymbol{\sigma})^k + i\theta\,\mathbf{n}\cdot\boldsymbol{\sigma}\sum_{\substack{k=1\\(k\text{ odd})}}^{\infty}\frac{1}{k!}(i\theta)^{k-1}(\mathbf{n}\cdot\boldsymbol{\sigma})^{k-1}.\end{aligned}$$

All powers of the product $\mathbf{n}\cdot\boldsymbol{\sigma}$ in both these sums are even, and hence, according to the solution of problem 4.4(i), are equal to the identity matrix I. Thus the above expression reduces to

$$\sum_{\substack{k=0\\(k \text{ even})}}^{\infty}\frac{1}{k!}(i\theta)^k + i\theta\,\mathbf{n}\cdot\boldsymbol{\sigma}\sum_{\substack{k=1\\(k \text{ odd})}}^{\infty}\frac{1}{k!}(i\theta)^{k-1}$$

$$= I\sum_{\substack{k=0\\(k \text{ even})}}^{\infty}\frac{(-1)^{k/2}}{k!}\theta^k + i\theta\,\mathbf{n}\cdot\boldsymbol{\sigma}\sum_{\substack{k=1\\(k \text{ odd})}}^{\infty}\frac{(-1)^{(k-1)/2}}{k!}\theta^{k-1}.$$

But the last two sums are the Taylor expansions of, respectively, the functions $\cos\theta$ and $\sin\theta$, near the point $\theta = 0$, so that, finally,

$$\exp\{i\theta\,\mathbf{n}\cdot\boldsymbol{\sigma}\} = \mathrm{I}\cos\theta + i\mathbf{n}\cdot\boldsymbol{\sigma}\sin\theta.$$

Of course, we could expect in advance that the matrix in question might be represented by a linear combination of the matrices I and $\mathbf{n}\cdot\boldsymbol{\sigma}$; however, it is remarkable how simple the coefficients in this combination are.

***Problem* 4.7.** Use the solution of problem 4.4(ii) to calculate $\exp\{\mathrm{A}\}$, where A is an arbitrary 2×2 matrix.

Solution: As was discussed in section 4.6 of the lecture notes, analytical functions of the linear operators (and hence of their matrices) are defined by their Taylor expansions, with the coefficients similar to those of c-number functions. In particular,

$$\exp\{\mathrm{A}\} \equiv \sum_{n=0}^{\infty}\frac{1}{n!}\mathrm{A}^n.$$

As was discussed in section 4.4 of the lecture notes, any 2×2 matrix may be represented in the form $(b\mathrm{I} + \mathbf{c}\cdot\boldsymbol{\sigma})$, where b is a scalar c-number, and $\mathbf{c}$ is a geometric vector with three c-number Cartesian components. Using such representation for the matrix A, we get

$$\exp\{\mathrm{A}\} = \sum_{n=0}^{\infty}\frac{1}{n!}(b\mathrm{I} + \mathbf{c}\cdot\boldsymbol{\sigma})^n.$$

Now using the result of problem 4.4(ii), we may write

$$\exp\{\mathrm{A}\} = \sum_{n=0}^{\infty}\frac{1}{n!}\left[\frac{(b+c)^n + (b-c)^n}{2}\mathrm{I} + \frac{(b+c)^n - (b-c)^n}{2}\frac{\mathbf{c}\cdot\boldsymbol{\sigma}}{c}\right]$$

$$= \frac{\mathrm{I}}{2}\left[\sum_{n=0}^{\infty}\frac{1}{n!}(b+c)^n + \sum_{n=0}^{\infty}\frac{1}{n!}(b-c)^n\right] + \frac{\mathbf{c}\cdot\boldsymbol{\sigma}}{2c}\left[\sum_{n=0}^{\infty}\frac{1}{n!}(b+c)^n - \sum_{n=0}^{\infty}\frac{1}{n!}(b-c)^n\right],$$

where $c \equiv \left(c_x^2 + c_y^2 + c_z^2\right)^{1/2}$. The sums in the last expression are just the 'usual' (c-number) exponents of $(b \pm c)$, so that we finally get

$$\exp\{\mathrm{A}\} = \frac{\mathrm{I}}{2}(\exp\{b+c\} + \exp\{b-c\}) + \frac{\mathbf{c}\cdot\boldsymbol{\sigma}}{2c}(\exp\{b+c\} - \exp\{b-c\})$$

$$\equiv e^b\left(\mathrm{I}\cosh c + \frac{\mathbf{c}\cdot\boldsymbol{\sigma}}{c}\sinh c\right) \equiv e^b\begin{pmatrix} \cosh c + \frac{c_z}{c}\sinh c & \frac{c_x - ic_y}{c}\sinh c \\ \frac{c_x + ic_y}{c}\sinh c & \cosh c - \frac{c_z}{c}\sinh c \end{pmatrix}.$$

As a sanity check, in the particular case when $\mathrm{A} = i\theta\mathbf{n}\cdot\boldsymbol{\sigma}$, i.e. if $b = 0$ and $\mathbf{c} = i\theta\mathbf{n}$ (i.e. $c = i\theta$, $c_x/c = \mathbf{n}_x$, etc), the last expression is reduced to

$$e^0\begin{pmatrix} \cosh(i\theta) + \mathbf{n}_z\sinh(i\theta) & (\mathbf{n}_x - i\mathbf{n}_y)\sinh(i\theta) \\ (\mathbf{n}_x + i\mathbf{n}_y)\sinh(i\theta) & \cosh(i\theta) - \mathbf{n}_z\sinh(i\theta) \end{pmatrix} \equiv \mathrm{I}\cosh(i\theta) + \mathbf{n}\cdot\boldsymbol{\sigma}\sinh(i\theta)$$

$$\equiv \mathrm{I}\cos\theta + i\mathbf{n}\cdot\boldsymbol{\sigma}\sin\theta,$$

i.e. to the result that was already obtained in the solution of the previous problem.

***Problem* 4.8.** Express the elements of the matrix $\mathrm{B} = \exp\{\mathrm{A}\}$ explicitly via those of the 2×2 matrix A. Spell out your result for the following matrices:

$$\mathrm{A} = \begin{pmatrix} a & a \\ a & a \end{pmatrix}, \quad \mathrm{A}' = \begin{pmatrix} i\varphi & i\varphi \\ i\varphi & i\varphi \end{pmatrix},$$

with real a and φ.

Solution: An explicit form of the solution of the previous problem,

$$\mathrm{B} = \exp\{\mathrm{A}\} = e^b\begin{pmatrix} \cosh c + \frac{c_z}{c}\sinh c & \frac{c_x - ic_y}{c}\sinh c \\ \frac{c_x + ic_y}{c}\sinh c & \cosh c - \frac{c_z}{c}\sinh c \end{pmatrix},$$

means that the elements of the matrix B are

$$B_{11} = e^b\left(\cosh c + \frac{c_z}{c}\sinh c\right), \qquad B_{12} = e^b\frac{c_x - ic_y}{c}\sinh c,$$
$$B_{21} = e^b\frac{c_x + ic_y}{c}\sinh c, \quad B_{22} = e^b\left(\cosh c - \frac{c_z}{c}\sinh c\right). \qquad (*)$$

In that problem, the argument matrix A was taken in the form

$$\mathrm{A} = b\mathrm{I} + \mathbf{c}\cdot\boldsymbol{\sigma} \equiv b\begin{pmatrix}1 & 0\\ 0 & 1\end{pmatrix} + c_x\begin{pmatrix}0 & 1\\ 1 & 0\end{pmatrix} + c_y\begin{pmatrix}0 & -i\\ i & 0\end{pmatrix} + c_z\begin{pmatrix}1 & 0\\ 0 & -1\end{pmatrix}$$
$$\equiv \begin{pmatrix} b + c_z & c_x - ic_y\\ c_x + ic_y & b - c_z\end{pmatrix}.$$

Comparing each matrix element of the last form with its explicit notation,

$$\mathrm{A} = \begin{pmatrix} A_{11} & A_{12}\\ A_{21} & A_{22}\end{pmatrix},$$

and solving the resulting simple systems of four linear equations, we get

$$b = \frac{A_{11} + A_{22}}{2}, \qquad c_x = \frac{A_{12} + A_{21}}{2}, \qquad c_y = i\frac{A_{12} - A_{21}}{2}, \qquad c_z = \frac{A_{11} - A_{22}}{2}, \qquad (**)$$

so that the combinations participating in Eq. (*) are:

$$c_x + ic_y = A_{21}, \qquad c_x - ic_y = A_{12},$$

$$c \equiv \left(c_x^2 + c_y^2 + c_z^2\right)^{1/2} = \frac{1}{2}[(A_{12} + A_{21})^2 - (A_{12} - A_{21})^2 + (A_{11} - A_{22})^2]^{1/2}$$
$$= \frac{1}{2}[4A_{12}A_{21} + (A_{11} - A_{22})^2]^{1/2}.$$

For the first particular matrix given in the problem's assignment (with all $A_{jj'} = a$), these relations give

$$b = c_x = a, \qquad c_y = c_z = 0, \qquad c = a,$$

so that Eqs. (*) yield

$$B_{11} = B_{22} = e^a \cosh a \equiv \frac{e^{2a} + 1}{2}, \qquad B_{12} = B_{21} = e^a \sinh a \equiv \frac{e^{2a} - 1}{2}.$$

Note that the off-diagonal elements of this simple matrix are 'boosted' by the exponential operation less than the diagonal ones.

For the second first particular matrix (with all $A'_{jj'} = i\varphi$), the general relations (**) give

$$b' = c'_x = i\varphi, \qquad c'_y = c'_z = 0, \qquad c' = i\varphi,$$

so that here Eqs. (*) yield

$$B'_{11} = B'_{22} = e^{i\varphi} \cos\varphi \equiv \frac{e^{2i\varphi} + 1}{2}, \qquad B'_{12} = B'_{21} = ie^{i\varphi} \sin\varphi \equiv \frac{e^{2i\varphi} - 1}{2}.$$

***Problem* 4.9.** Prove that for arbitrary square matrices A and B,

$$\text{Tr}\,(\text{AB}) = \text{Tr}\,(\text{BA}).$$

Is each diagonal element $(AB)_{jj}$ necessarily equal to $(BA)_{jj}$?

Solution: Using Eq. (4.96), and then Eq. (4.52) of the lecture notes, we get

$$\text{Tr}\,(\text{AB}) \equiv \sum_j (AB)_{jj} = \sum_{j,j'} A_{jj'}B_{j'j}.$$

This sum evidently does not depend on the operand order, i.e. does not change if we swap A and B.

However, single diagonal elements of the two products,

$$(AB)_{jj} = \sum_{j'} A_{jj'}B_{j'j}, \quad (BA)_{jj} = \sum_{j'} B_{jj'}A_{j'j},$$

are not necessarily equal. For example, looking at the Pauli matrix products calculated in problem 4.3, we see, e.g. that

$$(\sigma_x\sigma_y)_{11} = i, \quad \text{but } (\sigma_y\sigma_x)_{11} = -i.$$

***Problem* 4.10.** Calculate the trace of the following 2 × 2 matrix:

$$\text{A} \equiv (\mathbf{a}\cdot\boldsymbol{\sigma})(\mathbf{b}\cdot\boldsymbol{\sigma})(\mathbf{c}\cdot\boldsymbol{\sigma}),$$

where $\boldsymbol{\sigma}$ is the Pauli matrix vector, while **a**, **b**, and **c** are arbitrary *c*-number vectors.

Solution: From the definition of the Pauli matrices (see, e.g. Eq. (4.105) of the lecture notes), we get

$$\mathbf{a}\cdot\boldsymbol{\sigma} = a_x\sigma_x + a_y\sigma_y + a_z\sigma_z = \begin{pmatrix} a_z & a_x - ia_y \\ a_x + ia_y & -a_z \end{pmatrix}, \qquad (*)$$

and absolutely similar expressions for two other operands. A straightforward multiplication of the last two operands yields

$$\begin{aligned}(\mathbf{b}\cdot\boldsymbol{\sigma})(\mathbf{c}\cdot\boldsymbol{\sigma}) &= \begin{pmatrix} b_z & b_x - ib_y \\ b_x + ib_y & -b_z \end{pmatrix}\cdot\begin{pmatrix} c_z & c_x - ic_y \\ c_x + ic_y & -c_z \end{pmatrix} \\ &= \begin{pmatrix} b_zc_z + (b_x - ib_y)(c_x + ic_y) & b_z(c_x - ic_y) - (b_x - ib_y)c_z \\ (b_x + ib_y)c_z - b_z(c_x + ic_y) & (b_x + ib_y)(c_x - ic_y) + b_zc_z \end{pmatrix}.\end{aligned} \qquad (**)$$

For the calculation of $\text{Tr(A)} \equiv A_{11} + A_{22}$, we need only the diagonal terms of the product of the matrix (*) by the matrix (**); the multiplication yields

$$A_{11} = a_z\left[b_zc_z + (b_x - ib_y)(c_x + ic_y)\right] + (a_x - ia_y)\left[(b_x + ib_y)c_z - b_z(c_x + ic_y)\right],$$
$$A_{22} = (a_x + ia_y)\left[b_z(c_x - ic_y) - (b_x - ib_y)c_z\right] - a_z\left[(b_x + ib_y)(c_x - ic_y) + b_zc_z\right].$$

Now summing these two elements, opening parentheses, and collecting/canceling similar terms, we finally get

$$\begin{aligned}\mathrm{Tr(A)} &= 2i\left[a_z(b_xc_y - b_yc_x) + b_z(c_xa_y - c_ya_x) + c_z(a_xb_y - a_yb_x)\right] \\ &\equiv 2i\begin{vmatrix} a_x \; a_y \; a_z \\ b_x \; b_y \; b_z \\ c_x \; c_y \; c_z \end{vmatrix}.\end{aligned} \qquad (***)$$

A good sanity check is the invariance of the result with respect to the loop replacement of the operand vectors: $\mathbf{a} \to \mathbf{b} \to \mathbf{c} \to \mathbf{a}$, etc; indeed, this feature might be predicted in advance by applying to this trace the property Tr(AB) = Tr(BA), whose proof was the subject of the previous problem:

$$\mathrm{Tr}[(\mathbf{a}\cdot\boldsymbol{\sigma})(\mathbf{b}\cdot\boldsymbol{\sigma})(\mathbf{c}\cdot\boldsymbol{\sigma})] = \mathrm{Tr}[(\mathbf{c}\cdot\boldsymbol{\sigma})(\mathbf{a}\cdot\boldsymbol{\sigma})(\mathbf{b}\cdot\boldsymbol{\sigma})] = \mathrm{Tr}[(\mathbf{b}\cdot\boldsymbol{\sigma})(\mathbf{c}\cdot\boldsymbol{\sigma})(\mathbf{a}\cdot\boldsymbol{\sigma})].$$

Note also that according to Eq. (***), the trace vanishes if any two of the three vectors **a**, **b**, and **c** coincide.

***Problem* 4.11.** Prove that the matrix trace of an arbitrary operator does not change at its arbitrary unitary transformation.

Solution: Using the notation of section 4.4 of the lecture notes for the 'old' basis u_j and 'new' basis v_j, for the trace in the new basis we have, by definition,

$$\mathrm{Tr}\hat{A}|_{\text{in } v} \equiv \mathrm{Tr\,A}|_{\text{in } v} \equiv \sum_j A_{jj}|_{\text{in } v}.$$

Now we can use the general law (4.92) of the matrix element transformation, and then change the summation order:

$$\sum_j A_{jj}|_{\text{in } v} = \sum_{j,k,k'} U^\dagger_{jk} A_{kk'}|_{\text{in } u}\, U_{k'j} \equiv \sum_{k,k'} A_{kk'}|_{\text{in } u} \sum_j U^\dagger_{jk} U_{k'j}.$$

In the last sum we can use the explicit expressions (4.82)–(4.84) for the unitary matrix elements (valid in any basis):

$$\sum_j U^\dagger_{jk} U_{k'j} = \sum_j \langle v_j|u_k\rangle\langle u_{k'}|v_j\rangle.$$

This expression may be readily simplified by swapping these short brackets using Eq. (4.15), and then employing the completeness relation (4.44) for the identity operator:

$$\begin{aligned}\sum_j \langle v_j|u_k\rangle\langle u_{k'}|v_j\rangle &= \left(\sum_j \langle u_k|v_j\rangle\langle v_j|u_{k'}\rangle\right)^* \equiv \left(\langle u_k|\sum_j |v_j\rangle\langle v_j|\,|u_{k'}\rangle\right)^* \\ &= \langle u_k|u_{k'}\rangle^* = \delta_{kk'}.\end{aligned}$$

As a result,

$$\sum_j A_{jj}|_{\text{in } v} = \sum_{k,k'} A_{kk'}|_{\text{in } u}\delta_{kk'} = \sum_k A_{kk}|_{\text{in } u},$$

which is just the trace in the 'old' basis u_j. This is why the frequently used expression 'the trace of an operator' (without mentioning the basis) is quite meaningful.

***Problem* 4.12.** Prove that for any two full and orthonormal bases u_j, v_j of the same Hilbert space,

$$\text{Tr}\,(|u_j\rangle\langle v_{j'}|) = \langle v_{j'}|u_j\rangle.$$

Solution: By the definition of the trace,

$$\text{Tr}\,(|u_j\rangle\langle v_{j'}|) \equiv \sum_{j''}(|u_j\rangle\langle v_{j'}|)_{j''j''}.$$

Since the trace of any operator is basis-independent (see the previous problem), we may calculate these matrix elements in any basis, e.g. in u_j:

$$\begin{aligned}\sum_{j''}(|u_j\rangle\langle v_k|)_{j''j''} &= \sum_{j''}\langle u_{j''}|\,(|u_j\rangle\langle v_{j'}|)\,|u_{j''}\rangle \equiv \sum_{j''}\langle u_{j''}|u_j\rangle\langle v_{j'}|u_{j''}\rangle \\ &= \sum_{j''}\delta_{j''j}\langle v_{j'}|u_{j''}\rangle = \langle v_{j'}|u_j\rangle,\end{aligned}$$

q.e.d.[3]

***Problem* 4.13.** Is the 1D scattering matrix S, defined by Eq. (2.124) of the lecture notes, unitary? What about the 1D transfer matrix T defined by Eqs. (2.125)?

Solution: It is beneficial to use Eq. (2.127*a*) of the lecture notes (whose proof was the subject of problem 2.11), which takes into account the matrix element symmetry:

$$\text{S} = e^{i\theta}\begin{pmatrix} re^{i\varphi} & t \\ t & -re^{-i\varphi}\end{pmatrix}.$$

From here,

$$\text{SS}^\dagger = e^{i\theta}\begin{pmatrix} re^{i\varphi} & t \\ t & -re^{-i\varphi}\end{pmatrix} e^{-i\theta}\begin{pmatrix} re^{-i\varphi} & t \\ t & -re^{i\varphi}\end{pmatrix} = \begin{pmatrix} r^2+t^2 & 0 \\ 0 & r^2+t^2\end{pmatrix}.$$

But according to Eq. (2.127*b*), $r^2 + t^2 = 1$, so that the product is just the identity matrix I, i.e. the scattering matrix is unitary (reflecting the so-called *reciprocity principle*.)

[3] Just in case if the reader has not run into this famous acronym yet: q.e.d. = *quod erat demonstratum*, Latin for 'what had to be demonstrated'.

For the transfer matrix T, we may use a similarly structured expression obtained in the model solution of problem 2.12:

$$\mathrm{T} = \frac{1}{t}\begin{pmatrix} e^{i\theta} & -re^{-i\varphi} \\ -re^{i\varphi} & e^{-i\theta} \end{pmatrix},$$

with the similar relation between the real coefficients r and t. From here,

$$\mathrm{T}\mathrm{T}^{\dagger} = \frac{1}{t}\begin{pmatrix} e^{i\theta} & -re^{-i\varphi} \\ -re^{i\varphi} & e^{-i\theta} \end{pmatrix}\frac{1}{t}\begin{pmatrix} e^{-i\theta} & -re^{-i\varphi} \\ -re^{i\varphi} & e^{i\theta} \end{pmatrix} = \frac{1}{t^2}\begin{pmatrix} 1 + r^2 & -2re^{i(\theta-\varphi)} \\ -2re^{-i(\theta-\varphi)} & r^2 + 1 \end{pmatrix}.$$

Besides the trivial case $r = 0$ (and hence $t = 1$), this product is different from I, so that the matrix T is *not* unitary. Note, however, that

$$\mathrm{T}\sigma_z\mathrm{T}^{\dagger} = \frac{1}{t}\begin{pmatrix} e^{i\theta} & -re^{-i\varphi} \\ -re^{i\varphi} & e^{-i\theta} \end{pmatrix}\begin{pmatrix} 1 & 0 \\ 0 & -1 \end{pmatrix}\frac{1}{t}\begin{pmatrix} e^{-i\theta} & -re^{-i\varphi} \\ -re^{i\varphi} & e^{i\theta} \end{pmatrix} = \frac{1}{t^2}\begin{pmatrix} 1 - r^2 & 0 \\ 0 & r^2 - 1 \end{pmatrix} = \sigma_z;$$

the matrices that obey this relation (including T) are called *anti-unitary*.

***Problem* 4.14.** Calculate the trace of the following matrix:

$$\exp\{i\mathbf{a}\cdot\boldsymbol{\sigma}\}\exp\{i\mathbf{b}\cdot\boldsymbol{\sigma}\},$$

where $\boldsymbol{\sigma}$ is the Pauli matrix vector, while $\mathbf{a}$ and $\mathbf{b}$ are c-number geometric vectors.

Solution: Since we may always take $\mathbf{a} = a\mathbf{n}$, where $\mathbf{n}$ is the unit vector in the same direction, the solution of problem 4.6 (with θ replaced with a) may be rewritten as

$$\exp\{i\mathbf{a}\cdot\boldsymbol{\sigma}\} = \mathrm{I}\cos a + i\frac{\mathbf{a}\cdot\boldsymbol{\sigma}}{a}\sin a$$

(and similarly for the exponent including $\mathbf{b}$), so that

$$\begin{aligned}\exp\{i\mathbf{a}\cdot\boldsymbol{\sigma}\}\exp\{i\mathbf{b}\cdot\boldsymbol{\sigma}\} &= \left(\mathrm{I}\cos a + i\frac{\mathbf{a}\cdot\boldsymbol{\sigma}}{a}\sin a\right)\left(\mathrm{I}\cos b + i\frac{\mathbf{b}\cdot\boldsymbol{\sigma}}{b}\sin b\right) \\ &\equiv \mathrm{I}\cos a\cos b + i\left(\frac{\mathbf{a}\cdot\boldsymbol{\sigma}}{a}\sin a\cos b + \frac{\mathbf{b}\cdot\boldsymbol{\sigma}}{b}\cos a\sin b\right) \quad (*) \\ &\quad - \frac{(\mathbf{a}\cdot\boldsymbol{\sigma})(\mathbf{b}\cdot\boldsymbol{\sigma})}{ab}\sin a\sin b.\end{aligned}$$

The first term in the last expression is a diagonal matrix, with equal diagonal elements, so that its trace equals $2\cos a\cos b$. Next, since by the definition (4.105) of the Pauli matrices,

$$\mathbf{a}\cdot\boldsymbol{\sigma} = \begin{pmatrix} a_z & a_x - ia_y \\ a_x + ia_y & -a_z \end{pmatrix}, \quad \text{and} \quad \mathbf{b}\cdot\boldsymbol{\sigma} = \begin{pmatrix} b_z & b_x - ib_y \\ b_x + ib_y & -b_z \end{pmatrix}, \qquad (**)$$

the traces of these matrices, and hence that of the whole second term of Eq. (*), equal zero. Finally, the trace of the matrix product in the last term of Eq. (*) may be readily calculated, for example, directly from Eq. (**):

$$\mathrm{Tr}[(\mathbf{a}\cdot\boldsymbol{\sigma})(\mathbf{b}\cdot\boldsymbol{\sigma})] = \mathrm{Tr}\begin{pmatrix} a_z b_z + (a_x - ia_y)(b_x + ib_y) & \dots \\ \dots & (a_x + ia_y)(b_x - ib_y) + a_z b_z \end{pmatrix}$$

$$= 2(a_x b_x + a_y b_y + a_z b_z) \equiv 2\mathbf{a}\cdot\mathbf{b},$$

so that, finally,

$$\mathrm{Tr}(\exp\{i\mathbf{a}\cdot\boldsymbol{\sigma}\}\exp\{i\mathbf{b}\cdot\boldsymbol{\sigma}\}) = 2\left(\cos a \cos b - \frac{\mathbf{a}\cdot\mathbf{b}}{ab}\sin a \sin b\right).$$

***Problem* 4.15.** Prove the following vector-operator identity:

$$(\boldsymbol{\sigma}\cdot\hat{\mathbf{r}})(\boldsymbol{\sigma}\cdot\hat{\mathbf{p}}) = \mathrm{I}\,\hat{\mathbf{r}}\cdot\hat{\mathbf{p}} + i\boldsymbol{\sigma}\cdot(\hat{\mathbf{r}}\times\hat{\mathbf{p}}),$$

where I is the 2×2 identity matrix.

Hint: Take into account that the vector operators $\hat{\mathbf{r}}$ and $\hat{\mathbf{p}}$ are defined in the orbital-motion Hilbert space, independent of that of the Pauli vector-matrix $\boldsymbol{\sigma}$, and hence commute with it—even though they do not commute with each other.

Solution: Perhaps the simplest way to prove this identity is first to use the definitions of the scalar and vector products to express each of its parts via the Cartesian components of the matrices and operators. In particular, the left-hand side is

$$(\boldsymbol{\sigma}\cdot\hat{\mathbf{r}})(\boldsymbol{\sigma}\cdot\hat{\mathbf{p}}) = \sum_{j'}\sigma_{j'}\hat{r}_{j'}\sum_{j''}\sigma_{j''}\hat{p}_{j''} \equiv \sum_{j',j''}\sigma_{j'}\hat{r}_{j'}\sigma_{j''}\hat{p}_{j''}.$$

where all sums (and all those below in this solution) run from 1 to 3. Since the Pauli matrices commute with the coordinate and momentum operators, we may swap them and continue as

$$(\boldsymbol{\sigma}\cdot\hat{\mathbf{r}})(\boldsymbol{\sigma}\cdot\hat{\mathbf{p}}) = \sum_{j',j''}(\sigma_{j'}\sigma_{j''})\,\hat{r}_{j'}\hat{p}_{j''}.$$

Now using the first of Eqs. (*) in the model solution of problem 4.3, we get

$$(\boldsymbol{\sigma}\cdot\hat{\mathbf{r}})(\boldsymbol{\sigma}\cdot\hat{\mathbf{p}}) = \sum_{j',j''}(i\sigma_j\varepsilon_{jj'j''} + \mathrm{I}\delta_{j'j''})\,\hat{r}_{j'}\hat{p}_{j''} \equiv i\sum_{j',j''}\sigma_j\,\hat{r}_{j'}\hat{p}_{j''}\varepsilon_{jj'j''} + \mathrm{I}\sum_{j'}\hat{r}_{j'}\hat{p}_{j'}.$$

Due to the Levi-Civita symbol, which kills all terms with $j' = j''$, of the $3 \times 3 = 9$ terms of the first (double) sum, only six terms, with $j' \neq j''$, are different from 0; they may be regrouped into a single sum over the index $j \neq j', j''$:

$$(\boldsymbol{\sigma}\cdot\hat{\mathbf{r}})(\boldsymbol{\sigma}\cdot\hat{\mathbf{p}}) = i\sum_{j}\sigma_j(\hat{r}_{j'}\hat{p}_{j''} - \hat{r}_{j''}\hat{p}_{j'})\,\varepsilon_{jj'j''} + \mathrm{I}\sum_{j'}\hat{r}_{j'}\hat{p}_{j'}.$$

But this expression exactly coincides with the right-hand side of the equality we are proving, similarly spelled out via its Cartesian components:

$$\mathrm{I}\,\hat{\mathbf{r}}\cdot\hat{\mathbf{p}} + i\boldsymbol{\sigma}\cdot(\hat{\mathbf{r}}\times\hat{\mathbf{p}}) = \mathrm{I}\sum_{j'}\hat{r}_{j'}\hat{p}_{j'} + i\sum_{j}\sigma_j(\hat{r}_{j'}\hat{p}_{j''} - \hat{r}_{j''}\hat{p}_{j'})\varepsilon_{jj'j''},$$

so that the equality is indeed valid.

***Problem* 4.16.** Let A_j be the eigenvalues of some operator $\hat{A}$. Express the following two sums,

$$\Sigma_1 \equiv \sum_j A_j, \qquad \Sigma_2 \equiv \sum_j A_j^2,$$

via the matrix elements $A_{jj'}$ of this operator in an arbitrary basis.

Solution: According to Eq. (4.98) of the lecture notes, in the basis in that the operator's matrix is diagonal, the first sum is just the trace of the matrix:

$$\Sigma_1 = \sum_j A_{jj} = \mathrm{Tr(A)}. \qquad (*)$$

But according to the statement whose proof was the subject of problem 4.11, the operator's trace does not depend on the choice of the basis, so that Eq. (*) is valid in an arbitrary basis as well, and we may write

$$\Sigma_1 = \mathrm{Tr}(\hat{A}).$$

For the calculation of the sum Σ_2, we may notice that since the operators $\hat{A}$ and $\hat{A}^2$ commute, they share their eigenstates α_j, so that we may write

$$\hat{A}^2|\alpha_j\rangle \equiv \hat{A}\hat{A}|\alpha_j\rangle = \hat{A}A_j|\alpha_j\rangle = A_j\hat{A}|\alpha_j\rangle = A_j^2|\alpha_j\rangle.$$

This means that the eigenvalues of $\hat{A}^2$, corresponding to these states, are just A_j^2, so that Σ_2 is the trace of the operator $\hat{A}^2$, and hence may be calculated, in an arbitrary basis, as

$$\Sigma_2 = \mathrm{Tr}(\hat{A}^2) \equiv \sum_j (\mathrm{A}^2)_{jj} = \sum_{j,j'} A_{jj'}A_{j'j}.$$

Note that if the operator $\hat{A}$ is Hermitian, then $A_{j'j} = A_{jj'}^*$, and the last expression may be further simplified:

$$\Sigma_2 = \sum_{j,j'} A_{jj'}A_{jj'}^* = \sum_{j,j'} |A_{jj'}|^2.$$

***Problem* 4.17.** Calculate $\langle\sigma_z\rangle$ of a spin-½ in a quantum state with the following ket-vector:

$$|\alpha\rangle = \mathrm{const}\times(|\uparrow\rangle + |\downarrow\rangle + |\rightarrow\rangle + |\leftarrow\rangle),$$

where ($\uparrow$, $\downarrow$) and ($\rightarrow$, $\leftarrow$) are the eigenstates of the Pauli matrices σ_z and σ_x, respectively.

Hint: Double-check whether the solution you are giving is general.

Solution: A superficial solution to this problem stems from Eqs. (4.122) of the lecture notes, which yield

$$|\rightarrow\rangle + |\leftarrow\rangle = \sqrt{2}|\uparrow\rangle,$$

so that

$$|\alpha\rangle = \text{const} \times [(1 + \sqrt{2})|\uparrow\rangle + |\downarrow\rangle],$$

and after the normalization,

$$W_\uparrow = \frac{(1+\sqrt{2})^2}{(1+\sqrt{2})^2 + 1^2} \equiv \frac{3+2\sqrt{2}}{4+2\sqrt{2}}, \qquad W_\downarrow = 1 - W_\uparrow = \frac{1}{4+2\sqrt{2}}.$$

we get

$$\langle\sigma_z\rangle = (+1)\, W_\uparrow + (-1)\, W_\downarrow \equiv W_\uparrow - W_\downarrow = \frac{1}{\sqrt{2}} \approx 0.707. \qquad (*)$$

However, this is *not* the general solution of our problem. Indeed, during the discussion of the diagonalization of matrix σ_x in section 4.4, it was emphasized that the coefficients U_{11} and U_{21} may be multiplied by the same phase factor $\exp\{i\varphi_+\}$, with an arbitrary (real) φ_+. The same is evidently true for the second pair of coefficients, U_{22} and U_{12}, whose phase φ_- may be independent of φ_+. As a result, the most general form of that unitary matrix is

$$\mathrm{U}_x = \frac{1}{\sqrt{2}}\begin{pmatrix} e^{i\varphi_+} & e^{i\varphi_-} \\ e^{i\varphi_+} & -e^{i\varphi_-} \end{pmatrix}.$$

From this, Eqs. (4.122) have to be generalized as

$$|\rightarrow\rangle = \frac{\exp\{i\varphi_+\}}{\sqrt{2}}(|\uparrow\rangle + |\downarrow\rangle), \qquad |\leftarrow\rangle = \frac{\exp\{i\varphi_-\}}{\sqrt{2}}(|\uparrow\rangle - |\downarrow\rangle).$$

These expressions show that $\varphi_\pm$ are just the (arbitrary) phases of the states $\rightarrow$ and $\leftarrow$.

It is easy to verify that these phases don't affect not only the probabilities of these states, but also the discussion of all Stern–Gerlach experiments in section 4.4 of the lecture notes, so that there we could take $\varphi_+ = \varphi_- = 0$ quite legitimately. However, the state α being discussed in the current problem is affected by these phases:

$$|\alpha\rangle = C\left[\left(1 + \frac{e^{i\varphi_+}}{\sqrt{2}} + \frac{e^{i\varphi_-}}{\sqrt{2}}\right)|\uparrow\rangle + \left(1 + \frac{e^{i\varphi_+}}{\sqrt{2}} - \frac{e^{i\varphi_-}}{\sqrt{2}}\right)|\downarrow\rangle\right].$$

In particular, its normalization yields

$$|C|^{-2} = \left|1 + \frac{e^{i\varphi_+}}{\sqrt{2}} + \frac{e^{i\varphi_-}}{\sqrt{2}}\right|^2 + \left|1 + \frac{e^{i\varphi_+}}{\sqrt{2}} - \frac{e^{i\varphi_-}}{\sqrt{2}}\right|^2 \equiv 4 + 2\sqrt{2}\cos\varphi_+,$$

so that finally,

$$\langle\sigma_z\rangle = W_\uparrow - W_\downarrow = |C|^2\left[\left|1 + \frac{e^{i\varphi_+}}{\sqrt{2}} + \frac{e^{i\varphi_-}}{\sqrt{2}}\right|^2 - \left|1 + \frac{e^{i\varphi_+}}{\sqrt{2}} - \frac{e^{i\varphi_-}}{\sqrt{2}}\right|^2\right]$$
$$= \frac{\cos(\varphi_+ - \varphi_-) + \sqrt{2}\cos\varphi_-}{2 + \sqrt{2}\cos\varphi_+}.$$

For $\varphi_+ = \varphi_- = 0$, this expression is reduced to Eq. (*), but generally, depending on the choice of the phases φ_+ and φ_-, the calculated average $\langle\sigma_z\rangle$ may range all the way from (+1) to (−1).

***Problem* 4.18.** A spin-½ is fully polarized in the positive z-direction. Calculate the probabilities of the alternative outcomes of a perfect Stern–Gerlach experiment with the magnetic field oriented in an arbitrary different direction, performed on a particle in this spin state.

Solution: As was discussed in section 4.1 of the lecture notes, the Stern–Gerlach experiment measures the probabilities $W_\pm$ of the particle's magnetic moment $\mathbf{m}$ being oriented in the direction of the magnetic field of the SG apparatus, and opposite to it—in our current case, along the field's direction $\mathbf{n}$ and against it. Hence $W_\pm$ should be calculated using Eq. (4.120) of the lecture notes,

$$W_\pm = |\langle\alpha|m_\pm\rangle|^2, \qquad (*)$$

where $m_\pm$ are the normalized eigenstates of the corresponding scalar operator $\hat{m}_n \equiv \mathbf{n}\cdot\hat{\mathbf{m}}$, and α is the given state of the system—in our case, the polarized up-spin state $\uparrow$. As Eqs. (4.115)–(4.117) show, in the usual z-basis the matrix of the operator $\hat{m}_n$ is proportional to, and hence has the same eigenstates as the matrix $\mathrm{P}_n \equiv \mathbf{n}\cdot\boldsymbol{\sigma}$, where $\boldsymbol{\sigma}$ is the Pauli matrix vector with the Cartesian components given by Eq. (4.105):

$$\sigma_x \equiv \begin{pmatrix} 0 & 1 \\ 1 & 0 \end{pmatrix}, \qquad \sigma_y \equiv \begin{pmatrix} 0 & -i \\ i & 0 \end{pmatrix}, \qquad \sigma_z \equiv \begin{pmatrix} 1 & 0 \\ 0 & -1 \end{pmatrix}.$$

Taking the Cartesian components of the vector $\mathbf{n}$ in the usual polar-coordinate form,

$$n_x = \sin\theta\cos\varphi, \qquad n_y = \sin\theta\sin\varphi, \qquad n_z = \cos\theta,$$

where θ and φ are the usual polar angles of the vector $\mathbf{n}$, we get

$$\begin{aligned}
\mathrm{P}_n &\equiv \mathbf{n}\cdot\boldsymbol{\sigma} \equiv n_x\sigma_x + n_y\sigma_y + n_z\sigma_z \\
&= \sin\theta\cos\varphi\begin{pmatrix}0 & 1\\ 1 & 0\end{pmatrix} + \sin\theta\sin\varphi\begin{pmatrix}0 & -i\\ i & 0\end{pmatrix} + \cos\theta\begin{pmatrix}1 & 0\\ 0 & -1\end{pmatrix} \\
&\equiv \begin{pmatrix}\cos\theta & \sin\theta\, e^{-i\varphi}\\ \sin\theta\, e^{i\varphi} & -\cos\theta\end{pmatrix}.
\end{aligned}$$

Now we may use the procedure discussed in section 4.4 to find the eigenstates $P_\pm$ of this matrix, i.e. the coefficients $\langle\uparrow|P_\pm\rangle$ and $\langle\downarrow|P_\pm\rangle$ of their expansion over the z-basis states $\uparrow$ and $\downarrow$. For that, first, we may use the characteristic equation (4.103), in our case

$$\begin{vmatrix}\cos\theta - P_\pm & \sin\theta\, e^{-i\varphi}\\ \sin\theta\, e^{i\varphi} & -\cos\theta - P_\pm\end{vmatrix} = 0, \qquad (**)$$

to calculate the eigenvalues of the matrix: $P_\pm = \pm 1$.[4] Now plugging these values, one by one, into any of Eqs. (4.101), in our particular case having the form

$$\begin{aligned}
&(\cos\theta - P_\pm)\langle\uparrow|P_\pm\rangle + \sin\theta\, e^{-i\varphi}\langle\downarrow|P_\pm\rangle = 0,\\
&\sin\theta\, e^{i\varphi}\langle\uparrow|P_\pm\rangle + (-\cos\theta - P_\pm)\langle\downarrow|P_\pm\rangle = 0,
\end{aligned}$$

we get

$$\begin{aligned}
\langle\uparrow|P_+\rangle &= \langle\downarrow|P_+\rangle\frac{\sin\theta}{1-\cos\theta}e^{-i\varphi} \equiv \langle\downarrow|P_+\rangle\frac{\cos(\theta/2)}{\sin(\theta/2)}e^{-i\varphi},\\
\langle\uparrow|P_-\rangle &= \langle\downarrow|P_-\rangle\frac{1-\cos\theta}{\sin\theta}e^{-i\varphi} \equiv \langle\downarrow|P_-\rangle\frac{\sin(\theta/2)}{\cos(\theta/2)}e^{-i\varphi}.
\end{aligned}$$

Now performing the normalization, i.e. requiring $|\langle\uparrow|P_\pm\rangle|^2 + |\langle\downarrow|P_\pm\rangle|^2$ to equal 1, we get a very simple result:

$$\begin{aligned}
|\langle\uparrow|P_+\rangle| &= \cos\frac{\theta}{2}, \qquad |\langle\downarrow|P_+\rangle| = \sin\frac{\theta}{2},\\
|\langle\uparrow|P_-\rangle| &= \sin\frac{\theta}{2}, \qquad |\langle\downarrow|P_-\rangle| = \cos\frac{\theta}{2}.
\end{aligned}$$

But according to Eq. (*), with α being the state $\uparrow$, the squares of the two left expressions give us the required probabilities:

$$W_+ = \cos^2\frac{\theta}{2}, \qquad W_- = \sin^2\frac{\theta}{2}. \qquad (***)$$

[4] This result could be anticipated, because as we know from section 4.4 of the lecture notes, all Pauli matrices have the same eigenvalues, and they should not be affected by any rotation of the coordinate axes.

This is a very nontrivial[5] and important result (which will be used, in particular, for the discussion of Bell inequalities in the concluding chapter 10 of the lecture notes), so it makes sense to pass it through all possible sanity checks. The first (the most elementary) check is that $W_+ + W_- = 1$, as it has to be with the sum of all possible experimental outcomes.

Next, let us examine the most important particular cases. If $\theta = 0$ (i.e. we measure the spin along the axis of its prior polarization), then $W_+ = 1$ and $W_- = 0$—quite naturally; while if $\theta = \pi/2$ (the magnetic field is oriented along the axis x), we get $W_+ = W_- = ½$—the result which was already obtained in the lecture notes—see the first of Eqs. (4.123) and its discussion.

Finally, we may use Eqs. (***), and the calculated eigenvalues $P_\pm = \pm 1$ to find the expectation value of the observable $P_n \propto m_n$ from the general Eq. (1.37):

$$\langle P_n \rangle = P_+ W_+ + P_- W_- = \cos^2 \frac{\theta}{2} - \sin^2 \frac{\theta}{2} \equiv \cos\theta.$$

But the same expectation value may be found more simply, from Eq. (4.125), with the long bracket calculated directly in the z-basis:

$$\langle P_n \rangle = \langle \uparrow | \hat{P} | \uparrow \rangle = (1,\, 0) \begin{pmatrix} \cos\theta & \sin\theta\, e^{-i\varphi} \\ \sin\theta\, e^{i\varphi} & -\cos\theta \end{pmatrix} \begin{pmatrix} 1 \\ 0 \end{pmatrix} = \cos\theta.$$

The similarity of these results is perhaps the best confirmation of Eqs. (***). A forthcoming discussion in section 5.1 of the lecture notes will allow us to re-derive this important result in a much simpler way—see problem 5.1.

***Problem* 4.19.** In a certain basis, the Hamiltonian of a two-level system is described by the matrix

$$\mathrm{H} = \begin{pmatrix} E_1 & 0 \\ 0 & E_2 \end{pmatrix}, \qquad \text{with } E_1 \neq E_2,$$

while the operator of some observable A of this system, by the matrix

$$\mathrm{A} = \begin{pmatrix} 1 & 1 \\ 1 & 1 \end{pmatrix}.$$

For the system's state with the energy definitely equal to E_1, find the possible results of measurements of the observable A and the probabilities of the corresponding measurement outcomes.

Solution: Let us calculate the eigenvectors and eigenvalues of the operator of observable A in the given basis. For it, the system of equations (4.101)–(4.102) is

$$\begin{aligned} (1 - A_j)\, U_{1j} + U_{2j} &= 0, \\ U_{1j} + (1 - A_j)\, U_{2j} &= 0, \end{aligned} \qquad (*)$$

[5] Naively, one might expect the probability W_+ to be equal to the square of the z-component, n_z, of the vector **n**, i.e. to $\cos^2\theta$. The result given by Eqs. (***) is, of course, very different.

so that the characteristic equation (4.103) of their consistency is

$$\begin{vmatrix} 1 - A_j & 1 \\ 1 & 1 - A_j \end{vmatrix} \equiv (1 - A_j)^2 - 1 = 0.$$

Its (easy :-) solution yields two roots A_j: $A_1 = 0$ and $A_2 = 2$; these are the possible results of measurements of the observable A.

Now plugging these eigenvalues, one by one, into any equation of the system (*), for the unitary matrix elements we get

$$U_{11} = -U_{21}, \qquad U_{12} = U_{22},$$

so that after the proper normalization (4.104), and setting the inconsequential phase shifts to zero, we get

$$U_{11} = -U_{21} = \frac{1}{\sqrt{2}}, \qquad U_{21} = U_{22} = \frac{1}{\sqrt{2}}.$$

(Note that these results coincide with those obtained in section 4.4 of the lecture notes, for the Pauli matrix σ_x in the usual z-basis. This is not surprising, because the given matrix A is just the sum of σ_x and the identity matrix I, and the latter matrix is diagonal in any basis and thus cannot affect the results of diagonalization.)

Since, according to Eqs. (4.82) and (4.83), the obtained unitary transfer matrix elements U_{kj} (with indices k and j taking values 1 and 2) are just the short brackets $\langle u_k|a_j\rangle$ participating in the expansion,

$$|a_j\rangle = \sum_k \langle u_k|a_j\rangle |u_j\rangle,$$

of the eigenstates of the operator $\hat{A}$ over the states u_1, u_2 of the given basis, our result means that

$$|a_1\rangle = \frac{1}{\sqrt{2}}(|u_1\rangle - |u_2\rangle), \qquad |a_2\rangle = \frac{1}{\sqrt{2}}(|u_1\rangle + |u_2\rangle).$$

Now since the matrix H is diagonal in the given basis, and its energy E equals one of its eigenvalues (E_1), the system is the corresponding eigenstate, u_1. Hence the above expansion reduces to

$$|a_1\rangle = \frac{1}{\sqrt{2}}|u_1\rangle, \qquad |a_2\rangle = \frac{1}{\sqrt{2}}|u_1\rangle,$$

so that the probability of measuring each eigenvalue of A is 50%.

***Problem* 4.20.** Certain states $u_{1,2,3}$ form an orthonormal basis of a system with the following Hamiltonian:

$$\hat{H} = -\delta(|u_1\rangle\langle u_2| + |u_2\rangle\langle u_3| + |u_3\rangle\langle u_1|) + \text{h.c.},$$

where δ is a real constant, and h.c. means the Hermitian conjugate of the previous expression. Calculate its stationary states and energy levels. Can you relate this system with any other(s) discussed earlier in the course?

Solution: According to Eq. (4.159) of the lecture notes, the stationary states $a_{1,2,3}$ and the energy levels $E_{1,2,3}$ of the system are just the eigenstates and eigenvalues of the system's Hamiltonian, so that the problem is reduced to the diagonalization of the Hamiltonian matrix H. According to the assignment, and Eq. (4.59), in the $u_{1,2,3}$ basis the matrix has the form

$$\mathrm{H} = -\delta \begin{pmatrix} 0 & 1 & 1 \\ 1 & 0 & 1 \\ 1 & 1 & 0 \end{pmatrix},$$

so that its characteristic equation (4.103) is

$$\begin{vmatrix} -\varepsilon & -1 & -1 \\ -1 & -\varepsilon & -1 \\ -1 & -1 & -\varepsilon \end{vmatrix} = 0, \quad \text{where} \quad \varepsilon \equiv \frac{E_j}{\delta}, \quad \text{with} \quad j = 1, 2, 3.$$

This equality gives a cubic equation for ε:

$$f(\varepsilon) \equiv \varepsilon^3 - 3\varepsilon + 2 = 0;$$

fortunately, the equation is so simple that its roots may be readily guessed (and then verified by their substitution the guess into the equation) just by looking either at the equation itself, or at the plot of the function on its left-hand side—see the figure below:

$$\varepsilon_1 = -2, \quad \varepsilon_{2,3} = +1.$$

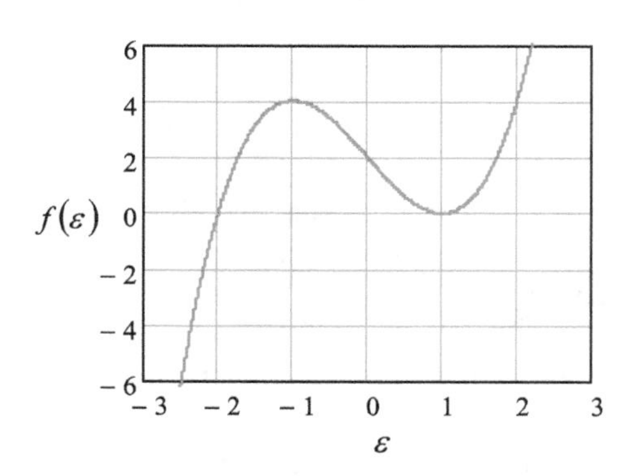

In order to calculate the corresponding eigenstates $a_{1,2,3}$ of the Hamiltonian, i.e. the coefficients in the expansions of its ket-vectors in the basis $u_{1,2,3}$:

$$\begin{aligned} |a_1\rangle &= U_{11}|u_1\rangle + U_{21}|u_2\rangle + U_{31}|u_3\rangle, \\ |a_2\rangle &= U_{12}|u_1\rangle + U_{22}|u_2\rangle + U_{32}|u_3\rangle, \\ |a_3\rangle &= U_{13}|u_1\rangle + U_{23}|u_2\rangle + U_{33}|u_3\rangle, \end{aligned} \qquad (*)$$

we need to plug in the calculated values $\varepsilon_{1,2,3}$, one by one, into the system of Eq. (4.101) for our matrix H. For the first root, $\varepsilon_1 = -2$, the system is

$$\begin{aligned} +2U_{11} - U_{21} - U_{31} = 0, \\ -U_{11} + 2U_{22} - U_{31} = 0, \\ -U_{11} - U_{22} + 2U_{31} = 0. \end{aligned}$$

It is evidently satisfied with any set of equal coefficients, $U_{11} = U_{21} = U_{31}$. Thus, requiring the set to be normalized in accordance with Eq. (4.104), we may spell out the first of Eqs. (*) as follows[6]:

$$|a_1\rangle = \frac{1}{\sqrt{3}}(|u_1\rangle + |u_2\rangle + |u_3\rangle). \qquad (**)$$

For the equal 2nd and 3rd roots, $\varepsilon_{2,3} = +1$, all equations of the system (*) look similarly:

$$-U_{1j} - U_{2j} - U_{3j} = 0, \quad \text{with } j = 2, 3. \qquad (***)$$

Due to the symmetry of these equations with respect to rotation of the first indices of the coefficients U_{kj}, it is natural to assume that the solutions differ only by a constant phase multiplier, i.e. to look for the solutions in the form

$$U_{kj} = U_0 \exp\{ik\alpha\}. \qquad (****)$$

Evidently, the change of k by 3 has to give the initial value U_{kj}, perhaps besides an inconsequential phase multiplier $\exp\{i2\pi n\}$, with any integer n. This requirement immediately yields just three physically distinguishable values of α: 0, and $\pm 2\pi/3$. The first of them does not satisfy Eq. (***),[7] but the two others do, because

$$U_{1j} + U_{2j} + U_{3j} = U_0 \exp\left\{\pm i\frac{2\pi}{3}\right\} + U_0 \exp\left\{\pm i2\frac{2\pi}{3}\right\} + U_0 \exp\left\{\pm i3\frac{2\pi}{3}\right\} \equiv 0,$$

for any U_0. This means that the two eigenstates corresponding to the degenerate eigenvalue $\varepsilon_{2,3} = 1$, may be taken, for example, in the form[8]

[6] As a reminder: according to Eq. (4.104), all coefficients of the string U_{12}, U_{21}, U_{31}, and hence the ket-vector $|a1\rangle$ as a whole, may be multiplied by coefficients $\exp\{i\varphi_1\}$, with an arbitrary real phase shift φ_1. The same is true for each of the two other eigenkets; for each of them the phase shift may be individual.

[7] Note, however, that solution (****) with such $\alpha = 0$ does describe our first eigenstate (**).

[8] Since these stationary states correspond to the same eigenenergy $E = 2\delta$, any their linear combination is also a stationary state. Remember, however, that for applications, it is important to keep such linear combinations orthogonal, i.e. their inner product (4.11) equals zero. (It is straightforward to verify that the above states $a_{1,2,3}$ do satisfy this requirement.)

$$|a_2\rangle = \frac{1}{\sqrt{3}}\left(|u_1\rangle + \exp\left\{i\frac{2\pi}{3}\right\}|u_2\rangle + \exp\left\{-i\frac{2\pi}{3}\right\}|u_3\rangle\right)$$
$$\equiv \frac{1}{\sqrt{3}}\left(|u_1\rangle + \frac{-1+i\sqrt{3}}{2}|u_2\rangle + \frac{-1-i\sqrt{3}}{2}|u_3\rangle\right),$$
$$|a_3\rangle = \frac{1}{\sqrt{3}}\left(|u_1\rangle + \exp\left\{-i\frac{2\pi}{3}\right\}|u_2\rangle + \exp\left\{+i\frac{2\pi}{3}\right\}|u_3\rangle\right)$$
$$\equiv \frac{1}{\sqrt{3}}\left(|u_1\rangle + \frac{-1-i\sqrt{3}}{2}|u_2\rangle + \frac{-1+i\sqrt{3}}{2}|u_3\rangle\right).$$

To summarize, all three eigenenergies of this system

$$E_1 = -2\delta, \qquad E_{2,3} = +\delta,$$

correspond to the states of the type (****), with different phase shifts α: $\alpha_1 = 0$ and $\alpha_{2,3} = \pm 2\pi/3$. But this is exactly the result which was obtained in problem 3.11 for three similar, and similarly coupled potential wells (say, located in the vertices of an equilateral triangle) within the tight-binding limit. In this particular physical implementation of our Hamiltonian, $u_{1,2,3}$ have the physical sense of particle's localized states inside the corresponding wells.

***Problem* 4.21.** Guided by Eq. (2.203) of the lecture notes, and by the solutions of problems 3.11 and 4.20, suggest a Hamiltonian describing particle's dynamics in an infinite 1D chain of similar potential wells in the tight-binding approximation, in the bra–ket formalism. Verify that its eigenstates and eigenvalues correspond to those discussed in section 2.7.

Solution: Inspired by the identity of solutions of problems 3.11 and 4.20, which give the similar sets of eigenstates and eigenvalues, we may readily guess the effective Hamiltonian describing the system in the vicinity of the nth localized level:

$$\hat{H}_n = E_n \sum_k |u_k\rangle\langle u_k| - \delta_n \left(\sum_k |u_k\rangle\langle u_{k+1}| + \text{h.c.}\right),$$

where the state u_k corresponds to the particle's position in kth well. Indeed, in the basis of these states, the (formally, infinite) matrix of this Hamiltonian is *tri-diagonal*:

$$\mathrm{H} = \begin{pmatrix} \cdots & \cdots & \cdots & \cdots & \cdots & \cdots & \cdots \\ \cdots & E_n & -\delta_n & 0 & 0 & 0 & \cdots \\ \cdots & -\delta_n & E_n & -\delta_n & 0 & 0 & \cdots \\ \cdots & 0 & -\delta_n & E_n & -\delta_n & 0 & \cdots \\ \cdots & 0 & 0 & -\delta_n & E_n & -\delta_n & \cdots \\ \cdots & 0 & 0 & 0 & -\delta_n & E_n & \cdots \\ \cdots & \cdots & \cdots & \cdots & \cdots & \cdots & \cdots \end{pmatrix},$$

so that for its eigenstates a_j (i.e. the coefficients U_{kj} of their expansion in the series

$$|a_j\rangle = \sum_k \langle u_k|a_j\rangle|u_k\rangle \equiv \sum_k U_{kj}|u_k\rangle$$

over the basis states u_k) and eigenvalues E_n, Eqs. (4.101) and (4.102) of the lecture notes give the following infinite system of equations,

$$-\delta_n U_{(k-1)j} + (E_n - E)\, U_{kj} - \delta_n U_{(k+1)j} = 0. \qquad (*)$$

Looking for the solution of this system in the Bloch-wave form[9],

$$U_{kj} = U_0 \exp\{ik\Delta\varphi\},$$

we immediately get the dispersion relation (2.206):

$$E = E_n - 2\delta_n \cos\Delta\varphi.$$

In this infinite system, the phase shift (i.e. the dimensionless quasi-momentum) $\Delta\varphi$ may take any real values.

***Problem* 4.22.** Calculate the eigenvectors and the eigenvalues of the following matrices:

$$\mathrm{A} = \begin{pmatrix} 0 & 1 & 0 \\ 1 & 0 & 1 \\ 0 & 1 & 0 \end{pmatrix}, \qquad \mathrm{B} = \begin{pmatrix} 0 & 0 & 0 & 1 \\ 0 & 0 & 1 & 0 \\ 0 & 1 & 0 & 0 \\ 1 & 0 & 0 & 0 \end{pmatrix}.$$

Solutions: Solving the characteristic equation (Eq. (4.103) of the lecture notes) for the matrix A,

$$\begin{vmatrix} -A & 1 & 0 \\ 1 & -A & 1 \\ 0 & 1 & -A \end{vmatrix} = 0, \qquad \text{giving } A^3 - 2A = 0,$$

we are getting three roots (the matrix eigenvalues) A_j: $A_1 = 0$ and $A_{2,3} = \pm\sqrt{2}$. Now we should plug these values, one by one, into the system of Eq. (4.101) for the elements $U_{kj} \equiv \langle u_k|a_j\rangle$ of the unitary matrix U performing the transformation from the vector of basis states $u_{1,2,3}$ to the corresponding eigenvector (number j):

$$|a_j\rangle = U_{1j}|u_1\rangle + U_{2j}|u_2\rangle + U_{3j}|u_3\rangle.$$

For our matrix A, the system is

$$\begin{aligned} -A_j U_{1j} + U_{2j} &= 0, \\ U_{1j} - A_j U_{2j} + U_{3j} &= 0, \\ U_{2j} - A_j U_{3j} &= 0. \end{aligned} \qquad (*)$$

[9] If necessary, please revisit Eq. (2.193) of the lecture notes and its discussion in section 2.7.

For the first eigenvalue, $A_j = A_1 = 0$, the top and the bottom equations immediately yield $U_{21} = 0$, so that the middle one yields $U_{11} = -U_{31}$. The requirement of the vector U_{k1} to be normalized (see Eq. (4.104) of the lecture notes), may be satisfied by setting $U_{11} = 1/\sqrt{2}$, $U_{31} = -1/\sqrt{2}$.

For the second and third eigenvalues, $A_{2,3} = \pm\sqrt{2}$, which differ only by the sign, Eqs. (*) are reduced to two very similar systems of equations:

$$\begin{aligned} &-\sqrt{2}\,U_{12} + U_{22} = 0, && \sqrt{2}\,U_{13} + U_{23} = 0,\\ &U_{12} - \sqrt{2}\,U_{22} + U_{32} = 0, && U_{13} + \sqrt{2}\,U_{23} + U_{33} = 0,\\ &-U_{22} - \sqrt{2}\,U_{32} = 0, && -U_{23} + \sqrt{2}\,U_{33} = 0, \end{aligned}$$

The top and the bottom equations of these systems immediately yield $U_{12} = -U_{32} = U_{22}/\sqrt{2}$, and $U_{33} = -U_{13} = U_{23}/\sqrt{2}$, so that the normalization conditions for these two strings may be satisfied by taking $U_{12} = U_{33} = 1/\sqrt{2}$, $U_{12} = U_{33} = ½$, and $U_{32} = U_{13} = -½$. So, the eigenvectors of the matrix are:

$$|a_1\rangle = \frac{1}{\sqrt{2}}(|u_1\rangle - |u_2\rangle), \quad |a_2\rangle = \frac{1}{2}\left(|u_1\rangle + \sqrt{2}\,|u_2\rangle - |u_3\rangle\right),$$
$$|a_3\rangle = \frac{1}{2}\left(- |u_1\rangle + \sqrt{2}\,|u_2\rangle + |u_3\rangle\right).$$

These eigenkets are defined to the phase factors $\exp\{i\varphi_j\}$, with arbitrary real φ_j.

For the matrix B, the characteristic equation is

$$\begin{vmatrix} -B & 0 & 0 & 1\\ 0 & -B & 1 & 0\\ 0 & 1 & -B & 0\\ 1 & 0 & 0 & -B \end{vmatrix} = 0, \quad \text{giving } B^4 - 2B^2 + 1 = 0. \qquad (**)$$

This is just a quadratic equation for B^2, with two equal roots $B^2 = 1$,[10] so that the four eigenvalues B_j of the matrix are

$$B_{1,2} = +1, \quad B_{3,4} = -1.$$

The system of Eq. (4.101), corresponding to the matrix B, is

$$\begin{aligned} -BU_{1j} + U_{4j} &= 0,\\ -BU_{2j} + U_{3j} &= 0,\\ U_{2j} - BU_{3j} &= 0,\\ U_{1j} - BU_{4j} &= 0, \end{aligned} \qquad (***)$$

where $U_{kj} \equiv \langle u_k|a_j\rangle$, with the indices k and j taking values from 1 to 4, are the elements of the unitary matrix performing the transformation from the vector of basis states u_k to the eigenvector corresponding to eigenvalue B_j:

[10] The (easy) task of solving the quadratic equation may be simplified even further by noticing that Eq. (**) may be rewritten as $(B^2 - 1)^2 = 0$.

$$|b_j\rangle = U_{1j}|u_1\rangle + U_{2j}|u_2\rangle + U_{3j}|u_3\rangle + U_{4j}|u_4\rangle.$$

Plugging the first value, $B_1 = +1$, into Eq. (***) we get $U_{11} = U_{41}$ and $U_{21} = U_{31}$. The result for the equal eigenvalue B_2 is of course similar: $U_{12} = U_{42}$ and $U_{22} = U_{32}$, so that the first two linearly independent eigenkets may be constructed, for example, by taking $U_{21} = U_{31} = 0$ and $U_{12} = U_{42} = 0$, giving (after the elementary normalization),

$$|b_1\rangle = \frac{1}{\sqrt{2}}(|u_1\rangle + |u_4\rangle), \quad |b_2\rangle = \frac{1}{\sqrt{2}}(|u_2\rangle + |u_3\rangle).$$

(Due to the degeneracy of these eigenvalues, $B_1 = B_2$, not only these kets, multiplied by arbitrary phase factors, but also any pair of their linearly-independent superpositions is also a legitimate solution.)

For the two remaining eigenvalues $B_{3,4} = -1$, the possible eigenstates are similar, with just the opposite signs of the nonvanishing matrix elements, so that we may take

$$|b_3\rangle = \frac{1}{\sqrt{2}}(|u_1\rangle - |u_4\rangle), \quad |b_4\rangle = \frac{1}{\sqrt{2}}(|u_2\rangle - |u_3\rangle),$$

with similar alternative options.

***Problem* 4.23.** A certain state γ is an eigenstate of each of two operators $\hat{A}$ and $\hat{B}$. What can be said about the corresponding eigenvalues a and b, if the operators anticommute?

Solution: Let us use the definition of the anticommutator[11] to calculate the following ket-vector:

$$\{\hat{A}, \hat{B}\}\,|\gamma\rangle \equiv \hat{A}\hat{B}\,|\gamma\rangle + \hat{B}\hat{A}\,|\gamma\rangle.$$

Since γ is an eigenstate of each of the operators, we may use Eqs. (4.68) and then (4.19) to proceed as follows:

$$\begin{aligned}\{\hat{A}, \hat{B}\}\,|\gamma\rangle &= \hat{A}b\,|\gamma\rangle + \hat{B}a\,|\gamma\rangle = b\hat{A}\,|\gamma\rangle + a\hat{B}\,|\gamma\rangle \\ &= ba\,|\gamma\rangle + ab\,|\gamma\rangle = 2ab\,|\gamma\rangle.\end{aligned} \qquad (*)$$

But the anticommutation of operators $\hat{A}$ and $\hat{B}$ means that

$$\{\hat{A}, \hat{B}\} = \hat{0},$$

where $\hat{0}$ is the null-operator, which is defined by Eq. (4.35), i.e. turns each state it acts upon into the null-state; in particular,

$$\{\hat{A}, \hat{B}\}\,|\gamma\rangle = |0\rangle. \qquad (**)$$

[11] See, e.g. Eq. (4.34) of the lecture notes.

Since the assignment implies that γ is a usual (not a null-) state, Eqs. (*) and (**) may be reconciled only if

$$ab = 0,$$

i.e. if at least one of these eigenvalues equals zero.

***Problem* 4.24.** Derive a differential equation for the time evolution of the expectation value of an observable, using both the Schrödinger picture and the Heisenberg picture of quantum dynamics.

Solution: Let us differentiate the basic Eq. (4.125) of the lecture notes over time, restoring, for clarity, the time arguments of the participating states and operators:

$$\frac{d}{dt}\langle A\rangle(t) = \left\langle \frac{\partial \alpha(t)}{\partial t} \right| \hat{A}(t)\,|\alpha(t)\rangle + \langle \alpha(t)|\,\frac{\partial \hat{A}(t)}{\partial t}\,|\alpha(t)\rangle + \langle\alpha(t)|\,\hat{A}(t)\left|\frac{\partial\alpha(t)}{\partial t}\right\rangle$$
$$= \left\langle\frac{\partial A}{\partial t}\right\rangle(t) + \left[\frac{\partial}{\partial t}\langle\alpha(t)|\right]\hat{A}(t)\,|\alpha(t)\rangle + \langle\alpha(t)|\,\hat{A}(t)\left[\frac{\partial}{\partial t}\,|\alpha(t)\rangle\right],$$

where the partial time derivative of A is over the explicit time dependence of its operator (if any)[12].

In the *Schrödinger picture*, the time evolution of the bra- and ket vectors of the state is described by Eq. (4.158) and its Hermitian conjugate (with $\hat{H}^\dagger = \hat{H}$),

$$-i\hbar\frac{\partial}{\partial t}\langle\alpha(t)| = \langle\alpha(t)|\,\hat{H}(t),$$

so that we get

$$i\hbar\frac{d}{dt}\langle A\rangle(t) - i\hbar\left\langle\frac{\partial A}{\partial t}\right\rangle(t) = -\,\langle\alpha(t)|\,\hat{H}(t)\hat{A}(t)\,|\alpha(t)\rangle + \langle\alpha(t)|\,\hat{A}(t)\hat{H}(t)\,|\alpha(t)\rangle$$
$$\equiv \langle\alpha(t)|\,[\hat{A}(t), \hat{H}(t)]\,\,|\alpha(t)\rangle.$$

Since this result does not include the initial time t_0, it is usually represented in a simpler form, with the time argument t just implied:

$$i\hbar\frac{d}{dt}\langle A\rangle = i\hbar\left\langle\frac{\partial A}{\partial t}\right\rangle + \langle[\hat{A}, \hat{H}]\rangle. \qquad (*)$$

In the Heisenberg picture, the derivation of this result is much simpler: it is sufficient to average both sides of Eq. (4.199)—as usual in this picture, over the statistical ensemble of the initial states of the system.

In the particular but very common case when the operator $\hat{A}$ does not depend on time explicitly, the first term on the right-hand side of Eq. (*) vanishes. If, in

[12] For example, this derivative vanishes for such time-independent operators as $\hat{\mathbf{p}} = -i\hbar\nabla$, ∇^2, $U(\mathbf{r})$, etc, even if the observables they describe do evolve in time.

addition, the commutator of this observable commutes with the Hamiltonian (at the same time instant), $[\hat{A}, \hat{H}] = 0$, then $\langle A \rangle$ remains constant in time. This is the quantum-mechanical analog of the classical integrals of motion, with the commutator playing the role of the Poisson brackets[13].

***Problem* 4.25.** At $t = 0$, a spin-½ system whose interaction with an external field is described by the Hamiltonian

$$\hat{H} = \mathbf{c} \cdot \hat{\boldsymbol{\sigma}} \equiv c_x\hat{\sigma}_x + c_y\hat{\sigma}_y + c_z\hat{\sigma}_z,$$

(where $c_{x,y,z}$ are real and time-independent c-numbers, and $\hat{\sigma}_{x,y,z}$ are the Pauli operators), was in the state $\uparrow$, one of the two eigenstates of the operator $\hat{\sigma}_z$. In the Schrödinger picture, calculate the time evolution of:

(i) the ket-vector $|\alpha\rangle$ of the system (in any time-independent basis you like),
(ii) the probabilities to find the system in the states $\uparrow$ and $\downarrow$, and
(iii) the expectation values of all three Cartesian components ($\hat{S}_x$, etc) of the spin operator $\hat{\mathbf{S}} = (\hbar/2)\hat{\boldsymbol{\sigma}}$.

Analyze and interpret the results for the particular case $c_y = c_z = 0$.

Hint: Think about the best basis to use for the solution.

Solutions: The problem is similar to the spin precession problem that was solved in section 4.6 of the lecture notes as an illustration, but is a bit more complex, because the z-basis is not the eigenbasis for our present Hamiltonian (unless $c_x = c_y = 0$, returning us to the simple precession problem). This is why in the general case we cannot directly use Eqs. (4.161), at least in the z-basis. There are two simple ways to solve the problem, and their comparison is rather instructive.

Approach 1 is to stay in the z-basis (of the $\uparrow$ and $\downarrow$ states), using the fact that in this basis the Hamiltonian operator may be represented with a simple matrix:

$$\mathrm{H} = c_x\sigma_x + c_y\sigma_y + c_z\sigma_z \equiv \begin{pmatrix} c_z & c_- \\ c_+ & -c_z \end{pmatrix}, \quad \text{where } c_\pm \equiv c_x \pm ic_y.$$

Let us start from the calculation of the time-evolution operator $\hat{u}$ of the system. In the z-basis, Eq. (4.157*b*) becomes the matrix equation

$$i\hbar\frac{d}{dt}\begin{pmatrix} u_{11} & u_{12} \\ u_{21} & u_{22} \end{pmatrix} = \begin{pmatrix} c_z & c_- \\ c_+ & -c_z \end{pmatrix}\begin{pmatrix} u_{11} & u_{12} \\ u_{21} & u_{22} \end{pmatrix} \equiv \begin{pmatrix} c_z u_{11} + c_- u_{21} & c_z u_{12} + c_- u_{22} \\ c_+ u_{11} - c_z u_{21} & c_+ u_{12} - c_z u_{22} \end{pmatrix}.$$

This system of four linear ordinary differential equations is actually a set of two independent (and similar) systems of two equations each: one for u_{11} and u_{21}, and another one for u_{12} and u_{22}. Their general solution is straightforward if a bit bulky; but

[13] See Eqs. (4.203)–(4.205) of the lecture notes–or the discussion in the end of *Part CM* section 10.1.

for the initial conditions (4.178) at $t = t_0 = 0$, i.e. for $u_{11} = u_{22} = 1$, and $u_{12} = u_{21} = 0$ at $t = 0$, the solution simplifies to

$$\mathrm{u}(t,\,0) = \begin{pmatrix} \cos\dfrac{ct}{\hbar} - i\dfrac{c_z}{c}\sin\dfrac{ct}{\hbar} & -i\dfrac{c_-}{c}\sin\dfrac{ct}{\hbar} \\ -i\dfrac{c_+}{c}\sin\dfrac{ct}{\hbar} & \cos\dfrac{ct}{\hbar} + i\dfrac{c_z}{c}\sin\dfrac{ct}{\hbar} \end{pmatrix}, \qquad (*)$$

where the scalar c is the length of the geometric vector $\mathbf{c}$:

$$c^2 \equiv c_x^2 + c_y^2 + c_z^2 \equiv c_+c_- + c_z^2.$$

(By the way, noting that the matrix (*) may be represented in the form

$$\mathrm{u}(t,\,0) = \begin{pmatrix} 1 & 0 \\ 0 & 1 \end{pmatrix}\cos\frac{ct}{\hbar} - \frac{i}{c}\begin{pmatrix} c_z & c_- \\ c_+ & -c_z \end{pmatrix}\sin\frac{ct}{\hbar} \equiv \mathrm{I}\cos\frac{ct}{\hbar} - \frac{i}{c}\mathbf{c}\cdot\boldsymbol{\sigma}\sin\frac{ct}{\hbar}$$

$$\equiv \mathrm{I}\cos\frac{ct}{\hbar} - \frac{i}{c}\mathrm{H}\sin\frac{ct}{\hbar},$$

we can guess that the same equality is also valid for the corresponding operators:

$$\hat{u}\,(t,\,0) = \hat{I}\,\cos\frac{ct}{\hbar} - \frac{i}{c}\hat{H}\,\sin\frac{ct}{\hbar},$$

independently of the basis. This fact (which is not used in this solution, but may be useful in other cases, including the next problem) may be indeed proved in two ways: either by the direct substitution of the last relation into Eq. (4.157*b*), for our particular Hamiltonian, or by the Taylor expansion of its solution (4.175), with $t_0 = 0$. My strong recommendation to the reader is to work out the both proofs, as an additional exercise.)

Now the remaining calculations are easy, using the fact that in this basis the initial state is represented by a simple row/column matrix (1, 0).

(i) According to Eq. (4.157*a*), the ket-vector of the system evolves as

$$|\,\alpha(t)\rangle = \hat{u}\,(t,\,0)|\,\alpha(0)\rangle,$$

where in our case $|\alpha(0)\rangle = |\uparrow\rangle$, so that in the z-basis,

$$\begin{pmatrix} \alpha_\uparrow(t) \\ \alpha_\downarrow(t) \end{pmatrix} = \mathrm{u}(t,\,0)\begin{pmatrix} \alpha_\uparrow(0) \\ \alpha_\downarrow(0) \end{pmatrix} = \begin{pmatrix} \cos\dfrac{ct}{\hbar} - i\dfrac{c_z}{c}\sin\dfrac{ct}{\hbar} & -i\dfrac{c_-}{c}\sin\dfrac{ct}{\hbar} \\ -i\dfrac{c_+}{c}\sin\dfrac{ct}{\hbar} & \cos\dfrac{ct}{\hbar} + i\dfrac{c_z}{c}\sin\dfrac{ct}{\hbar} \end{pmatrix}\begin{pmatrix} 1 \\ 0 \end{pmatrix}$$

$$\equiv \begin{pmatrix} \cos\dfrac{ct}{\hbar} - i\dfrac{c_z}{c}\sin\dfrac{ct}{\hbar} \\ -i\dfrac{c_+}{c}\sin\dfrac{ct}{\hbar} \end{pmatrix}.$$

(ii) According to Eq. (4.120), the probabilities to find the spin in ↑ and ↓ states are, respectively,

$$W_\uparrow = |\alpha_\uparrow(t)|^2 = \cos^2\frac{ct}{\hbar} + \frac{c_z^2}{c^2}\sin^2\frac{ct}{\hbar}, \qquad W_\downarrow = |\alpha_\downarrow(t)|^2 = \frac{c_x^2 + c_y^2}{c^2}\sin^2\frac{ct}{\hbar},$$

readily withstanding the sanity check: $W_\uparrow(t) + W_\downarrow(t) \equiv 1$. Notice that the time evolution matrix elements, and hence the state vectors, oscillate with the frequency $c/\hbar$, which is twice lower than the classical precession frequency $\Omega \equiv 2c/\hbar$, while all the observables oscillate with the full frequency Ω.

(iii) Since the operator $\hat{S}_z$ is diagonal in the z-basis, we can readily find its expectation value using the first form of Eq. (4.124):

$$\langle S_z\rangle = \left(+\frac{\hbar}{2}\right)W_\uparrow + \left(-\frac{\hbar}{2}\right)W_\downarrow = \frac{\hbar}{2}\left(\cos^2\frac{ct}{\hbar} + \frac{c_z^2 - c_x^2 - c_y^2}{c^2}\sin^2\frac{ct}{\hbar}\right)$$
$$\equiv \frac{\hbar}{2}\left(1 - \frac{2c_+c_-}{c^2}\sin^2\frac{ct}{\hbar}\right).$$

However, for other spin components we better use the general rule (4.125). Actually, for S_x most of the calculation has already been carried out in section 4.6 of the lecture notes—see the first form of Eq. (4.171):

$$\langle S_x\rangle = \frac{\hbar}{2}\left[\alpha_\uparrow(t)\alpha_\downarrow^*(t) + \alpha_\downarrow(t)\alpha_\uparrow^*(t)\right] = \hbar\left(\frac{c_y}{c}\cos\frac{ct}{\hbar} + \frac{c_xc_z}{c^2}\sin\frac{ct}{\hbar}\right)\sin\frac{ct}{\hbar}.$$

Carrying out a similar calculation for S_y, we get

$$\langle S_y\rangle = \langle\alpha(t)|\,\hat{S}_y\,|\,\alpha(t)\rangle = \frac{\hbar}{2}\begin{pmatrix}\alpha_\uparrow^* & \alpha_\downarrow^*\end{pmatrix}\begin{pmatrix}0 & -i\\ i & 0\end{pmatrix}\begin{pmatrix}\alpha_\uparrow\\ \alpha_\downarrow\end{pmatrix}$$
$$= \frac{\hbar}{2}i\left[\alpha_\uparrow(t)\alpha_\downarrow^*(t) - \alpha_\downarrow(t)\alpha_\uparrow^*(t)\right] = \hbar\left(-\frac{c_x}{c}\cos\frac{ct}{\hbar} + \frac{c_yc_z}{c^2}\sin\frac{ct}{\hbar}\right)\sin\frac{ct}{\hbar}.$$

For the particular case specified in the end of the assignment, $c_y = c_z = 0$, the precession's half-frequency is determined by the only nonvanishing Cartesian component of the field: $\Omega/2 = c_x/\hbar$, and the above results acquire a very simple form:

$$\mathrm{u}(t, 0) = \begin{pmatrix}\cos\dfrac{\Omega t}{2} & -i\sin\dfrac{\Omega t}{2}\\ -i\sin\dfrac{\Omega t}{2} & \cos\dfrac{\Omega t}{2}\end{pmatrix};$$

$$\langle S_x\rangle = 0, \qquad \langle S_y\rangle = -\frac{\hbar}{2}\sin\Omega t, \qquad \langle S_z\rangle = \frac{\hbar}{2}\cos\Omega t.$$

Comparing these formulas with Eqs. (4.173) and (4.174) of the lecture notes, we see that our results describe another particular case of the same spin precession about the direction of the field vector $\mathbf{c} = \{c_x, c_y, c_z\}$—now oriented along the x-axis rather than the z-axis. (A generalization of this result to the arbitrary field direction will be discussed in section 5.1 of the lecture notes.)

Approach 2 is to use the relations discussed in section 4.4 of the lecture notes to perform the unitary transform[14] from the z-basis to the eigenbasis of the Hamiltonian (at $t = 0$), then use the fact of the very simple time evolution (4.161) of the system in that 'new' basis, and finally perform the reciprocal transformation (at arbitrary $t \geqslant 0$). Though such a program may look rather involved, the calculations are in fact simpler than those of approach 1.

Indeed, let us call the Hamiltonian's eigenstates a_1 and a_2, keeping our usual notation, $\uparrow$ and $\downarrow$, for the states of the 'old' z-basis, and 'find' the eigenstates of the spin-½ in the field, i.e. calculate the coefficients of their expansion in the z-basis[15]:

$$|a_1\rangle = \langle\uparrow|a_1\rangle\,|\uparrow\rangle + \langle\downarrow|a_1\rangle\,|\downarrow\rangle, \qquad |a_2\rangle = \langle\uparrow|a_2\rangle\,|\uparrow\rangle + \langle\downarrow|a_2\rangle\,|\downarrow\rangle. \qquad (**)$$

For that, we need to solve the linear system of Eq. (4.101) (with $A_{kk'}$ replaced with $H_{kk'}$, and the 'old' basis $\{u_k\}$ now consisting of just two states $\uparrow$ and $\downarrow$), for each of two energy eigenvalues E_j: $E_1 = +c$ and $E_2 = -c$. For the first of them (in the notation of section 4.4, $j = 1$), the system is

$$\begin{aligned} k = 1: &\quad (c_z - c)\langle\uparrow|a_1\rangle + c_-\langle\downarrow|a_1\rangle = 0, \\ k = 2: &\quad c_+\langle\uparrow|a_1\rangle + (-c_z - c)\langle\downarrow|a_1\rangle = 0. \end{aligned}$$

Since c is an eigenvalue of this system, these two equations are compatible, and we can use any of them, for example, the second one, giving

$$\langle\downarrow|a_1\rangle = \frac{c_+}{c + c_z}\langle\uparrow|a_1\rangle.$$

Besides this linear relation, these two coefficients (which are of course nothing other than elements of the unitary transform matrix—see Eqs. (4.82)–(4.84) of the lecture notes), should also satisfy the normalization relation (4.104):

$$|\langle\uparrow|a_1\rangle|^2 + |\langle\downarrow|a_1\rangle|^2 = 1.$$

Solving these two equations together, selecting the arbitrary phase factor in the simplest way, and taking into account that $|c_\pm|^2 = c_+c_- = c_x^2 + c_y^2 = c^2 - c_z^2$, and $c \geqslant c_z$, so that

$$|c_\pm|^2 + |c \pm c_z|^2 = \left(c_x^2 + c_y^2\right) + \left(c^2 \pm 2cc_z + c_z^2\right) = 2c^2 \pm 2cc_z = 2c(c \pm c_z),$$

we readily get

$$\langle\uparrow|a_1\rangle = \frac{c + c_z}{[2c(c + c_z)]^{1/2}}, \qquad \langle\downarrow|a_1\rangle = \frac{c_+}{[2c(c + c_z)]^{1/2}}.$$

[14] The time-independent unitary matrix U of this transform should not be confused with the time-dependent unitary matrix u(t, 0) of the time-evolution operator.

[15] Here, 'for variety' (actually, for training the reader to use this popular alternative notation), I use short brackets to denote the elements of the unitary transform matrix U—see Eqs. (4.82) and (4.83) of the lecture notes.

Performing an absolutely similar calculation for $E_2 = -c$ ($j = 2$), we get a very similar result for the two remaining coefficients of the transform matrix:

$$\langle\uparrow|a_2\rangle = -\frac{c_-}{[2c(c+c_z)]^{1/2}}, \quad \langle\downarrow|a_2\rangle = \frac{c+c_z}{[2c(c+c_z)]^{1/2}}.$$

The first good sanity check is that in the z-oriented 'field' $\mathbf{c}$ (with $c_x = c_y = 0$, and $c_z = \pm c$), the transform matrix becomes diagonal, because each of the eigenstates $a_{1,2}$ coincides with one of the z-states, either $\uparrow$ or $\downarrow$, depending on the sign of c_z, i.e. on the field direction. Second, in a 'horizontal' field (with, say $c_y = c_z = 0$, and $c = c_x$), all the matrix elements are equal to $\pm 1/\sqrt{2}$, in accordance with Eq. (4.113).

Now we can use Eqs. (**), which are valid for any time moment, to calculate the initial state of the system in the basis of the eigenstates $a_{1,2}$. Since the system was initially entirely in the spin-up state ($\uparrow$), we get

$$\langle a_1|\alpha(0)\rangle = \langle a_1|\uparrow\rangle = \langle\uparrow|a_1\rangle^* = \frac{c+c_z}{[2c(c+c_z)]^{1/2}},$$

$$\langle a_2|\alpha(0)\rangle = \langle a_2|\uparrow\rangle = \langle\uparrow|a_2\rangle^* = -\frac{c_+}{[2c(c+c_z)]^{1/2}}.$$

According to Eq. (4.161), the time evolution of these matrix elements is reduced to their multiplication by simple phase factors $\exp\{-iE_j t/\hbar\} = \exp\{\mp ict/\hbar\}$:

$$\langle a_1|\alpha(t)\rangle \equiv \alpha_1(t) = \frac{c+c_z}{[2c(c+c_z)]^{1/2}}\exp\left\{-i\frac{ct}{\hbar}\right\},$$

$$\langle a_2|\alpha(t)\rangle \equiv \alpha_2(t) = -\frac{c_+}{[2c(c+c_z)]^{1/2}}\exp\left\{i\frac{ct}{\hbar}\right\}.$$

Now we may return to the z-basis, using the unitary transform reciprocal to Eqs. (**):

$$\begin{aligned}\alpha_\uparrow(t) \equiv \langle\uparrow|\alpha(t)\rangle &= \langle\uparrow|a_1\rangle\langle a_1|\alpha(t)\rangle + \langle\uparrow|a_2\rangle\langle a_2|\alpha(t)\rangle \\ &= \frac{(c+c_z)^2}{2c(c+c_z)}\exp\left\{-i\frac{ct}{\hbar}\right\} + \frac{c_+c_-}{2c(c+c_z)}\exp\left\{i\frac{ct}{\hbar}\right\} \\ &\equiv \cos\frac{ct}{\hbar} - i\frac{c_z}{c}\sin\frac{ct}{\hbar},\end{aligned}$$

$$\begin{aligned}\alpha_\downarrow(t) \equiv \langle\downarrow|\alpha(t)\rangle &= \langle\downarrow|a_1\rangle\langle a_1|\alpha(t)\rangle + \langle\downarrow|a_2\rangle\langle a_2|\alpha(t)\rangle \\ &= \frac{c_+}{[2c(c+c_z)]^{1/2}}\frac{c+c_z}{[2c(c+c_z)]^{1/2}}\exp\left\{-i\frac{ct}{\hbar}\right\} \\ &\quad - \frac{c-c_z}{[2c(c+c_z)]^{1/2}}\frac{c_+}{[2c(c+c_z)]^{1/2}}\exp\left\{i\frac{ct}{\hbar}\right\} \equiv -i\frac{c_+}{c}\sin\frac{ct}{\hbar}.\end{aligned}$$

This is exactly the same result as has been obtained using approach 1. Now we can carry out tasks (ii) and (iii) exactly as was done above.

***Problem* 4.26.** For the same system as in the previous problem, use the Heisenberg picture to calculate the time evolution of:

(i) all three Cartesian components of the Heisenberg spin operator $\hat{\mathbf{S}}_H(t)$, and
(ii) the expectation values of the spin components.

Compare the latter results with those of the previous problem.

Solutions:

(i) With the solutions of the previous problem on hand, one way to proceed is to use the result for time evolution matrix in the z-basis:

$$\mathrm{u}(t,\ 0) = \begin{pmatrix} \cos\frac{ct}{\hbar} - i\frac{c_z}{c}\sin\frac{ct}{\hbar} & -i\frac{c_-}{c}\sin\frac{ct}{\hbar} \\ -i\frac{c_+}{c}\sin\frac{at}{\hbar} & \cos\frac{ct}{\hbar} + i\frac{c_z}{c}\sin\frac{ct}{\hbar} \end{pmatrix},$$

$$\mathrm{u}^\dagger(t,\ 0) = \begin{pmatrix} \cos\frac{ct}{\hbar} + i\frac{c_z}{c}\sin\frac{ct}{\hbar} & i\frac{c_-}{c}\sin\frac{ct}{\hbar} \\ i\frac{c_+}{c}\sin\frac{ct}{\hbar} & \cos\frac{ct}{\hbar} - i\frac{c_z}{c}\sin\frac{ct}{\hbar} \end{pmatrix},$$

where

$$c_\pm \equiv c_x \pm ic_y, \quad c \equiv \left(c_x^2 + c_y^2 + c_z^2\right)^{1/2},$$

and Eqs. (4.190) of the lecture notes, to write

$$\mathbf{S}_H(t) = \mathrm{u}^\dagger(t,\ 0)\mathbf{S}_S\mathrm{u}(t,\ 0) \equiv \mathrm{u}^\dagger(t,\ 0)\mathbf{S}_H(\mathbf{0})\mathrm{u}(t,\ 0)$$

$$= \begin{pmatrix} \cos\frac{ct}{\hbar} + i\frac{c_z}{c}\sin\frac{ct}{\hbar} & i\frac{c_-}{c}\sin\frac{ct}{\hbar} \\ i\frac{c_+}{c}\sin\frac{ct}{\hbar} & \cos\frac{ct}{\hbar} - i\frac{c_z}{c}\sin\frac{ct}{\hbar} \end{pmatrix}\frac{\hbar}{2}\boldsymbol{\sigma}\begin{pmatrix} \cos\frac{ct}{\hbar} - i\frac{c_z}{c}\sin\frac{ct}{\hbar} & -i\frac{c_-}{c}\sin\frac{ct}{\hbar} \\ -i\frac{c_+}{c}\sin\frac{ct}{\hbar} & \cos\frac{ct}{\hbar} + i\frac{c_z}{c}\sin\frac{ct}{\hbar} \end{pmatrix},$$

where $\boldsymbol{\sigma}$ is the Pauli matrix vector—see Eq. (4.117). A straightforward though a bit bulky matrix multiplication yields

$$(S_x)_{11} = \frac{\hbar}{2}\sin\frac{ct}{\hbar}\left(\frac{2c_xc_z}{c^2}\sin\frac{ct}{\hbar} + \frac{2c_y}{c}\cos\frac{ct}{\hbar}\right),$$

$$(S_y)_{11} = \frac{\hbar}{2}\sin\frac{ct}{\hbar}\left(\frac{2c_z c_y}{c^2}\sin\frac{ct}{\hbar} - \frac{2c_x}{c}\cos\frac{ct}{\hbar}\right),$$

$$(S_z)_{11} = \frac{\hbar}{2}\left(1 - \frac{2c_+c_-}{c^2}\sin^2\frac{ct}{\hbar}\right),$$

and similarly for two other matrix elements.

Alternatively, we could use the fact (see the remark in the model solution of the previous problem) that the evolution matrix in this problem may be represented as

$$\mathrm{u}(t, 0) = \mathrm{I}\cos\frac{ct}{\hbar} - i\frac{\mathbf{c}\cdot\boldsymbol{\sigma}}{c}\sin\frac{ct}{\hbar},$$

and make the whole calculation in the vector form, using the multiplication and commutation rules for Pauli matrices, which we already know from the solution of problem 4.3.

One more option to derive the same result is by solving the differential equation (4.199) for the spin operator evolution, just as was done in the lecture notes for the simple particular case $c_x = c_y = 0$—see Eqs. (4.200)–(4.202).

(ii) Since the initial state of the system was ↑, according to Eq. (4.191) the expectation value of the spin is

$$\langle \mathbf{S}\rangle(t) = (1,\ 0)\begin{pmatrix}\mathbf{S}_{11}(t) & \mathbf{S}_{12}(t)\\ \mathbf{S}_{21}(t) & \mathbf{S}_{22}(t)\end{pmatrix}\begin{pmatrix}1\\0\end{pmatrix} \equiv \mathbf{S}_{11}(t).$$

As a result, we get:

$$\langle \mathbf{S}\rangle = \frac{\hbar}{2}\left[\mathbf{n}_x(S_x)_{11} + \mathbf{n}_y(S_y)_{11} + \mathbf{n}_z(S_z)_{11}\right]$$

$$= \frac{\hbar}{2}\left[\begin{matrix}\mathbf{n}_x 2\sin\frac{ct}{\hbar}\left(\frac{c_z c_x}{c^2}\sin\frac{ct}{\hbar} + \frac{c_y}{c}\cos\frac{ct}{\hbar}\right)\\ +\mathbf{n}_y 2\sin\frac{ct}{\hbar}\left(\frac{c_z c_y}{c^2}\sin\frac{ct}{\hbar} - \frac{c_x}{c}\cos\frac{ct}{\hbar}\right) + \mathbf{n}_z\left(1 - \frac{2c_+c_-}{c^2}\sin^2\frac{ct}{\hbar}\right)\end{matrix}\right],$$

i.e. (very fortunately :-) the same result as from the Schrödinger picture—see the solution of the previous problem.

***Problem* 4.27.** For the same system as in the two previous problems, calculate the matrix elements of the operator $\hat{\sigma}_z$ in the basis of the stationary states of the system.

Solution: The calculation is most straightforward in the z-basis {↑, ↓}, in which the operator $\hat{\sigma}_z$ has the simplest form—see the last of Eqs. (4.105) of the lecture notes:

$$(\sigma_z)_{11} \equiv \langle a_1|\hat{\sigma}_z|a_1\rangle = (\langle\uparrow|a_1\rangle^*,\ \langle\downarrow|a_1\rangle^*)\begin{pmatrix}1 & 0\\ 0 & -1\end{pmatrix}\begin{pmatrix}\langle\uparrow|a_1\rangle\\ \langle\downarrow|a_1\rangle\end{pmatrix} \equiv |\langle\uparrow|a_1\rangle|^2 - |\langle\downarrow|a_1\rangle|^2.$$

According to Eqs. (4.82)–(4.84), the brackets participating in this expression are just the elements of the unitary matrix of transfer between the stationary-state basis $\{a_1, a_2\}$ and the z-basis, and had already been calculated using approach 2 in the model solution of problem 4.25. Plugging them in, we get a very simple expression:

$$(\sigma_z)_{11} = \left|\frac{c + c_z}{[2c(c + c_z)]^{1/2}}\right|^2 - \left|\frac{c_+}{[2c(c + c_z)]^{1/2}}\right|^2 \equiv \frac{c + c_z}{2c} - \frac{c_+c_-}{2c(c + c_z)} \equiv \frac{c_z}{c},$$

where $c \equiv (c_x^{\ 2} + c_y^{\ 2} + c_z^{\ 2})^{1/2}$, and $c_\pm \equiv c_x \pm ic_y$, so that $c_+c_- = c_x^{\ 2} + c_y^{\ 2} = c^2 - c_z^{\ 2}$.

This result readily passes two key sanity checks. First, if $c_z = \pm c$ (i.e. $c_x = c_y = 0$), then $(\sigma_z)_{11} = \pm 1$, as it should be, because in this case the basis $\{a_1, a_2\}$ coincides either with the z-basis $\{\uparrow, \downarrow\}$, so that the answer is given by one of the diagonal elements of familiar matrix σ_z in the z-basis. Second, if only c_x is different from zero, i.e. $c_z = 0$, our result yields $(\sigma_z)_{11} = 0$. This is also what we should have, because the stationary states $\{a_1, a_2\}$ in such a 'horizontal field' **c** are the eigenstates of the corresponding 'horizontal operator' $\hat{\sigma}_x$ with the eigenstates described by Eqs. (4.122). In this basis, $(\sigma_z)_{11}$ evidently should vanish.

The expression for the second diagonal matrix element is similar,

$$(\sigma_z)_{22} = |\langle\uparrow|a_2\rangle|^2 - |\langle\downarrow|a_2\rangle|^2 = -\frac{c_z}{c},$$

and those for the off-diagonal elements may be also calculated absolutely similarly:

$$(\sigma_z)_{12} = (\langle\uparrow|a_1\rangle^*, \langle\downarrow|a_1\rangle^*)\begin{pmatrix}1 & 0\\ 0 & -1\end{pmatrix}\begin{pmatrix}\langle\uparrow|a_2\rangle\\ \langle\downarrow|a_2\rangle\end{pmatrix} = \langle\uparrow|a_1\rangle^*\langle\uparrow|a_2\rangle - \langle\downarrow|a_1\rangle^*\langle\downarrow|a_2\rangle$$

$$= -\frac{c + c_z}{[2c(c + c_z)]^{1/2}}\frac{c_-}{[2c(c + c_z)]^{1/2}} - \frac{c_-}{[2c(c + c_z)]^{1/2}}\frac{c + c_z}{[2c(c + c_z)]^{1/2}} \equiv -\frac{c_-}{c},$$

$$(\sigma_z)_{21} = (\sigma_z)_{12}^* = -\frac{c_+}{c}.$$

(The last step has used the fact that the operator $\hat{\sigma}_z$ is Hermitian.)

Note that because of the arbitrary phase factors in the unitary matrices, the expressions for the off-diagonal elements may be simultaneously multiplied by complex-conjugate factors $\exp\{\pm i\varphi\}$ with an arbitrary real phase φ. Note also that in contrast to the two previous problems, the results of this solution do not depend on the initial state of the system.

***Problem* 4.28.** In the Schrödinger picture of quantum dynamics, certain three operators satisfy the following commutation relation:

$$[\hat{A}, \hat{B}] = \hat{C}.$$

What is their relation in the Heisenberg picture, at a certain time instant t?

Solution: Using the definition of the commutator, then Eq. (4.190) of the lecture notes, and finally the unitary property (4.76) of the time-evolution operator, for the Heisenberg-picture operators (indicated by their dependence on time) we may write

$$\begin{aligned}
[\hat{A}(t), \hat{B}(t)] &\equiv \hat{A}(t)\hat{B}(t) - \hat{B}(t)\hat{A}(t) \\
&= \hat{u}^\dagger(t, t_0)\hat{A}(t_0)\hat{u}(t, t_0)\hat{u}^\dagger(t, t_0)\hat{B}(t_0)\hat{u}(t, t_0) \\
&\quad - \hat{u}^\dagger(t, t_0)\hat{B}(t_0)\hat{u}(t, t_0)\hat{u}^\dagger(t, t_0)\hat{A}(t_0)\hat{u}(t, t_0) \\
&= \hat{u}^\dagger(t, t_0)\hat{A}(t_0)\hat{B}(t_0)\hat{u}(t, t_0) - \hat{u}^\dagger(t, t_0)\hat{B}(t_0)\hat{A}(t_0)\hat{u}(t, t_0) \\
&\equiv \hat{u}^\dagger(t, t_0)[\hat{A}(t_0), \hat{B}(t_0)]\, \hat{u}(t, t_0)\,.
\end{aligned}$$

But by the definition (4.190) of the Heisenberg operators, for $t = t_0$, they are just the Schrödinger-picture operators, so that

$$\begin{aligned}
[\hat{A}(t), \hat{B}(t)] &= \hat{u}^\dagger(t, t_0)[\hat{A}, \hat{B}]\, \hat{u}(t, t_0) = \hat{u}^\dagger(t, t_0)\hat{C}\hat{u}(t, t_0) \equiv \hat{u}^\dagger(t, t_0)\hat{C}(t_0)\hat{u}(t, t_0) \\
&\equiv \hat{C}(t).
\end{aligned}$$

Hence, the commutation relation between operators does not depend on the picture they are all considered in. (This conclusion is valid for the interaction picture as well.) Note, however, that this is always true only if the time arguments of the operators coincide—see, e.g. a counter-example in a footnote in section 4.6 of the lecture notes.

***Problem* 4.29.** Prove the Bloch theorem, given by either Eq. (3.107) or Eq. (3.108) of the lecture notes.

Hint: Consider the *translation operator* $\hat{\mathcal{T}}_{\mathbf{R}}$, defined by the following result of its action on an arbitrary function $f(\mathbf{r})$:

$$\hat{\mathcal{T}}_{\mathbf{R}} f(\mathbf{r}) = f(\mathbf{r} + \mathbf{R}),$$

for the case when $\mathbf{R}$ is an arbitrary vector of the Bravais lattice (3.106). In particular, analyze the commutation properties of this operator, and apply them to an eigenfunction $\psi(\mathbf{r})$ of the stationary Schrödinger equation of a particle moving in the 3D periodic potential described by Eq. (3.105).

Solution: Let us act by the translation operator on the function $\hat{H}(\mathbf{r})f(\mathbf{r})$, there the Hamiltonian is $\mathbf{R}$-periodic in the sense of Eq. (3.105), and the function $f(\mathbf{r})$ is so far arbitrary, i.e. may or may not be periodic. The result is

$$\hat{\mathcal{T}}_{\mathbf{R}}\hat{H}(\mathbf{r})f(\mathbf{r}) = \hat{\mathcal{T}}_{\mathbf{R}}\big(\hat{H}(\mathbf{r})f(\mathbf{r})\big) = \hat{H}(\mathbf{r} + \mathbf{R})f(\mathbf{r} + \mathbf{R}) = \hat{H}(\mathbf{r})f(\mathbf{r} + \mathbf{R}) = \hat{H}(\mathbf{r})\hat{\mathcal{T}}_{\mathbf{R}} f(\mathbf{r}).$$

Since $f(\mathbf{r})$ is arbitrary, the comparison of the first and the last forms of this transformation chain shows that the operators $\hat{\mathcal{T}}_{\mathbf{R}}$ and $\hat{H}$ commute; hence, according to the discussion in section 4.5 of the lecture notes, they have to share their eigenfunctions $\psi(\mathbf{r})$:

$$\hat{H}\psi = E\psi, \qquad \hat{\mathcal{T}}_{\mathbf{R}}\psi = \tau(\mathbf{R})\psi,$$

where E and $\tau(\mathbf{R})$ are the corresponding eigenvalues. Without any restrictions, $\tau(\mathbf{R})$ may be taken in the form

$$\tau(\mathbf{R}) = e^{iF(\mathbf{R})}, \qquad (*)$$

where $F(\mathbf{R})$ may be a complex function.

Now, a successive action of two translation operators on any function evidently results in the net shift of its argument:

$$\hat{\mathcal{T}}_{\mathbf{R}}\hat{\mathcal{T}}_{\mathbf{R}'}f(\mathbf{r}) = \hat{\mathcal{T}}_{\mathbf{R}'}\hat{\mathcal{T}}_{\mathbf{R}}f(\mathbf{r}) = \hat{\mathcal{T}}_{\mathbf{R}+\mathbf{R}'}f(\mathbf{r}). \qquad (**)$$

Applying Eqs. (*) and (**) to a joint eigenfunction $\psi(\mathbf{r})$ of the operators $\hat{\mathcal{T}}_{\mathbf{R}}$ and $\hat{H}$, we may get any of the following two expressions:

$$\hat{\mathcal{T}}_{\mathbf{R}+\mathbf{R}'}\psi(\mathbf{r}) = \begin{cases} \hat{\mathcal{T}}_{\mathbf{R}}\hat{\mathcal{T}}_{\mathbf{R}'}\psi(\mathbf{r}) = \hat{\mathcal{T}}_{\mathbf{R}}e^{iF(\mathbf{R}')}\psi(\mathbf{r}) = e^{iF(\mathbf{R}')}\hat{\mathcal{T}}_{\mathbf{R}}\psi(\mathbf{r}) = e^{iF(\mathbf{R}')}e^{iF(\mathbf{R})}\psi(\mathbf{r}), \\ e^{iF(\mathbf{R}'+\mathbf{R})}\psi(\mathbf{r}) . \end{cases}$$

Hence the function $F(\mathbf{R})$ has to satisfy the following relation:

$$e^{iF(\mathbf{R}')}e^{iF(\mathbf{R})} = e^{iF(\mathbf{R}'+\mathbf{R})},$$

for an arbitrary choice of **R** and **R'** (from the Bravais lattice set). This is only possible if it is a linear function of each Cartesian component of **R**, with certain coefficients q_j:

$$F(\mathbf{R}) = q_x X + q_y Y + q_z Z = \mathbf{q}\cdot\mathbf{R}, \quad \text{i.e. if } \hat{\mathcal{T}}_{\mathbf{R}}\psi(\mathbf{r}) \equiv \psi(\mathbf{r}+\mathbf{R}) = \psi(\mathbf{r})\, e^{i\mathbf{q}\cdot\mathbf{R}}.$$

This is the Bloch theorem in the form (3.107). The only detail to add is to notice that the particle's probability density $\psi(\mathbf{r})\psi^*(\mathbf{r})$ should be periodic in the sense of Eq. (3.105), giving the requirement

$$\psi(\mathbf{r}+\mathbf{R})\psi^*(\mathbf{r}+\mathbf{R}) \equiv \psi(\mathbf{r})e^{i\mathbf{q}\cdot\mathbf{R}}\psi^*(\mathbf{r})e^{-i\mathbf{q}^*\cdot\mathbf{R}} = \psi(\mathbf{r})\psi^*(\mathbf{r}).$$

This gives the following condition: $\exp\{i(\mathbf{q}-\mathbf{q}^*)\cdot\mathbf{R}\} = 1$, which may be satisfied (for arbitrary **R**) only if the quasi-momentum is real: $\mathbf{q}^* = \mathbf{q}$.[16]

***Problem* 4.30.** A constant force F is applied to an (otherwise free) 1D particle of mass m. Calculate the eigenfunctions of the problem in:

(i) the coordinate representation, and
(ii) the momentum representation.

Discuss the relation between the results.

[16] This conclusion may be also made from the explicit expressions (3.111) for the primitive vectors of the reciprocal Bravais lattice (3.109).

Solutions:

(i) In the *coordinate* representation, the Hamiltonian of the particle,

$$\hat{H} = \frac{\hat{p}^2}{2m} - F\hat{x}, \qquad (*)$$

may be rewritten as

$$\hat{H} = -\frac{\hbar^2}{2m}\frac{\partial^2}{\partial x^2} - Fx.$$

so that the corresponding stationary Schrödinger equation is

$$\left(-\frac{\hbar^2}{2m}\frac{d^2}{dx^2} - Fx\right)\psi = E\psi.$$

This is just the Airy equation, which was discussed in section 2.4 of the lecture notes, and may be reduced to its canonical form (2.101),

$$\frac{d^2\psi}{d\zeta^2} - \zeta\psi = 0,$$

using the following dimensionless variable[17]

$$\zeta \equiv -\left(\frac{2mF}{\hbar^2}\right)^{1/3}\left(x + \frac{E}{F}\right). \qquad (**)$$

As was discussed in section 2.4, the fundamental solutions of this differential equation are the Airy functions $\mathrm{Ai}(\zeta)$ and $\mathrm{Bi}(\zeta)$—see figure 2.9a, reproduced in the figure below. Of them, only $\mathrm{Ai}(\zeta)$ is finite at all ζ, and since our wavefunction $\psi(\zeta)$ is defined on the whole axis $-\infty < \zeta < +\infty$, it has to be proportional to this function alone:

$$\psi(\zeta) = C_x\mathrm{Ai}(\zeta),$$

so that

$$\psi(x) = C_x\mathrm{Ai}\left[\left(\frac{2m}{\hbar^2F^2}\right)^{1/3}(E + Fx)\right]. \qquad (***)$$

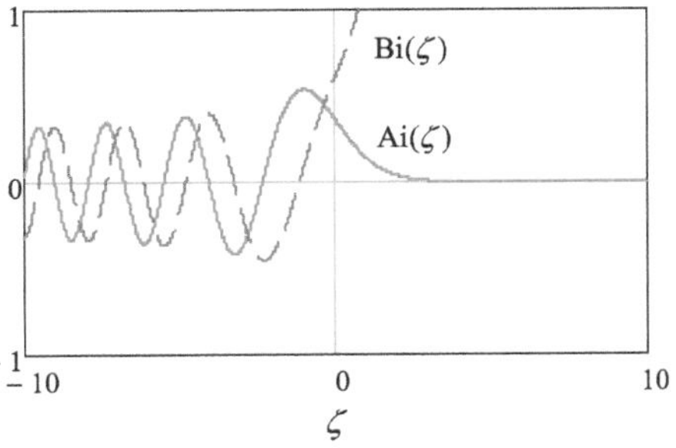

[17] Cf. Eq. (2.100) of the lecture notes, which differs only by the sign of $dU/dx = -F$ and hence by the opposite sign of x, selected to keep the expression under the cubic root positive.

(ii) In the *momentum* representation, we may use the first of Eqs. (4.269) to rewrite Eq. (*) as

$$\hat{H} = \frac{p^2}{2m} - i\hbar F\frac{\partial}{\partial p}.$$

The corresponding stationary Schrödinger equation for the wavefunction $\varphi(p)$, defined by Eq. (4.266), is

$$\left(\frac{p^2}{2m} - i\hbar F\frac{d}{dp}\right)\varphi = E\varphi.$$

Rewriting this equation in the variable-separated form,

$$i\hbar F\frac{d\varphi}{\varphi} = \left(\frac{p^2}{2m} - E\right)dp,$$

we see that it may be readily integrated, giving

$$i\hbar F \ln \varphi = \int\left(\frac{p^2}{2m} - E\right)dp = \left(\frac{p^3}{6m} - Ep\right) + \text{const},$$
$$\text{so that} \quad \varphi(p) = C_p \exp\left\{-\frac{i}{\hbar F}\left(\frac{p^3}{6m} - Ep\right)\right\}. \qquad (****)$$

Now let us verify that the functions (***) and (****) are indeed related with the Fourier transform (see Eq. (4.264) of the lecture notes):

$$\psi(x) = \frac{1}{(2\pi\hbar)^{1/2}}\int_{-\infty}^{+\infty} \varphi(p)\exp\left\{i\frac{px}{\hbar}\right\}dp.$$

Plugging Eq. (****) into this integral, and then introducing the dimensionless integration variable $\xi \equiv (2m\hbar F)^{-1/3}p$, we get

$$\begin{aligned}\psi(x) &= \frac{C_p}{(2\pi\hbar)^{1/2}}\int_{-\infty}^{+\infty} \exp\left\{i\left[\frac{px}{\hbar} - \frac{1}{\hbar F}\left(\frac{p^3}{6m} - Ep\right)\right]\right\}dp \\ &\equiv \frac{C_p(2m\hbar F)^{1/3}}{(2\pi\hbar)^{1/2}}2\pi\left[\frac{1}{\pi}\int_0^{+\infty} \cos\left(\frac{\xi^3}{3} + \zeta\xi\right)d\xi\right],\end{aligned}$$

where $\xi \equiv p/(2m\hbar F)^{1/3}$, and ζ is defined by Eq. (**).

But as was mentioned in section 2.4 of the lecture notes, the expression in the last square brackets is just the integral form of the function $\mathrm{Ai}(\zeta)$, so that this function $\psi(x)$ is proportional to the one given by Eq. (***). Hence, at the proper relation between their normalization coefficients C_x and C_p, the functions $\psi(x)$ and $\varphi(p)$ are indeed just the spatial Fourier images of each other.

***Problem* 4.31.** Use the momentum representation to re-solve the problem discussed in the beginning of section 2.6 of the lecture notes, i.e. calculate the eigenenergy of a 1D particle of mass m, localized in a very short potential well, of the 'area' $\mathscr{W}$.

Solution: The Hamiltonian of the system is

$$\hat{H} = \frac{\hat{p}^2}{2m} + U(\hat{x}), \quad \text{where } U(x) = -\mathscr{W}\,\delta(x), \quad \text{with } \mathscr{W} > 0. \qquad (*)$$

As was discussed in section 4.7 of the lecture notes, the momentum representation of its first term is just $p^2/2m$, but in contrast to the previous problem, the direct use of Eq. (4.269) to find the representation of the second term is problematic due to the highly nonlinear character of the delta-function, making its Taylor expansion non-trivial. This is why it is more prudent to start at square one, namely the general form of the eigenproblem:

$$\hat{H}\,|\alpha\rangle = E\,|\alpha\rangle.$$

Proceeding just as was done in section 4.7 for the coordinate representation, i.e. inner-multiplying both parts of this equation by $\langle p|$, and then using the closure relation analogous to Eq. (4.252),

$$\int dp' |p'\rangle\langle p'| = \hat{I},$$

we get

$$\langle p|\,\hat{H}\,|\,\alpha\rangle \equiv \int dp' \langle p|\,\hat{H}\,|\,p'\rangle\langle p'|\alpha\rangle = E\langle p|\alpha\rangle, \text{ i.e. } \int dp' \langle p|\,\hat{H}\,|\,p'\rangle\varphi(p') = E\varphi(p),$$

where $\varphi(p) \equiv \langle p|\alpha\rangle$ is the eigenfunction in the momentum representation. Now using the general Eq. (4.272) for the Hamiltonian bracket, for our particular potential we get

$$\frac{1}{2\pi\hbar}\int dp' \int dx \exp\left\{-i\frac{px}{\hbar}\right\}\left[\frac{\hat{p}^2}{2m} - \mathscr{W}\delta(x)\right]\exp\left\{i\frac{p'x}{\hbar}\right\}\varphi(p') = E\varphi(p). \qquad (**)$$

The integral of the first term gives just $(p^2/2m)\varphi(p)$, as it should:

$$\begin{aligned}
&\frac{1}{2\pi\hbar}\int dp' \int dx \exp\left\{-i\frac{px}{\hbar}\right\}\frac{\hat{p}^2}{2m}\exp\left\{i\frac{p'x}{\hbar}\right\}\varphi(p') \\
&\equiv \frac{1}{2\pi\hbar}\int dp' \frac{p'^2}{2m}\varphi(p') \int dx \exp\left\{i\frac{(p'-p)x}{\hbar}\right\} \\
&\equiv \int dp' \frac{p'^2}{2m}\varphi(p')\delta(p'-p) = \frac{p^2}{2m}\varphi(p),
\end{aligned}$$

while that of the second one is just

$$-\frac{\mathscr{W}}{2\pi\hbar}\int dp' \int dx \exp\left\{-i\frac{px}{\hbar}\right\}\delta(x)\exp\left\{i\frac{p'x}{\hbar}\right\} \equiv -\frac{\mathscr{W}}{2\pi\hbar}\int \varphi(p')dp',$$

i.e. a p-independent constant.

Hence Eq. (**) is reduced to

$$\frac{p^2}{2m}\varphi(p) - \frac{\mathscr{W}}{2\pi\hbar}\int \varphi(p')dp' = E\varphi(p),$$

immediately giving the implicit solution

$$\varphi(p) = \frac{\mathscr{W}/2\pi\hbar}{p^2/2m - E}\int \varphi(p')dp'.$$

Integrating both sides of this equality over p,[18] we get

$$\int \varphi(p)dp = \int \varphi(p')dp' \int \frac{\mathscr{W}/2\pi\hbar}{p^2/2m - E}dp \equiv \int \varphi(p')dp' \frac{\mathscr{W}}{2\pi\hbar}\left(\frac{m}{-2E}\right)^{1/2}\int_{-\infty}^{+\infty}\frac{d\xi}{1+\xi^2}$$
$$\equiv \int \varphi(p')dp' \frac{\mathscr{W}}{\hbar}\left(\frac{m}{-2E}\right)^{1/2}.$$

Requiring the integrals on both sides of this equality to be equal, we get exactly the same eigenenergy,

$$E = -\frac{m\mathscr{W}^2}{2\hbar^2},$$

as was obtained in section 2.6 using the wave mechanics, i.e. the coordinate representation.

***Problem* 4.32.** In the momentum representation, an operator of the 1D orbital motion equals p^{-1}. Find its coordinate representation.

Solution: Let us call the operator in question $\hat{\lambda}$, then according to the assignment, if

$$\hat{\lambda}\,|\alpha\rangle = |\alpha_\lambda\rangle,$$

where α is an arbitrary 1D orbital state of the particle, then

$$\frac{1}{p}\varphi(p) = \varphi_\lambda(p), \qquad (*)$$

[18] As was discussed in section 3.6 of the lecture notes, the eigenenergy E of the localized state has to be negative, so that the denominator is positive for all p, and we may use the standard table integral A.32a.

where $\varphi(p)$ and $\varphi_\lambda(p)$ are the momentum representations of the states α and α_λ, respectively. Hence,

$$\varphi(p) = p\varphi_\lambda(p). \qquad (**)$$

Now we may proceed using either of the following alternative approaches.

Approach 1. Let us consider the following ket-vector

$$|\alpha'\rangle \equiv \hat{p}|\alpha_\lambda\rangle.$$

In the momentum representation, this relation takes the form

$$\varphi'(p) = p\varphi_\lambda(p),$$

where $\varphi'(p)$ is the momentum representation of the state α'. The comparison of this expression with Eq. (**) shows that $\varphi'(p) = \varphi(p)$ for any state α, i.e. this state and α' have identical momentum representations. Since the set of the momentum basis states p, defined by Eq. (4.257) of the lecture notes, is full, this may be only if the states α and α' are identical, i.e.

$$|\alpha'\rangle = |\alpha\rangle, \qquad \text{so that} \quad \hat{p}|\alpha_\lambda\rangle \equiv \hat{p}\hat{\lambda}\,|\alpha\rangle = |\alpha\rangle,$$

for any state α. This means that (as we could expect from the very beginning),

$$\hat{p}\hat{\lambda} = \hat{I}.$$

(A similar calculation in the opposite order shows that $\hat{\lambda}\hat{p} = \hat{I}$ as well, so that the operators $\hat{p}$ and $\hat{\lambda}$ commute[19].) Hence the action of this operator product on an arbitrary coordinate-representation wavefunction cannot change it:

$$\hat{p}\hat{\lambda}\psi(x) = \hat{I}\psi(x) = \psi(x).$$

In the coordinate representation, this relation is

$$-i\hbar\frac{\partial}{\partial x}[\hat{\lambda}]_{\text{ in }x}\psi(x) = \psi(x).$$

Integrating both parts of this relation, rewritten for $x \to x'$, over the interval $[-\infty, x]$, and multiplying them by $i/\hbar$, we get:

$$[\hat{\lambda}]_{\text{ in }x}\psi(x) = \frac{i}{\hbar}\int_{-\infty}^{x}\psi(x')dx'. \qquad (***)$$

This equality[20], valid for any function $\psi(x)$, is the required coordinate representation of the operator $\hat{\lambda}$.

[19] Such operators are frequently called the *inverse* of each other, and one may run into symbolic equalities like $\hat{\lambda} = \hat{p}^{-1}$. However, such expressions should be treated with utmost care; in this course they are avoided.

[20] Actually, as follows from our derivation, the lower limit of this integral may be an arbitrary constant.

Approach 2. Let us use Eq. (4.264), then Eq. (*), and finally Eq. (4.265), to calculate the coordinate-representation wavefunction corresponding to the state α_λ:

$$\psi_\lambda(x) = \frac{1}{(2\pi\hbar)^{1/2}} \int_{-\infty}^{+\infty} \varphi_\lambda(p) \exp\left\{i\frac{px}{\hbar}\right\} dp \equiv \frac{1}{(2\pi\hbar)^{1/2}} \int_{-\infty}^{+\infty} \varphi(p) \exp\left\{i\frac{px}{\hbar}\right\} \frac{dp}{p}$$

$$= \frac{1}{(2\pi\hbar)^{1/2}} \int_{-\infty}^{+\infty} \left[\frac{1}{(2\pi\hbar)^{1/2}} \int_{-\infty}^{+\infty} \psi(x') \exp\left\{-i\frac{px'}{\hbar}\right\} dx'\right] \exp\left\{i\frac{px}{\hbar}\right\} \frac{dp}{p} \qquad (****)$$

$$\equiv \frac{1}{2\pi\hbar} \int_{-\infty}^{+\infty} \psi(x')\, dx'\, I_p, \quad \text{where} \quad I_p \equiv \int_{-\infty}^{+\infty} \exp\left\{i\frac{p(x-x')}{\hbar}\right\} \frac{dp}{p}.$$

The inner integral I_p may be readily worked out by comparing it with similar integrals over closed contours on the plane of the complex variable $\boldsymbol{p} \equiv \mathrm{Re}\,\boldsymbol{p} + i\mathrm{Im}\,\boldsymbol{p}$, where $\mathrm{Re}\,\boldsymbol{p} \equiv p$. Indeed, the exponential function under such an integral,

$$\exp\left\{i\frac{\boldsymbol{p}(x-x')}{\hbar}\right\} = \exp\left\{i\frac{(\mathrm{Re}\,\boldsymbol{p} + i\,\mathrm{Im}\,\boldsymbol{p})(x-x')}{\hbar}\right\}$$

$$= \exp\left\{i\frac{\mathrm{Re}\,\boldsymbol{p}\,(x-x')}{\hbar}\right\} \exp\left\{-\frac{\mathrm{Im}\,\boldsymbol{p}\,(x-x')}{\hbar}\right\},$$

is analytic for any $\boldsymbol{p}$. If $x - x' > 0$, it tends to zero exponentially at $\mathrm{Im}\,\boldsymbol{p} \to +\infty$. Hence I_p may be replaced with the integral over the contour C_+ shown on the left panel of the figure below, which bypasses the pole point $\boldsymbol{p} = 0$ via an infinitesimal semicircle, and returns to the initial point $p = -\infty$ along another semi-circle, this one with an infinite radius.

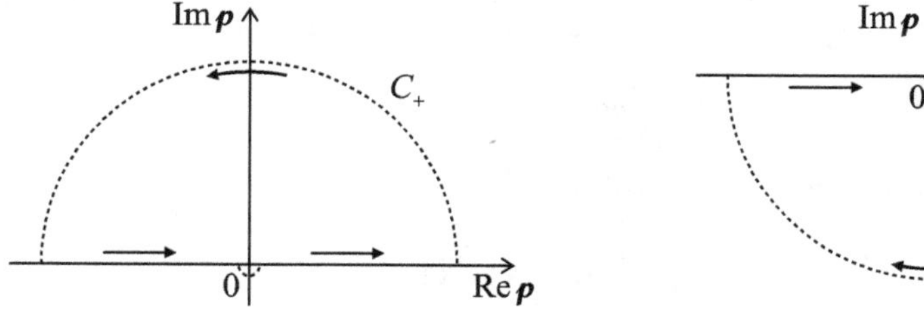

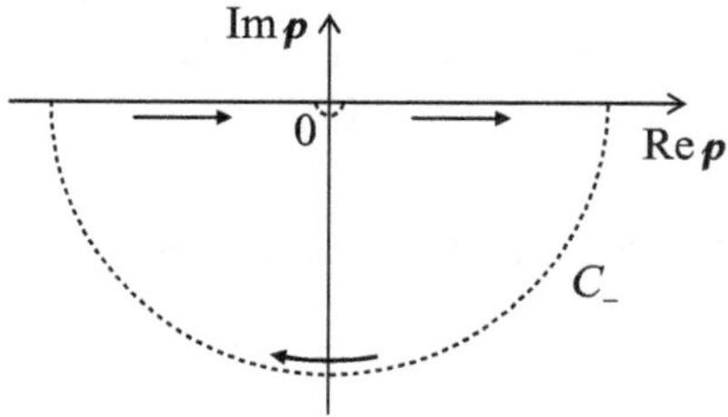

Applying the Cauchy integral formula to the exponential function alone[21], we get

$$I_p = \oint_{C+} \exp\left\{i\frac{\boldsymbol{p}(x-x')}{\hbar}\right\} \frac{d\boldsymbol{p}}{\boldsymbol{p}} = 2\pi i, \qquad \text{for } (x-x') > 0.$$

On the other hand, if $x - x' < 0$, the exponential function tends to zero at $\mathrm{Im}\,\boldsymbol{p} \to -\infty$. In this case, I_p may be replaced with the integral over the contour C_- shown in the right panel of the figure above, inside which the whole function under the integral, including the $1/\boldsymbol{p}$ factor, is analytic. Here the Cauchy integral theorem, applied to this composite function[22], yields

[21] See, e.g. Eq. (A.92) with $z = \boldsymbol{p}$, $z' = \boldsymbol{p}' = 0$, and $f(z) = \exp\{i\boldsymbol{p}(x-x')/\hbar\}$, so that $f(z') = 1$.
[22] See, e.g. Eq. (A.91) with $z = \boldsymbol{p}$, and $f(z) = \exp\{i\boldsymbol{p}(x-x')/\hbar\}/\boldsymbol{p}$.

$$I_p = \oint_{C-} \exp\left\{i\frac{p\,(x-x')}{\hbar}\right\}\frac{dp}{p} = 0, \qquad \text{for } (x-x') < 0.$$

Plugging these results into Eq. (****), we get

$$\psi_\lambda(x) \equiv [\hat{\lambda}]_{\text{ in }x}\psi(x) = \frac{1}{2\pi\hbar}\int_{-\infty}^{+\infty}\psi(x')\,dx' \times \begin{Bmatrix} 2\pi i, & \text{for } x' < x \\ 0, & \text{for } x' > x \end{Bmatrix} = \frac{i}{\hbar}\int_{-\infty}^{x}\psi(x')dx',$$

i.e. arrive at Eq. (***) again.

Finally note that acting absolutely similarly, for another operator (say, $\hat{\Lambda}$), defined by its coordinate representation as

$$[\hat{\Lambda}]_{\text{in }x} = \frac{1}{x},$$

we may readily get the following momentum representation:

$$[\hat{\Lambda}]_{\text{ in }p}\varphi(p) = -\frac{i}{\hbar}\int_{-\infty}^{p}\varphi(p')dp'.$$

This relation is similar to Eq. (***), besides the opposite sign—the change which may be expected already from the comparison of Eqs. (4.264) and (4.265) of the lecture notes.

Problem* 4.33. For a particle moving in a 3D periodic potential, develop the bra–ket formalism for the **q**-representation, in which a complex amplitude similar to a_q in Eq. (2.234) of the lecture notes (but generalized to 3D and all energy bands) plays the role of the wavefunction. In particular, calculate the operators **r** and **v** in this representation, and use the result to prove Eq. (2.237) for the 1D case in the low-field limit.

Solution: Let us consider the 3D orbital motion of a particle in a periodic potential $U(\mathbf{r})$, described by the Hamiltonian

$$\hat{H}_0 = \frac{\hat{p}^2}{2m} + U(\hat{\mathbf{r}}), \qquad \text{with } U(\mathbf{r}+\mathbf{R}) = U(\mathbf{R}), \qquad (*)$$

where $\mathbf{R}$ is an arbitrary vector of the Bravais lattice—see section 3.4. According to the Bloch theorem (3.108), in the coordinate representation the wavefunction of an arbitrary orbital state of the particle may be expressed as an expansion over the eigenfunctions of the Hamiltonian (*):

$$\psi(\mathbf{r}) = \sum_n \int a_{n,\mathbf{q}}u_{n,\mathbf{q}}(\mathbf{r})e^{i\mathbf{q}\cdot\mathbf{r}}d^3q\,, \qquad u_{n,\mathbf{q}}(\mathbf{r}+\mathbf{R}) = u_{n,\mathbf{q}}(\mathbf{r}),$$

which is a natural generalization of Eq. (2.234) to the 3D case and an arbitrary number of energy bands (numbered by the integer index n).

The radius-vector operator acts on this wavefunction as

$$\hat{\mathbf{r}}\psi(\mathbf{r}) = \sum_n \int a_{n,\mathbf{q}}\mathbf{r}u_{n,\mathbf{q}}(\mathbf{r})e^{i\mathbf{q}\cdot\mathbf{r}}d^3q\ . \qquad (**)$$

Now let us calculate the gradient of the last product under the integral, multiplied by $(-i)$, in the reciprocal $\mathbf{q}$-space of the quasi-momentum[23]:

$$\nabla_q(-iu_{n,\mathbf{q}}e^{i\mathbf{q}\cdot\mathbf{r}}) = \mathbf{r}u_{n,q}e^{i\mathbf{q}\cdot\mathbf{r}} - ie^{i\mathbf{q}\cdot\mathbf{r}}\nabla_q u_{n,\mathbf{q}}.$$

Expressing from this relation the first term on its right-hand side, and plugging it into Eq. (**), we get

$$\hat{\mathbf{r}}\psi = -i\sum_n \int a_{n,\mathbf{q}}\nabla_q(u_{n,\mathbf{q}}e^{i\mathbf{q}\cdot\mathbf{r}})\, d^3q + i\sum_n \int a_{n,\mathbf{q}}e^{i\mathbf{q}\cdot\mathbf{r}}\nabla_q u_{n,\mathbf{q}}d^3q.$$

Now let us integrate the first term on the right-hand side by parts over a volume so large that on its surface we may take $\psi = 0$, and in the second term, expand the $\mathbf{q}$-gradient of the function $u_{n,\mathbf{q}}$ into the series over the full set of these (mutually orthogonal) functions in all energy bands:

$$\nabla_q u_{n,\mathbf{q}} = -i\sum_{n'}\langle n,\mathbf{q}|\,\hat{\mathbf{\Omega}}\,|n',\mathbf{q}\rangle\, u_{n',\mathbf{q}},$$

where $|n, \mathbf{q}\rangle$ are the eigenstates of the Hamiltonian (*), with the Bloch wavefunctions

$$\psi_{n,q}(\mathbf{r}) = \langle \mathbf{r}|n,\mathbf{q}\rangle = u_{n,\mathbf{q}}e^{i\mathbf{q}\cdot\mathbf{r}},$$

and the vector operator $\hat{\mathbf{\Omega}}$ is, at this stage, defined just by its (vector!) matrix elements participating in the above expansion. The result of the integration is

$$\hat{\mathbf{r}}\psi = i\sum_n \int u_{n,\mathbf{q}}e^{i\mathbf{q}\cdot\mathbf{r}}\nabla_q a_{n,\mathbf{q}} + \sum_{n,n'}\langle n,\mathbf{q}|\,\hat{\mathbf{\Omega}}\,|n',\mathbf{q}\rangle\, a_{n,\mathbf{q}}u_{n',\mathbf{q}}e^{i\mathbf{q}\cdot\mathbf{r}}d^3q.$$

Swapping the indices n and n' in the last term, we may rewrite our result as

$$\hat{\mathbf{r}}\psi = \sum_n \int \left[i\nabla_q a_{n,\mathbf{q}} + \sum_{n'}\langle n',\mathbf{q}|\,\hat{\mathbf{\Omega}}\,|n,\mathbf{q}\rangle\, a_{n',\mathbf{q}} \right] u_{n,\mathbf{q}}e^{i\mathbf{q}\cdot\mathbf{r}}d^3q. \qquad (***)$$

On the other hand, in the $\mathbf{q}$-representation, the amplitude function $a_{n,\mathbf{q}}$ plays the role of the wavefunction—just as in the usual momentum representation, which was discussed in detail in section 4.7 for the 1D case, the state with the Schrödinger wavefunction (4.264) is described by the function $\varphi(p)$. In this representation, any operator should be described by its action on that wavefunction. This is why in the $\mathbf{q}$-representation, Eq. (***) may be re-written as

$$\hat{\mathbf{r}} = i\nabla_q + \hat{\mathbf{\Omega}}. \qquad (****)$$

[23] I am using this convenient term for $\mathbf{q}$, despite the fact (discussed in sections 2.7–2.8 and 3.4 of the lecture notes), that the genuine quasi-momentum equals $\hbar\mathbf{q}$.

In the basis of the $\{n, \mathbf{q}\}$ eigenstates, the matrix of the first term of this sum is diagonal in n. On the other hand, the matrix of the second operator is, by its definition, diagonal in the $\mathbf{q}$-subspace, and hence commutes with the Hamiltonian (*). Let us use this fact and Eq. (4.199) to calculate the velocity operator in the Heisenberg picture:

$$\hat{\mathbf{v}} \equiv \dot{\hat{\mathbf{r}}} = \frac{1}{i\hbar}[\hat{\mathbf{r}}, \hat{H}_0] = \frac{1}{i\hbar}[\, i\nabla_q, \hat{H}_0] + \dot{\hat{\mathbf{\Omega}}} \equiv \hat{\mathbf{v}}_\mathbf{q} + \dot{\hat{\mathbf{\Omega}}}.$$

As we know from the band theory, the Hamiltonian (*) is diagonalized by states $|n, \mathbf{q}\rangle$, with

$$\langle n, \mathbf{q}|\, \hat{H}_0 \,|n, \mathbf{q}\rangle = E_n(\mathbf{q}),$$

so that the first contribution, $\hat{\mathbf{v}}_q$, to the velocity operator is diagonal in both the n- and $\mathbf{q}$-spaces, with the diagonal elements that act on the $\mathbf{q}$-wavefunction $a_{n,\mathbf{q}}$ as

$$\hat{\mathbf{v}}_\mathbf{q} a_{n,\mathbf{q}} = \frac{1}{i\hbar}[i\nabla_q, E_n(\mathbf{q})]\, a_{s,\mathbf{q}} \equiv \frac{1}{\hbar}[\nabla_q E_n(\mathbf{q}) a_{n,\mathbf{q}} - E_n(\mathbf{q})\nabla_q a_{n,\mathbf{q}}] \equiv a_{n,\mathbf{q}} \nabla_q \frac{E_n(\mathbf{q})}{\hbar}.$$

For a wave packet localized in just one energy band n, with a narrow distribution of quasi-momentum about some central value q_0, the action of this operator is reduced to the multiplication by the expression

$$\mathbf{v}_{\text{gr}} = \nabla_q \frac{E_n(\mathbf{q})}{\hbar} \equiv \nabla_q \omega_n(\mathbf{q}),$$

in which we may immediately recognize the natural 3D generalization of Eq. (2.235) for the group velocity of the particle. Hence we may expect that at least in this case, only the first term in Eq. (****) is important.

To confirm this guess, let us discuss the second component, $\dot{\hat{\mathbf{\Omega}}}$, of the velocity operator $\hat{\mathbf{v}}$. Its matrix is diagonal in the $\mathbf{q}$-space, and we may use the same Heisenberg equation Eq. (4.199) to describe the dynamics of its nonvanishing elements:

$$\begin{aligned}\langle n, \mathbf{q}|\, \dot{\hat{\mathbf{\Omega}}}\, |n', \mathbf{q}\rangle &= \langle n, \mathbf{q}|\, \frac{1}{i\hbar}[\hat{\mathbf{\Omega}}, \hat{H}_0]\, |n', \mathbf{q}\rangle = \langle n, \mathbf{q}|\, \frac{1}{i\hbar}(\hat{\mathbf{\Omega}}\hat{H}_0 - \hat{H}_0\hat{\mathbf{\Omega}})\, |n', \mathbf{q}\rangle \\ &= \frac{E_{n'}(\mathbf{q}) - E_n(\mathbf{q})}{\hbar}\langle n, \mathbf{q}|\, \hat{\mathbf{\Omega}}\, |n', \mathbf{q}\rangle.\end{aligned}$$

Due to the definition of the matrix elements $\langle n, \mathbf{q}|\hat{\Omega}\, |n', \mathbf{q}\rangle$, it is clear that they should be finite (because for any finite $U(\mathbf{r})$, the functions $u_{n,q}(\mathbf{r})$ should be finite and continuous)[24]. Hence the second contribution to the velocity operator vanishes for $n = n'$, so that the operator $\hat{\Omega}$ is indeed important 'only'[25] for interband transitions,

[24] A good additional exercise for the interested reader: calculate these matrix elements explicitly for a 1D periodic potential $U(x)$, both in the tight-binding limit and the weak-potential approximation.

[25] Actually, this operator is of key importance for interband transitions in semiconductors, with such important applications as semiconductor lasers, photovoltaic cells, and detectors of radiation. For the reader interested in a detailed discussion of such transitions I can recommend, for example, the monograph [1]. (See also section 7.6 of the lecture notes.)

and in the absence of such transitions, does not affect the dynamics of a wave packet limited to just one energy band[26].

In order to analyze such dynamics in the presence of an external classical force $\mathbf{F}(t)$, we should add to the potential energy $U(\mathbf{q})$ an additional term with the spatial gradient equal to $-\mathbf{F}(t)$. For the simplest 1D case, the total Hamiltonian becomes

$$\hat{H} = \hat{H}_0 - F(t)\hat{x}.$$

If the scale F of the force magnitude is small in the sense of Eq. (2.236), $Fa \ll \Delta E_n$, Δ_n, it cannot cause interband transitions, so that if the system was initially localized in one energy band, we may ignore the second term in Eq. (****). The remaining first term of that relation obeys the commutation relations similar to the Heisenberg commutator (2.14); in the 1D case its only nonvanishing component is

$$[\hat{x}, \hat{q}_x] = \left[i\frac{\partial}{\partial q_x}, q_x\right] = i\hat{I}.$$

On the other hand, since the Hamiltonian (*) is diagonal in the q-representation, it commutes with the operator of the quasi-momentum q_x. Now we may use these commutation relations to spell out the Heisenberg equation of motion (4.199) for that operator:

$$\dot{\hat{q}}_x = \frac{1}{i\hbar}[\hat{q}_x, \hat{H}] = \frac{1}{i\hbar}\left[q_x, \hat{H}_0 - iF(t)\frac{\partial}{\partial q_x}\right] \equiv \frac{1}{\hbar}F(t).$$

For a narrow wave packet $a_{s,\mathbf{q}}$, with the center at point q_0, this equation is equivalent to the quasi-classical equation (2.237), thus (finally!) justifying all its applications discussed in section 2.8 of the lecture notes.

***Problem* 4.34.** A uniform, time-independent magnetic field $\mathscr{B}$ is induced in one semi-space, while the other semi-space is field-free, with a sharp, plane boundary between these two regions. A monochromatic beam of electrically-neutral spin-½ particles with a nonvanishing gyromagnetic ratio γ,[27] in a certain spin state, and with a kinetic energy E, is incident on this boundary, from the field-free side, under angle θ—see the figure below. Calculate the coefficient of particle reflection from the boundary.

[26] This fact is the mathematical expression of the vague statement 'the quasi-momentum is the average momentum' made in section 2.8. Indeed, an elementary 3D generalization of the first of Eqs. (4.269) shows that in the momentum ($\mathbf{p}$-) representation, $\hat{\mathbf{r}} = i\nabla_k$(where $\mathbf{k} = \mathbf{p}/\hbar$), i.e. coincides with the first component, $i\nabla_q$, of that operator in the $\mathbf{q}$-representation.

[27] The fact that γ may be different from zero even for electrically-neutral particles, such as neutrons, is explained by the Standard Model of elementary particles, in which a neutron 'consists' (in a broad sense of the word) of three electrically-charged quarks.

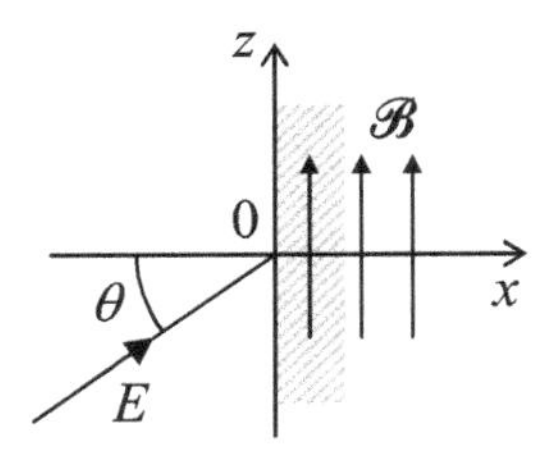

Solution: We may represent the spin ket-vector of the incident particles as a linear superposition of z-polarized states:

$$|\alpha\rangle = c_\uparrow |\uparrow\rangle + c_\downarrow |\downarrow\rangle,$$

where the z-axis is directed along the magnetic field—see the figure above. (The expression 'certain spin state' in the problem's assignment means that the coefficients $c_\uparrow$ and $c_\downarrow$ are known—up to a common phase factor.) Since the states ↑ and ↓ are the eigenstates of the Pauli Hamiltonian (4.163), in the field region (in the figure above, at $x > 0$), this operator may be replaced with its eigenvalues in this state basis: $E_\uparrow = -\gamma\mathscr{B}\hbar/2$ and $E_\downarrow = +\gamma\mathscr{B}\hbar/2$, respectively. Hence we may describe the orbital motion of the beam by a linear superposition of two wavefunctions[28], $\psi_\uparrow(\mathbf{r})$ and $\psi_\downarrow(\mathbf{r})$, which obey different Schrödinger equations:

$$-\frac{\hbar^2}{2m}\nabla^2\psi_\uparrow + U_\uparrow(\mathbf{r})\psi_\uparrow = E\psi_\uparrow, \quad \text{with } U_\uparrow(\mathbf{r}) = \begin{cases} 0, & \text{for } x < 0, \\ E_\uparrow, & \text{for } 0 < x. \end{cases}$$

$$-\frac{\hbar^2}{2m}\nabla^2\psi_\downarrow + U_\downarrow(\mathbf{r})\psi_\downarrow = E\psi_\downarrow, \quad \text{with } U_\downarrow(\mathbf{r}) = \begin{cases} 0, & \text{for } x < 0, \\ E_\downarrow, & \text{for } 0 < x. \end{cases}$$

But each of these independent[29] boundary problems is exactly similar to that of a spinless particle's reflection from a potential step—see problem 3.1. According to its solution, with the replacement of the parameter U_0 with either $E_\uparrow$ or $E_\downarrow$, respectively (provided that $\gamma > 0$), the reflection coefficients $\mathscr{R}$ of these two waves (referred to the initial intensity of the whole beam) are

$$\mathscr{R}_\uparrow = |c_\uparrow|^2 \times \left|\frac{\cos\theta - (\cos^2\theta + \gamma\mathscr{B}\hbar/2E)^{1/2}}{\cos\theta + (\cos^2\theta + \gamma\mathscr{B}\hbar/2E)^{1/2}}\right|^2,$$

[28] As will be discussed later in the course, such a set of two wavefunctions is frequently called a *spinor*.

[29] Still note that for the nearly monochromatic incident particles, and in the absence of dephasing (see chapter 7 of the lecture notes for its discussion), the de Broglie waves $\psi_\uparrow$ and $\psi_\downarrow$ remain coherent, and if their beams are recombined, they may exhibit quantum interference phenomena—see, e.g. section 3.1 of the lecture notes.

$$\mathscr{R}_{\downarrow} = |c_{\downarrow}|^2 \times \begin{cases} \left[\dfrac{\cos\theta - (\cos^2\theta - \gamma\mathscr{B}\hbar/2E)^{1/2}}{\cos\theta + (\cos^2\theta - \gamma\mathscr{B}\hbar/2E)^{1/2}}\right]^2, & \text{for } \dfrac{\gamma\mathscr{B}\hbar}{2E} < \cos^2\theta, \\ 1, & \text{for } \cos^2\theta < \dfrac{\gamma\mathscr{B}\hbar}{2E}. \end{cases}$$

(For $\gamma < 0$, there expressions should be interchanged.)

In particular, if the spin orientation energy $\gamma\mathscr{B}\hbar/2$ is larger than the incident particle energy E, the reflection of one de Broglie wave ($\psi_{\downarrow}$) is total for any incidence angle θ, so that only the other wave ($\psi_{\uparrow}$) propagates deep into the field region $x > 0$. Hence such a setup may be used for getting a beam of spin-polarized particles from an incident beam of unpolarized ones.

Reference

[1] Ridley B 2000 *Quantum Processes in Semiconductors* 4th edn (Oxford: Oxford University Press)

IOP Publishing

Quantum Mechanics
Problems with solutions
Konstantin K Likharev

Chapter 5

Some exactly solvable problems

***Problem* 5.1.** Use the discussion in section 5.1 of the lecture notes to find an alternative solution of problem 4.18.

Solution: According to the discussion in section 5.1, the expectation value of the z-component of spin-½, for the state polarized in the direction $\mathbf{n} = \mathbf{n}_x \sin\theta\cos\varphi + \mathbf{n}_y \sin\theta\sin\varphi + \mathbf{n}_x \cos\theta$, is given by the last of Eqs. (5.12):

$$\langle S_z \rangle = \frac{\hbar}{2}\cos\theta \equiv \frac{\hbar}{2}\left(2\cos^2\frac{\theta}{2} - 1\right). \qquad (*)$$

But according to Eq. (1.37), and the fact that in the z-basis the operator $\hat{S}_z$ has just two eigenstates, ↑ and ↓, with the eigenvalues $\pm\hbar/2$, this expectation value may be also represented as

$$\langle S_z \rangle = \left(+\frac{\hbar}{2}\right)W_\uparrow + \left(-\frac{\hbar}{2}\right)W_\downarrow \equiv \frac{\hbar}{2}(W_\uparrow - W_\downarrow). \qquad (**)$$

where $W_\uparrow$ and $W_\downarrow$ are the probabilities of the corresponding states. Since the sum of these two probabilities has to equal 1, Eq. (**) may be rewritten as

$$\langle S_z \rangle = \frac{\hbar}{2}[W_\uparrow - (1 - W_\uparrow)] \equiv \frac{\hbar}{2}(2W_\uparrow - 1).$$

Requiring this expression to give the same result as Eq. (*), we get

$$W_\uparrow = \cos^2\frac{\theta}{2}, \quad \text{and hence } W_\downarrow = 1 - W_\downarrow = 1 - \cos^2\frac{\theta}{2} \equiv \sin^2\frac{\theta}{2}.$$

Now we may argue that due to the isotropy of the free space, this result has to be independent of the absolute directions of the two axes (that of the initial polarization of the spin, and the direction of the magnetic field in the Stern–Gerlach apparatus),

doi:10.1088/2053-2563/aaf3a6ch5

and may depend only on the angle θ between these axes. Thus we confirm the solution of problem 4.18.

***Problem* 5.2.** A spin-½ is placed into an external magnetic field, with a time-independent orientation, its magnitude $\mathscr{B}(t)$ being an arbitrary function of time. Find explicit expressions for the Heisenberg operators and the expectation values of all three Cartesian components of the spin, as functions of time, in a coordinate system of your choice.

Solution: In the coordinate system with the z-axis directed along the applied magnetic field (evidently, the easiest choice), the Hamiltonian of our system is given by Eq. (4.163) of the lecture notes:

$$\hat{H} = -\gamma\mathscr{B}(t)\hat{S}_z \equiv -\gamma\mathscr{B}(t)\frac{\hbar}{2}\hat{\sigma}_z,$$

where γ is the gyromagnetic ratio of the particle, so that its matrix, in the z-basis, is

$$\mathrm{H} = -\gamma\mathscr{B}(t)\frac{\hbar}{2}\sigma_z \equiv -\gamma\mathscr{B}(t)\frac{\hbar}{2}\begin{pmatrix} 1 & 0 \\ 0 & -1 \end{pmatrix}.$$

This expression coincides with Eq. (5.3) of the lecture notes, with $b = 0$, and[1]

$$\mathbf{c} = -\gamma\mathscr{B}(t)\frac{\hbar}{2}\mathbf{n}_z.$$

Hence we may use Eq. (5.19), rewritten for the spin operator $\hat{\mathbf{S}} \equiv (\hbar/2)\hat{\boldsymbol{\sigma}}$,

$$\dot{\hat{\mathbf{S}}} = \boldsymbol{\Omega}(t) \times \hat{\mathbf{S}}, \quad \text{with} \quad \boldsymbol{\Omega}(t) \equiv \frac{2}{\hbar}\mathbf{c}(t) = -\gamma\mathscr{B}(t)\mathbf{n}_z,$$

so that in the Cartesian components,

$$\dot{\hat{S}}_x = -\Omega(t)\hat{S}_y, \qquad \dot{\hat{S}}_y = \Omega(t)\hat{S}_x, \qquad \dot{\hat{S}}_z = 0. \qquad (*)$$

The last equation immediately yields

$$\hat{S}_z(t) = \hat{S}_z(0) = \text{const},$$

while the easiest way to solve the system of the first two equations (*) is to introduce (just as was done for the orbital momentum in section 5.6 of the lecture notes), the *spin-ladder operators*[2]

$$\hat{S}_\pm \equiv \hat{S}_x \pm i\hat{S}_y,$$

[1] Alternatively, this expression follows from Eq. (5.13) with $\mathscr{B} = \mathscr{B}(t)\mathbf{n}_z$.

[2] Implicitly, these operators (or rather their expectation values) have already been used in the lecture notes—see Eqs. (4.172) and (4.173).

with the reciprocal relations

$$\hat{S}_x = \frac{\hat{S}_+ + \hat{S}_-}{2}, \qquad \hat{S}_y(t) = \frac{\hat{S}_+ - \hat{S}_-}{2i}. \qquad (**)$$

Using these relations, and Eqs. (*), to calculate the time derivatives of the ladder operators, we get two independent equations,

$$\dot{\hat{S}}_\pm = \pm i\Omega(t)\hat{S}_\pm,$$

with elementary solutions:

$$\hat{S}_\pm(t) = \hat{S}_\pm(0)\exp\{\pm i\varphi(t)\} \equiv [\hat{S}_x(0) \pm i\hat{S}_y(0)]\exp\{\pm i\varphi(t)\},$$

where

$$\varphi(t) \equiv \int_0^t \Omega(t')dt'. \qquad (***)$$

Now using Eqs. (**) to return the real Cartesian coordinates of the spin operator, we finally get

$$\hat{S}_x(t) = \hat{S}_x(0)\cos\varphi(t) - \hat{S}_y(0)\sin\varphi(t), \qquad \hat{S}_y(t) = \hat{S}_y(0)\cos\varphi(t) + \hat{S}_x(0)\sin\varphi(t).$$

In the particular case of a time-independent field, the argument (***) of the trigonometric functions in these expressions is just Ωt, so that our results are reduced to Eqs. (4.194) and (4.195) of the lecture notes (written there in the z-basis, but evidently valid, as any operator relations, in an arbitrary basis as well).

Since in the Heisenberg formalism the expectation value of any observable is calculated with time-independent bra- and ket-vectors, the relations between the expectation values of the spin components exactly replicate the above operator relations. Again, for the case of a constant magnetic field, they coincide with Eqs. (4.170)–(4.174).

***Problem* 5.3.** A two-level system is in a quantum state α described by the ket-vector $|\alpha\rangle = \alpha_\uparrow|\uparrow\rangle + \alpha_\downarrow|\downarrow\rangle$, with given (generally, complex) c-number coefficients $\alpha_{\uparrow\downarrow}$. Prove that we can always select a three-component vector $\mathbf{c} = \{c_x, c_y, c_z\}$ of real c-numbers, such that α is an eigenstate of the operator $\mathbf{c}\cdot\hat{\boldsymbol{\sigma}}$, where $\hat{\boldsymbol{\sigma}}$ is the Pauli vector-operator. Find all possible values of $\mathbf{c}$ satisfying this condition, and the second eigenstate (orthogonal to α) of the operator $\mathbf{c}\cdot\hat{\boldsymbol{\sigma}}$. Give a Bloch-sphere interpretation of your result.

Solution: In the z-basis, the operator $\mathbf{c}\cdot\hat{\boldsymbol{\sigma}}$ has the following matrix:

$$\mathbf{c}\cdot\boldsymbol{\sigma} = c_x\sigma_x + c_y\sigma_y + c_z\sigma_z$$
$$= c_x\begin{pmatrix} 0 & 1 \\ 1 & 0 \end{pmatrix} + c_y\begin{pmatrix} 0 & -i \\ i & 0 \end{pmatrix} + c_z\begin{pmatrix} 1 & 0 \\ 0 & -1 \end{pmatrix} = \begin{pmatrix} c_z & c_- \\ c_+ & -c_z \end{pmatrix}, \quad \text{where} \quad c_\pm \equiv c_x \pm ic_y,$$

so that the system of Eq. (4.102) for its eigenstates and eigenvalues has the form

$$\begin{pmatrix} c_z - \lambda_j & c_- \\ c_+ & -c_z - \lambda_j \end{pmatrix}\begin{pmatrix} \alpha_\uparrow \\ \alpha_\downarrow \end{pmatrix} = 0. \qquad (*)$$

The condition of compatibility of these equations yields the characteristic equation

$$\begin{vmatrix} c_z - \lambda_j & c_- \\ c_+ & -c_z - \lambda_j \end{vmatrix} = 0,$$

which has two roots $\lambda_\pm = \pm c$, where

$$c \equiv \left(c_z^2 + c_+c_-\right)^{1/2} \equiv \left(c_x^2 + c_y^2 + c_z^2\right)^{1/2} \equiv |\mathbf{c}|.$$

Plugging these roots back into the initial system of Eq. (*), we get

$$c_-\frac{\alpha_\downarrow}{\alpha_\uparrow} = \pm c - c_z, \qquad c_+\frac{\alpha_\uparrow}{\alpha_\downarrow} = \pm c + c_z.$$

With the characteristic equation satisfied, any one of these relations contains the same information. Moreover, up to this point this solution repeats Approach 2 in the solution of problem 4.25. However, since $c_-^* \equiv c_+$, while c and c_z are real, for our current purposes it is convenient to use the second equation together with the complex conjugate of the first equation,

$$c_+\frac{\alpha_\downarrow^*}{\alpha_\uparrow^*} = \pm c - c_z.$$

Indeed, subtracting the two equations, we may eliminate c (which is an inconvenient, nonlinear combination of the c-vector components):

$$c_+\left(\frac{\alpha_\uparrow}{\alpha_\downarrow} - \frac{\alpha_\downarrow^*}{\alpha_\uparrow^*}\right) = 2c_z. \qquad (**)$$

This is the *only* equation for c_+ we need to satisfy to make the given quantum state the eigenstate of our Hamiltonian, so that we may select *any* real c_z, then use Eq. (**) to calculate c_+,

$$c_+ = 2c_z\left(\frac{\alpha_\uparrow}{\alpha_\downarrow} - \frac{\alpha_\downarrow^*}{\alpha_\uparrow^*}\right)^{-1} = 2c_z\frac{\alpha_\downarrow\alpha_\uparrow^*}{|\alpha_\uparrow|^2 - |\alpha_\downarrow|^2},$$

and then calculate c_- as the complex conjugate of c_+:

$$c_- = 2c_z\frac{\alpha_\uparrow\alpha_\downarrow^*}{|\alpha_\uparrow|^2 - |\alpha_\downarrow|^2}.$$

Finally, we may return to c_x and c_y:

$$c_x = \frac{c_+ + c_-}{2} = c_z \frac{\alpha_\downarrow \alpha_\uparrow^* + \alpha_\uparrow \alpha_\downarrow^*}{|\alpha_\uparrow|^2 - |\alpha_\downarrow|^2}, \qquad c_y = \frac{c_+ - c_-}{2i} = c_z \frac{1}{i} \frac{\alpha_\downarrow \alpha_\uparrow^* - \alpha_\uparrow \alpha_\downarrow^*}{|\alpha_\uparrow|^2 - |\alpha_\downarrow|^2}.$$

In the Bloch-sphere representation given by Eq. (5.11) of the lecture notes, the above result takes a much nicer form:

$$c_\pm = c \sin\theta e^{\pm i\varphi} \quad \left(\text{i.e.} \quad c_x = c \sin\theta \cos\varphi, \quad c_y = c \sin\theta \sin\varphi\right), \quad c_z = c \cos\theta,$$

where c may be arbitrary, while θ and φ are the spherical angles of the state-representing point:

$$\theta = 2\cos^{-1}|\alpha_\uparrow|, \qquad \varphi = \arg \frac{\alpha_\downarrow |\alpha_\uparrow|}{\alpha_\uparrow (1 - |\alpha_\uparrow|^2)^{1/2}}.$$

This result has a very simple interpretation: in order for a state α, represented by a certain point on the Bloch sphere, to be an eigenstate of the operator $\mathbf{c} \cdot \hat{\boldsymbol{\sigma}}$, it is sufficient to have the 'field' vector $\mathbf{c}$ directed, from the origin, to this point. This is just an alignment of the (average) spin's direction with the field; the degree of freedom we have in the solution corresponds just to field's strength c.

The Bloch sphere graphics also gives a simple way to find the eigenstate α' orthogonal to the given one (α): their representing points should be diametrically opposite. Hence, to find the coefficients $\alpha'_\uparrow$ and $\alpha'_\downarrow$, it is sufficient to make the following replacements: $\theta \to (\pi - \theta)$, and $\varphi \to (\varphi + \pi)$ in Eqs. (5.10), getting

$$\begin{aligned} \alpha'_\uparrow &= \sin\frac{\theta}{2} e^{i(\gamma+\gamma')} \equiv \frac{\alpha_\uparrow |\alpha_\downarrow|}{|\alpha_\uparrow|} e^{i\gamma'}, \\ \alpha'_\downarrow &= -\cos\frac{\theta}{2} e^{i(\varphi+\gamma+\gamma')} \equiv -\frac{\alpha_\downarrow |\alpha_\uparrow|}{|\alpha_\downarrow|} e^{i\gamma'}, \end{aligned} \qquad (***)$$

with an arbitrary common phase γ'. It is straightforward to verify that this replacement also changes the sign of the eigenvalue λ: from $+c$ to $-c$ and vice versa.

Problem* 5.4. Analyze the statistics of the spacing $S \equiv E_+ - E_-$ between the energy levels of a two-level system, assuming that all elements $H_{jj'}$ of its Hamiltonian matrix (5.2) are independent random numbers, with equal and constant probability densities within the energy interval of interest. Compare the result with that for a purely diagonal matrix, with the similar probability distribution of the random diagonal elements.

Solution: According to Eq. (5.6) of the lecture notes,

$$S \equiv E_+ - E_- = 2\left[\left(\frac{H_{11} - H_{22}}{2}\right)^2 + H_{12}H_{21}\right]^{1/2}.$$

Since the Hamiltonian has to have real eigenvalues, its matrix has to be Hermitian, i.e. its diagonal elements H_{11} and H_{22} have to be real, while the off-diagonal elements H_{12} and H_{21} have to be complex-conjugate, so that we may rewrite the expression for S as

$$S = 2(X^2 + Y^2)^{1/2},$$

where $X \equiv (H_{11} - H_{22})/2$ and $Y \equiv |H_{12}| = |H_{21}|$ are independent random real numbers with a certain probability density w_0, so that the probability for them to be within an elementary interval $dXdY$ (i.e. an elementary area of the $[X, Y]$ plane) is

$$dw = w_0^2 dXdY.$$

Introducing the polar coordinates ρ and φ on that plane by standard relations

$$X = \rho \cos \varphi, \qquad Y = \rho \sin \varphi,$$

we get $dXdY = \rho d\rho d\varphi$ and $S = 2\rho$. Hence the probability for the system to be within a small interval $d\rho$, regardless of the angle φ (which does not affect S) is

$$dW = w_0^2 2\pi\rho d\rho = \frac{\pi}{2} w_0^2 S dS.$$

On the other hand, by the definition of the probability density of a variable (in our case, S), dW should be equal to $w(S)dS$, so that

$$w(S) = \frac{\pi}{2} w_0^2 S. \qquad (*)$$

However, this result may be taken for the genuine probability density of the interlevel spacing only for small S, because it does not reflect the fact that each two-level system has only one value of S. (For example, if integrated from 0 to some growing value S, Eq. (*) would eventually give a probability larger than 1.) The standard general way to correct this deficiency is to say that as S is increased, the probability $W(S)$ of having the spacing at the interval $[0, S]$ increases as

$$\frac{dW(S)}{dS} = w(S)[1 - W(S)].$$

Here the additional factor $[1 - W(S)]$ reflects the fact that only if the system had not had the spacing on the interval $[0, S]$, it may have it at larger values of the spacing. Integrating this simple differential equation, with the obvious boundary condition $W(0) = 0$, we get

$$W(S) = 1 - \exp\left\{-\int_0^S w(S')dS'\right\}.$$

Now the genuine probability density of the interlevel spacing may be calculated as

$$\tilde{w}(S) \equiv \frac{dW(S)}{dS} = w(S)\exp\left\{-\int_0^S w(S')dS'\right\}. \qquad (**)$$

At $S \to 0$, this probability density coincides with the 'seed density' $w(S)$, but at $S \to \infty$ it drops exponentially, ensuring the automatic convergence of the aggregate probability,

$$W(\infty) = \int_0^\infty \tilde{w}(S)dS,$$

to have the spacing at *some* value S, to the appropriate (unit) value, for any seed function $w(S)$:

$$\begin{aligned} W(\infty) &= \int_0^\infty w(S) \exp\left\{-\int_0^S w(S')dS'\right\}dS \\ &= \int_0^\infty \exp\left\{-\int_0^S w(S')dS'\right\}d\left[\int_0^S w(S')dS'\right] \\ &= \int_0^\infty \exp\{-\xi\}d\xi = 1. \end{aligned}$$

For our particular case, we may plug Eq. (*) into the general Eq. (**) to get

$$\tilde{w}(S) = \frac{\pi}{2}w_0^2 S \exp\left\{-\frac{\pi}{2}w_0^2\int_0^S S'dS'\right\} = \frac{\pi}{2}w_0^2 S \exp\left\{-\frac{\pi}{4}w_0^2 S^2\right\}. \qquad (***)$$

This result may be used, in particular, to calculate the average level spacing[3],

$$\begin{aligned} \langle S \rangle &= \int_0^\infty S\ \tilde{w}(S)dS = \frac{\pi}{2}w_0^2\int_0^\infty S^2 \exp\left\{-\frac{\pi}{4}w_0^2 S^2\right\}dS \\ &\equiv \frac{4}{\pi^{1/2}w_0}\int_0^\infty \xi^2 \exp\{-\xi^2\}d\xi = \frac{4}{\pi^{1/2}w_0}\frac{\pi^{1/2}}{4} \equiv \frac{1}{w_0}. \end{aligned}$$

The last relation allows one to express the parameter w_0 as $1/\langle S \rangle$, and rewrite Eq. (***) in its canonical form, as the probability density of the normalized spacing $s \equiv S/\langle S \rangle$:[4]

$$\tilde{w}(s) = \frac{dS}{ds}\tilde{w}(S) = \langle S \rangle \tilde{w}(S) = \frac{\pi}{2}s \exp\left\{-\frac{\pi}{4}s^2\right\}. \qquad (****)$$

This result (called the *Wigner surmise*) shows that the probability of having the two eigenenergies of a random two-level system very close to each other is vanishingly small. This fact, called the *level repulsion*, is perhaps the best known qualitative result of the field called the *Random Matrix Theory* (*RMT*)[5].

---

[3] Using the well-known Gaussian integral—see, e.g. Eq. (A.36*c*).

[4] The first of these equalities follows from the invariance of the elementary probability $dW = \tilde{w}(S)dS = \tilde{w}(s)ds$.

[5] The de-facto Bible of the RMT (whose founding father set notably includes E Wigner and F Dyson) is the monograph by M Mehta [1]. The field was inspired by the experimental observations, in the 1940–50s, of pseudo-random energy spectra of atomic nuclei, but is applicable to many other systems with uncontrollable parameters, for example solid-state 'quantum dots'—see, e.g. [2]. The general RMT shows that Eq. (****) is valid for the so-called *orthogonal ensemble*—just one of three major statistical ensemble types. (The reader interested in this classification and other details of the RMT is referred to the cited sources.)

In order to appreciate the nontrivial nature of Eq. (****), let us compare it with the statistics of spacing between independent energy values, which may be formally described by Eq. (5.2) with $H_{12} = H_{21} = 0$, so that

$$S = |H_{11} - H_{22}| \equiv 2|X|.$$

With the probability density of X equal to $w_0 = \text{const}$, we have $dW = w(S)dS = 4w_0 dX$ (with the additional factor of 2 coming from the contributions from two branches of the function $S(X)$, for positive and negative X), so that the seed density is

$$w(S) = 4w_0 = \text{const}.$$

Plugging this expression into the general Eq. (**), we get the purely exponential distribution,

$$\tilde{w}(S) = 4w_0 \exp\left\{-4w_0 \int_0^S dS'\right\} = 4w_0 \exp\{-4w_0 S\},$$

which is also usually recast into that for the normalized spacing s:[6]

$$\tilde{w}(s) = \exp\{-s\},$$

where $s \equiv S/\langle S\rangle$, in this case with

$$\langle S\rangle = \int_0^\infty S\tilde{w}(S)dS = \frac{1}{4w_0}\int_0^\infty \xi \exp\{-\xi\}d\xi = \frac{1}{4w_0}.$$

This exponential distribution is shown, together with the one given by Eq. (****), in the figure below. The difference is rather spectacular; the Wigner surmise may be interpreted as the direct result of the level repulsion at their anticrossing—see, e.g. figure 5.1 of the lecture notes.

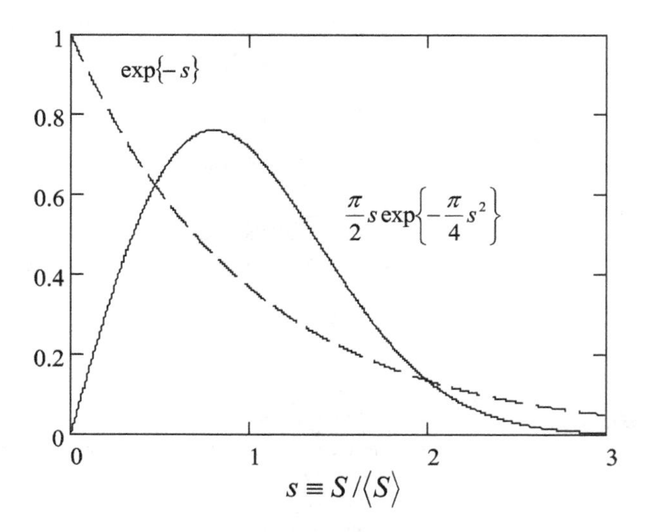

[6] In this field, such exponential distribution is frequently called 'Poissonian', but one should remember that it is just a limiting case (for $\langle n\rangle \ll 1$) of the genuine Poisson distribution (5.135).

Let me also note one more interesting aspect of the RMT: its application to the statistics of energy specta of classically chaotic Hamiltonian systems[7]. Namely, in classically chaotic systems, the level spacing distribution is closer to the Wigner surmise (****), while in non-chaotic ('integrable') systems it is closer to the 'Poissonian' (exponential) one. To the best of my knowledge, this *Bohigas–Giannoni–Schmit conjecture* has not received any general proof (yet), but it was confirmed by many numerical calculations for particular systems[8].

***Problem* 5.5.** For a periodic motion of a single particle in a confining potential $U(\mathbf{r})$, the *virial theorem* of non-relativistic classical mechanics[9] is reduced to the following equality:

$$\bar{T} = \frac{1}{2}\overline{\mathbf{r}\cdot\nabla U},$$

where T is the particle's kinetic energy, and the top bar means averaging over the time period of motion. Prove the quantum-mechanical version of the theorem for an arbitrary stationary quantum state, in the absence of spin effects:

$$\langle T\rangle = \frac{1}{2}\langle \mathbf{r}\cdot\nabla U\rangle,$$

where the angular brackets mean the expectation values of the observables.

Hint: Mimicking the proof of the classical virial theorem, consider the time evolution of the following operator:

$$\hat{G} \equiv \hat{\mathbf{r}}\cdot\hat{\mathbf{p}}.$$

Solution: According to Eq. (4.199), in the Heisenberg picture of quantum dynamics

$$i\hbar\dot{\hat{G}} = [\hat{G}, \hat{H}].$$

where, in the absence of spin effects, the particle's Hamiltonian may be taken in the form (4.237), so that

$$\dot{\hat{G}} = \frac{1}{i\hbar}[\hat{G}, \hat{H}] = \frac{1}{i\hbar}\left[\hat{\mathbf{r}}\cdot\hat{\mathbf{p}}, \frac{\hat{p}^2}{2m}\right] + \frac{1}{i\hbar}[\hat{\mathbf{r}}\cdot\hat{\mathbf{p}}, \hat{U}(\mathbf{r})].$$

Now using, for each Cartesian coordinate, the commutation relation (5.27), which yields the following vector relation:

$$[\hat{\mathbf{r}}, \hat{p}^2] = 2i\hbar\hat{\mathbf{p}},$$

[7] For a brief discussion of the classical deterministic chaos see, e.g. *Part CM* section 9.3; for a brief remark on quantum dynamics of classically chaotic systems, see a footnote in the beginning of section 3.5 of the lecture notes.

[8] See, e.g. [3].

[9] See, e.g. *Part CM* problem 1.12.

and also Eqs. (5.33) and (5.34), giving, in the coordinate representation, the following vector relation

$$[\hat{\mathbf{p}}, \hat{U}(\mathbf{r})] = -i\hbar \nabla U, \qquad (*)$$

we get

$$\dot{\hat{G}} = \frac{\hat{p}^2}{m} - \hat{\mathbf{r}} \cdot \nabla U \equiv 2\left(\hat{T} - \frac{1}{2}\hat{\mathbf{r}} \cdot \nabla U\right).$$

Averaging both parts of this equation over the initial quantum state, we get the following relation for the expectation values of the involved variables

$$\langle \dot{G} \rangle = 2\left(\langle T \rangle - \frac{1}{2}\langle \mathbf{r} \cdot \nabla U \rangle\right). \qquad (**)$$

Since, by its definition, for any localized motion G is a limited variable, the expectation value of its time derivative has to vanish in any stationary state of the system. Hence the right-hand side of Eq. (**) has to equal zero, thus proving the quantum virial theorem.

Note that, just as in classical mechanics, the theorem may be generalized as follows,

$$\langle T \rangle = \frac{1}{2}\left\langle \sum_{k=1}^{N} \hat{\mathbf{r}}_k \cdot \nabla_k U \right\rangle, \qquad (***)$$

for the motion of a system of N particles, described by the spin-independent Hamiltonian

$$\hat{H} = \sum_{k=1}^{N} \frac{\hat{p}_k^2}{2m_k} + \hat{U}(\hat{\mathbf{r}}_1, \hat{\mathbf{r}}_2, \ldots \hat{\mathbf{r}}_N).$$

Indeed, the commutators $[\hat{\mathbf{r}}_k, \hat{p}_{k'}]$ for all particle pairs ($k \neq k'$) vanish, because these operators belong to different Hilbert spaces, and Eq. (*), in the form

$$[\hat{\mathbf{p}}_k, \hat{U}] = -i\hbar \nabla_k U,$$

is valid even if the potential energy depends on positions of other particles as well, making the proof of Eq. (***) similar to that of its single-particle version.

***Problem* 5.6.** Calculate, in the WKB approximation, the transparency $\mathscr{T}$ of tunneling of a 2D particle with energy $E < U_0$ through a saddle-shaped potential 'pass'

$$U(x, y) = U_0\left(1 + \frac{xy}{a^2}\right),$$

where $U_0 > 0$ and a are real constants.

Solution: Equipotential lines of this potential profile, corresponding to $U(x, y) = E$, are hyperbolas symmetric with respect to the straight-line diagonals $y = \pm x$—see the figure below. Hence the quasi-classical instanton trajectory that provides the minimum of the tunneling exponent given by Eq. (5.56) of the lecture notes,

$$I \equiv \int_{\mathbf{r}_0}^{\mathbf{r}} \boldsymbol{\kappa} \cdot d\mathbf{r}, \quad \text{where} \quad \frac{\hbar^2\kappa^2}{2m} \equiv U(\mathbf{r}) - E,$$

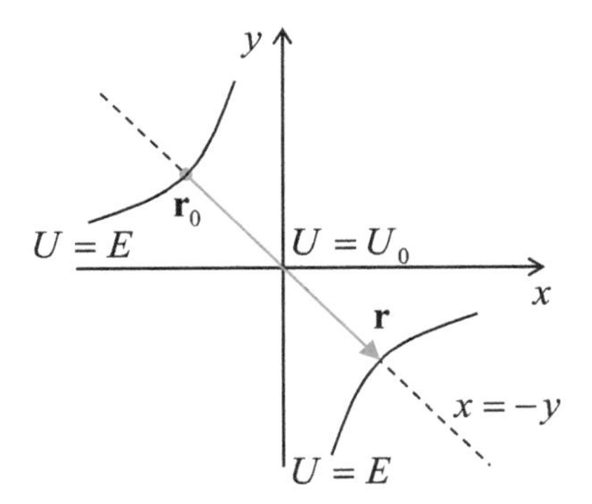

cannot deviate from one of these symmetry axes. (If it did, which side would that be?) If $U_0 > 0$, the diagonal connecting classically allowed regions with energies $E < U(0, 0) = U_0$ is $y = -x$ (see the arrow in the figure above). Plugging this relation into the function $U(x, y)$, and the latter into the above expression for $\boldsymbol{\kappa}$, we get

$$I = \frac{1}{\hbar}\int_{-l_c}^{+l_c} \left\{2m\left[U_0\left(1 - \frac{x^2}{a^2}\right) - E\right]\right\}^{1/2} dl,$$

with $dl = \sqrt{2}dx$, $l_c = \sqrt{2}x_c$, and $\pm x_c$ are the x-coordinates of the classical turning points $\mathbf{r}_0$ and $\mathbf{r}$, determined by the condition

$$U(x, y) - E \equiv U_0\left(1 - \frac{x_c^2}{a^2}\right) - E = 0.$$

Now a simple integration yields

$$I = \frac{\pi a}{\hbar}\left(\frac{m}{U_0}\right)^{1/2}(U_0 - E),$$

giving the following WKB transparency (5.58):

$$\mathscr{T} = \exp\{-2I\} = \exp\left\{-\frac{2\pi a}{\hbar}\left(\frac{m}{U_0}\right)^{1/2}(U_0 - E)\right\}.$$

***Problem* 5.7.** Calculate the so-called *Gamow factor*[10] for the alpha decay of atomic nuclei, i.e. the exponential factor in the transparency of the potential barrier resulting from the following simple model of the alpha-particle's potential energy as a function of its distance from the nuclear center:

$$U(r) = \begin{cases} U_0 < 0, & \text{for } r < R, \\ \dfrac{ZZ'e^2}{4\pi\varepsilon_0 r}, & \text{for } R < r, \end{cases}$$

(where $Ze = 2e > 0$ is the charge of the particle, $Z'e > 0$ is that of the nucleus after the decay, and R is the nucleus' radius), in the WKB approximation.

Solution: Evidently, such tunneling is possible only for energies $E > U(\infty) = 0$. Due to the spherical symmetry of the potential, the instanton trajectory, which minimizes the functional I given by Eq. (5.56) of the lecture notes, is a straight radial line, so that

$$I \equiv \int_R^{R'} \kappa(r)\, dr = \frac{1}{\hbar}\int_R^{R'} \{2m[U(r) - E]\}^{1/2} dr = \frac{1}{\hbar}\int_R^{R'} \left\{2m\left[\frac{ZZ'e^2}{4\pi\varepsilon_0 r} - E\right]\right\}^{1/2} dr,$$

where $R' = R'(E)$ is the radius at which the particle of energy E comes out from under the barrier:

$$U(R') \equiv \frac{ZZ'e^2}{4\pi\varepsilon_0 R'} = E.$$

Using this definition to simplify the function under the integral, and then the variable substitution $r \equiv R' \sin^2\xi$ (so that $dr = 2R' \sin\xi \cos\xi\, d\xi$), we get

$$\begin{aligned} I &= \frac{(2mE)^{1/2}}{\hbar}\int_R^{R'} \left(\frac{R'}{r} - 1\right)^{1/2} dr \\ &= \frac{2(2mE)^{1/2}R'}{\hbar}\int_{\sin^{-1}(R/R')^{1/2}}^{\pi/2} \left(\frac{1}{\sin^2\xi} - 1\right)^{1/2} \sin\xi \cos\xi\, d\xi \\ &\equiv \frac{2(2mE)^{1/2}R'}{\hbar}\int_{\sin^{-1}(R/R')^{1/2}}^{\pi/2} \cos^2\xi\, d\xi \\ &\equiv \frac{2(2mE)^{1/2}R'}{\hbar}\int_{\sin^{-1}(R/R')^{1/2}}^{\pi/2} \frac{1 + \cos 2\xi}{2}\, d\xi \\ &= \frac{(2mE)^{1/2}R'}{\hbar}\left[\frac{\pi}{2} - \sin^{-1}\left(\frac{R}{R'}\right)^{1/2} - \left(\frac{R}{R'}\right)^{1/2}\left(1 - \frac{R}{R'}\right)^{1/2}\right]. \end{aligned}$$

It is convenient to introduce the natural energy unit,

[10] Named after G Gamow, who made this calculation as early as in 1928.

$$E_0 \equiv U(R) = \frac{ZZ'e^2}{4\pi\varepsilon_0 R} \quad \left(\text{so that } \frac{R}{R'} = \frac{E}{E_0}\right),$$

to finally get

$$\mathscr{T} = e^{-2I}$$
$$= \exp\left\{-\frac{2(2mE_0)^{1/2}R}{\hbar}\left[\frac{\pi}{2}\left(\frac{E_0}{E}\right)^{1/2} - \left(\frac{E_0}{E}\right)^{1/2}\sin^{-1}\left(\frac{E}{E_0}\right)^{1/2} - \left(1 - \frac{E}{E_0}\right)^{1/2}\right]\right\}.$$

Just for the reader's reference, the typical energies E of emitted alpha particles are much smaller than E_0, so that the experimental data are reasonably well described by a simpler expression, following from the above result after dropping the two last terms in the square brackets:

$$\mathscr{T} \approx \exp\left\{-\frac{\sqrt{2}\,\pi m^{1/2}E_0 R}{\hbar E^{1/2}}\right\}$$
$$\equiv \exp\left\{-\frac{\sqrt{2}\,\pi m^{1/2}}{\hbar E^{1/2}}\frac{ZZ'e^2}{4\pi\varepsilon_0}\right\} \equiv \exp\left\{-ZZ'\,\alpha\left(\frac{2\pi^2 mc^2}{E}\right)^{1/2}\right\},$$

where α is the fine structure constant. Actually, the last expression gives reasonable results not only for alpha particles, but for many other nuclear reactions, provided that Z and Z' are taken equal to the numbers of protons in the reacting nuclei, and the particle's mass m is replaced with its reduced mass[11]:

$$m \rightarrow \frac{mm'}{m + m'}.$$

(For the relatively light alpha particles, this mass renormalization is not essential.)

***Problem* 5.8.** Use the WKB approximation to calculate the average time of ionization of a hydrogen atom, initially in its ground state, made metastable by the application of an additional weak, uniform, constant electric field $\mathscr{E}$. Formulate the conditions of validity of your result.

Solution: The net potential energy of the electron's motion in the atom is

$$U(\mathbf{r}) = -\frac{q^2}{4\pi\varepsilon_0 r} - q\mathscr{E}z,$$

where z is the electric field's direction. If the field is sufficiently weak,

$$q|\mathscr{E}|r_B \ll E_H, \tag{*}$$

[11] See, e.g. *Part CM* section 3.4 of the lecture notes, in particular Eq. (3.35). As will be discussed in chapter 8 below, this mass renormalization is valid in quantum mechanics as well.

where r_B is the Bohr radius and E_H is the Hartree energy given, respectively, by Eqs. (1.10) and (1.13) of the lecture notes, then the $\mathscr{E}$-induced correction to the ground-state energy E of the electron in atom is negligible, and we may use for it the unperturbed value (1.12) with $n = 1$:

$$E = -\frac{E_H}{2}.$$

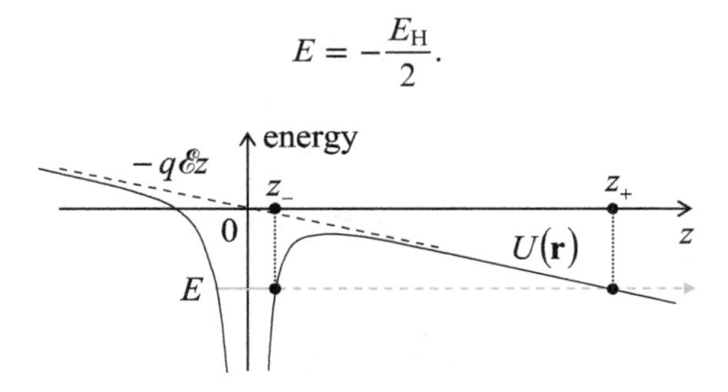

The green dashed arrow in the figure above shows the electron tunneling process leading to the atom's (positive) ionization. Due to the axial symmetry of the potential $U(\mathbf{r})$ and the ground-state wavefunction ψ_{100}, the instanton trajectory that minimizes the tunnel integral I given by Eq. (5.56), is a straight line along the z-axis (i.e. along the direction shown in the sketch[12]), so that

$$I = \int_{z_-}^{z_+} \kappa(z)dz, \quad \text{with} \quad \frac{\hbar^2\kappa^2(z)}{2m} = U(\mathbf{n}_z z) - E \equiv -\frac{q^2}{4\pi\varepsilon_0 z} - q\mathscr{E}z + \frac{E_H}{2}, \qquad (**)$$

where $z_\pm$ are the classical turning points that are defined by the condition $U(\mathbf{n}_z z_\pm) = E$ (see the figure above):

$$-\frac{q^2}{4\pi\varepsilon_0 z_\pm} - q\mathscr{E}z_\pm + \frac{E_H}{2} = 0.$$

In the limit $\mathscr{E} \to 0$, there is no need to solve this quadratic equation exactly, because evidently

$$z_- \approx \frac{q^2}{4\pi\varepsilon_0}\Big/\frac{E_H}{2} \equiv 2r_B, \qquad z_+ \approx \frac{E_H}{2}\Big/q\mathscr{E} \gg z_-,$$

and the integral I may be well approximated by neglecting the Coulomb term in Eq. (**) completely:

$$I \approx \frac{(2m)^{1/2}}{\hbar}\int_0^{z_+}\left(-q\mathscr{E}z + \frac{E_H}{2}\right)^{1/2} dz.$$

[12] Strictly speaking, for the electron's charge $q = -e < 0$, all z in the figure and all following relations should read $-z$, but the final results are still valid, with the replacement $q \to +e > 0$.

This is exactly the same (easy) integral as at the WKB approach to the Fowler–Nordheim tunneling problem (see problem 2.10, and also problem 2.14), with the replacements $U_0 - E \to E_H/2$ and $\mathbf{F} \to q\mathscr{E}$, and yields

$$\mathscr{T}_{\text{WKB}} = \exp\left\{-\frac{4}{3}\frac{(2m)^{1/2}}{\hbar}\frac{(E_H/2)^{3/2}}{q\mathscr{E}}\right\} \equiv \exp\left\{-\frac{2}{3}\frac{E_H}{q\mathscr{E}r_B}\right\}.$$

Since this expression does not include a possible pre-exponential factor[13], it does not make much sense to calculate the attempt time t_a exactly. Taking it at a reasonable value $\hbar/|E| \equiv 2\hbar/E_H$, from the general Eq. (2.153) of the lecture notes we get the following estimate of the lifetime of the metastable ground state, i.e. of the average ionization time:

$$\tau \equiv \frac{t_a}{\mathscr{T}} \sim \frac{2\hbar}{E_H}\exp\left\{\frac{2}{3}\frac{E_H}{q\mathscr{E}r_B}\right\}. \qquad (***)$$

Note that in the limit $\mathscr{E} \to 0$, the main, exponential factor in this result is quantitatively correct. If the condition (*) of our result's validity is fulfilled, this factor is much larger than 1. This condition is only violated in extremely high electric fields either of the order of, or above $E_H/er_B \sim 10^{12}$ V m^{-1}.

***Problem* 5.9.** For a 1D harmonic oscillator with mass m and frequency ω_0, calculate:

(i) all matrix elements $\langle n|\,\hat{x}^3|n'\rangle$, and
(ii) the diagonal matrix elements $\langle n|\,\hat{x}^4|n\rangle$,

where n and n' are arbitrary Fock states.

Solutions:

(i) Breaking $\hat{x}^3$ into the product of $\hat{x}$ by $\hat{x}^2$, and using the closure condition, we may write

$$\langle n'|\hat{x}^3|n\rangle \equiv \langle n|\hat{x}\hat{x}^2|n'\rangle = \sum_{n''=0}^{\infty}\langle n'|\hat{x}|n''\rangle\langle n''|\hat{x}^2|n\rangle.$$

Now we may use Eqs. (5.92) and (5.94) of the lecture notes to get

$$\langle n'|\,\hat{x}^3\,|n\rangle = \frac{x_0^3}{\sqrt{8}}\Big\{[n(n-1)(n-2)]^{1/2}\delta_{n',n-3}$$
$$+ 3n^{3/2}\delta_{n',n-1} + 3(n+1)^{3/2}\delta_{n',n+1} + [(n+1)(n+2)(n+3)]^{1/2}\delta_{n',n+3}\Big\}.$$

[13] The calculation of this factor (which in this 3D problem is different from that for the 1D Fowler–Nordheim tunneling, discussed in the solution of problem 2.10) is a good additional exercise, recommended to the reader.

(ii) Here it is simpler to factor $\hat{x}^4$ into the product of $\hat{x}^2$ by $\hat{x}^2$:

$$\langle n|\,\hat{x}^4\,|n\rangle \equiv \langle n|\,\hat{x}^2\hat{x}^2\,|n\rangle = \sum_{n'=0}^{\infty}\langle n|\,\hat{x}^2\,|n'\rangle\langle n'|\,\hat{x}^2\,|n\rangle,$$

and then again use Eq. (5.94), keeping only the partial products with equal and opposite differences between indices n and n' (because all other products vanish). The result is

$$\langle n|\,\hat{x}^4\,|n\rangle = \frac{3}{4}x_0^4(2n^2 + 2n + 1).$$

Note that for $n = 0$, this result may be readily calculated in the wave-mechanics style as well, using Eq. (2.275) for the ground state wavefunction, and the table integral (A.36*d*):

$$\begin{aligned}\langle 0|\,\hat{x}^4\,|0\rangle &= \int \psi_0^* x^4 \psi_0 dx \\ &= \frac{1}{\pi^{1/2}x_0}\int_{-\infty}^{+\infty} x^4 \exp\left\{-\frac{x^2}{x_0^2}\right\}dx \\ &\equiv \frac{2x_0^4}{\pi^{1/2}}\int_0^{+\infty} \xi^4 \exp\{-\xi^2\}\, d\xi = \frac{3}{4}x_0^4,\end{aligned}$$

but for higher values of n, such calculations are harder, because of the much more involved Eq. (2.284) for ψ_n.

***Problem* 5.10.** Calculate the sum (over all $n > 0$) of the so-called *oscillator strengths*,

$$f_n \equiv \frac{2m}{\hbar^2}(E_n - E_0)|\langle n|\,\hat{x}\,|0\rangle|^2,$$

(i) for a 1D harmonic oscillator, and
(ii) for a 1D particle confined in an arbitrary stationary potential well.

Solutions:

(i) According to Eq. (5.92) of the lecture notes, for a harmonic oscillator only one of the oscillator strengths is nonvanishing:

$$\langle n|\,\hat{x}\,|0\rangle = \left(\frac{\hbar}{2m\omega_0}\right)^{1/2}\delta_{1,n},$$

while $E_1 - E_0 = \hbar\omega_0$, so that

$$f_n = \frac{2m}{\hbar^2}\hbar\omega_0\frac{\hbar}{2m\omega_0}\delta_{1,n} \equiv \delta_{1,n},$$

and the sum in question equals 1.

(ii) According to Eqs. (4.191) and (4.199) of the lecture notes for the Heisenberg picture of quantum evolution, the time evolution of a matrix element of an operator $\hat{A}$ that does not depend on time explicitly, in the basis of stationary states n of an arbitrary system, is described by the following equation:

$$\begin{aligned} i\hbar \dot{A}_{nn'} &\equiv \langle n|\, i\hbar\dot{\hat{A}}(t)\, |n'\rangle = \langle n|\, [\hat{A}, \hat{H}]\, |n'\rangle \\ &\equiv \langle n|\, \hat{A}\hat{H}\, |n'\rangle - \langle n|\, \hat{H}\hat{A}\, |n'\rangle \\ &= \langle n|\, \hat{A}\, |n'\rangle E_{n'} - E_n \langle n|\, \hat{A}\, |n'\rangle \equiv A_{nn'}(E_{n'} - E_n). \end{aligned}$$

For the matrix element of our current interest, $x_{n0} \equiv \langle n|\hat{x}|0\rangle$, and its complex conjugate x_{0n}, this means

$$i\hbar\dot{x}_{n0} = (E_0 - E_n)x_{n0} \equiv -(E_n - E_0)x_{n0}, \qquad i\hbar\dot{x}_{0n} = (E_n - E_0)x_{0n}.$$

Now we may use these expressions to rewrite the oscillator strengths sum's definition,

$$\sum_{n>0} f_n \equiv \frac{2m}{\hbar^2}\sum_{n>0}(E_n - E_0)|x_{n0}|^2 \equiv \frac{2m}{\hbar^2}\sum_{n>0}(E_n - E_0)x_{0n}x_{n0},$$

in two different forms: either as

$$\sum_{n>0} f_n = \frac{2m}{\hbar^2}\sum_{n>0}(i\hbar\dot{x}_{0n})x_{n0} \equiv i\frac{2m}{\hbar}\sum_{n>0}\langle 0|\, \dot{\hat{x}}\, |n\rangle\langle n|\, \hat{x}\, |0\rangle,$$

or as

$$\sum_{n>0} f_n = -\frac{2m}{\hbar^2}\sum_{n>0}x_{0n}(i\hbar\dot{x}_{n0}) \equiv -i\frac{2m}{\hbar}\sum_{n>0}\langle 0|\, \hat{x}\, |n\rangle\langle n|\, \dot{\hat{x}}\, |0\rangle.$$

Taking the arithmetic average of these two expressions, and using the closure condition (4.44) for the orthonormal, full set of stationary states n, we get

$$\sum_{n>0} f_n = i\frac{m}{\hbar}\langle 0|(\dot{\hat{x}}\hat{x} - \hat{x}\dot{\hat{x}})|0\rangle. \qquad (*)$$

But for a particle with the time-independent Hamiltonian (4.237), we may use Eq. (5.29),

$$\dot{\hat{x}} = \frac{\hat{p}_x}{m},$$

and the Heisenberg's commutation relation (4.238), to transform Eq. (*) as

$$\sum_{n>0} f_n = \frac{i}{\hbar}\langle 0|(\hat{p}_x\hat{x} - \hat{x}\hat{p}_x)|0\rangle \equiv -\frac{i}{\hbar}\langle 0|[\hat{x}, \hat{p}_x]|0\rangle = \langle 0|\, \hat{I}\, |0\rangle = 1.$$

Thus, the oscillator strengths' sum equals 1 even for 1D systems that are far from the harmonic oscillator. This *Thomas–Reiche–Kuhn sum rule* is important for applications (see, e.g. section 9.2, and also *Part EM* section 7.2), because the coefficients f_n describe the intensity of dipole quantum transitions between the nth energy level and the ground state.

Just for reader's reference, this is only one of a broad family of very similar sum rules, which may be also proved similarly. Of those that would not be used in this course, perhaps the most useful rule is

$$\sum_{n'}(E_{n'} - E_n)^2 \; |\langle n| \hat{x} |n'\rangle|^2 = \frac{\hbar^2}{m^2}\langle n| \hat{p}^2 |n\rangle;$$

see also the next problem.

***Problem* 5.11.** Prove the so-called *Bethe sum rule*,

$$\sum_{n'}(E_{n'} - E_n) \; |\langle n| e^{ik\hat{x}} |n'\rangle|^2 = \frac{\hbar^2 k^2}{2m},$$

valid for a 1D particle moving in an arbitrary time-independent potential $U(x)$, and discuss its relation with the Thomas–Reiche–Kuhn sum rule whose derivation was the subject of the previous problem.

Hint: Calculate the expectation value, in a stationary state n, of the following double commutator,

$$\hat{D} \equiv [[\hat{H}, e^{ik\hat{x}}] , e^{-ik\hat{x}}],$$

in two ways—first, just spelling out both commutators, and, second, using the commutation relations between the operators $\hat{p}$ and $e^{ik\hat{x}}$, and compare the results.

Solution: Spelling out the commutators, we get

$$\begin{aligned}\hat{D} &= (\hat{H}e^{ik\hat{x}} - e^{ik\hat{x}}\hat{H})\, e^{-ik\hat{x}} - e^{-ik\hat{x}}(\hat{H}e^{ik\hat{x}} - e^{ik\hat{x}}\hat{H}) \\ &\equiv (\hat{H} - e^{ik\hat{x}}\hat{H}e^{-ik\hat{x}}) - (e^{-ik\hat{x}}\hat{H}e^{ik\hat{x}} - \hat{H}) \\ &\equiv 2\hat{H} - (e^{ik\hat{x}}\hat{H}e^{-ik\hat{x}} + e^{-ik\hat{x}}\hat{H}e^{ik\hat{x}}).\end{aligned}$$

Since the Hamiltonian operator is Hermitian, the second term inside the last parentheses is just the Hermitian conjugate of the first one, their expectation values are complex conjugates of each other, so that

$$D_n \equiv \langle n| \hat{D} |n\rangle = 2E_n - (\langle n| e^{ik\hat{x}}\hat{H}e^{-ik\hat{x}} |n\rangle + \text{c.c.}).$$

Inserting the identity operator on any side of the Hamiltonian operator, and then using the closure relation (4.44) in the stationary state basis, we may rewrite this relation as

$$\begin{aligned}
D_n &= 2E_n - (\langle n|\, e^{ik\hat{x}}\hat{H}\hat{I}e^{-ik\hat{x}}\, |n\rangle + \text{c.c.}) \\
&= 2E_n - \sum_{n'}(\langle n|\, e^{ik\hat{x}}\hat{H}\, |n'\rangle\langle n'|\, e^{-ik\hat{x}}\, |n\rangle + \text{c.c.}) \\
&= 2E_n - \sum_{n'}E_{n'}(\langle n|\, e^{ik\hat{x}}\, |n'\rangle\langle n'|\, e^{-ik\hat{x}}\, |n\rangle + \text{c.c.}) \qquad (*) \\
&= 2\left(E_n - \sum_{n'}E_{n'}\, |\langle n|\, e^{ik\hat{x}}\, |n'\rangle|^2\right).
\end{aligned}$$

Since the eigenenergy E_n is a c-number, we may formally represent it as a similar sum:

$$\begin{aligned}
E_n &= E_n\langle n|n\rangle = E_n\langle n|\, e^{ik\hat{x}}e^{-ik\hat{x}}\, |n\rangle \\
&= E_n\langle n|\, e^{ik\hat{x}}\hat{I}e^{-ik\hat{x}}\, |n\rangle = \sum_{n'}E_n\langle n|\, e^{ik\hat{x}}\, |n'\rangle\langle n'|\, e^{-ik\hat{x}}\, |n\rangle \\
&= \sum_{n'}E_n\, |\langle n|\, e^{ik\hat{x}}\, |n'\rangle|^2,
\end{aligned}$$

and then use this expression to recast Eq. (*) as

$$D_n = 2\sum_{n'}(E_n - E_{n'})\, |\langle n|\, e^{ik\hat{x}}\, |n'\rangle|^2 \equiv -2\sum_{n'}(E_{n'} - E_n)\, |\langle n|\, e^{ik\hat{x}}\, |n'\rangle|^2. \qquad (**)$$

On the other hand, the same double commutator given in the Hint may be calculated using the explicit form of the particle's Hamiltonian:

$$\hat{H} = \frac{\hat{p}^2}{2m} + U(\hat{x}),$$

where $\hat{p} \equiv \hat{p}_x$ for brevity. Since the operators $U(\hat{x})$ and $e^{\pm ik\hat{x}}$ are all functions of the coordinate operator, and hence commute with each other, the double commutator is reduced to

$$\hat{D} = \frac{1}{2m}[[\hat{p}^2, e^{ik\hat{x}}], e^{-ik\hat{x}}].$$

Reviewing in Eqs. (5.32)–(5.35) of the lecture notes, in which the commutator of the operators $\hat{p}$ and $U(\hat{x})$ was calculated, we may see that this calculation is valid for any function $f(\hat{x})$:

$$[\hat{p}, f(\hat{x})] = -i\hbar\frac{df(\hat{x})}{dx},$$

so that for the particular case $f(x) = e^{\pm ikx}$, we get

$$[\hat{p}, e^{\pm ik\hat{x}}] = -i\hbar\,(\pm ike^{ik\hat{x}}) \equiv \pm\hbar ke^{ik\hat{x}}, \quad \text{i.e. } \hat{p}e^{\pm ik\hat{x}} = e^{\pm ik\hat{x}}\hat{p} \pm \hbar ke^{\pm ik\hat{x}}. \qquad (***)$$

Applying this rule, with the plus sign, twice to the inner commutator in $\hat{D}$, we get

$$\begin{aligned}[\hat{p}^2, e^{ik\hat{x}}] &\equiv \hat{p}\hat{p}e^{ik\hat{x}} - e^{ik\hat{x}}\hat{p}\hat{p} = \hat{p}(e^{ik\hat{x}}\hat{p} + \hbar k e^{ik\hat{x}}) - e^{ik\hat{x}}\hat{p}\hat{p} \\ &= \hat{p}e^{ik\hat{x}}\hat{p} + \hbar k\hat{p}e^{ik\hat{x}} - e^{ik\hat{x}}\hat{p}\hat{p} \\ &= (e^{ik\hat{x}}\hat{p} + \hbar k e^{ik\hat{x}})\hat{p} + \hbar k\hat{p}e^{ik\hat{x}} - e^{ik\hat{x}}\hat{p}\hat{p} \\ &\equiv \hbar k\ (e^{ik\hat{x}}\hat{p} + \hat{p}e^{ik\hat{x}}).\end{aligned}$$

With this, our double commutator becomes

$$\hat{D} = \frac{\hbar k}{2m}[(e^{ik\hat{x}}\hat{p} + \hat{p}e^{ik\hat{x}}), e^{-ikx}] = \frac{\hbar k}{2m}(e^{ik\hat{x}}\hat{p}e^{-ik\hat{x}} - e^{-ik\hat{x}}\hat{p}e^{ik\hat{x}}).$$

Now applying Eq. (***) again, with the corresponding sign, to each of the two terms, we finally get

$$\hat{D} = \frac{\hbar k}{2m}(\hat{p} - \hbar k - \hat{p} - \hbar k) = -\frac{\hbar^2 k^2}{m}.$$

So, our double commutator is actually just a c-number, so that its expectation value in any state, including the stationary state n, is the same:

$$D_n \equiv \langle n|\ \hat{D}\ |n\rangle = -\frac{\hbar^2 k^2}{m}.$$

Requiring this expression to give the same result as Eq. (**), we get the Bethe sum rule:

$$\sum_{n'}(E_{n'} - E_n)\ |\langle n|\ e^{ik\hat{x}}\ |n'\rangle|^2 = \frac{\hbar^2 k^2}{2m}.$$

This relation, which is also valid in higher dimensionalities (with the replacement $e^{ikx} \to e^{i\mathbf{k}\cdot\mathbf{r}}$), is especially useful for the solid state theory[14].

Now, in the limit at $k \to 0$, we may expand the exponent inside the matrix element into the Taylor series, and keep only two leading terms:

$$\begin{aligned}\sum_{n'}(E_{n'} - E_n)\ |\langle n|\ e^{ik\hat{x}}\ |n'\rangle|^2 &\to \sum_{n'}(E_{n'} - E_n)\ |\langle n|\ (1 + ik\hat{x})\ |n'\rangle|^2 \\ &= \sum_{n'}(E_{n'} - E_n)\ |\delta_{nn'} + ik\langle n|\ \hat{x}\ |n'\rangle|^2 \\ &= \sum_{n'}(E_{n'} - E_n)\left(\delta_{nn'}^2 + k^2\langle n|\ \hat{x}\ |n'\rangle^2\right) \\ &= k^2\sum_{n'}(E_{n'} - E_n)\langle n|\ \hat{x}\ |n'\rangle^2,\end{aligned}$$

so that the Bethe sum rule is reduced to

[14] See, e.g. section 5.6 in the classical monograph by [4], and its later re-printings.

$$\sum_{n'}(E_{n'} - E_n)\ |\langle n|\ \hat{x}\ |n'\rangle|^2 = \frac{\hbar^2}{2m}.$$

In the particular case $n = 0$, and with the replacement $n' \to n$, this result is reduced to the Thomas–Reiche–Kuhn sum rule for the oscillator strength, proved in the previous problem:

$$\sum_n f_n \equiv \sum_{n'} \frac{2m}{\hbar^2}\ |\langle n|\ \hat{x}\ |0\rangle|^2 = 1,$$

so that this solution may be considered as its alternative (albeit longer) proof.

***Problem* 5.12.** Given Eq. (5.116) of the lecture notes, prove Eq. (5.117), using the hint given in the accompanying footnote.

Solution: Following the hint, we can write

$$\hat{f}(\lambda) = \hat{f}\,|_{\lambda=0} + \frac{\lambda}{1!}\frac{\partial \hat{f}}{\partial \lambda}\bigg|_{\lambda=0} + \frac{\lambda^2}{2!}\frac{\partial^2 \hat{f}}{\partial \lambda^2}\bigg|_{\lambda=0} + \dots. \qquad (*)$$

Let us calculate the derivatives participating in this expression, for our case

$$\hat{f}(\lambda) = \exp\{+\lambda\hat{A}\}\hat{B}\exp\{-\lambda\hat{A}\}.$$

Since both exponents $\exp\{\pm\lambda\hat{A}\}$ are defined by their Taylor expansions,

$$\exp\{\pm\lambda\hat{A}\} = \hat{I} \pm \frac{\lambda}{1!}\hat{A} + \frac{\lambda^2}{2!}\hat{A}\hat{A} \pm \frac{\lambda^3}{3!}\hat{A}\hat{A}\hat{A} + \dots,$$

their differentiation over the parameter λ gives

$$\frac{\partial}{\partial\lambda}\exp\{\pm\lambda\hat{A}\} = \pm\hat{A} + \frac{\lambda}{1!}\hat{A}\hat{A} \pm \frac{\lambda^2}{2!}\hat{A}\hat{A}\hat{A}\dots$$
$$\equiv \pm\hat{A}\left(\hat{I} \pm \frac{\lambda}{1!}\hat{A} + \frac{\lambda^2}{2!}\hat{A}\hat{A} + \dots\right) = \pm\hat{A}\exp\{\pm\lambda\hat{A}\}.$$

As a result, the differentiation of the operator $\hat{f}$ over λ yields

$$\frac{\partial \hat{f}}{\partial \lambda} = \hat{A}\hat{f} - \hat{f}\hat{A} = [\hat{A}, \hat{f}],$$
$$\frac{\partial^2 \hat{f}}{\partial \lambda^2} \equiv \frac{\partial}{\partial\lambda}\left(\frac{\partial \hat{f}}{\partial \lambda}\right) = \frac{\partial}{\partial\lambda}[\hat{A}, \hat{f}] = [\hat{A}, [\hat{A}, \hat{f}]], \dots$$

so that at $\lambda = 0$, when $\hat{f} = \hat{B}$, we may use Eq. (5.116) to get

$$\frac{\partial \hat{f}}{\partial \lambda}\Big|_{\lambda=0} = [\hat{A}, \hat{B}] = \mu\hat{I}, \quad \frac{\partial^2 \hat{f}}{\partial \lambda^2}\Big|_{\lambda=0} = [\hat{A}, [\hat{A}, \hat{B}]] = [\hat{A}, \mu\hat{I}] = \hat{0}, \quad \dots$$

Plugging this result into Eq. (*), we get (for *arbitrary* λ):

$$\hat{f}(\lambda) = \hat{B} + \frac{\lambda}{1!}\mu\hat{I},$$

for the particular case $\lambda = 1$ giving Eq. (5.117).

***Problem* 5.13.** Use Eqs. (5.116) and (5.117) of the lecture notes to simplify the following operators:

(i) $\exp\{+ia\hat{x}\}\ \hat{p}_x \exp\{-ia\hat{x}\}$, and
(ii) $\exp\{+ia\hat{p}_x\}\hat{x}\exp\{-ia\hat{p}_x\}$,

where a is a c-number.

Solutions:

(i) Let us apply Eq. (5.117) of the lecture notes,

$$\exp\{+\hat{A}\}\hat{B}\exp\{-\hat{A}\} = \hat{B} + \mu\hat{I}, \qquad (*)$$

where μ is the c-number coefficient in the commutation relation (5.116),

$$[\hat{A}, \hat{B}] = \mu\hat{I}, \qquad (**)$$

to the following operators: $\hat{A} = ia\hat{x}$ and $\hat{B} = \hat{p}_x$. Since, according to the Heisenberg uncertainty relation (4.238), for these two operators,

$$[\hat{A}, \hat{B}] \equiv [ia\hat{x}, \hat{p}_x] \equiv ia[\hat{x}, \hat{p}_x] = ia(i\hbar\hat{I}) \equiv -a\hat{I},$$

i.e. Eq. (**) is valid with $\mu = -a$, Eq. (*) yields

$$\exp\{+ia\hat{x}\}\ \hat{p}_x \exp\{-ia\hat{x}\} = \hat{p}_x - a\hat{I}.$$

(ii) Now applying Eq. (*) to the operators $\hat{A} = ia\hat{p}_x$ and $\hat{B} = \hat{x}$, for whom

$$[\hat{A}, \hat{B}] \equiv [ia\hat{p}_x, \hat{x}] \equiv ia[\hat{p}_x, \hat{x}] = ia\ (-i\hbar\hat{I}) = a\hat{I},$$

i.e. $\mu = +a$, we get

$$\exp\{+ia\hat{p}_x\}\ \hat{x}\exp\{-ia\hat{p}_x\} = \hat{x} + a\hat{I}.$$

***Problem* 5.14.** For a 1D harmonic oscillator, calculate:

(i) the expectation value of energy, and
(ii) the time evolution of the expectation values of the coordinate and momentum,

provided that in the initial moment ($t = 0$) it was in the state described by the following ket-vector:

$$|\alpha\rangle = \frac{1}{\sqrt{2}}(|31\rangle + |32\rangle),$$

where $|n\rangle$ are the ket-vectors of the stationary (Fock) states of the oscillator.

Solutions:

(i) In this Hamiltonian system, the total energy is conserved, so we may calculate it in the initial moment:

$$\langle E\rangle = \langle\alpha(0)|\ \hat{H}\ |\ \alpha(0)\rangle = \frac{1}{2}(\langle 31| + \langle 32\ |)\hat{H}(|31\rangle + |32\rangle).$$

Using the fact that the Hamiltonian is diagonal in the basis of Fock states n, with the diagonal elements equal to $E_n = \hbar\omega_0(n + ½)$, we get

$$\langle E\rangle = \frac{1}{2}(\langle 31|\ \hat{H}\ |31\rangle + \langle 32|\ \hat{H}\ |32\rangle) = \frac{\hbar\omega_0}{2}\left[\left(31 + \frac{1}{2}\right) + \left(32 + \frac{1}{2}\right)\right] = 32\hbar\omega_0.$$

(ii) The time evolution of the expectation values of x and p may be obtained, for example, from Eqs. (5.36) of the lecture notes, with $U = m\omega_0{}^2x^2/2$:

$$\langle\dot{x}\rangle = \frac{\langle p\rangle}{m}, \qquad \langle\dot{p}\rangle = -m\omega_0^2\langle x\rangle. \qquad (*)$$

These equations (which coincide with the classical equations of motion), have the well-known solution[15]

$$\begin{aligned}\langle x\rangle(t) &= \langle x\rangle(0)\cos\omega_0 t + \frac{\langle p\rangle(0)}{m\omega_0}\sin\omega_0 t,\\ \langle p\rangle(t) &= \langle p\rangle(0)\cos\omega_0 t - m\omega_0\langle x\rangle(0)\sin\omega_0 t,\end{aligned} \qquad (**)$$

so that the only thing still to be done is to find the expectation values of these observables at $t = 0$. This may be accomplished exactly as has been done above for energy, but with a little bit more care, because the matrices elements of the coordinate and momentum operators, in the Fock state basis, are *not* diagonal—see Eqs. (5.92)–(5.193) of the lecture notes:

[15] As a (hopefully, unnecessary) reminder: one can, for example, differentiate one of Eqs. (*) over time again, and plug the counterpart equation into the result, getting the standard second-order differential equation $\ddot{\xi} + \omega_0^2\xi = 0$, with ξ being either $\langle x\rangle$ or $\langle p\rangle$. The general solution of this equation is $\xi(t) = C_c\cos\omega_0 t + C_s\sin\omega_0 t$. Now calculating the constants C_c and C_s from initial conditions, we arrive at Eqs. (**).

$$\langle x\rangle(0) = \frac{1}{2}((\langle 31| + \langle 32|)\,\hat{x}\,(|31\rangle + |32\rangle))$$
$$= \frac{1}{2}(\langle 31|\ \hat{x}\ |32\rangle + \langle 32|\ \hat{x}\ |31\rangle)$$
$$= \frac{1}{2}\frac{x_0}{\sqrt{2}}(\sqrt{32} + \sqrt{32}) = 4x_0,$$
$$\langle p\rangle(0) = \frac{1}{2}((\langle 31| + \langle 32|)\hat{p}(|31\rangle + |32\rangle))$$
$$= \frac{1}{2}(\langle 31|\ \hat{p}\ |32\rangle + \langle 32|\ \hat{p}\ |31\rangle)$$
$$= \frac{1}{2}\frac{m\omega_0 x_0}{\sqrt{2}}(\sqrt{32} - \sqrt{32}) = 0.$$

Plugging these expressions into Eq. (**), we get

$$\langle x\rangle = 4x_0 \cos \omega_0 t, \quad \langle p\rangle = -4m\omega_0 x_0 \sin \omega_0 t.$$

Note that the exact answer could be different if there was a phase shift between the component Fock states—see, e.g. the solution of problem 4.17. However, even in this case the expectation values of the coordinate and momentum would oscillate with the oscillator's frequency ω_0.

Problem* 5.15. Re-derive the London dispersion force's potential of interaction of two isotropic 3D harmonic oscillators (already calculated in problem 3.16), using the language of mutually-induced polarization.

Solution[16]: The solution of problem 3.16, based on the calculation of the ground-state energy of the system, somewhat obscures the physical nature of the force. A more transparent physical picture of this effect is that the 'quantum fluctuations' (the term frequently used for more vivid description of quantum uncertainties) of the dipole moments $\mathbf{d}_{1,2}$ of the interacting oscillators, cause proportional fluctuating electric fields $\mathscr{E}_{1,2}(\mathbf{r}) \propto \mathbf{d}_{1,2}$ in their vicinity, including the location of the counterpart oscillator. Each of these fields induces a small additional polarization, $\tilde{\mathbf{d}}_2 \propto \mathscr{E}_1 \propto \mathbf{d}_1$, and $\tilde{\mathbf{d}}_1 \propto \mathscr{E}_2 \propto \mathbf{d}_2$, of the counterpart oscillator, on the top of its spontaneous fluctuations. In contrast to the mutually independent, spontaneous fluctuations $\mathbf{d}_{1,2}$, the induced parts $\tilde{\mathbf{d}}_{1,2}$ of the polarization are correlated (coherent) with their sources: $\tilde{\mathbf{d}}_2$ with $\mathbf{d}_1$, and $\tilde{\mathbf{d}}_1$ with $\mathbf{d}_2$, so that their interaction has a nonvanishing average component, resulting in a mutual attraction of the oscillators.

Let us make this argumentation quantitative, using the isotropic single-particle model already used for the model solution of problem 3.16: $\mathbf{d}_{1,2} = q\mathbf{s}_{1,2}$, where q is the electric charge of the effective oscillator's particle and $\mathbf{s}_{1,2}$ its displacement from

[16] This explanation of the long-range interaction between electroneutral molecules was suggested by P Debye in 1921, and quantified by F London in 1937.

the origin. Since the classical electric field $\mathscr{E}$ of a dipole is proportional to its moment **d**, the relation between their Heisenberg-picture operators in quantum mechanics is the same, namely[17]

$$\hat{\mathscr{E}}_k = \frac{1}{4\pi\varepsilon_0}\frac{3\mathbf{r}(\mathbf{r}\cdot\hat{\mathbf{d}}_k) - \hat{\mathbf{d}}_k r^2}{r^5} \equiv \frac{1}{4\pi\varepsilon_0 r^3}(-\hat{d}_{kx}\mathbf{n}_x - \hat{d}_{ky}\mathbf{n}_y + 2\hat{d}_{kz}\mathbf{n}_z) \equiv \frac{q}{4\pi\varepsilon_0 r^3}(-\hat{x}_k\mathbf{n}_x - \hat{y}_k\mathbf{n}_y + 2\hat{z}_k\mathbf{n}_z), \qquad (*)$$

where $k = 1, 2$ is the dipole number, r is the distance between the dipoles, and $\{x_k, y_k, z_k\}$ are the Cartesian components of the displacement vector $\mathbf{s}_k$, with axis z directed along the line connecting the dipoles, so that $\mathbf{r} = \{0, 0, r\}$. According to Eq. (5.141) of the lecture notes, in a 1D harmonic oscillator, the coordinate operators change with the oscillator's eigenfrequency, for example

$$\begin{aligned}\hat{x}_k(t) &= \left(\frac{\hbar}{2m\omega_k}\right)^{1/2}[\hat{a}_k(t) + \hat{a}_k^\dagger(t)]\\ &= \left(\frac{\hbar}{2m\omega_k}\right)^{1/2}[\hat{a}_k(0)\exp\{-i\omega_k t\} + \hat{a}_k^\dagger(0)\exp\{-i\omega_k t\}]\\ &\equiv \hat{X}_{k\omega}\exp\{-i\omega_k t\} + \hat{X}_{k\omega}^\dagger\exp\{i\omega_k t\}\\ &\equiv \hat{X}_{k\omega}\exp\{-i\omega_k t\} + \text{h.c.}, \quad \text{with } \hat{X}_{k\omega} \equiv \left(\frac{\hbar}{2m\omega_k}\right)^{1/2}\hat{a}(0),\end{aligned}$$

and similarly for two other Cartesian components[18], so that the electric fields (*) they induce have similar time dependences.

Next, since the classical equations of motion of the harmonic oscillator are linear, the Heisenberg equations of motion are also linear, the complex amplitudes of the Fourier components of the induced dipole moment operators may be also calculated using the classical relation[19,20]

$$\hat{\mathbf{d}}_{k\omega} \equiv q(\mathbf{n}_x\hat{X}_{k\omega} + \mathbf{n}_y\hat{Y}_{k\omega} + \mathbf{n}_z\hat{Z}_{k\omega}) = \frac{q^2\hat{\mathscr{E}}_{k'\omega}}{m\left(\omega_k^2 - \omega_{k'}^2\right)}, \qquad (**)$$

where the index k', which may be formally defined as $(3 - k)$, is used for the notation of the counterpart oscillator. Since these expressions diverge at $\omega_k \to \pm\omega_{k'}$, i.e. at

[17] See, e.g. *Part EM* section 3.1, in particular Eq. (3.13), which uses a different notation (**p**) for the electric dipole vector.

[18] Due to the assumed oscillator's isotropy, the frequencies ω_k are the same for all three coordinates.

[19] If this formula is not immediately evident, see, e.g. *Part CM* section 5.1.

[20] This relation between d_ω and $\mathscr{E}_\omega$ defines the complex electric permittivity $\varepsilon(\omega)$ of the continuous medium of such dipoles, which in turn determines the dispersion of electromagnetic wave propagating in it (see, e.g. *Part EM* section 7.2). This fact was the origin of the term *dispersion force*, coined by F London for the dipole–dipole interaction we are calculating.

$\omega_1 \to \pm\omega_2$, let us assume for a while that the oscillator eigenfrequencies are not exactly equal.

The energy of interaction of a dipole $\mathbf{d}_k$ with an external electric field $\mathscr{E}_{k'}$ is proportional to the scalar product $\mathbf{d}_k \cdot \mathscr{E}_{k'}$. In our case, the dipole moment of each oscillator is the sum of two parts: the spontaneous fluctuations $\mathbf{d}_k$ and the externally-induced polarization $\tilde{\mathbf{d}}_k$. Only the latter part is correlated with $\mathscr{E}_{k'}$,[21] so that only it contributes to a nonvanishing average interaction energy. Since $\tilde{\mathbf{d}}_k \propto \mathscr{E}_{k'}$, the energy needs the factor ½ before the product $-\tilde{\mathbf{d}}_k \cdot \mathscr{E}_{k'}$,[22] so that the expectation value of the full interaction potential may be calculated as

$$\langle U \rangle = -\frac{1}{2}\left\langle \overline{\hat{\mathscr{E}}_1 \cdot \hat{\tilde{\mathbf{d}}}_2} \right\rangle - \frac{1}{2}\left\langle \overline{\hat{\mathscr{E}}_2 \cdot \hat{\tilde{\mathbf{d}}}_1} \right\rangle, \qquad (***)$$

where the top bar means the time average. Let us spell out, for example, the first average of this sum, using Eq. (**) first, and then Eq. (*):

$$\begin{aligned}
\left\langle \overline{\hat{\mathscr{E}}_1 \cdot \hat{\tilde{d}}_2} \right\rangle &\equiv \left\langle \overline{\hat{\mathscr{E}}_{1x}(t)\, \hat{\tilde{d}}_{2x}(t) + \hat{\mathscr{E}}_{1y}(t)\, \hat{\tilde{d}}_{2y}(t) + \hat{\mathscr{E}}_{1z}(t)\, \hat{\tilde{d}}_{2z}(t)} \right\rangle \\
&= \frac{q^2}{m\left(\omega_2^2 - \omega_1^2\right)} \left\langle \hat{\mathscr{E}}_{1x\omega}\hat{\mathscr{E}}^\dagger_{1x\omega} + \hat{\mathscr{E}}_{1y\omega}\hat{\mathscr{E}}^\dagger_{1y\omega} + \hat{\mathscr{E}}_{1z\omega}\hat{\mathscr{E}}^\dagger_{1z\omega} \right\rangle \\
&= \frac{q^2}{m\left(\omega_2^2 - \omega_1^2\right)} \left(\frac{q}{4\pi\varepsilon_0 r^3}\right)^2 \left\langle \hat{X}_{1\omega}\hat{X}^\dagger_{1\omega} + \hat{Y}_{1\omega}\hat{Y}^\dagger_{1\omega} + 4\hat{Z}_{1\omega}\hat{Z}^\dagger_{1\omega} \right\rangle.
\end{aligned}$$

Due to the assumed isotropy of the oscillators, all coordinate product averages are equal, so that we get

$$\begin{aligned}
\left\langle \overline{\hat{\mathscr{E}}_1 \cdot \hat{\tilde{\mathbf{d}}}_2} \right\rangle &= \frac{q^2}{m\left(\omega_2^2 - \omega_1^2\right)} \left(\frac{q}{4\pi\varepsilon_0 r^3}\right)^2 6\left\langle \hat{X}_{1\omega}\hat{X}^\dagger_{1\omega} \right\rangle \\
&\equiv \frac{q^2}{m\left(\omega_2^2 - \omega_1^2\right)} \left(\frac{q}{4\pi\varepsilon_0 r^3}\right)^2 6\frac{\hbar}{2m\omega_1} \langle \hat{a}(0)\, \hat{a}^\dagger(0) + \text{h.c.} \rangle.
\end{aligned}$$

The last average may be treated as that of the Schrödinger-picture operators, and according to Eqs. (5.70) and (5. 75) of the lecture notes, equals 1 in the ground state of the oscillator, so that, finally,

$$\left\langle \overline{\hat{\mathscr{E}}_1 \cdot \hat{\tilde{\mathbf{d}}}_2} \right\rangle = \frac{6q^2}{m\left(\omega_2^2 - \omega_1^2\right)} \left(\frac{q}{4\pi\varepsilon_0 r^3}\right)^2 \frac{\hbar}{2m\omega_1} \equiv \frac{3q^4\hbar}{(4\pi\varepsilon_0)^2 m^2 r^6} \frac{1}{\omega_1\left(\omega_2^2 - \omega_1^2\right)}.$$

[21] Mathematically, the absence of mutual correlation of the main (spontaneous) part $\mathbf{d}_k$ with $\mathscr{E}_{k'} \propto \mathbf{d}_{k'}$ at $\omega_k \neq \omega_{k'}$ is expressed as the difference of their Heisenberg-picture operator frequencies, leading to averaging out of all terms $\exp\{\pm i(\omega_k \pm \omega_{k'})t\}$ of the $\mathbf{d}_k \cdot \mathbf{E}_{k'}$ products. The special case $\omega_k = \omega_{k'}$ (i.e. $\omega_1 = \omega_2$) needs a little bit more subtle analysis (based on random phase differences), which leads to the same conclusion.

[22] See, e.g. *Part EM* Eq. (3.15*b*).

The second term in Eq. (***) is absolutely similar, with swapped indices 1 and 2, so that we finally get[23]

$$\langle U \rangle = -\frac{3q^4\hbar}{2(4\pi\varepsilon_0)^2 m^2 r^6}\left[\frac{1}{\omega_1\left(\omega_2^2 - \omega_1^2\right)} + \frac{1}{\omega_2\left(\omega_1^2 - \omega_2^2\right)}\right]$$
$$\equiv -\frac{3q^4\hbar}{2(4\pi\varepsilon_0)^2 m^2 r^6}\frac{1}{\omega_1\omega_2(\omega_1 + \omega_1)}.$$

Now it is safe to consider the most important case of similar oscillators, $\omega_1 = \omega_2 \equiv \omega_0$, and verify that the result,

$$U = -\frac{3}{4}\frac{q^4\hbar}{(4\pi\varepsilon_0)^2 m^2 r^6 \omega_0^3} \equiv -\frac{3}{4}\mu^2\hbar\omega_0, \quad \text{with } \mu \equiv \frac{q^2}{4\pi\varepsilon_0 r^3 m\omega_0^2} \ll 1,$$

exactly coincides with the one obtained in the model solution of problem 3.16, by simpler means. However, using the current, more lengthy derivation, we not only have obtained a more clear physical picture of the London dispersion force, but also paved the way toward generalizations of this result to a more general model of the interacting atoms/molecules (see problem 6.18) and to their thermally-equilibrium state at a non-zero temperature (see problem 7.6).

Finally, let me note that one more popular form of the final result may be obtained by expressing it via the static *atomic polarizability* α, which may be defined by the relation[24]

$$\hat{\mathbf{d}}_\omega = \alpha\hat{\mathscr{E}}_\omega, \quad \text{at } \omega \to 0;$$

according to Eq. (**), in our simple oscillator model $\alpha = q^2/m\omega_0^2$, so that

$$U = -\frac{3}{4}\frac{\alpha^2\hbar\omega_0}{(4\pi\varepsilon_0)^2 r^6}.$$

However, I believe that this form conceals the resonant nature of the London dispersion force, which is so evident from the above calculation.

***Problem* 5.16.** An external force pulse $F(t)$, of a finite time duration $\mathcal{T}$, has been exerted on a 1D harmonic oscillator, initially in its ground state. Use the Heisenberg-picture equations of motion to calculate the expectation value of the oscillator's energy at the end of the pulse.

Solution: Plugging the system's Hamiltonian, which is a straightforward generalization of Eq. (5.62),

[23] To my personal taste, this 'miraculous' cancellation of the two divergences, which allows one to pursue the limit $\omega_1 \to \omega_2$ without mathematical complications, is one of the most beautiful results of quantum mechanics.

[24] See, e.g. *Part EM* Eq. (3.48).

$$\hat{H} = \frac{\hat{p}^2}{2m} + \frac{m\omega_0^2\hat{x}^2}{2} - F(t)\hat{x},$$

into the Heisenberg equations of motion (4.199), we get

$$i\hbar\dot{\hat{x}} = [\hat{x}, \hat{H}] = \frac{1}{2m}[\hat{x}, \hat{p}^2], \qquad i\hbar\dot{\hat{p}} = [\hat{p}, \hat{H}] = \frac{m\omega_0^2}{2}[\hat{p}, \hat{x}^2] - F(t)\,[\hat{p}, \hat{x}].$$

(The index H is implied.) The right-hand side of the first of these equations, and the first term in the second equation were already spelled out in section 5.2 of the lecture notes, and the last term in the second equation is just the product of the *c*-number function $F(t)$ by the basic commutator (4.238). As a result, we get the operator equations,

$$\dot{\hat{x}} = \frac{\hat{p}}{m}, \qquad \dot{\hat{p}} = -m\omega_0^2\hat{x} + F(t),$$

which have the same form as the classical equations of motion for the corresponding observables. Due to the linearity of these equations, they are satisfied by the following linear superposition:

$$\hat{x}(t) = \hat{x}_0(t) + X(t), \qquad \hat{p}(t) = \hat{p}_0(t) + P(t).$$

Here the index 0 marks the solution for $F(t) = 0$, which satisfies Eqs. (5.139) and the initial conditions, while X and P are the *c*-number additions due to the classical external force $F(t)$, which satisfy the similar equation of motion:

$$\dot{X} = \frac{P}{m}, \qquad \dot{P} = -m\omega_0^2 X + F(t),$$

with zero initial conditions: $X(0) = P(0) = 0$, where $t = 0$ is the moment of the beginning of the force pulse. The solution of this system of equations is well known from classical mechanics[25]; for $t = \mathcal{T}$ it gives

$$X(\mathcal{T}) = \frac{1}{m\omega_0}\int_0^{\mathcal{T}} F(t)\sin\omega_0(\mathcal{T} - t)\,dt, \quad P(\mathcal{T}) = \int_0^{\mathcal{T}} F(t)\cos\omega_0(\mathcal{T} - t)\,dt. \quad (*)$$

Now we may calculate the Heisenberg 'value' of the Hamiltonian at $t = \mathcal{T} + 0$, i.e. immediately after the end of the force pulse $F(t)$:

$$\begin{aligned}\hat{H}(\mathcal{T}) &= \frac{\hat{p}^2(\mathcal{T})}{2m} + \frac{m\omega_0^2\hat{x}^2(\mathcal{T})}{2}\\ &= \left[\frac{\hat{p}_0^2(\mathcal{T})}{2m} + \frac{m\omega_0^2\hat{x}_0^2(\mathcal{T})}{2}\right] + \left[\frac{P^2(\mathcal{T})}{2m} + \frac{m\omega_0^2X^2(\mathcal{T})}{2}\right] \qquad (**)\\ &\quad + \frac{1}{m}\hat{p}_0(\mathcal{T})P(\mathcal{T}) + m\omega_0^2\hat{x}_0(\mathcal{T})X(\mathcal{T})\,.\end{aligned}$$

[25] See, e.g. CM Eqs. (5.27) and (5.34), with $\delta = 0$, so that $\omega_0' = \omega_0$.

Since initially the oscillator was in its ground state, even at $t = \mathcal{T}$ the force-independent operators $\hat{x}_0$ and $\hat{p}_0$ still describe this state, with zero expectation values of the coordinate and momentum, and the energy equal to $\hbar\omega_0/2$. As a result, at the statistical (not time!) averaging of Eq. (**), the two last terms on its right-hand side vanish, and we get

$$\langle E\rangle \equiv \langle \hat{H}\rangle = \left\langle \frac{\hat{p}_0^2(\mathcal{T})}{2m} + \frac{m\omega_0^2\hat{x}_0^2(\mathcal{T})}{2}\right\rangle + \frac{P^2(\mathcal{T})}{2m} + \frac{m\omega_0^2 X^2(\mathcal{T})}{2} = \frac{\hbar\omega_0}{2} + E_F,$$

where E_F is the classical energy acquired from the force's pulse, which may be readily calculated from Eqs. (*):[26]

$$\begin{aligned} E_F &= \frac{P^2(\mathcal{T})}{2m} + \frac{m\omega_0^2 X^2(\mathcal{T})}{2} \\ &= \frac{1}{2m}\left\{\left[\int_0^{\mathcal{T}} F(t)\cos\omega_0(\mathcal{T}-t)dt\right]^2 + \left[\int_0^{\mathcal{T}} F(t)\sin\omega_0(\mathcal{T}-t)dt\right]^2\right\}. \end{aligned}$$

This result is in full correspondence with the physical picture of the final (Glauber) state, which may be described as the ground state of the oscillator, with its center translated, by the force pulse, into the point $\{X(\mathcal{T}), P(\mathcal{T})/m\omega_0\}$ on the phase plane—see section 5.5 of the lecture notes, in particular figure 5.8.

***Problem* 5.17.** Use Eqs. (5.144) and (5.145) of the lecture notes to calculate the uncertainties δx and δp of a squeezed ground state, and in particular prove Eqs. (5.143) for the case $\theta = 0$.

Solution: Let us represent the squeezed annihilation operator (5.144) in the following form,

$$\hat{b} = \mu\,\hat{a} + \nu\,\hat{a}^\dagger, \tag{*}$$

more convenient for calculations. According to Eq. (5.144),

$$\mu \equiv \cosh r, \qquad \nu \equiv e^{i\theta}\sinh r,$$

where r and θ are the real c-numbers describing the complex parameter $\zeta = re^{i\varphi}$ of its eigenstate. Since the parameter μ is real, the Hermitian conjugate of Eq. (*) is

$$\hat{b}^\dagger = \mu\,\hat{a}^\dagger + \nu^*\hat{a}. \tag{**}$$

Now solving the system of two linear equations (*) and (**), we get the reciprocal relations

[26] Alternatively, E_F may be calculated by integrating over time, from $t = 0$ to $t = T$, the instant power $\mathscr{P}(t) = F(t)V(t) = F(t)P(t)/m$ of the external force, using the second of Eqs. (**), rewritten for an arbitrary time t within the interval $[0, \mathcal{T}]$. (See, e.g. the model solution of *Part CM* problem 5.4.)

$$\hat{a} = \mu\,\hat{b} - \nu\,\hat{b}^\dagger, \qquad \hat{a}^\dagger = \mu\,\hat{b}^\dagger - \nu^*\hat{b}.$$

Since, according to Eqs. (5.66) of the lecture notes,

$$\hat{x} = \frac{x_0}{\sqrt{2}}(\hat{a} + \hat{a}^\dagger), \qquad \hat{p} = \frac{m\omega_0 x_0}{\sqrt{2}i}(\hat{a} - \hat{a}^\dagger),$$

we may express these operators via the squeezed creation–annihilation operators:

$$\hat{x} = \frac{x_0}{\sqrt{2}}\left[(\mu - \nu^*)\,\hat{b} + (\mu - \nu)\hat{b}^\dagger\right], \qquad \hat{p} = \frac{m\omega_0 x_0}{\sqrt{2}i}\left[(\mu + \nu^*)\,\hat{b} + (-\mu - \nu)\hat{b}^\dagger\right].$$

Let us use the first expression to calculate the expectation values of the coordinate and its square in a state ζ:

$$\langle x\rangle \equiv \langle\zeta|\,\hat{x}\,|\zeta\rangle = \frac{x_0}{\sqrt{2}}[(\mu - \nu^*)\langle\zeta|\;\hat{b}\;|\zeta\rangle + (\mu - \nu)\langle\zeta|\;\hat{b}^\dagger\;|\zeta\rangle],$$

$$\langle x^2\rangle = \frac{x_0^2}{2}\left[\begin{array}{l}(\mu - \nu^*)^2\langle\zeta|\;\hat{b}^2\;|\zeta\rangle + (\mu - \nu)^2\langle\zeta|\,\hat{b}^{\dagger 2}\;|\zeta\rangle \\ +(\mu - \nu^*)\,(\mu - \nu)\left(\langle\zeta|\;\hat{b}\hat{b}^\dagger\;|\zeta\rangle + \langle\zeta|\;\hat{b}^\dagger\hat{b}\;|\zeta\rangle\right)\end{array}\right]. \qquad (***)$$

According to Eqs. (5.144) and (5.145) of the lecture notes, for the general squeezed state,

$$\hat{b}\;|\zeta\rangle = (\alpha\mu + \alpha^*\nu)\;|\zeta\rangle, \quad \text{so that} \quad \langle\zeta|\;\hat{b}^\dagger = \langle\zeta|\;(\alpha^*\mu + \alpha\nu^*).$$

However, our task is to discuss only the special, *ground* squeezed state ζ, with $\alpha = 0$. For this state, the last relations are reduced to

$$\hat{b}\;|\zeta\rangle = 0, \qquad \langle\zeta|\;\hat{b}^\dagger = 0.$$

With these equalities, the first of Eqs. (***) immediately gives $\langle x\rangle = 0$,[27] while in the second of these expressions, only one average survives, giving

$$\langle x^2\rangle = \frac{x_0^2}{2}(\mu - \nu^*)\,(\mu - \nu)\,\langle\zeta|\;\hat{b}\hat{b}^\dagger\;|\zeta\rangle\,.$$

In order to evaluate this expression, let us use the fact that $\mu^2 - \nu\nu^* \equiv \cosh^2 r - \sinh^2 r = 1$, to verify that the squeezed creation–annihilation operators satisfy the same commutation relation as the usual creation–annihilation operators—see Eq. (5.68):

$$\begin{aligned}[\hat{b}, \hat{b}^\dagger] &= [(\mu\hat{a} + \nu\hat{a}^\dagger), (\mu\hat{a}^\dagger + \nu^*\hat{a})] \\ &= \mu^2[\hat{a}, \hat{a}^\dagger] + \nu\nu^*[\hat{a}^\dagger, \hat{a}] \\ &= (\mu^2 - \nu\nu^*)\;[\hat{a}, \hat{a}^\dagger] = \hat{I}.\end{aligned}$$

[27] This result could be readily anticipated from the physical sense of the squeezed ground state—see, e.g. its image in figure 5.8 of the lecture notes.

This relation may be rewritten as a convenient operator identity

$$\hat{b}\hat{b}^\dagger = \hat{b}^\dagger\hat{b} + \hat{I}; \qquad (****)$$

plugging it into the last expression for $\langle x^2 \rangle$, we get

$$\langle x^2 \rangle = \frac{x_0^2}{2}(\mu - \nu^*)(\mu - \nu)(\langle\zeta|\, \hat{b}^\dagger\hat{b}\, |\zeta\rangle + \langle\zeta|\, \hat{I}\, |\zeta\rangle)$$
$$= \frac{x_0^2}{2}(\mu - \nu^*)(\mu - \nu)\langle\zeta|\zeta\rangle.$$

According to its definition (5.142*b*), the squeezing operator $\hat{S}_\zeta$ is unitary, and hence the squeezed ground states are normalized to 1:

$$\langle\zeta|\zeta\rangle \equiv \langle 0|\, \hat{S}_\zeta^\dagger \hat{S}_\zeta\, |0\rangle = \langle 0|\, \hat{I}\, |0\rangle = 1,$$

so that, finally,

$$\langle x^2 \rangle = \frac{x_0^2}{2}(\mu - \nu^*)\,(\mu - \nu),$$

and we may use the general Eqs. (1.33) and (1.34) to get

$$\delta x = (\langle x^2 \rangle - \langle x \rangle^2)^{1/2} = \frac{x_0}{\sqrt{2}}[(\mu - \nu^*)\,(\mu - \nu)]^{1/2}$$
$$\equiv \frac{x_0}{\sqrt{2}}(\cosh^2 r + \sinh^2 r - 2\sinh r \cosh r \cos\theta)^{1/2}.$$

This general result depends on the phase θ, i.e. on time (reflecting the rotation of the squeezed state's image on the phase plane shown in figure 5.8), but for particular instants when θ is equal to 0 (plus any multiple of π), i.e. $\cos\theta = 1$, it takes the minimum value stated in the first of Eqs. (5.143):

$$\delta x = \frac{x_0}{\sqrt{2}}(\cosh^2 r + \sinh^2 r - 2\sinh r \cosh r)^{1/2} \equiv \frac{x_0}{\sqrt{2}}(\cosh r - \sinh r) \equiv \frac{x_0}{\sqrt{2}}e^{-r}.$$

The proof of the second of Eqs. (5.143), valid for the same moments of time, is absolutely similar.

***Problem* 5.18.** Calculate the energy of a harmonic oscillator in the squeezed ground state ζ.

Solution: Let us re-use the expressions

$$\hat{a} = \mu\hat{b} - \nu\hat{b}^\dagger, \qquad \hat{a}^\dagger = \mu\hat{b}^\dagger - \nu^*\hat{b},$$

derived in the solution of the previous problem. Plugging them into Eqs. (5.73) of the lecture notes, let us calculate the average $\langle N \rangle$, which determines the state's energy $E \equiv \langle H \rangle = \hbar\omega_0(\langle N \rangle + ½)$, in a squeezed state ζ:

$$\begin{aligned}\langle N\rangle \equiv \langle \hat{N}\rangle &= \langle \hat{a}^\dagger \hat{a}\rangle = \langle \zeta|\, \hat{a}^\dagger \hat{a}\, |\zeta\rangle \\ &= \langle \zeta|(\mu\hat{b}^\dagger - \nu^*\hat{b})(\mu\hat{b} - \nu\hat{b}^\dagger)|\zeta\rangle \\ &= \mu^2\langle \zeta|\, \hat{b}^\dagger\hat{b}\, |\zeta\rangle + \nu\nu^*\langle \zeta|\, \hat{b}\hat{b}^\dagger\, |\zeta\rangle - \mu\nu\langle \zeta|\, \hat{b}^\dagger\hat{b}^\dagger\, |\zeta\rangle - \mu\nu^*\langle \zeta|\, \hat{b}\hat{b}\, |\zeta\rangle.\end{aligned} \quad (*)$$

Our task is to discuss only the special, ground squeezed state ζ, with $\alpha = 0$. As was discussed in the previous problem's solution, it has the following properties:

$$\hat{b}\,|\zeta\rangle = 0, \quad \text{and hence} \quad \langle \zeta|\,\hat{b}^\dagger = 0,$$

Due to these properties, all terms in the last form of Eq. (*), besides the second one, are equal to zero, so that

$$\langle N\rangle = \nu\nu^*\langle \zeta|\, \hat{b}\hat{b}^\dagger\, |\zeta\rangle.$$

Applying, to this expression, the relations derived in the solution of the previous problem,

$$\hat{b}\hat{b}^\dagger = \hat{b}^\dagger\hat{b} + \hat{I}, \quad \text{and } \langle \zeta|\zeta\rangle = 1;$$

we finally get

$$\begin{aligned}\langle N\rangle &= \nu\nu^*\langle \zeta|\, \hat{b}\hat{b}^\dagger\, |\zeta\rangle = \nu\nu^*\langle \zeta|(\hat{b}^\dagger\hat{b} + \hat{I})|\zeta\rangle \\ &= \nu\nu^*\langle \zeta|\zeta\rangle = \nu\nu^* \equiv \sinh^2 r,\end{aligned}$$

so that

$$E = \hbar\omega_0\left(\sinh^2 r + \frac{1}{2}\right).$$

Proceeding to discussion of this result, note first of all that it is independent of the parameter θ. Actually, this fact could be predicted from the physical sense of that parameter as the (double) angle that determines the squeezing direction—see figure 5.8 of the lecture notes. Next, at $r \to 0$ (no squeezing), $E \to \hbar\omega_0/2$, which is the correct energy of the Fock/Glauber ground state. However, at $r \gg 1$, E is much larger than $\hbar\omega_0/2$, so that the adjective 'ground' in the name of this squeezed state should not be taken too literally. The same is true for the term *squeezed vacuum*, used for electromagnetic field oscillators (see section 9.1) in the ground state ζ; actually, such a 'vacuum' may have a lot of energy in it!

Problem* 5.19. Prove that the squeezed ground state, described by Eqs. (5.142), (5.144) and (5.145) of the lecture notes, may be sustained by a sinusoidal modulation of a harmonic oscillator's parameter, and calculate the squeezing factor r as a function of the parameter modulation depth, assuming that the depth is small, and the oscillator's damping is negligible.

Solution: The classical van der Pol analysis of a dissipation-free harmonic oscillator, of frequency ω_0, with one of its parameters weakly modulated with frequency

$2\omega \approx 2\omega_0$, gives[28] the following equation of motion of the complex amplitude a of the oscillations, defined by the relation $x(t) = \text{Re}[a(t)\exp\{-i\omega t\}]$:

$$\dot{a} = i\xi a - ima^*, \qquad (*)$$

where $\xi \equiv \omega - \omega_0$ is called the detuning, and m is proportional to the parameter modulation depth. (Eq. (*) is strictly valid only if both m and $|\xi|$ are much smaller than ω.) A straightforward analysis of this linear differential equation shows that the parametric excitation, i.e. an exponential growth of $|a(t)|$, takes place if m exceeds the following critical value:

$$m_c = |\xi|;$$

because of this, we will focus on the case $m < m_c$.

Due to the similarity of the equations of motion of observables in classical mechanics and the corresponding Heisenberg operators in quantum mechanics (see section 5.2 of the lecture notes), we may mimic Eq. (*) as the so-called RWA equation of the parametric oscillator:

$$\dot{\hat{a}} = i\xi\hat{a} - im\hat{a}^\dagger, \qquad \text{and hence} \qquad \dot{\hat{a}}^\dagger = -i\xi\hat{a}^\dagger + im\hat{a}.$$

Note that these $\hat{a}^\dagger$ and $\hat{a}$ are not exactly the creation–annihilation operators defined by Eqs. (5.65) of the lecture notes, in two aspects: first, they are not necessarily properly normalized (which does not matter for this linear system), and second, they include additional factors $\exp\{\pm i\omega t\}$; however, the latter difference also does not affect the forthcoming calculation.

Now, transferring to the mixed operator $\hat{b}$, defined by Eq. (5.144) of the lecture notes, and its Hermitian conjugate, in the form used in the solutions of the two previous problems,

$$\hat{b} = \mu\hat{a} + \nu\hat{a}^\dagger, \qquad \hat{b}^\dagger = \mu\hat{a}^\dagger + \nu^*\hat{a}, \qquad \text{where} \qquad \mu \equiv \cosh r, \qquad \nu \equiv e^{i\theta}\sinh r, \quad (**)$$

we get the following equation of motion of the operator $\hat{b}$:

$$\dot{\hat{b}} = i\,[\xi(\mu^2 + \nu\nu^*) + m\mu(\nu + \nu^*)]\,\hat{b} - i\,[m(\mu^2 + \nu^2) + 2\xi\mu\nu]\,\hat{b}^\dagger, \qquad (***)$$

and its Hermitian conjugate for the operator $\hat{b}^\dagger$. A time-independent[29] squeezed state corresponds to the parameters μ and ν (and hence r and θ) satisfying two conditions: the first square bracket in Eq. (***) should be purely real, and the second square bracket should completely vanish. Upon the substitution of the above expressions for μ and ν, these two conditions give the same results: $e^{2i\theta} = 1$,[30] and

$$\tanh 2r = \pm\frac{m}{\xi}.$$

[28] See, e.g. *Part CM* Eq. (5.78) with $m \equiv \mu\omega/4$ and $\delta = 0$.

[29] Again, besides its rotation, on the phase plane, with a constant frequency close to ω and ω_0—see figure 5.8.

[30] In the duality of the solution for the angle θ: $\theta_1 = 0$ and $\theta_2 = \pi$ (plus any multiple of 2π), we may readily recognize two possible (and equivalent) phases of the degenerate parametric excitation—see, e.g. *Part CM* section 5.5.

The last result shows that in the absence of the parameter modulation ($m = 0$), r vanishes, so that according to Eq. (**), the operators $\hat{a}$ and $\hat{b}$ (and hence their eigenstates) coincide, and the squeezed state turns into the Glauber state—as it should. On the other hand, as m approaches its critical value $m_c = |\xi|$, the squeezing factor $|r|$ tends to ∞, i.e. the squeezing becomes infinitely strong. According to the solution of the previous problem, this also means that the energy of the squeezed state tends to infinity, even if its center is located at the phase plane's origin, i.e. if the parameter α in Eq. (5.145) equals zero.

***Problem* 5.20.** Use Eqs. (5.148) of the lecture notes to prove that the operators $\hat{L}_j$ and $\hat{L}^2$ commute with the Hamiltonian of a spinless particle placed in any central potential field.

Solution: The Hamiltonian in question may be represented as

$$\hat{H} = \frac{\hat{p}^2}{2m} + U(\hat{\mathbf{r}}) \equiv \frac{\hat{p}^2}{2m} + f(\hat{r}^2), \quad \text{where} \quad \hat{r}^2 = \sum_{j=1}^{3} \hat{r}_j^2, \quad \text{and } \hat{p}^2 = \sum_{j=1}^{3} \hat{p}_j^2. \quad (*)$$

Let us first calculate the commutators of $\hat{L}_j$ with $\hat{r}_{j'}^2$ and $\hat{p}_{j'}^2$, for arbitrary j and j'. For the first of them, we may use Eq. (5.148) of the lecture notes, rewritten as

$$\hat{L}_j \hat{r}_{j'} = \hat{r}_{j'} \hat{L}_j + i\hbar \hat{r}_{j''} \varepsilon_{jj'j''}.$$

Using this relation twice, we get

$$\begin{aligned}
\left[\hat{L}_j, \hat{r}_{j'}^2\right] &\equiv \hat{L}_j \hat{r}_{j'}^2 - \hat{r}_{j'}^2 \hat{L}_j = \hat{L}_j \hat{r}_{j'} \hat{r}_{j'} - \hat{r}_{j'} \hat{r}_{j'} \hat{L}_j \\
&= (\hat{L}_j \hat{r}_{j'})\, \hat{r}_{j'} - \hat{r}_{j'} \hat{r}_{j'} \hat{L}_j = (\hat{r}_{j'} \hat{L}_j + i\hbar \hat{r}_{j''} \varepsilon_{jj'j''})\, \hat{r}_{j'} - \hat{r}_{j'} \hat{r}_{j'} \hat{L}_j \\
&= \hat{r}_{j'} (\hat{L}_j \hat{r}_{j'}) - \hat{r}_{j'} \hat{r}_{j'} \hat{L}_j + i\hbar \hat{r}_{j''} \hat{r}_{j'} \varepsilon_{jj'j''} \\
&= \hat{r}_{j'} (\hat{r}_{j'} \hat{L}_j + i\hbar \hat{r}_{j''} \varepsilon_{jj'j''}) - \hat{r}_{j'} \hat{r}_{j'} \hat{L}_j + i\hbar \hat{r}_{j''} \hat{r}_{j'} \varepsilon_{jj'j''} = 2i\hbar \hat{r}_{j'} \hat{r}_{j''} \varepsilon_{jj'j''}.
\end{aligned} \quad (**)$$

(For the last step, I have used the fact that all Cartesian coordinate operators do commute, and may be swapped at will.)

Now we may use this result to calculate

$$[\hat{L}_j, \hat{r}^2] \equiv \left[\hat{L}_j, \sum_{j'=1}^{3} \hat{r}_{j'}^2\right] = \sum_{j'=1}^{3} \left[\hat{L}_j, \hat{r}_{j'}^2\right] = 2i\hbar \sum_{j'=1}^{3} \hat{r}_{j'} \hat{r}_{j''} \varepsilon_{jj'j''}.$$

According to the definition of the Levi-Civita symbol, one term of the last sum, with $j' = j''$, equals zero, while two other terms are equal and opposite. Hence,

$$[\hat{L}_j, \hat{r}^2] = 0, \quad \text{so that} \quad [\hat{L}_j, f(\hat{r}^2)] = 0.$$

Since all three operators $\hat{L}_j$ commute with the function $f(\hat{r}^2)$, so do operators $\hat{L}_j^2$, and hence the operator $\hat{L}^2$, which is just their sum—see Eq. (5.150).

Next, due to the full similarity of the first and the second of Eqs. (5.149),

$$[\hat{L}_j, \hat{r}_{j'}] = i\hbar \hat{r}_{j''}\varepsilon_{jj'j''}, \qquad [\hat{L}_j, \hat{p}_{j'}] = i\hbar \hat{p}_{j''}\varepsilon_{jj'j''},$$

we may immediately reuse Eq. (**), just replacing $r_{j'}$ with $p_{j'}$:

$$[\hat{L}_j, \hat{p}_{j'}^2] = 2i\hbar \hat{p}_{j''}\hat{p}_{j'}\varepsilon_{jj'j''},$$

so that the summation over all j' immediately yields the similar result:

$$[\hat{L}_j, \hat{p}^2] = \sum_{j'=1}^{3}[\hat{L}_j, \hat{p}_{j'}^2] = 2i\hbar \sum_{j'=1}^{3} \hat{p}_{j''}\hat{p}_{j'}\varepsilon_{jj'j''} = 0.$$

Again, since all three operators $\hat{L}_j$ commute with $\hat{p}^2$, so do operators $\hat{L}_j^2$, and hence the operator $\hat{L}^2$.

Combining these results with Eq. (*), we finally get

$$[\hat{L}_j, \hat{H}] = \frac{1}{2m}[\hat{L}_j, \hat{p}^2] + [\hat{L}_j, f(\hat{r}^2)] = 0,$$
$$[\hat{L}^2, \hat{H}] = \frac{1}{2m}[\hat{L}^2, \hat{p}^2] + [\hat{L}^2, f(\hat{r}^2)] = 0.$$

According to Eq. (4.199) of the lecture notes, these equalities guarantee that the Heisenberg operators of L_j and L^2 do not change in time during the particle's motion in the central field; this quantum-mechanical fact corresponds to the classical-mechanical fact of the conservation of these observables.

***Problem* 5.21.** Use Eqs. (5.149), (5.150) and (5.153) of the lecture notes to prove Eqs. (5.155).

Solution: Let us spell out the following operator product:

$$\hat{L}_+\hat{L}_- \equiv (\hat{L}_x + i\hat{L}_y)(\hat{L}_x - i\hat{L}_y)$$
$$= \hat{L}_x^2 + \hat{L}_y^2 - i(\hat{L}_x\hat{L}_y - \hat{L}_y\hat{L}_x) \equiv \hat{L}_x^2 + \hat{L}_y^2 - i[\hat{L}_x, \hat{L}_y].$$

But according to the last of Eqs. (5.149) of the lecture notes, the last commutator equals $i\hbar\hat{L}_z$, so that we get the equality

$$\hat{L}_+\hat{L}_- = \hat{L}_x^2 + \hat{L}_y^2 + \hbar\hat{L}_z. \qquad (*)$$

Now, from the definition of the operator $\hat{L}^2$, given by Eq. (5.150), we get

$$\hat{L}_x^2 + \hat{L}_y^2 = \hat{L}^2 - \hat{L}_z^2.$$

Plugging the last relation into Eq. (*), we get the first of Eqs. (5.155). The second of these relations (which was already used in section 5.6) may be proved by the absolutely similar transformation of the product $\hat{L}_-\hat{L}_+$.

***Problem* 5.22.** Derive Eq. (5.164) of the lecture notes, using any of the prior formulas.

Solution: According to Eqs. (5.159) of the lecture notes and their discussion, the action of the ladder operators on the common eigenkets $|l, m\rangle$ of the operators $\hat{L}^2$ and $\hat{L}_z$ may be described as

$$\hat{L}_\pm |l, m\rangle = L_\pm^{(m)} |l, m \pm 1\rangle. \qquad (*)$$

(Note that according to this equality, $L_\pm^{(m)}$ are the only nonvanishing elements of the ladder operator matrices in the basis of the $|l, m\rangle$ states,

$$(L_\pm)_{m,m'} \equiv \langle l, m| \hat{L}_\pm |l, m'\rangle = L_\pm^{(m')}\delta_{m'\pm 1,m},$$

with a fixed orbital quantum number l.)

Let us calculate the coefficients $L_\pm^{(m)}$, assuming that the eigenstates are normalized: $\langle l, m|l, m\rangle = 1$. For that, first of all, let us notice that we are speaking essentially about finding just one, rather than two coefficient sets. Indeed, let us use the general rule bra–ket rule (4.25) to write

$$\langle l, m| (\hat{L}_+)^\dagger |l, m + 1\rangle = \langle l, m + 1| \hat{L}_+ |l, m\rangle^*.$$

Since, according to their definition (5.153), the ladder operators are the Hermitian conjugates of each other, this equality takes the following form:

$$\langle l, m| \hat{L}_- |l, m + 1\rangle = \langle l, m + 1| \hat{L}_+ |l, m\rangle^*.$$

Using Eq. (*) and the state normalization condition, the last equality yields

$$L_-^{(m+1)} = \left(L_+^{(m)}\right)^*, \qquad (**)$$

so that the problem is reduced to finding just one of these coefficient sets, say $L_+^{(m)}$.

This may be done, for example, by using the second of Eqs. (5.155) for writing a relation similar to the initial form of Eq. (5.163), but for an eigenstate with an arbitrary m:

$$\hat{L}^2 |l, m\rangle = \hbar\hat{L}_z |l, m\rangle + \hat{L}_z^2 |l, m\rangle + \hat{L}_-\hat{L}_+ |l, m\rangle.$$

Using the eigenvalues calculated in section 5.6 to evaluate the first three terms, and the definition (*) of the coefficients $L_\pm^{(m)}$ in the last term, we get

$$\begin{aligned}\hbar^2 l(l+1)|l, m\rangle &= \hbar^2 m\, |l, m\rangle + \hbar^2 m^2\, |l, m\rangle + \hat{L}_- L_+^{(m)}\, |l, m+1\rangle \\ &= [\hbar^2 m(m+1) + L_-^{(m+1)} L_+^{(m)}]\; |l, m\rangle.\end{aligned}$$

For all existing eigenstates (with $|m| \leqslant l$), this equality may be true only if the c-number factors in its first and last forms are equal. Together with Eq. (**), this gives us the final answer:

$$|L_+^{(m)}| = |L_-^{(m+1)}| = \hbar[\, l(l+1) - m(m+1)]^{1/2}.$$

The other two frequently used forms of the same result are

$$\left|L_\pm^{(m)}\right| = \hbar[l(l+1) - m(m \pm 1)]^{1/2} = \hbar[(l \pm m + 1)(l \mp m)]^{1/2};$$

the first of these forms, together with Eq. (*), gives Eq. (5.164) of the lecture notes.

As a sanity check, the coefficient $L_+^{(m)}$ turns into zero at $m = l$ (thus making the next state $|l, l+1\rangle$ impossible), while $L_-^{(m)}$ does the same at $m = -l$, thus terminating the state ladder on both sides—see figure 5.11.

***Problem* 5.23.** In the basis of the common eigenstates of the operators $\hat{L}_z$ and $\hat{L}^2$, described by kets $|l, m\rangle$:

(i) calculate the matrix elements $\langle l, m_1|\hat{L}_x|l, m_2\rangle$ and $\langle l, m_1|\hat{L}_x^2|l, m_2\rangle$;
(ii) spell out your results for the diagonal matrix elements (with $m_1 = m_2$) and their y-axis counterparts; and
(iii) calculate the diagonal matrix elements $\langle l, m|\hat{L}_x\hat{L}_y|l, m\rangle$ and $\langle l, m|\hat{L}_y\hat{L}_x|l, m\rangle$.

Solution:

(i) From the definition (5.153) of the ladder operators, we may readily get the reciprocal relations

$$\hat{L}_x = \frac{\hat{L}_+ + \hat{L}_-}{2}, \qquad \hat{L}_y = \frac{\hat{L}_+ - \hat{L}_-}{2i}, \qquad (*)$$

so that using Eq. (5.164) of the lecture notes (whose proof was the subject of the previous problem), we may calculate the matrix elements of the operator $\hat{L}_x$ in two equivalent forms:

$$\begin{aligned}\langle l, m_1|\hat{L}_x|l, m_2\rangle &= \frac{1}{2}\langle l, m_1|\hat{L}_+|l, m_2\rangle + \frac{1}{2}\langle l, m_1|\hat{L}_-|l, m_2\rangle \\ &= \frac{\hbar}{2}\Big\{ [(l + m_2 + 1)(l - m_2)]^{1/2}\delta_{m_2, m_1-1} \\ &\quad + [(l - m_2 + 1)(l + m_2)]^{1/2}\delta_{m_2, m_1+1}\Big\} \qquad (**) \\ &\equiv \frac{\hbar}{2}\Big\{ [(l + m_1)(l - m_1 + 1)]^{1/2}\delta_{m_1, m_2+1} \\ &\quad + [(l - m_1)(l + m_1 + 1)]^{1/2}\delta_{m_1, m_2-1}\Big\}.\end{aligned}$$

For the calculation of the matrix elements of $\hat{L}_x^2$, it is instrumental to represent this operator as the product $\hat{L}_x\hat{L}_x$, and then act by the first of them (a Hermitian operator!) upon the bra-vector, and with the second one, upon the ket-vector, using Eq. (**) twice—each time in the more convenient form:

$$\begin{aligned}\langle l, m_1|\hat{L}_x^2|l, m_2\rangle &\equiv \langle l, m_1|\hat{L}_x\hat{L}_x|l, m_2\rangle \\ &= \frac{\hbar}{2}\langle l, m_1 - 1|\{ [(l + m_1)(l - m_1 + 1)]^{1/2} \\ &\quad + \langle l, m_1 + 1|[(l - m_1)(l + m_1 + 1)]^{1/2}\} \\ &\quad \times \frac{\hbar}{2}\Big\{ [(l + m_2 + 1)(l - m_2)]^{1/2}|l, m_2 + 1\rangle \\ &\quad + [(l - m_2 + 1)(l + m_2)]^{1/2}|l, m_2 - 1\rangle\Big\} \\ &= \frac{\hbar^2}{4}\times\left\{\begin{array}{l} \Big\{[(l + m_1)(l - m_1 + 1)(l - m_2 + 1)(l + m_2)]^{1/2} \\ +[(l - m_1)(l + m_1 + 1)(l + m_2 + 1)(l - m_2)]^{1/2}\Big\}\ \delta_{m_1, m_2} \\ +[(l + m_1)(l - m_1 + 1)(l + m_2 + 1)(l - m_2)]^{1/2}\delta_{m_1-1, m_2+1} \\ +[(l - m_1)(l + m_1 + 1)(l - m_2 + 1)(l + m_2)]^{1/2}\delta_{m_1+1, m_2-1}\end{array}\right\}.\end{aligned}$$

These expressions show that the operator $\hat{L}_x^2$ 'connects' only the states whose magnetic quantum numbers either do not differ, or differ by ±2, in a clear analogy with the operator $\hat{x}^2$ for a harmonic oscillator—see Eq. (5.94).

(ii) For the diagonal matrix elements (with $m_1 = m_2 \equiv m$), these general formulas yield:

$$\begin{aligned}\langle l, m\,|\hat{L}_x|\,l, m\rangle &= 0, \\ \langle l, m\,|\hat{L}_x^2|\,l, m\rangle &= \frac{\hbar^2}{4}[(l + m)(l - m + 1) + (l - m)(l + m + 1)] \\ &\equiv \frac{\hbar^2}{2}[l(l + 1) - m^2]\,.\end{aligned}$$

Carrying out absolutely similar calculations for the operator $\hat{L}_y$ and its square, we get similar results:

$$\langle l, m_1|\hat{L}_y|l, m_2\rangle = \frac{1}{2i}\langle l, m_1|\hat{L}_+|l, m_2\rangle - \frac{1}{2i}\langle l, m_1|\hat{L}_-|l, m_2\rangle$$
$$= \frac{\hbar}{2i}\Big\{[(l + m_2 + 1)(l - m_2)]^{1/2}\delta_{m_2, m_1-1}$$
$$- [(l - m_2 + 1)(l + m_2)]^{1/2}\delta_{m_2, m_1+1}\Big\}$$
$$\equiv \frac{\hbar}{2i}\Big\{[(l + m_1)(l - m_1 + 1)]^{1/2}\delta_{m_1, m_2+1}$$
$$- [(l - m_1)(l + m_1 + 1)]^{1/2}\delta_{m_1, m_2-1}\Big\},$$
$$\langle l, m\, |\hat{L}_y|\, l, m\rangle = 0, \quad \langle l, m\, |\hat{L}_y^2|\, l, m\rangle = \frac{\hbar^2}{2}[l(l + 1) - m^2]\,.$$

(iii) Using exactly the same approach as for the calculation of the matrix elements of $\hat{L}_x^2$ and $\hat{L}_y^2$, for the mixed product of the component operators we get

$$\langle l, m|\, \hat{L}_x\hat{L}_y\, |l, m\rangle = \frac{i}{2}\hbar^2 m, \qquad \langle l, m|\, \hat{L}_y\hat{L}_x\, |l, m\rangle = -\frac{i}{2}\hbar^2 m. \qquad (***)$$

As a sanity check, let us verify this result by calculating the diagonal matrix elements of the commutator given by the last of Eqs. (5.149) of the lecture notes, using Eq. (5.158):[31]

$$\langle l, m|[\hat{L}_x, \hat{L}_y]\, |l, m\rangle = i\hbar\langle l, m|\, \hat{L}_z\, |l, m\rangle = i\hbar\langle l, m|\, \hbar m\, \, |l, m\rangle = i\hbar^2 m.$$

But this is exactly the result following from the subtraction of two Eqs. (***) from each other.

***Problem* 5.24.** For the state described by the common eigenket $|l, m\rangle$ of the operators $\hat{L}_z$ and $\hat{L}^2$ in a reference frame $\{x, y, z\}$, calculate the expectation values $\langle L_{z'}\rangle$ and $\langle L_{z'}^2\rangle$ in the reference frame whose axis z' forms angle θ with the axis z.

Solutions: The elementary trigonometry tells us that if a classical geometric vector $\mathbf{L}$ has Cartesian components L_x, L_y, L_z in a certain reference frame $\{x, y, z\}$, its projection to axis z' equals[32]

$$L_{z'} = L_x \sin\theta \cos\varphi + L_y \sin\theta \sin\varphi + L_z \cos\theta,$$

[31] An alternative way to get the same result is to use Eq. (5.164) of the lecture notes (whose proof was the task of problem 5.21). This approach is recommended to the reader as an additional exercise.

[32] This formula may be readily derived by representing $L_{z'}$ as the scalar product $\mathbf{L}\cdot\mathbf{n}_{z'}$, with $\mathbf{L} = \mathbf{n}_x L_x + \mathbf{n}_y L_y + \mathbf{n}_z L_z$and $\mathbf{n}_{z'} = \mathbf{n}_x \sin\theta\cos\varphi + \mathbf{n}_y \sin\theta\sin\varphi + \mathbf{n}_z\cos\theta$. (The last relation follows either from Eq. (A.59) with $\mathbf{r}\to\mathbf{n}_{z'}$, i.e. with $r = 1$, or just directly from figure below, taking into account that the length of the unit vector $\mathbf{n}'$ equals 1.)

where the angles θ and φ are defined as at the usual introduction of the spherical coordinates, with the unit vector $\mathbf{n}_{z'}$ replacing the usual radius-vector $\mathbf{r}$—see the figure below. As was discussed in section 1.2 of the lecture notes, all quantum-mechanical vector-operators, by definition, follow the same geometric relations as the classical geometric vectors, so that we may write

$$\hat{L}_{z'} = \hat{L}_x \sin\theta \cos\varphi + \hat{L}_y \sin\theta \sin\varphi + \hat{L}_z \cos\theta. \qquad (*)$$

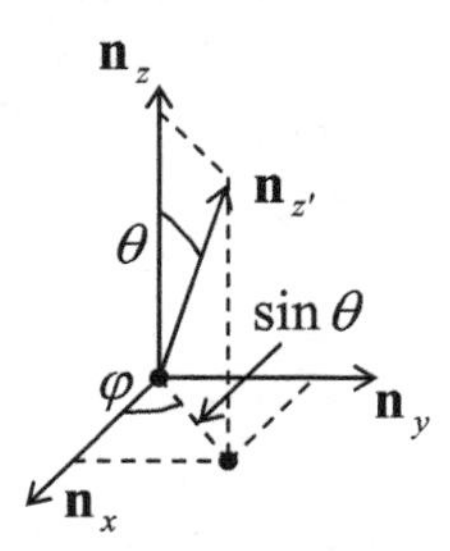

In order to find $\langle L_{z'}\rangle$, it is sufficient to calculate the expectation value of the right-hand side of Eq. (*):

$$\begin{aligned}\langle L_{z'}\rangle &\equiv \langle l, m|\, \hat{L}_{z'}\, |l, m\rangle \\ &= \langle l, m|\, \hat{L}_x\, |l, m\rangle \sin\theta \cos\varphi + \langle l, m|\, \hat{L}_y\, |l, m\rangle \sin\theta \sin\varphi \\ &\quad + \langle l, m|\, \hat{L}_z\, |l, m\rangle \cos\theta,\end{aligned}$$

and take into account that according to the solution of the previous problem, the first two matrix elements on the right-hand side of this expression equal zero, while according to Eq. (5.158) of the lecture notes, the last of them is equal to $\hbar m$. Hence,

$$\langle L_{z'}\rangle = \hbar m \cos\theta.$$

Now using Eq. (*), we may write (being careful not to swap non-commuting operators):

$$\begin{aligned}\hat{L}_{z'}^2 \equiv \hat{L}_{z'}\hat{L}_{z'} &= (\hat{L}_x \sin\theta \cos\varphi + \hat{L}_y \sin\theta \sin\varphi + \hat{L}_z \cos\theta) \\ &\quad (\hat{L}_x \sin\theta \cos\varphi + \hat{L}_y \sin\theta \sin\varphi + \hat{L}_z \cos\theta) \\ &= \hat{L}_x^2 \sin^2\theta \cos^2\varphi + \hat{L}_y^2 \sin^2\theta \sin^2\varphi + \hat{L}_z^2 \cos^2\theta \\ &\quad + (\hat{L}_x\hat{L}_y + \hat{L}_y\hat{L}_x)\, \sin^2\theta \sin\varphi \cos\varphi \\ &\quad + (\hat{L}_x\hat{L}_z + \hat{L}_z\hat{L}_x) \sin\theta \cos\theta \cos\varphi + (\hat{L}_y\hat{L}_z + \hat{L}_z\hat{L}_y) \sin\theta \cos\theta \sin\varphi.\end{aligned}$$

The expectation values of the first two operators participating in the last expression, were calculated in the previous problem:

$$\langle L_x^2\rangle = \left\langle L_y^2\right\rangle = \frac{\hbar^2}{2}[l(l+1) - m^2],$$

while that of the $\hat{L}_z^2 \equiv \hat{L}_z\hat{L}_z$ may be readily calculated using the fact that, according to Eq. (5.158) of the lecture notes, $|l, m\rangle$ is an eigenket ket of operator $\hat{L}_z$, with the eigenvalue $\hbar m$:

$$\langle L_z^2\rangle \equiv \langle l, m|\ \hat{L}_z\hat{L}_z\ |l, m\rangle = \langle l, m|\ \hat{L}_z\hbar m\ |l, m\rangle = \hbar m\langle l, m|\ \hat{L}_z\ |l, m\rangle = \hbar^2 m^2.$$

The expectation values of all other operator combinations vanish, as follows from the other results of the previous problem, and (in the last two cases) the same Eq. (5.158):

$$\begin{aligned}
\langle L_xL_y + L_yL_x\rangle &\equiv \langle l, m|\ \hat{L}_x\hat{L}_y\ |l, m\rangle + \langle l, m|\ \hat{L}_y\hat{L}_x\ |l, m\rangle \\
&= \frac{i}{2}\hbar^2 m - \frac{i}{2}\hbar^2 m = 0, \\
\langle L_xL_z + L_zL_x\rangle &\equiv \langle l, m|\ \hat{L}_x\hat{L}_z\ |l, m\rangle + \langle l, m|\ \hat{L}_z\hat{L}_x\ |l, m\rangle \\
&= \hbar m\langle l, m|\ \hat{L}_x\ |l, m\rangle + \langle l, m|\ \hat{L}_x\ |l, m\rangle = 0, \\
\langle L_yL_z + L_zL_y\rangle &\equiv \langle l, m|\ \hat{L}_y\hat{L}_z\ |\ l, m\rangle + \langle l, m|\ \hat{L}_z\hat{L}_y\ |l, m\rangle \\
&= \hbar m\langle l, m|\ \hat{L}_y\ |l, m\rangle + \hbar m\langle l, m|\ \hat{L}_y\ |l, m\rangle = 0,
\end{aligned}$$

so that finally we get

$$\begin{aligned}
\langle L_{z'}^2\rangle &= \frac{\hbar^2}{2}[l(l+1) - m^2]\sin^2\theta\cos^2\varphi \\
&+ \frac{\hbar^2}{2}[l(l+1) - m^2]\sin^2\theta\sin^2\varphi + \hbar^2 m^2\cos^2\theta \\
&\equiv \hbar^2\left[\frac{l(l+1) - m^2}{2}\sin^2\theta + m^2\cos^2\theta\right].
\end{aligned}$$

Note that the azimuthal angle φ shown in the figure above, i.e. the direction of the axes x and y of the initial reference frame, as well as the direction of the axes x' and y' of the 'primed' reference frame (at fixed axes z and z') do not affect the result. In hindsight, this looks very natural, and means that the above solution might be simplified by taking, from the very beginning, the azimuthal angle to be a constant—say, zero.

***Problem* 5.25.** Write down the matrices of the following angular momentum operators: $\hat{L}_x$, $\hat{L}_y$, $\hat{L}_z$, and$\hat{L}_\pm$, in the z-basis of the $\{l, m\}$ states with $l = 1$.

Solution: Since, according to Eqs. (5.158), (5.159), and (5.164) of the lecture notes, the action of all these operators on the ket- (or bra-) vectors $|l, m\rangle$ does not change the orbital quantum number l, their matrices consist of the elements with the same value of l (in our particular case, $l = 1$):

$$A_{mm'} = \langle l = 1, m|\ \hat{A}\ |l = 1, m'\rangle.$$

Since for $l = 1$ there are three possible values of the quantum number m (+1, 0 and −1), these are 3 × 3 matrices. Of them, the matrix of operator $\hat{L}_z$ is the simplest, because according to Eq. (5.158) of the lecture notes, it has only diagonal elements equal to $\hbar m$, so that numbering the states in the order accepted above, we may write

$$\mathrm{L}_z = \hbar \begin{pmatrix} 1 & 0 & 0 \\ 0 & 0 & 0 \\ 0 & 0 & -1 \end{pmatrix}.$$

Next, according to Eq. (5.164),

$$(L_\pm)_{mm'} = L_\pm^{(m')} \delta_{m'\pm 1,\, m}, \quad \text{with } \left| L_\pm^{(m)} \right| = \hbar[l(l+1) - m(m \pm 1)]^{1/2},$$

so that in our case $l = 1$ the only two nonvanishing matrix elements of each operator have the following magnitudes:

$$|(L_+)_{1,0}| = \hbar[1{\cdot}2 - 0{\cdot}1]^{1/2} = \sqrt{2}\,\hbar, \qquad |(L_+)_{0,-1}| = \hbar[1{\cdot}2 - (-1){\cdot}0]^{1/2} = \sqrt{2}\,\hbar,$$
$$|(L_-)_{0,1}| = \hbar[1{\cdot}2 - 1{\cdot}0]^{1/2} = \sqrt{2}\,\hbar, \qquad |(L_-)_{-1,0}| = \hbar[1{\cdot}2 - 0{\cdot}(-1)]^{1/2} = \sqrt{2}\,\hbar.$$

Note that the elements may be multiplied by phase factors $\exp\{i\varphi\}$, but the phases φ need to be related to keep the operators of observable momentum components,

$$\hat{L}_x = \frac{\hat{L}_+ + \hat{L}_-}{2} \quad \text{and } \hat{L}_y = \frac{\hat{L}_+ - \hat{L}_-}{2i}, \qquad (*)$$

Hermitian, i.e. Eqs. (4.65) of the lecture notes satisfied for their matrix elements. An elementary calculation using Eqs. (*) shows that this requires

$$(L_+)_{m'm} = (L_-)^*_{mm'}, \quad \text{i.e. } (L_+)_{1,0} = (L_-)^*_{0,1}, \quad \text{and } (L_+)_{0,-1} = (L_-)^*_{-1,0}.$$

This requirement is satisfied, for example, for the simplest choice $\varphi = 0$. In this case, we may represent our result as

$$\mathrm{L}_+ = \sqrt{2}\,\hbar \begin{pmatrix} 0 & 1 & 0 \\ 0 & 0 & 1 \\ 0 & 0 & 0 \end{pmatrix}, \qquad \mathrm{L}_- = \sqrt{2}\,\hbar \begin{pmatrix} 0 & 0 & 0 \\ 1 & 0 & 0 \\ 0 & 1 & 0 \end{pmatrix}.$$

Finally, using the operator relations (*) (which have to be followed by all matrix elements), we get

$$\mathrm{L}_x = \frac{\hbar}{\sqrt{2}} \begin{pmatrix} 0 & 1 & 0 \\ 1 & 0 & 1 \\ 0 & 1 & 0 \end{pmatrix}, \qquad \mathrm{L}_y = \frac{\hbar}{\sqrt{2}} \begin{pmatrix} 0 & -i & 0 \\ i & 0 & -i \\ 0 & i & 0 \end{pmatrix}.$$

***Problem* 5.26.** Calculate the angular factor of the orbital wavefunction of a particle with a definite value of L^2, equal to $6\hbar^2$, and the largest possible definite value of L_x. What is this value?

Solution: Let us introduce a new Cartesian coordinate system $\{x', y', z'\}$, with the same origin as the initial one $\{x, y, z\}$, but rotated relative to it as the figure below shows, so that

$$x' \equiv y, \quad y' \equiv z, \quad z' \equiv x, \quad \text{so that } r' = r. \qquad (*)$$

In this coordinate system, the state we are looking for has the same fixed value of L^2, and the largest definite value of $L_{z'}$. According to Eqs. (5.158) and (5.163) of the lecture notes, such a state is described by the eigenket $|l', m'\rangle$ with $m' = l' = 2$, and corresponds to $L_{z'} = m'\hbar = 2\hbar$ and $L^2 = \hbar^2 l'(l' + 1) = 6\hbar^2$. But according to the last of Eqs. (3.176) (due to the space isotropy, there formulas are valid for the primed coordinates (*) as well), the angular wavefunction corresponding to such a state is

$$\psi = \left(\frac{15}{32\pi}\right)^{1/2} \sin^2\theta' e^{2i\varphi'},$$

where the primed angles are related with the primed Cartesian coordinates (*) in the usual way[33]:

$$\sin\theta' \cos\varphi' = \frac{x'}{r'}, \qquad \sin\theta' \sin\varphi' = \frac{y'}{r'}, \qquad \cos\theta' = \frac{z'}{r'}. \qquad (**)$$

In order to use these relations, let us express ψ in terms of trigonometric functions of the single angles:

$$\begin{aligned}
\psi &\equiv \left(\frac{15}{32\pi}\right)^{1/2} \sin^2\theta'(\cos 2\varphi' + i \sin 2\varphi') \\
&\equiv \left(\frac{15}{32\pi}\right)^{1/2} \sin^2\theta'(\cos^2\varphi' - \sin^2\varphi' + 2i \sin\varphi' \cos\varphi') \\
&\equiv \left(\frac{15}{32\pi}\right)^{1/2} (\sin^2\theta' \cos^2\varphi' - \sin^2\theta' \sin^2\varphi' + 2i \sin^2\theta' \sin\varphi' \cos\varphi').
\end{aligned}$$

Now plugging Eqs. (**) into this result, and then using Eq. (*) to replace the coordinates back to the initial Cartesian ones, and finally to the initial (non-primed) spherical angles, we get

[33] See, e.g. Eq. (A.59).

$$\psi = \left(\frac{15}{32\pi}\right)^{1/2} \frac{1}{r'^2}(x'^2 - y'^2 + 2ix'y')$$
$$\equiv \left(\frac{15}{32\pi}\right)^{1/2} \frac{1}{r^2}(y^2 - z^2 + 2iyz)$$
$$\equiv \left(\frac{15}{32\pi}\right)^{1/2} (\sin^2\theta \sin^2\varphi - \cos^2\theta + 2i\sin\theta\sin\varphi\cos\theta).$$

Naturally, the coordinate replacement does not change the value $L_x = L_{z'} = 2\hbar$.

***Problem* 5.27.** For the state with the wavefunction $\psi = Cxye^{-\lambda r}$, with a real, positive λ, calculate:

(i) the expectation values of the observables L_x, L_y, L_z and L^2, and
(ii) the normalization constant C.

Solutions:

(i) Rewriting the given wavefunction in the spherical coordinates:

$$\psi = Cr^2e^{-\lambda r}\sin^2\theta\sin\varphi\cos\varphi = Cr^2e^{-\lambda r}\sin^2\theta\frac{\sin 2\varphi}{2} = Cr^2e^{-\lambda r}\sin^2\theta\frac{e^{2i\varphi} - e^{-2i\varphi}}{4i},$$

and comparing the result with the top and bottom lines of Eq. (3.176), we see that

$$\psi = \mathcal{R}(r) \times \frac{1}{\sqrt{2}}[Y_2^2(\theta, \varphi) + Y_2^{-2}(\theta, \varphi)], \qquad (*)$$

where

$$\mathcal{R}(r) = \frac{2}{i}\left(\frac{\pi}{15}\right)^{1/2} Cr^2e^{-\lambda r}. \qquad (**)$$

As Eq. (*) shows, the state is a linear superposition, with equal amplitudes (and hence equal probabilities $W_+ = W_- = ½$), of two angularly-orthogonal states: one with $l = 2$ and $m = 2$, and another one with $l = 2$ and $m = -2$. Hence, according to Eq. (5.158), the expectation value of L_z is

$$\langle L_z \rangle = \hbar(+2)W_+ + \hbar(-2)W_- = 0,$$

while according to Eq. (5.163), that of L^2 is

$$\langle L^2 \rangle = \hbar^2 2(2+1)W_+ + \hbar^2 2(2+1)W_- = 6\hbar^2.$$

Finally, according to Eq. (5.164) of the lecture notes, the expectation values of $L_\pm$, and hence of both $L_x = (L_+ - L_-)/2$ and $L_y = (L_x - L_y)/2i$, are equal to zero in any of these eigenstates, and hence in their linear superposition (*) as well.

(ii) Since all spherical harmonics are already normalized (see Eq. (3.173) of the lecture notes), so is the whole angular factor of the wavefunction (*), and it is sufficient to require the normalization of its radial part:

$$\int_0^\infty \mathcal{R}^*(r)\mathcal{R}(r)\, r^2 dr = 1.$$

With the function $\mathcal{R}(r)$given by Eq. (**), this equality gives the following condition on the constant C:

$$|C|^{-2} = \frac{4\pi}{15}\int_0^\infty r^6 e^{-2\lambda r} dr \equiv \frac{4\pi}{15}\left(\frac{1}{2\lambda}\right)^7 \int_0^\infty \xi^6 e^{-\xi} d\xi.$$

This dimensionless integral[34] equals $6! \equiv 720$, so that we finally get

$$|C| = \left[\frac{4\pi}{15}\left(\frac{1}{2\lambda}\right)^7 720\right]^{-1/2} \equiv \left(\frac{2}{3\pi}\right)^{1/2} \lambda^{7/2}.$$

(Just as a reminder, the normalization constant is defined to a phase multiplier $e^{i\varphi}$ with any real φ.)

***Problem* 5.28.** The angular state of a spinless particle is described by the following ket-vector:

$$|\alpha\rangle = \frac{1}{\sqrt{2}}(|l = 3, m = 0\rangle + |l = 3, m = 1\rangle).$$

Calculate the expectation values of the x- and y-components of its angular momentum. Is the result sensitive to a possible phase shift between the two component eigenkets?

Solution: Let us start from calculating the expectation values of the ladder operators:

$$\begin{aligned}\langle L_\pm \rangle &= \langle \alpha | \hat{L}_\pm |\alpha\rangle \\ &= \frac{1}{2}(\langle 3,0| \hat{L}_\pm |3,0\rangle + \langle 3,0| \hat{L}_\pm |3,1\rangle + \langle 3,1| \hat{L}_\pm |3,0\rangle + \langle 3,1| \hat{L}_\pm |3,1\rangle).\end{aligned}$$

Eq. (5.164) of the lecture notes shows that the diagonal matrix elements vanish, while each of the off-diagonal terms contributes to only one expectation value:

[34] See, e.g. Eq. (A.34*d*) with $n = 6$.

$$\langle L_+\rangle = \frac{1}{2}\langle 3,1|\,\hat{L}_+\,|3,0\rangle$$
$$= \frac{\hbar}{2}\langle 3,1|\,[l(l+1) - m(m+1)]^{1/2}_{l=3,\,m=0}\,|3,1\rangle = \sqrt{3}\,\hbar,$$
$$\langle L_-\rangle = \frac{1}{2}\langle 3,0|\,\hat{L}_-\,|3,1\rangle$$
$$= \frac{\hbar}{2}\langle 3,0|\,[l(l+1) - m(m-1)]^{1/2}_{l=3,\,m=1}\,|3,0\rangle = \sqrt{3}\,\hbar,$$

so that

$$\langle L_x\rangle = \frac{1}{2}(\langle L_+\rangle + \langle L_-\rangle) = \sqrt{3}\,\hbar, \quad \langle L_y\rangle = \frac{1}{2i}(\langle L_+\rangle - \langle L_-\rangle) = 0.$$

However, this result is valid only if the phase shift between the two components of the linear superposition is exactly zero. For an arbitrary phase shift, for example

$$|\alpha\rangle = \frac{1}{\sqrt{2}}\big(|3,0\rangle + e^{i\varphi}\,|3,1\rangle\big), \quad \text{so that } \langle\alpha| = \frac{1}{\sqrt{2}}\big(\langle 3,0\,| + e^{-i\varphi}\langle 3,1|\big),$$

the result becomes:

$$\langle L_+\rangle = \sqrt{3}\,\hbar e^{-i\varphi}, \quad \langle L_-\rangle = \sqrt{3}\,\hbar e^{i\varphi},$$

so that

$$\langle L_x\rangle = \sqrt{3}\,\hbar\cos\varphi, \quad \langle L_y\rangle = -\sqrt{3}\,\hbar\sin\varphi$$

—the situation to compare with the solutions of problems 4.17 and 5.14.

***Problem* 5.29.** A particle is in the state α with the orbital wavefunction proportional to the spherical harmonic $Y_1^1(\theta, \varphi)$. Find the angular dependence of the wavefunctions corresponding to the following ket-vectors:

(i) $\hat{L}_x|\alpha\rangle$,
(ii) $\hat{L}_y|\alpha\rangle$,
(iii) $\hat{L}_z|\alpha\rangle$,
(iv) $\hat{L}_+\hat{L}_-|\alpha\rangle$, and
(v) $\hat{L}^2|\alpha\rangle$.

Solution: According to the discussion of section 5.6, the given ket $|\alpha\rangle$ is the shared eigenket $|l, m\rangle$ of the operators $\hat{L}^2$ and $\hat{L}_z$, with $l = 1$ and $m = +1$, i.e. $|\alpha\rangle \propto |l = 1, m = 1\rangle \equiv |1, 1\rangle$. Hence these two operators, $\hat{L}^2$ and $\hat{L}_z$, acting upon the ket-vector of the state, do not change the angular dependence of its wavefunction, which is proportional to $\sin\theta\,\exp\{i\varphi\}$—see, e.g. the last line of Eq. (3.175) of the lecture notes. The same is true for the operator product $\hat{L}_+\hat{L}_-$, because according to Eq. (5.164), its right operand changes the initial state to $|1, 0\rangle$ (multiplied by a

c-number coefficient), but the left operand, acting next, returns the ket to its initial form $|1, 1\rangle$.

The results of action of the two remaining operators, $\hat{L}_x$ and $\hat{L}_y$, may be most simply obtained by expressing them via the ladder operators $\hat{L}_\pm$. From Eq. (5.153) we readily get

$$\hat{L}_x = \frac{1}{2}(\hat{L}_+ + \hat{L}_-), \quad \hat{L}_y = \frac{1}{2i}(\hat{L}_+ - \hat{L}_-).$$

According to the same Eq. (5.164) (see also figure 5.11),

$$\hat{L}_+ |1, 1\rangle = 0, \quad \hat{L}_- |1, 1\rangle = \text{const} \times |1, 0\rangle,$$

so that the wavefunctions corresponding to $\hat{L}_x|\alpha\rangle$ and $\hat{L}_y|\alpha\rangle$ are both proportional to the spherical harmonic with $l = 1$ and $m = 0$, i.e. to $Y_1^0(\theta, \varphi) \propto \cos\theta$, albeit with different coefficients.

***Problem* 5.30.** A charged, spinless 2D particle of mass m is trapped in a soft in-plane potential well $U(x, y) = m\omega_0^2(x^2 + y^2)/2$. Calculate its energy spectrum in the presence of a uniform magnetic field $\mathscr{B}$, normal to the plane.

Solution: Due to the evident axial symmetry of the problem, it may be most simply solved by the selection of the vector-potential not in the Landau form (3.44), but in the axially-symmetric form

$$\mathbf{A} = \frac{1}{2}\mathscr{B} \times \boldsymbol{\rho} \equiv \mathbf{n}_\varphi \frac{\mathscr{B}\rho}{2},$$

where $\boldsymbol{\rho} \equiv \{x, y\}$ is the 2D radius-vector. (Indeed, using the expression for curl of a vector in the cylindrical coordinates[35], it easy to check that this expression does satisfy the vector-potential's definition, $\nabla \times \mathbf{A} = \mathscr{B}$, with $\mathscr{B} = \mathbf{n}_z\mathscr{B} = \text{const}$.) The Cartesian components of this vector are

$$A_x = -\frac{\mathscr{B}}{2}y, \quad A_y = \frac{\mathscr{B}}{2}x,$$

so that the 2D form of the Hamiltonian (3.26), with the due replacement $q\phi \to U(x, y)$, is

$$\hat{H} = -\frac{\hbar^2}{2m}\left[\mathbf{n}_x\left(\frac{\partial}{\partial x} + i\frac{q\mathscr{B}}{2\hbar}y\right) + \mathbf{n}_y\left(\frac{\partial}{\partial y} - i\frac{q\mathscr{B}}{2\hbar}x\right)\right]^2 + \frac{m\omega_0^2}{2}(x^2 + y^2).$$

Squaring the brackets, and using the cyclotron frequency definition (3.48), but with the definite sign: $\omega_c \equiv -q\mathscr{B}/m$, in order for the vector $\boldsymbol{\omega}_c = \mathbf{n}_z\omega_c$ to have the correct direction (corresponding to the direction of the classical Lorentz force $\mathbf{F}_L = q\mathbf{v} \times \mathscr{B}$), we get

[35] See, e.g. Eq. (A.63).

$$\hat{H} = -\frac{\hbar^2}{2m}\left[\frac{\partial^2}{\partial x^2} + \frac{\partial^2}{\partial y^2} + \frac{m\omega_c}{\hbar^2} \times i\hbar\left(y\frac{\partial}{\partial x} - x\frac{\partial}{\partial y}\right)\right]$$
$$+ \frac{m\omega^2}{2}(x^2 + y^2), \quad \text{with} \quad \omega^2 \equiv \omega_0^2 + \frac{1}{4}\omega_c^2.$$

But according to Eq. (5.152) of the lecture notes, the factor following the × sign in this expression is just the coordinate representation of the operator $\hat{L}_z$, so that

$$\hat{H} = \hat{H}_{\text{osc}} + \frac{\omega_c}{2}\hat{L}_z,$$

where

$$\hat{H}_{\text{osc}} \equiv -\frac{\hbar^2}{2m}\left(\frac{\partial^2}{\partial x^2} + \frac{\partial^2}{\partial y^2}\right) + \frac{m\omega^2}{2}(x^2 + y^2)$$
$$\equiv \frac{1}{2m}\left(\hat{p}_x^2 + \hat{p}_y^2\right) + \frac{m\omega^2}{2}(\hat{x}^2 + \hat{y}^2). \qquad (*)$$

Using the commutation relations (5.149), it is straightforward to verify that the operator $\hat{L}_z$ commutes with the operators $(\hat{p}_x^2 + \hat{p}_y^2)$ and $(\hat{x}^2 + \hat{y}^2)$, and hence with the operator $\hat{H}_{\text{osc}}$. As a result, these operators share mutual eigenstates, which are also the eigenstates of the full Hamiltonian. Let us denote the ket-vectors of these states $|n, m\rangle$, by the reason which will be clear in a second[36]. The first of the component operators, $\hat{H}_{\text{osc}}$, is just that of an effective field-free 2D (planar) harmonic oscillator (with the frequency ω renormalized by the field, though), whose energy spectrum is given by Eq. (3.124) with $d = 2$, so that

$$\hat{H}_{\text{osc}}\,|n, m\rangle = \hbar\omega(n + 1)\,|n, m\rangle, \quad \text{with} \quad n = 0, 1, 2, \ldots .$$

On the other hand, according to Eq. (5.158),

$$\hat{L}_z\,|n, m\rangle = \hbar m\,|n, m\rangle, \quad \text{with } m \text{ integer},$$

so that

$$\hat{H}\,|n, m\rangle \equiv \left(\hat{H}_{\text{osc}} + \frac{\omega_c}{2}\hat{L}_z\right)|n, m\rangle = E_{n,m}\,|n, m\rangle,$$

with

$$E_{n,m} = \hbar\omega(n + 1) + \frac{\hbar\omega_c}{2}m \equiv \hbar\left(\omega_0^2 + \frac{\omega_c^2}{4}\right)^{1/2}(n + 1) + \frac{\hbar\omega_c}{2}m. \qquad (**)$$

[36] Note that preparing for this moment, I again (just as in the last sections of chapter 3) used a fancy font to denote the particle's mass m.

The physical sense of the last term, linear in $\omega_c \propto \mathscr{B}$ (and hence defining the main, linear response ΔE of the system to a weak field), is simple:

$$\Delta E = \frac{\omega_c}{2} L_z \equiv -\frac{q\mathscr{B}}{2m} L_z = -\mathbf{m} \cdot \mathscr{B},$$

where $\mathbf{m} = q\mathbf{L}/2m$ is the classical orbital magnetic moment of the particle—or even of any system of particles with the same q/m ratios[37].

In the opposite, high field limit, when $\omega \to \omega_c/2$, Eq. (**) tends to the Landau's result (3.50):

$$E_N \to \hbar|\omega_c|\left(N + \frac{1}{2}\right), \qquad \text{with } N = 0,\ 1,\ \ldots,$$

with the integer N in that relation equal to $(n + m')/2$, with $m' \equiv m\ \mathrm{sgn}(\omega_c)$, and the sign factor depending on that of the product $q\mathscr{B}$. The last correspondence means, in particular, that the sum $(n + m')$ has to be non-negative and even, i.e. that for fixed n, the magnetic quantum number may take only the following values: $m' = -n, -n + 2, -n + 4, \ldots$ From the physical sense of m, discussed above, we may expect this spectrum to be centered to zero, so that with $m'_{\min} = -n$, it gives $g = 2n/2 + 1 = n + 1$ values of m' (and hence of m).

A simpler way to get the last result, for an arbitrary field, is to say that since according to Eq. (*), the field-free 2D oscillator is just a system of two similar, independent 1D oscillators, each with the energy spectrum (5.86), the quantum number n is just the sum $n_x + n_y$, with both component quantum numbers taking integer values from 0 up. Hence the nth energy level of the 2D oscillator is $(n + 1)$-degenerate, with any of the component quantum numbers taking values $0, 1, \ldots n$. The applied magnetic field lifts this degeneracy, so that there should be $g = n + 1$ possible different values on m.[38]

In particular, this means that in the ground state of the oscillator (with $n = 0$), the magnetic quantum number m may have only one value, $m = 0$, so that the system does not have its own magnetic moment. This is very natural, because according to Eq. (3.125) of the lecture notes, with $d = 2$, its ground state-wavefunction is axially symmetric:

$$\psi_g = \frac{1}{\pi^{1/2} x_0} \exp\left\{-\frac{x^2 + y^2}{2x_0^2}\right\} \equiv \frac{1}{\pi^{1/2} x_0} \exp\left\{-\frac{\rho^2}{2x_0^2}\right\},$$

and cannot have any spontaneous magnetic moment $\mathbf{m}$. (If it did, what would be the sign of its only possible, z-component?)

[37] See, e.g. *Part EM* Eq. (5.95).

[38] In this context, the $\Delta m = \pm 2$ step of the magnetic quantum number in the high-field limit becomes somewhat less counter-intuitive, because any change of n_x or n_y by ± 1 has to be accompanied by an equal and opposite change of its counterpart, to keep n fixed.

***Problem* 5.31.** Solve the previous problem for a spinless 3D particle, placed (in addition to a uniform magnetic field $\mathscr{B}$) into a spherically-symmetric potential well $U(\mathbf{r}) = m\omega_0^2 r^2/2$.

Solution: Directing the axis z along the applied magnetic field, let us select its vector-potential just as was done in the (very similar) previous problem,

$$\mathbf{A} = \frac{1}{2}\mathscr{B} \times \boldsymbol{\rho},$$

where $\boldsymbol{\rho} \equiv \{x, y\}$ is the 2D radius-vector. The Cartesian components of this vector are

$$A_x = -\frac{\mathscr{B}}{2}y, \quad A_y = \frac{\mathscr{B}}{2}x,$$

so that Eq. (3.27), with the due replacement $q\phi \to U(\mathbf{r})$, takes the form

$$\hat{H} = -\frac{\hbar^2}{2m}\left[\mathbf{n}_x\left(\frac{\partial}{\partial x} + i\frac{q\mathscr{B}}{\hbar}y\right) + \mathbf{n}_y\left(\frac{\partial}{\partial y} - i\frac{q\mathscr{B}}{\hbar}x\right) + \mathbf{n}_z\frac{\partial}{\partial z}\right]^2$$
$$+ \frac{m\omega_0^2}{2}(x^2 + y^2 + z^2) \equiv \hat{H}_{x,y} + \hat{H}_z.$$

Let us represent this Hamiltonian as a sum:

$$\hat{H} = \hat{H}_{x,y} + \hat{H}_z,$$

where $\hat{H}_{x,y}$ is the Hamiltonian discussed in the solution of the previous problem,

$$\hat{H}_{x,y} = -\frac{\hbar^2}{2m}\left[\mathbf{n}_x\left(\frac{\partial}{\partial x} + i\frac{q\mathscr{B}}{\hbar}y\right) + \mathbf{n}_y\left(\frac{\partial}{\partial y} - i\frac{q\mathscr{B}}{\hbar}x\right)\right]^2 + \frac{m\omega_0^2}{2}(x^2 + y^2),$$

with the energy spectrum given by Eq. (**) of that solution:

$$E_{n,m} = \hbar\left(\omega_0^2 + \frac{\omega_c^2}{4}\right)^{1/2}(n + 1) + \frac{\hbar\omega_c}{2}m,$$
$$\text{where} \quad n = 0, 1, 2,..., \quad \text{and} \quad m = -n, -n + 2,..., +n,$$

while $\hat{H}_z$ is the Hamiltonian of a 1D harmonic oscillator:

$$\hat{H}_z = -\frac{\hbar^2}{2m}\frac{\partial^2}{\partial z^2} + \frac{m\omega_0^2}{2}z^2,$$

with eigenenergies $E_z = \hbar\omega_0(n_z + ½)$, where $n_z = 0, 1, 2,...$. These two Hamiltonians describe two completely independent systems, with different Hilbert spaces, so that their eigenvalues (i.e. energies) just add up. As a result, the full energy spectrum of the system is

$$E = E_{n,m} + E_z$$
$$= \hbar\left(\omega_0^2 + \frac{\omega_c^2}{4}\right)^{1/2}(n + 1) + \frac{\hbar\omega_c}{2}m + \hbar\omega_0\left(n_z + \frac{1}{2}\right). \quad (*)$$

In the limit of vanishing magnetic field ($\omega_c \to 0$), this result tends to Eq. (3.124) of the lecture notes, with $d = 3$, for an 3D oscillator

$$E \to \hbar\omega_0(n + 1) + \hbar\omega_0\left(n_z + \frac{1}{2}\right) \equiv \hbar\omega_0\left(N + \frac{3}{2}\right), \qquad \text{with } N \equiv n + n_z = 0,\ 1,\ 2, \ldots$$

In the opposite limit $\omega_0 \ll \omega_c$, the relative smallness of the coefficient $\hbar\omega_0$ in the last term of may be compensated by (possibly very large) values of the quantum number n_z, so that the result may be fairly represented as the sum,

$$E = \hbar\omega_0\left(n' + \frac{1}{2}\right) + E_z,$$

of the discrete Landau-level spectrum (3.50), with $n' \equiv n + m$, and the quasi-continuous energy E_z of the essentially-free motion along the z-axis.

***Problem* 5.32.** Calculate the spectrum of rotational energies of an axially-symmetric, rigid body.

Solution: According to classical mechanics[39], the rotational energy of an axially-symmetric rigid body, frequently called the *symmetric top*[40], is related to the principal-axis components of its angular momentum as

$$E = \frac{L_x^2 + L_y^2}{2I_1} + \frac{L_z^2}{2I_3},$$

where I_3 is the principal moment of inertia for rotation about the axis of symmetry (taken here for axis z), and $I_1 \equiv I_x = I_y$ is that for rotation about any axis perpendicular to z. According to the correspondence principle, this means that in quantum mechanics, the rotation may be described by the similar Hamiltonian:

$$\hat{H} = \frac{\hat{L}_x^2 + \hat{L}_y^2}{2I_1} + \frac{\hat{L}_z^2}{2I_3}.$$

According to Eq. (5.150) of the lecture notes, this Hamiltonian may be rewritten as

$$\hat{H} = \frac{\hat{L}^2 - \hat{L}_z^2}{2I_1} + \frac{\hat{L}_z^2}{2I_3} \equiv \frac{\hat{L}^2}{2I_1} + \frac{1}{2}\left(\frac{1}{I_3} - \frac{1}{I_1}\right)\hat{L}_z^2.$$

Since, according to Eq. (5.151), the operator $\hat{L}^2$ commutes with the operator $\hat{L}_z$, and hence with its square, they share the eigenstates described by the ket-vectors $|l, m\rangle$ which were discussed in section 5.6 of the lecture notes. As a result, we may use the corresponding eigenvalues given by Eqs. (5.158) and (5.163) to immediately write the eigenvalues of our Hamiltonian, i.e. the energy spectrum of the system:

[39] See, e.g. *Part CM* section 4.2, in particular Eqs. (4.25)–(4.26).

[40] Note that the set of symmetric tops, with two equal principal moments of inertia, is not limited to the axially-symmetric bodies; for example, any uniform equilateral triangle (see, e.g. *Part CM* figure 4.3) also belongs to this class. Hence, the energy spectrum calculated in this solution is even more general than the problem's assignment specifies.

$$E_{l,m} = \frac{\hbar^2 l(l+1)}{2I_1} + \frac{1}{2}\left(\frac{1}{I_3} - \frac{1}{I_1}\right)\hbar^2 m^2, \quad \text{with} \quad l = 0, 1, 2, \ldots; \; -l \leqslant m \leqslant +l.$$

Note that the second term of this expression may be either positive (for bodies stretched along the symmetry axis z, and hence having with $I_1 > I_3$—see *Part CM* Eq. (4.24) of the lecture notes), or equal zero (for a particular case of a *spherical top*, with all principal moment of inertia equal), or even negative (at $I_1 < I_3$). However, due to the condition $-l \leqslant m \leqslant +l$, even in the latter case all energies $E_{l,m}$ are still non-negative.

Note also that this calculation assumes that the body is 'rigid' indeed, in particular without internal degrees of freedom, and with distinguishable rotations by angles different from 2π. As a result, it may be only partially valid for real axially-symmetric molecules (such as the diatomic molecules H_2, N_2, etc); for their discussion see chapter 8.

***Problem* 5.33.** Simplify the following double commutator: $[\hat{r}_j, [\hat{L}^2, \hat{r}_{j'}]]$.

Solution: Using Eq. (5.150) of the lecture notes, we may write

$$\hat{A}_{jj'} \equiv [\hat{r}_j, [\hat{L}^2, \hat{r}_{j'}]] \equiv \left[\hat{r}_j, \left[\sum_{j''=1}^{3} \hat{L}_{j''}^2, \hat{r}_{j'}\right]\right] = \sum_{j''=1}^{3}\left[\hat{r}_j, [\hat{L}_{j''}\hat{L}_{j''}, \hat{r}_{j'}]\right]. \quad (*)$$

Let us start with spelling out the internal commutator in this expression. The calculations may be shortened a bit using the easily provable operator identity:

$$[\hat{B}\hat{C}, \hat{D}] = \hat{B}[\hat{C}, \hat{D}] + [\hat{B}, \hat{D}]\hat{C}. \quad (**)$$

Indeed, taking $\hat{B} = \hat{C} = \hat{L}_{j''}$ and $\hat{D} = \hat{r}_{j'}$, we get

$$[\hat{L}_{j''}\hat{L}_{j''}, \hat{r}_{j'}] = \hat{L}_{j''}[\hat{L}_{j''}, \hat{r}_{j'}] + [\hat{L}_{j''}, \hat{r}_{j'}]\hat{L}_{j''}.$$

Applying Eq. (5.148) in the form,

$$[\hat{L}_{j''}, \hat{r}_{j'}] = i\hbar\hat{r}_{j'''}\varepsilon_{j''j'j'''}, \quad (***)$$

to both commutators on the right-hand side, we get

$$[\hat{L}_{j''}\hat{L}_{j''}, \hat{r}_{j'}] = \hat{L}_{j''}i\hbar\hat{r}_{j'''}\varepsilon_{j''j'j'''} + i\hbar\hat{r}_{j'''}\varepsilon_{j''j'j'''}\hat{L}_{j''} \equiv i\hbar\varepsilon_{j''j'j'''}(\hat{L}_{j''}\hat{r}_{j'''} + \hat{r}_{j'''}\hat{L}_{j''}).$$

With this, the external commutator in Eq. (*) becomes

$$[\hat{r}_j, [\hat{L}_{j'}\hat{L}_{j''}, \hat{r}_{j'}]] = i\hbar\varepsilon_{j''j'j'''}\left([\hat{r}_j, \hat{L}_{j''}\hat{r}_{j'''}] + [\hat{r}_j, \hat{r}_{j'''}\hat{L}_{j''}]\right)$$
$$\equiv i\hbar\varepsilon_{j''j'''j'}\left([\hat{L}_{j''}\hat{r}_{j'''}, \hat{r}_j] + [\hat{r}_{j'''}\hat{L}_{j''}, \hat{r}_j]\right).$$

(Note the swap of the indices j' and j''' in the Levi-Civita symbol in the last expression, which compensates the sign reversal due to the swap of operands in both commutators.) Now we may apply the identity (**) again to both commutators on the right-hand side: in the first case, with $\hat{B} = \hat{L}_{j''}$, $\hat{C} = \hat{r}_{j'''}$, and $\hat{D} = \hat{r}_j$, and in the second case, with $\hat{B} = \hat{r}_{j'''}$, $\hat{C} = \hat{L}_{j''}$, and $\hat{D} = \hat{r}_j$. This gives

$$\left[\hat{r}_j, [\hat{L}_{j'}\hat{L}_{j''}, \hat{r}_{j'}]\right] = i\hbar\varepsilon_{j''j'''j'}\left(\hat{L}_{j''}[\hat{r}_{j'''}, \hat{r}_j] + [\hat{L}_{j''}, \hat{r}_j]\,\hat{r}_{j'''} + \hat{r}_{j'''}[\hat{L}_{j''}, \hat{r}_j] + [\hat{r}_{j'''}, \hat{r}_j]\hat{L}_{j''}\right),$$
$$= i\hbar\varepsilon_{j''j'''j'}\left([\hat{L}_{j''}, \hat{r}_j]\,\hat{r}_{j'''} + \hat{r}_{j'''}[\hat{L}_{j''}, \hat{r}_j]\right),$$

because the Cartesian coordinate operators commute regardless of their indices.

Let us apply to the commutator on the right-hand side of the last expression the basic Eq. (***) again, now in the form

$$[\hat{L}_{j''}, \hat{r}_j] = i\hbar\hat{r}_{j''''}\varepsilon_{j''jj''''},$$

where the index j'''' is complementary to the indices j and j' (it does not necessarily coincide with either j' or j'''). The result is

$$[\hat{r}_j, [\hat{L}_{j'}\hat{L}_{j''}, \hat{r}_{j'}]] = i\hbar\varepsilon_{j''j'''j'}(i\hbar\hat{r}_{j''''}\varepsilon_{j''jj''''}\hat{r}_{j'''} + \hat{r}_{j'''}i\hbar\hat{r}_{j''''}\varepsilon_{j'''jj''''})$$
$$\equiv -2\hbar^2\varepsilon_{j''j'''j'}\varepsilon_{j''jj''''}\hat{r}_{j'''}\hat{r}_{j''''},$$

so that the operator in question is reduced to

$$\hat{A}_{jj'} = -2\hbar^2\sum_{j''=1}^{3}\varepsilon_{j''j'''j'}\varepsilon_{j''jj''''}\hat{r}_{j'''}\hat{r}_{j''''}. \qquad (****)$$

Generally, the Levi-Civita product sums over a similar index may be calculated using the so-called 'contracted epsilon identity'[41], but in our case it is more prudent to use a more pedestrian way, because both indices j''' and j'''' depend on the summation index j''. Indeed, due to the index cycling symmetry, we need to consider not all nine possible combinations of the indices j and j', but only three different cases—of which two may be analyzed in one shot (and give the same result).

(i) If $j' = j$, then, by their definition, the indices j''' and j'''' are equal to each other for any j'', so that the Levi-Civita index product in sum (****) equals (−1) in both terms with $j'' \neq j$. (In the term with $j'' = j$, the product vanishes.) In the remaining two terms, the common index of the Cartesian coordinates takes both values not equal to j (and hence to j'), so that the result may be conveniently represented as

[41] See, e.g. Eq. (A.84*b*).

$$\hat{A}_{jj} = 2\hbar^2\left(\hat{r}^2 - \hat{r}_j^2\right).$$

(ii) If $j = j' \pm 1$,[42] the sum in Eq. (****) has not two, but just one nonvanishing term, because both the Levi-Civita symbols kill the terms with both $j'' = j'$ and $j'' = j = j' \pm 1$. In the only remaining term of the sum, $j'' = j' \mp 1$, and hence, by its definition as the complementary one to j' and j'', the index j''' is equal to $j' \pm 1 = j$, while the index j'''', by its definition as the complementary one to j and j'', equals j'. As the result, Eq. (****) is reduced to

$$\hat{A}_{jj'} = -2\hbar^2\varepsilon_{(j'\mp1)(j'\pm1)j'}\varepsilon_{(j'\mp1)(j'\pm1)j'}\hat{r}_j\hat{r}_{j'} = -2\hbar^2\hat{r}_j\hat{r}_{j'}, \quad \text{for } j \neq j'.$$

Both above results may be summarized as

$$\hat{A}_{jj'} = 2\hbar^2(\hat{r}^2\delta_{jj'} - \hat{r}_j\hat{r}_{j'}).$$

The reader may have noticed an interesting analogy between this final result and the well-known classical expression for the contribution of an elementary mass dm of a rigid body to its inertia tensor[43]:

$$dI_{jj'} = dm(r^2\delta_{jj'} - r_jr_{j'}).$$

***Problem* 5.34.** Prove the following commutation relation:

$$[\hat{L}^2, [\hat{L}^2, \hat{r}_j]] = 2\hbar^2(\hat{r}_j\hat{L}^2 + \hat{L}^2\hat{r}_j).$$

Solution: We may start from the following by-product of the model solution of the previous problem (with the number of primes in each j-index reduced by one):

$$\left[\hat{L}_{j'}^2, \hat{r}_j\right] = i\hbar(\hat{L}_{j'}\hat{r}_{j''} + \hat{r}_{j''}\hat{L}_{j'})\,\varepsilon_{j'jj''} \equiv -i\hbar\varepsilon_{jj'j''}(\hat{L}_{j'}\hat{r}_{j''} + \hat{r}_{j''}\hat{L}_{j'}).$$

Due to the Levi-Civita symbol's vanishing at $j = j'$, we get only two non-zero contributions to the full inner commutator

$$\begin{aligned}[\hat{L}^2, \hat{r}_j] &= \sum_{j'=1}^{3}\left[\hat{L}_{j'}^2, \hat{r}_j\right] \\ &= -i\hbar\sum_{j'=1}^{3}(\hat{L}_{j'}\hat{r}_{j''} + \hat{r}_{j''}\hat{L}_{j'})\varepsilon_{jj'j''} \\ &= -i\hbar\left[(\hat{L}_{j+1}\hat{r}_{j+2} + \hat{r}_{j+2}\hat{L}_{j+1}) - (\hat{L}_{j+2}\hat{r}_{j+1} + \hat{r}_{j+1}\hat{L}_{j+2})\right],\end{aligned} \quad (*)$$

[42] If $j' + 1 = 4$, such a sum means 1, and if $j' - 1 = 0$, the difference means 3. (In math speak, we define the combinations $j' \pm 1$ 'modulo 3'.)
[43] See, e.g. *Part CM* Eq. (4.16).

where the sums $(j + 1)$ and $(j + 2)$ are understood by modulo 3, i.e. the sums $(3 + 1)$ and $(2 + 2)$ are taken for 1, and $(3 + 2)$ for 2. Now using Eq. (5.148) of the lecture notes to write

$$\hat{L}_{j+1}\hat{r}_{j+2} = \hat{r}_{j+2}\hat{L}_{j+1} + i\hbar\hat{r}_j, \quad \text{and} \quad \hat{L}_{j+2}\hat{r}_{j+1} = \hat{r}_{j+1}\hat{L}_{j+2} - i\hbar\hat{r}_j,$$

we may rewrite Eq. (*) in a shorter form:

$$[\hat{L}^2, \hat{r}_j] = -2i\hbar(\hat{r}_{j+2}\hat{L}_{j+1} - \hat{r}_{j+1}\hat{L}_{j+2} + i\hbar\hat{r}_j). \tag{**}$$

Plugging this expression for the internal commutator into the left-hand side of the identity to be proved, and taking into account that according to Eq. (5.151) of the lecture notes, the operator of L^2 commutes with the operators of all L_j, we get

$$\begin{aligned}[\hat{L}^2, [\hat{L}^2, \hat{r}_j]] &= -2i\hbar[\hat{L}^2, (\hat{r}_{j+2}\hat{L}_{j+1} - \hat{r}_{j+1}\hat{L}_{j+2} + i\hbar\hat{r}_j)] \\ &= -2i\hbar\left\{[\hat{L}^2, \hat{r}_{j+2}]\hat{L}_{j+1} - [\hat{L}^2, \hat{r}_{j+1}]\hat{L}_{j+2} + i\hbar[\hat{L}^2, \hat{r}_j]\right\}.\end{aligned}$$

Now using Eq. (**) again for the first two commutators (with the corresponding replacement of the indices j), we get

$$\begin{aligned}[\hat{L}^2, [\hat{L}^2, \hat{r}_j]] &= -2i\hbar\{[-2i\hbar(\hat{r}_{j+1}\hat{L}_j - \hat{r}_j\hat{L}_{j+1} + i\hbar\hat{r}_{j+2})]\hat{L}_{j+1} \\ &\quad - [-2i\hbar(\hat{r}_j\hat{L}_{j+2} - \hat{r}_{j+2}\hat{L}_j + i\hbar\hat{r}_{j+1})]\hat{L}_{j+2} + i\hbar[\hat{L}^2, \hat{r}_j]\} \\ &\equiv 4\hbar^2(-\hat{r}_{j+1}\hat{L}_j\hat{L}_{j+1} + \hat{r}_j\hat{L}_{j+1}\hat{L}_{j+1} - i\hbar\hat{r}_{j+2}\hat{L}_{j+1} \\ &\quad + \hat{r}_j\hat{L}_{j+2}\hat{L}_{j+2} - \hat{r}_{j+2}\hat{L}_j\hat{L}_{j+2} + i\hbar\hat{r}_{j+1}\hat{L}_{j+2}) + 2\hbar^2(\hat{L}^2\hat{r}_j - \hat{r}_j\hat{L}^2).\end{aligned}$$

Adding and subtracting to the right-hand side of this relation the following expression:

$$4\hbar^2\hat{r}_j\hat{L}^2 \equiv 4\hbar^2\hat{r}_j(\hat{L}_j\hat{L}_j + \hat{L}_{j+1}\hat{L}_{j+1} + \hat{L}_{j+2}\hat{L}_{j+2}),$$

with the plus sign in the first case and the minus sign in the second case, we get

$$\begin{aligned}[\hat{L}^2, [\hat{L}^2, \hat{r}_j]] &= 4\hbar^2[-(\hat{r}_{j+1}\hat{L}_j + i\hbar\hat{r}_{j+2})\hat{L}_{j+1} - (\hat{r}_{j+2}\hat{L}_j - i\hbar\hat{r}_{j+1})\hat{L}_{j+2} - \hat{r}_j\hat{L}_j\hat{L}_j] \\ &\quad +2\hbar^2(\hat{L}^2\hat{r}_j + \hat{r}_j\hat{L}^2).\end{aligned}$$

With the commutation relation (5.148) applied to each term in the square brackets,

$$\hat{r}_{j+1}\hat{L}_j + i\hbar\hat{r}_{j+2} = \hat{L}_j\hat{r}_{j+1}, \qquad \hat{r}_{j+2}\hat{L}_j - i\hbar\hat{r}_{j+1} = \hat{L}_j\hat{r}_{j+2}, \qquad \hat{r}_j\hat{L}_j = \hat{L}_j\hat{r}_j,$$

this expression is reduced to

$$\begin{aligned}[\hat{L}^2, [\hat{L}^2, \hat{r}_j]] &= 4\hbar^2(-\hat{L}_j\hat{r}_{j+1}\hat{L}_{j+1} - \hat{L}_j\hat{r}_{j+2}\hat{L}_{j+2} - \hat{L}_j\hat{r}_j\hat{L}_j) \\ &\quad + 2\hbar^2(\hat{L}^2\hat{r}_j + \hat{r}_j\hat{L}^2) \\ &\equiv -4\hbar^2\hat{L}_j(\hat{r}_j\hat{L}_j + \hat{r}_{j+1}\hat{L}_{j+1} + \hat{L}_j\hat{r}_{j+2}\hat{L}_{j+2}) \\ &\quad + 2\hbar^2(\hat{L}^2\hat{r}_j + \hat{r}_j\hat{L}^2).\end{aligned}$$

But according to Eq. (5.147), the sum in the first parentheses equals zero:

$$\hat{r}_j\hat{L}_j + \hat{r}_{j+1}\hat{L}_{j+1} + \hat{r}_{j+2}\hat{L}_{j+2} \equiv \hat{\mathbf{r}} \cdot \hat{\mathbf{L}} \equiv \hat{\mathbf{r}} \cdot (\hat{\mathbf{r}} \times \hat{\mathbf{p}}) = 0,$$

so that we indeed get

$$[\hat{L}^2, [\hat{L}^2, \hat{r}_j]] \equiv 2\hbar^2(\hat{L}^2\hat{r}_j + \hat{r}_j\hat{L}^2).$$

***Problem* 5.35.** Use the commutation relation proved in the previous problem, and Eq. (5.148) of the lecture notes, to prove the orbital electric-dipole selection rules mentioned in section 5.6 of the lecture notes.

Solution: First, let us calculate the matrix elements of both sides of the identity proved in the previous problem,

$$[\hat{L}^2, [\hat{L}^2, \hat{r}_j]] = 2\hbar^2(\hat{r}_j\hat{L}^2 + \hat{L}^2\hat{r}_j), \qquad (*)$$

in the basis of the $\{l, m\}$ states discussed in section 5.6 of the lecture notes, i.e. the common eigenstates of the operators $\hat{L}^2$ and $\hat{L}_z$. For the left-hand side we get

$$\langle l, m|[\hat{L}^2, [\hat{L}^2, \hat{r}_j]\,]\,|l', m'\rangle \equiv \langle l, m|\left(\hat{L}^2[\hat{L}^2, \hat{r}_j]\,|l', m'\rangle - \langle l, m|[\hat{L}^2, \hat{r}_j]\,\hat{L}^2\right)\,|l', m'\rangle.$$

According to Eq. (5.163) of the lecture notes, the action of the rightmost operator $\hat{L}^2$ on the right ket-vector gives the same vector, but multiplied by $\hbar^2 l'(l' + 1)$, while the action of the similar (Hermitian!) operator in the leftmost position on the ket-vector gives the same vector, but multiplied by $\hbar^2 l(l + 1)$, so that we may continue as follows:

$$\begin{aligned}\langle l, m|[\hat{L}^2, [\hat{L}^2, \hat{r}_j]\,]\,|l', m'\rangle &= \hbar^2[l(l + 1) - l'(l' + 1)]\langle l, m|[\hat{L}^2, \hat{r}_j]\,|l', m'\rangle \\ &\equiv \hbar^2[l(l + 1) - l'(l' + 1)]\langle l, m|(\hat{L}^2\hat{r}_j - \hat{r}_j\hat{L}^2)\,|l', m'\rangle.\end{aligned}$$

Now the similar second step operation gives

$$\langle l, m|[\hat{L}^2, [\hat{L}^2, \hat{r}_j]\,]\,|l', m'\rangle = \hbar^4[l(l + 1) - l'(l' + 1)]^2\langle l, m|\,\hat{r}_j\,\,|l', m'\rangle.$$

The matrix elements of the right-hand side of Eq. (*) may be calculated in the similar way:

$$\begin{aligned}\langle l, m|\,2\hbar^2(\hat{r}_j\hat{L}^2 + \hat{L}^2\hat{r}_j)\,|l', m'\rangle &\equiv 2\hbar^2\left[\langle l, m|\,\hat{r}_j\hat{L}^2\,\,|l', m'\rangle + \langle l, m|(\hat{L}^2\hat{r}_j)\,|l', m'\rangle\right] \\ &= 2\hbar^4\langle l, m|\,\hat{r}_j\,\,|l', m'\rangle[l'(l' + 1) + l(l + 1)].\end{aligned}$$

Due to the identity (*), these matrix elements have to be equal, giving the result

$$\langle l, m|\, \hat{r}_j\, |l', m'\rangle f(l, l') = 0, \quad \text{where}$$
$$f(l, l') \equiv [l(l+1) - l'(l'+1)]^2 - 2[l'(l'+1) + l(l+1)].$$

This means that the matrix element of the jth coordinate has to vanish, unless $f(l, l')$ equals zero. Rewriting this function as

$$f(l, l') \equiv [(l + l' + 1)(l - l')]^2 - [(l + l' + 1)^2 + (l' - l)^2 - 1]$$
$$\equiv [(l + l' + 1)^2 - 1]\,[(l - l')^2 - 1],$$

and taking into account that l and l' cannot be negative, we see that $f(l, l')$ equals zero only if either $l = l' = 0$ (when the first square bracket of the last expression vanishes), or if $(l - l')^2 = 1$, i.e. if

$$l' = l \pm 1.$$

Since, according to Eq. (3.174), the angular wavefunction of the state with $l = 0$ (and hence $m = 0$) is a constant, the matrix elements

$$\langle l = 0, m = 0|\hat{r}_j|l = 0, m = 0\rangle \propto \oint \hat{r}_j d\Omega,$$

corresponding to the first case ($l = l' = 0$), vanish due to symmetry, the above equality $l' = l \pm 1$, gives the necessary condition to have at least some matrix element (or elements) $\langle l, m|\hat{r}_j\,|l', m'\rangle$ different from zero.

In order to get the second necessary condition, let us calculate the similar matrix elements of both parts of the vector equality (5.148), written for the operator of $L_3 \equiv L_z$,

$$[\hat{L}_3, \hat{r}_j] = i\hbar \hat{r}_{j'}\varepsilon_{3jj'}. \qquad (**)$$

For the left-hand side, we may write

$$\langle l, m|\, [\hat{L}_3, \hat{r}_j]\, |l', m'\rangle \equiv \langle l, m|\, \hat{L}_3\hat{r}_j\, |l', m'\rangle - \langle l, m|\, \hat{r}_j\hat{L}_3\, |l', m'\rangle.$$

According to Eq. (5.158) of the lecture notes, the action of the rightmost operator $\hat{L}_3 \equiv \hat{L}_z$ on the right ket-vector gives the same vector, but multiplied by $\hbar m'$, while the action of the similar (Hermitian!) operator in the leftmost position on the ket-vector gives the same vector, but multiplied by $\hbar m$, so that

$$\langle l, m|\, [\hat{L}_3, \hat{r}_j]\, |l', m'\rangle \equiv \hbar(m - m')\langle l, m|\, \hat{r}_j\, |l', m'\rangle. \qquad (***)$$

The similar matrix element of the right-hand side of Eq. (**) is just

$$\langle l, m|\, i\hbar\hat{r}_{j'}\varepsilon_{3jj'}\, |l', m'\rangle \equiv i\hbar\langle l, m|\, \hat{r}_{j'}\, |l', m'\rangle\varepsilon_{3jj'},$$

so that according to Eq. (**),

$$(m - m')\langle l, m|\, \hat{r}_j\, |l', m'\rangle = i\langle l, m|\, \hat{r}_{j'}\, |l', m'\rangle\varepsilon_{3jj'}. \qquad (****)$$

First, let us consider the case when the index j, and hence j', are not equal to 3, so that we may rewrite Eq. (****) for the second possible index $j' \neq 3$:

$$(m - m')\langle l, m| \hat{r}_{j'} \ |l', m'\rangle = i\langle l, m| \hat{r}_j \ |l', m'\rangle\varepsilon_{3j'j},$$

$$\text{i.e.} \quad \langle l, m| \hat{r}_{j'} \ |l', m'\rangle = \frac{i\varepsilon_{3j'j}}{m - m'}\langle l, m| \hat{r}_j \ |l', m'\rangle.$$

Plugging the last relation into the right-hand side of Eq. (***), and taking into account that by the Levi-Civita symbol's definition, $\varepsilon_{3jj'}\varepsilon_{3j'j} = -1$, we get

$$[(m - m')^2 - 1]\langle l, m| \hat{r}_j \ |l', m'\rangle = 0.$$

From here, the second necessary condition to have at least some matrix element(s) $\langle l, m|\hat{r}_j \ |l', m'\rangle$, with $j \neq 3$, different from zero is $(m - m')^2 = 1$, i.e.

$$m' = m \pm 1.$$

Finally, let us consider the case $j = 3$. In this case, the commutator (**) equals zero, so that Eq. (***) yields

$$(m - m')\langle l, m| \hat{r}_3 \ |l', m'\rangle = 0.$$

So, we may conclude that this particular matrix element may be different from 0 if

$$m = m',$$

thus completing the proof of the selection rules formulated in section 5.6.

As will be discussed in section 9.3, these rules, applied to the emission/absorption of electric-dipole radiation coupled to orbital motion of a quantum system, express the conservation of the total angular momentum, including that of the emitted/absorbed photon.

Note that these selection rules may be also obtained in a purely wave-mechanical way, using the following recurrent relations for the spherical functions (which are given here just for the reader's reference):

$$\cos\theta \ Y_l^m(\theta, \varphi) = \left[\frac{(l + 1 + m)(l + 1 - m)}{(2l + 1)(2l + 3)}\right]^{1/2} Y_{l+1}^m(\theta, \varphi) + \left[\frac{(l + m)(l - m)}{(2l + 1)(2l - 1)}\right]^{1/2} Y_{l-1}^m(\theta, \varphi),$$

$$\sin\theta \ Y_l^m(\theta, \varphi) = \left\{-\left[\frac{(l + 1 - m)(l + 2 - m)}{(2l + 1)(2l + 3)}\right]^{1/2} Y_{l+1}^{m-1}(\theta, \varphi) + \left[\frac{(l + m)(l - 1 + m)}{(2l + 1)(2l - 1)}\right]^{1/2} Y_{l-1}^{m+1}(\theta, \varphi)\right\}e^{i\varphi}.$$

***Problem* 5.36.** Express the commutators listed in Eq. (5.179) of the lecture notes, $[\hat{J}^2, \hat{L}_z]$ and $[\hat{J}^2, \hat{S}_z]$, via $\hat{L}_j$ and $\hat{S}_j$.

Solution: Using Eq. (5.181), and then the second of Eqs. (5.176), we may transform the first commutator as

$$\begin{aligned}[\hat{J}^2, \hat{L}_z] &= [\hat{L}^2 + \hat{S}^2 + 2\hat{\mathbf{L}} \cdot \hat{\mathbf{S}}, \hat{L}_z] \\ &= 2[\hat{\mathbf{L}} \cdot \hat{\mathbf{S}}, \hat{L}_z] \equiv 2[\hat{L}_x\hat{S}_x + \hat{L}_y\hat{S}_y + \hat{L}_z\hat{S}_z, \hat{L}_z] \\ &= 2\hat{S}_x[\hat{L}_x, \hat{L}_z] + 2\hat{S}_y[\hat{L}_y, \hat{L}_z].\end{aligned}$$

Now using the first of Eqs. (5.176), we finally get

$$[\hat{J}^2, \hat{L}_z] = 2i\hbar(-\hat{S}_x\hat{L}_y + \hat{S}_y\hat{L}_x).$$

Acting absolutely similarly (just swapping the operators of S and L everywhere), we get

$$[\hat{J}^2, \hat{S}_z] = 2i\hbar(\hat{S}_x\hat{L}_y - \hat{S}_y\hat{L}_x).$$

We see that indeed, neither of these commutators vanishes (though their sum, equal to $[\hat{J}^2, \hat{J}_z]$, does).

***Problem* 5.37.** Find the operator $\hat{\mathcal{T}}_\phi$ describing a quantum state's rotation by angle ϕ about a certain axis, using the similarity of this operation with the shift of a Cartesian coordinate, discussed in section 5.5 of the lecture notes. Then use this operator to calculate the probabilities of measurements of spin-½ components of a beam of particles with z-polarized spin, by a Stern–Gerlach instrument turned by angle θ within the $[z, x]$ plane, where y is the axis of particle propagation—see figure 4.1.[44]

Solution: In the course of our discussion of the Glauber states in section 5.5, we have proved that operator, defined by Eq. (5.111):

$$\hat{\mathcal{T}}_X = \exp\left\{-i\frac{\hat{p}_x}{\hbar}X\right\},$$

provides the wavefunction's translation by distance X along axis x. From section 5.6 we know that at a planar rotation about the axis z, the product $L_z\varphi$ plays the same role as the product $p_x x$ at the 1D motion along axis x. Hence, the operator

$$\hat{\mathcal{T}}_\varphi = \exp\left\{-i\frac{\hat{L}_z}{\hbar}\varphi\right\}$$

rotates any orbital wavefunction by the angle φ about axis z. It is straightforward to generalize this relation to the rotation by angle ϕ about an arbitrary axis with the unit vector $\mathbf{n}$:

[44] Note that the last task is just a particular case of problem 4.18 (see also problem 5.1).

$$\hat{\mathcal{T}}_\phi = \exp\left\{-i\frac{\hat{\mathbf{L}}\cdot\mathbf{n}}{\hbar}\phi\right\}.$$

Since all the commutation properties of the operator $\hat{\mathbf{S}}$ of spin are identical to those of $\hat{\mathbf{L}}$ (see section 5.7), the spin rotation should be described by a similar operator,

$$\hat{\mathcal{T}}_\phi = \exp\left\{-i\frac{\hat{\mathbf{S}}\cdot\mathbf{n}}{\hbar}\phi\right\}$$

so that for spin-½ particles, with $\hat{\mathbf{S}} = (\hbar/2)\hat{\boldsymbol{\sigma}}$,

$$\hat{\mathcal{T}}_\phi = \exp\left\{-i\hat{\boldsymbol{\sigma}}\cdot\mathbf{n}\frac{\phi}{2}\right\}.$$

The last relation may be recast into a simpler form by the expansion of the exponent into the Taylor series, separating the odd and even terms:

$$\begin{aligned}\hat{\mathcal{T}}_\phi &= \exp\left\{-i\hat{\boldsymbol{\sigma}}\cdot\mathbf{n}\frac{\phi}{2}\right\}\\ &= \sum_{k=2m}\frac{1}{k!}\left(-i\hat{\boldsymbol{\sigma}}\cdot\mathbf{n}\frac{\phi}{2}\right)^{2m} - i\hat{\boldsymbol{\sigma}}\cdot\mathbf{n}\frac{\phi}{2}\sum_{k=2m+1}\frac{1}{k!}\left(-i\hat{\boldsymbol{\sigma}}\cdot\mathbf{n}\frac{\phi}{2}\right)^{2m}.\end{aligned}$$

Since, as we know from chapter 4,

$$(\hat{\boldsymbol{\sigma}}\cdot\mathbf{n})^2 = (\hat{\sigma}_x n_x)^2 + (\hat{\sigma}_x n_y)^2 + (\hat{\sigma}_z n_z)^2 = \left(n_x^2 + n_y^2 + n_z^2\right)\hat{I} = \hat{I},$$

we may write

$$(\hat{\boldsymbol{\sigma}}\cdot\mathbf{n})^{2m} = \hat{I}, \qquad \text{so that } \hat{\mathcal{T}}_\varphi = \hat{I}\sum_{k=2m}\frac{1}{k!}(-1)^m\left(\frac{\phi}{2}\right)^k - i\hat{\boldsymbol{\sigma}}\cdot\mathbf{n}\sum_{k=2m+1}\frac{1}{k!}(-1)^m\left(\frac{\phi}{2}\right)^k.$$

But these sums are just the Taylor expansions of $\cos(\varphi/2)$ and $\sin(\varphi/2)$, respectively, so that

$$\hat{\mathcal{T}}_\varphi = \hat{I}\cos\frac{\phi}{2} - i\hat{\boldsymbol{\sigma}}\cdot\mathbf{n}\sin\frac{\phi}{2}.$$

If the unit vector $\mathbf{n}$ of rotation is directed along the particle's propagation axis (in our particular case, the y-axis), then $\hat{\boldsymbol{\sigma}}\cdot\mathbf{n} = \hat{\sigma}_y$, and the operator's matrix in the z-basis becomes very simple:

$$\mathrm{T}_\phi = \begin{pmatrix}\cos(\phi/2) & -\sin(\phi/2)\\ \sin(\phi/2) & \cos(\phi/2)\end{pmatrix},$$

where ϕ is now the angle between the final direction of state's rotation and its initial direction. If the latter is the z-axis, then i.e. ϕ is just the polar angle θ of the final direction. This means that if the initial and final states (let us call them 0 and θ) are expressed in the z-basis as, respectively,

$$|0\rangle = 0_\uparrow \,|\uparrow\rangle + 0_\downarrow \,|\downarrow\rangle, \quad \text{and} \quad |\theta\rangle = \theta_\uparrow \,|\uparrow\rangle + \theta_\downarrow \,|\downarrow\rangle,$$

then their probability amplitudes in the z-basis are related as

$$\begin{pmatrix} \theta_\uparrow \\ \theta_\downarrow \end{pmatrix} = \begin{pmatrix} \cos(\theta/2) & -\sin(\theta/2) \\ \sin(\theta/2) & \cos(\theta/2) \end{pmatrix} \begin{pmatrix} 0_\uparrow \\ 0_\downarrow \end{pmatrix}.$$

The rotation of a Stern–Gerlach instrument is evidently equivalent to the spin rotation by the same but opposite angle, so that in its basis, the z-polarized state is

$$|\uparrow\rangle = \cos\frac{\theta}{2}\,|\theta_\uparrow\rangle + \sin\frac{\theta}{2}\,|\theta_\downarrow\rangle$$

Hence the probabilities of the two outcomes of the SG measurements of the polarized particles are

$$W(\theta_\uparrow) = \cos^2\frac{\theta}{2}, \qquad W(\theta_\downarrow) = \sin^2\frac{\theta}{2},$$

—the same result as was obtained earlier in the solutions of problems 4.18 and 5.1.

***Problem* 5.38.** The rotation ('angle translation') operator $\hat{\mathcal{T}}_\phi$ analyzed in the previous problem, and the coordinate translation operator $\hat{\mathcal{T}}_X$ discussed in section 5.5 of the lecture notes, have a similar structure:

$$\hat{\mathcal{T}}_\lambda = \exp\left\{-i\frac{\hat{C}\lambda}{\hbar}\right\},$$

where λ is a real c-number characterizing the translation, and $\hat{C}$ is a Hermitian operator, which does not explicitly depend on time.

(i) Prove that such operators $\hat{\mathcal{T}}_\lambda$ are unitary.
(ii) Prove that if the shift by λ, induced by the operator $\hat{\mathcal{T}}_\lambda$, leaves the Hamiltonian of some system unchanged for any λ, then $\langle C\rangle$ is a constant of motion for any initial state of the system.
(iii) Discuss what does the last conclusion mean for the particular operators $\hat{\mathcal{T}}_X$ and $\hat{\mathcal{T}}_\phi$.

Solutions:

(i) As was repeatedly discussed in the lecture notes, the exponent of an operator is defined by its Taylor expansion. For our current case, such expansion is

$$\hat{\mathcal{T}}_\lambda \equiv \exp\left\{-i\frac{\hat{C}\lambda}{\hbar}\right\} = \sum_{k=0}^{\infty}\frac{1}{k!}\left(-i\frac{\hat{C}\lambda}{\hbar}\right)^k \equiv \sum_{k=0}^{\infty}\frac{1}{k!}(-i)^k\left(\frac{\lambda}{\hbar}\right)^k\hat{C}^k.$$

The Hermitian conjugate of this expression is

$$\hat{\mathcal{T}}_\lambda^\dagger \equiv \sum_{k=0}^{\infty}\frac{1}{k!}[(-i)^k]^*\left[\left(\frac{\lambda}{\hbar}\right)^k\right]^*(\hat{C}^k)^\dagger \equiv \sum_{k=0}^{\infty}\frac{1}{k!}i^k\left(\frac{\lambda}{\hbar}\right)^k\hat{C}^k, \qquad (*)$$

because the Hermitian conjugation of a c-number is equivalent to its complex conjugation, the ratio $\lambda/\hbar$ is real, and since the operator $\hat{C}$ is Hermitian, i.e. $\hat{C}^\dagger = \hat{C}$, so are all operators $\hat{C}^k$, with $k = 0, 1, 2, \ldots$ Indeed, applying the relation

$$(\hat{A}\hat{B})^\dagger = \hat{B}^\dagger\hat{A}^\dagger,$$

whose proof was the subject of problem 4.1(iii), to the operators $\hat{A} \equiv \hat{C}^{k-1}$ and $\hat{B} \equiv \hat{C}$, we get

$$(\hat{C}^k)^\dagger \equiv (\hat{C}^{k-1}\hat{C})^\dagger = \hat{C}^\dagger(\hat{C}^{k-1})^\dagger = \hat{C}(\hat{C}^{k-1})^\dagger.$$

Repeating this operation $(k - 1)$ more times, we get $(\hat{C}^k)^\dagger = \hat{C}^k$.

Now returning to Eq. (*), we may see that its last form is just the Taylor expansion of the operator

$$\exp\left\{+i\frac{\hat{C}\lambda}{\hbar}\right\} = \sum_{k=1}^{\infty}\frac{1}{k!}\left(i\frac{\hat{C}\lambda}{\hbar}\right)^k \equiv \sum_{k=1}^{\infty}\frac{1}{k!}\,i^k\left(\frac{\lambda}{\hbar}\right)^k\hat{C}^k.$$

Thus we have proved that

$$\hat{\mathcal{T}}^\dagger = \exp\left\{+i\frac{\hat{C}\lambda}{\hbar}\right\}, \quad \text{i.e.} \quad \exp\left\{+i\frac{\hat{C}\lambda}{\hbar}\right\} = \left(\exp\left\{-i\frac{\hat{C}\lambda}{\hbar}\right\}\right)^\dagger.$$

Let us apply to these two operators the general Eq. (5.117) of the lecture notes, with $\hat{A} \equiv i\hat{C}\lambda/\hbar$ and $\hat{B} \equiv \hat{I}$. Since the identity operator commutes with any other operator, Eq. (5.116) is valid with $\mu = 0$, so that Eq. (5.117) becomes

$$\exp\left\{+i\frac{\hat{C}\lambda}{\hbar}\right\}\exp\left\{-i\frac{\hat{C}\lambda}{\hbar}\right\} = \hat{I}, \quad \text{i.e. } \hat{\mathcal{T}}_\lambda^\dagger\hat{\mathcal{T}}_\lambda = \hat{I},$$

i.e. the translation operator $\hat{\mathcal{T}}_\lambda$ is indeed unitary[45].

[45] Note that Eq. (5.119), proved similarly in section 5.5 of the lecture notes, is just a particular case of this result for the operator defined by Eq. (5.113).

(ii) Let us spell out the commutator $[\hat{H}, \hat{\mathcal{T}}_\lambda]$, using the Taylor expansion of the latter operator:

$$[\hat{H}, \hat{\mathcal{T}}_\lambda] = \left[\hat{H}, \exp\left\{-i\frac{\hat{C}\lambda}{\hbar}\right\}\right] = \left[\hat{H}, \sum_{k=1}^{\infty}\frac{1}{k!}\left(-i\frac{\hat{C}\lambda}{\hbar}\right)^k\right] \equiv \sum_{k=1}^{\infty}\alpha_k\lambda^k, \qquad (**)$$

where

$$\alpha_k \equiv \frac{1}{k!}\left(-i\frac{1}{\hbar}\right)^k [\hat{H}, \hat{C}^k]$$

According to Eq. (4.93) of the lecture notes, the Hamiltonian's invariance under a unitary transform may be expressed as

$$\hat{U}^\dagger \hat{H} \hat{U} = \hat{H}.$$

Acting by the operator $\hat{U}$ on both sides of this relation, and taking into account that $\hat{U}\hat{U}^\dagger\hat{H} \equiv \hat{I}\hat{H} \equiv \hat{H}$, we see that this relation is equivalent to

$$\hat{H}\hat{U} = \hat{U}\hat{H}, \quad \text{i.e.} \quad [\hat{H}, \hat{U}] = 0.$$

Since we have proved that the operator $\hat{\mathcal{T}}_\lambda$ is unitary, the commutator $[\hat{H}, \hat{\mathcal{T}}_\lambda]$ equals zero (i.e. is a null operator), identically in the parameter λ. But according to Eq. (**), this is only possible when each coefficient α_k in the Taylor expansion of this expression equals zero, including α_1. This requirement gives

$$[\hat{H}, \hat{C}] = 0. \qquad (***)$$

Since, by the problem's conditions, the operator $\hat{C}$ does not depend on time explicitly, $\partial\hat{C}/\partial t = 0$, we may use the key Eq. (4.199) of the lecture notes, together with Eq. (***), to calculate its full time derivative in the Heisenberg picture of quantum dynamics:

$$i\hbar\frac{d\hat{C}}{dt} = [\hat{C}, \hat{H}] = 0.$$

According to the basic relation of the Heisenberg picture, Eq. (4.191), this means that for any initial state of the system, $\langle C\rangle$ does not depend on time.

(iii) For the linear translation operator $\hat{\mathcal{T}}_X$, defined by Eq. (5.111),

$$\hat{\mathcal{T}}_X \equiv \exp\left\{-i\frac{\hat{p}X}{\hbar}\right\},$$

the observable C is the linear momentum p (or rather its Cartesian component in the shift's direction x). Hence, the invariance of the Hamiltonian of a system with respect to such a shift means the conservation of $\langle p\rangle$ during an arbitrary motion of the system—the result well known in classical mechanics.

Similarly, if the rotations of a system about some axis $\mathbf{n}$, by an arbitrary angle ϕ, described by the operators

$$\hat{\mathcal{T}}_\phi \equiv \exp\left\{-i\frac{\hat{\mathbf{L}} \cdot \mathbf{n}}{\hbar}\phi\right\} \quad \text{and/or} \quad \hat{\mathcal{T}}_\phi \equiv \exp\left\{-i\frac{\hat{\mathbf{S}} \cdot \mathbf{n}}{\hbar}\phi\right\}$$

do not alter the Hamiltonian of a system (in particular, this is valid for any system symmetric with respect to rotation about the axis **n**), then the expectation values of the components of the orbital and/or spin angular momenta along axis **n** are conserved. For the orbital momentum vector **L**, this conclusion is also well known in classical mechanics (see, e.g. *Part CM* section 1.4), and it is only natural that the same fact holds for the spin vector **S**, because it is defined by similar commutation relations—cf. Eqs. (5.168) and (5.176).

***Problem* 5.39.** A particle with spin s is in a state with definite quantum numbers l and j. Prove that the observable $\mathbf{L} \cdot \mathbf{S}$ also has a definite value, and calculate it.

Solution: According to Eq. (5.177), in any state with a definite value of l, the variable L^2 always has a definite value equal to

$$L^2 = \hbar^2 l(l+1). \qquad (**)$$

(This is true even if the state does not have a definite value of m_l,[46] because such a state may be always represented as a linear superposition of states with the same l, but different m_l, and according to Eq. (5.177), the operator $\hat{L}^2$, acting upon such a superposition, multiplies each of its components by the same number (**).)

Absolutely similarly (and because the spin s of any particle is always fixed), Eqs. (5.169) and (5.175) show that for the specified state, the observables J^2 and S^2 also have definite values, respectively:

$$J^2 = \hbar^2 j(j+1), \qquad \text{and } S^2 = \hbar^2 s(s+1).$$

Hence, according to Eq. (5.181) of the lecture notes,

$$2\hat{\mathbf{L}} \cdot \hat{\mathbf{S}} = (\hat{J}^2 - \hat{L}^2 - \hat{S}^2),$$

in our state the scalar product $\mathbf{L} \cdot \mathbf{S}$ also has a definite value:

$$\mathbf{L} \cdot \mathbf{S} = \frac{\hbar^2}{2}[j(j+1) - l(l+1) - s(s+1)].$$

***Problem* 5.40.** For a spin-½ in a state with definite quantum numbers l, m_l, and m_s, calculate the expectation value of the observable J^2, and the probabilities of all its possible values. Interpret your results in the terms of the Clebsch–Gordan coefficients (5.190).

[46] As it is the case, for example, for the base states j, m_j of the coupled representation—see, e.g. figure 5.12 and Eq. (5.183).

Solution: Averaging the first form of Eq. (5.181) of the lecture notes, we get

$$\langle J^2\rangle = \langle L^2\rangle + \langle S^2\rangle + 2\langle \mathbf{L}\cdot\mathbf{S}\rangle \equiv \langle L^2\rangle + \langle S^2\rangle + 2\langle L_xS_x\rangle + 2\langle L_yS_y\rangle + 2\langle L_zS_z\rangle.$$

According to the second of Eqs. (5.177), a state with a definite quantum number l is an eigenstate of the operator $\hat{L}^2$, so that the first term on the right-hand side of the last explession is equal to the corresponding eigenvalue, $\hbar^2 l(l+1)$. Since the spin quantum number s of a particle is always fixed (definite), Eq. (5.169) allows us to make the similar conclusion about the second term: $\langle S^2\rangle = \hbar^2 s(s+1) = \hbar^2\cdot½\cdot(½+1) = (3/4)\,\hbar^2$.

Next, the fact that not only the squares of the vectors $\mathbf{L}$ and $\mathbf{S}$, but also their z-components, $L_z = \hbar m_l$ and $S_z = \hbar m_s$ have definite values in this quantum state, means that these vectors are uncoupled[47], so that we may write $\langle L_xS_x\rangle = \langle L_x\rangle\langle S_x\rangle$, etc, where the first average is over the Hilbert space of the orbital states, while the second average is over that of the spin states. But as we know from section 4.5 of the lecture notes (see, e.g. Eq. (4.134), which is valid, in the nomenclature of section 4.7, for $m_s = +½$), in a state with a definite m_s, the averages $\langle S_x\rangle$ and $\langle S_y\rangle$ equal zero, so that the only nonvanishing product of the component averages is $\langle L_z\rangle\langle S_z\rangle$, and we finally get

$$\langle J^2\rangle = \hbar^2\left[l(l+1) + \frac{3}{4} + 2m_l m_s\right]. \qquad (*)$$

Now, according to Eq. (5.189) of the lecture notes, in the quantum-statistical ensemble with a definite l (and $s = ½$), the quantum number j, and hence the variable $J^2 = \hbar^2 j(j+1)$, may take only two values each:

$$j_\pm = l \pm \frac{1}{2},$$
$$J_\pm^2 = \hbar^2 j_\pm(j_\pm + 1) = \hbar^2\left(l \pm \frac{1}{2}\right)\left(l + 1 \pm \frac{1}{2}\right) \qquad (**)$$
$$\equiv \hbar^2\left[l(l+1) + \frac{1}{4} \pm \frac{2l+1}{2}\right].$$

Hence there are only two nonvanishing probabilities, W_+ and $W_- = 1 - W_+$, to calculate, and the general Eq. (1.37) takes the form

$$\langle J^2\rangle = W_+J_+^2 + W_-J_-^2 = W_+J_+^2 + (1 - W_+)J_-^2 \equiv J_-^2 + W_+(J_+^2 - J_-^2),$$

so that

$$W_+ = \frac{\langle J^2\rangle - J_-^2}{J_+^2 - J_-^2}, \quad \text{and} \quad W_- = 1 - W_+ = \frac{\langle J^2\rangle - J_+^2}{J_-^2 - J_+^2}.$$

[47] More strictly, in the terms of section 5.7 of the lecture notes, such states are parts of the uncoupled-representation basis—see, e.g. figure 5.12 and the first line of Eq. (5.182).

Plugging into these expressions the relations (*) and (**), we get

$$W_+ = \frac{l + 1 + 2m_l m_s}{2l + 1}, \quad W_- = \frac{l - 2m_l m_s}{2l + 1}, \quad \text{i.e. } W_\pm = \frac{l + ½ \pm ½ \pm 2m_l m_s}{2l + 1}.$$

Finally, note that using the relation proved in section 5.7, $m_l = m_j - m_s$, and the fact that for spin-½ particles $(m_s)^2 = (\pm ½)^2 = 1/4$, we may recast the last formula in the form

$$W_\pm = \frac{l \pm 2m_j m_s + ½}{2l + 1} \equiv \frac{1}{2l + 1} \times \begin{cases} l \pm m_j + ½, & \text{for } m_s = +½, \\ l \mp m_j + ½, & \text{for } m_s = -½. \end{cases}$$

Comparing this formula with Eqs. (5.190), we see that $W_\pm$ are just the squared moduli of the Clebsch–Gordan coefficients—as they should be. So, if we need only the moduli of these coefficients (as we do for most applications), this solution presents a simple alternative way to calculate them.

***Problem* 5.41.** Derive the general recurrence relations for the Clebsch–Gordan coefficients.

Hint: Using the similarity of the commutation relations discussed in section 5.7, write the relations similar to Eqs. (5.164) of the lecture notes, for other components of the angular momentum, and apply them to Eq. (5.170).

Solution: The definition (5.170) of the total momentum implies that the ladder operators of its components, defined similarly to Eq. (5.153), are related simply as

$$\hat{J}_\pm = \hat{L}_\pm + \hat{S}_\pm.$$

Let us act by these operators on the corresponding parts of Eq. (5.183), keeping in mind that the brackets $\langle m_l, m_s | j, m_j \rangle$ in that relation are just the *c*-numbers (the Clebsch–Gordan coefficients), not affected by the operator action:

$$\hat{J}_\pm | j, m_j \rangle = \sum_{m_l, m_s} \langle m_l, m_s | j, m_j \rangle (\hat{L}_\pm + \hat{S}_\pm) | m_l, m_s \rangle, \qquad (*)$$

where the common quantum numbers l and s are implied for both representations.

In order to spell out the left-hand and right-hand sides of this equality, let us recall that Eq. (5.164) of the lecture notes could have been derived directly from the commutation relations (5.176), without any appeal to the wave-mechanics form of the orbital angular momentum operator. (See the model solution of problem 5.22.) Since, according to Eqs. (5.168) and (5.174), the commutation relations for the operators $\hat{\mathbf{S}}$ and $\hat{\mathbf{J}}$ are similar to those of $\hat{\mathbf{L}}$, we may repeat all the arguments to get similar formulas for the similarly defined ladder operators. In the notation of section 5.6, Eq. (5.164) takes the form

$$\hat{L}_\pm | m_l, m_s \rangle = \hbar [l(l + 1) - m_l(m_l \pm 1)]^{1/2} | m_l \pm 1, m_s \rangle$$
$$\equiv \hbar [(l \pm m_l + 1)(l \mp m_l)]^{1/2} | m_l \pm 1, m_s \rangle,$$

In the same uncoupled-presentation basis, we get[48]

$$\hat{S}_{\pm}|m_l, m_s\rangle = \hbar[(s \pm m_s + 1)(s \mp m_s)]^{1/2}|m_l, m_s \pm 1\rangle, \qquad (**)$$

while for the coupled-representation eigenstates,

$$\hat{J}_{\pm}|j, m_j\rangle = \hbar\left[(j \pm m_j + 1)(j \mp m_j)\right]^{1/2}|j, m_j \pm 1\rangle.$$

Plugging these expressions into Eq. (*), with the notation replacements $m_l \to \mu_l$ and $m_s \to \mu_s$ (the reason for them will be clear in a moment), we get:

$$\begin{aligned}&\left[(j \pm m_j + 1)(j \mp m_j)\right]^{1/2}|j, m_j \pm 1\rangle \\ &= \sum_{\mu_l, \mu_s}\left\langle \mu_l, \mu_s|j, m_j\right\rangle \\ &\times \left\{\left[(l \pm \mu_l + 1)(l \mp \mu_l)\right]^{1/2}|\mu_l \pm 1, \mu_s\rangle + \left[(s \pm \mu_s + 1)(s \mp \mu_s)\right]^{1/2}|\mu_l, \mu_s \pm 1\rangle\right\}.\end{aligned}$$

Now let us inner-multiply both sides of this relation by the bra-vector $\langle m_l, m_s|$. Since the vectors of any basis (in our case, of the uncoupled representation) are assumed to be orthonormal, the first term on the right-hand side gives a nonvanishing result only for $\mu_l = m_l \mp 1$ and $\mu_s = m_s$, and the second term, for $\mu_l = m_l$ and $\mu_s = m_s \mp 1$. As a result, the summation on that side is reduced to just two terms, and we get

$$\begin{aligned}&\left[(j \pm m_j + 1)(j \mp m_j)\right]^{1/2}\left\langle m_l, m_s|j, m_j \pm 1\right\rangle \\ &= [(l \pm m_l)(l \mp m_l + 1)]^{1/2}\left\langle m_l \mp 1, m_s|j, m_j\right\rangle \\ &\quad + [(s \pm m_s)(s \mp m_s + 1)]^{1/2}\left\langle m_l, m_s \mp 1|j, m_j\right\rangle.\end{aligned} \qquad (***)$$

The red and blue arrows in the figure below show the sets of the uncoupled-representation states, related by Eqs. (***) with, respectively, the upper and lower signs, on the rectangular lattice similar to the one that was shown in figure 5.14 of the lecture notes (in that case, for $s = ½$). These relations enable one to derive explicit formulas for the Clebsch–Gordan coefficients, similar to Eqs. (5.190), for an arbitrary spin s, starting from one of the two particular 'corner' states with $m_l = \pm l$ and $m_s = \pm s$, which may be represented by the same, single ket-vectors in both the uncoupled and coupled representations.

[48] For the most important case of spin-½ (with $s = ½$, and $m_s = \pm ½$), Eq. (**) is much simplified:

$$\hat{S}_{+}|m_l, m_s = +½\rangle = \hat{S}_{-}|m_l, m_s = -½\rangle = 0, \qquad \hat{S}_{\pm}|m_l, m_s = \mp ½\rangle = \hbar|m_l, m_s = \pm ½\rangle.$$

Verbally, this means that the ladder spin-½ operators of the proper sign just flip the spin's orientation. The orbital quantum number m_l aside, this is exactly the result we would get for these spin-½ operators $\hat{S}_{\pm} \equiv \hat{S}_x \pm i\hat{S}_y$ from Eqs. (4.128).

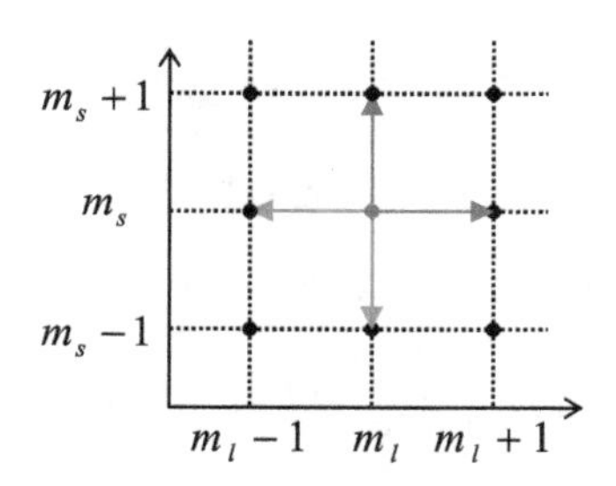

***Problem* 5.42.** Use the recurrence relations derived in the previous problem to prove Eqs. (5.190) of the lecture notes for the spin-½ Clebsch–Gordan coefficients.

Solution: Eqs. (5.190) is the set of four relations, corresponding to two independent signs in the relations $m_s = \pm½$ and $j - l = \pm ½$; as an example, let us consider the case of two upper signs:

$$m_s = +½, \quad j = l + ½.$$

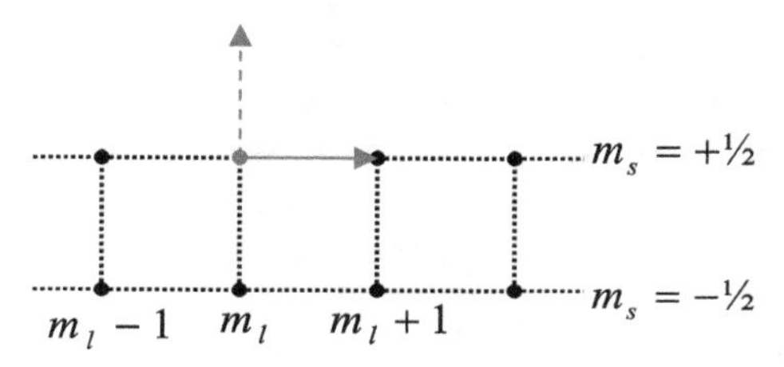

For our case $s = ½$, the rectangular diagram on the plane $[m_l, m_s]$ has the form shown in figure 5.14 of the lecture notes—see its simplified version in the figure above. If we apply the recurrence relations, derived in the solution of the previous problem, with the lower signs, to an arbitrary point $\{m_l, +½\}$ of the upper row, we may expect that one of the coefficients in these relations (corresponding to the dashed arrow on the diagram) will vanish, and we will get a direct recurrent relation between two adjacent Clebsch–Gordan coefficients of the upper row (in the figure above, corresponding to the points connected by the solid arrow).

Indeed, plugging the values $m_s = ½, j = l + ½$, and $m_j = m_l + ½ + 1$, into Eq. (***) of the solution of the previous problem, with the lower signs, we get

$$\begin{aligned}&[(l - m_l)(l + m_l + 2)]^{1/2}\langle m_l, +½|l + ½, m_l + ½\rangle \\ &= [(l - m_l)(l + m_l + 1)]^{1/2}\langle m_l + 1, +½|l + ½, m_l + ½ + 1\rangle \\ &\quad + [(0) \cdot (2)]^{1/2}\langle m_l, +½ + 1|l + ½, m_l + ½ + 1\rangle,\end{aligned}$$

so that the coefficient before the last bracket (which would describe an impossible state with $m_s = 3/2 > s$) vanishes as it has to, and we get a direct, simple recurrence relation,

$$\langle m_l, +½|l + ½, m_l + ½\rangle = \left(\frac{l + m_l + 1}{l + m_l + 2}\right)^{1/2} \langle m_l + 1, +½|l + ½, m_l + ½ + 1\rangle, \quad (*)$$

valid for all m_l that give a positive factor $l + m_l + 2$ in the denominator, i.e. for any point of the upper row of the rectangular diagram ($-l \leqslant m_l \leqslant +l$). As was discussed in the lecture notes, for the rightmost point of this row, i.e. for $m_l = l$, the Clebsch–Gordan coefficient should equal 1:

$$\langle l, +\tfrac{1}{2}|l + \tfrac{1}{2}, l + \tfrac{1}{2}\rangle = 1,$$

so that applying Eq. (*) sequentially to the points further and further left, we get:

$$\text{for } m_l = l - 1: \quad \langle l - 1, +\tfrac{1}{2}|l + \tfrac{1}{2}, l + \tfrac{1}{2} - 1\rangle$$
$$= \left(\frac{2l}{2l + 1}\right)^{1/2} \langle l, +\tfrac{1}{2}|l + \tfrac{1}{2}, l + \tfrac{1}{2}\rangle = \left(\frac{2l}{2l + 1}\right)^{1/2},$$
$$\text{for } m_l = l - 2: \quad \langle l - 2, +\tfrac{1}{2}|l + \tfrac{1}{2}, l + \tfrac{1}{2} - 2\rangle$$
$$= \left(\frac{2l - 1}{2l}\right)^{1/2} \langle l - 1, +\tfrac{1}{2}|l + \tfrac{1}{2}, l + \tfrac{1}{2} - 1\rangle = \left(\frac{2l}{2l + 1} \cdot \frac{2l - 1}{2l}\right)^{1/2},$$

etc, with both the numerator and denominator decreasing by 1 for each next fraction. Continuing this sequence to an arbitrary m_l, we finally get

$$\langle m_l, +\tfrac{1}{2}|l + \tfrac{1}{2}, m_l + \tfrac{1}{2}\rangle = \left(\frac{2l}{2l + 1} \cdot \frac{2l - 1}{2l} \cdot \cdot \frac{2l - 2}{2l - 1} \cdot \ldots \cdot \frac{l + m_l + 2}{l + m_l + 3} \cdot \frac{l + m_l + 1}{l + m_l + 2}\right)^{1/2}$$
$$\equiv \left(\frac{l + m_l + 1}{2l + 1}\right)^{1/2},$$

because all intermediate factors in the numerator and denominator cancel. Now rewriting this result in terms of $m_j = m_l + m_s = m_l + \tfrac{1}{2}$, we get

$$\left\langle m_j - \tfrac{1}{2}, +\tfrac{1}{2}\middle|l + \tfrac{1}{2}, m_j\right\rangle = \left(\frac{l + m_j + \tfrac{1}{2}}{2l + 1}\right)^{1/2},$$

i.e. the shorthand of the first line of Eqs. (5.190), for the upper sign of $(j - l)$.

The proof of these relations for other signs of m_s and $(j - l)$ is completely similar.

***Problem* 5.43.** A spin-½ particle is in a state with definite values of L^2, J^2, and J_z. Find all possible values of the observables S^2, S_z, and L_z, the probability of each listed value, and the expectation value for each of these observables.

Solution: Evidently, such a state is a common eigenstate of the operators $\hat{L}^2$, $\hat{J}^2$, and $\hat{J}_z$, with definite values of the corresponding quantum numbers l, j, and m_j, such that

$$L^2 = \hbar^2 l(l + 1), \qquad J^2 = \hbar^2 j(j + 1), \qquad \text{and} \quad J_z = \hbar m_j.$$

According to Eqs. (5.175), (5.177), and (5.189) of the lecture notes, these numbers should satisfy the following conditions:

$$l \geqslant 0, \qquad j = l \pm \tfrac{1}{2}, \qquad -j \leqslant m_j \leqslant +j.$$

As was discussed in section 5.7 of the lecture notes (see, e.g. figure 5.12), such a state is one of the basis states of the *coupled* representation, which may be represented by the ket-vector $|j, m_j\rangle$, where the definite quantum numbers l and s are implied. Using Eqs. (5.190) of the lecture notes, any such ket may be may be expressed via the kets $|m_l, m_s\rangle$ of the uncoupled representation:

$$|j = l \pm ½, m_j\rangle = \pm\left(\frac{l \pm m_j + ½}{2l+1}\right)^{1/2} |m_l = m_j - ½, m_s = +½\rangle$$
$$+ \left(\frac{l \mp m_j + ½}{2l+1}\right)^{1/2} |m_l = m_j + ½, m_s = -½\rangle. \qquad (*)$$

Now we are ready to start answering the problem's questions. First of all, according to the second of Eqs. (5.169), for the fixed $s = ½$, the observable S^2 may have only one value, $S^2 = \hbar^2 s(s + ½) = (3/4)\hbar^2$, so that the probability of this value is 100% and

$$\langle S^2\rangle = \frac{3}{4}\hbar^2,$$

regardless of the quantum state of the particle.

Next, due to the first of Eqs. (5.169), the possible values of S_z are $\hbar m_s = \pm\hbar/2$, and for the state (*), their probabilities are given by the squares of the corresponding (Clebsch–Gordan) coefficients:

$$W_\uparrow = \frac{l \pm m_j + ½}{2l+1}, \qquad W_\downarrow = \frac{l \mp m_j + ½}{2l+1}, \qquad (**)$$

giving the expectation value

$$\langle S_z\rangle = \frac{\hbar}{2}\frac{l \pm m_j + ½}{2l+1} - \frac{\hbar}{2}\frac{l \mp m_j + ½}{2l+1} \equiv \pm\frac{\hbar m_j}{2l+1} \qquad (***)$$

where the sign, as in Eq. (*), is determined by that in the relation $j = l \pm ½$. In any orbital s-state ($l = 0$), this result is reduced to the obvious formula $\langle S_z\rangle = \pm\hbar m_j = \pm\hbar/2$, because in this case the total orbital moment J is due to the particle's spin S alone.

Finally, the observable L_z may take values $\hbar m_l$. According to Eq. (*), for fixed j and m_j, the sign in the relation $m_l = m_j \mp ½$ is always opposite to that in relation $m_s = \pm ½$. This means that the probabilities (**) also describe those of these two possible values m_l:

$$W_{m_l = m_j - ½} = \frac{l \pm m_j + ½}{2l+1}, \qquad W_{m_l = m_j + ½} = \frac{l \mp m_j + ½}{2l+1},$$

and the expectation value of the observable L_z is

$$\langle L_z\rangle = \hbar\left(m_j - \frac{1}{2}\right)\frac{l \pm m_j + ½}{2l+1} + \hbar\left(m_j + \frac{1}{2}\right)\frac{l \mp m_j + ½}{2l+1} \equiv \hbar m_j\left(1 \mp \frac{1}{2l+1}\right).$$

Alternatively, the last formula may be obtained from averaging the operator equality (5.171):

$$\langle L_z \rangle = \langle J_z \rangle - \langle S_z \rangle = \hbar m_j - \langle S_z \rangle,$$

and plugging Eq. (***) for $\langle S_z \rangle$ into it.

***Problem* 5.44.** Re-solve the Landau-level problem, discussed in section 3.2 of the lecture notes, for a spin-½ particle. Discuss the result for the particular case of an electron, with the g-factor equal to 2.

Solution: The problem may be described by the Hamiltonian that is the sum of that of the orbital motion (see Eq. (3.26) of the lecture notes, with $\phi = 0$ and $\partial/\partial z = 0$) and the Pauli Hamiltonian (4.163), describing the interaction between its spin and the field:

$$\hat{H} = -\frac{\hbar^2}{2m}\left(\mathbf{n}_x\frac{\partial}{\partial x} + \mathbf{n}_y\frac{\partial}{\partial y} - i\frac{q}{\hbar}\mathbf{A}\right)^2 - \gamma\hat{\mathbf{S}}\cdot\mathbf{B},$$

Since the orbital and spin states are defined in different Hilbert spaces, and in this case do not interact, the total energy of the system is just the sum of the independent contributions from these two parts of the Hamiltonian—the first one given by Eq. (3.50), and the another one, by Eq. (4.167):

$$E = \hbar\omega_c\left(n + \frac{1}{2}\right) \pm \frac{\hbar\Omega}{2}, \quad \text{with } n = 0, 1, 2, \ldots.$$

This expression shows that generally the spin interaction with the field splits each Landau level into two sub-levels, with different spin orientation. However, as was noted in section 4.6, for an electron the frequencies $\omega_c \equiv e\mathscr{B}/m_e$ and $|\Omega| \equiv |\gamma\mathscr{B}| = |g_e e\mathscr{B}/2m_e|$ virtually[49] coincide, so that the above result may be recast as

$$E = \hbar\omega_c\left(n + \frac{1}{2} \pm \frac{1}{2}\right),$$

i.e. the energy takes only integer multiple values of $\hbar\omega_c$. This expression shows that besides the ground state, with energy $E = E_g = 0$, all other energy levels are doubly degenerate, with the two states differing not only by the spin direction, but also by the adjacent values of the orbital quantum number n.

***Problem* 5.45.** In the Heisenberg picture of quantum dynamics, find an explicit expression for the operator of acceleration,

$$\hat{\mathbf{a}} \equiv \frac{d\hat{\mathbf{v}}}{dt},$$

[49] Besides a very small correction due to the deviation of g_e from 2, which was mentioned in section 4.6 of the lecture notes.

of a spin-½ particle with electric charge q, moving in an arbitrary external electromagnetic field. Compare the result with the corresponding classical expression.

Hint: For the orbital motion's description, you may use Eq. (3.26) of the lecture notes.

Solution: The Hamiltonian of the particle in the field may be composed as the sum of the orbital Hamiltonian (3.26) and the Pauli Hamiltonian (4.163), corresponding to the magnetic field $\mathscr{B} = \nabla \times \mathbf{A}$:

$$\hat{H} = \frac{1}{2m}(\hat{\mathbf{P}} - q\hat{\mathbf{A}})^2 + q\hat{\phi} - \gamma\hat{\mathbf{S}} \cdot (\nabla \times \hat{\mathbf{A}}).$$

Here the operator signs over the scalar and vector potentials are spelled out, as reminders that the potentials are generally functions of not only time, but also of the particle's position, which in this calculation should be treated as an operator. Hence the potentials should be also treated as linear operators (commuting with $\hat{\mathbf{r}}$), even if the quantum properties of the electromagnetic field itself are negligible.

The orbital operators $\hat{\mathbf{r}}$ and $\hat{\mathbf{P}}$ are defined in a Hilbert space different from that of the spin operator and hence commute with it. On the other hand, similar Cartesian components of the orbital operators do not commute. In particular, the commutation relation Eq. (4.238), valid for an arbitrary jth component,

$$[\hat{r}_j, \hat{p}_j] = i\hbar, \qquad (*)$$

is not affected by the addition, to the kinetic momentum operator $\hat{\mathbf{p}}$, of the field part $q\hat{\mathbf{A}}(\hat{\mathbf{r}}, t)$, because the latter commutes with the coordinate operator $\hat{\mathbf{r}}$, so that

$$[\hat{r}_j, \hat{P}_j] = i\hbar. \qquad (**)$$

As a result, the equation of motion of the coordinate operator components is given by Eq. (5.29) of the lecture notes even in the field. In the vector form,

$$\hat{\mathbf{v}} \equiv \frac{d\hat{\mathbf{r}}}{dt} = \frac{\hat{\mathbf{p}}}{m} = \frac{\hat{\mathbf{P}} - q\hat{\mathbf{A}}}{m}, \qquad (***)$$

so that the Hamiltonian may be rewritten in the form[50]

$$\hat{H} = \frac{m\hat{v}^2}{2} + q\hat{\phi} - \gamma\hat{\mathbf{S}} \cdot (\nabla \times \hat{\mathbf{A}}).$$

Plugging these two expressions into the general Eq. (4.199), we may calculate the acceleration operator:

$$\hat{\mathbf{a}} \equiv \frac{d\hat{\mathbf{v}}}{dt} = \frac{1}{i\hbar}[\hat{\mathbf{v}}, \hat{H}] = \frac{1}{i\hbar}\frac{1}{m}[\hat{\mathbf{p}}, q\hat{\phi}(\hat{\mathbf{r}}, t)] + \frac{1}{i\hbar}\frac{m}{2}[\hat{\mathbf{v}}, \hat{v}^2].$$

[50] Note also that Eq. (***) allows us to rewrite Eq. (3.28) in a very simple and natural form: $\mathbf{j} = \psi^*\hat{\mathbf{v}}\psi$.

Due to the similarity of Eqs. (*) and (**), the Cartesian components of the first term may be calculated exactly as was done in section 5.2 of the lecture notes, giving Eq. (5.35) with the replacement

$$-\frac{\partial}{\partial \hat{r}_j}U(\hat{\mathbf{r}}) \to -q\frac{\partial}{\partial \hat{r}_j}\hat{\phi}(\hat{\mathbf{r}}, t) = q\hat{\mathscr{E}}_j(\hat{\mathbf{r}}, t),$$

where $\hat{\mathscr{E}} = -\nabla\hat{\phi}$ is the vector-operator of the electric field. Next, calculating the jth Cartesian component of the second commutator,

$$[\hat{\mathbf{v}}, \hat{v}^2]_j \equiv \left[\hat{\mathbf{v}}, \sum_{j'=1}^{3}\hat{v}_{j'}^2\right]_j \equiv \sum_{j'=1}^{3}\mathbf{n}_j\left[\hat{v}_j, \hat{v}_{j'}^2\right],$$

we should remember that the jth component of the vector-potential $\mathbf{A}$ may be (and most typically is) a function of *all* Cartesian coordinates $r_{j'}$, and hence (in contrast with the situation with the canonical momentum operator $\hat{\mathbf{P}}$), different components of the kinetic momentum operator $\hat{\mathbf{p}} = \hat{\mathbf{P}} - q\hat{\mathbf{A}}$, and hence of the velocity operator (***), do not commute:

$$\begin{aligned}
[\hat{v}_j, \hat{v}_{j'}] &= \frac{1}{m^2}\left[(\hat{P} - q\hat{A})_j, (\hat{P} - q\hat{A})_{j'}\right] \\
&= \frac{1}{m^2}\left[\left(-i\hbar\frac{\partial}{\partial r_j} - q\hat{A}_j\right), \left(-i\hbar\frac{\partial}{\partial r_{j'}} - q\hat{A}_{j'}\right)\right] \\
&= \frac{i\hbar q}{m^2}\left(\left[\frac{\partial}{\partial r_j}, \hat{A}_{j'}\right] + \left[\hat{A}_j, \frac{\partial}{\partial r_{j'}}\right]\right) \\
&= \frac{i\hbar q}{m^2}\left(\frac{\partial \hat{A}_{j'}}{\partial r_j} - \frac{\partial \hat{A}_j}{\partial r_{j'}}\right),
\end{aligned}$$

where, in the last expression, the derivatives act only upon the components of the vector-potential $\mathbf{A}$, but not upon the following wavefunction. But the last combination of derivatives is just (plus or minus) the j'th component of the magnetic field $\mathscr{B} = \nabla \times \mathbf{A}$, so that

$$[\hat{v}_j, \hat{v}_{j'}] = \frac{i\hbar q}{m^2}\hat{\mathscr{B}}_{j''}\varepsilon_{jj'j''}, \quad \text{i.e.} \quad \hat{v}_j\hat{v}_{j'} = \hat{v}_{j'}\hat{v}_j + \frac{i\hbar q}{m^2}\hat{\mathscr{B}}_{j''}\varepsilon_{jj'j''},$$

where $\varepsilon_{jj'j''}$ is the Levi-Civita symbol. Applying the last relation twice, we get

$$[\hat{\mathbf{v}}, \hat{v}^2]_j \equiv \sum_{j'=1}^{3}\left[\hat{v}_j, \hat{v}_{j'}^2\right] = \sum_{j'=1}^{3}(\hat{v}_j\hat{v}_{j'}\hat{v}_{j'} - \hat{v}_{j'}\hat{v}_{j'}\hat{v}_j)$$

$$= \sum_{j'=1}^{3}\left(\hat{v}_{j'}\hat{v}_j\hat{v}_{j'} + \frac{i\hbar q}{m^2}\hat{\mathscr{B}}_{j''}\varepsilon_{jj'j''}\hat{v}_{j'} - \hat{v}_{j'}\hat{v}_{j'}\hat{v}_j\right)$$

$$= \sum_{j'=1}^{3}\left(\hat{v}_{j'}\hat{v}_{j'}\hat{v}_j + \hat{v}_{j'}\frac{i\hbar q}{m^2}\hat{\mathscr{B}}_{j''}\varepsilon_{jj'j''} + \frac{i\hbar q}{m^2}\hat{\mathscr{B}}_{j''}\varepsilon_{jj'j''}\hat{v}_{j'} - \hat{v}_{j'}\hat{v}_{j'}\hat{v}_j\right)$$

$$\equiv \frac{i\hbar q}{m^2}\sum_{j'=1}^{3}(\hat{v}_{j'}\hat{\mathscr{B}}_{j''} + \hat{\mathscr{B}}_{j''}\hat{v}_{j'})\varepsilon_{jj'j''}\,.$$

But the last sum is just the jth component of the vector-operator $\hat{\mathbf{v}} \times \hat{\mathscr{B}} - \hat{\mathscr{B}} \times \hat{\mathbf{v}}$, so that, merging the above expressions in the vector form, we finally get

$$\hat{\mathbf{a}} = \frac{q}{m}\left(\hat{\mathscr{E}} + \frac{\hat{\mathbf{v}} \times \hat{\mathscr{B}} - \hat{\mathscr{B}} \times \hat{\mathbf{v}}}{2}\right). \qquad (****)$$

Perhaps the most important feature of these results is that the operators of velocity and acceleration of the electron (and hence its orbital motion) are not affected by the particle's spin. (This conclusion is valid for particles with any spin, but only if the relativistic effect of the spin–orbit interaction[51] is ignored.) Also, Eq. (****) formally coincides with the well-known formula for the Lorentz-force-induced acceleration in the classical electromagnetism, rewritten in the form

$$\mathbf{a} = \frac{\mathbf{F}_\mathrm{L}}{m}, \qquad \mathbf{F}_\mathrm{L} = q(\mathscr{E} + \mathbf{v} \times \mathscr{B}) \equiv q\left(\mathscr{E} + \frac{\mathbf{v} \times \mathscr{B} - \mathscr{B} \times \mathbf{v}}{2}\right),$$

though since the quantum operators $\hat{\mathbf{v}}$ and $\hat{\mathscr{B}}$ do not commute, making the transition from these classical variables to their operators in the first form of the classical relation for $\mathbf{F}_\mathrm{L}$ would give a wrong result.

***Problem* 5.46.** A byproduct of the solution of problem 5.41 is the following relation for the spin operators (valid for any spin s):

$$\langle m_s|\hat{S}_\pm|m_s \pm 1\rangle = \hbar[(s \pm m_s + 1)(s \mp m_s)]^{1/2}.$$

Use this result to spell out the matrices S_x, S_y, S_z, and S^2 of a particle with $s = 1$, in the z-basis—defined as the basis in which the matrix S_z is diagonal.

[51] See, e.g. sections 6.3 and 9.7 of the lecture notes.

Solution: According to Eqs. (5.169), the matrices S_z and S^2 are diagonal in the basis of the states with definite quantum numbers m_s. For $s = 1$, this is the three-functional basis with $m_s = +1$, 0, and −1. In this basis, these relations (with the implied constant s) give

$$S_z = \hbar\begin{pmatrix}1 & 0 & 0\\ 0 & 0 & 0\\ 0 & 0 & -1\end{pmatrix}, \qquad S^2 = 2\hbar^2\begin{pmatrix}1 & 0 & 0\\ 0 & 1 & 0\\ 0 & 0 & 1\end{pmatrix},$$

while the expression given in the assignment yields

$$S_- = \hbar\begin{pmatrix}0 & 0 & 0\\ \sqrt{2} & 0 & 0\\ 0 & \sqrt{2} & 0\end{pmatrix}, \qquad S_+ = \hbar\begin{pmatrix}0 & \sqrt{2} & 0\\ 0 & 0 & \sqrt{2}\\ 0 & 0 & 0\end{pmatrix}.$$

Since the operators $\hat{S}_\pm$ are defined via the Cartesian component operators as $\hat{S}_\pm \equiv \hat{S}_x \pm i\hat{S}_y$, for the Cartesian component matrices we get

$$S_x \equiv \frac{S_+ + S_-}{2} = \frac{\hbar}{\sqrt{2}}\begin{pmatrix}0 & 1 & 0\\ 1 & 0 & 1\\ 0 & 1 & 0\end{pmatrix}, \qquad S_y \equiv \frac{S_+ - S_-}{2i} = \frac{\hbar}{\sqrt{2}}\begin{pmatrix}0 & -i & 0\\ i & 0 & -i\\ 0 & i & 0\end{pmatrix}.$$

Note that the analogy between these 3 × 3 matrices and the corresponding 2 × 2 matrices for $s = ½$, following from Eqs. (4.105) and (4.116) of the lecture notes,

$$S_x = \frac{\hbar}{2}\begin{pmatrix}0 & 1\\ 1 & 0\end{pmatrix}, \qquad S_y = \frac{\hbar}{2}\begin{pmatrix}0 & -i\\ i & 0\end{pmatrix},$$

is evident—and beautiful.

Problem* 5.47. For a particle with an arbitrary spin s, find the quantum numbers m_j and j that are necessary to describe, in the coupled-representation basis:

(i) all states with a definite quantum number l, and
(ii) a state with definite values of not only l, but also m_l and m_s.

Give an interpretation of your results in terms of a classical geometric vector diagram (figure 5.13 of the lecture notes).

Solutions: For arbitrary l and s,[52] the 'rectangular diagram' of the basis states (see figure 5.14 of the lecture notes), looks as shown in the figure below. Here each point corresponds to one of $N_m = (2l + 1) \times (2s + 1)$ basis states of the uncoupled representation, each with definite quantum numbers m_l and m_s, while each tilted

[52] For clarity, I had to draw the diagram for certain values of l and s, but none of the expressions below uses these values, besides the assumed restriction $s \leqslant l$; the opposite case will be discussed below. Also, note that all the formulas below are valid whether s is integer or half-integer—as in figure 5.14 of the lecture notes.

straight line symbolizes one or several basis states of the coupled representation, with fixed quantum numbers j and $m_j = m_l + m_s$.

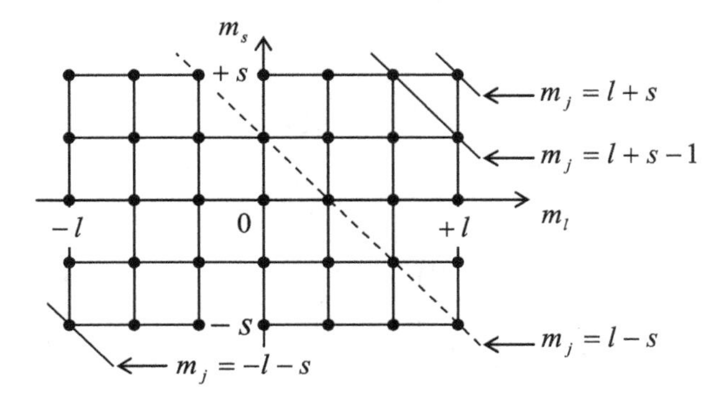

(i) The diagram immediately shows that for the set of quantum numbers m_s and m_l, following from Eqs. (5.169) and (5.177), namely $-s \leqslant m_s \leqslant +s$ and $-l \leqslant m_l \leqslant +l$, the range of possible numbers m_j is

$$-l - s \leqslant m_j \leqslant l + s,$$

but the range of possible numbers j is a bit less evident. In order to calculate it, let us use the fact that in order to have a unique set of linearly-independent relations (5.183),

$$|j, m_j\rangle = \sum_{m_l, m_s} |m_l, m_s\rangle \langle m_l, m_s | j, m_j \rangle,$$

the number of the basis ket vectors (and hence the basis states) in both representation should be equal.

Let us count the number of states of the coupled representation, starting from the top right corner of the diagram. Evidently, there may be only one such state, corresponding to the single uncoupled-representation state with quantum numbers $m_l = +l$ and $m_s = +s$, i.e. with $m_j = (m_j)_{\max} \equiv (m_l + m_s)_{\max} = l + s$. According to the last of Eqs. (5.175) of the lecture notes, $-j \leqslant m_j \leqslant j$, the quantum number j has to be exactly equal to this maximum value of m_j, so that

$$j_{\max} = l + s. \qquad (*)$$

According to the same relation between j and m_j, the full set of the basis states of the coupled representation should include $n(j_{\max}) = 2(m_j)_{\max} + 1 \equiv 2l + 2s + 1$ states corresponding to $j = j_{\max}$—see the rightmost vertical line in the figure below—in which we will gradually plot the number of the coupled-representation basis states as a function of j.

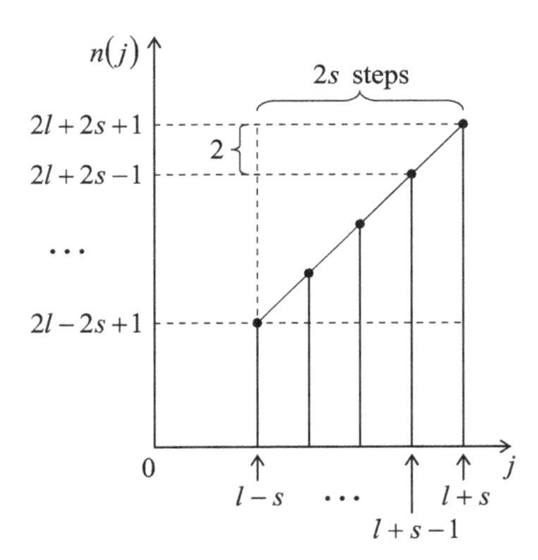

Moving on to the next value, $m_j = l + s - 1$, the rectangular diagram above shows two $\{m_l, m_s\}$ states, both with $m_j \equiv m_l + m_s = (m_j)_{\max} - 1 \equiv l + s - 1$, which need, for their linear representation, two different $\{j, m_j\}$ states. Since both these states have the same m_j, they cannot be described by the same $j = j_{\max}$. Hence we need one more value of j; repeating the discussion above, we see that this value has to be equal to the current value of m_j, i.e. to $l + s - 1$, by one less than $j_{\max} = l + s$. The overall number of m_j-states, corresponding to this new value $j = j_{\max} - 1$ is $n(j_{\max} - 1) = 2(l + s - 1) + 1 \equiv 2l + 2s - 1$, i.e. by two less than $n(j_{\max})$—see the second vertical line from the right in the state counting diagram above.

Repeating this process again and again, we move to the left on both diagrams by unit horizontal steps, each time reducing m_j by one, and getting one more new value of j, which is also less than the smallest previous value by one, and has the number of m_j-states lower by two—see the figure above. The process breaks only after $2s$ steps, when the tilted line of equal m_j, on the rectangular diagram, hits its lower right corner—with $m_l = l$, $m_s = -s$, i.e. with $m_j \equiv m_l + m_s = l - s$, and the lowest value of j equal to

$$j_{\min} = j_{\max} - 2s \equiv l - s \geqslant 0, \qquad (**)$$

responsible for $n(j_{\min}) = 2l - 2s - 1$ states with different values of m_j.

Let us count the total number N_j of these basis states of the coupled representation, corresponding to the already covered values of j, within the interval $j_{\min} \leqslant j \leqslant j_{\max}$. From the number-of-state diagram above we get

$$N_j \equiv \sum_{j=j_{\min}}^{j_{\max}} n(j) = \sum_{j=j_{\min}}^{j_{\max}} \left[n(j_{\min}) + 2(j - j_{\min})\right]$$

$$= \sum_{j=l-s}^{l+s} [(2l - 2s + 1) + 2(j - l + s)] \equiv \sum_{j=l-s}^{l+s} (2j + 1).$$

This is just the sum of $(2s + 1)$ unit terms, plus a difference of two standard arithmetic progressions, so that using the well-known formula for the progression[53], we get

$$N_j = (2s + 1) + 2\sum_{j=1}^{l+s} j - 2\sum_{j=1}^{l-s-1} j$$
$$= (2s + 1) + 2\frac{(l + s)(l + s + 1)}{2} - 2\frac{(l - s - 1)(l - s)}{2}$$
$$\equiv (2s + 1)(2l + 1).$$

But this number is exactly equal to N_m; hence we have reached the state number equality in both representations without involving lower values of j. (The states on the left of the dashed tilted line on the rectangular diagram are covered by the already counted values of j; for example, the left bottom corner state with $m_j = -s - l \equiv -j_{\max}$ is evidently described by $j = j_{\max}$.)

Hence Eqs. (*) and (**) indeed give the boundaries of the range of j for all $N \equiv N_m = N_j$ states of the system in the case $s \leqslant l$; in the opposite case $l \leqslant s$ it is evidently sufficient to repeat all the above arguments after transposing the rectangular diagram, getting the same result besides the reversal $s \leftrightarrow l$, so that $j_{\min} = s - l$ is again non-negative. Both cases may be summarized as follows:

$$|l - s| \leqslant j \leqslant l + s. \qquad (***)$$

(ii) In a system in which the quantum numbers m_l and m_s are fixed, and hence $m_j \equiv m_l + m_s$ is fixed as well, we may repeat all the process of motion from the top right corner of the rectangular diagram, discussed above. However, if m_j is positive and larger than $|l - s|$, i.e. the point $\{m_l, m_s\}$ is located to the right of the dashed tilted line on the diagram, we may stop the process as soon as we have reached this point, because the state in question has been, by construction, covered by the already accounted values of j. Similarly, if m_j is negative, and its magnitude is larger than $|l - s|$, we may repeat the process starting from the lower left corner, and again stop it when the tilted line of fixed m_j has reached the given point $\{m_l, m_s\}$. Summarizing these cases, we get the following range of j:

$$\max[|l - s|, |m_l + m_s|] \leqslant j \leqslant l + s. \qquad (****)$$

These results may be readily interpreted using the classical vector diagrams, such as shown in figure 5.13 of the lecture notes. As a reminder, on such diagrams, the products $\hbar m_l$, $\hbar m_s$, and $\hbar m_j$ are associated with the z-components, L_z, S_z, and J_z of the angular momenta vectors **L**, **S**, and **J**, respectively, while $\hbar l$, $\hbar s$, and $\hbar j$, with the lengths L, S, and J of these vectors, i.e. the differences between l and $[l(l + 1)]^{1/2}$, etc (which are of a purely quantum origin) are ignored.

[53] See, e.g. Eq. (A.8*a*).

In the case (i), when the z-components of the angular momentum vectors are not fixed, the largest length of vector $\mathbf{J} \equiv \mathbf{L} + \mathbf{S}$ corresponds to the parallel alignment of vectors $\mathbf{L}$ and $\mathbf{S}$: $J_{\max} = L + S$, while its smallest value, to their antiparallel alignment: $J_{\max} = |L - S|$ (see the figure below), in agreement with Eq. (***).

L S L

$J_{\max} = L + S$ $J_{\min} = |L - S|$ S

In the case (ii), i.e. when the exact values of L_z and S_z are fixed (in addition to the fixation of L and S), the classical picture shows an additional limitation imposed at J from below: $J_{\min} = |L_z + S_z|$, if $|L_z + S_z| > |L - S|$, in agreement with the left inequality in Eq. (****)—see the first panel in the figure below. However, in the classical picture the exact parallel alignment of the vectors $\mathbf{L}$ and $\mathbf{S}$, necessary to reach the maximum value $J_{\max} = L + S$, implied by the right part of Eq. (****), is possible only at a certain exact proportion between their z-components and lengths: $S_z/L_z = S/L$. In quantum mechanics, there is no such exact restriction[54]—the fact which again emphasizes the limitations of the classical model.

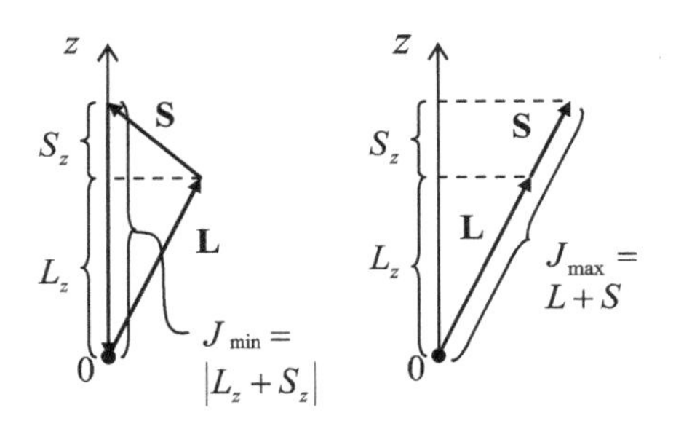

***Problem* 5.48.** A particle of mass m, with electric charge q and spin s, free to move along a plane ring of radius R, is placed into a constant, uniform magnetic field $\mathscr{B}$, directed normally to the ring's plane. Calculate the energy spectrum of the system. Explore and interpret the particular form the result takes when the particle is an electron with the g-factor $g_e = 2$.

Solution: Directing the z-axis parallel to the field, i.e. normally to ring's plane, we may describe the problem by a Hamiltonian which is the sum of that of the rotator's kinetic energy in the presence of magnetic field (see Eq. (3.131) of the lecture notes, with the convenient, axially-symmetric choice (3.132) of the vector potential), and

[54] In the limit $l, s \to \infty$, the asymptotic correspondence to classical mechanics is achieved by the Clebsch–Gordan coefficients $\langle m_l, m_s | j, m_j \rangle$ tending to zero everywhere at the rectangular diagram, besides very close to its diagonal, i.e. at $m_s/m_l \approx s/l$, thus tending to the classical restriction.

the Pauli Hamiltonian (4.163) of the interaction between the particle's spin and the field:

$$\hat{H} = -\frac{1}{2m}\left(-i\frac{\hbar}{R}\frac{\partial}{\partial\varphi} - qA_\varphi\right)^2 - \gamma\mathscr{B}\hat{S}_z \equiv \frac{1}{2m}\left(\frac{\hat{L}_z}{R} - q\frac{\mathscr{B}R}{2}\right)^2 - \gamma\mathscr{B}\hat{S}_z,$$

where γ is the gyromagnetic factor of the particle.

As the Hamiltonian shows, in this system the orbital and spin degrees of freedom interact with the external magnetic field, but not with each other. Moreover, each of the operators $\hat{L}_z$ and $\hat{S}_z$ commutes with the Hamiltonian, and hence their eigenstates (in the corresponding Hilbert spaces) are eigenstates of the total Hamiltonian as well. As a result, the total energy of the system is just the sum of the independent contributions from these two parts of the Hamiltonian, with the observables S_z and L_z quantized independently in accordance with Eqs. (5.169) and (5.177):

$$E = \frac{1}{2m}\left(\frac{\hbar m_l}{R} - q\frac{\mathscr{B}R}{2}\right)^2 - \gamma\mathscr{B}\hbar m_s,$$
$$\text{with} \quad m_l = 0, \pm 1, \pm 2, \ldots, \quad \text{and} - s \leqslant m_s \leqslant +s.$$

This expression shows that each orbital level (with a particular magnetic quantum number m_l) is split into $(2s + 1)$ equidistant spin sub-levels, numbered by the spin magnetic number m_s.

For an electron, $q = -e$, $\gamma = \gamma_e \equiv qg_e/2m_e = -e/m_e$, and $m_s = \pm\frac{1}{2}$, so that the result is reduced to

$$E = \frac{1}{2m_e}\left(\frac{\hbar m_l}{R} + \frac{e\mathscr{B}R}{2}\right)^2 \pm \frac{e\mathscr{B}\hbar}{2m_e}$$
$$\equiv \frac{\hbar^2 m_l^2}{2m_e R^2} + \frac{e\hbar}{2m_e}(m_l \pm 1)\mathscr{B} + \frac{e^2\mathscr{B}^2R^2}{8m_e}.$$

The first term of the last expression is the electron's quantized kinetic energy in the absence of magnetic field, while the next two terms describe the field's effects. The second term, linear in $\mathscr{B}$ and hence dominating in weak fields (with the magnetic flux $\Phi = \pi|\mathscr{B}|R^2$ through the rotator's ring much smaller than the 'normal' field quantum $\Phi_0' \equiv 2\pi\hbar/e$), may be interpreted as a result of the field's interaction with the pre-existing magnetic moment of the system, with the normal (z-) component

$$m_z = -\frac{e\hbar}{2m_e}(m_l \pm 1) \equiv -\frac{e}{2m_e}(L_z + 2S_z).$$

This double contribution of the electron's spin to its magnetic moment, and hence to its interaction with the external magnetic field, is responsible for all the complications of the Zeeman effect—see section 6.4 of the lecture notes.

Finally, the last term, quadratic in $\mathscr{B}$ and hence dominating at $\Phi \gg \Phi_0'$, describes the essentially classical (and hence independent of both quantum numbers) effect of the orbital diamagnetism of the system, i.e. the energy of the field's interaction with the magnetic moment it has induced[55].

References

[1] Mehta M 2004 *Random Matrices* (Elsevier/Academic Press)
[2] Beenakker C 1997 *Rev. Mod. Phys.* **69** 731
[3] McDonald S and Kaufmann A 1979 *Phys. Rev. Lett.* **42** 1189
[4] Ziman J 1979 *Principles of the Theory of Solids* 2nd edn (Cambridge: Cambridge University Press)

[55] See, e.g. *Part EM* section 5.5—in particular the model solution of problem 5.14(i), with the appropriate replacement $\langle x^2\rangle + \langle y^2\rangle \to R^2$.

IOP Publishing

Quantum Mechanics
Problems with solutions
Konstantin K Likharev

Chapter 6

Perturbative approaches

***Problem* 6.1.** Use Eq. (6.14) of the lecture notes to prove the following general form of the Hellmann–Feynman theorem (whose proof in the wave-mechanics domain was the task of problem 1.5):

$$\frac{\partial E_n}{\partial \lambda} = \langle n|\frac{\partial \hat{H}}{\partial \lambda}|n\rangle,$$

where λ is an arbitrary c-number parameter.

Solution: In the basic Eq. (6.1), let us take

$$\hat{H}^{(0)} = \hat{H}|_{\lambda=\lambda_0}, \quad \hat{H}^{(1)} = \left.\frac{\partial \hat{H}}{\partial \lambda}\right|_{\lambda=\lambda_0}(\lambda - \lambda_0),$$

where λ_0 is arbitrary, and the difference $(\lambda - \lambda_0)$ is small. In the first approximation in this difference, the eigenenergy perturbation may be also expressed by the linear term of the Taylor series:

$$E_n(\lambda) - E_n(\lambda_0) = E^{(1)} = \left.\frac{\partial E_n}{\partial \lambda}\right|_{\lambda=\lambda_0}(\lambda - \lambda_0).$$

Plugging these relations into Eq. (6.14), with $n^{(0)}$ being the eigenstate n at $\lambda = \lambda_0$, we get the Hellmann–Feynman theorem at this value of λ (which, again, is arbitrary).

As was shown in the solution of problem 3.34, this theorem may be used, in particular, for a very simple proof of the first of Eqs. (3.211). The remaining two formulas of Eq. (3.211) may be proved in a similar way[1].

[1] See, e.g. [1].

doi:10.1088/2053-2563/aaf3a6ch6

***Problem* 6.2.** Establish a relation between Eq. (6.16) of the lecture notes and the result of the classical theory of weakly anharmonic ('nonlinear') oscillations for negligible damping.

Hint: Use the N Bohr's reasoning, discussed in problem 1.1.

Solution: Following the N Bohr's arguments, let us use Eqs. (2.262) and (6.16) to calculate the frequency of quantum transitions between the adjacent high energy levels of the anharmonic oscillator ($n \gg 1$):

$$\omega \equiv \frac{1}{\hbar}(E_{n+1} - E_n) \approx \frac{1}{\hbar}\frac{dE_n}{dn} \approx \frac{1}{\hbar}\frac{d\left(E_n^{(0)} + E_n^{(1)}\right)}{dn} = \omega_0 + \frac{3}{2}\frac{\beta x_0^4}{\hbar}(2n + 1).$$

Since the correction to the frequency ω_0 is already proportional to the small parameter β, we may combine this result with the expression for the effective real amplitude A of the sinusoidal oscillations, defined as

$$\frac{A^2}{2} \equiv \langle x^2 \rangle,$$

using the unperturbed expression (5.95). This gives $A^2 = x_0^2(2n + 1)$, so that we may write

$$\omega \approx \omega_0 + \frac{3}{2}\frac{\beta x_0^2}{\hbar}A^2 \equiv \omega_0 + \frac{3}{2}\frac{\beta}{m\omega_0}A^2. \qquad (*)$$

On the other hand, the classical theory of weakly nonlinear oscillations at negligible damping, described by the differential equation[2]

$$\ddot{x} + \omega_0^2 x = \alpha_{\text{cl}} x^3, \qquad (**)$$

gives the following approximate (but at $\alpha_{\text{cl}} \to 0$, asymptotically correct) expression for the oscillation frequency as a function of the amplitude A of the nearly-sinusoidal oscillations $x(t)$:[3]

$$\omega \approx \omega_0 - \frac{3}{8}\frac{\alpha_{\text{cl}}}{\omega_0}A^2. \qquad (***)$$

In order to reveal the relation between the coefficients α_{cl} and β, let us write the classical Lagrangian function corresponding to the Hamiltonian (6.2) with $\alpha = 0$:

$$L \equiv T - U = \frac{m\dot{x}^2}{2} - \left(\frac{m\omega_0^2}{2}x^2 + \beta x^4\right).$$

The corresponding Lagrange equation of motion is

[2] See, e.g. *Part CM* Eq. (5.43) with $\delta = 0, f_0 = 0$, and $\omega \approx \omega_0$, so that $\xi \equiv \omega - \omega_0 \approx (\omega - \omega_0)^2/2\omega_0$. (The index 'cl' is attached to the constant α just to avoid any chance of its confusion with that participating in Eq. (6.2) of the lecture notes of this course.)

[3] See, e.g. *Part CM* Eq. (5.49).

$$\frac{d}{dt}\frac{\partial L}{\partial \dot{x}} - \frac{\partial L}{\partial x} = 0, \qquad \text{giving} \quad m\ddot{x} + m\omega_0^2 x + 4\beta x^3 = 0.$$

The comparison of this equation and Eq. (**) shows that they coincide if $\alpha_{cl} = -4\beta/m$. But with this substitution, Eq. (***) exactly coincides with Eq. (*).

So, in the limit $n \to \infty$, the quantum and classical theories yield the same result—as they should by the correspondence principle.

***Problem* 6.3.** A weak, time-independent additional force F is exerted on a 1D particle that was placed into a hard-wall potential well

$$U(x) = \begin{cases} 0, & \text{for } 0 < x < a, \\ +\infty, & \text{otherwise.} \end{cases}$$

Calculate, sketch, and discuss the 1st-order perturbation of its ground-state wavefunction.

Solution: The unperturbed wavefunctions and energy levels of the problem have been calculated in section 1.7 of the lecture notes—see Eqs. (1.84) and (1.85); in the notation of section 6.1:

$$\psi_n^{(0)} = \left(\frac{2}{a}\right)^{1/2} \sin\frac{\pi n x}{a}, \qquad E_n^{(0)} = \frac{\pi^2\hbar^2}{2m}n^2, \qquad \text{with } n = 1, 2, \ldots.$$

The first-order perturbation of the ground-state wavefunction may be calculated using the coordinate representation of Eq. (6.18) with $n = 1$ and the notation change $n' \to n$:

$$\psi_g^{(1)} \equiv \psi_1^{(1)} = \sum_{n>1}^{\infty} \frac{H_{1,n}^{(1)}}{E_1^{(0)} - E_n^{(0)}} \psi_n^{(0)}, \qquad (*)$$

where $H_{1,n}^{(1)}$ are the matrix elements (6.8) of the perturbation Hamiltonian $\hat{H}^{(1)} = -Fx$:

$$H_{1,n}^{(1)} \equiv \int \psi_1^{(0)*}(x)\hat{H}^{(1)}\psi_n^{(0)}(x)dx = -\frac{2F}{a}\int_0^a \sin\frac{\pi x}{a}\, x\, \sin\frac{\pi n x}{a} dx = -\frac{2F}{a}\left(\frac{a}{\pi}\right)^2 I,$$

where

$$I \equiv \int_0^{\pi} \xi \sin\xi \, \sin n\xi \; d\xi, \qquad \text{and } \xi \equiv \frac{\pi x}{a}.$$

The last integral may be re-written using Eq. (A.18*c*):

$$I \equiv \int_0^{\pi} \xi \sin\xi \, \sin n\xi \; d\xi = \frac{I_- - I_+}{2}, \quad \text{where } I_\pm \equiv \int_0^{\pi} \xi \cos(n \pm 1)\xi \; d\xi.$$

Now we can work out the integrals $I_\pm$ by parts,

$$I_\pm = \int_{\xi=0}^{\xi=\pi} \xi \; \frac{d[\sin(n \pm 1)\xi]}{(n \pm 1)} = \left[\xi \; \frac{\sin(n \pm 1)\xi}{(n \pm 1)}\right]_{\xi=0}^{\xi=\pi} - \frac{1}{(n \pm 1)} \int_{\xi=0}^{\xi=\pi} \sin(n \pm 1)\xi \; d\xi$$

$$= \frac{1}{(n \pm 1)^2}[\cos(n \pm 1)\pi - 1],$$

so that

$$I = \frac{1}{2}\left\{\frac{1}{(n-1)^2}[\cos(n-1)\pi - 1] - \frac{1}{(n+1)^2}[\cos(n+1)\pi - 1]\right\}$$

$$= -\frac{4n}{(n^2-1)^2} \times \begin{cases} 1, & \text{for } n \text{ even,} \\ 0, & \text{for } n \text{ odd.} \end{cases}$$

As a result, Eq. (*) yields

$$\psi_g^{(1)} = -\frac{8}{\pi^2}\frac{Fa}{E_1^{(0)}} \sum_{n=2,4,\ldots}^{\infty} \frac{n}{(n^2-1)^3}\psi_n^{(0)}. \qquad (**)$$

The numerical fraction under the sum is a rapidly decreasing function of n (its first value, for $n = 2$, is $2/27 \approx 0.074$, while the next nonvanishing value, for $n = 4$, is already $4/512 \approx 0.0012$), so that a very good approximation of the result is given by the first term alone:

$$\psi_g^{(1)} \approx -\frac{8}{\pi^2}\frac{Fa}{E_1^{(0)}}\frac{2}{(2^2-1)^3}\psi_2^{(0)} = -\frac{32}{27\pi^4}\frac{Fa}{\hbar^2/m}\left(\frac{2}{a}\right)^{1/2}\sin\frac{2\pi x}{a}.$$

The red line in the figure below shows this function, together with the exact result (**) shown with the blue line (these lines virtually operlap), while the dashed line shows (not to scale!) the unperturbed ground state wavefunction $\psi_g^{(0)} \equiv \psi_1^{(0)}$. Sketching their sum $\psi_g^{(0)} + \psi_g^{(1)}$, we may see that the external force F shifts the total wavefunction slightly in its direction (if $F > 0$, then to the right)—as it should. (A good additional exercise for the reader: use the above results to calculate the resulting shift of the expectation value $\langle x \rangle$ from its unperturbed values $a/2$.)

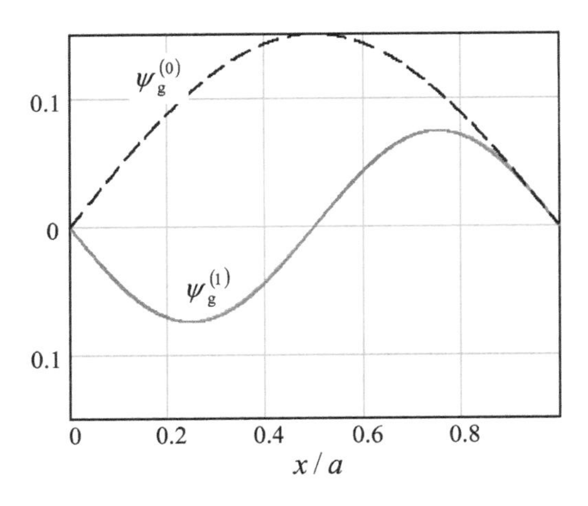

One more additional exercise: calculate the corresponding shift of the ground state energy.

***Problem* 6.4.** A time-independent force $\mathbf{F} = \mu(\mathbf{n}_x y + \mathbf{n}_y x)$, where μ is a small constant, is applied to a 3D harmonic oscillator of mass m and frequency ω_0. Calculate, in the first order of the perturbation theory, the effect of the force upon the ground state energy of the oscillator, and its lowest excited energy level. How small should the constant μ be for your results to be quantitatively correct?

Solution: Any potential force $\mathbf{F} = \mathbf{F}(\mathbf{r})$ may be described by an additional potential energy $U(\mathbf{r})$, such that $\mathbf{F} = -\nabla U$. In our particular case, independent integrations of the force $\mathbf{F}$ along each of three coordinates yield the results

$$U = -\int F_x dx + f_1(y, z) = -\int \mu y dx + f_1(y, z) = -\mu xy + f_1(y, z),$$

$$U = -\int F_y dy + f_2(x, z) = -\int \mu x dy + f_2(y, z) = -\mu xy + f_2(x, z),$$

$$U = -\int F_z dz + f_3(x, y) = -\int 0 dz + f_3(x, y) = f_3(x, y),$$

which are compatible only if $f_3(x,y) = -\mu xy + \text{const}$, and $f_1(y,z) = f_2(x,z) = \text{const}$. Dropping this inconsequential constant, we may use the resulting U as the perturbation Hamiltonian:

$$\hat{H}^{(1)} = -\mu xy. \qquad (*)$$

As was discussed in the beginning of section 3.5 of the lecture notes, unperturbed eigenstates of the 3D oscillator may be described by the products of eigenfunctions of 1D similar harmonic oscillators:

$$\psi_{klm}^{(0)}(\mathbf{r}) = \psi_k(x)\,\psi_l(y)\,\psi_m(z),$$

with each of the indices $\{k, l, m\}$ taking independent integer values 0, 1, 2, …. The corresponding unperturbed energies are

$$E_{klm}^{(0)} = \hbar\omega_0\left(k + l + m + \frac{3}{2}\right),$$

so that there is only one ground state with the wavefunction ψ_{000} and energy $E_0 = (3/2)\hbar\omega_0$, and 3 lowest excited states with wavefunctions ψ_{100}, ψ_{010}, and ψ_{001}, all with the same energy $E_1 = (5/2)\hbar\omega_0$.

According to Eq. (2.275), the non-degenerate ground state of the oscillator is described by an even function of all its arguments:

$$\psi_{000}^{(0)}(\mathbf{r}) = \psi_0(x)\,\psi_0(y)\,\psi_0(z) \propto \exp\left\{-\frac{x^2 + y^2 + z^2}{2x_0^2}\right\}, \qquad \text{with } x_0 \equiv \left(\frac{\hbar}{m\omega_0}\right)^{1/2},$$

so that the first-order shift (6.14) of its energy,

$$E_{000}^{(1)} = \int_{-\infty}^{+\infty} dx \int_{-\infty}^{+\infty} dy \int_{-\infty}^{+\infty} dz \;\psi_{000}^{(0)*}(\mathbf{r})\hat{H}^{(1)}\;\psi_{000}^{(0)}(\mathbf{r}),$$

vanishes for the perturbation (*), which is an odd function of x and y.

Similarly, all *diagonal* matrix elements (6.8) for the lowest excited states vanish as well, because ψ_{100}, ψ_{010}, and ψ_{001} are odd functions of one coordinate, but even functions of the other two coordinates. For example, according to Eqs. (2.282) and (2.284),

$$\psi_{100}(\mathbf{r}) = \psi_1(x)\,\psi_0(y)\,\psi_0(z) = \frac{1}{\sqrt{2}\,\pi^{3/4}x_0^{3/2}}\frac{2x}{x_0}\exp\left\{-\frac{x^2+y^2+z^2}{2x_0^2}\right\},$$

so that

$$E_{100}^{(1)} = \int_{-\infty}^{+\infty} dx \int_{-\infty}^{+\infty} dy \int_{-\infty}^{+\infty} dz \;\psi_{100}^{(0)*}(\mathbf{r})\hat{H}^{(1)}\;\psi_{100}^{(0)}(\mathbf{r}) \propto \int_{-\infty}^{+\infty} y\exp\left\{-\frac{y^2}{x_0^2}\right\}dy = 0,$$

and similarly for two other eigenstates.

However, since the excited eigenstates are degenerate, their mixed matrix elements are also important, and one pair of these elements (not involving the eigenfunction ψ_{001}) is different from zero:

$$\begin{aligned}
H_{100,010}^{(1)} &= \int_{-\infty}^{+\infty} dx \int_{-\infty}^{+\infty} dy \int_{-\infty}^{+\infty} dz \;\psi_{100}^{(0)*}(\mathbf{r})\hat{H}^{(1)}\;\psi_{010}^{(0)}(\mathbf{r}) \\
&= -\mu\int_{-\infty}^{+\infty} dx \int_{-\infty}^{+\infty} dy \int_{-\infty}^{+\infty} dz \;\psi_1^*(x)\,\psi_0^*(y)\,\psi_1^*(z)\,xy\;\psi_0(x)\,\psi_1(y)\,\psi_0(z) \\
&= -\mu\frac{1}{2\,\pi^{3/2}x_0^3}\int_{-\infty}^{+\infty} dx \int_{-\infty}^{+\infty} dy \int_{-\infty}^{+\infty} dz \;xy\,\frac{2x}{x_0}\frac{2y}{x_0}\exp\left\{-\frac{x^2+y^2+z^2}{x_0^2}\right\} \\
&= -\mu\frac{2x_0^2}{\pi^{3/2}}\left(\int_{-\infty}^{+\infty}\xi^2\exp\{-\xi^2\}d\xi\right)^2\left(\int_{-\infty}^{+\infty}\exp\{-\xi^2\}d\xi\right).
\end{aligned}$$

These are two standard Gaussian integrals[4] equal, respectively, to $\pi^{1/2}/2$ and $\pi^{1/2}$, so that

$$H_{100,010}^{(1)} = -\frac{\mu x_0^2}{2} \equiv -\frac{\mu\hbar}{2m\omega_0},$$

and absolutely similarly for $H^{(1)}{}_{010,100}$.

Hence the characteristic equation (6.26) for the states {100} and {010} has the form

$$\begin{vmatrix} H_{100,100}^{(1)} - E^{(1)} & H_{010,100}^{(1)} \\ H_{100,010}^{(1)} & H_{010,010}^{(1)} - E^{(1)} \end{vmatrix} \equiv \begin{vmatrix} -E^{(1)} & -\mu\hbar/2m\omega_0 \\ -\mu\hbar/2m\omega_0 & -E^{(1)} \end{vmatrix} = 0,$$

[4] See, e.g. Eq. (A.36).

and has two roots

$$E_{\pm}^{(1)} = \pm\frac{\mu\hbar}{2m\omega_0}. \qquad (**)$$

Their difference describes lifting the degeneracy of these two states, while the state {001} is not involved in the interaction, and its energy is not changed.

The above calculation is asymptotically exact if the energy correction (**) is much smaller than the basic level spacing $\hbar\omega_0$:

$$\left|\frac{\mu\hbar}{2m\omega_0}\right| \ll \hbar\omega_0, \qquad \text{i.e. if } |\mu| \ll 2m\omega_0^2.$$

Note the Planck constant has dropped from this condition, so it has a classical character. This happens because the perturbation (*) is a quadratic form of the coordinates, just as the unperturbed potential energy of the oscillator,

$$U^{(0)} = \frac{m\omega_0^2}{2}(x^2 + y^2 + z^2),$$

so that the relation of their magnitudes does not depend on the oscillator's quantum length scale x_0. This fact also allows an exact solution of this problem, similar to that of problem 3.16. The reader is challenged to carry out such solution, and compare the result with Eq. (**).

***Problem* 6.5.** A 1D particle of mass m is localized in a very short potential well that may be approximated with a delta-function:

$$U(x) = -\mathscr{w}\delta(x), \qquad \text{with } \mathscr{w} > 0.$$

Calculate the change of its ground state's energy by an additional weak, time-independent force F, in the first nonvanishing approximation of the perturbation theory. Discuss the limits of validity of this result, taking into account that at $F \neq 0$, the localized state of the particle is metastable.

Solution: As was discussed in the beginning of section 2.6 of the lecture notes (see also section 6.6), the unperturbed Hamiltonian of this system has just one localized state with a negative energy,

$$E_g^{(0)} = -\frac{\hbar^2\kappa^2}{2m} = -\frac{m\mathscr{w}^2}{2\hbar^2},$$

and an exponentially decaying wavefunction:

$$\psi_g^{(0)} = \kappa^{1/2}e^{-\kappa|x|}, \qquad \text{where } \kappa = \frac{m\mathscr{w}}{\hbar^2}.$$

Treating the force F as a perturbation, with

$$\hat{H}^{(1)} = -Fx, \tag{*}$$

we immediately see from Eq. (6.14) that the first-order perturbation of the ground state energy vanishes, because the function under the corresponding integral is odd:

$$E_g^{(1)} = \langle g^{(0)} | \hat{H}^{(1)} | g^{(0)} \rangle = \int_{-\infty}^{+\infty} \psi_g^{(0)*} \hat{H}^{(1)} \psi_g^{(0)} dx = -F\kappa \int_{-\infty}^{+\infty} x e^{-\kappa|x|} dx = 0.$$

Hence, we need to proceed to the second order perturbation (6.20), which includes all unperturbed states:

$$E_g^{(2)} \equiv E_0^{(2)} = \sum_{n \neq 0} \frac{|H_{n0}^{(1)}|^2}{E_0^{(0)} - E_n^{(0)}}. \tag{**}$$

At this point we need to notice that the force-unperturbed system also has a continuum of positive-energy states, with energies

$$E_n^{(0)} = \frac{\hbar^2 k_n^2}{2m}.$$

As was discussed in section 6.6 of the lecture notes, the only consequential states of this set, with eigenfunctions

$$\psi_n^{(0)} = \left(\frac{2}{l}\right)^{1/2} \sin k_n x,$$

where $l \gg 1/\kappa$, $1/k_n$ is the length of the artificial normalization segment, are not affected by the delta-functional potential of the well. The matrix elements we need,

$$H_{n0}^{(1)} \equiv \langle n^{(0)} | \hat{H}^{(1)} | g^{(0)} \rangle = \int_{-\infty}^{+\infty} \psi_n^{(0)*} \hat{H}^{(1)} \psi_g^{(0)} dx = -2F \left(\frac{2\kappa}{l}\right)^{1/2} \frac{2\kappa k_n}{\left(\kappa^2 + k_n^2\right)^2},$$

were also calculated in section 6.6. (The factor-of-2 difference between the above expression and Eq. (6.130) is due to the similar difference between the operator $\hat{H}^{(1)}$ defined by Eq. (*), and the operator $\hat{A}$ given by Eq. (6.123) of the lecture notes.) As a result, Eq. (**) yields

$$E_g^{(2)} = \sum_{n \neq 0} \frac{|H_{n0}^{(1)}|^2}{E_0^{(0)} - E_n^{(0)}} = -\frac{64 m F^2 \kappa^3}{\hbar^2 l} \sum_{n > 0} \frac{k_n^2}{\left(\kappa^2 + k_n^2\right)^5}.$$

In the limit $\kappa l \to \infty$, the distance between the adjacent values of k_n becomes much less than κ, so that using the density of the final states, also calculated in section 6.6:

$$\frac{dn}{dk_n} = \frac{l}{2\pi}, \qquad \rho_n \equiv \frac{dn}{dE_n} = \frac{lm}{2\pi\hbar^2 k_n}$$

the sum in may be approximated by the following integral[5]:

[5] This cancellation of the artificial length l is a necessary condition of the correctness of such normalization procedure.

$$E_{\mathrm{g}}^{(2)} = -\frac{64mF^2\kappa^3}{\hbar^2 l}\int_0^\infty \frac{k_n^2}{\left(\kappa^2 + k_n^2\right)^5}dn = -\frac{64mF^2\kappa^3}{\hbar^2 l}\int_0^\infty \frac{k_n^2}{\left(\kappa^2 + k_n^2\right)^5}\frac{dn}{dk_n}dk_n$$
$$= -\frac{32mF^2}{\pi\hbar^2\kappa^4}I, \qquad (***)$$

where

$$I \equiv \int_0^\infty \frac{\xi^2 d\xi}{(1+\xi^2)^5}, \qquad \text{and} \quad \xi \equiv \frac{k_n}{\kappa}.$$

The dimensionless integral I may be worked out, for example, by reducing it to a table one by *integration over a parameter*—the trick so simple and so frequently used that, as an exception, I will demonstrate it here—despite all my usual strong focus on physics rather than math. Let us define the following function:

$$J(\lambda) \equiv \int_0^\infty \frac{d\zeta}{(1+\lambda\zeta^2)^4}.$$

Then, on one hand, we may write

$$\frac{dJ}{d\lambda} = \int_0^\infty \frac{d}{d\lambda}\left[\frac{1}{(1+\lambda\zeta^2)^4}\right]d\zeta = -4\int_0^\infty \frac{\zeta^2 d\zeta}{(1+\lambda\zeta^2)^5}, \qquad \text{so that our } I = -\frac{1}{4}\frac{dJ}{d\lambda}\bigg|_{\lambda=1}.$$

On the other hand, introducing a new integration variable $\xi \equiv \lambda^{1/2}\zeta$, so that $d\zeta = \lambda^{-1/2}d\xi$ and $\lambda\zeta^2 = \xi^2$, $J(\lambda)$ may be readily reduced to a well-known table integral[6]:

$$J(\lambda) \equiv \int_0^\infty \frac{d\zeta}{(1+\lambda\zeta^2)^4} = \lambda^{-1/2}\int_0^\infty \frac{d\xi}{(1+\xi^2)^4} = \lambda^{-1/2}\frac{\pi}{2}\cdot\frac{1\cdot 3\cdot 5}{2\cdot 4\cdot 6} = \frac{5\pi}{32}\lambda^{-1/2},$$

so that

$$\frac{dJ}{d\lambda} = \frac{5\pi}{32}\left(-\frac{1}{2}\right)\lambda^{-3/2} \equiv -\frac{5\pi}{64}\lambda^{-3/2},$$

and, finally, the integral we need is

$$I = -\frac{1}{4}\frac{dJ}{d\lambda}\bigg|_{\lambda=1} = -\frac{1}{4}\cdot\left(-\frac{5\pi}{64}\right) = \frac{5\pi}{256}.$$

With this, Eq. (***) yields the following final result for the ground-state energy's shift:

$$E_{\mathrm{g}}^{(2)} = -\frac{32mF^2}{\pi\hbar^2\kappa^4}\cdot\frac{5\pi}{256} \equiv -\frac{5}{8}\frac{mF^2}{\hbar^2\kappa^4}.$$

As was discussed in section 6.1 of the lecture notes, the necessary condition of the perturbation theory's validity is that this shift is much smaller than the unperturbed

[6] See, e.g. Eq. (A.32c) with $n = 4$.

energy. In our case, this condition, apart from numerical factors of the order of 1, reads

$$\frac{mF^2}{\hbar^2\kappa^4} \ll \frac{\hbar^2\kappa^2}{m}, \quad \text{giving } \frac{|F|}{\kappa} \ll \frac{\hbar^2\kappa^2}{m},$$
$$\text{i.e. } |F|\,\Delta x \ll |E_g|, \quad \text{where } \Delta x \equiv \frac{1}{\kappa}. \qquad (****)$$

The physics of the last condition is very clear: the potential work of the force F moving the particle within the localization region, of the effective width Δx, has to be much smaller than the unperturbed energy.

Now let us discuss whether this condition is affected by one more feature of this system: as the figure below shows, for any $F \neq 0$, the localized state is separated from the continuum of states with $E > U_{\text{full}}(x)$ by a triangular potential barrier of the maximum height $|E_g|$ and the width $t = |E_g/F|$, so that the state is metastable, with a finite lifetime τ. The calculation of this lifetime, within the WKB approximation, was the subject of problem 2.14, and the result was

$$\tau_{\text{WKB}} = \frac{\hbar^3}{m\mathscr{w}^2}\exp\left\{\frac{2m^2\mathscr{w}^3}{3\hbar^4|F|}\right\}.$$

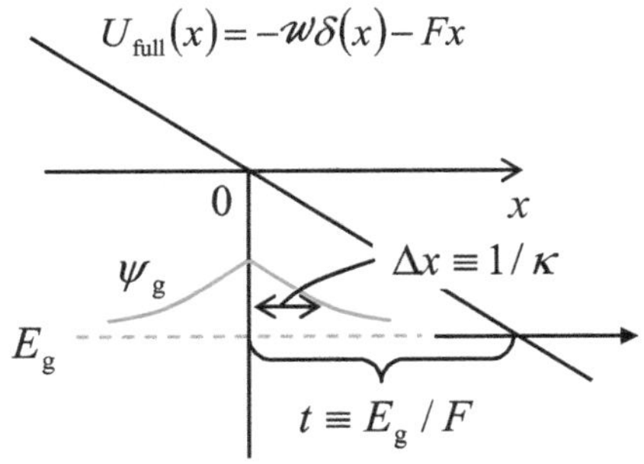

As was discussed in section 2.5 of the lecture notes, the very notion of the energy of such a metastable state lifetime is well defined only if its lifetime is much longer than the attempt time's scale $t_a \sim \hbar/|E_g| = 2\hbar^3/m\mathscr{w}^2$. This is correct if the above exponent is much larger than 1, i.e. if

$$|F| \ll \frac{m^2\mathscr{w}^3}{\hbar^4}.$$

But given the above relation between $\mathscr{w}$ and κ, this is the same condition as given by Eq. (****).[7]

Hence, the metastability of the ground state does not affect the above result for its energy—within the condition (****) of validity of the perturbative approach to the problem.

[7] Note that this condition may be also written in another, very intuitive form: $t \gg \Delta x$—see the figure above.

***Problem* 6.6.** Use the perturbation theory to calculate the eigenvalues of the observable L^2 (where $\mathbf{L}$ is the orbital angular momentum) in the limit $|m| \approx l \gg 1$, by purely wave-mechanical means.

Hint: You may like to use the following substitution: $\Theta(\theta) = f(\theta)/\sin^{1/2}\theta$.

Solution: According to Eqs. (5.146) and (5.166) of the lecture notes, in the coordinate representation the eigenproblem for the operator $\hat{L}^2$, may be reduced to the following equation,

$$\hbar^2\left[-\frac{1}{\sin\theta}\frac{d}{d\theta}\left(\sin\theta\frac{d\Theta}{d\theta}\right) + \frac{m^2}{\sin^2\theta}\Theta\right] = L^2\Theta,$$

for the polar-angle factor $\Theta(\theta)$ of the eigenfunction. Here L^2 is the (at this stage, unknown) eigenvalue of the operator, and m is the 'magnetic' quantum number.

Plugging in the substitution suggested in the Hint, and multiplying all terms by $(\sin^{1/2}\theta)/\hbar^2$, we get the following differential equation for the new function $f(\theta)$:

$$-\frac{d^2f}{d\theta^2} + u(\theta)f = \varepsilon f, \quad \text{where} \quad u(\theta) \equiv \left(\frac{m^2 - 1/4}{\sin^2\theta} - \frac{1}{4}\right), \quad \text{and} \quad \varepsilon \equiv \frac{L^2}{\hbar^2}. \quad (*)$$

The structure of this equation is identical to the Schrödinger equation for a 1D particle, with the function $u(\theta)$ playing the role of the effective potential energy. At $m^2 \gg 1$, this function has a deep minimum at $\theta \approx \pi/2$, so that the lowest-energy wavefunctions of the system are localized near this minimum. In order to calculate these eigenfunctions (and corresponding 'energies' ε), we may approximate the effective potential energy by Taylor-expanding the function $u(\theta)$ with respect to the small deviation

$$\tilde{\theta} \equiv \theta - \frac{\pi}{2},$$

and keeping only the leading terms of this expansion:

$$u(\theta) \equiv -\frac{1}{4} + \left(m^2 - \frac{1}{4}\right)\frac{1}{\cos^2\tilde{\theta}} \approx -\frac{1}{4} + \left(m^2 - \frac{1}{4}\right)\left(1 + \tilde{\theta}^2 + \frac{2}{3}\tilde{\theta}^4\right), \quad \text{for } |\tilde{\theta}| \ll 1.$$

In this approximation, Eq. (*) takes the form

$$-\frac{d^2f}{d\theta^2} + \left[\left(m^2 - \frac{1}{4}\right)\tilde{\theta}^2 + \frac{2}{3}\left(m^2 - \frac{1}{4}\right)\tilde{\theta}^4\right]f = \left(\varepsilon - m^2 + \frac{1}{2}\right)f,$$

so that it is similar to the Schrödinger equation[8],

[8] Let me hope that the subscript 'ef' excludes any possibility of confusion between the effective mass and the magnetic quantum number.

$$-\frac{\hbar^2}{2m_{\text{ef}}}\frac{d^2\psi}{dx^2}+\left[\frac{m_{\text{ef}}\omega_{\text{ef}}^2}{2}x^2+\alpha_{\text{ef}}x^3+\beta_{\text{ef}}x^4\right]\psi=E_{\text{ef}}\psi,$$

of the anharmonic oscillator described by the Hamiltonian (6.2), and coincides with it if we take

$$m_{\text{ef}}=\frac{\hbar^2}{2},\quad \frac{m_{\text{ef}}\omega_{\text{ef}}^2}{2}=m^2-\frac{1}{4}\approx m^2,\quad \text{so that } \hbar\omega_{\text{ef}}\approx 2\,|m|\,,$$
$$\alpha_{\text{ef}}=0,\quad \beta_{\text{ef}}=\frac{2}{3}\left(m^2-\frac{1}{4}\right)\approx\frac{2}{3}m^2,\quad \text{and } E_{\text{ef}}=\varepsilon-m^2+\frac{1}{2},\qquad (**)$$

where the approximations are justified by the initial assumption $m^2 \gg 1$. According to Eq. (6.16) of the lecture notes, with $x_0 \equiv (\hbar/m_{\text{ef}}\omega_{\text{ef}})^{1/2}$, in the first-order of the perturbation theory, the energy spectrum of such an oscillator is

$$E_{\text{ef}}=E_n^{(0)}+E_n^{(1)}=\hbar\omega_{\text{ef}}\left(n+\frac{1}{2}\right)+\frac{3}{4}\beta_{\text{ef}}\frac{\hbar^2}{m_{\text{ef}}^2\omega_{\text{ef}}^2}(2n^2+2n+1),$$

with n taking values 0, 1, 2,... Plugging in the effective oscillator parameters (**), we get a surprisingly simple result:

$$\varepsilon\equiv E_{\text{ef}}+m^2-\frac{1}{2}=m^2+2\,|m|\,n+|m|+n^2+n$$
$$\equiv(|m|+n)\,(|m|+n+1),$$

Now calling the sum $|m| + n$, which can take integer values $|m|$, $|m| + 1$, $|m| + 2$, ..., a new quantum number l, we get the result,

$$L^2\equiv\hbar^2\varepsilon=\hbar^2l(l+1),\qquad (***)$$

which coincides with Eq. (5.163).

The above simple derivation of this (exact) formula is valid only if the height, $\hbar\omega_{\text{ef}}(n + 1/2) = |m|(2n + 1)$, of the nth energy level of the effective oscillator over the bottom of the potential well $u(\theta)$ is much smaller than its depth $\sim m^2$, i.e. only if

$$1,\, n \ll |m|,$$

so that $l \equiv |m| + n$ is relatively close to $|m|$: $l - |m| \ll |m|$. As was shown in section 5.6 of the lecture notes, operator methods allow a derivation of Eq. (***) more easily, and for arbitrary m and l.

However, the above calculation has its value, at least because it illuminates, from one more standpoint, the notorious difference between the square of the largest eigenvalue of observable L_z, equal to $\hbar^2(m^2)_{\max} = \hbar^2 l^2$, and that of L^2, equal to $\hbar^2 l(l + 1) \geqslant \hbar^2 l^2$, which was repeatedly discussed in the lecture notes. Indeed, in the picture developed above, the difference is due to the nonvanishing spread, even at $n = 0$ (i.e. at $|m| = l \equiv m_{\max}$), of the function $f(\theta)$, and hence of the probability to find the system near the values $\theta = \pi/2$.[9] As a result of this spread, the angular

[9] Besides analytical results, this spread is clearly visible on the rightmost and leftmost plots in figure 3.20 of the lecture notes.

momentum vector **L** is never definitely directed along axis z, so that the expectation values of L_x^2 and L_y^2 do not vanish even at $m = \pm l$, giving $L^2 > L_z^2$.

***Problem* 6.7.** In the first nonvanishing order of the perturbation theory, calculate the shift of the ground-state energy of an electrically charged spherical rotator (i.e. a particle of mass m, free to move over a spherical surface of radius R) due to a weak, uniform, time-independent electric field $\mathscr{E}$.

Solution: As was discussed in section 6.2 of the lecture notes, in the coordinate representation, with the z-axis directed along the applied field $\mathscr{E}$, the Hamiltonian of the field-induced perturbation is proportional to the cosine of the polar angle θ, and independent of the azimuthal angle φ:

$$\hat{H}^{(1)} = -q\mathscr{E}z = -q\mathscr{E}R\cos\theta, \qquad (*)$$

where q is the rotator's electric charge. On the other hand, as was discussed in section 3.6 of the lecture notes, the unperturbed ground-state wavefunction of the rotator is independent of both angles:

$$\psi_g^{(0)}(\theta, \varphi) = Y_0^0(\theta, \varphi) = \frac{1}{(4\pi)^{1/2}} = \text{const.}$$

As a result, the first-order correction (6.14) to the ground-state energy vanishes,

$$E_g^{(1)} = H_{gg}^{(1)} \equiv \langle g^{(0)} | \hat{H}^{(1)} | g^{(0)} \rangle \equiv \int_0^{2\pi} d\varphi \int_0^{\pi} \sin\theta d\theta\, \psi_g^{(0)*} \hat{H}^{(1)} \psi_g^{(0)}$$
$$= -\frac{q\mathscr{E}R}{2} \int_0^{\pi} \sin\theta\cos\theta d\theta = 0.$$

In the expression for the second-order correction, which follows from Eq. (6.20),

$$E_g^{(2)} = \sum_{n\neq g} \frac{|H_{gn}^{(1)}|^2}{E_g^{(0)} - E_n^{(0)}}$$
$$\equiv \sum_{n\neq g} \frac{1}{E_g^{(0)} - E_n^{(0)}} \left| \frac{q\mathscr{E}R}{(4\pi)^{1/2}} \int_0^{2\pi} d\varphi \int_0^{\pi} \sin\theta d\theta\, \psi_n^{(0)*} \cos\theta \right|^2, \qquad (**)$$

(where, according to Eq. (3.163) with $l = 0$, $E_g^{(0)} = 0$), nonvanishing contributions may come only from the unperturbed wavefunctions $\psi_n^{(0)} = Y_l^m(\theta, \varphi)$ that have angular dependence similar to that given by Eq. (*), i.e. are proportional to the cosine of the polar angle and are independent of the azimuthal angle. Looking at Eqs. (3.174)–(3.176) of the lecture notes, it is evident that there is only one such function:

$$Y_1^0(\theta, \varphi) = \left(\frac{3}{4\pi}\right)^{1/2} \cos\theta.$$

corresponding to the eigenenergy (3.163) with $l = 1$:

$$E_1^{(0)} = \frac{\hbar^2}{mR^2}.$$

With these substitutions, Eq. (**) yields

$$\begin{aligned}
E_g^{(2)} &= \frac{1}{E_g^{(0)} - E_1^{(0)}} \left| \frac{q\mathscr{E}R}{(4\pi)^{1/2}} \int_0^{2\pi} d\varphi \int_0^{\pi} \sin\theta d\theta \; Y_1^0(\theta, \varphi) \cos\theta \right|^2 \\
&= \frac{1}{0 - \hbar^2/mR^2} \left| \frac{q\mathscr{E}R}{(4\pi)^{1/2}} \int_0^{2\pi} d\varphi \int_0^{\pi} \sin\theta d\theta \left[\left(\frac{3}{4\pi} \right)^{1/2} \cos\theta \right] \cos\theta \right|^2 \\
&= -\frac{3mR^2(q\mathscr{E}R)^2}{4\hbar^2} \left| \int_0^{\pi} \cos^2\theta \sin\theta d\theta \right|^2 \\
&= -\frac{3mR^2(q\mathscr{E}R)^2}{4\hbar^2} \left| \int_{-1}^{1} \cos^2\theta d(\cos\theta) \right|^2 \\
&= -\frac{3mR^2(q\mathscr{E}R)^2}{4\hbar^2} \left(\frac{2}{3} \right)^2 \equiv -\frac{q^2\mathscr{E}^2 mR^4}{3\hbar^2}.
\end{aligned}$$

Note that the correction is negative, as it should be the second-order correction to the ground state energy of any system—see the discussion in section 6.1 of the lecture notes.

***Problem* 6.8.** Use the perturbation theory to evaluate the effect of a time-independent, uniform electric field $\mathscr{E}$ on the ground-state energy E_g of a hydrogen atom. In particular:

(i) calculate the 2nd-order shift of E_g, neglecting the extended unperturbed states with $E > 0$, and bring the result to the simplest analytical form you can,
(ii) find the lower and the upper bounds on the shift, and
(iii) discuss the simplest manifestations of this *quadratic Stark effect*.

Solutions:

(i) The perturbation Hamiltonian is the same as in the previous problem:

$$\hat{H}^{(1)} = -q\mathscr{E}r\cos\theta \equiv e\mathscr{E}r\cos\theta.$$

As was shown in section 6.2 of the lecture notes, in this case the diagonal element of the perturbation matrix (6.8), corresponding to the ground state of the atom (with the quantum numbers $n = 1$, $l = 0$, and $m = 0$), equals zero—see Eq. (6.34). Hence, according to Eq. (6.14), the first-order correction to the ground-state energy vanishes. To calculate its shift ΔE_g in the lowest nonvanishing, second order, we may use Eq. (6.20) of the lecture notes. For the hydrogen atom, neglecting the extended unperturbed states with $E > 0$, the summation index n' on the right-hand side of this relation should be understood as a shorthand for the set of quantum numbers n, l, and m (with $0 \leqslant l \leqslant n-1$, and $-l \leqslant m \leqslant +l$), so that we may write

$$\Delta E_g = \sum_{n=2}^{\infty} \frac{1}{E_1^{(0)} - E_n^{(0)}} \sum_{l=0}^{n-1} \sum_{m=-l}^{+l} |\langle 1,\ 0,\ 0|\ e\mathscr{E}r \cos\theta\ |n,\ l,\ m\rangle|^2, \qquad (*)$$

where the unperturbed states are denoted with the sets of their quantum numbers. Neglecting the very small fine-structure effects discussed in section 6.3, for the unperturbed eigenenergies we may use Eq. (3.201) with $E_0 = E_\mathrm{H}$, where E_H is the Hartree energy (1.13),

$$E_n^{(0)} = -\frac{E_\mathrm{H}}{2n^2},$$

so that the first fraction in Eq. (*) is

$$\frac{1}{E_1^{(0)} - E_n^{(0)}} = -\frac{1}{E_\mathrm{H}\left(-\dfrac{1}{2} + \dfrac{1}{2n^2}\right)} = -\frac{2}{E_\mathrm{H}} \frac{n^2}{n^2 - 1}.$$

Plugging this expression, and using Eqs. (3.171) and (3.200) for the unperturbed wavefunctions of the atom, in the coordinate representation we get

$$\begin{aligned}\Delta E_\mathrm{g} = &-\frac{2e^2\mathscr{E}^2}{E_\mathrm{H}} \sum_{n=2}^{\infty} \frac{n^2}{n^2-1} \sum_{l=0}^{n-1} \left(\int_0^\infty \mathcal{R}_{n,l}(r)\, \mathcal{R}_{1,0}(r) r^3 dr \right)^2 \\ &\times \frac{(2l+1)}{4\pi} \sum_{m=-l}^{+l} \frac{(l-m)!}{(l+m)!} \left| \int_0^\pi P_l^m(\cos\theta) \cos\theta \sin\theta d\theta \int_0^{2\pi} e^{-im\varphi} d\varphi \right|^2,\end{aligned}$$

where it is taken into account that for the ground level, with $n = 1$ and $l = m = 0$, $P_l^m(\cos\theta) = 1$ and $e^{im\varphi} = 1$. This expression may be simplified, because the integral over φ equals 2π for $m = 0$ and vanishes for all other m, so that the sum over m gives just one term:

$$\begin{aligned}\Delta E_\mathrm{g} = &-\frac{2e^2\mathscr{E}^2}{E_\mathrm{H}} \sum_{n=2}^{\infty} \frac{n^2}{n^2-1} \sum_{l=0}^{n-1} \left(\int_0^\infty \mathcal{R}_{n,l}(r)\, \mathcal{R}_{1,0}(r) r^3 dr \right)^2 \\ &\times \frac{(2l+1)}{2} \left| \int_0^\pi P_l^0(\cos\theta) \cos\theta \sin\theta d\theta \right|^2 \\ \equiv &-\frac{2e^2\mathscr{E}^2}{E_\mathrm{H}} \sum_{n=2}^{\infty} \frac{n^2}{n^2-1} \sum_{l=0}^{n-1} \left(\int_0^\infty \mathcal{R}_{n,l}(r)\, \mathcal{R}_{1,0}(r) r^3 dr \right)^2 \\ &\times \frac{(2l+1)}{2} \left| \int_{-1}^{+1} P_l(\xi) \xi d\xi \right|^2.\end{aligned}$$

where $P_l(\xi) = P_l^0(\xi)$ are the Legendre polynomials (3.165). Since the polynomials are orthonormal in the sense of Eq. (3.167), and according to the second of Eqs. (3.166), ξ in the last integral may be considered as $P_1(\xi)$, and the sum over l is reduced to just one term, with $l = 1$,[10] giving

$$\Delta E_g = -\frac{2e^2\mathscr{E}^2}{E_H}\sum_{n=2}^{\infty}\frac{n^2}{n^2-1}\left(\int_0^{\infty}\mathcal{R}_{n,1}(r)\,\mathcal{R}_{1,0}(r)r^3dr\right)^2\frac{(2\cdot 1+1)}{2}\left|\int_{-1}^{+1}\xi^2 d\xi\right|^2$$
$$\equiv -\frac{4e^2\mathscr{E}^2}{3E_H}\sum_{n=2}^{\infty}\frac{n^2}{n^2-1}\left(\int_0^{\infty}\mathcal{R}_{n,1}(r)\,\mathcal{R}_{1,0}(r)r^3dr\right)^2.$$

(In accordance with a general remark made in section 6.1 of the lecture notes, the second-order correction to the ground state energy is negative.)

(ii) For finding the lower and upper bounds of the shift, we may notice that since the fraction under the sum changes only within a narrow interval,

$$1 < \left.\frac{n^2}{n^2-1}\right|_{n\geqslant 2} \leqslant \frac{4}{3}, \quad \text{i.e.}\ \frac{2}{E_H} < \left|\frac{1}{E_1^{(0)} - E_n^{(0)}}\right|_{n\geqslant 2} \leqslant \frac{4}{3}\frac{2}{E_H},$$

Eq. (*) may be used to get the following bounds for the energy shift magnitude:

$$\frac{2e^2\mathscr{E}^2}{E_H}\Sigma < \left|\Delta E_g\right| \leqslant \frac{4}{3}\frac{2e^2\mathscr{E}^2}{E_H}\Sigma,$$

where

$$\Sigma \equiv \sum_{1>0}|\langle 1,0,0|\,r\cos\theta\,|n,l,m\rangle|^2$$
$$= \sum_{n>1}\langle 1,0,0|\,r\cos\theta\,|n,l,m\rangle\langle n,l,m|\,r\cos\theta\,|1,0,0\rangle.$$

The sum Σ would not change if we add to it the similar term with $n = 1$ (with $l = m = 0$), because, as was discussed above, the matrix element in it equals zero. Hence we may write

$$\Sigma = \sum_{\text{all } n,l,m}\langle 1,0,0|\,r\cos\theta\,|n,l,m\rangle\langle n,l,m|\,r\cos\theta\,|1,0,0\rangle.$$

But due to the completeness of the unperturbed state set, we may apply to this sum the closure condition (4.44), getting the following simple expression:

$$\Sigma = \langle 1,0,0|\,(r\cos\theta)^2\,|1,0,0\rangle \equiv \frac{1}{4\pi}\int_0^{\infty}\mathcal{R}_{1,0}^2(r)r^4dr\int_0^{\pi}\cos^2\theta\sin\theta d\theta\int_0^{2\pi}d\varphi,$$

Now using Eq. (3.208), variable replacements $\zeta \equiv 2r/r_B$, $\xi \equiv \cos\theta$, and a table integral over ζ,[11] we get

---

[10] Such vanishing of all terms with $m \neq 0$ and $l \neq 1$ is one of manifestations of the general quantum transition selection rules, which were repeatedly discussed in this course—see, e.g. problem 5.35.

[11] See, e.g. Eq. (A.34d) with $n = 4$.

$$\Sigma = \frac{1}{2}\frac{4}{r_B^3}\int_0^\infty e^{-2r/r_B}r^4 dr \int_{-1}^{+1} \cos^2\theta \; d(\cos\theta)$$
$$= \frac{1}{2}\frac{4}{r_B^3}\left(\frac{r_B}{2}\right)^5 \int_0^\infty e^{-\zeta}\zeta^4 d\zeta \int_{-1}^{+1} \xi^2 d\xi \;=\frac{1}{2}4\frac{1}{2^5}4!\frac{2}{3}r_B^2 = r_B^2,$$

so that, finally, the second-order correction is confined to a narrow interval

$$2\frac{e^2\mathscr{E}^2 r_B^2}{E_H} < \left|\Delta E_g\right| \leqslant \frac{8}{3}\frac{e^2\mathscr{E}^2 r_B^2}{E_H}. \qquad (**)$$

Just for reader's reference, the exact theory[12] gives a value,

$$\Delta E_g = -\frac{9}{4}\frac{e^2\mathscr{E}^2 r_B^2}{E_H}, \qquad (***)$$

within the interval (**).

(iii) Thus, if temperature is not extremely high, so that an atom is reliably in its ground state, the change of its energy in electric field is negative and proportional to $\mathscr{E}^2$. This fact may be expressed in the following (traditional) form:

$$\Delta E = -\alpha\frac{\mathscr{E}^2}{2}.$$

The most significant manifestation of this effect is in that, according to the theory of electric polarization[13], the coefficient α in the last formula is just the *atomic* (or 'molecular') *polarizability*, which relates the induced dipole moment **d** of an atom/molecule to the applied field[14],

$$\mathbf{d} = \alpha\mathscr{E} \qquad \left(\text{so that } \Delta E = -\frac{\mathbf{d}\cdot\mathscr{E}}{2}\right),$$

and hence determines the electric susceptibility χ_e (and the dielectric constant $\kappa \equiv 1 + \chi_e$) of a medium (e.g. a gas) with a relatively low volumic density n of such atoms/molecules, making their interaction negligible[15]:

$$\chi_e = \frac{\alpha n}{\varepsilon_0}, \qquad \kappa = 1 + \frac{\alpha n}{\varepsilon_0}.$$

According to Eq. (1.13), the Hartree energy may be represented as $E_H = e^2/4\pi\varepsilon_0 r_B$, so that $e^2 r_B^2/E_H = 4\pi\varepsilon_0 r_B^3$, and Eq. (***) may be rewritten as

[12] See, e.g. [2]. This result may be also obtained using the second-order perturbation theory, but with the account of a contribution from the extended unperturbed states—which turns out to be relatively small.

[13] See, e.g. *Part EM* section 3.1, in particular Eqs. (3.15*b*) and (3.48). (In that part of my series, following tradition, the electric dipole moment is denoted as **p**, rather than **d** as in this part.)

[14] Alternatively, **d** may be calculated as the expectation value of $q\mathbf{r} = -e\mathbf{r}$, using the first-order approximation (6.18) for the state perturbations—giving the same result.

[15] See, e.g. *Part EM* Eqs. (3.44) and (3.50).

$$\Delta E_g = -4\pi\varepsilon_0\left(\frac{9}{2}r_B^3\right)\frac{\mathscr{E}^2}{2}.$$

This means that the atomic polarizability of the hydrogen atom is

$$\alpha = 4\pi\varepsilon_0\frac{9}{2}r_B^3 \approx 4\pi\varepsilon_0\,(1.651\ r_B)^3$$

—the result to be compared with $\alpha = 4\pi\varepsilon_0 R^3$ for a sphere of a hypothetical material that perfectly screens out the external electric field[16]. (Metals do that, but only if R is much larger than r_B—see, e.g. *Part EM* section 2.1.)

***Problem* 6.9.** A particle of mass m, with electric charge q, is in its ground s-state with a given energy $E_g < 0$, being localized by a very short-range, spherically-symmetric potential well. Calculate its static electric polarizability α.

Solution: As was discussed in the solution of the previous problem, the polarizability is related, as

$$\Delta E_g = -\alpha\frac{\mathscr{E}^2}{2},$$

to the quadratic shift of the ground-state energy of the system, caused by a weak external electric field $\mathscr{E}$. (The linear shift, given by Eq. (6.14) of the lecture notes,

$$E_g^{(1)} = \int \psi_g^*(\mathbf{r})\hat{H}^{(1)}\psi_g(\mathbf{r})d^3r \propto \mathscr{E},$$

evidently vanishes for our perturbation Hamiltonian (6.29),

$$\hat{H} = -q\mathscr{E}z = -q\mathscr{E}r\cos\theta,$$

due to the spherical symmetry of the ground-state wavefunction $\psi_g(\mathbf{r}) = \psi_g(r)$.) As Eq. (6.20) shows, in order to calculate $\Delta E_g = E_g^{(2)}$, we need to evaluate all matrix elements

$$H_{gn}^{(1)} = \int \psi_g^*(\mathbf{r})\hat{H}^{(1)}\psi_n(\mathbf{r})d^3r = -q\mathscr{E}\int \psi_g^*(r)\ r\cos\theta\psi_n(\mathbf{r})d^3r$$
$$= -\,q\mathscr{E}\oint d\Omega\int_0^\infty r^2dr\ \ \psi_g^*(r)\ r\cos\theta\psi_n(\mathbf{r}).$$

Since, according to Eqs. (3.166) and (3.171), $\cos\theta$ may be represented as $P_1^0(\cos\theta) \propto Y_1^0(\theta,\varphi)$, and all spherical harmonics are orthogonal in the sense of Eq. (3.173), in our case only the wavefunctions with $l = 1$ and $m = 0$,

$$\psi_n(r) = \mathcal{R}_{1,k}(r)Y_1^0(\theta,\varphi) \equiv \mathcal{R}_{1,k}(r)\left(\frac{3}{4\pi}\right)^{1/2}\cos\theta, \qquad (*)$$

give nonvanishing matrix elements of the perturbation Hamiltonian:

[16] See, e.g. *Part EM* Eq. (3.11).

$$
\begin{aligned}
H_{gn}^{(1)} &= -q\mathscr{E}\left(\frac{3}{4\pi}\right)^{1/2} \oint \cos^2\theta d\Omega \int_0^\infty r^2dr\ \psi_g^*(r)\, r\, \mathcal{R}_{1,k}(r) d^3r \\
&= -\ q\mathscr{E}\left(\frac{4\pi}{3}\right)^{1/2} \int_0^\infty r^2dr\ \psi_g^*(r)\, r\, \mathcal{R}_{1,k}(r)\ .
\end{aligned} \qquad (**)
$$

The radial eigenfunctions for $l = 1$ and $m = 0$ for a spherically-symmetric region with $U(r) = 0$ have been already discussed in section 3.6 (where they have been valid for $r < R$)—see Eq. (3.186)

$$
\mathcal{R}_{1,k}(r) = C_k j_1(kr) \equiv C_k\left(\frac{\sin kr}{k^2r^2} - \frac{\cos kr}{kr}\right).
$$

For the normalization of these functions, let us introduce (just as was done in the course repeatedly, starting from section 1.7) an auxiliary, sufficiently large volume, to which these wavefunctions will be confined. It is convenient, for example, to take a sphere of a radius $R \gg 1/k_0$, where k_0 is the characteristic scale of the wave numbers k—still to be determined. Then the normalization of wavefunctions with $k \sim k_0$, i.e. $kR \gg 1$, requires

$$
|C_k|^{-2} = \int_0^R \left(\frac{\sin kr}{k^2r^2} - \frac{\cos kr}{kr}\right)^2 r^2dr \approx \frac{1}{k^2}\int_0^R \cos^2 kr\, dr \approx \frac{R}{2k^2}, \quad \text{for } kR \gg 1,
$$

so that we may take[17]

$$
\mathcal{R}_{1,k}(r) = \left(\frac{2}{R}\right)^{1/2}\left(\frac{\sin kr}{kr^2} - \frac{\cos kr}{r}\right).
$$

Note that the confinement makes the spectrum of wave vectors k discrete; at $kR \gg 1$, the distance between the eigenvalues is constant:

$$
kR = \frac{\pi}{2} + n\pi, \qquad \text{i.e. } k = \frac{\pi}{2R} + n\frac{\pi}{R}, \qquad \text{for } n \gg 1. \qquad (***)
$$

Proceeding to the ground-state wavefunction ψ_g: as was discussed in the model solution of problem 3.24, it has the form

$$
\psi_g = C_g\frac{\exp\{-\kappa r\}}{r}, \qquad \text{with } \frac{\hbar^2\kappa^2}{2m} = -E_g > 0, \qquad \text{for } r \geqslant R';
$$

in our current limit of a very small well ($R' \to 0$), this form is valid for any $r \neq 0$, so that the normalization condition,

$$
\int \psi_g^*(\mathbf{r})\psi_g(\mathbf{r})d^3r \equiv 4\pi\int_0^\infty r^2dr\ \left|\psi_g\right|^2 \equiv 4\pi\left|C_g\right|^2\int_0^\infty e^{-2\kappa r}dr\ = 1,
$$

after an elementary integration, yields $|C_g|^2 = \kappa/2\pi$, and we may take

[17] Just as a usual reminder, we may always multiply a wavefunction by $\exp\{i\varphi\}$, where φ is any real constant, but in most stationary-function cases, the choice $\varphi = 0$ is the most convenient one.

$$\psi_g = \left(\frac{\kappa}{2\pi}\right)^{1/2} \frac{\exp\{-\kappa r\}}{r}.$$

Now we may, finally, calculate the nonvanishing matrix elements (**):

$$\begin{aligned}
H_{gn}^{(1)} &= -q\mathscr{E}\left(\frac{4\pi}{3}\right)^{1/2} \int_0^\infty r^2 dr \left(\frac{\kappa}{2\pi}\right)^{1/2} \frac{\exp\{-\kappa r\}}{r} r \left(\frac{2}{R}\right)^{1/2} \left(\frac{\sin kr}{kr^2} - \frac{\cos kr}{r}\right) \\
&\equiv -q\mathscr{E}\left(\frac{4\kappa}{3R}\right)^{1/2} \left[\frac{1}{k} \operatorname{Im} \int_0^\infty e^{(ik-\kappa)r} dr - \operatorname{Re} \int_0^\infty r e^{(ik-\kappa)r} dr\right] \\
&= -q\mathscr{E}\left(\frac{4\kappa}{3R}\right)^{1/2} \left[\frac{1}{k} \operatorname{Im} \frac{\exp\{(ik-\kappa)r\}}{(ik-\kappa)} + \operatorname{Re} \frac{\exp\{(ik-\kappa)r\}}{(ik-\kappa)^2}\right]_0^\infty \\
&\equiv -q\mathscr{E}\left(\frac{4\kappa}{3R}\right)^{1/2} \frac{2k^2}{(k^2+\kappa^2)^2}.
\end{aligned}$$

Note that according to the solution of problem 3.24, the unperturbed eigenenergy of the state (*) is just $\hbar^2 k^2/2m$. With this, and the above expressions for E_g and $H_{gn}^{(1)}$, for the ground state energy, Eq. (6.20) of the lecture notes takes the form

$$\Delta E_g = E_g^{(2)} = \sum_{n>0} \frac{\left|H_{gn}^{(1)}\right|^2}{E_n^{(0)} - E_g^{(0)}} = -\frac{32 q^2 \mathscr{E}^2 \kappa m}{3R\hbar^2} \sum_{n>0} \frac{k^4}{(k^2+\kappa^2)^5},$$

where k and n are related by Eq. (***). Since our result for $E_g^{(2)}$ is only valid for $kR \gg 1$, i.e. $n \gg 1$, we may ignore the first term in that relation, and transfer from the summation to integration over n, and then over k, with $dk = (\pi/R)dn$, i.e. $dn = (R/\pi)dk$:

$$\begin{aligned}
\Delta E_g &= -\frac{32 q^2 \mathscr{E}^2 \kappa m}{3R\hbar^2} \int_0^\infty \frac{k^4 dn}{(k^2+\kappa^2)^5} = -\frac{32 q^2 \mathscr{E}^2 \kappa m}{3\pi\hbar^2} \int_0^\infty \frac{k^4 dk}{(k^2+\kappa^2)^5} \\
&\equiv -\frac{32 q^2 \mathscr{E}^2 m}{3\pi\kappa^4\hbar^2} \int_0^\infty \frac{\xi^4 d\xi}{(1+\xi^2)^5},
\end{aligned}$$

with $\xi \equiv k/\kappa$.[18] The last integral may be readily worked out by recasting its numerator as a sum of three terms proportional to different powers of $(1 + \xi^2)$:

$$\xi^4 \equiv [(1+\xi^2) - 1]^2 = (1+\xi^2)^2 - 2(1+\xi^2) + 1,$$

and hence representing the integral as a sum of three terms, all proportional to integrals of the type Eq. (A.32*c*), but with different n (equal to 3, 4, and 5, respectively):

$$\begin{aligned}
\int_0^\infty \frac{\xi^4 d\xi}{(1+\xi^2)^5} &= \int_0^\infty \frac{d\xi}{(1+\xi^2)^3} - 2\int_0^\infty \frac{d\xi}{(1+\xi^2)^4} + \int_0^\infty \frac{d\xi}{(1+\xi^2)^5} \\
&= \frac{\pi}{2}\frac{1\cdot 3}{2\cdot 4}\left[1 - 2\frac{5}{6} + \frac{5\cdot 7}{6\cdot 8}\right] \equiv \frac{3\pi}{256}.
\end{aligned}$$

[18] Note that the auxiliary bounding radius $R \gg 1/k_0 \sim 1/\kappa$ has fallen out of the result, thus satisfying a necessary condition of the self-consistency of this state-counting procedure.

As a result, for the energy shift by the electric field we get[19]

$$\Delta E_g = -\frac{q^2\mathscr{E}^2 m}{8\kappa^4\hbar^2} \equiv -\frac{q^2\hbar^2}{16mE_g^2}\frac{\mathscr{E}^2}{2},$$

so that the electric polarizability of the system is

$$\alpha = \frac{q^2\hbar^2}{16mE_g^2} \equiv \frac{q^2 m}{4\kappa^4\hbar^2}.$$

The first of the expressions for α shows that the larger the unperturbed $|E_g|$ is, the smaller is the electric field effect. This is natural, because the stronger particle's confinement (the larger $|E_g|$), the smaller is the effective radius $r_{ef} = 1/\kappa \propto |E_g|^{-1/2}$ of the ground-state wavefunction ψ_g, and hence the smaller is the potential energy difference $\Delta E \sim q\mathscr{E}r_{ef}$ created by the external field for the localized particle.

***Problem* 6.10.** In some atoms, the charge-screening effect of other electrons on the motion of each of them may be reasonably well approximated by the replacement of the Coulomb potential (3.190), $U = -C/r$, with the so-called *Hulthén potential*

$$U = -\frac{C/a}{\exp\{r/a\} - 1} \to -C \times \begin{cases} 1/r, & \text{for } r \ll a, \\ \exp\{-r/a\}/a, & \text{for } a \ll r, \end{cases}$$

where a is the effective screening radius. Assuming that $a \gg r_0$, use the perturbation theory to calculate the energy spectrum in this model, in the lowest order needed to lift the l-degeneracy of the levels.

Solution: As was discussed in section 3.7 of the lecture notes, the radial extension of the eigenfunctions of an electron in a hydrogen-like atom/ion scales as n^2r_0, where n is the principal quantum number. Hence, if n is not too high ($n^2 \ll a/r_0$), we may consider the difference between the Hulthén and Coulomb potentials as a perturbation:

$$\hat{H}^{(1)} \equiv -\frac{C/a}{\exp\{r/a\} - 1} - \left(-\frac{C}{r}\right) \equiv \frac{C}{r}\left[1 - \frac{r/a}{\exp\{r/a\} - 1}\right] \ll \frac{C}{r}, \quad \text{for } r \ll a,$$

and limit the Taylor expansion of the function in the square brackets, in the parameter $\xi \equiv r/a \ll 1$,

$$1 - \frac{\xi}{e^{\xi} - 1} = \frac{\xi}{2} - \frac{\xi^2}{12} + \frac{\xi^4}{720} - \dots$$

to its two leading terms[20], so that

[19] Note that this expression may be represented in the form similar to Eq. (***) of the model solution of the previous problem: $\Delta E_g = -(q\mathscr{E}r_{ef})^2/8E_g$, where $r_{ef} \equiv 1/\kappa$ is the particle localization radius.

[20] If we kept just the first term, the resulting (constant) shift of the energy levels would not lift the l-degeneracy.

$$\hat{H}^{(1)} \approx C\left(\frac{1}{2a} - \frac{r}{12a^2}\right), \qquad \text{for } r \ll a.$$

With this perturbation, Eq. (6.14) of the lecture notes takes the form

$$E_{n,l}^{(1)} = \langle n, l, m \mid \hat{H}^{(1)} \mid n, l, m\rangle = \frac{C}{2a} - C\frac{\langle n, l, m \mid \hat{r} \mid n, l, m\rangle}{12a^2} \equiv \frac{C}{2a} - C\frac{\langle r\rangle_{n,l}}{12a^2},$$

where $\{n, l\}$ are the unperturbed states of the Bohr atom. Now using Eq. (3.210) of the lecture notes, we finally get

$$E_{n,l}^{(1)} = \frac{C}{2a} + \frac{Cr_0}{24a^2}[-3n^2 + l(l+1)].$$

This expression describes the sought-after lifting of the energy level degeneracy, with higher values of the orbital quantum number l giving higher energy—the effect already mentioned in section 3.7, and pertinent to virtually any realistic perturbation of the Coulomb potential—see, for example, section 6.3, in particular Eqs. (6.51) and (6.60), and figure 6.4.

***Problem* 6.11.** In the first nonvanishing order of the perturbation theory, calculate the correction to energies of the ground state and all lowest excited states of a hydrogen-like atom/ion, due to an electron's penetration into its nucleus, modeling it as a spinless, uniformly charged sphere of a radius $R \ll r_B/Z$.

Solution: The electrostatic potential ϕ inside a uniformly-charged sphere of radius R, with the total charge $Q = Ze$, may be readily calculated either using the Gauss law, or solving the corresponding Poisson equation. The result is[21]

$$\phi(r) = \frac{Ze}{4\pi\varepsilon_0 R}\left(\frac{3}{2} - \frac{r^2}{2R^2}\right), \qquad \text{for } r \leqslant R.$$

The arbitrary constant in this expression is selected so that at $r = R$ the potential coincides with the usual expression for the Coulomb potential outside the sphere (the same as of the point charge Ze):

$$\phi_0(r) = \frac{Ze}{4\pi\varepsilon_0 r}, \qquad (*)$$

which tends to zero at $r \to \infty$. Since the potential (*) has already been taken into account in the solution of the basic Bohr atom problem (see, e.g. section 3.6 of the lecture notes), the perturbation Hamiltonian of our current problem is due to their difference:

[21] See, e.g. *Part EM* Eq. (1.51).

$$\hat{H}^{(1)} = U^{(1)}(\mathbf{r}) = -e\big[\phi(r) - \phi_0(r)\big] = -\frac{Ze^2}{4\pi\varepsilon_0 R} \times \begin{cases} \left(\dfrac{3}{2} - \dfrac{r^2}{2R^2} - \dfrac{R}{r}\right), & \text{for } r \leqslant R, \\ 0, & \text{for } r \geqslant R. \end{cases}$$

Since the spatial extension scale of the unperturbed wavefunctions $\psi^{(0)}(\mathbf{r})$ of the atom/ion is given by the radius $r_0 = r_B/Z \ll R$, in the first order in the parameter $R/r_0 \ll 1$ (in actual atoms, as small as $\sim 10^{-5}$) we may approximate this potential with a 3D delta-function

$$U^{(1)}(\mathbf{r}) \approx \mathscr{W}\delta(\mathbf{r}),$$

with the weight

$$\begin{aligned}
\mathscr{W} &= \int U^{(1)}(\mathbf{r})\, d^3r = -\frac{Ze^2}{4\pi\varepsilon_0 R}\int_{r\leqslant R}\left(\frac{3}{2} - \frac{r^2}{2R^2} - \frac{R}{r}\right) d^3r \\
&= -\frac{Ze^2}{4\pi\varepsilon_0 R} 4\pi \int_0^R \left(\frac{3}{2} - \frac{r^2}{2R^2} - \frac{R}{r}\right) r^2 dr \\
&\equiv -\frac{Ze^2}{4\pi\varepsilon_0 R} 4\pi R^3 \int_0^1 \left(\frac{3}{2} - \frac{\xi^2}{2} - \frac{1}{\xi}\right)\xi^2 d\xi \\
&= -\frac{Ze^2R^2}{4\pi\varepsilon_0} 4\pi \left(\frac{3}{2}\cdot\frac{1}{3} - \frac{1}{2}\cdot\frac{1}{5} - \frac{1}{2}\right) \\
&\equiv \frac{Ze^2R^2}{4\pi\varepsilon_0} 4\pi \frac{1}{10}.
\end{aligned}$$

As a result, the matrix elements (6.8) of the perturbation may be calculated as

$$\hat{H}^{(1)}_{n'n''} \equiv \int \psi^{(0)*}_{n'}(\mathbf{r}) \hat{H}^{(1)} \psi^{(0)}_{n''}(\mathbf{r}) d^3r \approx \mathscr{W} \psi^{(0)*}_{n'}(0) \psi^{(0)}_{n''}(0), \qquad (**)$$

where each of the indices n encodes the whole appropriate set of quantum numbers—in the case of the hydrogen-like atom, n, l, and m—see section 3.7 of the lecture notes.

For the ground state ($n = 1$, $l = m = 0$), Eqs. (3.174), (3.200), and (3.208) yield

$$\psi^{(0)}_{g}(0) = \frac{1}{(4\pi)^{1/2}} \frac{2}{r_0^{3/2}}.$$

Since this state is non-degenerate, the first-order correction to its energy may be calculated using the simple Eq. (6.14):

$$\begin{aligned}
E^{(1)}_{g} = H^{(1)}_{gg} &= \mathscr{W} \psi^{(0)*}_{g}(0)\, \psi^{(0)}_{g}(0) = \frac{Ze^2R^2}{4\pi\varepsilon_0} 4\pi \frac{1}{10} \frac{1}{4\pi} \frac{4}{r_0^3} \\
&\equiv 0.4 \frac{Ze^2}{4\pi\varepsilon_0 r_0}\left(\frac{R}{r_0}\right)^2 \equiv 0.4 E_0 \left(\frac{R}{r_0}\right)^2,
\end{aligned}$$

where the energy scale E_0 is defined by Eq. (3.191) with $C = Ze^2/4\pi\varepsilon_0$, so that $E_0 = Z^2E_H$, and for the generic hydrogen atom, with $Z = 1$, this is just the Hartree energy $E_H \approx 27.2$ eV—see Eq. (1.13).

Generally, for the 4 lowest excited states (all with $n = 2$, but with either $l = m = 0$, or $l = 1$ and $m = 0, \pm1$) we should be more accurate, because in the absence of perturbation, they are degenerate, so that the perturbation that may lift their degeneracy should be treated using the approach discussed in the last part of section 6.1 of the lecture notes. However, according to the second of Eqs. (3.209), for all p-states with $l = 1$, $\psi^{(0)}(0) = 0$, so that within the approximation (**), all the matrix elements involving these states vanish. Since, according to the first of Eqs. (3.209), this is not true for the exited s-state (with $n = 2$, $l = 0$):

$$\psi_s^{(0)}(0) = \frac{1}{(4\pi)^{1/2}}\frac{2}{(2r_0)^{3/2}},$$

the only nonvanishing effect of this perturbation on the excited states is the s-state's energy shift by

$$E_s^{(1)} = H_{ss}^{(1)} = \mathscr{W}\psi_s^{(0)*}(0)\psi_s^{(0)}(0) = \frac{Ze^2R^2}{4\pi\varepsilon_0}4\pi\frac{1}{10}\frac{4}{4\pi(2r_0)^3}$$
$$= 0.05\frac{Ze^2}{4\pi\varepsilon_0 r_0}\left(\frac{R}{r_0}\right)^2 \equiv 0.05E_0\left(\frac{R}{r_0}\right)^2,$$

i.e. by the magnitude, 8 times lower than the ground state's shift. For the actual atoms, both shifts are very small, $\sim 10^{-10}E_H \sim 10^{-9}$ eV.

***Problem* 6.12.** Prove that the kinetic-relativistic correction operator (6.48) indeed has only diagonal matrix elements in the basis of unperturbed Bohr atom states (3.200).

Solution: In Eq. (6.49) of the lecture notes, let us act by the first of the two similar (Hermitian) operators $\hat{H}^{(0)}$ upon the bra-vector, and by the second one, on the ket-vector. Since these vectors describe eigenstates of $\hat{H}^{(0)}$, and the corresponding eigenvalues depend only on the principal quantum number n, we get

$$\langle nlm|\,\hat{H}^{(1)}\,|nl'm'\rangle = -\frac{1}{4mc^2}\langle nlm|(E_n - \hat{U}(r))\,(E_n - \hat{U}(r))\,|nl'm'\rangle$$
$$\equiv -\frac{1}{4mc^2}\Big[E_n^2\delta_{nn'} - 2E_n\langle nlm|\,\hat{U}(r)|n'l'm'\rangle \qquad (*)$$
$$+ \langle nlm|\,\hat{U}^2(r)|\,n'l'm'\rangle\Big].$$

The operators of the spherically-symmetric functions $U(r)$ and $U^2(r)$ can act only on the radial factors $\mathcal{R}_{n,l}(r)$ of the wavefunctions $\psi_{n,l,m}$. Hence, due to the orthogonality of the angular factors (spherical functions), all matrix elements (*) with either $m \neq m'$, or $l \neq l'$ (or both) do indeed vanish.

***Problem* 6.13.** Calculate the lowest-order relativistic correction to the ground-state energy of a 1D harmonic oscillator.

Solution: The perturbation Hamiltonian for this problem as the same as in Eq. (6.47) of the lecture notes:

$$\hat{H}^{(1)} = -\frac{1}{8m^3c^2}\hat{p}^4,$$

besides that in our current case, $\hat{p}$ should be understood as the 1D operator, so that it may be expressed via the creation—annihilation operators of the harmonic oscillator—see Eq. (5.66):

$$\hat{p} = \frac{(\hbar m\omega_0)^{1/2}}{\sqrt{2}i}(\hat{a} - \hat{a}^\dagger).$$

Since the energy levels of the unperturbed oscillator are non-degenerate, the first-order correction to the groundstate energy may be calculated using Eq. (6.14) of the lecture notes:

$$E_{\rm g}^{(1)} \equiv \langle 0|\, \hat{H}^{(1)}\, |0\rangle = -\frac{1}{8m^3c^2}\langle 0|\, \hat{p}^4\, |0\rangle \equiv -\frac{(\hbar\omega_0)^2}{32mc^2}I, \qquad (*)$$
$$\text{with } I \equiv \langle 0|\, (\hat{a} - \hat{a}^\dagger)^4\, |0\rangle.$$

The bracket I may be calculated either just as $\langle n|x^4|n\rangle$ was calculated in the model solution of problem 5.9(ii) (i.e. using Eq. (5.93) of the lecture notes), or even simpler—using the fact that according to Eqs. (5.89),

$$\hat{a}\,|0\rangle = 0, \qquad \text{so that } \langle 0|\,\hat{a}^\dagger = 0.$$

As a result, if in the last form of the expression

$$I = \langle 0|\,(\hat{a} - \hat{a}^\dagger)^2(\hat{a} - \hat{a}^\dagger)^2\,|0\rangle$$
$$\equiv \langle 0|\,(\hat{a}^2 - \hat{a}\hat{a}^\dagger - \hat{a}^\dagger\hat{a} + \hat{a}^{\dagger 2})(\hat{a}^2 - \hat{a}\hat{a}^\dagger - \hat{a}^\dagger\hat{a} + \hat{a}^{\dagger 2})|0\rangle,$$

we act by the operators in the first parentheses upon the bra-vector, and by those in the second parentheses, upon the ket-vector, two of each of the four terms vanish, giving

$$I = \langle 0|\,(\hat{a}^2 - \hat{a}\hat{a}^\dagger)(-\hat{a}\hat{a}^\dagger + \hat{a}^{\dagger 2})|0\rangle$$
$$\equiv \langle 0|\,(-\hat{a}^3\hat{a}^\dagger + \hat{a}^2\hat{a}^{\dagger 2} + \hat{a}\hat{a}^\dagger\hat{a}\hat{a}^\dagger - \hat{a}\hat{a}^{\dagger 3})|0\rangle.$$

According to Eqs. (5.89), the first and the last terms in the parentheses of the last expression yield zero contributions to the expectation value, because they have different powers of the creation and annihilation operators, so that their sequential action on, say, the ket-vector gives a non-ground ket-vector, orthogonal to the ground-state bra-vector. The remaining two terms may be calculated directly, using the same Eqs. (5.89):

$$I = \langle 0|(\hat{a}^2\hat{a}^{\dagger 2} + \hat{a}\hat{a}^\dagger\hat{a}\hat{a}^\dagger)|0\rangle \equiv \langle 0|\ \hat{a}(\hat{a}\hat{a}^\dagger + \hat{a}^\dagger\hat{a})\hat{a}^\dagger\ |0\rangle$$
$$= \langle 1|(\hat{a}\hat{a}^\dagger + \hat{a}^\dagger\hat{a})\ |1\rangle \equiv \langle 1|\ \hat{a}\hat{a}^\dagger\ |1\rangle + \langle 1|\ \hat{a}^\dagger\hat{a}\ |1\rangle$$
$$= \langle 2|\sqrt{2}\cdot\sqrt{2}\ |2\rangle + \langle 0\ |\ \sqrt{1}\cdot\sqrt{1}\ |0\rangle = 3\ ,$$

so that, finally, Eq. (*) yields

$$E_{\rm g}^{(1)} = -\frac{3}{32}\frac{(\hbar\omega_0)^2}{mc^2}.$$

This expression is quantitatively valid only if this correction is much smaller than the unperturbed ground-state energy $E_{\rm g}^{(0)} = \hbar\omega_0/2$, i.e. if $\hbar\omega_0 \ll mc^2$. Since, according to Eq. (5.97), $\hbar\omega_0$ also gives the scale of the kinetic energy T of the particle in this state, this requirement is essentially the same as was discussed in section 6.3 of the lecture notes: $T \ll mc^2$.

***Problem* 6.14.** Use the perturbation theory to calculate the contribution to the magnetic susceptibility $\chi_{\rm m}$ of a dilute gas, that is due to the orbital motion of a single electron inside each gas' particle. Spell out your result for a spherically-isotropic ground state of the electron, and give an estimate of the magnitude of this *orbital susceptibility*.

Solution: According to classical electrodynamics[22], the magnetic energy u per unit volume of a linear, isotropic medium may be expressed as

$$u = \frac{\mathscr{B}^2}{2\mu},$$

where $\mathscr{B}$ Bis the applied magnetic field, and μ is the magnetic permeability, related to the magnetic susceptibility as $\mu = \mu_0(1 + \chi_{\rm m})$. For a dilute gas, the susceptibility is small in comparison with 1; in this case we may separate the magnetic energy u into a sum of the energy $\mathscr{B}^2/2\mu_0$ of the magnetic field in free space, and a small correction $u_{\rm m}$ proportional to $\chi_{\rm m}$:

$$u_{\rm m} \equiv u - u_0 = \frac{\mathscr{B}^2}{2\mu} - \frac{\mathscr{B}^2}{2\mu_0} = \frac{\mathscr{B}^2}{2\mu_0}\left(\frac{1}{1+\chi_{\rm m}} - 1\right) \approx -\chi_{\rm m}\frac{\mathscr{B}^2}{2\mu_0},$$

On the other hand, for a medium of non-interacting atoms/molecules with a spatial density n, u_m should be equal to $n\langle E_{\rm m}\rangle$, where $E_{\rm m}$ is the change of the energy of one atom due to its magnetization. Comparing these two formulas, we see that in this case the susceptibility may be calculated as[23]

[22] See, e.g. *Part EM* section 5.5, in particular Eqs. (5.112) and (5.140).

[23] An alternative way to get the same formula (*) is to combine the classical expression $E_m = -\mathscr{m}\cdot\mathscr{B}/2$ for the energy of interaction between the magnetic field $\mathscr{B}$ and the magnetic dipole $\mathscr{m}$ it has induced, the definition of $\chi_{\rm m}$ as $\mathscr{M}/\mathscr{H}$, where the magnetization $\mathscr{M}$ of a dilute medium may be calculated as $\mathscr{M} = n\mathscr{m}$, and the fact that if $|\chi_{\rm m}| \ll 1$, then $\mathscr{H} \approx \mu_0\mathscr{B}$—see, e.g. *Part EM* section 5.5. (In quantum mechanics, $\mathscr{m}$, $\mathscr{M}$, and $E_{\rm m}$ in these relations have to be understood either as the operators of these observables, or as their averages over the ensemble of the corresponding quantum states.)

$$\chi_{\rm m} = -\frac{2\mu_0 \langle E_{\rm m}\rangle}{\mathscr{B}^2} n. \qquad (*)$$

In order to calculate the energy due to the orbital motion of a single electron in the confining potential, we may use the Hamiltonian (6.63), neglecting the term linear in the field[24], but keeping the term quadratic in the vector-potential $\mathbf{A}$, representing it in the convenient form (6.64):

$$\hat{H}_{\rm m} = \frac{e^2}{2m_{\rm e}}\hat{\mathbf{A}}^2 = \frac{e^2}{8m_{\rm e}}|\mathscr{B} \times \hat{\mathbf{r}}|^2 = \frac{e^2\mathscr{B}^2}{8m_{\rm e}}(\hat{x}^2 + \hat{y}^2),$$

where x and y are the Cartesian coordinates perpendicular to the field direction (taken for axis z). If the field is not too high, this Hamiltonian may be used as $\hat{H}^{(1)}$ in the relations of the perturbation theory. For the most important case of the atom in its ground state 0 (which is always non-degenerate), we may immediately use Eq. (6.14) to get

$$\langle E_{\rm m}\rangle = \frac{e^2\mathscr{B}^2}{8m_{\rm e}}\langle 0|\, \hat{x}^2 + \hat{y}^2 \,|0\rangle,$$

so that Eq. (*) yields

$$\chi_{\rm m} = -\frac{\mu_0 n e^2}{4m_{\rm e}}\langle 0|\, \hat{x}^2 + \hat{y}^2 \,|0\rangle.$$

We see that such $\chi_{\rm m}$ is always negative—the effect that is called the *orbital* (or 'Larmor') *diamagnetism*. For the motion in a central field, in which the ground state's wavefunction is spherically-symmetric (see section 3.6), this expression may be further simplified by noting that the averages of all Cartesian components squared have to be equal, so that

$$\langle 0|\, \hat{x}^2 \,|0\rangle = \langle 0|\, \hat{y}^2 \,|0\rangle = \frac{1}{3}\langle 0|\, \hat{r}^2 \,|0\rangle \equiv \frac{1}{3}\langle r^2\rangle, \quad \text{and} \quad \chi_{\rm m} = -\frac{\mu_0 n e^2}{6m_{\rm e}}\langle r^2\rangle. \qquad (**)$$

Remarkably, this formula coincides with the one calculated from a reasonable classical model—see, e.g. *Part EM* problem 5.14(i). It gives a good semi-quantitative description of experimental data for gases, and even liquids and solids of some multi-electron atoms (especially those with filled electron shells, whose net spontaneous orbital and spin momenta vanish), assuming that the contributions of all atom's electrons add up independently.

In order to estimate the magnitude of the effect described by Eq. (**), note that according to the definition of the electromagnetic constants ε_0 and μ_0, the latter of them equals $1/\varepsilon_0 c^2$, where c is the speed of light in free space, so that our result for the susceptibility (which is dimensionless by its definition) may be rewritten as

[24] That term, while contributing to the energy level splitting by the field, i.e. the Zeeman effect (see section 6.4), gives zero contribution to the energy of s-states with no spontaneous angular momentum, in particular of the ground state of an electron moving in a spherically-symmetric potential.

$$\chi_{\mathrm{m}} = -\frac{4\pi}{6} n \frac{e^2}{4\pi\varepsilon_0} \frac{1}{m_{\mathrm{e}}c^2} \langle r^2 \rangle \equiv -\frac{2\pi}{3} n\alpha^2 r_{\mathrm{B}} \langle r^2 \rangle,$$

where $\alpha \equiv e^2/4\pi\varepsilon_0\hbar c \approx 1/137 \ll 1$ is the fine structure constant (6.62), and r_{B} is the Bohr radius (1.10). Since r_{B} gives the fair scale of $\langle r^2 \rangle$ in atoms and (not very large) molecules, we may write the following estimate:

$$|\chi_{\mathrm{m}}| \sim n\alpha^2 r_{\mathrm{B}}^3 \ll n r_{\mathrm{B}}^3. \qquad (***)$$

This estimate shows that the orbital diamagnetism is so weak that it corresponds to pushing out of the atom only a tiny part of the order of $\alpha^2 \sim 10^{-4}$ of the magnetic field lines—while a perfect diamagnetic, for example a bulk superconductor, would push all of them out. Note that an uncompensated net spin of atoms/molecules may give them, due to the spin's polarization by the field, a different, paramagnetic contribution to the magnetic susceptibility—see, e.g. *Part EM* section 5.5 and problem 5.14(ii). At sufficiently low temperatures, this *spin paramagnetism* may be much higher than the orbital diamagnetism (***)—see, e.g. *Part SM* problems 2.4 and 3.10.

***Problem* 6.15.** How to calculate the energy level degeneracy lifting, by a time-independent perturbation, in the second order in the perturbation in $\hat{H}^{(1)}$, assuming that it is not lifted in the first order? Carry out such calculation for a plane rotator of mass m and radius R, carrying electric charge q, and placed into a weak, uniform, constant electric field $\mathscr{E}$.

Solution: If all first-order matrix elements $H_{n'n''}$ connecting degenerate states, i.e. participating in Eqs. (6.24)–(6.26) of the lecture notes, vanish in the first order of the perturbation theory, we need to calculate them in the second order. For that, it is sufficient to repeat the calculation that was used to derive Eq. (6.19), starting from the exact system of equations (6.7) and the expansions (6.9)–(6.10). With the assumption that $H^{(1)}_{n'n''} = 0$, so that according to Eq. (6.14), $E^{(1)}_n = 0$ as well, the balance of the terms $O(\mu^2)$ gives us a system of equations similar to Eq. (6.24), but with the very natural replacement

$$H^{(1)}_{n'n''} \to H^{(2)}_{n'n''} \equiv \langle n'^{(0)} | \hat{H}^{(1)} | n''^{(1)} \rangle,$$

where $n''^{(1)}$ is state n'', calculated in the first order. It is described by Eq. (6.18), so that renaming the quantum numbers in that formula ($n \to n''$, $n' \to n$), we get

$$H^{(2)}_{n'n''} \equiv \sum_{n \neq n''} \frac{H^{(1)}_{nn''}}{E^{(0)}_{n''} - E^{(0)}_n} \langle n'^{(0)} | \hat{H}^{(1)} | n^{(0)} \rangle \equiv \sum_{n \neq n''} \frac{H^{(1)}_{n'n} H^{(1)}_{nn''}}{E^{(0)}_{n''} - E^{(0)}_n}. \qquad (*)$$

(In the particular case $n' = n''$, this expression is reduced to Eq. (6.20) of the lecture notes.)

The basic properties of the unperturbed ($\mathscr{E} = 0$) plane rotator were discussed in section 3.5 of the lecture notes. In the coordinate representation, the rotator's Hamiltonian (3.126) is

$$\hat{H}^{(0)} = -\frac{1}{2mR^2}\frac{\partial^2}{\partial\varphi^2},$$

its eigenfunctions (3.129) are[25]

$$\psi_m^{(0)} = \frac{1}{(2\pi)^{1/2}}e^{im\varphi},$$

and the energy spectrum (3.130) is

$$E_m^{(0)} = m^2 E_1, \qquad \text{where } E_1 \equiv \frac{\hbar^2}{2mR^2}.$$

Note that all these energy values, besides the ground state ($m = 0$), are doubly degenerate:

$$E_{-m}^{(0)} = E_m^{(0)}.$$

This degeneracy is lifted by the perturbation created by the applied electric field,

$$\hat{H}^{(1)} = -q\mathscr{E}x \equiv -q\mathscr{E}R\cos\varphi, \qquad (**)$$

where $\mathscr{E}$ is the electric field's component within rotator's plane, and the x-axis is directed along that component. However, calculating the first-order matrix elements (6.8) of the perturbation (with the natural notation replacement $n \to m$):

$$H_{m'm''}^{(1)} = \int_0^{2\pi} \psi_{m'}^{(0)*}\hat{H}^{(1)}\psi_{m''}^{(0)}d\varphi \equiv \frac{1}{2\pi}\int_0^{2\pi} e^{-im'\varphi}(-q\mathscr{E}R\cos\varphi)\,e^{im''\varphi}d\varphi$$
$$= -\frac{q\mathscr{E}R}{2\pi}\int_0^{2\pi} e^{-im'\varphi}\frac{e^{i\varphi}+e^{-i\varphi}}{2}e^{im''\varphi}d\varphi = -\frac{q\mathscr{E}R}{2}(\delta_{m',m''+1} + \delta_{m',m''-1}),$$

we see that they connect only the states with the quantum numbers m different by ± 1, while the difference between these numbers for each degenerate state couple is $|2m| > 1$. Hence, in the first order of the perturbation theory the level degeneracy is not lifted[26], and, according to Eq. (6.14), the eigenenergies are not affected.

[25] See also Eq. (5.146). The mth unperturbed eigenfunction corresponds to the eigenvalue $(L_z)_m = m\hbar$—see Eq. (5.158).

[26] Note that this fact is not generally true for the rotator in an additional magnetic field. Indeed, as Eq. (3.134) and figure 3.18 of the lecture notes show, if the magnetic flux Φ piercing rotator's area is a half-multiple of the flux quantum $\Phi_0' \equiv 2\pi\hbar/q$, the lowest unperturbed energy levels $E_m = E_1(m - 1/2)^2 = E_1/4$ are equal for the states with $\Delta m = 1$, for example $m = 1$ and $m = 0$. This degeneracy is lifted by the electric field already in the first order of the perturbation theory. Note also the analogy between this problem and the 1D band theory (see section 2.7 of the lecture notes), where the role of magnetic field is played by the quasi-momentum $\hbar q$, and each Fourier harmonic of the periodic potential $U(x)$, with amplitude U_l, acts similarly to the sinusoidal perturbation (**). Indeed, the weak-potential limit explored in section 2.7(ii) of the lecture notes is just the first-order perturbation theory in small U_l.

Proceeding to the second-order perturbation theory, let us spell out Eq. (*) for two states of each degenerate pair ($n' = m'$, $n'' = \pm m'$):

$$H^{(2)}_{m',\pm m'} = \sum_{m\neq m'} \frac{H^{(1)}_{m',m} H^{(1)}_{m,\pm m'}}{E^{(0)}_{m'} - E^{(0)}_{m}} = \frac{1}{E_1}\left(\frac{q\mathscr{E}R}{2}\right)^2 \sum_{m\neq m'} \frac{(\delta_{m',m+1} + \delta_{m',m-1})(\delta_{m,\pm m'+1} + \delta_{m,\pm m'-1})}{m'^2 - m^2}.$$

This expression yields nonvanishing diagonal matrix elements for any m' (with the sum contributed by two terms, with $m = m' \pm 1$):

$$\begin{aligned} H^{(2)}_{m',m'} &= \frac{1}{E_1}\left(\frac{q\mathscr{E}R}{2}\right)^2 \left[\frac{1}{m'^2 - (m'+1)^2} + \frac{1}{m'^2 - (m'-1)^2}\right] \\ &\equiv \frac{1}{E_1}\left(\frac{q\mathscr{E}R}{2}\right)^2 \frac{2}{4m'^2 - 1}, \end{aligned} \qquad (***)$$

but off-diagonal matrix elements only for $m' = \pm 1$ (with the sum limited to only one term, with $m = 0$):

$$H^{(2)}_{1,-1} = H^{(2)}_{-1,1} = \frac{1}{E_1}\left(\frac{q\mathscr{E}R}{2}\right)^2. \qquad (****)$$

Hence all the energy levels with $m' > 1$ are just shifted by the amount given by Eq. (***), but for $m' = 1$ we get a system of two linear equations, whose matrix consists of two equal off-diagonal elements (****) and two equal diagonal elements (***):

$$H^{(2)}_{1,1} = H^{(2)}_{-1,-1} = \frac{1}{E_1}\left(\frac{q\mathscr{E}R}{2}\right)^2 \frac{2}{3}.$$

Solving the corresponding characteristic equation,

$$\begin{vmatrix} H^{(2)}_{1,1} - E^{(2)}_1 & H^{(2)}_{1,-1} \\ H^{(2)}_{-1,1} & H^{(2)}_{-1,-1} - E^{(2)}_1 \end{vmatrix} = 0, \qquad \text{which gives} \qquad \begin{vmatrix} \frac{2}{3} - \lambda & 1 \\ 1 & \frac{2}{3} - \lambda \end{vmatrix} = 0,$$

$$\text{for } \lambda \equiv E^{(2)}_1 \Big/ \frac{1}{E_1}\left(\frac{q\mathscr{E}R}{2}\right)^2,$$

we get two roots, $\lambda_\pm = 2/3 \pm 1$, showing that in the second order of the perturbation theory the double degeneracy is lifted:

$$\begin{aligned} (E_1)_\pm &= E^{(0)}_1 + E^{(2)}_1 = E_1 + \frac{1}{E_1}\left(\frac{q\mathscr{E}R}{2}\right)^2\left(\frac{2}{3} \pm 1\right) \\ &\equiv E_1\left[1 + \left(\frac{q\mathscr{E}R}{2E_1}\right)^2 \times \begin{Bmatrix} 5/3 \\ -1/3 \end{Bmatrix}\right]. \end{aligned}$$

This expression is quantitatively valid only at $q\mathscr{E}R/2 \ll E_1 \equiv \hbar^2/2m R^2$, i.e. in a sufficiently low applied electric field.

Problem* 6.16. The Hamiltonian of a quantum system is slowly changed in time.

(i) Develop a theory of quantum transitions in the system, and spell out its result in the first order in the speed of the change.
(ii) Use the first-order result to calculate the probability that a finite-time pulse of a slowly changing force $F(t)$ drives a 1D harmonic oscillator, initially in its ground state, into an excited state.
(iii) Compare the last result with the exact one.

Solutions:

(i) Let us solve the Schrödinger equation (4.158),

$$i\hbar\frac{\partial}{\partial t}|\alpha\rangle = \hat{H}|\alpha\rangle,$$

for a system described by a Hamiltonian $\hat{H}$ which is slowly changed in time. In the 0th approximation in the speed, when the change is adiabatic, its general solution may be represented as the sum

$$|\alpha\rangle = \sum_n a_n e^{i\varphi_n(t)}|n\rangle,$$

where

$$\hat{H}|n\rangle = E_n|n\rangle, \quad \text{and} \quad \varphi_n(t) \equiv -\frac{1}{\hbar}\int_0^t E_n(t')\,dt', \quad \text{i.e.} \quad \dot{\varphi}_n \equiv -E_n/\hbar, \qquad (*)$$

while a_n are some constant c-numbers. (The physical sense of each a_n is the probability amplitude of our system being, at time t, in the instant state $|n\rangle$ defined by Eq. (*) for the same t.) If the Hamiltonian changes in time, so are E_n and $|n\rangle$, and the above relations do not represent the *exact* solution of the Schrödinger equation, but we may look for such a solution in the same form, with E_n, $|n\rangle$, and φ_n defined by Eqs. (*) at each particular instant, but with the probability amplitudes a_n also being some slow functions of time.

Plugging this form into the Schrödinger equation, we get

$$i\hbar\sum_n\left(\dot{a}_n|n\rangle - ia_n\dot{\varphi}_n|n\rangle + a_n|\dot{n}\rangle\right)e^{i\varphi_n(t)} = \sum_n \hat{H}a_n|n\rangle e^{i\varphi_n(t)}.$$

According to the definitions (*), the right-hand side of this equality cancels with the second term on its left-hand side, and the equation, rewritten for the index n', reduces to

$$i\hbar\sum_{n'}(\dot{a}_{n'}|n'\rangle + a_{n'}|\dot{n}'\rangle)\,e^{i\varphi_{n'}(t)} = 0.$$

Now inner-multiplying the left-hand side by $\langle n|\exp\{-i\varphi_n\}$ with an arbitrary n, and using the orthonormality of the kets n at any instant t, we get the following equation for the time evolution of the coefficients a_n:

$$\dot{a}_n = -\sum_{n'} a_{n'} \langle n|\dot{n}'\rangle\, e^{i(\varphi_{n'}-\varphi_n)}.$$

In order to spell out the inner product on the right-hand side of this expression, let us take the partial derivative over time of the first relation in Eq. (*), also rewritten for the index n':

$$\dot{\hat{H}}\, |n'\rangle + \hat{H}\, |\dot{n}'\rangle = \dot{E}_{n'}\, |n'\rangle + E_{n'}\, |\dot{n}'\rangle,$$

where the dot over the Hamiltonian operator means its differentiation only over its explicit time dependence. Inner-multiplying both parts of this equation by $\langle n|$, we get

$$\dot{\hat{H}}_{nn'} + \langle n|\, \hat{H}\, |\dot{n}'\rangle = \langle n|\, \dot{E}_{n'}\, |n'\rangle + \langle n|\, E_{n'}\, |\dot{n}'\rangle, \qquad \text{where } \dot{\hat{H}}_{nn'} \equiv \langle n|\, \dot{\hat{H}}_{nn'}\, |n'\rangle.$$

Both $E_{n'}$ and its time derivative are just (time-dependent) c-numbers, and may be taken out of the corresponding long brackets, so that acting by the Hamiltonian (a Hermitian operator!) in the second term upon the bra-vector on its left, we may reduce this relation to

$$\dot{\hat{H}}_{nn'} + E_n\langle n|\dot{n}'\rangle = \dot{E}_n\langle n|n'\rangle + E_{n'}\langle n|\dot{n}'\rangle.$$

For any $n' \neq n$, the first term on the right-hand side vanishes due to the same orthonormality of the set n, and we get

$$\langle n|\dot{n}'\rangle = -\frac{\dot{\hat{H}}_{nn'}}{E_n - E_{n'}}, \qquad \text{for } n' \neq n.$$

On the other hand, for $n' = n$ we may differentiate over time the normalization condition $\langle n|n\rangle = 1$, getting

$$\langle n|\dot{n}\rangle + \langle \dot{n}|n\rangle \equiv \langle n|\dot{n}\rangle + \langle n|\dot{n}\rangle^* \equiv 2\,\mathrm{Re}\,\langle n|\dot{n}\rangle = 0. \qquad (**)$$

This equality means that $\langle n|\dot{n}\rangle$ is always purely imaginary, i.e. equal to $i\phi(t)$, with a real $\phi(t)$. But we may always select the phase of the ket $|n\rangle$, defined by Eqs. (*), arbitrarily for any time instant, in other words, make the replacement $|n\rangle \rightarrow |n\rangle\exp\{i\Phi(t)\}$ with arbitrary real $\Phi(t)$. At such a replacement, the inner product we are discussing changes as

$$\begin{aligned}\langle n|\dot{n}\rangle &\rightarrow \left(\langle n|\exp\{-i\Phi\}\right)\frac{\partial}{\partial t}\left(|n\rangle \exp\{i\Phi\}\right)\\ &\equiv \left(\langle n|\exp\{-i\Phi\}\right)\left(|\dot{n}\rangle + |n\rangle\, i\dot{\Phi}\right)\exp\{i\Phi\}\\ &\equiv \langle n|\dot{n}\rangle + i\dot{\Phi} \equiv i\left(\varphi + \dot{\Phi}\right).\end{aligned}$$

Hence, with the proper choice of the function $\Phi(t)$, we may always make the product $\langle n|\dot{n}\rangle$ real[27], so that Eq. (**) yields

$$\langle n|\dot{n}\rangle = 0.$$

As a result, we get

$$\dot{a}_n = \sum_{n'\neq n} a_{n'} \frac{\dot{\hat{H}}_{nn'}}{E_n - E_{n'}} e^{i(\varphi_{n'}-\varphi_n)}. \qquad (***)$$

So far, this is an exact result, exactly equivalent to the initial Schrödinger equation. (In this aspect, Eq. (***) is similar to Eqs. (6.84) of the 'usual' perturbation theory, though these two relations are based on different approaches, and each of them is more convenient in its own domain of applications.)

Now let us reduce Eq. (***) to an approximate form for the case when at $t < 0$ the system was definitely is its ground state ($n' = 0$). Then, in the first approximation in $\dot{\hat{H}}$, on the right-hand side of Eqs. (***) we may take $a_0 = 1$, and all other $a_{n'} = 0$, so that

$$\dot{a}_n = \frac{\dot{\hat{H}}_{n0}}{E_n - E_0} e^{i(\varphi_0-\varphi_n)} \equiv \frac{\dot{\hat{H}}_{n0}}{E_n - E_0} \exp\left\{\frac{i}{\hbar}\int_0^t [E_n(t') - E_0(t')]dt'\right\}.$$

(ii) For a 1D harmonic oscillator under the effect of an additional force $F(t)$,

$$\hat{H} = \frac{\hat{p}^2}{2m} + \frac{m\omega_0^2\hat{x}^2}{2} - F(t)\hat{x}, \quad \text{so that } \dot{\hat{H}} \equiv \frac{\partial}{\partial t}\hat{H} = -\dot{F}(t)\hat{x},$$

$$\text{i.e. } \dot{\hat{H}}_{n0} = -\dot{F}(t)x_{n0}.$$

As we know from Eq. (5.92) of the lecture notes, in the harmonic oscillator only one of these matrix elements, $x_{10} = x_0/\sqrt{2}$, is different from zero, so that in the first-order approximation, only the first excited state ($n = 1$) has a non-zero probability amplitude $a_n = a_1$. Taking into account also that for the harmonic oscillator $E_1 - E_0 = \hbar\omega_0 =$ const, we get

$$\dot{a}_1 = -\frac{\dot{F}(t)}{E_1 - E_0}\frac{x_0}{\sqrt{2}} \exp\left\{\frac{i}{\hbar}\int_0^t [E_1(t') - E_0(t')]dt'\right\} \equiv -\frac{x_0}{\sqrt{2}\hbar\omega_0}\dot{F}(t)e^{i\omega_0 t},$$

so that, taking into account that $x_0 = (\hbar/m\omega_0)^{1/2} =$ const, the probability to find the oscillator in the first excited state after the end of the force pulse is

[27] It would be, however, an error to say that $|n\rangle$ itself may be made real in any representation. For example, in the coordinate representation the corresponding wavefunction $\psi_n \equiv \langle \mathbf{r}|n\rangle$ cannot be made real in many cases—see, for example, the wavefunctions (3.129) with $m \neq 0$, or more generally any eigenstate with a nonvanishing density $\mathbf{j}$ of probability current—see Eq. (1.49).

$$W_1(t) \equiv |a_1(t)|^2 = \frac{x_0^2}{2\hbar^2\omega_0^2}\left[\int_0^t \dot{F}(t')\, e^{i\omega_0 t'}dt'\right]^2 \equiv \frac{1}{2\hbar m\omega_0^3}\left[\int_0^t \dot{F}(t')\, e^{i\omega_0 t'}dt'\right]^2.$$

For a finite-time pulse of force, we may always select the time interval $[0, t]$ so broad that $F(t) = 0$ outside of it, and it makes sense to work out the involved integral by parts:

$$\int_{-\infty}^{+\infty} \dot{F}(t)\, e^{i\omega_0 t}dt = \int_{t=-\infty}^{t=+\infty} e^{i\omega_0 t}d[F(t)] = [F(t)e^{i\omega_0 t}]_{t=-\infty}^{t=+\infty} - \int_{t=-\infty}^{t=+\infty} F(t)d(e^{i\omega_0 t})$$
$$\equiv 0 - i\omega_0\int_{t=-\infty}^{t=+\infty} F(t)e^{i\omega_0 t}dt,$$

and rewrite the expression for the final value of W_1 in a more convenient form:

$$W_1 = \frac{|I|^2}{2\hbar m\omega_0}, \qquad \text{with } I \equiv \int_{-\infty}^{+\infty} F(t)e^{i\omega_0 t}dt. \qquad (****)$$

By construction of the first-order approximation, this result is only valid if $W_1 \ll 1$.

(iii) For this particular system (the harmonic oscillator, initially in the ground state), the exact solution, valid for *any* W_1, is also possible. Indeed, as was discussed in the model solution of problem 5.16, if the oscillator was in the ground state (which is one of the Glauber states) initially, it remains in the Glauber state (5.107) at an arbitrary time t, with the time-dependent central point

$$X(t) = \frac{1}{m\omega_0}\int_0^t F(t')\sin\omega_0(t-t')dt', \quad P(t) = \int_0^t F(t')\cos\omega_0(t-t')dt',$$

i.e. with the dimensionless complex amplitude α (defined by Eq. (5.102) of the lecture notes) equal to

$$\alpha(t) \equiv \left(\frac{m\omega_0}{2\hbar}\right)^{1/2}\left[X(t) + i\frac{P(t)}{m\omega_0}\right] = i\left(\frac{1}{2m\hbar\omega_0}\right)^{1/2}\int_0^t F(t)\, e^{i\omega_0(t-t')}dt'$$
$$\equiv i\left(\frac{1}{2m\hbar\omega_0}\right)^{1/2} e^{i\omega_0 t}\int_0^t F(t)\, e^{-i\omega_0 t'}dt'.$$

According to this formula, the value of the Poisson distribution parameter $\langle n\rangle$, defined by Eq. (5.137), after the end of the pulse (formally, at $t = +\infty$) is

$$\langle n\rangle \equiv |\alpha(+\infty)|^2 = \frac{|I|^2}{2m\hbar\omega_0},$$

where I is the same integral as in Eq. (****). Now we may use Eq. (5.135) of the lecture notes to calculate the probability of the oscillator's transfer, by the pulse, into the first excited (Fock) state:

$$W_1 = \langle n\rangle e^{-\langle n\rangle} \equiv \frac{|I|^2}{2m\hbar\omega_0}\exp\left\{-\frac{|I|^2}{2m\hbar\omega_0}\right\}.$$

In the limit of a very small I, this (exact) result reduces to Eq. (****), confirming its correctness.

Problem* 6.17. Use the single-particle approximation to calculate the complex electric permittivity $\varepsilon(\omega)$ of a dilute gas of similar atoms, due to their induced electric polarization by a weak external ac field, for a field frequency ω very close to one of quantum transition frequencies $\omega_{nn'}$. Based on the result, calculate and estimate the absorption cross-section of each atom.

Hint: In the single-particle approximation, an atom's properties are determined by Z similar, non-interacting electrons, each moving in a similar static attracting potential, generally different from the Coulomb one, because it is contributed not only by the nucleus, but also by other electrons.

Solution: According to the complex electric permittivity's definition[28], in the approximation of non-interacting atoms it may be calculated as

$$\varepsilon(\omega) \equiv \frac{\mathscr{D}_\omega}{\mathscr{E}_\omega} = \frac{\varepsilon_0 \mathscr{E}_\omega + \mathscr{n} d_\omega}{\mathscr{E}_\omega} \equiv \varepsilon_0 + \mathscr{n}\frac{d_\omega}{\mathscr{E}_\omega},$$

where $\mathscr{n}$ is the number of atoms per unit volume[29], and d_ω is the complex amplitude of the time-dependent expectation value of the electric dipole moment of one atom,

$$\langle d \rangle = d_\omega e^{-i\omega t} + d_\omega^* e^{i\omega t},$$

induced by the applied weak, classical ac field[30]

$$\mathscr{E} = \mathscr{E}_\omega e^{-i\omega t} + \mathscr{E}_\omega^* e^{i\omega t}.$$

In the single-particle approximation, d may be calculated as Zqx, where $q = -e$ is the single electron charge, Z is the number of electrons per atom, and x is the electron's coordinate along the applied field's direction, so that its expectation value in the time-dependent quantum state $\alpha(t)$ (i.e. in the Schrödinger picture) may be calculated as

$$\langle d \rangle = Zq\langle \alpha(t)| \, \hat{x} \, |\alpha(t)\rangle. \qquad (*)$$

According to the discussion of section 6.5, if the field's frequency ω is very close to that of a quantum transition between two eigenstates, in the expansion of the bra- and ket-vectors of the state α in a series over eigenstate vectors, we may keep only these two terms. At weak applied fields and not extremely high temperatures, the atoms

[28] See, e.g. *Part EM* sections 3.3 and 7.2 (where the electric dipole moment is denoted as **p**).

[29] This fancy font is used here to avoid any possibility of confusion between the volumic density of atoms ($\mathscr{n}$) and the state number (n).

[30] Generally, $\mathscr{E}(t)$and $\langle \mathbf{d}(t)\rangle$ are vectors with different directions, not necessarily proportional to each other—see, e.g. *Part EM* sections 3.1–3.2. Dielectric properties of a medium may be described by a scalar, field-independent function $\varepsilon(\omega)$ only if these vectors are essentially parallel (which is true for any disordered matter) and proportional to each other (which is always true for sufficiently low fields, in the absence of spontaneous polarization). Since the problem assignment asks for such a scalar function, we may assume that these conditions are satisfied.

spend most time in their ground state (due to unavoidable energy relaxation—see chapter 7 for its discussion), so that one of the involved states, with the probability amplitude very close to 1, has to be the ground state of the system. As a result, using Eq. (6.82), we may approximate the state vectors as

$$|\alpha(t)\rangle = |0\rangle \exp\left\{-i\frac{E_0 t}{\hbar}\right\} + a(t)\,|n\rangle \exp\left\{-i\frac{E_n t}{\hbar}\right\},$$

$$\langle\alpha(t)| = \langle 0|\, \exp\left\{i\frac{E_0 t}{\hbar}\right\} + a^*(t)\langle n|\, \exp\left\{i\frac{E_n t}{\hbar}\right\},$$

where the coefficient $a(t)$ is proportional to the applied field, and hence small[31]. Plugging this expression into Eq. (*) and keeping only the terms proportional to a (and hence to $\mathscr{E}_\omega$), we get

$$\langle d\rangle \to Zq\left[a(t)\exp\left\{i\frac{E_0 - E_n}{\hbar}t\right\}x_{0n} + a^*(t)\exp\left\{i\frac{E_n - E_0}{\hbar}t\right\}x_{n0}\right],$$

where x_{0n} and x_{n0} are time-independent matrix elements of the Hermitian operator of the coordinate:

$$x_{0n} \equiv \langle 0|\, \hat{x}\, |n\rangle = \langle n|\, \hat{x}\, |0\rangle^* \equiv x^*_{n0}.$$

Generally, the time evolution of the probability amplitude $a(t)$ of the excited state n has to be found from the system of equations (6.88), with the Hamiltonian amplitude,

$$\hat{A}_\omega = -q\mathscr{E}_\omega\hat{x},$$

corresponding to the perturbation (6.29), with z replaced with x. However, in our resonant case when $\omega \approx \omega_{n0} \equiv (E_n - E_0)/\hbar$, we may reuse the approximate solution of the system, expressed by the first term of Eq. (6.90), with $n' = 0$ and $A_{nn'} = A_{n0} = -q\mathscr{E}_\omega x_{n0}$. As a result, we get

$$\begin{aligned}\langle d\rangle &= Zq[a(t)\exp\{-i\omega_{n0}t\}x_{0n} + a^*(t)\exp\{i\omega_{n0}t\}x_{n0}]\\ &\equiv \frac{ZqA_{n0}}{\hbar(\omega - \omega_{n0})}[\exp\{i(\omega_{0n} - \omega)t\} - 1]\exp\{-i\omega_{n0}t\}x_{0n} + \text{c.c.}\\ &\equiv -\frac{Zq^2\mathscr{E}_\omega}{\hbar(\omega - \omega_{n0})}x_{n0}x_{0n}[\exp\{-i\omega t\} - \exp\{-i\omega_{n0}t\}] + \text{c.c.}\end{aligned}$$

This means that the complex amplitude of the only frequency component we are interested in (changing in time as $\exp\{-i\omega t\}$, i.e. with the same frequency as the field, and hence giving a nonvanishing average contribution to the electric permittivity)[32] is

[31] We will assume $|a| \ll 1$—the relation which should be used to derive the qualitative condition of the electric field smallness, necessary for the validity of our result.

[32] Re-examining the analysis of section 6.5, we may see that the second term in this result, having a different frequency, is the artifact of the zero initial conditions assumed at the sharp turning on of the interaction Hamiltonian—see Eq. (6.86). In real systems with a nonvanishing (if very small) dissipation, this component decays with time, and the whole atomic response retains only one frequency (ω), representing the quantum version of the classical forced oscillations—see, e.g. *Part CM* section 5.1.

$$d_\omega = -\frac{Zq^2 \, |x_{n0}|^2}{\hbar(\omega - \omega_{n0})} \mathscr{E}_\omega = \frac{Zq^2 \, |x_{n0}|^2}{\hbar(\omega_{n0} - \omega)} \mathscr{E}_\omega.$$

Finally, summing the contributions from all quantum transitions (all excited states n), we get the following formal result:

$$\varepsilon(\omega) = \varepsilon_0 + \mathscr{n}Z\frac{q^2}{\hbar}\sum_{n>0}\frac{|x_{n0}|^2}{\omega_{n0} - \omega}, \qquad (**)$$

though it is strictly valid only in the vicinity of any of the transition frequencies ω_{n0} (namely, at $|\omega - \omega_{n0}| \ll |\omega_{n0}|$). It describes odd-resonant, diverging responses of the system near each quantum transition frequency ω_{n0}.

In the particular case when the atom may be modeled by a 1D harmonic oscillator of an eigenfrequency ω_0, we may use Eq. (5.92) to get $|x_{n0}|^2 = (\hbar/2m\omega_0)\delta_{n0}$, and hence the function $\varepsilon(\omega)$, at positive frequencies, has only one such singularity (*pole*):

$$\varepsilon(\omega) = \varepsilon_0 + \mathscr{n}Z\frac{q^2}{2m\omega_0(\omega_0 - \omega)}, \qquad \text{at } \omega \approx \omega_0. \qquad (***)$$

Remarkably, this result exactly coincides with the classical one for a set of $\mathscr{n}Z$ similar harmonic oscillators, with negligible damping, per unit volume[33], illustrating again that the harmonic oscillator is 'the most classical' of all spatially-confined quantum systems, due to the linearity of its Heisenberg equations of motion. (Due to this property, Eq. (***) might be obtained much more simply—just using the fact that due to the linearity of Eqs. (5.36) with $U = m\omega_0^2x^2/2 - q\mathscr{E}x$, the expectation value of the coordinate of the oscillator follows the classical equations of motion.)

Now returning to the general quantum-mechanical result (**), valid for an arbitrary confining potential, we may notice that it has a structure similar to Eq. (***), and may be rewritten in the form inspired by it:

$$\varepsilon(\omega) = \varepsilon_0 + \mathscr{n}Z\frac{q^2}{2m}\sum_{n>0}\frac{f_n}{\omega_{n0}(\omega_{n0} - \omega)},$$

$$\text{where} \quad f_n \equiv \frac{2m}{\hbar}\omega_{n0}\,|x_{n0}|^2 \equiv \frac{2m}{\hbar^2}(E_n - E_0)\,|x_{n0}|^2.$$

Due to the analogy with Eq. (***), the coefficient f_n is called the *oscillator strength* of the atomic excitation to the nth energy level. According to Eq. (***), for a harmonic oscillator only one such of these coefficients, $f_1 = 1$, is different from zero. For an arbitrary atomic system, this is not true, but the sum of all f_n still equals 1—see the solution of problem 5.10.

The above results may leave the impression that the complex electric permittivity dielectric constant is a purely real function of frequency. However, the *Kramers–Kronig dispersion relations*, based on very general causality arguments, and hence valid for the results of the quantum-mechanical analysis as well, show that the real

[33] See, e.g. *Part EM* Eq. (7.32) with n replaced with $\mathscr{n}Z$, $\delta = 0$ and $\omega \to \omega_0$.

and imaginary parts of the function $\varepsilon(\omega) = \varepsilon'(\omega) + i\varepsilon''(\omega)$ are in fact related[34]. In particular, each pole of $\varepsilon'(\omega)$ at a certain frequency corresponds to a proportional delta-function of the imaginary part $\varepsilon''(\omega)$ of this function, which characterizes energy loss in the medium[35]:

$$\frac{1}{\omega_{n0} - \omega} \quad \text{in } \varepsilon'(\omega) \quad \rightarrow \quad \pi\delta(\omega - \omega_{n0}) \quad \text{in } \varepsilon''(\omega).$$

For our particular case, this correspondence yields the following result (for $\omega > 0$):

$$\varepsilon''(\omega) = \varkappa Z\frac{\pi q^2}{2m}\sum_{n>0}\frac{f_n}{\omega_{n0}}\delta(\omega - \omega_{n0}), \qquad (****)$$

describing a series of infinitely narrow resonance peaks of the media's absorption at each quantum transition frequency.

This result may seem mysterious, because at our current description of the system by a Hamiltonian, it should not, apparently, have any energy loss. Indeed, this result, of the same class as the Golden Rule discussed in sections 6.6–6.7 of the lecture notes[36], is one of the deepest (and hence most beautiful) results of quantum mechanics, which essentially preempts the analysis of open quantum systems. Such analysis, to be discussed in chapter 7, describes the physics of energy loss by the system's coupling to its environment, and in particular shows that as the coupling tends to zero, the finite-width peaks of the dissipative functions similar to $\varepsilon''(\omega)$ become infinitely narrow, but retain their 'areas', i.e. the delta-functions' weights given by Eq. (****).

The result (****) may be readily re-calculated into the absorption cross-section σ of one atom, which may be defined by the following (hopefully, self-explanatory) relation for the time-averaged power of the incident wave per unit area of its front[37]:

$$\frac{d\overline{\mathscr{P}}}{dz} = -\varkappa\sigma\overline{\mathscr{P}},$$

where z is the direction of the wave propagation. Namely, elementary macroscopic electrodynamics[38] tells us that the same power gradient of a monochromatic wave of frequency ω may be calculated as

$$\frac{d\overline{\mathscr{P}}}{\overline{\mathscr{P}}} = -2k''(\omega)dz,$$

where $k''(\omega)$ is the imaginary part of the complex wave number $k(\omega) = \omega[\varepsilon(\omega)\mu(\omega)]^{1/2}$.

[34] See, e.g. *Part EM* section 7.2.

[35] See, e.g. *Part EM* Eqs. (7.55) and (7.56).

[36] Actually, all the essential physics of the relation between the reversible and irreversible quantum dynamics is described already by the simple metastable state model discussed in the end of section 2.5, and the reader having conceptual issues with this relation may be referred there for an additional thoughtful review.

[37] This expression implies incoherent addition of the energies absorbed by each atom, which is adequate for a diluted gas, with random interatomic distances.

[38] See, e.g. *Part EM* section 7.9, in particular Eqs. (7.215) and (7.216).

For a non-magnetic medium (with $\mu(\omega) = \mu_0$), with relatively weak absorption ($\varepsilon''(\omega) \ll \varepsilon'(\omega) = \varepsilon_0$), we may Taylor-expand this expression for k:

$$k \equiv k' + ik'' = \omega\left[(\varepsilon_0 + i\varepsilon'')\mu_0\right]^{1/2}$$

$$\equiv \omega\left[\varepsilon_0\left(1 + i\frac{\varepsilon''}{\varepsilon_0}\right)\mu_0\right]^{1/2} \approx \omega(\varepsilon_0\mu_0)^{1/2}\left(1 + i\frac{\varepsilon''}{2\varepsilon_0}\right),$$

thus getting

$$2k''(\omega) = \omega\left(\frac{\mu_0}{\varepsilon_0}\right)^{1/2}\varepsilon''(\omega) \equiv \frac{\omega}{\varepsilon_0 c}\varepsilon''(\omega).$$

Now comparing the two above expressions for the power loss per unit length, we get

$$\sigma = \frac{1}{n}\frac{\omega}{\varepsilon_0 c}\varepsilon''(\omega),$$

so that using Eq. (****), with $q = -e$, we finally obtain

$$\sigma = Z\frac{\pi q^2}{2m\varepsilon_0 c}\sum_{n>0} f_n \delta(\omega - \omega_{n0}) \equiv 4\pi^2 Z\alpha\omega\sum_{n>0}|x_{n0}|^2\,\delta(\omega - \omega_{n0}),$$

where $\alpha \equiv (e^2/4\pi\varepsilon_0)/\hbar c \approx 1/137$ is the fine-structure constant.

Since the square of the coordinate matrix element is typically of the order of the 'physical' cross-section σ_0 of the atom, this result shows that the frequency-averaged cross-section,

$$\bar{\sigma} \sim \frac{1}{\omega_{n0}}\int_{\Delta\omega\sim\omega_{n0}} \sigma d\omega,$$

is of the order of $\alpha Z\sigma_0$, i.e. is much smaller than σ_0 for not too heavy atoms. This relation gives one more illustration of the meaning of the fine-structure constant α as a measure of the electromagnetic interaction weakness in quantum mechanics.

***Problem* 6.18.** Use the solution of the previous problem to generalize the expression for the London dispersion force between two atoms (whose calculation in the harmonic oscillator model was the subject of problems 3.16 and 5.15) to the single-particle model with an arbitrary energy spectrum.

Solution: The result obtained in problem 5.15 for a single-oscillator model of each atom,

$$U = -\frac{3q^4\hbar}{2(4\pi\varepsilon_0)^2 m^2 r^6}\frac{1}{\omega_1\omega_2(\omega_1 + \omega_2)},$$

may be readily generalized to the modeling of each atom by a set of isotropic 3D harmonic oscillators, with eigenfrequencies ω_n and relative numbers f_n:

$$U = -\frac{3q^4\hbar}{2(4\pi\varepsilon_0)^2 m^2 r^6}\sum_{n,n'}\frac{f_n f_{n'}}{\omega_n \omega_{n'}(\omega_n + \omega_{n'})}.$$

Now using the fact that according to the solution of problem 5.15, this formula is based on the same resonance response as was discussed in the solution of the previous problem, we may generalize it further to describe the long-range interaction between two arbitrary (but still isotropic) single-particle systems:

$$U = -\frac{3q^4\hbar}{2(4\pi\varepsilon_0)^2 m^2 r^6}\sum_{n,n'>0}\frac{f_n f_{n'}}{\omega_{n0}\omega_{n'0}(\omega_{n0} + \omega_{n'0})},$$

where the frequency sets $\omega_{n0} \equiv (E_n - E_0)/\hbar$ and $\omega_{n'0} \equiv (E_{n'} - E_0)/\hbar$ describe the excitation spectra of the counterpart atoms, and f_n and $f_{n'}$ are oscillator strengths of the corresponding transitions:

$$f_n \equiv \frac{2m}{\hbar}\omega_{n0}\,|x_{n0}|^2.$$

(Due to the assumed isotropy, the matrix elements x_{n0} are the same for all coordinates.)

This was the main result obtained in 1930 by F London. Its main restriction is that it assumes the instant propagation of the dipole's electric field by the inter-atomic distance r. This assumption (which is clearly made in the model solutions of problems 3.16 and 5.15) is valid only if $k_{n0}r \ll 1$, where the wave number k_{n0} equals ω_{n0}/c in vacuum, and may be somewhat higher in a dense medium. Since the typical frequencies ω_{n0} for atoms and molecules are rather high ($\sim 10^{16}$ s^{-1}), noticeable deviations from London's result may start already from $r \sim 1$ μm. The later extensions of this theory to arbitrary values of $k_{n0}r$, notably by H Casimir and D Polder in 1948, and by E Lifshitz in 1956,[39] have shown a remarkable (and somewhat counter-intuitive) connection between the London dispersion force and the fundamental *Casimir effect*—for its discussion, see section 9.1 of the lecture notes.

***Problem* 6.19.** Use the solution of the previous problem to express the potential energy of interaction of two hydrogen atoms, both in their ground state and separated by distance $r \gg r_B$, in the simplest possible analytical form, and use the result to estimate the energy.

Solution: For the effective potential energy of the far-range interaction of two similar atoms, we may use the solution of the previous problem, with $f_n = f_{n'}$ and $\omega_{n0} = \omega_{n'0}$:

[39] For a comprehensive review of this topic see, e.g. [3].

$$U = -\frac{3}{4}\left(\frac{q^2}{4\pi\varepsilon_0}\right)^2 \frac{\hbar}{m^2 r^6} \sum_{n>0} \frac{f_n^2}{\omega_{n0}^3} \equiv -3\left(\frac{q^2}{4\pi\varepsilon_0}\right)^2 \frac{1}{\hbar r^6} \sum_{n>0} \frac{|x_{n0}|^4}{\omega_{n0}}.$$

For a hydrogen atom, $q = -e$, and $m = m_e$, so that according to the Eq. (1.9), $(q^2/4\pi\varepsilon_0)^2 = (e^2/4\pi\varepsilon_0)^2 = (\hbar^2/m_e)E_H$, where $E_H \approx 27.2$ eV is the Hartree energy unit, and we may rewrite U in a form more convenient for our purposes:

$$U = -E_H \frac{3\hbar}{m_e r^6} \sum_{n>0} \frac{|x_{n0}|^4}{\omega_{n0}}. \qquad (*)$$

Due to the atom's isotropy, we may replace the matrix elements x_{n0} with the elements[40]

$$z_{n1} \equiv \langle nlm|\, \hat{z}\, |100\rangle = \langle nlm|\, r\cos\theta\, |100\rangle = I_\Omega I_r,$$

with the same moduli, where

$$I_\Omega \equiv \oint d\Omega [Y_l^m(\theta, \varphi)]^* \cos\theta \;\; Y_0^0(\theta, \varphi), \quad \text{and } I_r \equiv \int_0^\infty r^2 dr\, \mathcal{R}_{nl}^*(r) r \mathcal{R}_{1,0}(r).$$

Since $\cos\theta \equiv P_1(\cos\theta)$, Eq. (3.170) of the lecture notes shows[41] that all integrals I_Ω vanish besides one, with $l = 1$ and $m = 0$:[42]

$$I_\Omega = 2\pi \int_0^\pi \sin\theta d\theta \left(\frac{3}{4\pi}\right)^{1/2} \cos\theta\, \cos\theta \left(\frac{1}{4\pi}\right)^{1/2}$$
$$= \frac{\sqrt{3}}{2} \int_{-1}^{+1} \cos^2\theta d(\cos\theta) = \frac{1}{\sqrt{3}}.$$

As a result, Eq. (*) is reduced to

$$U = -\frac{1}{3}\frac{\hbar^2}{m_e}\frac{1}{r^6} E_H \sum_{n>1} \frac{|I_n|^4}{\hbar\omega_{n0}}, \qquad \text{with} \quad I_n \equiv \int_0^\infty \mathcal{R}_{n,1}^*(r)\mathcal{R}_{1,0}(r)\, r^3 dr,$$

where, according to Eq. (1.8) of the lecture notes,[43]

$$\hbar\omega_{n,0} = E_H\left[-\frac{1}{2n^2} - \left(-\frac{1}{2\cdot 1^2}\right)\right] \equiv \frac{\hbar^2}{2m_e r_B^2}\left(1 - \frac{1}{n^2}\right),$$

so that, finally

[40] Note that due to the specific (traditional) choice of the principal quantum numbers n, equal to 1 rather than 0 for the ground state, the index '0' in the general formulas of the perturbation theory has to be understood as the set $\{n = 1, l = 0, m = 0\}$.

[41] This fact is even more obvious from the explicit form of the lowest-order spherical functions—see Eqs. (3.174)–(3.176).

[42] Actually, this is just one more manifestation of the selection rules for the electric-dipole transitions, $\Delta l = 1$, $\Delta m = 0, \pm 1$, which was repeatedly discussed in this course—see in particular problem 5.35.

[43] Alternatively, see Eq. (3.201) with $C = e^2/4\pi\varepsilon_0$.

$$U = -\frac{2}{3}E_H\left(\frac{r_B}{r}\right)^6 \sum_{n>1} \frac{(|I_n|/r_B)^4}{1 - 1/n^2}. \qquad (**)$$

For the first (and the largest) term of the sum, with $n = 2$, the integral I_n may be readily calculated using Eqs. (3.208)–(3.209) with $r_0 = r_B$, and the table integral (A.34*d*):

$$I_2 = \int_0^\infty \frac{1}{(2r_B)^{3/2}} \frac{r}{3^{1/2} r_B} \exp\left\{-\frac{r}{2r_B}\right\} \frac{2}{r_B^{3/2}} \exp\left\{-\frac{r}{r_B}\right\} r^3 dr$$
$$\equiv \frac{r_B}{6^{1/2}}\left(\frac{2}{3}\right)^5 \int_0^\infty \xi^4 e^{-\xi} d\xi = \frac{r_B}{6^{1/2}}\left(\frac{2}{3}\right)^5 4! \equiv \frac{2^{15/2}}{3^{9/2}} r_B,$$

so that the first term of the sum in Eq. (**) is $(2^{15/2}/3^{9/2})^4/(1 - 1/2^2) = 2^{32}/3^{20} \approx 3.69$.

In order to calculate I_n for higher values of n we would need to use the heavy artillery of Eqs. (3.195)–(3.197), but since the radial wavefunctions $\mathcal{R}_{n,l}$ in the integrals I_n are normalized in the sense of Eq. (3.194), and the scale of the spatial extension of the product $\mathcal{R}_{n,1}\mathcal{R}_{1,0}$ is $[n/(n + 1)]r_B$, all these integrals are of the order of r_B, decreasing fast with n—compare, for example, the blue-line plots for $\mathcal{R}_{n,1}$ with $n = 2$ and $n = 3$ in figure 3.22. As a result, the whole sum in Eq. (**) is not much larger than its first term, and hence, taking into account the front coefficient (2/3), the interaction energy may be fairly estimated as

$$U \approx -3E_H\left(\frac{r_B}{r}\right)^6.$$

Since our calculations are based on a perturbative approach, this result is only valid if $-U \ll E_H$, i.e. $r \gg r_B$.

***Problem* 6.20.** In a certain quantum system, distances between three lowest energy levels are slightly different—see the figure below ($|\xi| \ll \omega_{1,2}$). Assuming that the involved matrix elements of the perturbation Hamiltonian are known, and are all proportional to the external ac field's amplitude, find the time necessary to populate the first excited level almost completely (with a given precision $\varepsilon \ll 1$), using the Rabi oscillation effect, if at $t = 0$ the system is completely in its ground state. Spell out your result for a weakly anharmonic oscillator.

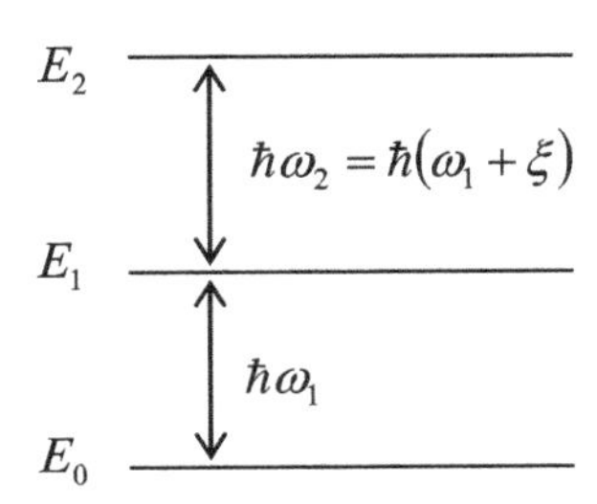

Solution: Since $\varepsilon = 1 - (W_1)_{\max} \ll 1$, the undesirable population of the second excited level, $W_2(t) < \varepsilon$, is small at all times, so that we may first calculate $W_1(t)$ ignoring the probability amplitude a_2, i.e. exactly as was done in the two-level approximation in section 6.5 of the lecture notes. For the precise tuning of the external excitation frequency, $\omega = \omega_1$ (which is necessary to approach $(W_1)_{\max} = 1$), i.e. for $\Delta \to \Delta_{10} = 0$, Eq. (6.99) with $A = A_{10}$ gives $\Omega = |A_{10}|/\hbar$, we may use Eq. (6.100) to get

$$b_1(t) = -i \sin \frac{|A_{10}|\, t}{\hbar} \equiv -i \sin \zeta t, \quad \text{where } \zeta \equiv \frac{|A_{10}|}{\hbar},$$

so that from the first of Eqs. (6.96), also with $\Delta = 0$, we obtain

$$a_1(t) = -i \sin \zeta t. \tag{*}$$

Now, neglecting the effect of the small amplitude $a_2(t)$ of the second excited state on the function a_1(t), we may plug Eq. (*) into the first of Eqs. (6.94) written for $n = 2$, $n' = 1$, and hence with $\Delta = \Delta_{21} \equiv \omega - \omega_2 = -\xi$, and $A = A_{21}$;[44]

$$\begin{aligned} i\hbar \dot{a}_2 &= a_1(t) A_{21} e^{i\xi t} = -iA_{21} \sin \zeta t\, e^{i\xi t} \\ &\equiv -\frac{A_{21}}{2}[\exp\{i(\xi + \zeta)t\} - \exp\{i(\xi - \zeta)t\}]. \end{aligned}$$

Its integration (with the initial condition $a_2(0) = 0$) is elementary, and yields

$$a_2(t) = \frac{A_{21}}{2\hbar}\left[\frac{\exp\{i(\xi + \zeta)\, t\} - 1}{\xi + \zeta} - \frac{\exp\{i(\xi - \zeta)\, t\} - 1}{\xi - \zeta}\right].$$

According to Eq. (*), the occupation $W_1(t) = |a_1(t)|^2 = |\sin \zeta t|^2$ of the first excited state reaches its maximum periodically, at the moments

$$t_m = \frac{\pi(m + ½)}{\zeta}, \qquad \text{when} \exp\{i\zeta t_m\} = \pm i, \tag{**}$$

where the sign depends on whether the integer m is even (+) or odd (−). At these moments,

$$a_2(t_m) = \frac{A_{21}}{\hbar}\frac{\pm i\xi \exp\{i\xi t_m\} + \zeta}{\xi^2 - \zeta^2},$$

From here, the second level's occupation at these moments is

$$W_2(t_m) = |a^2(t_m| = \frac{|A_{21}|^2}{\hbar^2}\frac{\xi^2 \mp 2\xi\zeta \sin \zeta t_m + \zeta^2}{|\xi^2 - \zeta^2|^2}.$$

[44] At $\xi \to 0$, the direct excitation of the second-level state from the initial one is negligible in comparison with a nearly-resonant excitation of that level from the first excited state.

Generally, the frequencies ξ and ζ are incommensurate, so that the sine function in this result may take any values between −1 and +1. We are interested in the *smallest* possible W_2, achieved at the lower limit, so that we may write

$$\min_m[W_2(t_m)] = \frac{|A_{21}|^2}{\hbar^2}\frac{\xi^2 - 2\xi\zeta + \zeta^2}{(\xi^2 - \zeta^2)^2} \equiv \frac{|A_{21}|^2}{\hbar^2}\frac{1}{|\xi + \zeta|^2}.$$

From the requirement for this expression is not larger than the given $\varepsilon \ll 1$, and the proportionality of both $|A_{10}|$ and $|A_{21}|$ to the same ac excitation amplitude, the minimum occupancy inversion time is achieved at $|\zeta| \ll |\xi|$:

$$(t_m)_{\min} = \frac{\pi\hbar}{2\,|A_{10}|_{\max}} = \frac{\pi}{2\varepsilon^{1/2}\,|\xi|}\left|\frac{A_{21}}{A_{10}}\right|. \qquad (***)$$

The time tends to infinity at $\varepsilon \to 0$ (the perfect inversion requirement) and/or $\xi \to 0$ (equidistant energy levels, such as in a harmonic oscillator). For a slightly anharmonic 1D oscillator, excited by an external classical force $F(t)$, the perturbation Hamiltonian is

$$\hat{H}^{(1)}(t) = -F(t)\hat{x},$$

and the matrix element ratio that participates in our result may be calculated from Eq. (5.92) of the lecture notes:

$$\left|\frac{A_{21}}{A_{10}}\right| = \left|\frac{x_{21}}{x_{10}}\right| = \left|\frac{\langle 2|\,\hat{x}\,|1\rangle}{\langle 1|\,\hat{x}\,|0\rangle}\right| = \sqrt{2},$$

while the detuning ξ may be evaluated using Eqs. (6.16) and (6.23):

$$\xi \equiv \omega_2 - \omega_1 \equiv \frac{(E_2 - E_1) - (E_1 - E_0)}{\hbar}$$

$$= 3\frac{\beta x_0^4}{\hbar} - \frac{15}{2}\frac{\alpha x_0^6}{\hbar^2\omega_0} \equiv 3\frac{\beta\hbar}{m^2\omega_0^2} - \frac{15}{2}\frac{\alpha\hbar}{m^3\omega_0^4},$$

where α and β are the anharmonicity coefficients defined by Eq. (6.2). Evidently, the smaller the coefficients (or rather the magnitude of their combination on the right-hand side of the last expression), the larger is the shortest population inversion time (***). This trend will be used in section 8.5 of the lecture notes to explain why adding external (linear) circuit elements to Josephson-junction qubits, while having a beneficial effect of decreasing their coupling to environment, and hence increasing the dephasing time, creates a problem of the quantum computing speed reduction.

Problem* 6.21. Analyze the possibility of a slow transfer of a system from one of its energy levels to another one (in the figure below, from level 1 to level 3), using the scheme shown in that figure, in which the monochromatic external excitation amplitudes A_+ and A_- may be slowly changed at will.

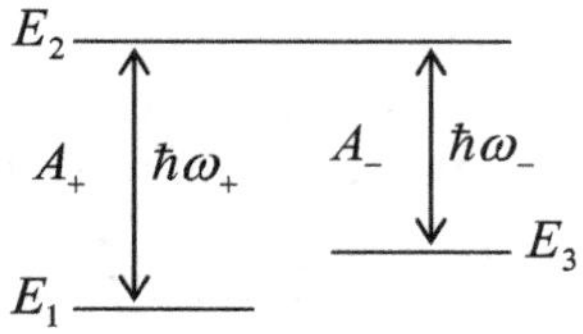

E_2, E_1, E_3, A_+, $\hbar\omega_+$, A_-, $\hbar\omega_-$

Solution: Assuming, for the sake of simplicity, the exact tuning of the excitation frequencies[45],

$$\hbar\omega_+ = E_2 - E_1, \quad \hbar\omega_- = E_2 - E_3,$$

and ignoring, for time being, the slow time evolution of the amplitudes A_+ and A_-, we may write the following evident generalization of Eqs. (6.97), with $\Delta = 0$, for this system:

$$i\hbar\dot{b}_1 = A_+ b_2, \qquad i\hbar\dot{b}_2 = A_+^* b_1 + A_-^* b_3, \qquad i\hbar\dot{b}_3 = A_- b_2. \qquad (*)$$

Looking for the partial solution of this system of three homogeneous linear differential equations in the usual form $\exp\{\lambda t\}$, we get the following characteristic equation:

$$\begin{vmatrix} -i\hbar\lambda & A_+ & 0 \\ A_+^* & -i\hbar\lambda & A_-^* \\ 0 & A_- & -i\hbar\lambda \end{vmatrix} = 0, \qquad \text{i.e. } -i\hbar\lambda[(-i\hbar\lambda)^2 - A_+A_+^* - A_-A_-^*] = 0.$$

This equation has three roots: two of them, $\lambda_\pm = \pm i\Omega$ corresponding to the Rabi oscillation half-frequency

$$\Omega = \frac{1}{\hbar}\left(A_+A_+^* + A_-A_-^*\right)^{1/2} \qquad (**)$$

(this expression is an evident generalization of Eq. (6.99) of the lecture notes, with $\Delta = 0$), and one more root, $\lambda_0 = 0$. The mathematical origin of the last result is clearly visible from the comparison of the first and the last equations (*): it shows that if $b_2 \neq 0$, then

$$\frac{\dot{b}_1}{A_+} = \frac{\dot{b}_3}{A_-},$$

i.e. that the system of equations has the following integral of motion:

$$A_- b_1 - A_+ b_3 = C = \text{const}. \qquad (***)$$

Physically, the origin of this relation is that in this three-level system, the probability amplitudes b_1 and b_3 may change only due to b_2, and at fixed $A_\pm$, these changes are proportional.

[45] The procedure discussed below is more tolerant to a small common detuning $\Delta \equiv E_2 - (E_1 + \hbar\omega_+) = E_2 - (E_3 + \hbar\omega_-)$, than to the relative detuning $\delta \equiv (E_1 + \hbar\omega_+) - (E_3 + \hbar\omega_-)$—see, e.g. figure 10 in the review by N Vitanov *et al* [4].

It is intuitively clear that Eqs. (*) have to hold if the excitation amplitudes A_+ and A_- are changed in time sufficiently slowly ('adiabatically')—actually, much slower than the Rabi frequency Ω. As a result, they may be used to describe the following counter-intuitive operation. First, starting with the one of the energy levels populated (say, $b_1 = 1$), but two other energy levels 2 and 3 empty ($b_2 = b_3 = 0$), let us turn on the external field of frequency ω_-, i.e. make A_-^* to equal some $A_0 \neq 0$, thus making the last two of Eqs. (*) valid—so far with $A_+ = 0$.[46] However, since $b_2 = 0$, Eq. (***) is still *not* valid. Indeed, as the first of Eqs. (*) shows, the state corresponding to the energy level E_1 is still fully uncoupled from the (now, Rabi-coupled) 'dark' quantum states of the levels E_2 and E_3.

Now let us slowly increase the amplitude A_+ of the external field ω_+, simultaneously decreasing A_-, so that the Rabi frequency (**) would stay constant, equal to $A_0/\hbar$; in this case, Eq. (***), now valid, may be conveniently rewritten as

$$b_1 \cos\theta - b_3 \sin\theta = c = \text{const}, \qquad \text{where } \cos\theta = \frac{A_-}{A_0}, \qquad \sin\theta = \frac{A_+}{A_0}. \qquad (****)$$

At some, rather small, value of A_+ (which depends on the rate of its increase) the Rabi effect makes the probability amplitude b_2 sufficiently different from zero to make Eq. (****) valid, thus establishing the integration constant c. Since at this moment b_1 is still very close to 1, and A_- to A_0, i.e. the angle θ to 0, this constant c is close to 1. According to Eq. (****), as this slow process is continued, with the parameter θ being moved from 0 to $\pi/2$, the variable $|b_3|$, and hence the occupancy $W_3 = |b_3|^2$ of the level E_3 gradually increases, while $W_1 = |b_1|^2$ decreases. Finally, at some small value $A_- \ll A_0$, i.e. at $\theta \approx \pi/2$, when Eq. (****) is still valid, it yields $b_3 \approx -1$, and $b_1 \approx 0$, i.e. $W_3 \approx 1$, while $W_1 \approx 0$. At this point, a full turn-off of the field of frequency ω_- completely isolates the now-occupied state on level E_3 from the other two states (which now form another pair of Rabi-coupled 'dark' states, with $W_{1,2} \ll 1$). After that, the excitation of frequency ω_+ may be turned off without any effect on these (zero) occupancies.

So, using the Rabi coupling of two initially empty ('dark') states *first*, and only *then* turning on the coupling of this pair with initially occupied state, we may perform a virtually complete, adiabatic transfer of the system from one energy level to another one[47]. In contrast to the direct π-pulse of Rabi-oscillations between the initial and final levels, this procedure does not require exact timing. Due to this advantage, the STIRAP process, initially just a quantum curiosity, is now finding more and more applications in atomic, molecular, and solid-state physics and chemistry—see, e.g. the already cited review by Vitanov *et al.* [4].

---

[46] Such Rabi-connected, but unoccupied energy levels are frequently called *dark states*.

[47] Because of its counter-intuitive nature, this *STIRAP* (*Stimulated Raman Adiabatic Passage*) procedure was invented, by U Gaubatz *et al*, only in 1988, i.e. only after four decades of studies of various Rabi-oscillation effects by many research groups. It is even more curious that the opposite, more apparent time sequence of the external ac field increase gives worse transfer results, in particular because of its higher sensitivity to unintentional (but unavoidable) coupling to environment, to be discussed in chapter 7.

***Problem* 6.22.** A weak external force pulse $F(t)$, of a finite time duration, is applied to a 1D harmonic oscillator that initially was in its ground state.

(i) Calculate, in the lowest nonvanishing order of the perturbation theory, the probability that the pulse drives the oscillator into its lowest excited state.
(ii) Compare the result with the exact solution of the problem.
(iii) Spell out the perturbative result for a Gaussian-shaped waveform,

$$F(t) = F_0 \exp\{-t^2/\tau^2\},$$

and analyze its dependence on the time scale τ of the pulse.

Solutions:

(i) The general approach to such problems is given by the set of (exact!) Eqs. (6.84) of the lecture notes. As was argued at the derivation of Eq. (6.90) (for a specific time dependence of the perturbation), in the lowest order of the perturbation theory we may leave, on the right-hand sides of these equations, only the terms whose probability amplitudes $a_{n'}$ are initially different from zero. In our current problem this is only the ground-state amplitude, $a_0 = 1$, so that the right-hand side of each Eq. (6.84), with $n' = 0$, is reduced to just one term:

$$i\hbar\dot{a}_n = H_{n0}^{(1)}(t)\, e^{i\omega_{n0}t}. \qquad (*)$$

Just as was repeatedly discussed in this course, a weak, coordinate-independent force $F(t)$ may be described by the following Hamiltonian's perturbation,

$$\hat{H}^{(1)}(t) = -F(t)\hat{x},$$

so that Eq. (*) takes the form

$$i\hbar\dot{a}_n = -F(t)x_{n0}e^{i\omega_{n0}t}, \qquad (**)$$

where x_{n0} are the matrix elements of the coordinate operator in the unperturbed brackets—see Eq. (6.8). As Eq. (5.92) of the lecture notes shows, for a harmonic oscillator all such matrix elements equal zero, with just one exception:

$$x_{n0} = \left(\frac{\hbar}{2m\omega_0}\right)^{1/2} \times \begin{cases} 1, & \text{for } n = 1, \\ 0, & \text{otherwise,} \end{cases}$$

so that only one of all Eqs. (*), with $n = 1$, has a nonvanishing right-hand side:

$$i\hbar\dot{a}_1 = -\left(\frac{\hbar}{2m\omega_0}\right)^{1/2} F(t)\, e^{i\omega_0 t},$$

where Eq. (5.75) has been used to spell out the quantum transition frequency: $\omega_{10} \equiv (E_1 - E_0)/\hbar = \omega_0$.

This differential equation for amplitude $a_1(t)$, with the initial condition $a_1(-\infty) = 0$, may be readily integrated to find its final value:

$$a_1(+\infty) = i\left(\frac{1}{2m\hbar\omega_0}\right)^{1/2} \int_{-\infty}^{+\infty} F(t)\, e^{i\omega_0 t} dt.$$

From here, the probability of the oscillator's excitation by the pulse is

$$W_1(+\infty) \equiv |a_1(+\infty)|^2 = \frac{|I|^2}{2m\hbar\omega_0}, \quad \text{where } I \equiv \int_{-\infty}^{+\infty} F(t)\, e^{i\omega_0 t} dt. \qquad (***)$$

This integral converges for any pulse of a finite duration, i.e. if $F(t) \to 0$ at $t \to \pm\infty$.

(ii) Due to our initial assumptions, the above formulas are only valid if $|a_1(t)|^2 \ll 1$ for all t, in particular, if $W_1(+\infty) \ll 1$, i.e. if

$$|I|^2 \ll 2m\hbar\omega_0. \qquad (****)$$

As was discussed in the model solution of problem 6.16, for this particular system (the harmonic oscillator, initially in the ground state), the exact solution, valid for *any* W_1, is also possible:

$$W_1 = \langle n \rangle e^{-\langle n \rangle} \equiv \frac{|I|^2}{2m\hbar\omega_0} \exp\left\{-\frac{|I|^2}{2m\hbar\omega_0}\right\}.$$

If the condition (****) is satisfied, this formula duly reduces to the perturbative result (***).

(iii) For the particular pulse shape given in the assignment, I is a standard Gaussian integral, which may be readily worked out as was discussed in section 2.2 of the lecture notes—see Eqs. (2.21)–(2.23):

$$\begin{aligned} I &= F_0 \int_{-\infty}^{+\omega} \exp\left\{-\frac{t^2}{\tau^2} + i\omega_0 t\right\} dt = F_0 \exp\left\{\left(\frac{-i\omega_0\tau}{2}\right)^2\right\} \\ &\times \int_{-\infty}^{+\infty} \exp\left\{-\left[\left(\frac{t}{\tau}\right)^2 + 2\left(\frac{t}{\tau}\right)\left(\frac{-i\omega_0\tau}{2}\right) + \left(\frac{-i\omega_0\tau}{2}\right)^2\right]\right\} dt \\ &= F_0 \exp\left\{-\frac{\omega_0^2\tau^2}{4}\right\} \int_{-\infty}^{+\infty} \exp\left\{-\left[\frac{t}{\tau} + \frac{-i\omega_0\tau}{2}\right]^2\right\} dt \\ &= \pi^{1/2} F_0 \tau \exp\left\{-\frac{\omega_0^2\tau^2}{4}\right\}, \end{aligned}$$

so that, finally, the probability of the oscillator's excitation is

$$W_1(+\infty) = \frac{|I|^2}{2m\hbar\omega_0} = \frac{\pi F_0^2\tau^2}{2m\hbar\omega_0} \exp\left\{-\frac{\omega_0^2\tau^2}{2}\right\}.$$

According to this formula, at fixed other parameters, the excitation is most effective if the pulse's duration is of the order of oscillator's period: $\tau_{\text{opt}} = \sqrt{2}/\omega_0$. This is natural, because a much shorter pulse does not give the system enough time to accomplish the interlevel quantum transition, while a very long pulse is just an adiabatic change of an oscillator's parameter (namely, of its equilibrium position $X_0(t) = F(t)/m\omega_0^2$) and, according to the discussion in the beginning of section 6.5 of the lecture notes (and in the solution of problem 6.16), leaves the system in the initial quantum state.

***Problem* 6.23.** A spatially-uniform, but time-dependent external electric field $\mathscr{E}(t)$ is applied, starting from $t = 0$, to a charged plane rotator, initially in its ground state.

(i) Calculate, in the lowest nonvanishing order in the field's strength, the probability that by time $t > 0$, the rotator is in its nth excited state.
(ii) Spell out and analyze your results for a constant-magnitude field rotating, with a constant angular velocity ω, within the rotator's plane.
(iii) Do the same for a monochromatic field of frequency ω, with a fixed polarization.

Solutions:

(i) Acting exactly as in section 6.5 of the lecture notes (see also the model solution of the previous problem), let us solve Eq. (6.84), with the appropriate notation replacement $n \to m$, in the first perturbation order, by taking all probability amplitudes $a_{m'}(t)$ equal to zero in the right hand part, except that of the ground state: $a_0(t) = 1$ for all t. The resulting (approximate) equation is

$$i\hbar\dot{a}_m = H^{(1)}_{m0}(t)\exp\{i\omega_{m0}t\}, \quad \text{with} \quad \omega_{m0} = \frac{E_m - E_0}{\hbar}, \quad \text{for } m \neq 0,$$

where $H^{(1)}_{m0}$ are the matrix elements of the perturbation created by the field, in the unperturbed-state basis of the rotator. This equation may be readily integrated, with zero initial conditions $a_m(0) = 0$, to give

$$a_m(t) = \frac{1}{i\hbar}\int_0^t H^{(1)}_{m0}(t')\exp\{i\omega_{m0}t'\}dt', \quad \text{for } m > 0.$$

Hence the probability to find the system, at a moment $t > 0$, in its mth excited state is

$$W_m(t) = \frac{1}{\hbar^2}\left|\int_0^t H^{(1)}_{m0}(t')\exp\{i\omega_{m0}t'\}dt'\right|^2. \qquad (*)$$

Now let us specify the matrix elements participating in this general expression. The perturbation Hamiltonian created by a uniform electric plane applied within the rotator's plane (say, [x, y]) is a natural generalization of Eq. (6.29):

$$\hat{H}^{(1)}(t) = -q\boldsymbol{\rho}\cdot\mathscr{E}(t) = -q\left[x\mathscr{E}_x(t) + y\mathscr{E}_y(t)\right] = -qR\left[\cos\varphi\;\mathscr{E}_x(t) + \sin\varphi\;\mathscr{E}_y(t)\right].$$

In the basis of the stationary states of the rotator, with the wavefunctions given by Eq. (3.129) of the lecture notes (with the evident normalization $|C_m|^2 = 1/2\pi$):

$$\psi_m = \frac{1}{(2\pi)^{1/2}} e^{im\varphi}, \quad \text{with} \quad m = 0, \pm 1, \pm 2, \dots, \qquad (**)$$

the matrix elements participating in Eq. (*) are

$$H^{(1)}_{mo}(t) = -qR\int_0^{2\pi} \psi^*_m\left[\cos\varphi\, \mathscr{E}_x(t) + \sin\varphi\, \mathscr{E}_y(t)\right]\psi_0 d\varphi$$

$$\equiv -\frac{qR}{2\pi}\int_0^{2\pi} e^{-im\varphi}\left[\frac{e^{i\varphi}+e^{-i\varphi}}{2}\mathscr{E}_x(t) + \frac{e^{i\varphi}-e^{-i\varphi}}{2i}\mathscr{E}_y(t)\right]d\varphi$$

$$\equiv -\frac{qR}{4\pi}\left\{\left[\mathscr{E}_x(t) - i\mathscr{E}_y(t)\right]\int_0^{2\pi} e^{i(1-m)\varphi}d\varphi + \left[\mathscr{E}_x(t) + i\mathscr{E}_y(t)\right]\int_0^{2\pi} e^{i(-1-m)\varphi}d\varphi\right\}.$$

This expression shows that only the matrix elements with $m = \pm 1$ are different from zero[48]:

$$H^{(1)}_{\pm 1,0}(t) = -\frac{qR}{2}\left[\mathscr{E}_x(t) \mp i\mathscr{E}_y(t)\right],$$

so that the pulse may drive the system only into any of its two lowest excited states, with the same energy, $E_{+1} = E_{-1} = \hbar^2/2mR^2$:

$$W_{\pm 1}(t) = \frac{q^2R^2}{4\hbar^2}\left|\int_0^t \left[\mathscr{E}_x(t') \mp i\mathscr{E}_y(t')\right]\exp\{i\omega_{10}t'\}dt'\right|^2, \qquad (***)$$
$$\text{with} \quad \omega_{10} = \frac{E_{\pm 1}}{\hbar} = \frac{\hbar}{2mR^2}.$$

(ii) This general expression shows that the probabilities of excitation of these two degenerate states are not always equal. The best example is that of a field rapidly rotating within the rotator's plane—for example, counterclockwise:

$$\mathscr{E}_x(t) = \mathscr{E}_0\cos(\omega t + \alpha), \quad \mathscr{E}_y(t) = \mathscr{E}_0\sin(\omega t + \alpha), \quad \text{for } t > 0,$$

where $\mathscr{E}_0$, ω, and α are constants, so that

$$\mathscr{E}_x(t) \mp i\mathscr{E}_y(t) = \mathscr{E}_0[\cos(\omega t + \alpha) \mp i\sin(\omega t + \alpha)] \equiv \mathscr{E}_0\exp\{\mp i(\omega t + \alpha)\}.$$

Plugging this expression into Eq. (***), we get

[48] This is again one of the manifestations of the selection rules in quantum transitions. Note that according to our calculation, in axially-symmetric 2D systems the rules require a change of the *magnetic* quantum number m by ± 1, while in the spherically-symmetric 3D systems, they require a similar change of the *orbital* quantum number l, while the magnetic number may either change by ± 1 or stay constant—see the footnote at the end of section 5.6, and problem 5.35.

$$W_{\pm 1}(t) = \frac{q^2R^2\mathscr{E}_0^2}{4\hbar^2}\left|\int_0^t \exp\{i[(\omega_{10} \mp \omega)t' - \alpha]\}dt'\right|^2$$
$$= \frac{q^2R^2\mathscr{E}_0^2}{4\hbar^2}\left|\frac{\exp\{i[(\omega_{10} \mp \omega)t]\} - 1}{i(\omega_{10} \mp \omega)}e^{-i\alpha}\right|^2.$$
$$\equiv \frac{q^2R^2\mathscr{E}_0^2}{2\hbar^2}\frac{1 - \cos(\omega_{10} \mp \omega)t}{(\omega_{10} \mp \omega)^2} \equiv \frac{q^2R^2\mathscr{E}_0^2}{\hbar^2(\omega_{10} \mp \omega)^2}\sin^2\frac{(\omega_{10} \mp \omega)t}{2}.$$

This expression shows that, as might be expected from the discussion in section 6.5 of the lecture notes, and in particular Eq. (6.90), both probabilities oscillate in time, generally with comparable but different amplitudes. However, if the external field frequency ω is very close to that (ω_{10}) of the potential interlevel transitions, the oscillations of the probability W_{+1} are much larger than those of its counterpart, W_{-1}.[49] This is very natural, because the enhanced wavefunction, corresponding to Eq. (**) with $m = +1$,

$$\Psi_{+1}(\varphi, t) \propto \psi_{+1}(\varphi)\exp\left\{-\frac{iE_{+1}t}{\hbar}\right\} \propto \exp\{i(\varphi - \omega_{10}t)\},$$

describes a de Broglie wave propagating in the same (counterclockwise) direction as the field. (For the opposite, clockwise direction of the field rotation, the probability of the opposite state, with $m = -1$, is similarly enhanced.)

This effect may be partly interpreted classically, by saying that the rotator's angular momentum L_z picks up a part of the momentum of the rotating field. However, the classical language does not describe the rotator momentum's quantization, $L_z = n\hbar$, and the related effect of its energy quantization, so it cannot describe the excitation's resonance at $\omega \to \omega_{10}$.

(iii) For a linearly-polarized ac field,

$$\mathscr{E}_x(t) = \mathscr{E}_0 \cos\varphi_0 \cos(\omega t + \alpha), \quad \mathscr{E}_y(t) = \mathscr{E}_0 \sin\varphi_0 \cos(\omega t + \alpha), \quad \text{for } t > 0,$$

where φ_0 is the angle between the field's direction and axis x, the result is different:

$$\mathscr{E}_x(t) \mp i\mathscr{E}_y(t) = \mathscr{E}_0[\cos\varphi_0 \mp i\sin\varphi_0]\cos(\omega t + \alpha)$$
$$= \mathscr{E}_0\exp\{\mp i\varphi_0\}\frac{\exp\{i(\omega t + \alpha)\} + \exp\{-i(\omega t + \alpha)\}}{2},$$

so that Eq. (***) yields

$$W_{\pm 1}(t) = \frac{q^2R^2\mathscr{E}_0^2}{4\hbar^2}\left|\int_0^t \exp\{i[(\omega_{10} + \omega)t' + \alpha]\}dt' + \int_0^t \exp\{i[(\omega_{10} - \omega)t' - \alpha]\}dt'\right|^2$$
$$= \frac{q^2R^2\mathscr{E}_0^2}{4\hbar^2}\left|\frac{\exp\{i[(\omega_{10} + \omega)t]\} - 1}{i(\omega_{10} + \omega)}e^{i\alpha} + \frac{\exp\{i[(\omega_{10} - \omega)t]\} - 1}{i(\omega_{10} - \omega)}e^{-i\alpha}\right|^2.$$

[49] In exploring this resonance behavior, we should not forget that this perturbative result is only valid if both probabilities $W_{\pm 1}$ are much smaller than 1.

This bulky result (which does not benefit much from its further processing) shows that the probabilities of the excitation of both states are equal, and do not depend on the field's inclination angle φ_0. Both these features may be explained by the fact that the linearly-polarized field may be always represented as the sum of two rotating fields (if we speak about waves, by two circularly-polarized waves) with equal amplitudes, regardless of the angle φ_0. This is why at $\omega \to \omega_{10}$, both probabilities exhibit the same resonance behavior as in the rotating field case.

***Problem* 6.24.** A spin-½ with a gyromagnetic ratio γ is placed into a magnetic field including a time-independent component $\mathscr{B}_0$, and a perpendicular field of a constant magnitude $\mathscr{B}_\text{r}$, rotated with a constant angular velocity ω. Can this *magnetic resonance* problem be reduced to one already discussed in chapter 6 of the lecture notes?

Solution: According to Eq. (4.163) of the lecture notes, we may describe this situation by the following time-dependent Pauli Hamiltonian:

$$\hat{H}(t) = -\frac{\hbar\gamma}{2}\hat{\boldsymbol{\sigma}} \cdot \mathscr{B}(t) = -\frac{\hbar\gamma}{2}\left[\hat{\sigma}_x\mathscr{B}_\text{r}\cos(\omega t + \varphi) - \hat{\sigma}_y\mathscr{B}_\text{r}\sin(\omega t + \varphi) + \hat{\sigma}_z\mathscr{B}_0\right],$$

where the axis z is directed along the field $\mathscr{B}_0$, and φ is a constant depending on the direction of the x- and y-axes (within the plane normal to z) and the time origin. Note that at $\omega > 0$, this expression describes the clockwise rotation of the field, but we may also use it to describe the opposite direction of rotation, by taking ω negative.

In the standard z-basis, the Hamiltonian's matrix is

$$\begin{aligned}\text{H}(t) &= -\frac{\hbar\gamma}{2}\left[\sigma_x\mathscr{B}_\text{r}\cos(\omega t + \varphi) - \sigma_y\mathscr{B}_\text{r}\sin(\omega t + \varphi) + \sigma_z\mathscr{B}_0\right] \\ &\equiv -\frac{\hbar\gamma}{2}\begin{pmatrix} \mathscr{B}_0 & \mathscr{B}_\text{r}e^{i(\omega t+\varphi)} \\ \mathscr{B}_\text{r}e^{-i(\omega t+\varphi)} & -\mathscr{B}_0 \end{pmatrix}.\end{aligned}$$

Using the same Schrödinger picture of the quantum evolution as was used in most of chapter 6, we may look for the ket-vector of the spin's state in the usual form (5.1),

$$|\alpha(t)\rangle = \alpha_\uparrow(t)\ |\uparrow\rangle + \alpha_\downarrow(t)\ |\downarrow\rangle.$$

Plugging the above expressions for $\text{H}(t)$ and $|\alpha(t)\rangle$ into the Schrödinger equation (4.158), we get the following system of equations for the probability amplitudes $\alpha_{\uparrow\downarrow}(t)$:

$$i\hbar\begin{pmatrix}\dot{\alpha}_\uparrow \\ \dot{\alpha}_\downarrow\end{pmatrix} = -\frac{\hbar\gamma}{2}\begin{pmatrix} \mathscr{B}_0 & \mathscr{B}_\text{r}e^{i(\omega t+\varphi)} \\ \mathscr{B}_\text{r}e^{-i(\omega t+\varphi)} & -\mathscr{B}_0 \end{pmatrix}\begin{pmatrix}\alpha_\uparrow \\ \alpha_\downarrow\end{pmatrix} \equiv -\frac{\hbar}{2}\begin{pmatrix} -\Omega\alpha_\uparrow + \gamma\mathscr{B}_\text{r}e^{i(\omega t+\varphi)}\alpha_\downarrow \\ \gamma\mathscr{B}_\text{r}e^{-i(\omega t+\varphi)}\alpha_\uparrow + \Omega\alpha_\downarrow \end{pmatrix}, \qquad (*)$$

where, in the last expression, $\Omega \equiv -\gamma\mathscr{B}_0$ is the frequency of spin's precession in the constant field $\mathscr{B}_0$—see Eq. (4.164) of the lecture notes. At $\mathscr{B}_\text{r} = 0$, the equations for $\alpha_{\uparrow\downarrow}(t)$ become uncoupled, and have simple solutions $\alpha_\uparrow = c_\uparrow\exp\{-i\Omega t/2\}$, $\alpha_\downarrow =$

$c_\downarrow \exp\{i\Omega t/2\}$, describing such precession, where $c_\uparrow$ and $c_\downarrow$ are some constants, defined by the initial state of the spin.

As a result, even in the case $\mathscr{B}_r \neq 0$ we may simplify the system (*) by looking for its solution in a similar form, but with time-dependent coefficients:

$$\alpha_\uparrow(t) = a_\uparrow(t)\, e^{-i\Omega t/2}, \qquad \alpha_\downarrow(t) = a_\downarrow(t)\, e^{i\Omega t/2}.$$

With this variable replacement, Eqs. (*) yield

$$i\hbar\dot{a}_\uparrow = a_\downarrow A e^{-i\Delta t}, \qquad i\hbar\dot{a}_\downarrow = a_\uparrow A^* e^{i\Delta t},$$

where

$$A \equiv -\frac{\hbar\gamma\mathscr{B}_r}{2}e^{i\varphi}, \quad \text{and} \quad \Delta \equiv \Omega - \omega.$$

This system of equations is identical to Eqs. (6.94) of the lecture notes[50], and we may use the results of its analysis in section 6.5. In particular, it describes the Rabi oscillations of the probabilities,

$$W_\uparrow(t) = |\alpha_\uparrow(t)|^2 = |a_\uparrow(t)|^2 \qquad \text{and } W_\downarrow(t) = |\alpha_\downarrow(t)|^2 = |a_\downarrow(t)|^2,$$

to find the spin in the eigenstates ↑ and ↓ of the 'unperturbed' ($\mathscr{B}_r = 0$) system. The oscillations have the frequency (6.99),

$$2\Omega = \left(\Delta^2 + 4\frac{|A|^2}{\hbar^2}\right)^{1/2} \equiv [(\Omega - \omega)^2 + (\gamma\mathscr{B}_r)^2]^{1/2},$$

and an amplitude which has a resonance dependence on the rotating field's frequency ω, peaking at $\omega = \Omega$, i.e. at the frequency of precession in the constant field $\mathscr{B}_0$.

This is exactly what is called the 'magnetic resonance'; it is crucial that at $\omega \to \Omega$, the probability oscillation amplitude may be very large (see, e.g. figure 6.9 of the lecture notes) even if the rotating field's amplitude is very small, $\mathscr{B}_r \ll \mathscr{B}_0$—as it is in practical applications of this effect, in particular in nuclear magnetic resonance (NMR) spectroscopy. In this limit, the field-induced FWHM bandwidth $\Delta\omega = 2\gamma\mathscr{B}_r$ of the resonance becomes negligible, and the experimentally observed bandwidth is defined by the effects of spin coupling to its environment—to be discussed in chapter 7 of this course.

***Problem* 6.25.** Develop the general theory of quantum excitations of the higher levels of a discrete-spectrum system, initially in the ground state, by a weak time-dependent perturbation, up to the second order. Spell out and discuss the result for the case of a monochromatic excitation, with a nearly perfect tuning of its frequency ω to *the half* of a certain quantum transition frequency $\omega_{n0} \equiv (E_n - E_0)/\hbar$.

[50] Note, however, that while Eqs. (6.94) are approximate, valid at the strong condition (6.93), the equations in this solutions are *exact*—the result of the simplicity of our current (two-level) system, and the special (rotating-field) way of its ac excitation.

Solution: Let us start from the general (exact) system of equations (6.84) for probability amplitudes a_n:

$$i\hbar\dot{a}_n = \sum_{n'} a_{n'} H_{nn'}^{(1)}(t) \exp\{i\omega_{nn'}t\}, \qquad \text{with } \omega_{nn'} \equiv \frac{E_n - E_{n'}}{\hbar}, \qquad (*)$$

where E_n are the energy levels of the system in the absence of perturbation, and look for its solution in the form of an expansion similar to Eqs. (6.9)–(6.10) of the lecture notes:

$$a_n(t) = a_n^{(0)}(t) + a_n^{(1)}(t) + a_n^{(2)}(t) + \dots, \qquad \text{with } a_n^{(k)}(t) \propto \mu^k,$$

where the small parameter μ is the scale of the perturbation Hamiltonian $\hat{H}^{(1)}(t)$. If by the beginning of the perturbation (the instant that we may take for $t = 0$), the system was in the ground state ($n = 0$), then we may take $a_n^{(0)}(t) = \delta_{n,0}$. Plugging these values into the right-hand side of Eqs. (*) with $n \neq 0$, in the first order we get simplified equation,

$$i\hbar\dot{a}_n = H_{n0}^{(1)}(t) \exp\{i\omega_{n0}t\},$$

with the elementary solution[51]

$$a_n^{(1)}(t) = \frac{1}{i\hbar}\int_0^t H_{n0}^{(1)}(t') \exp\{i\omega_{n0}t'\}\, dt', \qquad \text{for } n \neq 0. \qquad (**)$$

The resulting occupancy of the nth energy level is

$$W_n^{(1)}(t) = |a_n^{(1)}(t)|^2$$

If for some reason we need a more exact result (say, if $a_n^{(1)} \approx 0$ for a certain state we are interested in), we may proceed to the second-order approximation by plugging Eq. (**), rewritten for index n', into the right-hand side of Eqs. (*). This gives us the equation,

$$\begin{aligned} i\hbar\dot{a}_n^{(2)} &= \sum_{n'\neq 0} a_{n'}^{(1)} H_{nn'}^{(1)}(t) \exp\{i\omega_{nn'}t\} \\ &= \sum_{n'\neq 0} \frac{1}{i\hbar} H_{nn'}^{(1)}(t) \exp\{i\omega_{nn'}t\} \int_0^t H_{n'0}^{(1)}(t') \exp\{i\omega_{n'0}t'\} dt', \end{aligned}$$

whose solution may be also expressed in the integral form:

$$a_n^{(2)}(t) = -\frac{1}{\hbar^2}\sum_{n'\neq 0}\int_0^t dt' H_{nn'}^{(1)}(t') \exp\{i\omega_{nn'}t'\} \int_0^{t'} dt'' \; H_{n'0}^{(1)}(t'') \exp\{i\omega_{n'0}t''\}. \quad (***)$$

This probability amplitude determines the occupation of the levels not excited in the first order:

[51] Note that this relation was already used in the model solution of problem 6.23, and (in a more particular form) in that of problem 6.22.

$$W_n^{(2)}(t) = |a_n^{(2)}(t)|^2.$$

In the particular case of the monochromatic (sinusoidal) excitation, described by Eq. (6.86) of the lecture notes,

$$\hat{H}^{(1)}(t) = \begin{cases} 0, & \text{for } t < 0, \\ \hat{A}e^{-i\omega t} + \hat{A}^{\dagger}e^{+i\omega t}, & \text{for } t \geqslant 0, \end{cases}$$

the integrals in Eq. (***) may be readily worked out. The result is especially compact in the case of nearly perfect tuning, mentioned in the assignment, when $\xi \equiv 2\omega - \omega_{n0} \to 0$:

$$a_n^{(2)} \approx \frac{1}{\hbar^2 \xi} \sum_{n' \neq 0} \frac{A_{nn'}A_{n'0}}{\omega_{n'0} - \omega}(e^{i\xi t} - 1), \quad \text{so that}$$

$$W_n^{(2)} \approx \frac{4}{\hbar^4 \xi^2} \left| \sum_{n' \neq 0} \frac{A_{nn'}A_{n'0}}{\omega_{n'0} - \omega} \right|^2 \sin^2(\xi t/2).$$

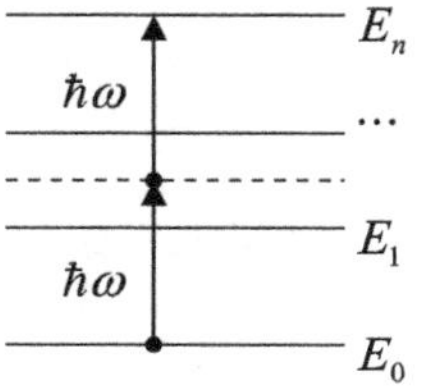

This expression (valid only if $W_n^{(2)} \ll 1$) formally diverges at $\xi \to 0$, i.e. describes a resonant excitation of the nth state, taking place even if the 'direct' matrix element A_{n0} vanishes, and even if the excitation frequency ω itself is not in an exact resonance with any of the quantum transition frequencies $\omega_{n'0}$, and as a result, no other states are strongly excited—see the figure above[52]. This is an example of what is usually, especially in quantum optics, called the *two-photon excitation*.

Another possible case of a two-photon (and more generally, *multi-photon*) excitation takes place when the perturbation Hamiltonian $\hat{H}^{(1)}$ is a nonlinear function of the applied field. (At the perturbative description of this case, it may be unnecessary to go beyond the first-order result (**), because the time dependence of the perturbation would contain higher harmonics of the field's frequency.)

***Problem* 6.26.** A heavy, relativistic particle, with electric charge $q = Ze$, passes by a hydrogen atom, initially in its ground state, with an impact parameter (the shortest distance) b within the range $r_B \ll b \ll r_B/\alpha$, where α is the fine structure

[52] Note that this situation differs from that considered in problem 6.20, where the involved energy levels are nearly equidistant, and the excitation of the second of them is conditioned by a very high occupation of the first one.

constant (6.62). Calculate the probabilities of atom's transition into its lowest excited states.

Solution: Due to the large mass of the flying-by particle, we may neglect the effects of its Coulomb interaction with the atom, as well as quantum-mechanical uncertainty, on its motion. In the absence of these effects, the motion is classical, uniform (acceleration-free), so that the electric and magnetic fields felt by the atom may be calculated classically. Such calculation yields[53]

$$\mathscr{E}_x(t) = 0, \quad \mathscr{E}_y(t) = -\frac{q}{4\pi\varepsilon_0}\frac{u\gamma t}{(u^2\gamma^2t^2 + b^2)^{3/2}}, \quad \mathscr{E}_z(t) = \frac{q}{4\pi\varepsilon_0}\frac{\gamma b}{(u^2\gamma^2t^2 + b^2)^{3/2}}, \qquad (*)$$
$$\mathscr{B}_x(t) = \frac{u}{c^2}\mathscr{E}_z(t), \quad \mathscr{B}_y(t) = 0, \quad \mathscr{B}_z(t) = 0.$$

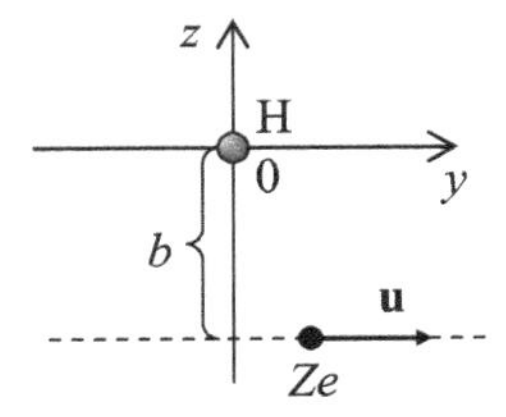

Here the y-axis is directed along the particle's velocity $\mathbf{u}$, z is the direction from the nearest-approach point toward the atom (see the figure above), time t, measured in the reference frame of the hydrogen atom, is referred to the instant of the nearest approach, and $\gamma \equiv (1 - u^2/c^2)^{-1/2} \geqslant 1$ is the relativistic parameter. Note that Eqs. (*) are exactly valid only at the center of the atom (from which the impact parameter is measured), but given the condition $r_B \ll b$, we may use them at all essential distances $r \sim r_B$ of the electron from the center.

Since according to Eqs. (1.13) of the lecture notes, the effective speed $v \sim \omega_0 r_B$ (where $\omega_0 \equiv E_H/\hbar$) of the electron inside the atom is of the order of $\alpha c \ll c$, Eqs. (*) shows that the magnetic component $-e\mathbf{v} \times \mathscr{B}$ of the Lorentz force acting on the atom's electron is much smaller than its electric component $-e\mathscr{E}$. Hence we may include only the latter force in the Hamiltonian of the particle–electron interaction:

$$\hat{H}_{\text{int}}(t) = e\mathscr{E}(t) \cdot \mathbf{r} \equiv e\left[\mathscr{E}_y(t)y + \mathscr{E}_z(t)z\right] \equiv er\left[\mathscr{E}_y(t)\sin\theta\sin\varphi + \mathscr{E}_z(t)\cos\theta\right].$$

Now using the first-order perturbation result (discussed, in particular, in the previous problem), for the total probability (at $t \gg b/u\gamma$) of the transition from the ground state (with the quantum numbers $n = 1$, $l = 0$, and $m = 0$) into the lowest excited states (with $n = 2$, and either $l = 0$ and $m = 0$, or $l = 1$ and $m = -1$, 0, or +1), we get

[53] See, e.g. *Part EM* section 9.5, in particular Eqs. (9.139) and (9.140), with the axis notation replacement $x \to y \to z \to x$.

$$W = \sum_{l,m} W_{l,m}, \qquad \text{where}$$

$$W_{l,m} = \frac{1}{\hbar^2}\left|\int_{-\infty}^{+\infty} \langle 1, 0, 0|\, \hat{H}_{\text{int}}(t)|\, 2, l, m\rangle \exp\{i\omega_{21}t\}\, dt\right|^2, \qquad (**)$$

where and l and m are the quantum numbers of the final state, while, according to Eq. (1.12),

$$\omega_{21} \equiv \frac{E_2 - E_1}{\hbar} = \frac{E_{\text{H}}}{\hbar}\left[-\frac{1}{2\cdot 2^2} - \left(-\frac{1}{2\cdot 1^2}\right)\right] \equiv \frac{3E_{\text{H}}}{8\hbar}.$$

As Eqs. (*) shows, for a relativistic particle, with $u \sim c$, the field pulse it induces at the atom's location, and hence the interaction Hamiltonian as the function of time, have the duration Δt of the order of $b/\gamma u \sim b/c$. The second of the conditions given in the assignment, $b \ll r_{\text{B}}/\alpha$, means that this duration is much less than $r_{\text{B}}/\alpha c$. On the other hand, the exponent in Eq. (**) changes with frequency $\omega_{12} \sim E_{\text{H}}/\hbar$, so that their product satisfies the condition

$$\omega_{12}\Delta t \ll \frac{E_{\text{H}}}{\hbar}\frac{r_{\text{B}}}{\alpha c} \equiv 1.$$

Hence the exponent in Eq. (**) cannot change significantly during the particle' passage, and we may take it from under the integral over time, getting

$$W_{l,m} = \frac{1}{\hbar^2}\left|\int_{-\infty}^{+\infty} \langle 1, 0, 0|\, \hat{H}_{\text{int}}(t)\, |2, l, m\rangle dt\right|^2$$

$$\equiv \frac{e^2}{\hbar^2}\left|\int_{-\infty}^{+\infty} \langle 1, 0, 0|\; r\left[\mathscr{E}_y(t)\sin\theta\,\sin\varphi + \mathscr{E}_z(t)\cos\theta\right]\, |2, l, m\rangle dt\right|^2.$$

Since, according to Eq. (*), $\mathscr{E}_y(t)$ is an odd function of time, the integral of the first term in the symmetric (infinite) limits vanishes, and we are left with

$$W_{l,m} = \frac{e^2}{\hbar^2}\, |I|^2 |\, \langle 1, 0, 0\, |\, r\cos\theta\, |2, l, m\rangle\, |^2, \quad \text{where } I \equiv \int_{-\infty}^{+\infty} \mathscr{E}_z(t)\, dt.$$

The integral I is readily reduced to a table one[54]:

$$I \equiv \int_{-\infty}^{+\infty} \mathscr{E}_z(t)dt = \frac{q}{4\pi\varepsilon_0}\gamma b\int_{-\infty}^{+\infty}\frac{dt}{(u^2\gamma^2t^2 + b^2)^{3/2}}$$

$$= \frac{q}{4\pi\varepsilon_0}\frac{2}{ub}\int_0^{+\infty}\frac{d\xi}{(\xi^2+1)^{3/2}} = \frac{q}{4\pi\varepsilon_0}\frac{2}{ub}.$$

Now we can make use of Eq. (3.200) of the lecture notes to spell out the matrix elements:

[54] See, e.g. Eq. (A.32b).

$$\langle 1, 0, 0|\, r\cos\theta\, |2, l, m\rangle = \int_0^{2\pi} d\varphi \int_0^{\pi} \sin\theta\, d\theta\ [Y_0^0(\theta,\varphi)]^* \cos\theta\ Y_l^m(\theta,\varphi)$$
$$\times \int_0^{\infty} r^2 dr\, \mathcal{R}_{1,0}(r) r\, \mathcal{R}_{2,l}(r).$$

Together with Eqs. (3.174) and (3.175), this expression shows, first of all, that the matrix element for the final 2s state (with $l = 0$, $m = 0$) vanishes because of the integral over the polar angle θ:

$$\int_0^{\pi} \sin\theta\, d\theta\ \ Y_0^0(\theta,\varphi)\cos\theta\ Y_0^0(\theta,\varphi) = \frac{1}{4\pi}\int_{-1}^{+1} \cos\theta\, d(\cos\theta) = 0.$$

For the two 2p states (with $l = 1$, and $m = \pm 1$) whose spherical harmonics are proportional to $e^{\pm i\varphi}$, the matrix elements vanish because of the integral over the azimuthal angle, and only for the 2p state with $m = 0$, the matrix element is different from zero:

$$\langle 1, 0, 0\, |\, r\cos\theta\, |2,1,0\rangle = \int_0^{2\pi} d\varphi \int_0^{\pi} \sin\theta d\theta\ [Y_0^0(\theta,\varphi)]^* \cos\theta\ Y_1^0(\theta,\varphi)$$
$$\times \int_0^{\infty} r^2 dr\, \mathcal{R}_{1,0}(r)\, r\, \mathcal{R}_{2,1}(r)$$
$$= \frac{\sqrt{3}}{2}\int_{-1}^{+1} \cos^2\theta d(\cos\theta) \int_0^{\infty} r^2 dr\, \mathcal{R}_{1,0}(r) r\, \mathcal{R}_{2,1}(r)$$
$$\equiv \frac{1}{\sqrt{3}}\int_0^{\infty} \mathcal{R}_{1,0}(r)\, \mathcal{R}_{2,1}(r)\, r^3 dr.$$

Now we may use Eq. (3.208) and the second of Eqs. (3.209), with $r_0 = r_B$, to calculate the remaining radial integral, by reducing it to a table one[55]:

$$\int_0^{\infty} \mathcal{R}_{1,0}(r)\, \mathcal{R}_{2,1}(r) r^3 dr = \int_0^{\infty} \frac{2}{r_B^{3/2}} e^{-r/r_B} \frac{r}{(2r_B)^{3/2}\sqrt{3}\, r_B} e^{-r/2r_B} r^3 dr$$
$$= \frac{1}{\sqrt{6}\, r_B^4}\int_0^{\infty} e^{-3r/2r_B} r^4 dr$$
$$= \frac{r_B}{\sqrt{6}}\left(\frac{2}{3}\right)^5 \int_0^{\infty} e^{-\xi}\xi^4 d\xi = \frac{2^5 r_B}{3^5\sqrt{6}} 4!\ ,$$

so that, finally (for $q = Ze$):

$$W = \frac{e^2}{\hbar^2}\left(\frac{q}{4\pi\varepsilon_0}\frac{2}{ub}\right)^2 \left(\frac{1}{\sqrt{3}}\frac{2^5 r_B}{3^5\sqrt{6}} 4!\right)^2 = \frac{2^{17}}{3^{10}}\left(\frac{Ze^2}{4\pi\varepsilon_0}\frac{r_B}{\hbar u b}\right)^2$$
$$\equiv \frac{2^{17}}{3^{10}}\left(\frac{Z\omega_0 r_B^2}{ub}\right)^2 \approx 2.22\left(Z\gamma \frac{\omega_0 b}{u\gamma}\frac{r_B^2}{b^2}\right)^2,$$

[55] See, e.g. Eq. (A.34d) with $n = 4$. Note also that this matrix element was already calculated in the solution of problem 6.19.

where $\omega_0 \equiv E_H/\hbar$.

As was discussed above, this result was derived for $\omega_{12}\Delta t \sim \omega_0 b/u\gamma \ll 1$, and $r_B \ll b$, so that two last fractions inside the last parentheses are much smaller than 1, meaning that for not extremely high values of Z and γ, the excitation probability is very low. (If the product $Z\gamma$ is so high that our result yields W of the order of 1 or higher, it should be revised, because the used version of the perturbation theory is only valid if it yields $W \ll 1$—see Eq. (6.89) of the lecture notes.)

***Problem* 6.27.** A particle of mass m is initially in the localized ground state, with the known energy $E_g < 0$, of a very small, spherically-symmetric potential well. Calculate the rate of its delocalization by an applied force $\mathbf{F}(t)$ with time-independent amplitude F_0, frequency ω, and direction $\mathbf{n}_F$.

Solution: This is essentially a generalization of the 1D problem solved in section 6.6 of the lecture notes, and may be solved similarly, with due respect to the 3D aspects. First, the wavefunction of the initial (ground) state is different—see the solution of problem 6.9:

$$\psi_{n'} = \left(\frac{\kappa}{2\pi}\right)^{1/2} \frac{\exp\{-\kappa r\}}{r}, \qquad \text{where } \frac{\hbar^2\kappa^2}{2m} = -E_g > 0.$$

Second, the perturbative Hamiltonian now includes a scalar product:

$$\hat{H}^{(1)} = -\mathbf{F}(t)\cdot\hat{\mathbf{r}} = -\mathbf{n}_F\cdot\hat{\mathbf{r}}\ F_0 \cos\omega t.$$

Since the rest of the system is isotropic, we may direct the z-axis along the (time-independent) direction $\mathbf{n}_F$ of the force, and thus reduce the Hamiltonian to the standard form (6.86):

$$\hat{H}^{(1)} = -\ F_0\hat{z}\cos\omega t \equiv \hat{A}e^{-i\omega t} + \hat{A}^\dagger e^{i\omega t},$$

$$\text{with} \qquad \hat{A} = -\frac{F_0 z}{2} \equiv -\frac{F_0 r\cos\theta}{2}, \qquad \text{for } t \geqslant 0,$$

where θ is the usual polar angle. Finally, we need to find the extended final-state wavefunctions ψ_n, describing the escaped (delocalized) particle[56], that would give non-zero values of the matrix elements

$$A_{nn'} = \int \psi_n^*(\mathbf{r})\hat{A}\psi_{n'}(\mathbf{r})d^3r = -\frac{F_0}{2}\left(\frac{\kappa}{2\pi}\right)^{1/2}\int_0^\infty re^{-\kappa r}dr \oint d\Omega\ \cos\theta\ \psi_n(\mathbf{r}),$$

which determine the Golden-Rule rate (6.111). Since $\cos\theta$ is proportional to one of the spherical harmonics (namely, $Y_1^0(\theta, \varphi)$—see Eq. (3.175) of the lecture notes), and all such harmonics are orthogonal, this integral does not vanish only for the final-state wavefunctions proportional to $\cos\theta$ and independent of φ.

[56] Of course, such states exist only for energies $E_n > 0$, and hence the 'ionization' is an effect with a low-frequency threshold: $\Gamma = 0$ if $\omega < \omega_{\min} \equiv |E_g|/\hbar$.

There are two natural ways to construct such wavefunctions. One is to use the solution of the problem solved at the end of section 3.6, giving such functions in the form (3.187) with $l = 1$:

$$\psi_n(\mathbf{r}) = C_n j_1(k_n r)\cos\theta,$$

where $j_1(\xi)$ is one of spherical Bessel functions, and k_n is related to the final-state energy as

$$E_n = \frac{\hbar^2 k_n^2}{2m}. \qquad (*)$$

However, let me leave using this approach for the next, similar problem, and use this occasion to illustrate an alternative approach based on plane de Broglie waves $\psi_n(\mathbf{r}) \propto \exp\{i\mathbf{k}_n\cdot\mathbf{r}\}$, where the magnitude of vector $\mathbf{k}_n$ satisfies the same Eq. (*). Indeed, since the localizing potential $U(\mathbf{r})$ differs from zero only at the origin, such wavefunctions are legitimate solutions of the Schrödinger equation at $r \neq 0$. Moreover, the spherically-symmetric potential $U(\mathbf{r})$, proportional to $Y_0^0(\theta, \varphi) =$ const, does not alter those of the waves that give nonvanishing contributions to $A_{nn'}$. The normalization coefficient may be found, as usual (see, e.g. section 1.7), requiring the particle to be confined within an artificial, very large volume $V \gg 1/k_n^3$, $1/\kappa^3$. Then the normalized wavefunctions are

$$\psi_n(r) = \frac{1}{V^{1/2}}\exp\{i\mathbf{k}_n\cdot\mathbf{r}\} \equiv \frac{1}{V^{1/2}}\exp\left\{i\left(k_x x + k_y y + k_z z\right)\right\},$$

and our matrix elements may be rewritten as

$$A_{nn'} = -\frac{F_0}{2}\left(\frac{\kappa}{2\pi V}\right)^{1/2}\int_V z\exp\left\{i\left(k_x x + k_y y + k_z z\right)\right\}\frac{\exp\{-\kappa r\}}{r}d^3r.$$

Perhaps the easiest way to calculate the integral in this expression is to notice that it is equal to $\partial I/\partial(ik_z)$, where I is a similar integral, but without the factor z before the exponent:

$$I \equiv \int_V \exp\left\{i\left(k_x x + k_y y + k_z z\right)\right\}\frac{\exp\{-\kappa r\}}{r}d^3r \equiv \int_V \exp\{i\mathbf{k}_n\cdot\mathbf{r}\}\frac{\exp\{-\kappa r\}}{r}d^3r.$$

Indeed, the integral I is evidently independent of the direction of the vector k_n, and we may take this direction for a new axis z' (independent of the direction z of the applied force!), and calculate it in these new spherical coordinates $\{r, \theta', \varphi'\}$:[57]

[57] Since the function under integral decays at distances $r \sim \kappa^{-1} \ll V^{1/3}$, the integration volume restriction may be ignored here.

$$I = \int_V \exp\{ik_n r \cos\theta '\} \frac{\exp\{-\kappa r\}}{r} d^3r$$
$$\equiv \int_0^\infty r^2 dr \int_0^\pi \sin\theta ' d\theta ' \int_0^{2\pi} d\varphi' \ \exp\{ik_n r \cos\theta '\} \frac{\exp\{-\kappa r\}}{r}$$
$$= 2\pi \int_0^\infty re^{-\kappa r} dr \int_{-1}^{+1} \exp\{ik_n r \xi\} d\xi$$
$$= -2\pi \int_0^\infty re^{-\kappa r} dr \frac{\exp\{-ik_n r\} - \exp\{+ik_n r\}}{ik_n r}$$
$$\equiv \frac{4\pi}{k_n} \int_0^\infty e^{-\kappa r} \sin k_n r \ dr = \frac{4\pi}{k_n} \text{Im} \int_0^\infty \exp\{(-\kappa + ik_n)r\} dr$$
$$= -\frac{4\pi}{k_n} \text{Im} \frac{1}{-\kappa + ik_n} \equiv \frac{4\pi}{\kappa^2 + k_n^2}.$$

From here,

$$A_{nn'} = -F_0 \left(\frac{2\pi\kappa}{V}\right)^{1/2} \frac{\partial I}{\partial(ik_z)} = -iF_0 \left(\frac{2\pi\kappa}{V}\right)^{1/2} \frac{2k_z}{\left(\kappa^2 + k_x^2 + k_y^2 + k_z^2\right)^2}$$
$$= -iF_0 \left(\frac{8\pi\kappa}{V}\right)^{1/2} \frac{k_n \cos\theta}{(\kappa^2 + k_n^2)^2}.$$

Since this matrix element depends not only on the magnitude, but also on the direction of the vector $\mathbf{k}_n$, we should use the state number counting rule (1.90) carefully, first applying it to the states with vectors $\mathbf{k}_n$ within a small solid angle $d\Omega \ll 4\pi$, with a virtually constant angle θ:

$$dN_n = \frac{V}{(2\pi)^3} k_n^2 dk_n d\Omega.$$

Combined with the derivative of Eq. (*), $dE_n = (\hbar^2/m)k_n dk_n$, this relation gives the 'directional' (angle-differential) density of states

$$d\rho_n \equiv \frac{dN_n}{dE_n} \equiv \frac{dN}{dk_n} \Big/ \frac{dE_n}{dk_n} = \frac{V}{(2\pi)^3} k_n^2 dk_n d\Omega \Big/ \frac{\hbar^2}{m} k_n dk_n = \frac{Vmk_n}{(2\pi)^3 \hbar^2} d\Omega,$$

so that the Golden Rule (6.111) yields[58]

$$d\Gamma = \frac{2\pi}{\hbar} |A_{nn'}|^2 \, d\rho_n = \frac{2\pi}{\hbar} \times F_0^2 \frac{8\pi\kappa}{V} \frac{k_n^2 \cos^2\theta}{\left(\kappa^2 + k_n^2\right)^4} \times \frac{Vmk_n}{(2\pi)^3 \hbar^2} d\Omega$$
$$\equiv \frac{2}{\pi} \frac{F_0^2 m}{\hbar^3} \frac{\kappa k_n^3}{\left(\kappa^2 + k_n^2\right)^4} \cos^2\theta d\Omega.$$

[58] Note that the artificial binding volume V has cancelled—an usual necessary condition of the legitimacy of this normalization procedure.

This partial rate is the probability of a particle's delocalization in unit time, with the condition that its final wave vector $\mathbf{k}_n$ is within the elementary solid angle $d\Omega$. Note that the angular distribution of the delocalized particles vanishes in all directions normal to that of the applied force (with $\theta = \pi/2$).

The total rate may now be calculated by the summation of such partial rates over the whole solid angle:

$$\Gamma = \oint d\Gamma = \frac{2}{\pi}\frac{F_0^2 m}{\hbar^3}\frac{\kappa k_n^3}{\left(\kappa^2 + k_n^2\right)^4} \times 2\pi \int_0^\pi \cos^2\theta \sin\theta d\theta$$

$$= \frac{8}{3}\frac{F_0^2 m}{\hbar^3}\frac{\kappa k_n^3}{\left(\kappa^2 + k_n^2\right)^4}.$$

In order to discuss the frequency dependence of this rate, it is useful to notice that due to the energy conservation (formally expressed by Eq. (6.93) of the lecture notes—see also figure 6.10),

$$\hbar\omega = E_n - E_g \equiv \frac{\hbar^2 k_n^2}{2m} - \left(-\frac{\hbar^2\kappa^2}{2m}\right) \equiv \frac{\hbar^2\left(k_n^2 + \kappa^2\right)}{2m},$$

our result may be rewritten in a simpler form:

$$\Gamma = \frac{8}{3}\frac{F_0^2 m}{\hbar^3}\frac{\kappa k_n^3}{(2m\omega/\hbar)^4} = \frac{1}{6}\frac{F_0^2\hbar}{m^3}\frac{\kappa k_n^3}{\omega^4} \propto \frac{(\omega - \omega_{\min})^{3/2}}{\omega^4}.$$

According to this formula, the rate first increases when ω exceeds its threshold value $\omega_{\min} \equiv |E_g|/\hbar$, and then (after $\omega = (8/5)\omega_{\min}$) decreases again, vanishing at $\omega \to \infty$. This behavior is qualitatively similar to, but quantitatively different from that in the 1D case—cf. Eq. (6.133) of the lecture notes.

***Problem* 6.28.** Calculate the rate of ionization of a hydrogen atom, initially in its ground state, by a classical, linearly polarized electromagnetic wave with a frequency ω within the range

$$\frac{\hbar}{m_e r_B^2} \ll \omega \ll \frac{c}{r_B},$$

where r_B is the Bohr radius. Recast your result in terms of the cross-section of this electromagnetic wave absorption process. Discuss briefly what changes of the theory would be necessary if each of the above two conditions had been violated.

Solution: Due to the second condition specified in the assignment, the electromagnetic wavelength, $\lambda = 2\pi c/\omega$, is much larger than the Bohr radius, i.e. the linear scale of the atom's wavefunctions. In such a 'quasi-static' approximation[59], the spatial variation of the field may be ignored, so that following Eq. (6.29) of the lecture notes, with $q = -e$, the perturbation Hamiltonian may be taken in the form

[59] See, e.g. *Part EM* section 6.3.

$$\hat{H}^{(1)} = e\mathscr{E}_0\hat{z}\cos\omega t = \hat{A}e^{-i\omega t} + \hat{A}^\dagger e^{+i\omega t}, \qquad \text{with} \quad \hat{A} = \hat{A}^\dagger = \frac{1}{2}e\mathscr{E}_0\hat{z},$$

where $\mathscr{E}_0$ is the wave field's amplitude, and the z-axis is directed along the electric field's polarization.

The first of the given conditions ensures that the electron's final state energy,

$$E_n = E_{n'} + \hbar\omega = -\frac{E_H}{2} + \hbar\omega, \quad \text{where } E_H = \frac{\hbar^2}{m_e r_B^2},$$

is much higher than the Hartree energy $E_H \approx 27$ eV. This condition allows us to neglect an atom's effect on the final state's wavefunctions, and take them in the form of free-particle de Broglie waves[60]. It is possible to take these wavefunctions in the form of plane waves, $\psi_n(\mathbf{r}) \propto \exp\{i\mathbf{k}_n\cdot\mathbf{r}\}$, with the magnitude of wave vector $\mathbf{k}_n$ related to the final state energy $E_n > 0$ as

$$\frac{\hbar^2 k_n^2}{2m_e} = E_n, \qquad (*)$$

but then the matrix elements $A_{nn'}$ that participate in the Golden Rule (6.111) would depend not only on the magnitude of vector $\mathbf{k}_n$ (i.e. on the final state energy E_n) but also on the direction of this vector, making the Golden Rule directly applicable only to each subset of final states within a small solid angle $d\Omega$, with a definite angle θ between the vectors $\mathbf{k}_n$ and $\mathbf{n}_z$.

Though this technical difficulty may be readily overcome (see the solution of the previous problem), let me use here an alternative approach, only mentioned in that solution. Let us look at the angular structure of the matrix elements, in the coordinate representation:

$$A_{nn'} = \int \psi_n^* \hat{A} \psi_{n'} d^3r.$$

While according to Eqs. (3.174), (3.200), and (3.208) of the lecture notes, with $r_0 = r_B$, the initial, ground-state wavefunction $\psi_{n'}$ is spherically-symmetric,

$$\psi_{n'} = Y_0^0(\theta, \varphi)\mathcal{R}_{1,0}(r) = \frac{1}{\pi^{1/2} r_B^{3/2}} \exp\left\{-\frac{r}{r_B}\right\},$$

the perturbation operator does have an angular dependence, and according to Eq. (3.175), may be represented as

$$\hat{A} = \frac{1}{2}e\mathscr{E}_0 z \equiv \frac{1}{2}e\mathscr{E}_0 r\cos\theta = \left(\frac{\pi}{3}\right)^{1/2} e\mathscr{E}_0 r Y_1^0(\theta, \varphi).$$

[60] Note that the previous, conceptually very similar problem could be solved analytically without imposing such a condition, because the very short-range binding potential $U(\mathbf{r})$, considered in it, does not disturb the relevant finite states, corresponding to $l = 1$, with any energy $E_n > 0$.

Since the spherical harmonics are orthogonal, the angular factor of the matrix element integral does not vanish only if the final state is also proportional to the same spherical harmonic—with the quantum numbers $l = 1$ and $m = 0$. As we know from the problem solved at the end of section 3.6 of the lecture notes, for a free particle with energy E_n such a function is

$$\psi_n = C_n j_1(k_n r) Y_1^0(\theta, \varphi), ,$$

where the wave number k_n satisfies the same Eq. (*), and the spherical Bessel function j_1 is very simple—see Eq. (3.186):

$$j_1(\xi) = \frac{\sin \xi}{\xi^2} - \frac{\cos \xi}{\xi}.$$

The normalization coefficient C_n may be readily obtained introducing an auxiliary confining sphere of a sufficiently large radius $R \gg 1/k_n$, and imposing the boundary condition $\psi_n(R) = 0$ on its surface. In this case, the second term in the expression for $j_1(\xi)$ dominates the normalization integral:

$$\int \psi_n \psi_n^* d^3r = \oint | Y_{10}(\theta, \varphi)|^2 \, d\Omega \int_0^R |j_1(k_n r)|^2 \, r^2 dr$$

$$\approx |C_n|^2 \int_0^R \left(\frac{\cos k_n r}{k_n r}\right)^2 r^2 dr \approx |C_n|^2 \frac{R}{2k_n^2} = 1,$$

(where we have taken into account that the spherical harmonics are orthonormal), so that we may take C_n in a real form

$$C_n = \left(\frac{2}{R}\right)^{1/2} k_n.$$

Now, assuming that this R is much larger than r_B as well (as we certainly may, due to the artificial character of R), we can calculate the required matrix element ignoring this upper bound:

$$A_{nn'} = \frac{1}{\pi^{1/2} r_B^{3/2}} \left(\frac{\pi}{3}\right)^{1/2} e\mathscr{E}_0 \left(\frac{2}{R}\right)^{1/2} k_n \oint | Y_1^0(\theta, \varphi)|^2 \, d\Omega \times I,$$

$$\text{with} \quad I \equiv \int_0^\infty \exp\left\{-\frac{r}{r_B}\right\} r \, j_1(k_n r) r^2 dr,$$

where the angular integral equals 1 again. It may look like that due to the first of the conditions given in the assignment, $k_n \gg 1/r_B$, in the radial integral I we can again keep only the second, more slowly decaying part of $j_1(k_n r)$. However, due to the exponential cut-off, this is not so; indeed:

$$I = \int_0^\infty \exp\left\{-\frac{r}{r_B}\right\} r \left(\frac{\sin k_n r}{k_n^2 r^2} - \frac{\cos k_n r}{k_n r}\right) r^2 dr$$

$$\equiv \frac{1}{k_n^2} \text{Im} \int_0^\infty \exp\{-\lambda r\} \, r dr - \frac{1}{k_n} \text{Re} \int_0^\infty \exp\{-\lambda r\} \, r^2 dr,$$

$$\text{with} \quad \lambda \equiv r_B^{-1} - ik_n r.$$

These are two well-known integrals[61], equal to $1/\lambda^2$ and $2/\lambda^3$, correspondingly, so that

$$I = \frac{1}{k_n^2}\mathrm{Im}\frac{1}{\left(r_\mathrm{B}^{-1} - ik_n\right)^2} - \frac{2}{k_n}\mathrm{Re}\frac{1}{\left(r_\mathrm{B}^{-1} - ik_n\right)^3}$$

$$= \frac{1}{k_n^2}\frac{2r_\mathrm{B}^{-1}k_n}{\left(r_\mathrm{B}^{-2} + k_n^2\right)^2} - \frac{2}{k_n}\frac{r_\mathrm{B}^{-3} - 3r_\mathrm{B}^{-1}k_n^2}{\left(r_\mathrm{B}^{-2} + k_n^2\right)^3} \to \frac{2r_\mathrm{B}^{-1}}{k_n^5} + \frac{6r_\mathrm{B}^{-1}}{k_n^5}.$$

We see that the contributions from both terms are comparable even in the limit $k_n \gg r_\mathrm{B}^{-1}$ used at the last step. As a result, we get

$$A_{nn'} = \frac{1}{\pi^{1/2}r_B^{3/2}}\left(\frac{\pi}{3}\right)^{1/2} e\mathscr{E}_0 \left(\frac{2}{R}\right)^{1/2} k_n \times \left(-\frac{8}{k_n^5 r_\mathrm{B}}\right)$$

$$\equiv -\left(\frac{128}{3}\right)^{1/2} \frac{e\mathscr{E}_0}{k_n^4 r_\mathrm{B}^{5/2} R^{1/2}}.$$

What remains is to calculate the density of final states. For $k_nR \gg 1$, the boundary condition $j_1(k_nR) = 0$ yields $k_nR \approx \pi n + \text{const}$, so that $dn/dk_n = R/\pi$. Combining this relation with the derivative of Eq. (*), $dE_n = (\hbar^2/m_\mathrm{e})k_n dk_n$, we may calculate the final state density:

$$\rho_n \equiv \frac{dn}{dE_n} \equiv \frac{dn}{dk_n}\Big/\frac{dE_n}{dk_n} = \frac{R}{\pi}\Big/\frac{\hbar^2 k_n}{m_\mathrm{e}} \equiv \frac{Rm_\mathrm{e}}{\pi\hbar^2 k_n}.$$

Now the Golden Rule (6.111) yields[62]

$$\Gamma = \frac{2\pi}{\hbar}\left[\left(\frac{128}{3}\right)^{1/2} \frac{e\mathscr{E}_0}{k_n^4 r_\mathrm{B}^{5/2} R^{1/2}}\right]^2 \frac{Rm_\mathrm{e}}{\pi\hbar^2 k_n} \equiv \frac{256}{3}\frac{e^2\mathscr{E}_0^2 m_\mathrm{e}}{\hbar^3 k_n^9 r_\mathrm{B}^5}.$$

(Notice that the artificial confinement radius R has cancelled, as it had to.)

In order to get a better feeling of the intensity of the ionization process, it is useful to consider it as the electromagnetic wave absorption (with the corresponding photoelectron emission), and borrow from the scattering theory the notion of its full cross-section σ. Creatively reworking Eq. (3.59) of the lecture notes, we may define this notion as

$$\sigma \equiv \frac{\text{average number of photons absorbed by atom per second}}{\text{incident flux of photons per unit area}}$$

$$= \frac{\Gamma}{S/\hbar\omega},$$

[61] See, e.g. Eq. (A.34*d*) with $n = 1$ and $n = 2$. Since the functions under the integrals are analytical, these formulas are valid even for complex λ, provided that Re $\lambda > 0$, so that the integrals converge at their upper limits.

[62] The reader is highly encouraged to re-calculate this result, using the alternative plane-wave approach discussed in the previous problem.

where S is the Pointing vector's magnitude, i.e. the electromagnetic wave's power per unit area of its front. For plane waves in vacuum[63]:

$$S = \frac{\mathscr{E}_0^2}{2Z_0}, \quad \text{where } Z_0 = \left(\frac{\mu_0}{\varepsilon_0}\right)^{1/2} \equiv \frac{1}{\varepsilon_0 c}.$$

Using our result for Γ, the cross-section may be represented in the form

$$\sigma = \frac{256}{3}\frac{e^2\mathscr{E}_0^2 m_e}{\hbar^3 k_n^9 r_B^5} \Bigg/ \frac{\varepsilon_0 c \mathscr{E}_0^2}{2\hbar\omega} = \frac{512}{3}\frac{k_{EM} e^2 m_e}{\varepsilon_0 \hbar^2 k_n^9 r_B^5} = \frac{2{,}048\pi}{3}\sigma_B \frac{k_{EM} r_B}{(k_n r_B)^9},$$

where $k_{EM} = \omega/c$ is the electromagnetic wave's wave number, and

$$\sigma_B \equiv \pi r_B^2 \approx 0.88 \times 10^{-20} \text{m}^2$$

is the effective 'physical' cross-section of the hydrogen atom. (The Bohr radius $r_B \equiv 4\pi\varepsilon_0\hbar^2/e^2 m_e$.)

Since the strong conditions given in the assignment may be rewritten as

$$k_{EM} r_B \ll 1 \ll k_n r_B,$$

we see that within our frequency range the cross-section of the ionization process is much smaller than σ_B. By the way, it is useful to estimate how broad this frequency range is. The ratio of its upper and lower bounds for ω, by the order of magnitude, is

$$\frac{c/r_B}{\hbar/m_e r_B^2} = \frac{4\pi\varepsilon_0 c\hbar}{e^2} \equiv \frac{1}{\alpha} \approx 137,$$

where α is the fine structure constant (6.62). So, the specified frequency range is broad, but not too broad: if we want to keep both inequalities strong indeed, it is about one order of magnitude—in electromagnetic wavelengths, from ~10 nm to ~100 nm.[64] In the middle of the range (at $k_{EM} r_B \sim 0.1$, $k_n r_B \sim 10$), we get $\sigma \sim 2\times10^{-5}\sigma_B \sim 2\times10^{-25}$ m^2.[65] This number means that if a sample has $n \sim 0.5\times10^{28}$ hydrogen atoms per m^3 (the number typical for organic condensed matter), the penetration length $\delta = 1/n\sigma$ of such radiation[66] due to the hydrogen ionization is about 10^{-3} m, i.e. 1 mm.

---

[63] See, e.g. *Part EM* section 7.1.

[64] Electromagnetic waves in this range are usually called the *extreme ultra-violet* radiation—EUV (or XUV) for short.

[65] Note that this σ of the *inelastic* photoionization process is still much larger than the electron's cross-section $\sigma_T = (8/3)\sigma_B\alpha^4 \approx 0.67\times10^{-29}$ m^2 of the *elastic* ('Thomson') scattering of electromagnetic waves—see, e.g. *Part EM* section 8.3, in particular Eq. (8.41).

[66] Let me hope that the reader has met the used expression, $\delta = 1/n\sigma$, earlier. If not: it may be readily obtained from the requirement that the expression for the decay $-dS$ of the incident radiation in a layer of thickness $dx \ll \delta$, calculated from the definition of the cross-section, $-dS = \sigma S dN = S\sigma n dx$, coincides with the one, $-dS = S dx/\delta$, following from the exponential decay law, $S \propto \exp\{-x/\delta\}$, where x is the direction of radiation's propagation. Note that a similar (though not identical) expression is valid for the particle mean free path—see, e.g. *Part SM* chapter 6.

Moreover, as may be readily proved by reviewing the above calculations, for hydrogen-like atoms/ions with atomic number $Z > 1$, our result is valid with the replacement $\sigma \rightarrow Z^5\sigma$. Though this model is rather crude for other atoms, such scaling still hints that the EUV penetration length into any condensed matter is very small. This is true indeed, causing many problems with its practical applications, including high-resolution photolithography—the main process used for integrated circuit patterning. (For one, no refractive optics is available for the EUV, due to such high absorption.)

Returning to the last assignment of the problem, in order to overcome the given lower frequency bound, we would need to modify the final-state wavefunction for interaction with the $1/r$-type potential profile of the hydrogen nuclei. (This may be done by studying the solutions of Eq. (3.193) of the lecture notes, with $l = 1$, for $\varepsilon > 0$.) On the other hand, at frequencies higher than the specified upper bound of the frequency range, we would need to take into account the spatial dependence of the electromagnetic field. Neither is too easy.

Problem* 6.29. Use the quantum-mechanical Golden Rule to derive the general expression for the electric current I through a weak tunnel junction between two conductors, biased with dc voltage V, via the matrix elements that describe the tunneling, treating the conductors as Fermi gases of electrons with negligible direct interaction. Simplify the result in the low-voltage limit.

Hint: The electric current flowing through a weak tunnel junction is so low that it does not substantially perturb the electron states inside each conductor.

Solution: In thermal equilibrium (which is not perturbed by weak tunnel current), the Fermi gas of non-interacting electrons in each conductor may be described by a dense (quasi-continuous) set of energy levels E_n, occupied by electrons in accordance with the Fermi distribution $f(E_n)$, which drops from 1 to 0 on the scale $\sim k_B T$, when the energy crosses some value E_F, called the *Fermi energy*[67].

The figure below shows (schematically) the energy diagram of a tunnel junction biased by a time-independent (dc) voltage V, which creates the Fermi levels' difference $E_F - E'_F = -eV$. Hence each energy level $E_{n'}$ in one conductor (referred, say, to its Fermi energy) becomes aligned with the level $E_n = E_{n'} + eV$ of another conductor. If the level $E_{n'}$ is fully occupied, and the level E_n is completely empty, the Golden Rule (6.137) for the rate of electron tunneling from left to right reads

$$\Gamma(E_{n'}) = \frac{2\pi}{\hbar} |T_{nn'}|^2 \rho_R(E_{n'} + eV),$$

[67] For a detailed discussion of the Fermi distribution and the Fermi gas see, e.g. *Part SM* sections 2.8 and 3.3; for our current purposes, the particular form of the function $f(E)$ is not important.

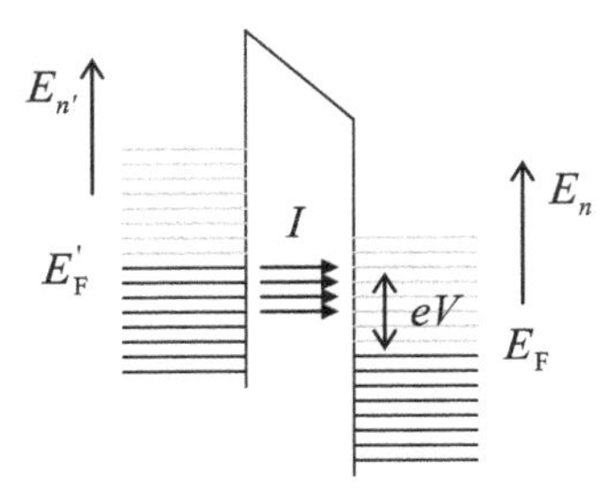

where $T_{nn'}$ is the matrix element of the electrode interaction, describing the tunneling barrier's transparency, and ρ_R is the density of states in the right electrode (which should take into account the two-fold spin degeneracy of each orbital energy level). The electric current due to this rate would be just $-e\Gamma(E_{n'})$; however, in order to account for an incomplete occupancy of the initial and final states of the electron, this expression has to be multiplied by $f(E_{n'})[1 - f(E_n)] = f(E_{n'})[1 - f(E_{n'} + eV)]$. (The second multiplier is due to the Pauli principle for Fermi particles: if a state is already occupied by an electron, tunneling of an additional electron into the same state is forbidden.) Thus the total current from the left conductor to the right one may be calculated as a sum over the quasi-continuous set of levels $E_{n'}$, with a density $\rho_L(E_{n'})$:

$$I_{\to} = -e\frac{2\pi}{\hbar}\sum_{n'}|T_{nn'}|^2\,\rho_R(E_{n'} + eV)f(E_{n'})[1 - f(E_{n'} + eV)]$$
$$\approx -\frac{2\pi e}{\hbar}\int \rho_L(E_{n'})dE_{n'}\,|T_{nn'}|^2\,\rho_R(E_{n'} + eV)f(E_{n'})[1 - f(E_{n'} + eV)]\,.$$

Now, we should also take into account the back current $I_{\leftarrow}$ due to reciprocal tunneling from the right to the left conductor, which differs from $I_{\to}$ only by the occupancy factors. As a result, for the net current we get

$$I(V) \equiv I_{\to} - I_{\leftarrow} = -\frac{2\pi e}{\hbar}\int \rho_L(E_{n'})dE_{n'}\,|T_{nn'}|^2\,\rho_R(E_{n'} + eV)$$
$$\times\begin{Bmatrix} f(E_{n'})[1 - f(E_{n'} + eV)] \\ -f(E_{n'} + eV)[1 - f(E_{n'})]\end{Bmatrix}$$
$$\equiv -\frac{2\pi e}{\hbar}\int |T_{nn'}|^2\,\rho_L(E_{n'})\rho_R(E_{n'} + eV)\big[f(E_{n'}) - f(E_{n'} + eV)\big]dE_{n'}.$$

This expression is broadly used in the theory of tunnel junctions. It may be readily simplified in the limit when eV is small in comparison with the scale of the energy dependence of the matrix elements $T_{nn'}$ and the densities of states (but not necessarily with the thermal excitation scale k_BT), so that $[f(E_{n'}) - f(E_{n'} + eV)] \approx -df/dE|_{En'=EF}\, eV$, and

$$I(V) \to GV, \quad \text{with } G = \frac{2\pi e^2}{\hbar}\langle |T_{nn'}|^2\rangle_{E=E_F}\rho_L(E_F)\rho_R(E_F),$$

where the averaging is over all states on the Fermi surface (which is dominated by the states with the group velocity virtually normal to the junction plane).

In tunnel junctions between most conductors (say, semiconductors or normal metals), this Ohm's law works very well at applied voltages up to several hundred mV. However, in some cases this is not so. For example, in superconductors the effective density of single-electron states is strongly suppressed within the so-called *superconducting energy gap* $\Delta \sim k_B T_c$, where T_c is the critical temperature, near the Fermi energy. As a result, the I–V curves of tunnel junctions between superconductors may be strongly nonlinear on the scale of a few meV, with the current rising sharply as the voltage reaches the threshold value $V_t = (\Delta_L + \Delta_R)/e$.[68]

Problem* 6.30. Generalize the result of the previous problem to the case when a weak tunnel junction is biased with voltage $V_0 + A\cos\omega t$, with $\hbar\omega$ generally comparable with eV_0 and eA.

Solution: Since weak tunneling does not perturb the electron states inside conductors, they remain stationary, i.e. their time dependence is reduced to $\exp\{-iE_n t/\hbar\}$, if the state energy E_n is referred to a fixed level inside the same conductor. However, if the voltage $V(t)$ between two conductors is changed in time, we should count their energies from the same reference level, for example the Fermi level of one of the conductors—say the 'left' one. Then for the states in that electrode, we may still use the same time dependence, $\exp\{-iE'_n t/\hbar\}$, but that in the 'right' conductor should be appropriately generalized.

To accomplish that, let us recall that the simple complex-exponential (i.e. sinusoidal) dependence results from applying the Schrödinger equation (1.25)[69] to the situation when the wavefunction may be factored into its spatial and temporal parts—see Eq. (1.57):

$$\Psi_n(\mathbf{r}, t) = a_n(t)\psi_n(\mathbf{r}).$$

We can evidently perform such partition in our case, because the stationary eigenfunctions do not depend on the energy reference level. However, the Hamiltonian of the right electrode now acquires a time-dependent component:

$$\hat{H}_R(t) = \hat{H}_R|_{V=0} - eV(t),$$

so that the equation for the complex amplitudes $a_n(t)$ is now more general than Eq. (1.61*a*):

[68] Historically, the experimental observation of this threshold in 1960 by I Giaever (which brought him a Nobel Prize in Physics) was a key confirmation of the BCS theory of superconductivity developed in 1956–57 by J Bardeen, L Cooper, and R Schrieffer.

[69] The absolutely similar separation follows from the bra–ket form, Eq. (4.158), of the Schrödinger equation.

$$i\hbar\dot{a}_n = [E_n - eV(t)]\, a_n,$$

This linear, ordinary differential equation may be readily integrated for an arbitrary time dependence $V(t)$, giving

$$a_n(t) = a_n(0)\exp\left\{-i\frac{E_n t}{\hbar} + i\frac{e}{\hbar}\int_0^t V(t')dt'\right\}. \qquad (*)$$

If the voltage is constant in time (say, $A = 0$), this result is reduced to the usual exponential dependence:

$$a_n(t)\bigg|_{A=0} = a_n(0)\exp\left\{-i\frac{E_n t}{\hbar} + i\frac{e}{\hbar}V_0 t\right\} = a_n(0)\exp\left\{-i\frac{(E_n - eV_0)\, t}{\hbar}\right\}.$$

In this case, the requirement for the matrix element $T_{nn'} \propto a_n^*(t)a_{n'}(t)$ to be time-independent leads to the level-alignment condition $E_n - eV = E_{n'}$ that was already used in the solution of the previous problem.

For the voltage including a sinusoidal ac component with a nonvanishing amplitude A as well, Eq. (*) yields

$$\begin{aligned} a_n(t) &= a_n(0)\exp\left\{-i\frac{(E_n - eV_0)\, t}{\hbar} + i\frac{eA}{\hbar}\int_0^t \cos\omega t' dt'\right\} \\ &= a_n(0)\exp\left\{-i\frac{(E_n - eV_0)\, t}{\hbar}\right\}\exp\left\{i\frac{eA}{\hbar\omega}\sin\omega t\right\}. \end{aligned}$$

Using the integral representation of the Bessel functions $J_k(\xi)$ of the first kind with integer indices[70], the last exponent may be represented as a simple Fourier series[71]:

$$\exp\left\{i\frac{eA}{\hbar\omega}\sin\omega t\right\} = \sum_{k=-\infty}^{+\infty} J_k\left(\frac{eA}{\hbar\omega}\right)e^{ik\omega t}.$$

Let us use this representation to rewrite the above result for $a_n(t)$ as

$$a_n(t) = a_n(0)\sum_{k=-\infty}^{+\infty} J_k\left(\frac{eA}{\hbar\omega}\right)\exp\left\{-\frac{i(E_n - eV_k)t}{\hbar}\right\}, \quad \text{with } V_k \equiv V_0 + k\frac{\hbar\omega}{e}.$$

Translated into plain English, this formula tells us that from the point of view of the left conductor (where our energy reference is now located), each eigenstate of the right conductor is reproduced at an infinite set of energies separated by equal intervals $\Delta E = \hbar\omega$, each with the amplitude multiplier $J_k(eA/\hbar\omega)$. Since within the region of validity of the Golden Rule, these states may be treated as independent (incoherent) ones, the tunneling currents into/from each state of this set just add up. Hence, repeating all calculations of the previous problem, and taking into account

[70] See, e.g. Eq. (A.42*a*). The basic properties and plots of the Bessel functions $J_n(\xi)$ may be found, e.g. in *Part EM* section 2.7.

[71] Such an expansion is of course possible for any periodic function of time, but for the sinusoidal argument it has especially simple coefficients.

that the component of $|T_{nn'}|^2 \propto a_n^*(t)a_{n'}(t)$, corresponding to the kth term of the sum, is now proportional to $J_k^2(eA/\hbar\omega)$, we may express the final result via that in the dc case (at $A = 0$):

$$I(\overline{V}, A) = \sum_{k=-\infty}^{\infty} J_k^2\left(\frac{eA}{\hbar\omega}\right) I(V_k, 0) \equiv \sum_{k=-\infty}^{\infty} J_k^2\left(\frac{eA}{\hbar\omega}\right) I\left(V_0 + k\frac{\hbar\omega}{e}, 0\right). \qquad (**)$$

This famous *Tien–Gordon formula*[72] shows that in the presence of an ac signal of amplitude A, the dc I–V curve may be represented as a sum of the original I–V curves (measured at $A = 0$), shifted along the dc voltage by intervals $k\hbar\omega/e$, with their currents factored by $J_k^2(eA/\hbar\omega)$. (Since these factors decrease, at large k, faster than $1/k$, the sum is always finite.) This result is especially spectacular if the original current (in the absence the ac voltage) is negligible until a certain threshold V_t and then makes a finite jump—as it does in the case of a tunnel junction between two superconductors—see the discussion at the end of the previous problem's solution. Then the external electromagnetic radiation creates similar jumps ('current steps') at lower voltages $V_t - k\hbar\omega/e$. This effect, called *photon-assisted tunneling*, may be interpreted as follows: the kth term of the sum (**) describes either absorption or emission of k quanta of the external radiation responsible for the ac part of the voltage $V(t)$. Let me hope that the reader appreciates how smartly does the quantum mechanics manage to smuggle in the notion of electromagnetic field quanta even when we try to describe the field in a completely classical way.

Let me finish with offering the reader two additional tasks:

(i) Use the properties of the Bessel functions to show that if the ac frequency is reduced so much that the voltage interval $\hbar\omega/e$ is much smaller than the voltage scale of the original I–V curve's nonlinearity, the Tien–Gordon formula (**) is reduced to the classical result

$$I(V_0, A) = \overline{I(V_0 + A\cos\omega t, 0)},$$

where the top bar means, as usual, time averaging—in our case, over the ac signal's period.

(ii) Explain why the Tien–Gordon formula differs from the solution of a similar problem for a Josephson junction—see problem 1.6.

Problem* 6.31. Use the quantum-mechanical Golden Rule to derive the Landau–Zener formula (2.257).

[72] It was derived in 1963 by P Tien and J Gordon to explain the spectacular (and initially very surprising) effects of microwave irradiation of tunnel junctions between two superconductors, observed a year earlier by A Dayem and R Martin.

Solution[73]: As was discussed in section 2.8 of the lecture notes, the Landau–Zener formula, which may be conveniently rewritten in the form (2.259),

$$W_{n'} = \exp\left\{-\frac{2\pi \,|U_{nn'}|^2}{\hbar u}\right\},$$

may be interpreted as the expression for the probability of the system's transfer to another branch of the anticrossing diagram (reproduced in the figure below), provided that it is dragged through the anticrossing with a constant 'energy speed'

$$u \equiv \frac{d}{dt}(E_n - E_{n'}),$$

where E_n and $E_{n'}$ are the unperturbed energies of the two states in the absence of their coupling described by the matrix coefficient $U_{nn'}$. Note that the Landau–Zener tunneling (indicated schematically with the solid, straight arrow in the figure below) corresponds to the *conservation* of the initial unperturbed state n', while its *change*, $n' \to n$ (shown with the dashed, curved arrow) corresponds to an adiabatic motion of the system along the same branch of the anticrossing diagram.

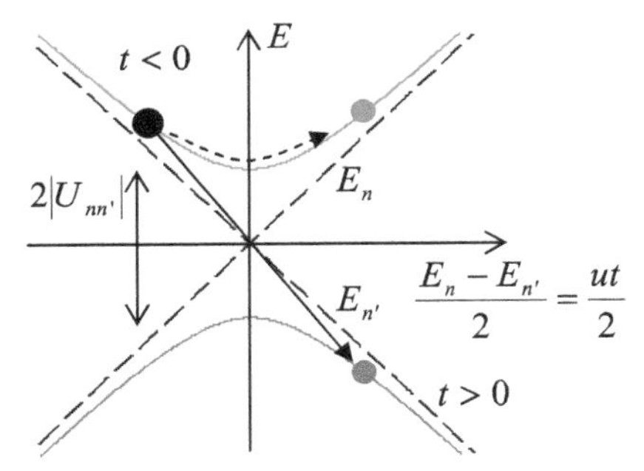

Of course, there is no explicit continuum of final states in this situation; however, from the point of view of the initial state (with energy $E_{n'}$), the second energy level E_n passes by the reference level with the 'velocity' u, and on the average may be considered as forming a pseudo-continuum of states. As was briefly discussed in section 6.6, a formal way to describe this fact is to replace the density of states ρ_n with the Dirac's delta-function $\delta(E_n - E_{n'})$. In our current case, it is equal to $\delta(ut)$, so that the Golden Rule gives a time-dependent transition rate

$$\Gamma(t) = \frac{2\pi}{\hbar}\,|U_{nn'}|^2\,\delta(ut). \qquad (*)$$

Now let us write the probability decay equation (6.113) (essentially the definition of the rate Γ),

[73] See, e.g. [5] and references therein.

$$\dot{W}_{n'} = -\Gamma(t)W_{n'}, \qquad \text{i.e.}\ \frac{dW_{n'}}{W_{n'}} = -\Gamma(t)dt,$$

where $W_{n'}$ is the probability of the particle to stay in the initial state n', i.e. in our case the probability of the Landau–Zener tunneling—see the figure above. This equation may be readily solved for an arbitrary function $\Gamma(t)$, provided that it has a limited time duration:

$$W_{n'}(t) = W_{n'}(-\infty)\exp\left\{-\int_{-\infty}^{t}\Gamma(t')dt'\right\}.$$

For our particular case (*), and $W_{n'}(t < 0) = 1$, the integration immediately yields

$$W_{n'}(t > 0) = \exp\left\{-\frac{2\pi}{\hbar}|U_{nn'}|^2\int_{t<0}^{t>0}\delta(ut')dt'\right\}$$
$$= \exp\left\{-\frac{2\pi}{\hbar u}|U_{nn'}|^2\int_{\xi<0}^{\xi>0}\delta(\xi)d\xi\right\} = \exp\left\{-\frac{2\pi}{\hbar u}|U_{nn'}|^2\right\},$$

thus giving us the Landau–Zener formula.

Actually, this result is to some extent due to good luck. Indeed, the standard derivation of the Golden Rule (see section 6.6) is based on ignoring the coherence of the partial final states with close but different energies E_n. At the Landau–Zener transition, the role of all these fixed-energy states is played by a single state n at different moments of time, and the coherence of its increments is effectively destroyed by the linear change of its eigenenergy in time. Nevertheless, this is a spectacular example of a good theory working beyond its anticipated limits.

References

[1] Shankar R 1980 *Quantum Mechanics* 2nd edn (Springer) pp 470–1
[2] Dalgarno A and Lewis J 1956 *Proc. Roy. Soc.* **A233** 70
[3] Dzyaloshinskii I *et al* 1961 *Sov. Phys. Uspekhi* **4** 153
[4] Vitanov N *et al* 2017 *Rev. Mod. Phys.* **89** 015006
[5] Amin M *et al* 2009 *Phys. Rev.* A **79** 022107

IOP Publishing

Quantum Mechanics
Problems with solutions
Konstantin K Likharev

Chapter 7

Open quantum systems

***Problem* 7.1.** Calculate the density matrix of a two-level system described by the Hamiltonian matrix

$$\mathrm{H} = \mathbf{c}\cdot\boldsymbol{\sigma} \equiv c_x\sigma_x + c_y\sigma_y + c_z\sigma_z,$$

where σ_k are the Pauli matrices, and c_j are c-numbers, in thermodynamic equilibrium at temperature T.

Solution: According to Eq. (7.24) of the lecture notes, in the thermodynamic equilibrium the density operator of our two-level system is diagonal in the energy eigenstate basis:

$$w_{nn'} = W_n\delta_{nn'}, \quad \text{with } W_n = \frac{1}{Z}\exp\left\{-\frac{E_n}{k_BT}\right\}, \quad Z = \sum_{n=1,2}\exp\left\{-\frac{E_n}{k_\mathrm{B}T}\right\}.$$

The state energies are given by Eq. (5.6) with $b = 0$:

$$E_{1,2} = \pm c, \quad \text{where } c \equiv \left(c_x^2 + c_y^2 + c_z^2\right)^{1/2} \geqslant 0.$$

As a result, the statistical sum is just

$$Z = \exp\left\{\frac{c}{k_\mathrm{B}T}\right\} + \exp\left\{-\frac{c}{k_\mathrm{B}T}\right\},$$

so that the state probabilities are

$$W_{1,2} = \frac{\exp\{\mp c/k_\mathrm{B}T\}}{\exp\{c/k_\mathrm{B}T\} + \exp\{-c/k_\mathrm{B}T\}} \equiv \frac{1}{1 + \exp\{\pm 2c/k_\mathrm{B}T\}}. \qquad (*)$$

Now we have to transfer the density operator from the basis of energy eigenstates (let us call them a_1 and a_2) to the z-basis of the states (say, $\uparrow$ and $\downarrow$), in which the

Hamiltonian matrix is given. This may be done using Eq. (4.93), valid for any operator and any two bases $\{a_1, a_2\}$ and $\{u_1, u_2\}$:

$$w_{jj'}\,|_{\text{in }u} = \sum_{n,n'=1,2} U_{jn} w_{nn'}\,|_{\text{in }a} U^{\dagger}_{n'j'} \equiv \sum_{n=1,2} U_{jn} W_n U^{\dagger}_{nj'}$$
$$= U_{j1} W_1 U^{\dagger}_{1j'} + U_{j2} W_2 U^{\dagger}_{2j'},$$

where U is the unitary matrix of transform between the bases, with the elements given by Eqs. (4.83) and (4.84):

$$U_{jn} = \langle u_j | a_n \rangle, \qquad U^{\dagger}_{n'j'} = \langle a_{n'} | u_{j'} \rangle = \langle u_{j'} | a_{n'} \rangle^*,$$

so that

$$w_{jj'}\,|_{\text{in }u} = \langle u_j | a_1 \rangle W_1 \langle u_{j'} | a_1 \rangle^* + \langle u_j | a_2 \rangle W_2 \langle u_{j'} | a_2 \rangle^*.$$

For our particular Hamiltonian, and the special basis $\{u_1, u_2\} \equiv \{\uparrow, \downarrow\}$, the transfer matrix coefficients may be borrowed from the model solution of problem 4.25 (approach 2):

$$\langle \uparrow | a_1 \rangle = \langle \downarrow | a_2 \rangle = \frac{c + c_z}{[2c(c + c_z)]^{1/2}},$$
$$\langle \uparrow | a_2 \rangle = -\langle \downarrow | a_1 \rangle^* = -\frac{c_-}{[2c(c + c_z)]^{1/2}},$$

where $c_\pm \equiv c_x \pm ic_y$, so that we finally get[1]

$$\mathrm{w} = \frac{1}{2c}\begin{pmatrix} (c + c_z)W_1 + (c - c_z)W_2 & c_-(W_1 - W_2) \\ c_+(W_1 - W_2) & (c - c_z)W_1 + (c + c_z)W_2 \end{pmatrix},$$

with $W_{1,2}$ given by Eq. (*).

This result emphasizes once again that even in thermal equilibrium, the density matrix is diagonal only in a certain (stationary) basis, but not in others.

***Problem* 7.2.** In the usual z-basis, spell out the density matrix of a spin-½ with gyromagnetic ratio γ:

(i) in the pure state with the spin definitely directed along the z-axis,
(ii) in the pure state with the spin definitely directed along the x-axis,
(iii) in the thermal equilibrium at temperature T, in a magnetic field directed along the z-axis, and
(iv) in the thermal equilibrium at temperature T, in a magnetic field directed along the x-axis.

[1] As a sanity check, if $c_x = c_y = 0$, i.e. $c = c_z$ and $c_\pm = 0$, the matrix becomes diagonal, as it should.

Solutions:

(i) In this case, the probability of the state ↑ is 100%, so that the basis of the states w_j in that the density matrix is diagonal, coincides with the z-basis. Hence we may use Eq. (7.18*a*) of the lecture notes to write

$$w_{jj'} = \delta_{j1}\delta_{j'1}, \quad \text{with } j, j' = 1, 2, \quad \text{i.e. } \mathrm{w} = \begin{pmatrix} 1 & 0 \\ 0 & 0 \end{pmatrix}.$$

(ii) In this case, the density matrix in diagonal in a different (x-) basis, so we need to use Eq. (7.18*b*),

$$w_{jj'} = U_{1j}^* U_{1j'}, \qquad (*)$$

where $U_{jj'} \equiv \langle x_j | z_{j'} \rangle$ is the unitary matrix of the transform from the x-basis to the z-basis, and the x-basis $w_{1,1}$ is taken for 1. (In the original Eq. (7.18*b*), describing a general quantum system, this state had the number j'). As we know from section 4.4 of the lecture notes, this matrix may be taken in the form[2]

$$U = U^\dagger = \frac{1}{\sqrt{2}}\begin{pmatrix} 1 & 1 \\ 1 & -1 \end{pmatrix}, \qquad (**)$$

As a result, in the z-basis we get

$$\mathrm{w} = \frac{1}{2}\begin{pmatrix} 1 & 1 \\ 1 & 1 \end{pmatrix}.$$

(iii) In the thermally-equilibrium state, the density matrix is diagonal in the basis of the stationary states n of the system, i.e. in the eigenbasis of its Hamiltonian—see Eqs. (7.23) and (7.24) of the lecture notes:

$$w_{nn'} = W_n \delta_{nn'}, \quad \text{with } W_n = \frac{1}{Z}\exp\left\{-\frac{E_n}{k_\mathrm{B}T}\right\}, \quad \text{and } Z \equiv \sum_n \exp\left\{-\frac{E_n}{k_\mathrm{B}T}\right\},$$

where E_n are the eigenvalues of system's energy. As was discussed in section 4.4 of the lecture notes, for a spin-½ particle with the gyromagnetic ratio γ in a magnetic field $\mathscr{B}$, the energy (referred to its field-free value) has two eigenvalues,

$$E_{1,2} = \mp\gamma\frac{\hbar}{2}\mathscr{B},$$

so that the two-term statistical sum and the 2 × 2 density matrix may be readily spelled out:

$$Z = e^+ + e^- \equiv 2\cosh\frac{\gamma\hbar\mathscr{B}}{2k_\mathrm{B}T}, \quad \text{where } e^\pm \equiv \exp\left\{\pm\frac{\gamma\hbar\mathscr{B}}{2k_\mathrm{B}T}\right\},$$

[2] As a reminder, the unitary matrix elements $U_{jj'}$ are defined to an arbitrary phase multiplier $\exp\{i\varphi_j\}$.

$$\mathrm{w} = \frac{1}{e^+ + e^-}\begin{pmatrix} e^+ & 0 \\ 0 & e^- \end{pmatrix}. \qquad (***)$$

In our current case of the field directed along axis z, the basis of stationary states coincides with the z-basis, i.e. the last formula gives the final answer to the posed question. In the limit of low temperatures, $k_B T \ll \gamma\hbar\mathscr{B}$, $e^+ \gg e^-$, and Eq. (***) is virtually reduced to the result of part (i), showing that the system is in a virtually pure state $\uparrow$.

(iv) In the case of the magnetic field directed along axis x, Eq. (***) is valid in the x-basis, and we may apply the general rule (4.93), with the unitary matrix (**), to the statistical operator, to calculate each element of the density matrix in the z-basis:

$$w_{jj'} = \sum_{n,n'=1}^{2} U_{jn} w_{nn'} \,|_{\text{in } x} U^{\dagger}_{n'j'} = \sum_{n=1}^{2} U_{jn} W_n U^{\dagger}_{nj'}.$$

A straightforward calculation of this sum yields

$$w = \frac{1}{2(e^+ + e^-)}\begin{pmatrix} e^+ + e^- & e^+ - e^- \\ e^+ - e^- & e^+ + e^- \end{pmatrix} = \frac{1}{2}\begin{pmatrix} 1 & \tanh(\gamma\hbar\mathscr{B}/2k_B T) \\ \tanh(\gamma\hbar\mathscr{B}/2k_B T) & 1 \end{pmatrix}.$$

This result shows that the diagonal elements of the density matrix, i.e. the probabilities of finding the spin oriented in both directions of axis z, are equal to 50% each. This is exactly what we could expect for the particle in an x-oriented field, with zero z-component. The result for the off-diagonal elements is somewhat less trivial. It shows that these elements, equal to each other, are always smaller than the diagonal ones, approaching them only in the low-temperature limit $k_B T \ll \gamma\hbar\mathscr{B}$, when the density matrix is reduced to the one calculated in part (ii).

Note that all these results may be obtained, in the corresponding limits, from the result of solution of problem 7.1, with the magnitude c of the vector $\mathbf{c}$ equal to $\gamma\hbar\mathscr{B}/2$.

***Problem* 7.3.** Calculate the Wigner functions of a harmonic oscillator:

(i) in the thermodynamic equilibrium at temperature T,
(ii) in the ground state, and
(iii) in the Glauber state with a dimensionless complex amplitude α.

Discuss the relation between the first of the results and the Gibbs distribution.

Solutions:

(i) In section 7.2 of the lecture notes, the following result for the density matrix of the system was derived—see Eq. (7.44),

$$w(x, x') = \frac{1}{Z}\left(\frac{m\omega_0}{2\pi\hbar \sinh[\hbar\omega_0/k_B T]}\right)^{1/2} \times \exp\left\{-\frac{m\omega_0[(x^2 + x'^2)\cosh[\hbar\omega_0/k_B T] - 2xx']}{2\hbar \sinh[\hbar\omega_0/k_B T]}\right\}, \qquad (*)$$

where the statistical sum Z is given by Eq. (7.25):

$$Z = \frac{\exp\{-\hbar\omega_0/2k_B T\}}{1 - \exp\{-\hbar\omega_0/k_B T\}}.$$

Let us use Eq. (7.50) to recalculate the matrix into the Wigner function

$$W(X, P) \equiv \frac{1}{2\pi\hbar}\int w\left(X + \frac{\tilde{X}}{2}, X - \frac{\tilde{X}}{2}\right)\exp\left\{-\frac{iP\tilde{X}}{\hbar}\right\}d\tilde{X}.$$

Leaving the normalization coefficient alone for a while, let us spell out the exponent in Eq. (*) as a function of the new variables:

$$x^2 + x'^2 = \left(X + \frac{\tilde{X}}{2}\right)^2 + \left(X - \frac{\tilde{X}}{2}\right)^2 = 2\left(X^2 + \frac{\tilde{X}^2}{4}\right),$$

$$2xx' = 2\left(X + \frac{\tilde{X}}{2}\right)\left(X - \frac{\tilde{X}}{2}\right) = 2\left(X^2 - \frac{\tilde{X}^2}{4}\right),$$

so that

$$\exp\left\{-\frac{m\omega_0[(x^2 + x'^2)\cosh[\hbar\omega_0/k_B T] - 2xx']}{2\hbar \sinh[\hbar\omega_0/k_B T]}\right\}$$

$$= \exp\left\{-\frac{m\omega_0}{\hbar}\left(X^2\frac{\cosh[\hbar\omega_0/k_B T] - 1}{\sinh[\hbar\omega_0/k_B T]} + \frac{\tilde{X}^2}{4}\frac{\cosh[\hbar\omega_0/k_B T] + 1}{\sinh[\hbar\omega_0/k_B T]}\right)\right\}$$

$$\equiv \exp\left\{-X^2\frac{m\omega_0}{\hbar}\tanh\frac{\hbar\omega_0}{2k_B T}\right\}\exp\left\{-\frac{\tilde{X}^2}{4}\frac{m\omega_0}{\hbar}\coth\frac{\hbar\omega_0}{2k_B T}\right\}.$$

The Fourier transform in the Wigner function's definition affects only the second exponent in the last expression, and is a standard Gaussian integral

$$\int_{-\infty}^{+\infty}\exp\left\{-\frac{\tilde{X}^2}{4}\frac{m\omega_0}{\hbar}\coth\frac{\hbar\omega_0}{2k_B T}\right\}\exp\left\{-\frac{iP\tilde{X}}{\hbar}\right\}d\tilde{X} \equiv \int_{-\infty}^{+\infty}\exp\left\{-\frac{\tilde{X}^2}{2x_c^2} - \frac{iP\tilde{X}}{\hbar}\right\}d\tilde{X},$$

where[3]

$$x_c^2 = \frac{2\hbar}{m\omega_0}\tanh\frac{\hbar\omega_0}{2k_B T}.$$

[3] Note that this parameter (essentially the correlation length of the oscillator) has a temperature dependence opposite to $\langle x^2 \rangle$ (the coordinate's variance—see Eq. (7.48) of the lecture notes): it *decreases* with the growth of temperature, at $k_B T \gg \hbar\omega_0$ approaching its value (7.37) for a free particle, with the qualitatively similar temperature behavior. Its discussion may be found in section 7.2 of the lecture notes, just after Eq. (7.37).

This integral may be worked out either as usual, by complementing the exponent to a full square, or just reusing the results of an absolutely similar integration performed in section 2.2. The final result is proportional to

$$\exp\left\{-\frac{1}{\hbar m\omega_0}\tanh\frac{\hbar\omega_0}{2k_BT}P^2\right\},$$

so that the Wigner function as the whole is

$$\begin{aligned} W(X, P) &= A\exp\left\{-\tanh\frac{\hbar\omega_0}{2k_BT}\left[\frac{P^2}{\hbar m\omega_0}+\frac{m\omega_0X^2}{\hbar}\right]\right\} \\ &\equiv A\exp\left\{-\frac{H(X, P)}{\langle E\rangle}\right\}, \end{aligned} \qquad (**)$$

where $H(X, P)$ is classical Hamiltonian function of the oscillator, taken of the arguments X and P,

$$H(X, P) \equiv \frac{P^2}{2m}+\frac{m\omega_0^2X^2}{2},$$

and $\langle E\rangle$ is the average energy of the oscillator—see Eq. (7.26) of the lecture notes:

$$\langle E\rangle = \frac{\hbar\omega_0}{2}\coth\frac{\hbar\omega_0}{2k_BT}.$$

So, we have proved Eq. (7.62), and simultaneously found the coefficient C participating in it. What remains is to calculate the normalization coefficient A. The easiest way to do this is to require that the integration of W over X and P gives 1. This procedure readily yields

$$A = \frac{\omega_0}{2\pi\langle E\rangle} = \frac{1}{\pi\hbar}\tanh\frac{\hbar\omega_0}{2k_BT},$$

so that, finally,

$$W(X, P) = \frac{\omega_0}{2\pi\langle E\rangle}\exp\left\{-\frac{H(X, P)}{\langle E\rangle}\right\}.$$

We see that the Wigner function looks much simpler than the corresponding density matrix (*), while carrying all the information contained in the latter. A very interesting feature of this result is that its functional dependence on $H(X, P)$ coincides with that of the Gibbs distribution (7.25), but with the energy normalized to $\langle E\rangle$ rather than k_BT. (The two coincide only in the high-temperature limit $k_BT \gg \hbar\omega_0$.)

(ii) The easiest way to calculate the Wigner function in the ground state is to use the result (**) in the low-temperature limit ($k_BT \ll \hbar\omega_0$, $\langle E\rangle \approx E_0 = \hbar\omega_0/2$):

$$W_0(X, P) = \frac{1}{\pi\hbar}\exp\left\{-\left(\frac{m\omega_0^2 X^2}{2} + \frac{P^2}{2m}\right)\bigg/\frac{\hbar\omega_0}{2}\right\} \equiv \frac{1}{\pi\hbar}\exp\left\{-\frac{H(X, P)}{E_0}\right\}.$$

This is the function plotted in panel (a) of figure 7.3 of the lecture notes.

Alternatively, this result may be readily obtained by the direct integration of the factorable density matrix of this pure quantum state:

$$W(X, P) \equiv \frac{1}{2\pi\hbar}\int \psi_0\left(X + \frac{\tilde{X}}{2}\right)\psi_0^*\left(X - \frac{\tilde{X}}{2}\right)\exp\left\{-\frac{iP\tilde{X}}{\hbar}\right\}d\tilde{X}.$$

(iii) According to Eqs. (5.107) of the lecture notes, the wavefunction of the Glauber state with the dimensionless complex amplitude α (5.102) may be obtained from the ground-state wavefunction $\psi_0(x)$ by the shift of its argument by $\sqrt{2}x_0 \operatorname{Re} \alpha$, and its multiplication by a phase exponent corresponding to a monochromatic wave with the momentum $\sqrt{2}\, m\omega_0 x_0 \operatorname{Im} \alpha$. Such shifts result in similar shifts of the arguments of the ground-state Wigner function by the listed amounts, so that it becomes

$$W(X, P) = \frac{1}{\pi\hbar}\exp\left\{-\left[\frac{m\omega_0^2(X - \sqrt{2}x_0 \operatorname{Re} \alpha)^2}{2} + \frac{(P - \sqrt{2}x_0 m\omega_0 \operatorname{Im} \alpha)^2}{2m}\right]\bigg/\frac{\hbar\omega_0}{2}\right\},$$

where $x_0 \equiv (\hbar/m\omega_0)^{1/2}$.

***Problem* 7.4.** Calculate the Wigner function of a harmonic oscillator, with mass m and frequency ω_0, in its first excited stationary state ($n = 1$).

Solution: Since this is a pure state, we may use Eq. (7.63) of the lecture notes:

$$W(X, P) = \frac{1}{2\pi\hbar}\int \psi_1\left(X + \frac{\tilde{X}}{2}\right)\psi_1^*\left(X - \frac{\tilde{X}}{2}\right)\exp\left\{-\frac{iP\tilde{X}}{\hbar}\right\}d\tilde{X},$$

where $\psi_1(x)$ is the wavefunction of the oscillator in its first excited Fock state. This function is given by Eqs. (2.282) and (2.284) with $n = 1$:

$$\psi_1(x) = \frac{1}{(2^1 \cdot 1!)^{1/2}\pi^{1/4}x_0^{1/2}}\exp\left\{-\frac{x^2}{2x_0^2}\right\}H_1\left(\frac{x}{x_0}\right) \equiv \frac{2^{1/2}x}{\pi^{1/4}x_0^{3/2}}\exp\left\{-\frac{x^2}{2x_0^2}\right\},$$

where $x_0 \equiv (\hbar/m\omega_0)^{1/2}$, so that

$$\psi_1\left(X + \frac{\tilde{X}}{2}\right)\psi_1^*\left(X - \frac{\tilde{X}}{2}\right) = 2\frac{(X + \tilde{X}/2)(X - \tilde{X}/2)}{\pi^{1/2}x_0^3}\exp\left\{-\frac{(X + \tilde{X}/2)^2 + (X - \tilde{X}/2)^2}{2x_0^2}\right\}$$

$$\equiv 2\frac{X^2 - \tilde{X}^2/4}{\pi^{1/2}x_0^3}\exp\left\{-\frac{X^2 + \tilde{X}^2/4}{x_0^2}\right\},$$

and

$$W(X, P) = \frac{1}{\pi\hbar}\int_{-\infty}^{+\infty} \frac{X^2 - \tilde{X}^2/4}{\pi^{1/2}x_0^3} \exp\left\{-\frac{X^2 + \tilde{X}^2/4}{x_0^2}\right\} \exp\left\{-\frac{iP\tilde{X}}{\hbar}\right\} d\tilde{X}$$

$$\equiv \frac{1}{\pi^{3/2}\hbar x_0^3} \exp\left\{-\frac{X^2}{x_0^2}\right\} \int_{-\infty}^{+\infty} \left(X^2 - \frac{\tilde{X}^2}{4}\right) \exp\left\{-\frac{\tilde{X}^2}{4x_0^2} - \frac{iP\tilde{X}}{\hbar}\right\} d\tilde{X}.$$

After the usual completion of the expression under the exponent to the full square,

$$-\frac{\tilde{X}^2}{4x_0^2} - \frac{iP\tilde{X}}{\hbar} = -\frac{\left(\tilde{X} + 2iPx_0^2/\hbar\right)^2}{4x_0^2} - \frac{P^2x_0^2}{\hbar^2} = -\frac{Z^2}{4x_0^2} - \frac{P^2x_0^2}{\hbar^2},$$

where $Z \equiv \tilde{X} + 2iPx_0^2/\hbar$, so that $\tilde{X} = Z - 2iPx_0^2/\hbar$, we get

$$W(X, P) = \frac{1}{\pi^{3/2}\hbar x_0} \exp\left\{-\frac{X^2}{x_0^2} - \frac{P^2x_0^2}{\hbar^2}\right\} I,$$

where

$$I \equiv \int_{-\infty}^{+\infty} \left(\frac{X^2}{x_0^2} - \frac{\left(Z - 2iPx_0^2/\hbar\right)^2}{4x_0^2}\right) \exp\left\{-\frac{Z^2}{4x_0^2}\right\} d\tilde{X}$$

$$= \left(\frac{X^2}{x_0^2} + \frac{P^2x_0^2}{\hbar^2}\right) \int_{-\infty}^{+\infty} \exp\left\{-\frac{Z^2}{4x_0^2}\right\} d\tilde{X}$$

$$+ \frac{iP}{\hbar}\int_{-\infty}^{+\infty} Z \exp\left\{-\frac{Z^2}{4x_0^2}\right\} d\tilde{X} - \frac{1}{4x_0^2}\int_{-\infty}^{+\infty} Z^2 \exp\left\{-\frac{Z^2}{4x_0^2}\right\} d\tilde{X}.$$

As was discussed in section 2.2 of the lecture notes, since the functions under these integrals are analytical, and tend to zero at $\tilde{X} \to \pm\infty$, the purely imaginary shift between Z and $\tilde{X}$ does not affect them, so that the second integral vanishes (because the function under it is asymmetric), while the first and the third ones are reduced, by substitution $\xi \equiv \tilde{X}/2x_0$, to table integrals[4]

$$\int_{-\infty}^{+\infty} e^{-\xi^2} d\xi = \pi^{1/2}, \qquad \int_{-\infty}^{+\infty} \xi^2 e^{-\xi^2} d\xi = \frac{\pi^{1/2}}{2}.$$

As a result, we finally get

$$W(X, P) = \frac{2}{\pi\hbar}\left(\frac{X^2}{x_0^2} + \frac{P^2x_0^4}{\hbar^2} - \frac{1}{2}\right) \exp\left\{-\frac{X^2}{x_0^2} - \frac{P^2x_0^2}{\hbar^2}\right\}$$

$$\equiv \frac{2}{\pi\hbar}\left(\rho^2 - \frac{1}{2}\right) \exp\{-\rho^2\},$$

[4] See, e.g. Eq. (A.36).

where

$$\rho \equiv \left(\frac{X^2}{x_0^2} + \frac{P^2 x_0^4}{\hbar^2}\right)^{1/2} \equiv \frac{H(X, P)}{E_0}, \quad \text{with } H = \frac{P^2}{2m} + \frac{m\omega_0^2 X^2}{2}, \text{ and } E_0 = \frac{\hbar\omega_0}{2},$$

is the normalized distance from the center of the $[X, P]$ plane. This function, whose plot is shown on panel (b) of figure 7.3, is negative at $\rho < 1/\sqrt{2}$, illustrating the impossibility to interpret the Wigner function as the probability density—see the conclusion of section 7.2 of the lecture notes.

Problem* 7.5. A harmonic oscillator is weakly coupled to an Ohmic environment.

(i) Use the rotating-wave approximation to write the reduced equations of motion for the Heisenberg operators of the complex amplitude of oscillations.
(ii) Calculate the expectation values of the correlators of the fluctuation force operators, participating in these equations, and express them via the average number $\langle n \rangle$ of thermally-induced excitations in equilibrium, given by the second of Eqs. (7.26*b*) of the lecture notes.

Solutions:

(i) Differentiating the definitions (5.65) of the creation–annihilation operators over time, then using the Heisenberg equations (7.144) of motion of an oscillator with Ohmic damping, and then the reciprocal relations (5.66), we get the following equations of motion of these operators:

$$\begin{aligned} \dot{\hat{a}} &= -i\omega_0 \hat{a} - \delta(\hat{a} - \hat{a}^\dagger) + \frac{i}{(2\hbar m\omega_0)^{1/2}} \hat{\tilde{F}}(t), \\ \dot{\hat{a}}^\dagger &= i\omega_0 \hat{a}^\dagger + \delta(\hat{a} - \hat{a}^\dagger) - \frac{i}{(2\hbar m\omega_0)^{1/2}} \hat{\tilde{F}}(t), \end{aligned} \qquad (*)$$

where $\delta \equiv \eta/2m$ is the damping constant. At negligible coupling to the environment, these equations are reduced to Eqs. (5.140) with just the first terms on their right-hand sides, and hence follow the oscillating solutions (5.141), so that the following operators,

$$\hat{\alpha} \equiv \hat{a}e^{i\omega_0 t}, \quad \hat{\alpha}^\dagger \equiv \hat{a}^\dagger e^{-i\omega_0 t}, \qquad (**)$$

do not evolve in time. Evidently, in the classical limit (e.g. for Glauber states with $|\alpha| \gg 1$), these operators have the meaning of the complex amplitude of the oscillations and its complex conjugate, so that even in the general case they may be interpreted as the Heisenberg operators of the amplitudes.

Coupling to the environment results in a variation of the amplitudes. The exact equations of this motion, that may be obtained by plugging Eqs. (*) into Eqs. (**) differentiated over time,

$$\dot{\tilde{a}} = (\dot{\hat{a}} + i\omega_0\hat{a})e^{i\omega_0 t} = -\delta(\tilde{a}e^{-i\omega_0 t} - \tilde{a}^\dagger e^{i\omega_0 t})e^{i\omega_0 t} + \frac{i}{(2\hbar m\omega_0)^{1/2}}\hat{\tilde{F}}(t)e^{i\omega_0 t},$$

$$\dot{\tilde{a}}^\dagger = (\dot{\hat{a}}^\dagger - i\omega_0\hat{a})e^{-i\omega_0 t} = \delta(\tilde{a}e^{-i\omega_0 t} - \tilde{a}^\dagger e^{i\omega_0 t})e^{-i\omega_0 t} - \frac{i}{(2\hbar m\omega_0)^{1/2}}\hat{\tilde{F}}(t)e^{-i\omega_0 t},$$

are coupled and have an analytical but rather cumbersome solution.

However, we may notice that all right-hand-side terms of these equations are proportional to the weak coupling to the environment, i.e. small, so that the time evolution of the operators (**) is slow. Hence, using the basic idea of the rotating-wave approximation (RWA) discussed in section 6.5,[5] we may average these right-hand parts over a relatively long time interval Δt, thus eliminating the rapidly-oscillating terms, because they have a small effect on the amplitude evolution. This first-order approximation yields the following 'reduced' (or 'RWA') Heisenberg–Langevin equations[6]:

$$\dot{\tilde{a}} = -\delta\tilde{a} + \hat{f}(t), \quad \dot{\tilde{a}}^\dagger = -\delta\tilde{a}^\dagger + \hat{f}^\dagger(t), \qquad (***)$$

where the reduced force operator is defined as

$$\hat{f}(t) \equiv \frac{i}{(2\hbar m\omega_0)^{1/2}}\overline{\hat{\tilde{F}}(t)e^{i\omega_0 t}} \equiv \frac{i}{(2\hbar m\omega_0)^{1/2}}\frac{1}{\Delta t}\int_{t-\Delta t/2}^{t+\Delta t/2}\hat{\tilde{F}}(t)e^{i\omega_0 t}dt,$$

and $\hat{f}^\dagger(t)$ as its Hermitian adjoint. Evidently, the first terms on the right-hand sides of the reduced equations (which dominate if $\langle\alpha\rangle$ is large enough to neglect fluctuations) try to decrease the amplitude exponentially, as $\exp\{-\delta t\}$, while the second terms, representing the fluctuation force, disturb such deterministic decay. In particular, these terms do not allow the operators $\tilde{a}$ and $\tilde{a}^\dagger$ to approach zero, because that would violate their commutation relation

$$[\tilde{a}, \tilde{a}^\dagger] = \hat{I},$$

which follows from Eq. (**) above and Eq. (5.68) of the lecture notes.

Note that since the Langevin operator $\hat{\tilde{F}}(t)$ is a random function of time, i.e. is not periodic, the averaging interval Δt should be chosen more carefully than in the deterministic case[7]. On one hand, it should be *much longer* than $2\pi/\omega_0$, in order to suppress the high-frequency components of the product under the integral. On the other hand, in order to describe the amplitude evolution correctly, Δt should be *much shorter* than the decay time constant $1/\delta$:

[5] See also the discussion of the van der Pol method (i.e. the classical version of the RWA) in *Part CM* section 5.3.

[6] Actually, such equations are also valid for a more general situation when the operator $\hat{\tilde{F}}(t)$ describes not only the fluctuations, but also an external force with a frequency ω close enough to the resonance frequency ω_0 of the oscillator. Quantum-mechanical solutions of several other problems (such as parametric excitation, see *Part CM* section 5.6) may be also obtained using straightforward generalizations of these equations.

[7] If the right-hand part is $2\pi/\omega$—periodic (where ω is close to, but not necessarily exactly equal to ω_0—see *Part CM* section 5.3–5.6), then a perfect averaging out of fast components may be achieved by taking Δt equal to that period.

$$\frac{2\pi}{\omega_0} \ll \Delta t \ll \frac{1}{\delta}.$$

Such a choice is of course possible only if the Q-factor $Q \equiv \omega_0/2\delta$ of the oscillator is much higher than 1, which is therefore the necessary condition of applicability of the rotating-wave approximation.

(ii) Let us calculate the statistical properties of the effective low-frequency forces $\hat{f}(t)$ and $\hat{f}^\dagger(t)$. Their statistical averages (in the language of mathematical statistics, their *first moments*), are proportional to those of the 'parent force' $\hat{\tilde{F}}(t)$ and hence are equal to zero—see the discussion of Eq. (7.92) in the lecture notes. Thus, let us start from the calculation of the following second moment of the forces, based on their *anticommutator*:

$$\left\langle \left\{\hat{f}(t_1), \hat{f}^\dagger(t_2)\right\} \right\rangle = \frac{1}{2\hbar m\omega_0}\frac{1}{(\Delta t)^2}\int_{t_1-\Delta t/2}^{t_1+\Delta t/2} dt' \\ \times \int_{t_2-\Delta t/2}^{t_2+\Delta t/2} dt'' \left\langle \left\{ \hat{\tilde{F}}(t'), \hat{\tilde{F}}(t'')\right\} \right\rangle \exp\{i\omega_0(t' - t'')\}.$$

With the account of Eqs. (7.110)–(7.111), the double integral in this relation is

$$\int_{t_1-\Delta t/2}^{t_1+\Delta t/2} dt' \int_{t_2-\Delta t/2}^{t_2+\Delta t/2} dt''\, 2K_F(t' - t'')\exp\{i\omega_0(t' - t'')\}.$$

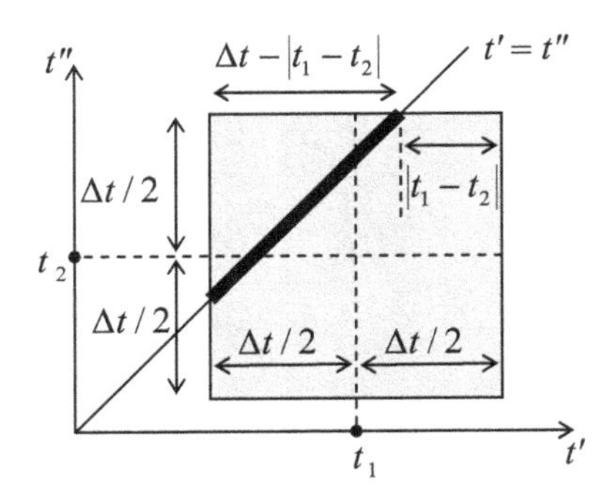

The averaging time Δt, which satisfies the above strong conditions, is much longer than the correlation time τ_c of the environment force fluctuations (i.e. the characteristic time of decay of its correlation function K_F), so that only a relatively narrow area of the $[t', t'']$ plane, shown schematically by the bold line segment in the figure above, gives a noticeable contribution to this double integral. (Note that at $|t_1 - t_2| > \Delta t$, the segment disappears, i.e. the integral vanishes.) Hence the internal integral may be formally taken in the infinite limits of the difference $t'' - t'$, and we may use the first form of Eq. (7.112) to get

$$\left\langle \left\{ \hat{f}(t_1), \hat{f}^\dagger(t_2) \right\} \right\rangle = \frac{2\pi S_F(\omega_0)}{\hbar m \omega_0} D(t_1 - t_2),$$

$$\text{with} \quad D(\tau) \equiv \frac{1}{(\Delta t)^2} \times \begin{cases} \Delta t - |\tau|, & \text{if } |\tau| < \Delta t, \\ 0, & \text{otherwise.} \end{cases}$$

On the time scale ($\sim 1/\delta$) of the amplitude evolution, the function $D(\tau)$ may be well approximated with a delta-function $C\delta(\tau)$ with 'area'

$$C = \frac{1}{(\Delta t)^2} \int_{-\Delta t}^{+\Delta t} (\Delta t - |\tau|) d\tau = \int_{-1}^{1} (1 - |\xi|) d\xi = 1,$$

so that finally, we may write

$$\left\langle \left\{ \hat{f}(t_1), \hat{f}^\dagger(t_2) \right\} \right\rangle = \frac{2\pi S_F(\omega_0)}{\hbar m \omega_0} \delta(t_1 - t_2).$$

Now the fluctuation–dissipation theorem (7.134) may be used to recast this result as

$$\left\langle \left\{ \hat{f}(t_1), \hat{f}^\dagger(t_2) \right\} \right\rangle = = \frac{\chi''(\omega_0)}{m\omega_0} \coth \frac{\hbar\omega_0}{2k_B T} \delta(t_1 - t_2),$$

so that at the Ohmic dissipation (7.138),

$$\left\langle \left\{ \hat{f}(t_1), \hat{f}^\dagger(t_2) \right\} \right\rangle = \frac{\eta}{m} \coth \frac{\hbar\omega_0}{2k_B T} \delta(t_1 - t_2) \equiv 2\delta \coth \frac{\hbar\omega_0}{2k_B T} \delta(t_1 - t_2).$$

Performing an absolutely similar calculation for the *commutator* of the two fluctuating force operators, with the only difference of using the Green–Kubo formula (7.109) rather than the fluctuation–dissipation theorem, we get

$$\left\langle \left[\hat{f}(t_1), \hat{f}^\dagger(t_2) \right] \right\rangle = 2\delta \times \delta(t_1 - t_2).$$

Now we may combine these two results to calculate the required correlation functions:

$$\begin{aligned} \left\langle \hat{f}(t_1) \hat{f}^\dagger(t_2) \right\rangle &\equiv \frac{1}{2} \left(\left\langle \left\{ \hat{f}(t_1), \hat{f}^\dagger(t_2) \right\} \right\rangle + \left\langle \left[\hat{f}(t_1), \hat{f}^\dagger(t_2) \right] \right\rangle \right) \\ &= 2\delta \left[\coth(\hbar\omega_0/k_B T) + 1 \right] \delta(t_1 - t_2), \\ \langle \hat{f}(t_1)^\dagger \hat{f}(t_2) \rangle &\equiv \frac{1}{2} \left(\left\langle \left\{ \hat{f}(t_1), \hat{f}^\dagger(t_2) \right\} \right\rangle - \left\langle \left[\hat{f}(t_1), \hat{f}^\dagger(t_2) \right] \right\rangle \right) \\ &= 2\delta \left[\coth(\hbar\omega_0/k_B T) - 1 \right] \delta(t_1 - t_2). \end{aligned}$$

Comparing these formulas with Eq. (7.26*b*), which may be rewritten as

$$\langle n \rangle = \frac{1}{2} \left(\coth \frac{\hbar\omega_0}{2k_B T} - 1 \right),$$

we finally get a very elegant couple of relations[8],

$$\left\langle \hat{f}(t_1)\hat{f}^\dagger(t_2) \right\rangle = 2\delta\left(\langle n \rangle + 1\right)\delta(t_1 - t_2),$$
$$\left\langle \hat{f}^\dagger(t_1)\hat{f}(t_2) \right\rangle = 2\delta\,\langle n \rangle\,\delta(t_1 - t_2). \qquad (****)$$

The result expressed by Eqs. (***) and (****), frequently with addition of other terms reflecting other forces acting upon the oscillator, is broadly used in quantum optics and electronics.

The physical reason why the noise source approximation by the delta-correlated functions in the RWA equations gives sufficient accuracy (i.e. the equations correct to the first order in H_{int}) is very simple. As was discussed in section 7.5, of the whole broad spectrum of the fluctuations (in our case, coming from the thermally-equilibrium environment), a high-Q oscillator is substantially affected only by the Fourier components very close to its resonance frequency, and we commit no substantial error if we replace the genuine spectral density $S_F(\omega)$ with a frequency-independent value equal to $S_F(\omega_0)$. However, as was discussed in section 7.4, a random process with a constant spectral density is delta-correlated. This argumentation might be used for a faster calculation of the average *anticommutator* of the fluctuation force; however, for the calculation of its *commutator* we still need the integration carried out above, based on the (more subtle) Green–Kubo formula.

***Problem* 7.6.** Calculate the average potential energy of long-range electrostatic interaction between two similar isotropic, 3D harmonic oscillators, each with the electric dipole moment $\mathbf{d} = q\mathbf{s}$, where $\mathbf{s}$ is the oscillator's displacement from its equilibrium position, at arbitrary temperature T.

Solution: For $T = 0$, this interaction (in the zero-temperature limit, called the London dispersion force) was calculated, by two different methods (both due to F London) in the solution of problems 3.16 and 5.15. Reviewing the latter solution[9], we see that the calculations are completely valid at $T \neq 0$, up to the following result,

$$\left\langle \overline{\hat{\mathscr{E}}_1 \cdot \hat{\mathbf{d}}_2} \right\rangle = \frac{3q^2}{m^2}\left(\frac{q}{4\pi\varepsilon_0 r^3}\right)^2 \frac{\hbar}{\omega_1\left(\omega_2^2 - \omega_1^2\right)} \langle \hat{a}(0)\hat{a}^\dagger(0) + \text{h.c.}\rangle_1, \qquad (*)$$

and the similar result, with swapped indices 1 and 2, for the second component,

$$\left\langle \overline{\hat{\mathscr{E}}_2 \cdot \hat{\mathbf{d}}_1} \right\rangle = \frac{3q^2}{m^2}\left(\frac{q}{4\pi\varepsilon_0 r^3}\right)^2 \frac{\hbar}{\omega_2\left(\omega_1^2 - \omega_2^2\right)} \langle \hat{a}(0)\hat{a}^\dagger(0) + \text{h.c.}\rangle_2, \qquad (**)$$

[8] To the best of my knowledge, they were first derived by M Lax in 1966.

[9] The reader is challenged to generalize the former solution.

of the average interaction between the oscillators,

$$\langle U\rangle = -\frac{1}{2}\left\langle\overline{\hat{\mathscr{E}}_1\cdot\hat{\mathbf{d}}_2}\right\rangle - \frac{1}{2}\left\langle\overline{\hat{\mathscr{E}}_2\cdot\hat{\mathbf{d}}_1}\right\rangle,$$

provided that the averaging on the right-hand sides of Eqs. (*) and (**) is understood not only in the quantum-mechanical, but also in the statistical-ensemble sense.

As Eq. (5.70) of the lecture notes shows, such an average is just that of the oscillator's Hamiltonian, i.e. its average energy $\langle E\rangle$, divided by the ground-state energy $\hbar\omega_k/2$ of the corresponding oscillator, so that for the system in thermal equilibrium at temperature T we may use Eq. (7.26) to immediately obtain the required answer:

$$\langle U\rangle = -\frac{3q^2}{2m^2}\left(\frac{q}{4\pi\varepsilon_0 r^3}\right)^2\left[\frac{1}{\omega_1\left(\omega_2^2-\omega_1^2\right)}\coth\frac{\hbar\omega_1}{2k_BT} + \frac{1}{\omega_2\left(\omega_1^2-\omega_2^2\right)}\coth\frac{\hbar\omega_2}{2k_BT}\right]. \qquad (***)$$

In the classical limit ($k_BT \gg \hbar\omega_{1,2}$), this expression is reduced to

$$\langle U\rangle = -\frac{3q^2}{m^2}\left(\frac{q}{4\pi\varepsilon_0 r^3}\right)^2\left[\frac{1}{\omega_1^2\left(\omega_2^2-\omega_2^2\right)} + \frac{1}{\omega_2^2\left(\omega_1^2-\omega_1^2\right)}\right]k_BT$$

$$\equiv -\frac{3q^2}{m^2}\left(\frac{q}{4\pi\varepsilon_0 r^3}\right)^2\frac{1}{\omega_1^2\omega_2^2}k_BT.$$

For the most important case of similar oscillators, of frequency $\omega_0 \equiv \omega_1 = \omega_2$, we get

$$\langle U\rangle = -\frac{3q^2}{m^2\omega_0^4}\left(\frac{q}{4\pi\varepsilon_0 r^3}\right)^2 k_BT \equiv -3\mu^2 k_BT, \quad \text{where } \mu \equiv \frac{q^2}{4\pi\varepsilon_0 r^3 m\omega_0^2} \ll 1.$$

Note that this expression differs from the result following from Eq. (***) at $T = 0$,[10]

$$\langle U\rangle = -\frac{3}{4}\mu^2\hbar\omega_0,$$

by a not-quite-trivial replacement $\hbar\omega_0/4 \to k_BT$ (instead of the more usual $\hbar\omega_0/2 \to k_BT$), because of the resonant behavior of both terms in the general result (***).

***Problem* 7.7.** A semi-infinite string with mass μ per unit length is attached to a wall, and stretched with a constant force (tension) $\mathscr{T}$. Calculate the spectral density of the transverse force exerted on the wall, in thermal equilibrium.

[10] As a reminder, this result was already obtained in the solutions of not only problem 5.15, but also (by different means) problem 3.16.

Solution: Classical mechanics says that the string may support transverse waves with two independent polarizations—for example, two mutually perpendicular linear polarizations[11]. If the waves are small, their dynamics is independent, so that in order to calculate one Cartesian coordinate of the transverse force (say, along axis x, normal to the string's direction z), it is sufficient to analyze the waves $x(z, t)$, with displacements in one plane, $[x, z]$.

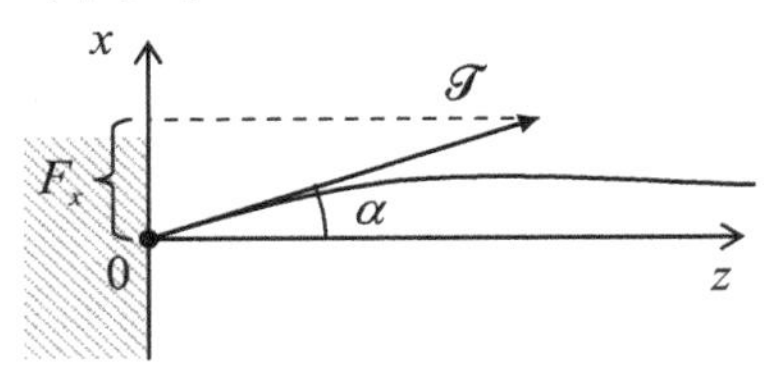

According to the discussion in section 7.4 of the lecture notes, in order to use Eq. (7.134) of the lecture notes for the spectral density of the force,

$$S_F(\omega) = \frac{\hbar}{2\pi}\text{Im}\,\chi(\omega)\coth\frac{\hbar\omega}{2k_\text{B}T}, \tag{*}$$

it is sufficient to calculate the generalized susceptibility $\chi(\omega)$, defined by Eq. (7.124),

$$F_\omega = \chi(\omega)x_\omega,$$

i.e. the ratio of complex amplitudes of two classical processes: externally imposed classical 1D oscillations $x(0, t)$ of the sting's end, and the resulting force F_x exerted on the oscillation source by the string.

From the system's geometry (see the figure above), with the wall's position taken for $z = 0$, in the small oscillation limit ($dx/dz \to 0$), this component of the force is

$$F_x = \mathscr{T}\sin\alpha \approx \mathscr{T}\tan\alpha \equiv \mathscr{T}\frac{\partial x}{\partial z}|_{z=+0}. \tag{**}$$

Very similarly, the net force exerted on a small arbitrary fragment dz of the string by its adjacent right and left parts is

$$dF_x \approx \mathscr{T}\left(\frac{\partial x}{\partial z}|_{z+dz/2} - \frac{\partial x}{\partial z}|_{z-dz/2}\right) \approx \mathscr{T}\left[\frac{\partial^2 x}{\partial^2 z}|_z\left(+\frac{dz}{2}\right) - \frac{\partial^2 x}{\partial^2 z}|_z\left(-\frac{dz}{2}\right)\right]$$
$$= \mathscr{T}\frac{\partial^2 x}{\partial^2 z}dz.$$

Since the mass of the fragment is μdz, the second Newton law for its motion in the direction x gives us the following equation:

$$\mu\frac{\partial^2 x}{\partial t^2} = \mathscr{T}\frac{\partial^2 x}{\partial^2 z}.$$

[11] See, e.g. *Part CM* section 6.4. Note that much of this solution just reproduces the classical analysis discussed in that part of the EAP series.

This is the well-known wave equation, with the general solution

$$x(z, t) = f_{\rightarrow}\left(t - \frac{z}{v}\right) + f_{\leftarrow}\left(t + \frac{z}{v}\right),$$

where

$$v \equiv \left(\frac{\mathscr{T}}{\mu}\right)^{1/2} > 0,$$

is the wave velocity, and $f_{\rightarrow}$ and $f_{\leftarrow}$ are some functions of a single argument, which are determined by initial and boundary conditions. If the wave on the string is excited, as in our case, only by the motion of its end (at $z = 0$), the wave may travel only to the right:

$$x(z, t) = f_{\rightarrow}\left(t - \frac{z}{v}\right), \quad \text{so that} \quad \frac{\partial x}{\partial z}(z, t) = -\frac{1}{v}\frac{\partial f_{\rightarrow}}{\partial t}\left(t - \frac{z}{v}\right), \quad \text{at } z \geqslant 0. \quad (***)$$

Plugging these expressions, for $z = +0$, into Eq. (**), we get

$$F_x(t) = \mathscr{T}\frac{\partial x}{\partial z}(0, t) = -\frac{\mathscr{T}}{v}\frac{\partial f_{\rightarrow}}{\partial t}(t) \equiv -\frac{\mathscr{T}}{v}\frac{\partial x}{\partial t}(0, t).$$

Comparing this relation with Eq. (7.137) of the lecture notes, we see that the action of the string (which carries the induced wave (***), together with the associated energy, to infinity) is equivalent to kinematic friction, with the drag coefficient[12]

$$\eta = \frac{\mathscr{T}}{v} \equiv (\mathscr{T}\mu)^{1/2}.$$

This means that we may immediately use Eq. (7.138),

$$\text{Im}\,\chi(\omega) = \eta\omega = (\mathscr{T}\mu)^{1/2}\omega,$$

so that Eq. (*) yields the required answer:

$$S_F(\omega) = \frac{\hbar\omega}{2\pi}(\mathscr{T}\mu)^{1/2}\coth\frac{\hbar\omega}{2k_BT}.$$

Note that this is a good example of a dissipative environment that may be described by a time-independent Hamiltonian. As a result, this (and other similar) models give a good opportunity to explore some difficult issues of the theory of open systems, by using reliable theoretical methods developed for Hamiltonian systems.

[12] Note that in our case this coefficient is just the *wave impedance* Z of the string—see, e.g. *Part CM* Eq. (6.48).

Problem* 7.8. Calculate the low-frequency spectral density of small fluctuations of the voltage V across a Josephson junction, shunted with an Ohmic conductor, and biased with a dc external current $I > I_c$.

Hint: You may use Eqs. (1.73)–(1.74) of the lecture notes to describe the junction's dynamics, and assume that the shunting conductor remains in thermal equilibrium.

Solution: In the Heisenberg–Langevin approach, we may use Eqs. (1.73)–(1.74) for the Heisenberg operators of the corresponding observables:

$$\hat{I}(t) = I_c \sin\hat{\varphi}, \qquad \hat{V} = \frac{\hbar}{2e}\frac{d\hat{\varphi}}{dt},$$

taking

$$\hat{I}(t) = \bar{I} - \hat{I}_G(t),$$

where I_G is the current flowing through the shunt—see the figure below. As we know from section 7.4 of the lecture notes (see in particular Eq. (7.92) and its discussion), we may represent this current as the sum of its average component, and the fluctuations that may be calculated in the absence of voltage[13]:

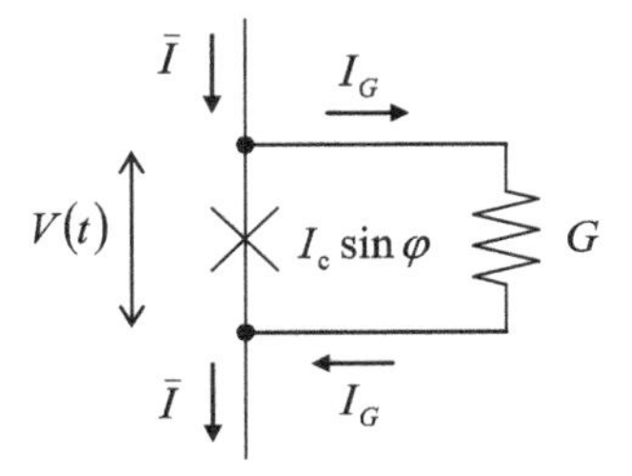

$$\hat{I}_G(t) = \langle\hat{I}_G\rangle - \tilde{I}(t) = G\hat{V} - \tilde{I}(t),$$

where G is the Ohmic conductance of the shunt, so that the resulting Heisenberg equation of motion of the Josephson phase φ is.

$$I_c \sin\hat{\varphi} + \frac{\hbar G}{2e}\frac{d\hat{\varphi}}{dt} = \bar{I} + \tilde{I}(t). \qquad (*)$$

Since we have been asked to analyze only small fluctuations, we may look for the solution of this equation in the form $\hat{\varphi} = \varphi_0 + \hat{\tilde{\varphi}}$, where φ_0 is the solution of the fluctuation-free classical equation

$$I_c \sin\varphi_0 + \frac{\hbar G}{2e}\frac{d\varphi_0}{dt} = \bar{I},$$

[13] Note that the statistical averaging in the first form of this relation is only over the degrees of freedom of the environment (i.e. the shunting conductor), so that from the point of view of the Josephson junction as such, the voltage in its second form is still an operator. Also, the sign before the fluctuating term is a matter of convention, and is taken negative here just for the compactness of the following calculations.

and the operator $\hat{\tilde{\varphi}}$ describes small fluctuations of the phase, obeying the linearized version of Eq. (*):

$$I_c \cos \varphi_0 \, \hat{\tilde{\varphi}} + \frac{\hbar G}{2e} \frac{d\hat{\tilde{\varphi}}}{dt} = \hat{\tilde{I}}(t).$$

The solution of these two differential equations may be made less bulky by introducing normalized variables[14]: current $\imath \equiv I/I_c$ and time $\tau \equiv t/(\hbar G/2eI_c)$. In these dimensionless variables,

$$\dot{\varphi}_0 + \sin \varphi_0 = \bar{\imath}, \qquad \dot{\hat{\tilde{\varphi}}} + \cos \varphi_0 \, \hat{\tilde{\varphi}} = \hat{\tilde{\imath}}(\tau), \qquad (**)$$

where (in this solution only!) the dot means the differentiation over τ, rather than t.

The first (classical but nonlinear) equation[15] may be readily solved analytically, because the separation of variables leads to a table integral[16]:

$$\frac{d\varphi_0}{\bar{\imath} - \cos \varphi_0} = d\tau, \qquad \text{so that } \tau = \int \frac{d\varphi_0}{\bar{\imath} - \cos \varphi_0}.$$

However, for our purposes the explicit form of the function $\varphi_0(t)$ is less important than the following expression for its time derivative, and hence for the unperturbed part of the voltage $V \propto d\varphi/dt \propto d\varphi/d\tau$ across the junction (see the figure above):

$$\dot{\varphi}_0 = \frac{\bar{v}^2}{\bar{\imath} - \cos(\bar{v}\tau + \text{const})}, \quad \text{where } \bar{v} \equiv (\bar{\imath}^2 - 1)^{1/2} > 0, \quad \text{for } \bar{\imath} > 1, \text{ i.e. } \bar{I} > I_c,$$

where the inconsequential constant depends on the selected time origin; for what follows, it is convenient to take it as equal to zero. This result shows that $\dot{\varphi}_0$ is a periodic (but not a sinusoidal!) function of τ, with the normalized frequency $\bar{v}$ approaching zero at $\bar{\imath} \rightarrow 1$. (This is just the same Josephson oscillations of frequency (1.75) that were briefly discussed at the end of section 1.6 of the lecture notes, besides that in our current case of a constant external current, they are the simultaneous oscillations of the supercurrent $I_c \sin\varphi$ and the voltage V.) For our calculation, we will need the Fourier expansion of this periodic function and its time derivative,

$$\dot{\varphi}_0 = \sum_{k=-\infty}^{\infty} v_k e^{-ik\bar{v}\tau}, \qquad \ddot{\varphi}_0 = \sum_{k=-\infty}^{\infty} v_k(-ik\bar{v})\, e^{-ik\bar{v}\tau},$$

$$\text{with} \quad v_k = \frac{1}{2\pi} \int_{-\pi}^{\pi} \dot{\varphi}_0 e^{ik\bar{v}\tau} d(\bar{v}\tau) \equiv \frac{\bar{v}^2}{\pi} \int_0^{\pi} \frac{\cos k\xi \, d\xi}{\bar{\imath} - \cos \xi}.$$

[14] Let me hope that the use of different fonts makes the difference between the normalized current ($\imath$) and the imaginary unity (i) sufficiently clear.

[15] This well-known equation (see, e.g. *Part CM* section 5.4) at $\bar{\imath} < 1$ has the stationary solution $\varphi_0 = \sin^{-1} \bar{\imath}$. The calculation of the phase and voltage fluctuations in this case is recommended to the reader as a (simple) additional exercise. The answer is: at $\omega \rightarrow 0$, $S_\varphi(\omega) \rightarrow S_I(\omega)/I_c^2 \cos^2 \varphi_0$; $S_V(\omega) \rightarrow (\hbar\omega/2e)^2 S_\varphi(\omega) \rightarrow 0$.

[16] See, e.g. Eq. (A.30*c*).

This is also a table integral[17], giving

$$v_k = \bar{v}\,(\bar{\imath} - \bar{v})^{|k|},$$

so that, in particular, $v_0 = \bar{v}$—hence the notation.

Now solving the second of Eqs. (**) by the standard method of variable coefficients (fully applicable to linear equations for operators), we readily get

$$\hat{\tilde{\varphi}}(\tau) = f(\tau)\int^{\tau} \frac{\hat{\tilde{\imath}}(\tau')}{f(\tau')}d\tau', \quad \text{where} \quad f(\tau) \equiv \exp\left\{-\int^{\tau} \cos\varphi_0(\tau'')d\tau''\right\}.$$

This result may be recast in a form more convenient for calculations. Indeed, differentiating the first of Eqs. (**) over τ, we get

$$\ddot{\varphi}_0 + \cos\varphi_0\,\dot{\varphi}_0 = 0, \quad \text{i.e.}\ \frac{d\dot{\varphi}_0}{d\tau} + \cos\varphi_0\,\dot{\varphi}_0 = 0, \quad \text{so that}\ \frac{d\dot{\varphi}_0}{\dot{\varphi}_0} = -\cos\varphi_0 d\tau.$$

Integrating this elementary differential equation, we see that the above function $f(\tau)$ is just $\dot{\varphi}(\tau)$, give or take a time-independent multiplier. As a result, we get,

$$\hat{\tilde{\varphi}}(\tau) = \dot{\varphi}_0(\tau)\int^{\tau} \frac{\hat{\tilde{\imath}}(\tau')}{\dot{\varphi}_0(\tau')}d\tau', \quad \text{so that} \quad \dot{\hat{\tilde{\varphi}}}(\tau) = \hat{\tilde{\imath}}(\tau) + \ddot{\varphi}_0(\tau)\int^{\tau} \frac{\hat{\tilde{\imath}}(\tau')}{\dot{\varphi}_0(\tau')}d\tau'.$$

The last formula describes the (normalized) fluctuations of the voltage V, i.e. exactly the subject of our interest. With the above explicit forms of $\dot{\varphi}_0$, and the Fourier expansion of its time derivative, it becomes

$$\dot{\hat{\tilde{\varphi}}}(\tau) = \hat{\tilde{\imath}}(\tau) - i\sum_{k=-\infty}^{\infty} k\bar{v}(\bar{\imath} - \bar{v})^{|k|}e^{-ik\bar{v}\tau}\int^{\tau}\left(\bar{\imath} - \frac{e^{i\bar{v}\tau'} + e^{-i\bar{v}\tau'}}{2}\right)\hat{\tilde{\imath}}(\tau')d\tau'. \qquad (***)$$

Now using the Fourier expansions of the (normalized) voltage and current fluctuations, similar to Eq. (7.115) of the lecture notes,

$$\dot{\hat{\tilde{\varphi}}}(\tau) = \int_{-\infty}^{+\infty} \hat{v}_{\omega}e^{-i\omega\tau}d\omega, \quad \hat{\tilde{\imath}}(\tau) = \int_{-\infty}^{+\infty} \hat{\imath}_{\omega}e^{-i\omega\tau}d\omega,$$

we may readily perform the integration in Eq. (***) explicitly. Requiring the complex amplitudes of all harmonics in both parts of the resulting equation to be equal, we get

$$\hat{v}_{\omega} = \sum_{k=-\infty}^{+\infty} z_k(\omega)\,\hat{\imath}_{\omega-k\bar{v}}, \quad \text{with}\ z_k(\omega) = \delta_{k,0} + \frac{k(\bar{\imath} - \bar{v})^{|k|}\bar{\imath}}{\omega - k\bar{v}} - \frac{1}{2}\sum_{\pm}\frac{(k \pm 1)(\bar{\imath} - \bar{v})^{|k\pm 1|}}{\omega - (k \pm 1)\bar{v}}.$$

The first of these expressions describes the so-called *mixing* of the current fluctuations with harmonics of the Josephson oscillations, due to the sinusoidal nonlinearity of the Josephson supercurrent[18]. According to Eq. (7.114) of the lecture notes, it

---

[17] See, e.g. Eqs. (2.5.16)–(2.5.22) in the manual by Prudnikov *et al*, cited in appendix A, section A.16.

[18] For a brief discussion of this (essentially, classical, or rather mathematical) effect see, e.g. *Part CM* section 6.7.

means that the symmetrized spectral densities of the voltage and current fluctuations are related as follows:

$$S_v(\omega) = \sum_{k=-\infty}^{+\infty} |z_k|^2\, S_\imath(\omega - k\bar{v}).$$

This is a very informative result, in particular (at $\omega \approx n\bar{v}$) describing the fluctuation-induced broadening of the Josephson oscillation harmonics. For our purposes, however, we need only its low-frequency limit, $\omega \to 0$. In this limit, thanks to the identity $(\bar{\imath} - \bar{v})^{-1} = (\bar{\imath} + \bar{v})$ following from the above formula for $\bar{v}$, the result simplifies, and includes only three (essentially, two) terms:

$$S_v(0) = \frac{1}{\bar{v}^2}\left[\bar{\imath}^2 S_\imath(0) + \frac{1}{4}S_\imath(\bar{v}) + \frac{1}{4}S_\imath(-\bar{v})\right] = \frac{\bar{\imath}^2}{\bar{v}^2}\left[S_\imath(0) + \frac{1}{2\bar{\imath}^2}S_\imath(\bar{v})\right].$$

The front factor in this expression is just the square of the dc differential ('dynamic') resistance $R_\mathrm{d} \equiv d\bar{V}/d\bar{I}$ of the system, in our normalized units:

$$r_\mathrm{d} = \frac{d\bar{v}}{d\bar{\imath}} = \frac{d(\bar{\imath}^2 - 1)^{1/2}}{d\bar{\imath}} = \frac{\bar{\imath}}{(\bar{\imath}^2 - 1)^{1/2}} = \frac{\bar{\imath}}{\bar{v}},$$

so that returning to the initial, dimensional units, we finally get

$$S_V(0) = R_\mathrm{d}^2\left[S_I(0) + \frac{I_\mathrm{c}^2}{2\bar{I}^2}S_I(\omega_\mathrm{J})\right], \qquad (****)$$

where

$$\omega_\mathrm{J} = \frac{2eI_\mathrm{c}}{\hbar G}\bar{v} = \frac{2e}{\hbar}\bar{V}$$

is the dimensional frequency of the Josephson oscillations—see Eq. (1.75). The first term in the square brackets of Eq. (****) describes the simple, and very natural transform of low-frequency fluctuations of the current in the shunt, to voltage fluctuations of the same frequency, at the differential resistance of the system: $\tilde{V} = R_\mathrm{d}\tilde{I}$. The second term is much less trivial: it describes the intensity of fluctuations induced by high-frequency ($\omega \approx \omega_\mathrm{J}$) current fluctuations, and transformed to nearly-zero frequencies due to their mixing with the Josephson self-oscillations.

Since the current fluctuations of the conductance G, remaining in the thermal equilibrium with temperature T, obey the fluctuation–dissipation theorem (7.134) with $\mathrm{Im}\,\chi(\omega)/\omega = G$:[19]

$$S_I(\omega) = \frac{\hbar\omega}{2\pi}G \coth\frac{\hbar\omega}{2k_\mathrm{B}T},$$

[19] See the footnote just before Eq. (7.139) of the lecture notes.

their spectral density at frequency ω_J does not vanish even at low temperatures:

$$S_I(\omega_J) \approx \frac{\hbar\omega_J}{2\pi}G, \quad \text{at } k_B T \ll \hbar\omega_J,$$

and represents purely quantum fluctuations—see the discussion following Eq. (7.152) of the lecture notes. Eq. (****) is exactly the theoretical result (circa 1972) that was later used by R Koch *at al* for comparison with their experimental results[20]. The good agreement of the data with this theory gave a firm evidence of the reality of the quantum fluctuations in an Ohmic environment, without any explicit oscillator at their frequency (ω_J).

***Problem* 7.9.** Prove that in the interaction picture of quantum dynamics, the expectation value of an arbitrary observable A may be indeed calculated using Eq. (7.167) of the lecture notes.

Solution: The basic Eq. (7.5) may be represented as

$$\langle A\rangle(t) = \text{Tr}[\hat{A}\hat{u}\hat{w}(0)\hat{u}^\dagger],$$

where $\hat{u}$ is the full evolution operator, which obeys Eq. (4.157*b*). Using Eqs. (4.209) and (4.210) of the interaction picture, we may express the expectation value via products of the partial evolution operators $\hat{u}_0$ and $\hat{u}_I$, and then group the operands as follows:

$$\langle A\rangle(t) = \text{Tr}\left[\hat{A}\hat{u}_0\hat{u}_I\hat{w}(0)\hat{u}_I^\dagger\hat{u}_0^\dagger\right] = \text{Tr}\left[\left(\hat{A}\hat{u}_0\hat{u}_I\hat{w}(0)\hat{u}_I^\dagger\right)\left(\hat{u}_0^\dagger\right)\right].$$

From chapter 4 we know that the trace of a product of two operators does not depend on their order, so that in the last relation, we may swap the parentheses and then regroup the terms:

$$\langle A\rangle(t) = \text{Tr}\left[\left(\hat{u}_0^\dagger\right)\left(\hat{A}\hat{u}_0\hat{u}_I\hat{w}(0)\hat{u}_I^\dagger\right)\right] = \text{Tr}\left[\left(\hat{u}_0^\dagger\hat{A}\hat{u}_0\right)\left(\hat{u}_I\hat{w}(0)\hat{u}_I^\dagger\right)\right].$$

But according to Eqs. (4.214) and (7.166), this relation is nothing but Eq. (7.167):

$$\langle A\rangle(t) = \text{Tr}[A_I(t)\hat{w}_I(t)].$$

***Problem* 7.10.** Show that the quantum-mechanical Golden Rule (6.149) and the master equation (7.196) give the same results for the rate of spontaneous quantum transitions $n' \to n$ in a system with a discrete energy spectrum, weakly coupled to a low-temperature heat bath ($k_B T \ll \hbar\omega_{nn'}$).

Hint: Establish a relation between the function $\chi''(\omega_{nn'})$ that participates in Eq. (7.196), and the density of states ρ_n, which participates in the Golden Rule formula, by considering a particular case of sinusoidal classical oscillations in the system of interest.

[20] For the reference, see chapter 7 of the lecture notes.

Solution: First, let us consider a system coupled with environment via the interaction Hamiltonian (7.90), in the particular case when its variable x performs sinusoidal classical oscillations, $x = x_0\cos\omega t \equiv (x_0/2)(\exp\{-i\omega t\} + \exp\{i\omega t\})$, with frequency $\omega = \omega_{nn'} \gg k_B T/\hbar$. Then the background occupation of the excited levels (n) of the environment is negligible, and we may apply to it the Golden Rule in the form of Eq. (6.111), with $A_{nn'} = (x_0/2)F_{nn'}$:

$$\Gamma = \frac{2\pi}{\hbar}\left|\frac{x_0 F_{nn'}}{2}\right|^2 \rho_n = \frac{\pi}{2\hbar}x_0^2\,|F_{nn'}|^2\,\rho_n.$$

The quantum transitions at this rate, transferring to the environment the energy $\hbar\omega$ each, result in the average power flow

$$\overline{\langle \mathscr{P} \rangle} = \Gamma\hbar\omega = \frac{\pi}{2}\omega x_0^2\,|F_{nn'}|^2\,\rho_n.$$

On the other hand, the same power may be expressed by Eq. (7.127) of the lecture notes:

$$\overline{\langle \mathscr{P} \rangle} = \frac{\omega x_0^2}{2}\chi''(\omega).$$

Comparing these two expressions for power, we find the connection we have been looking for:

$$\chi''(\omega_{nn'}) = \pi\,|F_{nn'}|^2\,\rho_n.$$

Since this expression includes only the characteristics of the environment, it should be valid for any process in the system of our interest—either classical or quantum. With this substitution, Eq. (7.196) applied to the spontaneous energy-reducing transition, i.e. to the case $E_{n'} > E_n$, and taken for in the low-temperature limit,

$$\Gamma_{n'\to n} = \frac{2}{\hbar}\,|x_{nn'}|^2\,\chi''(\omega_{nn'}),$$

reads

$$\Gamma_{n'\to n} = \frac{2\pi}{\hbar}\,|x_{nn'}|^2|F_{nn'}|^2\,\rho_n.$$

But this is exactly Eq. (6.149), taking into account the replacements $A \to \pm x$, and $B \to \mp F$, which follow from the comparison of the interaction Hamiltonians (6.145) and (7.90).

***Problem* 7.11.** For a harmonic oscillator with weak Ohmic dissipation, use Eqs. (7.208) and (7.209) of the lecture notes to find the evolution of the expectation value $\langle E \rangle$ of oscillator's energy in time at arbitrary initial state, and compare the result with that following from the Heisenberg–Langevin approach.

Solution: Spelling out Eqs. (7.208) and (7.209),

$$\dot{W}_n = \Gamma_{n+1\to n}W_{n+1} + \Gamma_{n-1\to n}W_{n-1} - (\Gamma_{n\to n+1} + \Gamma_{n\to n-1})W_n$$
$$= 2\delta\{(n+1)(n_e+1)W_{n+1} + nn_eW_{n-1} - [(n+1)n_e + n(n_e+1)]\}W_n,$$

and plugging them into the expression for the expectation value of the oscillator's energy (referred, for the calculation convenience, to its ground-state level $\hbar\omega_0/2$),

$$\langle E\rangle = \hbar\omega_0\sum_{n=1}^{\infty}nW_n, \quad \text{so that } \frac{d}{dt}\langle E\rangle = \hbar\omega_0\sum_{n=1}^{\infty}n\dot{W}_n.$$

we get

$$\frac{d}{dt}\langle E\rangle = 2\delta\hbar\omega_0\sum_{n=1}^{\infty}n\{(n+1)(n_e+1)W_{n+1} + nn_eW_{n-1} - [(n+1)n_e + n(n_e+1)]W_n\}$$
$$\equiv 2\delta\hbar\omega_0\left\{\sum_{n=1}^{\infty}n(n+1)(n_e+1)W_{n+1} + \sum_{n=1}^{\infty}n^2n_eW_{n-1} - \sum_{n=1}^{\infty}n[(n+1)n_e + n(n_e+1)]W_n\right\}.$$

Let us replace the summation index n with $(n-1)$ in the first sum, and with $(n+1)$ in the second sum. This gives

$$\frac{d}{dt}\langle E\rangle = 2\delta\hbar\omega_0\left\{\sum_{n=2}^{\infty}(n-1)n(n_e+1)W_n + \sum_{n=0}^{\infty}(n+1)^2n_eW_n\right.$$
$$\left. - \sum_{n=1}^{\infty}n[(n+1)n_e + n(n_e+1)]W_n\right\} \qquad (*)$$
$$\equiv 2\delta\hbar\omega_0\sum_{n=0}^{\infty}(n_e-n)W_n \equiv 2\delta(E_e - \langle E\rangle),$$

where E_e is the equilibrium value of the energy:

$$E_e \equiv \hbar\omega_0\sum_{n=0}^{\infty}n_eW_n = \hbar\omega_0 n_e\sum_{n=0}^{\infty}W_n = \hbar\omega_0 n_e.$$

Since this value is time-independent, Eq. (*) may be rewritten as

$$\frac{d}{dt}\tilde{E} = -2\delta\tilde{E}, \quad \text{where } \tilde{E} \equiv \langle E\rangle - E_e,$$

and has a simple solution,

$$\tilde{E}(t) = \tilde{E}(0)\exp\{-2\delta t\}, \quad \text{i.e. } \langle E\rangle(t) = E_e + [\langle E\rangle(0) - E_e]\exp\{-2\delta t\}, \qquad (**)$$

describing an exponential change of the energy from its initial value to the equilibrium one.

It is remarkable that this very simple result is valid for arbitrary initial distribution of the probabilities W_n (though their time evolution may be rather involved—see figure 7.8 of the lecture notes), and for arbitrary phase shifts of the initial Fock states (because according to Eq. (7.205), the evolution of the off-diagonal elements of the density matrix of the oscillator does not affect that of its diagonal elements, W_n.)

Now proceeding to the Heisenberg–Langevin approach to the same problem, we may start from Eq. (7.145) of the lecture notes:

$$m\ddot{\hat{x}} + \eta\dot{\hat{x}} + m\omega_0^2\hat{x} = \hat{F}(t). \qquad (***)$$

Looking for a solution of the corresponding homogeneous equation in the usual form $\exp\{\lambda t\}$, we get the following well-known[21] characteristic equation

$$\lambda^2 + \frac{\eta}{m}\lambda + \omega_0^2 = 0,$$

whose roots may be simplified in the low-damping limit:

$$\lambda_\pm \approx \pm i\omega_0 - \frac{\eta}{2m} \equiv \pm i\omega_0 - \delta.$$

This means that in the absence of the external force $F(t)$, all linear operators ($\hat{x}$, $\hat{p}$, $\hat{a}$, $\hat{a}^\dagger$, etc) of the system depend on time as $\exp\{(\pm i\omega_0 - \delta)t\}$, while their Hermitian quadratic forms (such as the energy) are proportional to the modulus square of this function, i.e. to $\exp\{-2\delta t\}$.

Due to the linearity of Eq. (***), its solution may be represented as the sum of stationary fluctuations induced by the force described by its right-hand side, and the exponentially decaying oscillations due to the initial conditions. Since these two processes are independent of each other (mutually incoherent), their energies may be just added up:

$$\langle E\rangle(t) = E_\mathrm{e} + C\exp\{-2\delta t\}.$$

Selecting the constant C in this expression so that at $t = 0$ it coincides with the initial value $\langle E\rangle(0)$, we get

$$\langle E\rangle(t) = \langle E\rangle(0)\exp\{-2\delta t\} + E_\mathrm{e}(1 - \exp\{-2\delta t\}),$$

i.e. the same result as follows from the density matrix approach—cf Eq. (**). We see that calculations using the Heisenberg–Langevin formalism are indeed much simpler—where they work, i.e. for linear systems.

***Problem* 7.12.** Derive Eq. (7.219) of the lecture notes in an alternative way, using an expression dual to Eq. (4.244).

Solution: We need to calculate the diagonal matrix element of the operator $[\hat{x}, [\hat{x}, \hat{w}]]$ in the momentum representation. First, let us rewrite this element, using the closure condition (4.220), written for the eigenstates p of the momentum operator, twice:

[21] See, e.g. *Part CM* section 5.1.

$$\begin{aligned}\langle p|\,[\hat{x},\,[\hat{x},\,\hat{w}]]\,|p\rangle &= \langle p|\,[\hat{x}(\hat{x}\hat{w} - \hat{w}\hat{x}) - (\hat{x}\hat{w} - \hat{w}\hat{x})\hat{x}]\,|p\rangle \\ &= \int dp'\langle p|\,\hat{x}\,|p'\rangle \int dp''[\langle p'|\,\hat{x}\,|p''\rangle w(p'',\,p) \\ &\quad - w(p',\,p'')\langle p''|\,\hat{x}\,|p\rangle] \qquad (*) \\ &\quad - \int dp''\langle p''|\,\hat{x}\,|p\rangle \int dp'[\langle p|\,\hat{x}\,|p'\rangle w(p',\,p'') \\ &\quad - w(p,\,p')\langle p'|\,\hat{x}\,|p''\rangle],\end{aligned}$$

thus expressing it via the matrix elements of the coordinate operator in this representation. On the other hand, reproducing the discussion starting from Eq. (4.240) for the momentum representation, we get a formula dual to Eq. (4.244):

$$\int dp'\langle \hat{p}|\,\hat{x}\,|\hat{p}'\rangle\varphi(p') = i\hbar\frac{\partial}{\partial p}\varphi(p), \qquad (**)$$

where $\varphi(p)$ is an arbitrary wavefunction in the momentum representation. In order to make calculations more compact, we may use the definition of the delta-function to represent Eq. (**) in a shorthand form

$$\langle \hat{p}|\,\hat{x}\,|\hat{p}'\rangle = i\hbar\frac{\partial}{\partial p}\delta(p - p').$$

This relation may look a bit intimidating, because it apparently requires one to differentiate the delta-function explicitly. However, the symmetric nature of the commutators to be evaluated eliminates the need to do that[22].

Indeed, let us start from spelling out the inner integral (over p'') in the first term on the right-hand side of the last form of Eq. (*):

$$\begin{aligned}&\int dp''[\langle p'|\,\hat{x}\,|p''\rangle w(p'',\,p) - w(p',\,p'')\langle p''|\,\hat{x}\,|p\rangle] \\ &= i\hbar\int dp''\left[\frac{\partial}{\partial p'}\delta(p' - p'')w(p'',\,p) - w(p',\,p'')\frac{\partial}{\partial p''}\delta(p'' - p)\right].\end{aligned}$$

Differentiating the product $\delta(p' - p'')w(p'' - p)$ by parts, and noticing that the derivative of the delta-function $\delta(p' - p'')$ is nonvanishing only at $p' \to p''$, we see that, after the external integration over p', the terms with such derivatives cancel, so that the inner integral (over p'') may be reduced to

$$i\hbar\int dp''\delta(p' - p'')\frac{\partial}{\partial p''}w(p'',\,p) = i\hbar\frac{\partial}{\partial p'}w(p',\,p).$$

The inner integral (over p') in the second term of Eq. (*) is similar (with the replacements $p' \to p$, and $p \to p''$), giving

[22] The reader scared of such high-riding is invited to reproduce the following calculation in a (longer) integral form, using Eq. (**).

$$\int dp'[\langle p|\,\hat{x}\,|p'\rangle w(p',p'') - w(p,p')\langle p'|\,\hat{x}\,|p''\rangle] = i\hbar\frac{\partial}{\partial p}w(p,p'').$$

Now let us change the notation from p'' to p' in the second term of Eq. (*), and merge both terms of the right-hand side:

$$\langle p|\,[\hat{x},[\hat{x},\hat{w}]]\,|p\rangle = \int dp'\left[\langle p|\,\hat{x}\,|p'\rangle i\hbar\frac{\partial}{\partial p'}w(p',p) - i\hbar\frac{\partial}{\partial p}w(p,p')\langle p'|\,\hat{x}\,|p\rangle\right]$$

$$= -\hbar^2\int dp'\left[\frac{\partial}{\partial p}\delta(p-p')\frac{\partial}{\partial p'}w(p',p) - \frac{\partial}{\partial p}w(p,p')\frac{\partial}{\partial p'}\delta(p-p')\right],$$

so that repeating the same calculation as in the inner integrals, we get a simple result,

$$\langle p|\,[\hat{x},[\hat{x},\hat{w}]]\,|p\rangle = -\hbar^2\int dp'\delta(p-p')\frac{\partial}{\partial p}\frac{\partial}{\partial p'}w(p',p) = -\hbar^2\frac{\partial^2}{\partial p^2}w(p),$$

which is equivalent to Eq. (7.219). (Its derivation described in section 7.6 of the lecture notes is arguably more elegant.)

***Problem* 7.13.** A particle in a system of two coupled potential wells (see, e.g. figure 7.4 in the lecture notes) is weakly coupled to an Ohmic environment.

(i) Derive equations describing time evolution of the density matrix elements.
(ii) Solve these equations in the low-temperature limit, when the energy level splitting is much larger than k_BT, to calculate the time evolution of the probability of finding the particle in one of the wells, after it had been placed there at $t = 0$.

Solutions:

(i) As follows from numerous discussions in the lecture notes (see, in particular, sections 5.1 and 7.3), the problem may be adequately described by the following simple Hamiltonian:

$$\hat{H} = \mathbf{c}\cdot\hat{\boldsymbol{\sigma}} - \hat{\sigma}_z\hat{f}\{\lambda\} + \hat{H}_e\{\lambda\},$$

where $\mathbf{c}$ is a c-number vector, with c_z describing the well level misalignment (see Eq. (7.69) and figure 7.4), and the complex numbers $c_\pm \equiv c_x \pm ic_y$, tunneling between the wells. Its first part, $\hat{H}_s \equiv \mathbf{c}\cdot\hat{\boldsymbol{\sigma}}$, describing an unperturbed two-level system (in our case, two coupled potential wells) was discussed in detail in chapter 4. In particular, the elements of the unitary matrix of transfer between the basis of the eigenstates $a_{1,2}$ of this Hamiltonian and the z-basis in that the Pauli operators have the standard matrices (4.105), were calculated in the model solution of problem 4.25 (approach 2),

$$\langle\uparrow|a_1\rangle = \langle\downarrow|a_2\rangle = \frac{c + c_z}{[2c(c + c_z)]^{1/2}},$$
$$\langle\downarrow|a_1\rangle = -\langle\uparrow|a_2\rangle^* = -\frac{c_+}{[2c(c + c_z)]^{1/2}}, \qquad (*)$$

(with $c \equiv (c_x^2 + c_y^2 + c_z^2)^{1/2}$), and then used in the solution of problem 4.27 to calculate the matrix elements of the operator $\hat{\sigma}_z$ in the eigenstate basis:

$$(\sigma_z)_{11} = -(\sigma_z)_{22} = \frac{c_z}{c}, \qquad (\sigma_z)_{12} = (\sigma_z)_{21}^* = -\frac{c_-}{c}e^{i\varphi},$$

with an arbitrary real c-number φ.

Now we may use the last result to calculate, from Eqs. (7.199) and (7.203) (with the replacement $\hat{x} \to \hat{\sigma}_z$), the rates[23]

$$\frac{1}{T_1} \equiv \Gamma_{1\to2} \approx \frac{2c}{\hbar}\eta\frac{c_+c_-}{c^2} \equiv \frac{2c}{\hbar}\eta\left(1 - \frac{c_z^2}{c^2}\right),$$
$$\frac{1}{T_2} \approx \frac{1}{2}\Gamma_{1\to2} + \frac{4k_BT}{\hbar^2}\eta\frac{c_z^2}{c^2},$$

that participate in Eqs. (7.194) and (7.201) of the density matrix element evolution in the interaction picture (in the low-temperature limit $k_BT \ll E_1 - E_2 \equiv 2c$, when $\Gamma_{2\to1} \ll \Gamma_{1\to2}$):

$$\dot{w}_{11} = -\dot{w}_{22} = -\frac{1}{T_1}w_{11},$$
$$\dot{w}_{12} = -\left(\frac{1}{T_2} + i\Delta_{12}\right)w_{12}, \qquad \dot{w}_{21} = -\left(\frac{1}{T_2} - i\Delta_{12}\right)w_{21}.$$

(ii) According to Eqs. (7.161) and (7.165), the transfer back to the usual Schrödinger picture may be accomplished by adding terms $\pm i2c/\hbar$ to the right-hand sides of the equations' off-diagonal matrix elements. With these additions (on whose background the small, interaction-due terms $\pm i\Delta_{12}$ may be ignored), the equations have the following solutions:

$$w_{11}(t) = w_{11}(0)\exp\left\{-\frac{t}{T_1}\right\},$$
$$w_{22}(t) = w_{22}(0) + w_{11}(0)\left(1 - \exp\left\{-\frac{t}{T_1}\right\}\right), \qquad (**)$$
$$w_{12}(t) = w_{12}(0)\exp\left\{i\frac{2ct}{\hbar} - \frac{t}{T_2}\right\}, \qquad w_{21}(t) = w_{12}^*(t).$$

[23] Note that in the limit of a very high barrier separating the wells, insurmountable for tunneling, $c_\pm \to 0$, the rate $1/T_1$ vanishes (as it should), while the result for $1/T_2$ is reduced to that given by Eq. (7.204) of the lecture notes. (This limit is the only motivation for keeping the second, usually much smaller, second term in the expression for $1/T_2$.)

The initial values of the matrix elements should be obtained from our specific initial condition of having the particle definitely in the one of the wells, in our current notation in the state $\uparrow$. We may use the coefficients (*) to write the ket-vector of this (pure) initial state as

$$|\alpha(0)\rangle = |\uparrow\rangle = \alpha_1(0)\,|a_1\rangle + \alpha_2(0)\,|a_2\rangle,$$

with

$$\alpha_1(0) = \langle a_1|\uparrow\rangle = \langle\uparrow|a_1\rangle^* = \frac{c + c_z}{[2c(c + c_z)]^{1/2}},$$
$$\alpha_2(0) = \langle a_2|\uparrow\rangle = \langle\uparrow|a_2\rangle^* = -\frac{c_+}{[2c(c + c_z)]^{1/2}},$$

and then use Eq. (7.20) to write

$$w_{11}(0) = |\alpha_1(0)|^2 = \left|\frac{c + c_z}{[2c(c + c_z)]^{1/2}}\right|^2 \equiv \frac{c + c_z}{2c},$$
$$w_{22}(0) = |\alpha_2(0)|^2 = \left|\frac{c_-}{[2c(c + c_z)]^{1/2}}\right|^2 \equiv \frac{c_x^2 + c_y^2}{2c(c + c_z)} = \frac{c - c_z}{2c},$$
$$|w_{12}(0)| = |w_{21}(0)| = |\alpha_1(0)\alpha_2^*(0)| = \left|\frac{c + c_z}{[2c(c + c_z)]^{1/2}}\frac{c_+}{[2c(c + c_z)]^{1/2}}\right| \equiv \frac{\left(c_x^2 + c_y^2\right)^{1/2}}{2c},$$

where I have used the fact that due to the definition of the parameter c, there is the following relations between the Cartesian components of the vector $\mathbf{c}$: $c_x^{\ 2} + c_y^{\ 2} \equiv c^2 - c_z^{\ 2} = (c - c_z)(c + c_z)$. The upper two lines show that the diagonal matrix elements automatically satisfy the normalization condition

$$w_{11}(0) + w_{22}(0) = \frac{c + c_z}{2c} + \frac{c - c_z}{2c} \equiv 1.$$

(The equations (**) of their time evolution immediately show that this condition, $w_{11}(t) + w_{22}(t) = 1$, is sustained for any $t > 0$.) With these initial conditions, the solution (**) takes the form

$$w_{11}(t) = \frac{c + c_z}{2c}\exp\left\{-\frac{t}{T_1}\right\}, \quad w_{22}(t) = 1 - \frac{c + c_z}{2c}\exp\left\{-\frac{t}{T_1}\right\},$$
$$w_{12}(t) = -\frac{\left(c_x^2 + c_y^2\right)^{1/2}}{2c}\exp\left\{i\frac{2ct}{\hbar} - \frac{t}{T_2}\right\}, \quad w_{21}(t) = w_{12}^*(t).$$

What remains now is to transform this result back into the z-basis, because only in that basis the required probability, say $W_L \equiv W_\uparrow$, is simply expressed via the diagonal element of the density matrix:

$$W_L(t) = W_\uparrow(t) = w_{\uparrow\uparrow}(t).$$

Since, due to dephasing, at $t > 0$ our two-level system is not in a pure state (though, at finite t, it still differs from a classical mixture!), we need to carry out the back transformation carefully, using the general Eq. (4.93), which is valid for any operator (in particular, the density operator) and any two bases a_n and u_j:

$$w_{jj'}\,|_{\text{in }u} = \sum_{n,n'} U_{jn} w_{nn'}\,|_{\text{in }a} U^{\dagger}_{n'j'}$$

where U is the unitary matrix of transform between the bases—see Eqs. (4.83) and (4.84):

$$U_{jn} = \langle u_j | a_n \rangle, \quad U^{\dagger}_{n'j'} = \langle a_{n'} | u_{j'} \rangle \equiv \langle u_{j'} | a_{n'} \rangle^{*}.$$

In our case, the u_j is just the two-functional z-basis $\{\uparrow, \downarrow\}$, so that for the only element of the density matrix we are interested in, we get

$$W_{\text{L}}(t) = \langle \uparrow | a_1 \rangle w_{11}(t) \langle \uparrow | a_1 \rangle^{*} + \langle \uparrow | a_1 \rangle w_{12}(t) \langle \uparrow | a_2 \rangle^{*} + \langle \uparrow | a_2 \rangle w_{21}(t) \langle \uparrow | a_1 \rangle^{*} + \langle \uparrow | a_2 \rangle w_{22}(t) \langle \uparrow | a_2 \rangle^{*}$$

$$\equiv |\langle \uparrow | a_1 \rangle|^2\, w_{11}(t) + |\langle \uparrow | a_2 \rangle|^2\, w_{22}(t) + 2\,\text{Re}[\langle \uparrow | a_1 \rangle w_{12}(t) \langle \uparrow | a_2 \rangle^{*}],$$

and plugging in the matrix elements from Eq. (*), we finally get

$$\begin{aligned} W_{\text{L}}(t) = &\left| \frac{c + c_z}{[2c(c + c_z)]^{1/2}} \right|^2 \frac{c + c_z}{2c} \exp\left\{-\frac{t}{T_1}\right\} \\ &+ \left| \frac{c_-}{[2c(c + c_z)]^{1/2}} \right|^2 \left(1 - \frac{c + c_z}{2c} \exp\left\{-\frac{t}{T_1}\right\}\right) \\ &+ 2\,\text{Re}\left[\frac{c + c_z}{[2c(c + c_z)]^{1/2}} \frac{\left(c_x^2 + c_y^2\right)^{1/2}}{2c} \exp\left\{i\frac{2ct}{\hbar} - \frac{t}{T_2}\right\} \frac{c_+}{[2c(c + c_z)]^{1/2}} \right] \\ \equiv &\frac{c - c_z}{2c} + \frac{c_z(c + c_z)}{2c^2} \exp\left\{-\frac{t}{T_1}\right\} + \frac{(c - c_z)(c + c_z)}{2c^2} \cos\frac{2ct}{\hbar} \exp\left\{-\frac{t}{T_2}\right\}. \end{aligned}$$

(As a sanity check, $W_{\text{L}}(0) = 1$, as it should be for our initial conditions.)

The last expression shows that the initial occupancy of the well not only relaxes to its equilibrium value with the time constant T_1 due to system's transfer to the lower energy level, i.e. due to the energy relaxation, but also performs quantum oscillations (which were discussed in section 2.6, then in section 5.1, and then again at the end of section 6.1 of the lecture notes) with their amplitude decaying with the time constant T_2, due to dephasing. The initial value, $(1 - c_z^2/c^2)/2$, of the oscillation amplitude is largest if $c_z \to \pm c$ (i.e. at low tunneling between the wells; $c_{\pm} \to 0$); note, however, that in this limit the rate $1/T_2$ is highest.

Finally, let us interpret these results for another important two-level system, spin-½—see sections 4.6 and 5.1. As was discussed there, in the absence of coupling to environment ($T_{1,2} \to \infty$) the sinusoidal process with frequency $\Omega = 2c/\hbar = -\gamma\mathscr{B}$ describes the torque-induced precession of the spin's expectation value about the direction of applied magnetic field $\mathscr{B}$. The dissipative coupling causes a gradual decay of the precession amplitude, with the expectation value of the magnetic moment eventually aligned with the field, thus minimizing the potential energy (4.162).

***Problem* 7.14.** A spin-½ with gyromagnetic ratio γ is placed into magnetic field $\mathscr{B}(t) = \mathscr{B}_0 + \tilde{\mathscr{B}}(t)$ with an arbitrary but small time-dependent component, and is also weakly coupled to a dissipative environment. Derive differential equations describing the time evolution of the expectation values of spin's Cartesian components, at arbitrary temperature.

Solution: At the specified conditions, the Hamiltonians of spin interaction with its environment and with the time-dependent component of the magnetic field may be both considered as small perturbations of the basic Pauli Hamiltonian given by Eq. (4.163) of the lecture notes:

$$\hat{H}^{(0)} = -\hat{\mathbf{m}} \cdot \mathscr{B}_0,$$

so that the contributions of these perturbations to the right-hand sides of the equation of motion of the vector $\langle \mathbf{S} \rangle$ may be calculated independently.

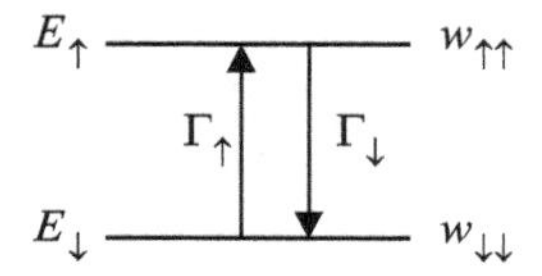

The contribution due to the environment coupling may be calculated from Eqs. (7.194) and (7.201) for the density matrix elements $w_{nn'}$, in the stationary-state basis of the unperturbed Hamiltonian. For our unperturbed two-level system (see the figure above), defined by the dc field component $\mathscr{B}_0$, this is the usual z-basis of stationary states $\uparrow$ and $\downarrow$, with energies $E_\downarrow$ and $E_\uparrow$, so there are only two master equations (7.194) for the diagonal components of the density matrix $w_{\uparrow\uparrow}$ and $w_{\downarrow\downarrow}$ (as we know, equal to the probabilities to find the particle in the corresponding state). In the low-temperature limit, discussed in the model solution of the previous problem, only the rate $\Gamma_\downarrow$ was important, but in our current case of arbitrary temperature, we need to use the full form of these equations:

$$\begin{aligned}\dot{w}_{\uparrow\uparrow} &= \Gamma_\uparrow w_{\downarrow\downarrow} - \Gamma_\downarrow w_{\uparrow\uparrow},\\ \dot{w}_{\downarrow\downarrow} &= -\Gamma_\uparrow w_{\downarrow\downarrow} + \Gamma_\downarrow w_{\uparrow\uparrow}.\end{aligned} \tag{*}$$

(For the environment in thermodynamic equilibrium at temperature T, the rates $\Gamma_{\uparrow\downarrow}$ are fundamentally related by Eq. (7.197):

$$\frac{\Gamma_\uparrow}{\Gamma_\downarrow} = \exp\left\{-\frac{\Delta}{k_\mathrm{B}T}\right\},$$

where $\Delta \equiv E_\uparrow - E_\downarrow = \hbar\gamma\mathscr{B}_0$ is the interlevel energy gap. Note also that the master equations (*) satisfy the relation

$$\dot{w}_{\uparrow\uparrow} = -\dot{w}_{\downarrow\downarrow},$$

which naturally sustains the conservation of the total probability $w_{\uparrow\uparrow} + w_{\downarrow\downarrow} = 1$, so that they are not independent.)

For the two off-diagonal components of the density matrix, due to the weakness of coupling to the environment, we may neglect, in Eq. (7.201), the small frequency corrections $\Delta_{nn'}$ in comparison with the basic energy gap Δ. As a result (as for any two-level system, in which the sums over other levels, on the right-hand side of Eq. (7.203), vanish) we get similar equations for both off-diagonal elements:

$$\dot{w}_{\uparrow\downarrow} = -\frac{1}{T_2}w_{\uparrow\downarrow}, \quad \dot{w}_{\downarrow\uparrow} = -\frac{1}{T_2}w_{\downarrow\uparrow}. \qquad (**)$$

Now let us spell out the fundamental relation (7.5) between the expectation value of any observable and the density matrix, for the Cartesian components of the spin, in the same z-basis of states $\uparrow$ and $\downarrow$, using the Schrödinger picture of quantum dynamics—in which all time evolution is delegated to the density matrix. For example, for the x-component we get

$$\begin{aligned}\langle S_x(t)\rangle &= \mathrm{Tr}\{\mathrm{S}_x\mathrm{w}(t)\} = \frac{\hbar}{2}\mathrm{Tr}\{\sigma_x\mathrm{w}(t)\} = \frac{\hbar}{2}\mathrm{Tr}\left\{\begin{pmatrix}0 & 1\\ 1 & 0\end{pmatrix}\begin{pmatrix}w_{\uparrow\uparrow}(t) & w_{\uparrow\downarrow}(t)\\ w_{\downarrow\uparrow}(t) & w_{\downarrow\downarrow}(t)\end{pmatrix}\right\} \\ &= \frac{\hbar}{2}\{w_{\uparrow\downarrow}(t) + w_{\downarrow\uparrow}(t)\}.\end{aligned} \qquad (***)$$

Differentiating both parts of this relation over time, and then using Eqs. (**), we get the following expression for environment's contribution to the spin derivative:

$$\langle \dot{S}_x\rangle_{\mathrm{env}} = -\frac{1}{T_2}\left(\dot{w}_{\uparrow\downarrow} + \dot{w}_{\downarrow\uparrow}\right) = -\frac{\hbar}{2}\frac{1}{T_2}(w_{\uparrow\downarrow} + w_{\downarrow\uparrow}).$$

Now using Eq. (***) again, we get a very simple relation describing spin dephasing:

$$\langle \dot{S}_x\rangle_{\mathrm{env}} = -\frac{1}{T_2}\langle S_x\rangle.$$

An absolutely similar calculation gives the similar result for the y-component of spin,

$$\langle \dot{S}_y\rangle_{\mathrm{env}} = -\frac{1}{T_2}\langle S_y\rangle.$$

However, the result for the z-component of spin, i.e. the its component along the direction of the base field $\mathscr{B}_0$, is different. Indeed[24]:

[24] Actually, Eq. (****) has already been derived and used in the lecture notes—see Eq. (7.75).

$$\langle S_z(t)\rangle = \frac{\hbar}{2}\mathrm{Tr}\{\sigma_z \mathrm{w}(t)\} = \frac{\hbar}{2}\mathrm{Tr}\left\{\begin{pmatrix}1 & 0\\ 0 & -1\end{pmatrix}\begin{pmatrix} w_{\uparrow\uparrow}(t) & w_{\uparrow\downarrow}(t)\\ w_{\downarrow\uparrow}(t) & w_{\downarrow\downarrow}(t)\end{pmatrix}\right\}$$
$$= \frac{\hbar}{2}\{w_{\uparrow\uparrow}(t) - w_{\downarrow\downarrow}(t)\}. \qquad (****)$$

Solving this equation, together with the normalization condition $w_{\uparrow\uparrow} + w_{\downarrow\downarrow} = 1$, for the two probabilities, we get

$$w_{\uparrow\uparrow} = \frac{1}{2}\left(1 + \frac{\langle S_z\rangle}{\hbar/2}\right), \quad w_{\downarrow\downarrow} = \frac{1}{2}\left(1 - \frac{\langle S_z\rangle}{\hbar/2}\right).$$

Now differentiating both sides of Eq. (****) over time, and then using Eqs. (*) and the above expressions for $w_{\uparrow\uparrow}$ and $w_{\downarrow\downarrow}$, we get the expression

$$\langle \dot{S}_z\rangle_{\mathrm{env}} = (\Gamma_\uparrow - \Gamma_\downarrow)\frac{\hbar}{2} - (\Gamma_\uparrow + \Gamma_\downarrow)\langle S_z\rangle,$$

which is usually recast in the following equivalent form:

$$\langle \dot{S}_z\rangle_{\mathrm{env}} = -\frac{\langle S_z\rangle - S_{\mathrm{e}}}{T_1},$$

where $1/T_1$ is the effective spin relaxation rate[25]:

$$\frac{1}{T_1} \equiv \Gamma_\uparrow + \Gamma_\downarrow,$$

while the constant

$$S_{\mathrm{e}} \equiv \frac{\hbar}{2}\frac{\Gamma_\uparrow - \Gamma_\downarrow}{\Gamma_\uparrow + \Gamma_\downarrow} \equiv \frac{\hbar}{2}\frac{\Gamma_\uparrow/\Gamma_\downarrow - 1}{\Gamma_\uparrow/\Gamma_\downarrow + 1} = \frac{\hbar}{2}\frac{\exp\{-\Delta/k_{\mathrm{B}}T\} - 1}{\exp\{-\Delta/k_{\mathrm{B}}T\} + 1} \equiv \frac{\hbar}{2}\tanh\frac{\Delta}{2k_{\mathrm{B}}T}$$

has the physical sense of the thermal-equilibrium value of $\langle S_z\rangle$.[26] Note that at high temperatures, $k_{\mathrm{B}}T \gg \Delta$, this value tends to zero, because both levels of the system, in equilibrium, are almost equally occupied.

On the other hand, as was discussed in section 5.1 of the lecture notes, the contribution of the time-dependent magnetic field into the spin dynamics may be merged with that of the base field $\mathscr{B}_0$, into one vector equation (5.22):

$$\langle \dot{\mathbf{S}}\rangle_{\mathrm{field}} = \gamma\langle \mathbf{S}\rangle \times \mathscr{B}(t).$$

Now merging all right-hand sides of the equations for the spin component derivatives, we finally get

[25] This expression is a natural generalization of the low-temperature case of (7.199), which is valid when $k_{\mathrm{B}}T \ll \Delta$ and hence $\Gamma_\uparrow \ll \Gamma_\downarrow$—see the model solution of the previous problem.

[26] Due to problem's axial symmetry, in the equilibrium, $\langle S_x\rangle = \langle S_y\rangle = 0$.

$$\langle \dot{S}_x \rangle = \gamma[\langle \mathbf{S} \rangle \times \mathscr{B}(t)]_x - \frac{\langle S_x \rangle}{T_2}, \quad \langle \dot{S}_y \rangle = \gamma[\langle \mathbf{S} \rangle \times \mathscr{B}(t)]_y - \frac{\langle S_y \rangle}{T_2},$$
$$\langle \dot{S}_z \rangle = \gamma[\langle \mathbf{S} \rangle \times \mathscr{B}(t)]_z - \frac{\langle S_z \rangle - S_e}{T_1}.$$

These equations are especially popular for the description of experiments with the sets of $N \gg 1$ similar (practically, non-interacting) nuclear spins in condensed matter samples, where they take the form of *Bloch equations*[27] for the Cartesian components of the nuclear magnetization $\mathbf{M} = n\gamma\langle \mathbf{S} \rangle$ of the sample (where $n \equiv N/V$ is the spin density):

$$\dot{M}_x = \gamma[\mathbf{M} \times \mathscr{B}(t)]_x - \frac{M_x}{T_2}, \quad \dot{M}_y = \gamma[\mathbf{M} \times \mathscr{B}(t)]_y - \frac{M_y}{T_2},$$
$$\dot{M}_z = \gamma[\mathbf{M} \times \mathscr{B}(t)]_z - \frac{M_z - M_e}{T_1}.$$

Since at $N \gg 1$ the quantum fluctuations of individual spins are effectively averaged out, in many cases the Bloch equations may be considered as classical equations of motion, similar to those describing torque-induced precession of a symmetric top[28], but with a very specific damping.

The most important effect described by these equations is the environment-induced broadening of the *magnetic resonance*, which was briefly discussed in chapters 5 and 6—see also the model solution of problem 6.23. As was shown in that solution, in the simplest case when $\tilde{\mathscr{B}}(t)$ is a field with a constant magnitude $\mathscr{B}_r$, rotating in the plane perpendicular to the time-independent field $\mathscr{B}_0$, with the angular velocity ω,[29] this effect obeys Eqs. (6.94) of the lecture notes, with $|A| = \gamma\mathscr{B}_r/2$, which describe the Rabi oscillations of the level occupancy, with an amplitude exhibiting a resonance at $\omega \approx \Delta/\hbar$. At negligible coupling of the spin to its dissipative environment, the FWHM bandwidth $\Delta\omega$ of this resonance is equal to $2\gamma\mathscr{B}_r$, but as the above Bloch equations imply, the environment provides an additional broadening of the resonance by $\Delta\omega \sim 1/T_{1,2}$; this broadening dominates at $\mathscr{B}_r \to 0$.

For the most important variety of this effect, the *nuclear magnetic resonance* (NMR), the nuclear spin interaction with the environment is typically very weak, with the times $T_{1,2}$ in some cases exceeding 1 s, while the resonance frequency $\Delta/2\pi\hbar$ may be of the order 100 MHz in practical magnetic fields $\mathscr{B}_0$ of a few tesla. As a result, the resonance may have a very small relative bandwidth, of the order of 10^{-8}, so that its detection allows experimental measurements of tiny local variations of $\mathscr{B}_0$. This effect has important applications in condensed matter physics, chemistry, and biomedicine[30].

[27] Their dissipation-free form (5.23) was briefly discussed in the end of section 5.1 of the lecture notes.

[28] See, e.g. *Part CM* section 4.5.

[29] Actually, the resonance takes place for any sinusoidal field of frequency $\omega \approx \Delta/\hbar$; just the quantitative description of its effect far from the resonance may be somewhat different.

[30] See, e.g. the monograph by J Keeler, cited in section 6.5 of the lecture notes.

IOP Publishing

Quantum Mechanics
Problems with solutions
Konstantin K Likharev

Chapter 8

Multiparticle systems

***Problem* 8.1.** Prove that Eq. (8.30) of the lecture notes indeed yields $E_g^{(1)} = (5/4)E_H$.

Solution: According to Eq. (8.30), we need to calculate

$$E_g^{(1)} = \frac{1}{\left(\pi r_0^3\right)^2}\int \exp\left\{-\frac{2r_1}{r_0}\right\}d^3r_1\int \exp\left\{-\frac{2r_2}{r_0}\right\}d^3r_2\frac{e^2}{4\pi\varepsilon_0\,|\mathbf{r}_1 - \mathbf{r}_2|}.$$

Instead of calculating this 6D integral directly, we may notice that this is just the classical energy of the Coulomb interaction[1] of two independent distributed electrostatic charges with a similar spherically-symmetric density:

$$\rho(r) = \frac{e}{\pi r_0^3}\exp\left\{-\frac{2r}{r_0}\right\}.$$

The (radially directed) electric field $\mathscr{E}$, induced by one of these charges, may be readily calculated by applying the Gauss law[2] to a sphere of radius r:

$$4\pi r^2\mathscr{E}(r) = \frac{Q}{\varepsilon_0} \equiv \frac{1}{\varepsilon_0}\int_{r'<r}\rho(r')d^3r' \equiv \frac{1}{\varepsilon_0}\frac{e}{\pi r_0^3}4\pi\int_0^r \exp\left\{-\frac{2r'}{r_0}\right\}r'^2dr'$$
$$\equiv \frac{e}{2\varepsilon_0}\int_0^{2r/r_0} \exp\{-\xi\}\xi^2d\xi,$$

where $\xi \equiv 2r'/r_0$. The last (dimensionless) integral may be readily worked out by parts; its indefinite form is equal to $\exp\{-\xi\}(-\xi^2 - 2\xi - 2)$, so that after the limit substitution, we get

$$\mathscr{E}(r) = \frac{e}{4\pi\varepsilon_0 r^2}\left[\exp\left\{-\frac{2r}{r_0}\right\}\left(-\frac{2r^2}{r_0^2} - \frac{2r}{r_0} - 1\right) + 1\right].$$

[1] See, e.g. *Part EM* Eqs. (1.38) and (1.55).
[2] See, e.g. *Part EM* Eq. (1.16).

doi:10.1088/2053-2563/aaf3a6ch8

The second charge creates an absolutely similar field, so that $\mathscr{E}_{\text{total}} = 2\mathscr{E}(r)$, and using the well-known expression for the electric field energy[3], the Coulomb interaction energy of these two distributed charges may be calculated as

$$E_{\text{int}} \equiv E_{\text{total}} - 2E_{\text{parial}} = \frac{\varepsilon_0}{2}\int[2\mathscr{E}(r)]^2 d^3r - 2\frac{\varepsilon_0}{2}\int[\mathscr{E}(r)]^2 d^3r$$

$$\equiv \varepsilon_0 \int[\mathscr{E}(r)]^2 d^3r = 4\pi\varepsilon_0 \int_0^\infty [\mathscr{E}(r)]^2 r^2 dr$$

$$= \frac{e^2}{4\pi\varepsilon_0}\int_0^\infty \left[\exp\left\{-\frac{2r}{r_0}\right\}\left(-\frac{2r}{r_0^2} - \frac{2}{r_0} - \frac{1}{r}\right) + \frac{1}{r}\right]^2 dr$$

$$\equiv \frac{e^2}{4\pi\varepsilon_0}\frac{1}{2r_0}\int_0^\infty \left[\exp\{-\xi\}\left(\xi + 2 + \frac{2}{\xi}\right) - \frac{2}{\xi}\right]^2 d\xi,$$

where $\xi \equiv 2r/r_0$ again. The last integral may be readily calculated by squaring the brackets, and integrating each of the resulting terms by parts[4]. The result equals 5/4, so that using the fact that for helium ($Z = 2$), $r_0 \equiv r_B/Z = r_B/2$, we finally get

$$E_g^{(1)} = E_{\text{int}} = \frac{e^2}{4\pi\varepsilon_0}\frac{1}{2r_0}\frac{5}{4} \equiv \frac{5}{4}\frac{e^2}{4\pi\varepsilon_0 r_B} \equiv \frac{5}{4}E_H,$$

i.e. the result used in section 8.2 of the lecture notes.

***Problem* 8.2.** For a diluted gas of helium atoms in their ground state, with n atoms per unit volume, calculate its:

(i) electric susceptibility χ_e, and
(ii) magnetic susceptibility χ_m,

and compare the results.

Hint: You may use the model solution of problems 6.8 and 6.14, and the results of the variational description of the helium atom's ground state in section 8.2 of the lecture notes.

Solutions:

(i) As was discussed in the model solution of problem 6.8, the atomic polarizability of the hydrogen atom, in its ground state, is

$$\alpha = 4\pi\varepsilon_0 \frac{9}{2} r_B^3, \qquad (*)$$

[3] See, e.g. *Part EM* Eq. (1.65).
[4] Alternatively, we may use for most of them the table integral (A.34*d*) with the corresponding values of n.

where r_B is the Bohr radius (1.13). Rescaling this result for a hydrogen-like 'atom' (or rather positive ion), with the nuclear charge $Q = Ze$ and one bound electron, we get

$$\alpha = 4\pi\varepsilon_0 \frac{9}{2}\frac{r_B^3}{Z^4}. \qquad (**)$$

Indeed, Eq. (*) is just a representation of the following result of the solution of problem 6.8, the change of atom's basic energy induced by the applied electric field $\mathscr{E}$:

$$\Delta E = -\frac{9}{2}\frac{r_B^2}{E_H}\frac{e^2\mathscr{E}^2}{2},$$

where E_H is the Hartree energy unit (1.9). As we know from section 3.6 of the lecture notes, for the Bohr-like atom/ion with $Q = Ze$, r_B should be replaced with $r_0 = r_B/Z$, and E_H with $E_0 = E_H Z^2$, so that instead of Eq. (***) we get

$$\Delta E = -\frac{9}{2}\frac{(r_B/Z)^2}{(E_H Z^2)}\frac{e^2\mathscr{E}^2}{2} \equiv -\frac{9}{2Z^4}\frac{r_B^2}{E_H}\frac{e^2\mathscr{E}^2}{2},$$

immediately giving Eq. (**).

Now taking Z equal to its variational-optimized value (8.34), $Z_{ef} = 2 - 5/16 \equiv 27/16$,[5] and adding these contributions from its two electrons, we get

$$\frac{\alpha_{He}}{4\pi\varepsilon_0} = 2 \cdot \frac{9}{2}\frac{r_B^3}{(27/16)^4} \approx 1.11 r_B^3.$$

In a diluted gas (with density $n \ll r_B^{-3}$), the atom interactions are negligible, so that their induced electric dipole moments $\mathbf{d} = \alpha\mathscr{E}$ just add up. According to the electrostatic basics[6], such a linear isotropic polarization, may be readily recalculated into the electric susceptibility:

$$(\chi_e)_{He} = \frac{n d_{He}}{\varepsilon_0 \mathscr{E}} = \frac{n\alpha_{He}}{\varepsilon_0} \approx 14 n r_B^3 \ll 1.$$

(ii) According to the discussion in section 8.2 of the lecture notes, the ground state of the helium atom is a spin singlet, so its net spin is zero: $S = M_S = 0$. As a result, the atom does not have spin paramagnetism, and its magnetic susceptibility is due to the orbital diamagnetism of the electrons[7]. Since, according to the discussion in

[5] As was mentioned in section 8.2 of the lecture notes, this variational approach describes experimental results with an accuracy better than 1%.

[6] See, e.g. *Part EM* section 3.3. Note again that the electric dipole moment, in that part of the series, is denoted as **p**.

[7] Strictly speaking, we also should consider the (very weak) spin paramagnetism of the two-proton *nucleus* of the helium atom, but since protons are also spin-½ fermions, the ground state of the nucleus may be also considered as a spin singlet (despite strong interaction of the protons), making its net spin equal to zero as well.

section 8.2 of the lecture notes, the orbital ground state of each electron is very close to that of a hydrogen-like atom, with the effective nuclear charge $Z_{ef} = 2 - 5/16 = 27/16$, we may use the solution of problem 6.14 for an arbitrary single-electron system,

$$(\chi_m)_{single} = -\frac{2\pi}{3} n\alpha^2 r_B \langle r^2 \rangle,$$

(where n is the number of atoms per unit volume, and $\alpha \approx 1/137$ is the fine structure constant) to write

$$(\chi_m)_{He} = -2 \cdot \frac{2\pi}{3} n\alpha^2 r_B \langle r_1^2 \rangle_{H \text{ with } Z \to Z_{ef}}.$$

This expectation value for a hydrogen-like atom, in its ground state, may be readily calculated using Eq. (3.208):[8]

$$\begin{aligned}\langle r_1^2 \rangle &= \oint_{4\pi} \left[Y_0^0(\theta, \varphi) \right]^2 d\Omega \int_0^\infty r^2 \,|\, \mathcal{R}_{1,0}(r)|^2 \, r^2 dr \\ &= \frac{4}{r_0^3} \int_0^\infty r^4 e^{-2r/r_0} dr \;=\; \frac{r_0^2}{8} \int_0^\infty \xi^4 e^{-\xi} d\xi,\end{aligned}$$

with $\xi \equiv 2r/r_0$. This is a table integral[9], equal to $\Gamma(5) = 4! = 24$, so that

$$\left\langle r_1^2 \right\rangle_H = 3r_0^2,$$

where, according to Eqs. (1.13) and (3.192), $r_0 = r_B/Z$. Now making the replacement $Z \to Z_{ef} = 27/16$, we get

$$(\chi_m)_{He} = -\frac{4\pi}{3} n\alpha^2 r_B \frac{3r_B^2}{Z_{ef}^2} = -4\pi \left(\frac{16}{27} \right)^2 \alpha^2 n r_B^3.$$

Comparing the results for χ_e and χ_m, we see that of these two dimensionless parameters (at the same n), the magnitude of the latter one is much (by a factor of $\sim \alpha^{-2} \sim 10^4$) smaller. This is very natural, since (as was repeatedly discussed in *Part EM* of this series) the orbital magnetism is a relativistic effect, very small for the effective velocities $v \sim 10^{-2} c$ of the quantum motion of electrons in atoms and molecules.

***Problem* 8.3.** Calculate the expectation values of the following observables: $\mathbf{s}_1 \cdot \mathbf{s}_2$, $S^2 \equiv (\mathbf{s}_1 + \mathbf{s}_2)^2$ and $S_z \equiv s_{1z} + s_{2z}$, for the singlet and triplet states of the system of two spins-½, defined by Eqs. (8.18) and (8.21) of the lecture notes, directly, without using

---

[8] As a reminder, the spherical harmonics and the radial wavefunctions (including $\mathcal{R}_{1,0}$), presented in chapter 3, are already normalized—see Eqs. (3.173) and (3.194).

[9] See, e.g. Eq. (A.34*a*) with $s = 5$.

the general rule (8.48) of spin addition. Compare the results with those for the system of two classical vectors of magnitude $\hbar/2$ each.

Solution: Let us calculate the action of the scalar product operator on the ket-vectors of all states of the uncoupled-representation basis of the system, by first spelling them out, and then returning to the shorthand notation. Starting from the simplest state with both spins up:

$$\begin{aligned}\hat{\mathbf{s}}_1 \cdot \hat{\mathbf{s}}_2 \,|\uparrow\uparrow\rangle &= (\hat{s}_{1x}\hat{s}_{2x} + \hat{s}_{1y}\hat{s}_{2y} + \hat{s}_{1z}\hat{s}_{2z})|\uparrow\uparrow\rangle \\ &\equiv (\hat{s}_{1x}\hat{s}_{2x} + \hat{s}_{1y}\hat{s}_{2y} + \hat{s}_{1z}\hat{s}_{2z})|\uparrow\rangle \otimes |\uparrow\rangle \\ &= \hat{s}_{1x}\,|\uparrow\rangle \otimes \hat{s}_{2x}\,|\uparrow\rangle + \hat{s}_{1y}\,|\uparrow\rangle \otimes \hat{s}_{2y}\,|\uparrow\rangle \\ &\quad + \hat{s}_{1z}\,|\uparrow\rangle \otimes \hat{s}_{2z}\,|\uparrow\rangle.\end{aligned}$$

As we know quite well by now (see, e.g. Eq. (4.128) of the lecture notes), for each particle

$$\hat{s}_x\,|\uparrow\rangle = \frac{\hbar}{2}\,|\downarrow\rangle, \quad \hat{s}_y\,|\uparrow\rangle = i\frac{\hbar}{2}\,|\downarrow\rangle, \quad \hat{s}_z\,|\uparrow\rangle = \frac{\hbar}{2}\,|\uparrow\rangle,$$

while the operators of one spin do not affect the ket-vectors of its counterpart, so that we get

$$\begin{aligned}\hat{\mathbf{s}}_1 \cdot \hat{\mathbf{s}}_2 \,|\uparrow\uparrow\rangle &= \left(\frac{\hbar}{2}\right)^2 (|\downarrow\rangle \otimes |\downarrow\rangle - |\downarrow\rangle \otimes |\downarrow\rangle + |\uparrow\rangle \otimes |\uparrow\rangle) \\ &= \left(\frac{\hbar}{2}\right)^2 |\uparrow\rangle \otimes |\uparrow\rangle \equiv \left(\frac{\hbar}{2}\right)^2 |\uparrow\uparrow\rangle.\end{aligned} \qquad (*)$$

So, the simple (factorable) triplet state $\uparrow\uparrow$ is indeed an eigenstate of the scalar product's operator, with eigenvalue $(\hbar/2)^2$. An absolutely similar calculation for the opposite simple triplet state, $\downarrow\downarrow$, taking into account that

$$\hat{s}_x\,|\downarrow\rangle = \frac{\hbar}{2}\,|\uparrow\rangle, \quad \hat{s}_y\,|\downarrow\rangle = -i\frac{\hbar}{2}\,|\uparrow\rangle, \quad \hat{s}_z\,|\downarrow\rangle = -\frac{\hbar}{2}\,|\downarrow\rangle,$$

yields the similar result:

$$\hat{\mathbf{s}}_1 \cdot \hat{\mathbf{s}}_2 \,|\downarrow\downarrow\rangle = \left(\frac{\hbar}{2}\right)^2 |\downarrow\downarrow\rangle, \qquad (**)$$

also corresponding to the initial state. These two results are in agreement with the classical picture of two aligned vectors of length $\hbar/2$ each.

However, the absolutely similar calculations for the oppositely directed spin states give quite different results:

$$\hat{\mathbf{s}}_1 \cdot \hat{\mathbf{s}}_2 \,|\uparrow\downarrow\rangle = \left(\frac{\hbar}{2}\right)^2 (2|\downarrow\uparrow\rangle - |\uparrow\downarrow\rangle), \quad \hat{\mathbf{s}}_1 \cdot \hat{\mathbf{s}}_2 \,|\downarrow\uparrow\rangle = \left(\frac{\hbar}{2}\right)^2 (2|\uparrow\downarrow\rangle - |\downarrow\uparrow\rangle),$$

showing that these states are *not* eigenstates of the scalar product's operator. However, their two linear superpositions, the entangled triplet (sign +) and singlet (sign −) states, defined by Eqs. (8.18) and (8.20),

$$|s_\pm\rangle = \frac{1}{\sqrt{2}}(|\uparrow\downarrow\rangle \pm |\downarrow\uparrow\rangle),$$

are eigenstates of this operator:

$$\hat{\mathbf{s}}_1 \cdot \hat{\mathbf{s}}_2 \, |s_\pm\rangle = \left(\frac{\hbar}{2}\right)^2 \frac{1}{\sqrt{2}} \times \begin{cases} (|\uparrow\downarrow\rangle + |\downarrow\uparrow\rangle) \\ (-3\,|\uparrow\downarrow\rangle - 3\,|\downarrow\uparrow\rangle) \end{cases} \equiv \left(\frac{\hbar}{2}\right)^2 \times \begin{cases} |s_+\rangle, \\ -3\,|s_-\rangle, \end{cases}$$

though with rather different eigenvalues:

$$\langle \hat{\mathbf{s}}_1 \cdot \hat{\mathbf{s}}_2 \rangle = \left(\frac{\hbar}{2}\right)^2 \times \begin{cases} (+1), & \text{for the triplet state } (s_+), \\ (-3), & \text{for the singlet state } (s_-). \end{cases} \qquad (***)$$

Note that according to Eqs. (*)–(***), the product's eigenvalue for the entangled triplet state is the same as for both factorable triplet states—a fact rather counter-intuitive for such a linear superposition of two states with oppositely directed spins as s_+. The same may be said about the result (***) for the singlet state, with its 'unnaturally' high modulus; it is obviously in a sharp contradiction with the classical prediction $\mathbf{s}_1 \cdot \mathbf{s}_2 = -(\hbar/2)^2$ for two equal and antiparallel vectors of magnitude $\hbar/2$ each.

Let us now consider the total spin operator (8.47),

$$\hat{\mathbf{S}} \equiv \hat{\mathbf{s}}_1 + \hat{\mathbf{s}}_2.$$

Its square may be calculated using the fact that the operators of the two component spins are defined in different Hilbert spaces, and hence commute:

$$\hat{S}^2 \equiv \hat{\mathbf{S}} \cdot \hat{\mathbf{S}} \equiv (\hat{\mathbf{s}}_1 + \hat{\mathbf{s}}_2) \cdot (\hat{\mathbf{s}}_1 + \hat{\mathbf{s}}_2) = \hat{\mathbf{s}}_1^2 + \hat{\mathbf{s}}_2^2 + 2\hat{\mathbf{s}}_1 \cdot \hat{\mathbf{s}}_2. \qquad (****)$$

As we know from chapter 4, squares of single-particle spin-½ operators are proportional to the identity operator:

$$\hat{s}_{1,2}^2 = \left(\hat{s}_x^2 + \hat{s}_y^2 + \hat{s}_z^2\right)_{1,2} = 3\left(\frac{\hbar}{2}\right)^2 \hat{I}.$$

Hence in any state of the system, the expectation value of the operator (****) is

$$\langle S^2 \rangle = 6\left(\frac{\hbar}{2}\right)^2 + 2\langle \mathbf{s}_1 \cdot \mathbf{s}_2 \rangle,$$

so that using Eqs. (*)–(***), we get the same Eq. (8.52),

$$\langle S^2 \rangle = \left(\frac{\hbar}{2}\right)^2 \times \begin{cases} 8, & \text{for all triplet states,} \\ 0, & \text{for the singlet state,} \end{cases}$$

which was obtained in section 8.2 from the general relations (8.48) (valid for particles with any spin), with the quantum number $S = 1$ for the triplet states and $S = 0$ for the singlet state—see also the 'rectangular diagram' in figure 8.2. (The counter-intuitive nature of this result for the entangled triplet state was already noted at its discussion in the lecture notes.)

However, for the z-component of the vector sum (8.47), the quantum-mechanical result coincides with what we could expect for classical vectors. Indeed, for example,

$$\begin{aligned}\hat{S}_z\,|\uparrow\downarrow\rangle &\equiv (\hat{s}_{1z} + \hat{s}_{2z})\,|\uparrow\rangle \otimes |\downarrow\rangle \\ &= \hat{s}_{1z}\,|\uparrow\rangle \otimes |\downarrow\rangle + |\uparrow\rangle \otimes \hat{s}_{2z}\,|\downarrow\rangle \\ &= |\uparrow\rangle \otimes |\downarrow\rangle - |\uparrow\rangle \otimes |\downarrow\rangle = 0,\end{aligned}$$

and similarly for the second component state, $\downarrow\uparrow$, and hence for both entangled states:

$$\hat{S}_z\,|s_\pm\rangle = 0.$$

This result is in again in full agreement with the general theory of the spin addition, because, as shown in figure 8.2, for both these states the 'magnetic' quantum number $M_S = (m_s)_1 + (m_s)_2$ is equal to zero. An absolutely similar calculation shows that the factorable states $\uparrow\uparrow$ and $\downarrow\downarrow$ are also the eigenstates of the operator $\hat{S}_z$, with the eigenstates, respectively, $+\hbar$ and $-\hbar$, corresponding to $M_S = \pm 1$—also in concord with Eq. (8.48) for the net spin $S = 1$, and classical expectations.

***Problem* 8.4.** Discuss the factors $\pm 1/\sqrt{2}$ that participate in Eqs. (8.18) and (8.20) of the lecture notes for the entangled states of the system of two spins-½, in terms of Clebsh–Gordan coefficients similar to those discussed in section 5.7.

Solution: As was discussed in section 8.2 of the lecture notes, the sum (8.47) of two spins has the same properties (8.48) as the sum (5.170) of the orbital and spin angular momenta of a single particle. Hence, for a system of two spins-½ we may repeat all the discussion of section 5.7 with the following replacements:

$$\begin{aligned}&\hat{\mathbf{L}} \rightarrow \hat{\mathbf{s}}_1, &&\hat{\mathbf{S}} \rightarrow \hat{\mathbf{s}}_2, &&\hat{\mathbf{J}} \equiv \hat{\mathbf{L}} + \hat{\mathbf{S}} \rightarrow \hat{\mathbf{S}} = \hat{\mathbf{s}}_1 + \hat{\mathbf{s}}_2,\\ &l \rightarrow s_1 = \tfrac{1}{2}, &&s \rightarrow s_2 = \tfrac{1}{2}, &&j \rightarrow S = s_1 \pm s_2 = 0,\ 1,\\ &m_l \rightarrow m_1 = \pm\tfrac{1}{2}, &&m_s \rightarrow m_2 = \pm\tfrac{1}{2}, &&m_j \rightarrow M_S = \begin{cases}0, & \text{for } S = 0,\\ 0,\ \pm 1, & \text{for } S = 1,\end{cases}\end{aligned} \qquad (*)$$

where the index s in the magnetic quantum numbers m_s of the component spins is just implied, to avoid excessively cluttered notation.

With these replacements, instead of the two state groups listed in Eq. (5.182), we get the following two possible bases, of four states each, available for representation of an arbitrary state of the composite system:

- the uncoupled-representation basis: states $|m_1, m_2\rangle$, and
- the coupled-representation basis: states $|S, M_S\rangle$.

In particular, as was discussed in section 8.2, and confirmed by the direct calculation in the solution of the previous problem, both entangled states (8.18) and (8.20) belong to the coupled-representation basis, both with $M_S = 0$, but with $S = 1$ for the triplet state, and $S = 0$ for the singlet state.

Now we may use the replacements (*) to write the following analogs of Eqs. (5.190) for the Clebsh–Gordan coefficients of the two-particle system:

$$\begin{aligned}\langle m_1 = M_S - \tfrac{1}{2}, m_2 = +\tfrac{1}{2}|S = \tfrac{1}{2} \pm \tfrac{1}{2}, M_S\rangle &= \pm\left(\frac{\tfrac{1}{2} \pm M_S + \tfrac{1}{2}}{2 \cdot \tfrac{1}{2} + 1}\right)^{1/2} \\ &\equiv \pm\frac{1}{\sqrt{2}}(1 \pm M_S)^{1/2}, \\ \langle m_1 = M_S + \tfrac{1}{2}, m_2 = -\tfrac{1}{2}|S = \tfrac{1}{2} \pm \tfrac{1}{2}, M_S\rangle &= +\left(\frac{\tfrac{1}{2} \mp M_S + \tfrac{1}{2}}{2 \cdot \tfrac{1}{2} + 1}\right)^{1/2} \\ &\equiv +\frac{1}{\sqrt{2}}(1 \mp M_S)^{1/2}.\end{aligned} \qquad (**)$$

These results are valid for states with any S and M_S of our list. In particular, for the factorable triplet state $\uparrow\uparrow$, with $S = 1$ and $M_S = +1$, these formulas are reduced to

$$\begin{aligned}\langle m_1 = +\tfrac{1}{2}, m_2 = +\tfrac{1}{2}|S = 1, M_S = +1\rangle &= 1, \\ \langle m_1 = +\tfrac{3}{2}, m_2 = -\tfrac{1}{2}|S = 1, M_S = +1\rangle &= 0,\end{aligned}$$

while for the factorable triplet state $\downarrow\downarrow$, with $S = 0$ and $M_S = -1$,

$$\begin{aligned}\langle m_1 = -\tfrac{3}{2}, m_2 = +\tfrac{1}{2}|S = 1, M_S = -1\rangle &= 0, \\ \langle m_1 = +\tfrac{1}{2}, m_2 = -\tfrac{1}{2}|S = 1, M_S = -1\rangle &= 1.\end{aligned}$$

The two non-trivial relations of this set mean simply that the factorable triplet states belong to both the coupled- and uncoupled representation:

$$\begin{aligned}|\uparrow\uparrow\rangle &\equiv |m_1 = +\tfrac{1}{2}, m_2 = +\tfrac{1}{2}\rangle = |S = 1, M_S = +1\rangle, \\ |\downarrow\downarrow\rangle &\equiv |m_1 = -\tfrac{1}{2}, m_2 = -\tfrac{1}{2}\rangle = |S = 1, M_S = -1\rangle,\end{aligned}$$

as was already discussed in section 8.2 of the lecture notes—see the top-right and bottom-left points in the 'rectangular diagram' shown in figure 8.2.

On the other hand, for the entangled states, both with $M_S = 0$, Eqs. (**) are reduced to:

$$\begin{aligned}\langle m_1 = -\tfrac{1}{2}, m_2 = +\tfrac{1}{2}|S = \tfrac{1}{2} \pm \tfrac{1}{2}, M_S = 0\rangle &\equiv \langle\downarrow\uparrow|S = \tfrac{1}{2} \pm \tfrac{1}{2}, M_S = 0\rangle \\ &= \pm\frac{1}{\sqrt{2}}, \\ \langle m_1 = +\tfrac{1}{2}, m_2 = -\tfrac{1}{2}|S = \tfrac{1}{2} \pm \tfrac{1}{2}, M_S = 0\rangle &\equiv \langle\uparrow\downarrow|S = \tfrac{1}{2} \pm \tfrac{1}{2}, M_S = 0\rangle \\ &= +\frac{1}{\sqrt{2}}.\end{aligned}$$

According to these relations, we may write

$$|S = 1, M_S = 0\rangle = \frac{1}{\sqrt{2}}(|\uparrow\downarrow\rangle + |\downarrow\uparrow\rangle) \equiv |s_+\rangle,$$
$$|S = 0, M_S = 0\rangle = \frac{1}{\sqrt{2}}(|\uparrow\downarrow\rangle - |\downarrow\uparrow\rangle) \equiv |s_-\rangle,$$

confirming once again the fact that the entangled states $s_\pm$ belong to the coupled-representation basis, with the listed quantum numbers of the net spin.

Thus the factors $\pm 1/\sqrt{2}$ that participate in the definitions (8.18) and (8.20) of these states may be considered just as particular cases of the Clebsh–Gordan coefficients in the linear relations between the basis vectors of the uncoupled and coupled representations of the two-spin-½ system.

Problem* 8.5. Use the perturbation theory to calculate the contribution to the so-called *hyperfine splitting* of the ground energy of the hydrogen atom[10], due to the interaction between spins of the nuclei (proton) and the electron.

Hint: The proton's magnetic moment operator is described by the same Eq. (4.115) of the lecture notes as the electron, but with a positive gyromagnetic factor $\gamma_p = g_p e/2m_p \approx 2.675 \times 10^8\ \text{s}^{-1}\ \text{T}^{-1}$, whose magnitude is much smaller than that of the electron ($|\gamma_e| \approx 1.761 \times 10^{11}\ \text{s}^{-1}\ \text{T}^{-1}$), due to the much higher mass, $m_p \approx 1.673 \times 10^{-27}$ kg $\approx 1{,}835\ m_e$. (The g-factor of the proton is also different, $g_p \approx 5.586$.[11])

Solution: The Hamiltonian perturbation due to the interaction between the magnetic dipole $\mathbf{m}_e$ of the electron and the magnetic field $\mathscr{B}_p$ induced by proton's magnetic moment,

$$\hat{\mathbf{m}}_p = \gamma_p \hat{\mathbf{S}}_p = g_p \frac{e}{2m_p} \hat{\mathbf{S}}_p,$$

may be taken in the usual Pauli form (see Eq. (4.163) of the lecture notes):

$$\hat{H}_{ss} = -\hat{\mathbf{m}}_e \cdot \hat{\mathscr{B}}_p, \quad \text{with} \quad \hat{\mathbf{m}}_e = \gamma_e \hat{\mathbf{S}}_e = -g_e \frac{e}{2m_e} \hat{\mathbf{S}}_e, \quad g_e \approx 2.0023.$$

Due to the proton's relatively large mass, its position's uncertainty (very small in comparison with the scale of electron's wavefunction spread, the Bohr radius r_B) may be disregarded, so that the relation between the operators $\mathscr{B}_p$ and $\mathbf{m}_p$ may be borrowed from the classical electrodynamics' result for the field of a point dipole[12]:

$$\hat{\mathscr{B}}_p(\mathbf{r}) = \frac{\mu_0}{4\pi}\left(\frac{3\mathbf{r}(\mathbf{r} \cdot \hat{\mathbf{m}}_p) - r^2 \hat{\mathbf{m}}_p}{r^5} + \frac{8\pi}{3} \hat{\mathbf{m}}_p \delta(\mathbf{r})\right).$$

[10] This effect was discovered experimentally by A Michelson in 1881, and explained theoretically by W Pauli in 1924.

[11] The anomalously large value of the proton's g-factor results from the composite quark-gluon structure of this particle. (An exact calculation of g_p remains a challenge for quantum chromodynamics.)

[12] See, e.g. *Part EM* section 5.4.

As a reminder, the delta-functional term in this expression provides a 'coarse-grain' description of the field source—in our case, the proton. In our case its use is legitimate due to the (very) strong relation $r_B \gg a_p$, where $a_p \sim 10^{-15}$ m is the effective spread of the proton's electric charge.

Combining these formulas, we may use Eq. (6.14) of the lecture notes to write the following first-order correction to the energy of the ground (and hence non-degenerate) state of the atom:

$$E_0^{(1)} = \langle 0|\, \hat{H}_{ss}\, |0\rangle$$
$$= \frac{\mu_0}{4\pi}\frac{g_e g_p e^2}{4m_e m_p}\left[\langle 0|\frac{3(\mathbf{r}\cdot\hat{\mathbf{s}}_e)(\mathbf{r}\cdot\hat{\mathbf{s}}_p)}{r^5}|0\rangle - \langle 0|\frac{r^2\hat{\mathbf{s}}_e\cdot\hat{\mathbf{s}}_p}{r^5}|0\rangle + \frac{8\pi}{3}\langle 0|\hat{\mathbf{s}}_e\cdot\hat{\mathbf{s}}_p\delta(\mathbf{r})|0\rangle\right], \qquad (*)$$

where 0 is the unperturbed ground state of the atom. Since the vector of this state is spin–orbit factorable in the sense of Eq. (8.12), each average inside the square brackets of the last expression may be calculated separately for the orbital and spin components. For example, the first average is

$$\langle s_{ep}| \otimes \langle o_0|\, \frac{3(\mathbf{r}\cdot\hat{\mathbf{s}}_e)(\mathbf{r}\cdot\hat{\mathbf{s}}_p)}{r^5}\, |o_0\rangle \otimes |s_{ep}\rangle$$
$$= \langle s_{ep}| \int \psi_{100}^*(\mathbf{r})\frac{3(\mathbf{r}\cdot\hat{\mathbf{s}}_e)(\mathbf{r}\cdot\hat{\mathbf{s}}_p)}{r^5}\psi_{100}(\mathbf{r})d^3r\, |s_{ep}\rangle,$$

(with the ground-state wavefunctions calculated in section 3.6), where the integral over the orbital motion's space may be taken by temporarily treating the electron and proton spin operators as usual (c-number) 3D vectors $\mathbf{s}_e$ and $\mathbf{s}_p$—because such treatment is valid for any matrix element of these operators.

Selecting the coordinate axis x within the common plane of these two vectors, bisecting the angle φ_0 between them (see the figure below), and the z-axis normal to this plane, in the usual polar coordinates (in which the x–y component $\boldsymbol{\rho}$ of the radius-vector $\mathbf{r}$ has the length $\rho = r\sin\theta$), we get

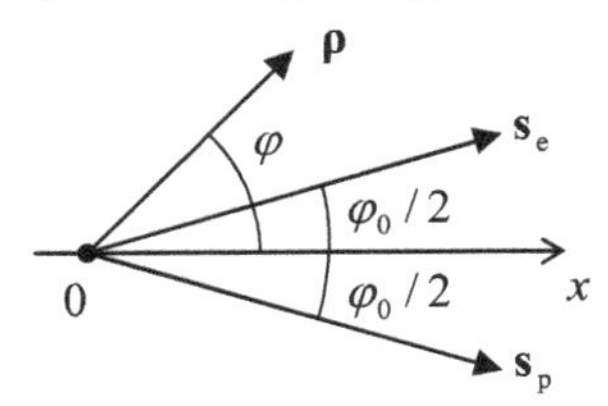

$$3(\mathbf{r}\cdot\hat{\mathbf{s}}_e)(\mathbf{r}\cdot\hat{\mathbf{s}}_p) = 3(\boldsymbol{\rho}\cdot\mathbf{s}_e)(\boldsymbol{\rho}\cdot\mathbf{s}_p)$$
$$= 3\left[\hat{s}_e r\sin\theta\cos\left(\varphi - \frac{\varphi_0}{2}\right)\right]\left[\hat{s}_p r\sin\theta\cos\left(\varphi + \frac{\varphi_0}{2}\right)\right]$$
$$\equiv \frac{3}{2}r^2\hat{s}_e\hat{s}_p\sin^2\theta(\cos\varphi_0 + \cos 2\varphi);$$
$$r^2(\hat{\mathbf{s}}_e\cdot\hat{\mathbf{s}}_p) = r^2\hat{s}_e\hat{s}_p\cos\varphi_0.$$

The wavefunction $\psi_{100}(\mathbf{r})$ (describing the ground 1s electronic state of the atom) is spherically-symmetric. With the account of this fact, and the above expressions, the integration over the full solid angle of the two first terms on the right-hand side of Eq. (*) gives similar results, so that their difference vanishes. As a result, the only contribution to the energy correction is given by the third, delta-functional term:

$$E_0^{(1)} = \frac{\mu_0}{4\pi}\frac{g_e g_p e^2}{4m_e m_p}\frac{8\pi}{3}\left\langle s_{ep}\right|\int \psi_{100}^*(\mathbf{r})\hat{\mathbf{s}}_e \cdot \hat{\mathbf{s}}_p \delta(\mathbf{r})\psi_{100}(\mathbf{r})d^3r\left|s_{ep}\right\rangle$$

$$\equiv \frac{\mu_0}{6}\frac{g_e g_p e^2}{m_e m_p}\left|\psi_{100}(0)\right|^2\left\langle s_{ep}\right|\hat{\mathbf{s}}_e \cdot \hat{\mathbf{s}}_p\left|s_{ep}\right\rangle.$$

(According to Eqs. (3.174) and (3.208), $|\psi_{100}(0)|^2 = 1/\pi r_B^3$.)

Now proceeding to the spin factor: since the proton is also a Fermi-particle with spin ½, the spin average in the above expression may be calculated exactly as in the solution of problem 8.3, giving $(\hbar/2)^2$ for any triplet state, and $-3(\hbar/2)^2$ for the singlet state, with the difference equal to $\hbar^2$.[13] As a result, the ground state energy splits to two hyperfine sublevels, with triplet states' energy higher than that of the singlet state by

$$\Delta E_{ss} = \frac{\mu_0}{6\pi}\frac{g_e g_p e^2\hbar^2}{m_e m_p r_B^3}. \qquad (**)$$

Plugging in the values of the participating constants, we get $\Delta E_{ss} \approx 5.884 \times 10^{-6}$ eV, almost 7 orders of magnitude smaller than the Hartree energy $E_0 \approx 27$ eV, thus giving an *a posteriori* justification of our perturbative treatment. Moreover, this splitting is much smaller than the fine structure of the energy due to spin–orbit interaction (see section 6.3 of the lecture notes)—hence the term *hyper*fine[14].

The hyperfine splitting of hydrogen levels is very important for astronomy, because due to the cosmic microwave background radiation, the effective temperature of the hydrogen gas in space is at least ~ 3 K, i.e. substantially higher than the minimum ($T_{ss} \sim \Delta E_{ss}/k_B \sim 0.1$ K) necessary for the spontaneous thermal excitation of the higher (triplet) states. After such thermal excitation, the hydrogen atom eventually returns to the genuine ground (singlet) state, emitting a microwave photon with frequency $\omega_{ss} = \Delta E_{ss}/\hbar \approx 0.8924 \times 10^{10}$ s^{-1} ($f_{ss} \equiv \omega_{ss}/2\pi \approx 1420.4$ MHz)[15] corresponding to the

[13] Note that according to the solution of the same problem 8.3, all triplet states have the net spin $S = 1$, while the singlet state, $S = 0$. The change of S at the quantum transition between the hyperfine sub-levels may be interpreted by saying that the spin balance is carried away by the emitted circularly-polarized photon with spin 1—the notion to be discussed in chapter 9.

[14] The splitting (**) affects each sub-level of the fine structure. Note also that in more complex atoms and molecules, several other mechanisms, most notably including the interaction between the quadrupole electric moment of the nucleus (see, e.g. *Part EM* section 8.9) with the electrons' electric field gradient, also make comparable contributions to the hyperfine structure of the energy levels.

[15] These experimental values (measured to the 13th decimal place!) differ from the above theoretical value by ~ 0.2%, due to quantum-electrodynamic effects, ignored in the above treatment.

wavelength $\lambda_{ss} = c/f_{ss} \approx 21.11$ cm. This famous *21-cm line* gives radioastronomy one of the most important tools for the study of the spatial distribution of the Universe's most ubiquitous atoms. (In particular, it was used to discover the spiral structure of our galaxy.)

***Problem* 8.6.** In the simple case of just two similar spin-interacting particles, distinguishable by their spatial location, the famous *Heisenberg model* of ferromagnetism[16] is reduced to the following Hamiltonian:

$$\hat{H} = -J\hat{\mathbf{s}}_1 \cdot \hat{\mathbf{s}}_2 - \gamma\mathscr{B} \cdot (\hat{\mathbf{s}}_1 + \hat{\mathbf{s}}_2),$$

where J is the spin interaction constant, γ is the gyromagnetic ratio of each particle, and $\mathscr{B}$ is the external magnetic field. Find the stationary states and eigenenergies of this system for spin-½ particles.

Solution: According to the solution of problem 8.3, all three triplet states (8.21), and the singlet state (8.18) are the eigenstates of both terms of this Hamiltonian, and hence are the stationary states of this system, with the following energies:

$$E = \begin{cases} -J(\hbar/2)^2 - \hbar\gamma\mathscr{B}, & \text{for the state } \uparrow\uparrow, \\ -J(\hbar/2)^2 + \hbar\gamma\mathscr{B}, & \text{for the state } \downarrow\downarrow, \\ -J(\hbar/2)^2, & \text{for the entangled triplet state,} \\ +3J(\hbar/2)^2, & \text{for the (entangled) singlet state.} \end{cases}$$

If the magnetic field effect is negligibly small, the energy spectrum is reduced to just two levels: the triple-degenerate level: $E_{\text{triplet}} \equiv -J(\hbar/2)^2$, and the singlet level $E_{\text{singlet}} \equiv 3J(\hbar/2)^2$. So, the singlet and triplet spin states may naturally form even if two similar particles are distinguishable (in our current case, by their fixed spatial positions), due to their explicit interaction.

On the other hand, a substantial magnetic field, with $\gamma\mathscr{B} \sim \hbar J$, lifts the triplet level's degeneracy. (Note a substantial similarity of this effect with that for the excited states of the ^{4}He atom, discussed at the end of section 8.2 of the lecture notes.)

***Problem* 8.7.** Two particles, both with spin-½, but different gyromagnetic ratios γ_1 and γ_2, are placed into external magnetic field $\mathscr{B}$. In addition, their spins interact as in the Heisenberg model:

$$\hat{H}_{\text{int}} = -J\hat{\mathbf{s}}_1 \cdot \hat{\mathbf{s}}_2.$$

Find the eigenstates and eigenenergies of the system[17].

[16] It was suggested in 1926, independently by W Heisenberg and P Dirac. A discussion of temperature effects on this and other similar systems (especially the Ising model of ferromagnetism) may be found in *Part SM* chapter 4.

[17] For similar particles (in particular, with $\gamma_1 = \gamma_2$) the problem is evidently reduced to the previous one.

Solution: In the usual z-basis for each particle, with the z-axis directed along the applied magnetic field, the total Hamiltonian of the system has the following matrix:

$$\mathrm{H} = -J\left(\frac{\hbar}{2}\right)^2(\sigma_{1x}\sigma_{2x} + \sigma_{1y}\sigma_{2y} + \sigma_{1z}\sigma_{2z}) - \gamma_1\mathscr{B}\frac{\hbar}{2}\sigma_{1z} - \gamma_2\mathscr{B}\frac{\hbar}{2}\sigma_{2z}.$$

Using this expression, and the solution of problem 8.3, we may readily calculate the effects of the Hamiltonian on the each of four states of the uncoupled-representation basis of the system:

$$\begin{aligned}
\hat{H}\,|\uparrow\uparrow\rangle &= \left(-J\frac{\hbar^2}{4} - \gamma_1\mathscr{B}\frac{\hbar}{2} - \gamma_2\mathscr{B}\frac{\hbar}{2}\right)|\uparrow\uparrow\rangle,\\
\hat{H}\,|\downarrow\downarrow\rangle &= \left(-J\frac{\hbar^2}{4} + \gamma_1\mathscr{B}\frac{\hbar}{2} + \gamma_2\mathscr{B}\frac{\hbar}{2}\right)|\downarrow\downarrow\rangle,\\
\hat{H}\,|\uparrow\downarrow\rangle &= \left(J\frac{\hbar^2}{4} - \gamma_1\mathscr{B}\frac{\hbar}{2} + \gamma_2\mathscr{B}\frac{\hbar}{2}\right)|\uparrow\downarrow\rangle - 2J\frac{\hbar^2}{4}\,|\downarrow\uparrow\rangle,\\
\hat{H}\,|\downarrow\uparrow\rangle &= \left(J\frac{\hbar^2}{4} + \gamma_1\mathscr{B}\frac{\hbar}{2} - \gamma_2\mathscr{B}\frac{\hbar}{2}\right)|\downarrow\uparrow\rangle - 2J\frac{\hbar^2}{4}\,|\uparrow\downarrow\rangle.
\end{aligned} \qquad (*)$$

The first two of these formulas show that the factorable triplet states $\uparrow\uparrow$ and $\downarrow\downarrow$ are always the eigenstates of this system, with eigenenergies, respectively,

$$E_{\uparrow\uparrow} = -J\frac{\hbar^2}{4} - (\gamma_1 + \gamma_2)\mathscr{B}\frac{\hbar}{2}, \quad \text{and} \quad E_{\downarrow\downarrow} = -J\frac{\hbar^2}{4} + (\gamma_1 + \gamma_2)\mathscr{B}\frac{\hbar}{2}. \qquad (**)$$

On the other hand, Eqs. (*) mean that the two remaining eigenstates, which will be denoted ($\pm$), are not the simple (factorable) states $\uparrow\downarrow$ and $\downarrow\uparrow$, but rather some of their linear superpositions:

$$|+\rangle = a_+\,|\uparrow\downarrow\rangle + b_+\,|\downarrow\uparrow\rangle, \qquad |-\rangle = a_-\,|\uparrow\downarrow\rangle + b_-\,|\downarrow\uparrow\rangle.$$

The coefficient pairs $\{a_\pm, b_\pm\}$ in these relations may be found (to a common multiplier, which should be calculated from the normalization condition) as the eigenvectors of the following partial (2×2) matrix of the Hamiltonian, written in the basis of two uncoupled-representation states, $\uparrow\downarrow$ and $\downarrow\uparrow$:

$$\begin{pmatrix} J\frac{\hbar^2}{4} - (\gamma_1 - \gamma_2)\mathscr{B}\frac{\hbar}{2} & -2J\frac{\hbar^2}{4} \\ -2J\frac{\hbar^2}{4} & J\frac{\hbar^2}{4} + (\gamma_1 - \gamma_2)\mathscr{B}\frac{\hbar}{2} \end{pmatrix}. \qquad (***)$$

As usual, we should find the corresponding eigenvalues (i.e. the energy levels) first, as the roots $E_\pm$ of the consistency equation (4.103), in our case

$$\begin{vmatrix} J\frac{\hbar^2}{4} - (\gamma_1 - \gamma_2)\mathscr{B}\frac{\hbar}{2} - E & -2J\frac{\hbar^2}{4} \\ -2J\frac{\hbar^2}{4} & J\frac{\hbar^2}{4} + (\gamma_1 - \gamma_2)\mathscr{B}\frac{\hbar}{2} - E \end{vmatrix} = 0.$$

A straightforward calculation yields:

$$E_{\mp} = J\frac{\hbar^2}{4} \pm \left[\left(2J\frac{\hbar^2}{4}\right)^2 + \left((\gamma_1 - \gamma_2)\mathscr{B}\frac{\hbar}{2}\right)^2\right]^{1/2}. \qquad (****)$$

According to Eqs. (**) and (****), at negligible magnetic fields, $\gamma_{1,2}\mathscr{B} \ll \hbar J$, the energy spectrum of the system is reduced to just two levels: a triplet level: $E_{\uparrow\uparrow} \to E_{\downarrow\downarrow} \to E_+ \to E_{\text{triplet}} \equiv -J\hbar^2/4$, and a singlet level, $E_- \to E_{\text{singlet}} \equiv 3J\hbar^2/4$. On the other hand, at very high fields, when $(\gamma_1 \pm \gamma_2)\mathscr{B} \gg \hbar J$, all four energies are different:

$$E_{\uparrow\uparrow} \approx (-\gamma_1 - \gamma_2)\mathscr{B}\frac{\hbar}{2}, \quad E_{\downarrow\downarrow} \approx (\gamma_1 + \gamma_2)\mathscr{B}\frac{\hbar}{2},$$
$$E_- \approx (\gamma_1 - \gamma_2)\mathscr{B}\frac{\hbar}{2}, \quad E_+ \approx (-\gamma_1 + \gamma_2)\mathscr{B}\frac{\hbar}{2},$$

and may be interpreted as algebraic sums of individual spin energies in the magnetic field. Thus at very high fields the spin coupling is unimportant, and the spins behave independently—as could be expected.

Now plugging the results (****), one by one, into the system of equations for the coefficients $\{a_\pm, b_\pm\}$, that corresponds to the matrix (***), we get similar expressions for their ratios:

$$\frac{b_+}{a_+} = -\frac{a_-}{b_-} = \frac{(\gamma_1 - \gamma_2)\mathscr{B}}{\hbar J} + \left[1 + \left(\frac{(\gamma_1 - \gamma_2)\mathscr{B}}{\hbar J}\right)^2\right]^{1/2}.$$

For either similar particles, i.e. for $\gamma_1 - \gamma_2 = 0$, or for any $\gamma_{1,2}$ but in very low magnetic fields, these formulas yield $b_+ = a_+$ and $b_- = -a_-$, so that the eigenstate (+) is the usual mixed triplet state s_+ (see, e.g. Eq. (8.20) of the lecture notes), while the eigenstate (−) is the singlet state (8.18), in agreement with the previous problem's solution. However, at $\gamma_1 \neq \gamma_2$, a substantial magnetic field lifts the triplet degeneracy, and makes the linear superpositions (±) different from the entangled states of the coupled-representation basis. In the limit of very large fields, $(\gamma_1 - \gamma_2)\mathscr{B} \gg \hbar J$, the coefficient a_+ becomes much smaller than b_+, and the coefficient b_- much less than a_-, meaning that the eigenstate (+) tends to the state $\uparrow\downarrow$ of the uncoupled-representation basis, and the eigenstate (−) tends to the opposite state $\downarrow\uparrow$ of this basis.

***Problem* 8.8.** Two similar spin-½ particles, with the gyromagnetic ratio γ, localized at two points separated by distance a, interact via the field of their magnetic dipole moments. Calculate the spin eigenstates and eigenvalues of the system.

Solution: In classical electrodynamics, the energy of interaction of magnetic dipoles $\mathbf{m}_1$ and $\mathbf{m}_2$, separated by distance a is[18]

$$U = -\frac{\mu_0}{4\pi a^3}\frac{3\mathbf{a}(\mathbf{a}\cdot\mathbf{m}_1) - a^2\mathbf{m}_1}{a^5}\cdot\mathbf{m}_2$$
$$\equiv \frac{\mu_0}{4\pi a^3}(m_{1x}m_{2x} + m_{1y}m_{2y} - 2m_{1z}m_{2z}),$$

where the z-axis is directed along the vector $\mathbf{a}$ connecting the dipole positions. According to the correspondence principle of quantum mechanics, the interaction is described by the Hamiltonian that is similarly expressed via the Cartesian components of the magnetic moment operators given by Eq. (4.115) of the lecture notes:

$$\hat{H} = \frac{\mu_0}{4\pi a^3}(\hat{m}_{1x}\hat{m}_{2x} + \hat{m}_{1y}\hat{m}_{2y} - 2\hat{m}_{1z}\hat{m}_{2z}), \quad \text{where } \hat{\mathbf{m}}_j = \gamma\hat{\mathbf{s}}_j,$$

and $\hat{\mathbf{s}}_j$ is the spin vector-operator of the jth particle. For spin-½ particles, in the standard z-basis, the operator is described by Eqs. (4.116) and (4.117), so that the Hamiltonian matrix is

$$\mathrm{H} = E_0(\sigma_{1x}\sigma_{2x} + \sigma_{1y}\sigma_{2y} - 2\sigma_{1z}\sigma_{2z}), \quad \text{with} \quad E_0 \equiv \frac{\mu_0}{4\pi a^3}\left(\gamma\frac{\hbar}{2}\right)^2,$$

where $\sigma_{1,2}$ are the Pauli matrices (4.105), acting on spin vectors of the corresponding particles. (Note that this Hamiltonian substantially differs from that in the Heisenberg model.)

Using this expression and acting just as in problem 8.3, we may readily calculate the result of action of the Hamiltonian on each of four states of the uncoupled-representation basis of the two-spin system. The result is

$$\hat{H}\,|\uparrow\uparrow\rangle = -2E_0\,|\uparrow\uparrow\rangle, \qquad \hat{H}\,|\downarrow\downarrow\rangle = -2E_0\,|\downarrow\downarrow\rangle,$$
$$\hat{H}\,|\uparrow\downarrow\rangle = 2E_0(|\uparrow\downarrow\rangle + |\downarrow\uparrow\rangle), \quad \hat{H}\,|\downarrow\uparrow\rangle = 2E_0(|\uparrow\downarrow\rangle + |\downarrow\uparrow\rangle);$$

it means that the spin-aligned states, ↑↑ and ↓↓, are stationary states of the system, with the same energy,

$$E_{\uparrow\uparrow} = E_{\downarrow\downarrow} = -2E_0.$$

On the other hand, the spin-misaligned states ↑↓ and ↓↑, while not mixing with the spin-aligned states, interact with each other, forming a partial (2 × 2) matrix

$$\mathrm{H} = 2E_0\begin{pmatrix}1 & 1\\ 1 & 1\end{pmatrix}. \qquad (*)$$

Solving the characteristic equation (4.103) for this matrix,

$$\begin{vmatrix}1-\lambda_\pm & 1\\ 1 & 1-\lambda_\pm\end{vmatrix} = 0, \quad \text{i.e. } \lambda_\pm^2 - 2\lambda_\pm = 0,$$

[18] See, e.g. *Part EM* Eqs. (5.99) and (5.100).

we get $\lambda_+ = 2$, $\lambda_- = 0$, giving us two more (non-degenerate) energy levels:

$$E_+ = 2E_0\lambda_+ = 4E_0, \qquad E_- = 2E_0\lambda_- = 0.$$

Now plugging these values, one by one, into any Eq. (4.102) used for the diagonalization of the matrix (*),

$$(1 - \lambda_\pm)U_{1\pm} + U_{2\pm} = 0, \qquad U_{1\pm} + (1 - \lambda_\pm)U_{2\pm} = 0,$$

we get $U_{1+} = U_{2+}$, $U_{1-} = -U_{2-}$, meaning that the eigenstates corresponding to the energy levels $E_\pm$ correspond, respectively, to the familiar entangled triplet and singlet states—see Eqs. (8.18) and (8.20):

$$|s_\pm\rangle \equiv \frac{1}{\sqrt{2}}(|\uparrow\downarrow\rangle \pm |\downarrow\uparrow\rangle).$$

Hence, in this system the energies of the triplet states are different even at the absence of the external magnetic field: the spin-aligned (factorable) states have an energy ($-2E_0$) lower that that ($+4E_0$) of the entangled state, in accordance with the classical trend of the magnetic moment self-alignment due to their dipole interaction.

***Problem* 8.9.** Consider the permutation of two identical particles, each of spin s. How many different symmetric and antisymmetric spin states can the system have?

Solution: As was discussed in section 5.7 of the lecture notes, each of the particles of spin s may have $(2s + 1)$ different, linearly-independent spin states—for example, the states with definite and different 'magnetic' quantum numbers m_s on the interval $-s \leqslant m_s \leqslant +s$—see, e.g. Eq. (5.169) of the lecture notes. Thus, generally, there are $(2s + 1)^2$ different direct products,

$$|m_s\rangle \otimes |m_s'\rangle, \qquad (*)$$

with arbitrary m_s and m_s' of the above list. Of them, $(2s + 1)$ products have $m_s = m_s'$, and hence are symmetric with respect to the particle permutations, while

$$(2s + 1)^2 - (2s + 1) \equiv 2s(2s + 1)$$

states have $m_s \neq m_s'$. Of the latter states, one can form $s(2s + 1)$ symmetric and the same number of asymmetric combinations of the type

$$|m_s\rangle \otimes |m_s'\rangle \pm |m_s'\rangle \otimes |m_s\rangle. \qquad (**)$$

Since all these states are linearly-independent, they form a full system of $(2s + 1)^2$ spin states, and combining them to symmetric and antisymmetric independent linear combinations does not change their total numbers. Hence, the total number of different symmetric states is

$$N_s = (2s + 1) + s(2s + 1) \equiv (s + 1)(2s + 1),$$

while the number of different antisymmetric states is only

$$N_a = s(2s + 1) < N_s.$$

As the trivial example, for two spin-free particles ($s = 0$) we get $N_s = 1$, $N_a = 0$—fine, because the only possible spin(-free) state of this system is symmetric with respect to particle permutation. As the next simplest example, $s = ½$, our results yield $N_a = ½\cdot(2\cdot½ + 1) = 1$ and $N_s = (½ + 1)\cdot(2\cdot½ + 1) = 3$, corresponding to the three triplet states (8.21) and the only singlet state (8.18).

Note, however, that for systems of two particles with $s > ½$, not all basis vectors of the coupled representation, i.e. not all simultaneous eigenkets of the operators $\hat{S}^2$ and $\hat{S}_z$ have the simple form (*) or (**), and their more complex linear superpositions may be needed—see, for example, the next problem.

***Problem* 8.10.** For a system of two identical particles with $s = 1$:

(i) List all possible spin states in the uncoupled-representation basis.
(ii) List all possible pairs $\{S, M_S\}$ of the quantum numbers describing the states of the coupled-representation basis—see Eq. (8.48) of the lecture notes.
(iii) Which of the $\{S, M_S\}$ pairs describe the states symmetric, and which the states antisymmetric, with respect to the particle permutation?

Solutions:

(i) The state vectors of the uncoupled representation are given by Eq. (*) of the solution of the previous problem, with each of quantum numbers m_s and m_s' taking one of three values of the set $\{-1, 0, +1\}$. In the shorthand notation, used in sections 8.1 and 8.2 of the lecture notes, their ket-vectors are

$$|+1, +1\rangle, \quad |+1, 0\rangle, \quad |+1, -1\rangle, \quad |0, +1\rangle, \quad |0, 0\rangle,$$
$$|0, -1\rangle, \quad |-1, +1\rangle, \quad |-1, 0\rangle, \quad |-1, -1\rangle.$$

For what follows, it is instrumental to represent these states as the points at the 'rectangular diagram' shown in the figure below. (It similar to that for $s = ½$, shown in figure 8.2 of the lecture notes.)

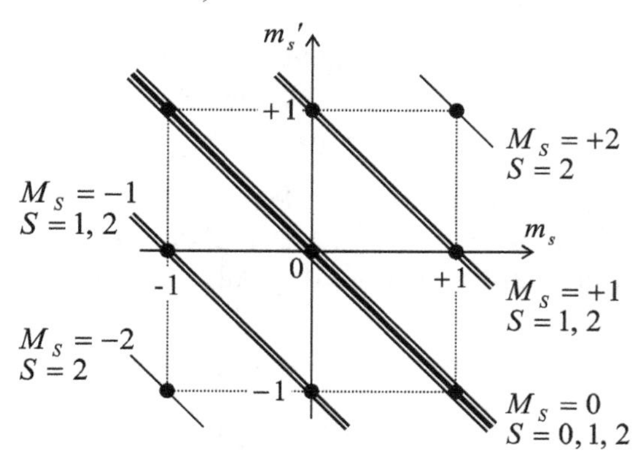

(ii) As was discussed in section 5.7 of the lecture notes for the addition of the angular momentum operators $\hat{\mathbf{L}}$ and $\hat{\mathbf{S}}$, and reproduced in section 8.2 for the addition of two spin operators $\hat{\mathbf{s}}_1$ and $\hat{\mathbf{s}}_2$, the kets of the coupled-representation basis, i.e. the common eigenkets of the operators $\hat{S}^2 \equiv (\hat{\mathbf{s}}_1 + \hat{\mathbf{s}}_2)^2$ and $\hat{S}_z = \hat{s}_{1z} + \hat{s}_{2z}$, are linear combinations of the uncoupled-representation kets with the fixed sum

$$m_s + m_s' = M_S,$$

and hence may be represented, on the rectangular diagram, by the straight lines of the slope (−1), that connect the corresponding $\{m_s, m_s'\}$ points—see the figure above. Hence, the assignment of the numbers M_S is elementary, and is shown in that figure.

For several coupled-representation states, the assignment of the quantum number S is also straightforward. For example, let us consider the top-right and bottom-left points of the rectangular diagram, representing the states with $m_s = m_s' = +1$ and $m_s = m_s' = -1$. Since each of the lines representing coupled-representation states, which passes through one of these points, does not pass through any other point, the corresponding states belong also to the coupled-representation basis, with $M_S = \pm 2$. They, evidently have the largest possible value of M_S, namely $|M_S|_{\max} = 2$. According to Eq. (8.48), this value serves as the corresponding number S, so that we may write

$$\begin{aligned} |m_s = +1, m_s' = +1\rangle &= |S = 2, M_S = +2\rangle, \\ |m_s = -1, m_s' = -1\rangle &= |S = 2, M_S = -2\rangle. \end{aligned} \qquad (*)$$

Now going, from the corners, one step toward the center of the diagram, i.e. to the states with $M_S = \pm 1$, we may notice, first of all, that since there are two uncoupled-representation states for each of these values, there should be also two of their linear superpositions giving different coupled-representation states, and they cannot have the same number S, because otherwise they would not be linearly-independent. Next, as was discussed in section 5.7 for the addition of vectors $\hat{\mathbf{L}}$ and $\hat{\mathbf{S}}$, the quantum number characterizing the square of their sum (there j, and in our current case, S) can change, at such step, only by ±1. Hence the two coupled-representation states, for each of these M_S, should have $S = 1$ and $S = 2$. Since, due to the symmetry of the rectangular diagram with respect to the axis $m_s = m_s'$, the moduli of the weights of the two uncoupled-representation kets, participating in each of such superpositions, have to be equal, the only possible superpositions have the familiar form (give or take an arbitrary phase multiplier):

$$\begin{aligned} \left|S = \begin{Bmatrix} 2 \\ 1 \end{Bmatrix}, M_S = +1\right\rangle &= \frac{1}{\sqrt{2}}(|m_s = +1, m_s' = 0\rangle \pm |m_s = 0, m_s' = +1\rangle), \\ \left|S = \begin{Bmatrix} 2 \\ 1 \end{Bmatrix}, M_S = -1\right\rangle &= \frac{1}{\sqrt{2}}(|m_s = -1, m_s' = 0\rangle \pm |m_s = 0, m_s' = -1\rangle), \end{aligned} \qquad (**)$$

—cf. Eqs. (8.18) and (8.20) of the lecture notes.

Finding the similarly explicit forms for the three remaining coupled-representation states with $M_S = 0$, corresponding to the lines passing through the origin of the rectangular diagram, in more complex[19], but this is not necessary to fulfill our task. Indeed, since all these kets have the same M_S, they all should have different S, in order to be linearly-independent. But according to Eq. (8.50), spelled out for our case $S_{\max} = s_1 + s_2 = 2$, the list of possible values of S is limited to $S = 0$, 1 and 2, thus giving us the final answer—see the labels in the figure above.

(iii) The two states (*) are evidently symmetric with respect to a particle's permutation, and so are two of the four states (**), corresponding to $S = 2$. So, besides the three states with $M_S = 0$, four states are symmetric, and two antisymmetric. But according to the solution of the previous problem, for $s = 1$ we must have six symmetric and three antisymmetric states, so that of the three states with $M_S = 0$, two have to be symmetric and one antisymmetric. Since the state with $S = 0$ and $M_S = 0$ has to be symmetric for any s, this means that the state with $S = 1$ and $M_S = 0$ has to be antisymmetric, while the state with $S = 2$ and $M_S = 0$, has to be symmetric.

***Problem* 8.11.** Represent the operators of the total kinetic energy and the total orbital angular momentum of a system of two particles, with masses m_1 and m_2, as combinations of terms describing the center-of-mass motion and the relative motion. Use the results to calculate the energy spectrum of the so-called *positronium*—a metastable 'atom'[20] consisting of one electron and its positively charged antiparticle, the positron.

Solution: The operators in question are the sums of the single-particle operators, defined by Eqs. (1.27) and (5.147) of the lecture notes:

$$\hat{T} = \hat{T}_1 + \hat{T}_2 \equiv \frac{\hat{p}_1^2}{2m_1} + \frac{\hat{p}_2^2}{2m_2}, \qquad \hat{\mathbf{L}} = \hat{\mathbf{L}}_1 + \hat{\mathbf{L}}_2 \equiv [\hat{\mathbf{r}}_1 \times \hat{\mathbf{p}}_1] + [\hat{\mathbf{r}}_2 \times \hat{\mathbf{p}}_2].$$

Following the classical mechanics lead[21], let us introduce two new radius-vector operators:

$$\hat{\mathbf{R}} \equiv \frac{m_1\hat{\mathbf{r}}_1 + m_2\hat{\mathbf{r}}_2}{M}, \qquad \hat{\mathbf{r}} \equiv \hat{\mathbf{r}}_1 - \hat{\mathbf{r}}_2, \qquad (*)$$

[19] This may be done, for example, using the Clebsh–Gordan coefficients for $s_1 = s_2 = 1$—which, in turn, may be derived from the recurrent relations that were derived (for the $\hat{\mathbf{J}} = \hat{\mathbf{L}} + \hat{\mathbf{S}}$ addition), in the solution of problem 5.41. Just for the reader's reference,

$$|S = 2,\ M_S = 0\rangle = (|m_s = +1,\ m_s' = -1\rangle + |m_s = -1,\ m_s' = +1\rangle + 2\,|m_s = 0,\ m_s' = 0\rangle)/\sqrt{6},$$
$$|S = 1,\ M_S = 0\rangle = (|m_s = +1,\ m_s' = -1\rangle - |m_s = -1,\ m_s' = +1\rangle)/\sqrt{2},$$
$$|S = 0,\ M_S = 0\rangle = (|m_s = +1,\ m_s' = -1\rangle + |m_s = -1,\ m_s' = +1\rangle - |m_s = 0,\ m_s' = 0\rangle)/\sqrt{3}.$$

These expressions may be readily verified, for example, by combining Eq. (5.164) of the lecture notes, with $l = 1$, duly translated to the spin language, and Eq. (****) of the solution of problem 8.3. This verification is highly recommended to the reader as an additional task.

[20] Its lifetime (either 0.124 ns or 138 ns, depending on the parallel or antiparallel configuration of the components spins), is limited by the weak interaction of its components, which causes their annihilation with the emission of several gamma-ray photons.

[21] See, e.g. *Part CM* Eq. (3.32).

where $M \equiv m_1 + m_2$ is the total mass of the system. Considering these definitions as a system of two linear equations for the initial radius-vector operators, we may readily solve it to get reciprocal relations

$$\hat{\mathbf{r}}_1 = \hat{\mathbf{R}} + \frac{m_2}{M}\hat{\mathbf{r}}, \quad \hat{\mathbf{r}}_2 = \hat{\mathbf{R}} - \frac{m_1}{M}\hat{\mathbf{r}}. \qquad (**)$$

Now let us define two new momentum operators as

$$\hat{\mathbf{P}} \equiv \hat{\mathbf{p}}_1 + \hat{\mathbf{p}}_2, \quad \hat{\mathbf{p}} \equiv \frac{m_2}{M}\hat{\mathbf{p}}_1 - \frac{m_1}{M}\hat{\mathbf{p}}_2, \qquad (***)$$

with the reciprocal relations

$$\hat{\mathbf{p}}_1 = \frac{m_1}{M}\hat{\mathbf{P}} + \hat{\mathbf{p}}, \quad \hat{\mathbf{p}}_2 = \frac{m_2}{M}\hat{\mathbf{P}} - \hat{\mathbf{p}}. \qquad (****)$$

Using the fact that the operators of different particles are defined in their separate Hilbert spaces and hence commute, we may verify that the Cartesian components of the new operators satisfy the Heisenberg commutation relations (2.14), while commuting with each other. For example (with the index x of the particle momentum operators just implied):

$$\begin{aligned}
[\hat{X}, \hat{P}_x] &= \left[\left(\frac{m_1}{M}\hat{x}_1 + \frac{m_2}{M}\hat{x}_2\right), (\hat{p}_1 + \hat{p}_2)\right] = \frac{m_1}{M}[\hat{x}_1, \hat{p}_1] + \frac{m_2}{M}[\hat{x}_2, \hat{p}_2] \\
&= \frac{m_1}{M}i\hbar + \frac{m_2}{M}i\hbar = i\hbar, \\
[\hat{x}, \hat{p}_x] &= \left[(\hat{x}_1 - \hat{x}_2), \left(\frac{m_2}{M}\hat{p}_1 - \frac{m_1}{M}\hat{p}_2\right)\right] = \frac{m_2}{M}[\hat{x}_1, \hat{p}_1] + \frac{m_1}{M}[\hat{x}_2, \hat{p}_2] \\
&= \frac{m_2}{M}i\hbar + \frac{m_1}{M}i\hbar = i\hbar, \\
[\hat{X}, \hat{p}_x] &= \left[\left(\frac{m_1}{M}\hat{x}_1 + \frac{m_2}{M}\hat{x}_2\right), \left(\frac{m_2}{M}\hat{p}_1 - \frac{m_1}{M}\hat{p}_2\right)\right] \\
&= \frac{m_1 m_2}{M^2}[\hat{x}_1, \hat{p}_1] - \frac{m_2 m_1}{M^2}[\hat{x}_2, \hat{p}_2] = 0, \\
[\hat{x}, \hat{P}_x] &= [(\hat{x}_1 - \hat{x}_2), (\hat{p}_1 + \hat{p}_2)] = [\hat{x}_1, \hat{p}_1] - [\hat{x}_2, \hat{p}_2] = 0,
\end{aligned}$$

with similar relations for two other Cartesian components. Hence, the operators (***) are the legitimate operators of the momenta corresponding to the radius-vector operators (*).[22]

Now plugging Eqs. (**) and (****) into the operators of interest, we readily get

$$\hat{T} = \frac{\hat{P}^2}{2M} + \frac{\hat{p}^2}{2m}, \quad \hat{\mathbf{L}} = [\hat{\mathbf{R}} \times \hat{\mathbf{P}}] + [\hat{\mathbf{r}} \times \hat{\mathbf{p}}],$$

[22] Another, perhaps less convincing way to prove this fact is to plug Eqs. (**) into the coordinate-representation form of the particle momentum operators, $\hat{\mathbf{p}}_{1,2} = -i\hbar\nabla_{1,2}$ to get Eqs. (****) with $\hat{\mathbf{P}} = -i\hbar\nabla_{\mathbf{R}}$, $\hat{\mathbf{p}} = -i\hbar\nabla_{\mathbf{r}}$, and then argue that the relation between the operators should not depend on their particular representation.

where m is the same *reduced mass* that participates in classical dynamics of two-particle systems:

$$m = \frac{m_1 m_2}{M}, \quad \text{i.e.} \quad \frac{1}{m} = \frac{1}{m_1} + \frac{1}{m_2}, \qquad \text{so that } m \leqslant m_1, m_2.$$

Evidently, the uppercase operators (and the total mass M) describe the motion of the center of mass of the system of two particles, while the lowercase operators (and the reduced mass m) describe their mutual motion. If we are not interested in the motion of the atom as a whole, in the case of the Coulomb interaction of the electron and positron, with equal and opposite charges $\pm e$, i.e. neglecting the very small fine-structure effects discussed in section 6.3, we may take the Hamiltonian of the system in the form

$$\hat{H} = \hat{T} + \hat{U} = \frac{\hat{\mathbf{p}}^2}{2m} - \frac{e^2}{4\pi\varepsilon_0 r}.$$

In the particular case of the positronium, $m = m_e/2$, so that the energy spectrum is similar to that of the hydrogen atom, but with the twice smaller mass. This change leaves the functional form of Bohr's theory result (1.12) intact, but reduces the effective value of the Hartree energy E_H (and hence of all eigenenergies) by a factor of two:

$$E_n = -\frac{(E_H/2)}{2n^2} = -\frac{E_H}{4n^2}, \qquad \text{with } E_H \equiv \left(\frac{e^2}{4\pi\varepsilon_0 \hbar c}\right)^2 m_e c^2 \equiv \alpha^2 m_e c^2 \approx 27.2 \text{ eV},$$

so that the ground-state energy of the positronium is $-E_H/4 \approx -6.8$ eV.

***Problem* 8.12.** Two particles with similar masses $\mathscr{m}$ and charges q are free to move along a round, plane ring of radius R. In the limit of strong Coulomb interaction of the particles, find the lowest eigenenergies of the system, and sketch the system of its energy levels. Discuss possible effects of particle indistinguishability.

Solution: On the basis of our discussion of single-particle rotation in sections 3.5 and 5.6, it is straightforward to write the system's Hamiltonian:

$$\hat{H} = \hat{T} + \hat{U}, \quad \hat{T} = -\frac{1}{2\mathscr{m}R^2}\frac{\partial^2}{\partial\varphi_1^2} + \frac{1}{2\mathscr{m}R^2}\frac{\partial^2}{\partial\varphi_2^2}, \quad \hat{U} = \frac{q^2}{4\pi\varphi_0 r_{12}}.$$

Here r_{12} is the distance between the particles, which may be readily expressed via the difference,

$$\varphi \equiv \varphi_1 - \varphi_2, \tag{*}$$

of their angular coordinates $\varphi_{1,2}$ modulo 2π—see the figure below:

$$r_{12} = 2R \sin \left| \frac{\varphi}{2} \right|.$$

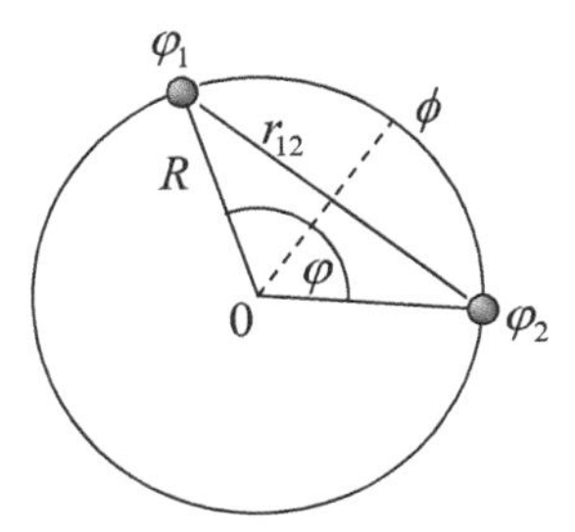

Despite the apparent simplicity of the Hamiltonian, its general analysis is rather complex, but in the strong interaction limit,

$$\frac{q^2}{4\pi\varepsilon_0 R} \gg \frac{\hbar^2}{2mR^2},$$

the lowest eigenenergies may be readily found analytically, using clues from the classical properties of the system. Indeed, classically, the lowest energies of the system correspond to the particles pressed, by their Coulomb repulsion, to the opposite ends of the same ring's diameter $D = 2R$, so that the potential energy U of the system equals $U_0 \equiv q^2/4\pi\varepsilon_0 D = q^2/8\pi\varepsilon_0 R$. The interaction does not prevent the pair from its free joint rotation around the ring with an arbitrary angular velocity Ω.

One more contribution to system's energy may come from small oscillations of the particles near the equilibrium position $r_{12} = 2D$. In order to find the oscillation frequency, let us introduce, besides the difference angle φ defined by Eq. (*), another, independent linear combination of the particles' coordinates:

$$\phi = \frac{\varphi_1 + \varphi_2}{2}, \qquad (**)$$

characterizing their joint rotation—just as is done in with vector coordinates in the previous problem (and in classical mechanics). Solving Eqs. (*) and (**) for $\varphi_{1,2}$, and plugging the result,

$$\varphi_1 = \phi + \frac{\varphi}{2}, \quad \varphi_2 = \phi - \frac{\varphi}{2},$$

into the classical expression for the kinetic energy of the system,

$$T = \frac{m(R\dot{\varphi}_1)^2}{2} + \frac{m(R\dot{\varphi}_2)^2}{2},$$

we get

$$T = T_{\text{rot}} + T_{\text{osc}}, \quad \text{with } T_{\text{rot}} = \frac{I}{2}\dot{\phi}^2, \quad \text{and} \quad T_{\text{osc}} = \frac{I'}{2}\dot{\varphi}^2,$$

where $I \equiv (2m)R^2$ is the total moment of inertia of the system, while $I' \equiv m_{ef}R^2 = (m/2)R^2$ is the 'reduced moment of inertia', similar to the 'reduced mass' $m_{ef} = m/2$ in problems with equal particle masses. Now we may Taylor-expand the potential energy near its minimum value:

$$U = \frac{q^2}{4\pi\varepsilon_0 r_{12}} = \frac{q^2}{8\pi\varepsilon_0 R \sin|\varphi/2|} = U_0 + \tilde{U}, \quad \text{with}$$

$$\tilde{U} \approx \kappa_{ef}\frac{\tilde{\varphi}^2}{2}, \quad \text{where } \tilde{\varphi} \equiv \varphi - \pi,$$

to calculate the effective spring constant κ_{ef} of the small oscillations $\tilde{\varphi}$ of the difference angle φ. A straightforward calculation yields

$$\kappa_{ef} \equiv \frac{d^2U}{d\varphi^2}\Big|_{\varphi=\pi} = \frac{q^2}{4\pi\varepsilon_0 R}\frac{1}{8}.$$

so that the total energy of small oscillations is

$$E_{osc} = T_{osc} + \tilde{U} = \frac{I'}{2}\dot{\tilde{\varphi}}^2 + \frac{\kappa_{ef}}{2}\tilde{\varphi}^2 \equiv \frac{I'}{2}\dot{\tilde{\varphi}}^2 + \frac{I'\omega_0^2}{2}\tilde{\varphi}^2, \quad \text{with}$$

$$\omega_0 \equiv \left(\frac{\kappa_{ef}}{I'}\right)^{1/2} = \left(\frac{q^2}{4\pi\varepsilon_0}\frac{1}{mR^3}\right)^{1/2}.$$

As we know from classical mechanics, ω_0 so defined is the small oscillations frequency.

Now proceeding to quantum mechanics, Eqs. (*) and (**), and the replacement $U \to \tilde{U}$ allow us to rewrite the initial Hamiltonian in the form of the sum of two independent components[23]:

$$\hat{H} = \hat{H}_{rot} + \hat{H}_{osc}, \quad \text{with } \hat{H}_{rot} = -\frac{\hbar^2}{4mR^2}\frac{\partial^2}{\partial\phi^2},$$

$$\hat{H}_{osc} = -\frac{\hbar^2}{2I'}\frac{\partial^2}{\partial\tilde{\varphi}^2} + \frac{I'\omega_0^2}{2}\tilde{\varphi}^2,$$

whose eigenenergies may be calculated separately, and then added. The small-oscillation part, which was repeatedly discussed in this course, immediately yields

$$E_{osc} = \hbar\omega_0\left(n + \frac{1}{2}\right) \equiv \hbar\left(\frac{q^2}{4\pi\varepsilon_0}\frac{1}{mR^3}\right)^{1/2}\left(n + \frac{1}{2}\right), \quad \text{with } n = 0, 1, 2, \ldots.$$

The eigenfunctions of the rotational Hamiltonian, which has only a simple kinetic-energy component, have been also discussed several times, starting from section 3.5:

[23] An alternative way to get these expressions is to use the solution of problem 8.8, taking $\hat{\mathbf{L}} = \mathbf{n}_z\hat{L}_z$, where $\mathbf{n}_z$ is the unit vector perpendicular to the ring's plane.

$$\psi(\phi) = \frac{1}{(2\pi)^{1/2}} e^{im\phi}. \qquad (***)$$

However, at this point we need to be careful, because now this wavefunction describes the joint rotation of a *two-particle* system, and their permutation properties are important. If the particles are distinguishable (either by their nature, or by the state of some internal degrees of freedom), $\psi(\phi)$ should be invariant to system's rotation by $\Delta\phi = 2\pi$ (and its multiples), just as in the case of a single-particle rotator. This condition immediately gives the condition $\exp\{im2\pi\} = 1$, i.e. forces m to be integer, giving the energy spectrum[24]

$$E_{\text{rot}} = \frac{\hbar^2}{4\mathscr{m}R^2} m^2, \quad \text{with } m = 0, \pm 1, \pm 2, \ldots \qquad (****)$$

The figure below shows the scheme of the system's lowest energy levels for this case. Note the double rotational degeneracy of all levels (besides those with $m = 0$), and the level hierarchy due to the strong relation between the gaps between adjacent rotational and oscillational energies,

$$\Delta E_{\text{osc}} = \hbar\omega_0 = \hbar\left(\frac{q^2}{4\pi\varepsilon_0}\frac{1}{\mathscr{m}R^3}\right)^{1/2}$$
$$= \left(\frac{q^2}{8\pi\varepsilon_0 R}\frac{2\hbar^2}{\mathscr{m}R^2}\right)^{1/2} \sim (U_0 \Delta E_{\text{rot}})^{1/2} \gg \Delta E_{\text{rot}},$$

in our limit of strong particle interaction ($U_0 \gg \Delta E_{\text{rot}}$).

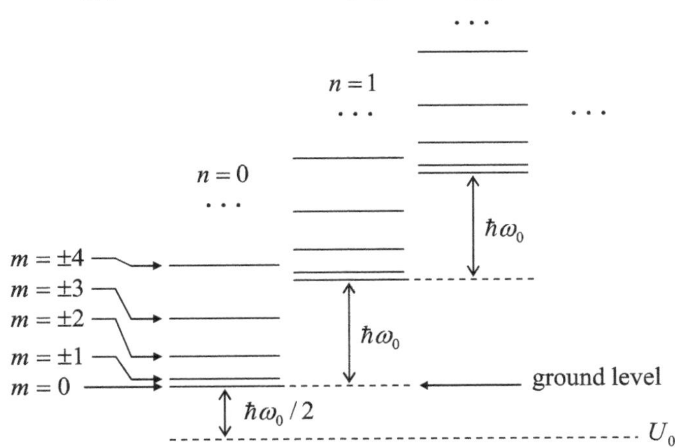

However, if the particles are indistinguishable, the situation is more complicated. Indeed, let the background (spin plus oscillation) state of the system be symmetric with respect to the particle permutation. (This is true, for example, if the particles are

[24] Let me hope that the difference between the fonts used for the 'magnetic' quantum number (m), and the particles' mass ($\mathscr{m}$) is sufficiently clear.

spinless bosons, and the system is in an even vibrational state with $n = 2n'$, with integer n', for example in the ground state with $n = 0$.) Then the system's rotation by just $\pm\pi$ should result in an identical quantum state, so that we should require that the wavefunction (***) satisfies the requirement $\psi(\phi \pm \pi) = \psi(\phi)$, i.e. that $\exp\{im\pi\} = 1$. This requirement allows only *even* values of m, thus decimating the rotational energy spectrum (****).

On the other hand, in an odd vibrational state of the bosonic system, with $n = 2n' + 1$, the oscillator's wavefunction changes sign at the replacement $\varphi \to -\varphi$. According to the definition of the angles φ and ϕ (see, e.g. the figure above), such replacement is equivalent, modulo 2π, to the replacement $\phi \to \phi \pm \pi$, so that the sign change of the total wavefunction may be compensated by that of $\psi(\phi)$, but only if $\exp\{im\pi\} = -1$, i.e. if m is even. Hence, now the rotational spectrum (****) is decimated again, but now by forbidding the energy levels with even m, notably including the ground state with $m = 0$.

This effect becomes even more involved in the case of two indistinguishable spin-½ fermions (e.g. electrons or protons). As was discussed in section 8.2 of the lecture notes, the necessary antisymmetry of the system's wavefunction to particle permutation may be achieved either via the spin asymmetry (in the singlet spin state) or the orbital asymmetry, with a symmetric (triplet) spin state, with $S = 1$. For our system this means, for example, that in the ground vibrational state ($n = 0$), there are two sets of spin-rotational states: spin-singlet states with m even, and spin-triplet states with m odd.

Rather counter-intuitively, such symmetry effects may affect readily observable properties of real systems (for example, such important diatomic molecules as O_2 and N_2) even when they are imposed by *nuclear* spins, despite their extremely week interaction with other degrees of freedom—see the next problem.

***Problem* 8.13.** Low-energy spectra of many diatomic molecules may be well described modeling the molecule as a system of two particles connected with a light and elastic, but very stiff spring. Calculate the energy spectrum of a molecule in this approximation. Discuss possible effects of nuclear spins on the spectra of so-called *homonuclear* molecules, formed by two similar atoms.

Solution: In the specified model, the system's Hamiltonian is

$$\hat{H} = \frac{\hat{p}_1^2}{2m_1} + \frac{\hat{p}_2^2}{2m_2} + \frac{\kappa(r-a)^2}{2}, \qquad \text{with } \mathbf{r} \equiv \mathbf{r}_1 - \mathbf{r}_2,$$

where a is the classical-equilibrium distance between the atoms. Using the solution of problem 8.11 to transform the kinetic-energy part of the Hamiltonian, we may rewrite it as

$$\hat{H} = \frac{\hat{P}^2}{2M} + \frac{\hat{p}^2}{2m} + \frac{\kappa(r-a)^2}{2}, \qquad (*)$$

where $M \equiv m_1 + m_2$ is the total mass of the particles, m is the reduced mass of their mutual motion,

$$m = \frac{m_1 m_2}{M} = \left(\frac{1}{m_1} + \frac{1}{m_2}\right)^{-1},$$

while the momenta operators,

$$\hat{\mathbf{P}} \equiv \hat{\mathbf{p}}_1 + \hat{\mathbf{p}}_2, \quad \text{and} \quad \hat{\mathbf{p}} \equiv \frac{m_2}{M}\hat{\mathbf{p}}_1 - \frac{m_1}{M}\hat{\mathbf{p}}_2,$$

are related in the usual ('canonical') way with, respectively, the radius-vector $\mathbf{R}$ of the center of mass of the particles, and the relative radius-vector $\mathbf{r}$.

As Eq. (*) shows, the system's spectrum is the sum of two independent contributions: the kinetic energy of the molecule as a whole, with the mass M, free to move in space, and an 'effective' single particle, with the reduced mass m, moving in spherically-symmetric potential $U = \kappa(r - a)^2/2$. As we know well, the former motion is simple, giving a continuous energy spectrum (1.89), while the latter is generally complex, and is simplified only if the spring is rather stiff:

$$\kappa a^2 \gg \frac{\hbar^2}{ma^2}. \qquad (**)$$

In this limit, we may separate the general 3D motion of the effective particle into rotation at the fixed distance $r = a$, and small radial (1D) oscillations about this point, and use Eqs. (2.262) and (3.163) to write the total energy spectrum as the sum

$$E = \frac{\hbar^2 k^2}{2M} + \frac{\hbar^2 l(l+1)}{2ma^2} + \hbar\omega_0\left(n + \frac{1}{2}\right), \quad \text{with } \omega_0 \equiv \left(\frac{\kappa}{m}\right)^{1/2}, \qquad (***)$$

where the wave vector $\mathbf{k}$ may take arbitrary values[25], while l and n are integer quantum numbers, which may take values from the same set {0, 1, 2,...}.

For most diatomic molecules, the rotational constant $B \equiv \hbar^2/ma^2$ is in the range from 10^{-5} to 10^{-2} eV. (Even in the extreme case of the hydrogen molecule H_2, with its very light nuclei, $m_{1,2} \approx 1.7 \times 10^{-27}$ kg, and small equilibrium distance $a \approx 0.074$ nm $\approx 1.4\, r_B$ between them, $B \approx 7.6$ meV.) As a result, at room temperatures, $T \sim 300$ K, with $k_BT \sim 25$ meV, thermal fluctuations are sufficient to excite many lower rotational levels, with $l > 0$, of all such molecules.

On the other hand, the vibration frequencies $\omega_0/2\pi$ are between 10^{12} and 10^{14} Hz, so that even the lowest oscillation energies are in the much higher range from $\sim 10^{-2}$ to ~ 1 eV. (In the ultimate case of H_2, $\hbar\omega_0 \approx 0.54$ eV.) As a result, at room temperatures the vibrational levels with $n > 0$ are virtually *not* populated. The electronic state excitations have comparable or even higher energies, in the 1 eV ballpark, because they do not involve the motion of relatively heavy nuclei.

[25] The quantum transitions used for most spectral measurements are unaffected by the translational motion, described by the first term in Eq. (***), so that this component will be ignored in the forthcoming discussion.

This hierarchy justifies the model explored in this problem, and in particular the stiff-spring condition (**).

This is essentially the whole story for *heteronuclear* molecules, such as CO, NO, or HCl, consisting of different, and hence distinguishable atoms. On the other hand, for *homonuclear* molecules, such as H_2, O_2, or N_2, the indistinguishability effects may be important even at room temperatures, because (as was discussed above) most molecules are in their ground electronic and vibrational states, and hence are symmetric or antisymmetric with respect to atom permutation, depending on their total spin, including not only its electronic, but also the nuclear component.

For example, the ^{16}O nucleus (of the oxygen isotope most abundant at natural conditions) has the total spin equal to zero, while the total (orbital + spin) ground-state electronic wavefunction of the molecule is *antisymmetric* with respect to atoms' swap. As a result, just as was discussed in the solution of the previous problem, the rotational wavefunction has to be antisymmetric with respect to the replacement $\mathbf{r} \rightarrow -\mathbf{r}$, i.e. the simultaneous replacement $\theta \rightarrow \pi - \theta$, and $\varphi \rightarrow \varphi + \pi$ (modulo 2π), where θ and φ are the usual polar angles of the effective radius-vector $\mathbf{r}$. As we know from the properties of the spherical harmonics (see, e.g. Eqs. (3.168) and (3.171), or just figure 3.20 of the lecture notes), this condition may be only fulfilled for odd values of the quantum number l. This means, rather counter-intuitively, that the ground state of the O_2 molecule corresponds to $l = 1$, i.e. essentially, to its rotation:

$$E_g = (E_g)_{\text{electronic}} + \frac{\hbar^2}{2ma^2} + \frac{\hbar\omega_0}{2}.$$

Even more interesting (and historically, more important) are the properties of the nitrogen molecule N_2, with the nitrogen atoms of the (dominating) isotope ^{14}N. The nucleus of such an atom has 7 protons and 7 neutrons, and the ground-state spin $I = 1$, while the total ground electronic state of the covalent-bound N_2 molecule is *symmetric* with respect to the nuclei permutation[26]. As a result, the lowest rotational energy of the molecule depends on the net spin I_Σ of two nuclei. Since the operator of the total spin is a vector sum similar to those discussed in sections 5.7 and 8.2 of the lecture notes:

$$\hat{\mathbf{I}}_\Sigma = \hat{\mathbf{I}}_1 + \hat{\mathbf{I}}_2, \quad \text{with } I_1 = I_2 = 1,$$

and obeys the relations similar to those given by Eq. (8.48), we may use the solution of problem 8.9, with $s = 1$, to conclude that the nuclear spin system has six

[26] Let me emphasize once again that such *symmetry* of the ground electronic state does not contradict its fundamental *asymmetry* with respect to the permutation of any two electrons. For example, in the ground state of the simplest H_2 molecule, its only two electrons are in the simple spin-singlet state (8.24), which is of course antisymmetric with respect to the swap of the *electrons*—in this particular case, because of the spin factor. However, the same Eq. (8.24) clearly shows that the swap of the H_2 *nuclei*, which is equivalent to the simultaneous change of the signs of both radius-vectors $\mathbf{r}_1$ and $\mathbf{r}_2$, does not change the total sign of the ket vector of this electron state. Such symmetry differences are possible because each electron in this state is a 'common property' of both nuclei, and its number cannot be associated with that of a particular nucleus.

symmetric states (allowing only even values of l) and three asymmetric states (allowing only odd values of l). Since for most molecules, at room temperature, $k_BT \gg B \equiv \hbar^2/ma^2$, so that many lower rotational states are thermally excited, the number of such states with even values of l is twice larger than that for odd values of l. The selection rules discussed in section 5.6 of the lecture notes (see also the solution of problem 5.35) enable distinguishing the parity of l by measuring which quantum (in particular, optical) transitions from this state to a fixed final state are allowed.

In the late 1920s, i.e. before the experimental discovery of neutrons in 1932, the observation of this ratio in experimental molecular spectra of N_2 molecules (by L Ornstein) had helped to establish the fact that the spin I of the ^{14}N nucleus is indeed equal to 1, and hence to discard the then-plausible model in which the nucleus would consist of 14 protons and 7 electrons, giving it the observed mass $m \approx 14m_p$ and the net electric charge $Q = 7e$. (In that model, the ground-state value of I would be semi-integer.)[27]

***Problem* 8.14.** Two indistinguishable spin-½ particles are attracting each other at contact:

$$U(x_1, x_2) = -\mathscr{w}\,\delta(x_1 - x_2), \quad \text{with } \mathscr{w} > 0,$$

but are otherwise free to move along the x-axis. Find the energy and the wavefunction of the ground state of the system.

Solution: The system's Hamiltonian is

$$\hat{H} = \hat{T} + \hat{U} = \frac{\hat{p}_1^2}{2m} + \frac{\hat{p}_2^2}{2m} - \mathscr{w}\,\delta(x_1 - x_2)$$
$$= -\frac{\hbar^2}{2m}\left(\frac{\partial^2}{\partial x_1^2} + \frac{\partial^2}{\partial x_2^2}\right) - \mathscr{w}\,\delta(x_1 - x_2),$$

so that the corresponding Schrödinger equation is satisfied, at $x_1 \neq x_2$, with exponential orbital wavefunctions

$$\psi_{\kappa_1,\kappa_2} = \text{const} \times \exp\{\kappa_1 x_1 + \kappa_2 x_2\}, \qquad (*)$$

with any (real or complex) c-numbers $\kappa_{1,2}$, corresponding to the following energy:

$$E_{\kappa_1,\kappa_2} = -\frac{\hbar^2}{2m}\left(\kappa_1^2 + \kappa_2^2\right).$$

Since the potential energy U depends only on the distance $x \equiv x_1 - x_2$ between the particles, it is natural to replace the variables, introducing, besides x, another independent

[27] The author is grateful to P vanNieuwenhuizen for sharing this historic note.

linear combination of coordinates $x_{1,2}$, namely the position $X \equiv (x_1 + x_2)/2$ of their center of mass. Plugging the reciprocal relation between the old and new variables[28],

$$x_1 = X + \frac{x}{2}, \quad x_2 = X - \frac{x}{2},$$

into Eq. (*), we get

$$\psi_{\kappa_1,\kappa_2} = \text{const} \times \exp\{\kappa x + \mathcal{K}X\}, \quad \text{with } \kappa \equiv \frac{\kappa_1 - \kappa_2}{2}, \quad \text{and} \quad \mathcal{K} \equiv \kappa_1 + \kappa_2,$$

so that

$$\kappa_1 = \frac{\mathcal{K}}{2} + \kappa, \quad \kappa_2 = \frac{\mathcal{K}}{2} - \kappa, \quad \text{and} \quad E_{\kappa_1,\kappa_2} = -\frac{\hbar^2}{2m}\left(\frac{\mathcal{K}^2}{2} + 2\kappa^2\right).$$

The potential energy is independent of the coordinate X, so that the constant $\mathcal{K}$ cannot have any real part (otherwise the wavefunction would diverge at either $X \to +\infty$ or $X \to -\infty$), i.e. has to be purely imaginary, $\mathcal{K} = iK$. But this would give a positive contribution, $\hbar^2 K^2/m$, to the energy, so that the lowest eigenstates correspond to $K = 0$, physically meaning that in its ground state the system as a whole is at rest (at a completely uncertain location X), and its wavefunction may be marked with just one index, κ:

$$\psi_\kappa = \text{const} \times \exp\{\kappa x\} \equiv \text{const} \times \exp\{\kappa(x_1 - x_2)\},$$
$$\text{with} \quad E_\kappa = -\frac{\hbar^2\kappa^2}{m} \equiv -\frac{\hbar^2\kappa^2}{2(m/2)}. \qquad (**)$$

(The last, apparently awkward form of the expression for E_κ corresponds better to its physical sense: this is the kinetic energy of the effective single particle with the reduced mass $m_{ef} = m/2$, which always appears in two-body problems—either quantum or classical.)

Now we need to use Eq. (**) to compose eigenfunctions that would satisfy:

(i) the boundary conditions at $x \to \pm\infty$,
(ii) the fermionic permutation rule—see Eqs. (8.14) of the lecture notes, and
(iii) the boundary condition at $x = 0$.

The first of these conditions, that the wavefunction should not diverge at $x \to \pm\infty$, may be readily satisfied by assigning to κ opposite signs in the regions $x > 0$ and $x < 0$, i.e. with taking the wavefunction in one of the following two forms:

$$\psi = \text{const} \times \begin{cases} \exp\{-\kappa\,|x|\} \equiv \exp\{-\kappa\,|x_1 - x_2|\}, \\ \text{sgn}\,(x)\exp\{-\kappa\,|x|\} \equiv \text{sgn}\,(x)\exp\{-\kappa\,|x_1 - x_2|\}, \end{cases} \qquad (***)$$

both with positive κ. (The sign of κ does not affect E_κ.)

[28] This transformation of coordinate and momenta is, of course, just a particular case of that discussed in problem 8.11, for the particular case $m_1 = m_2 = m$, so that the reduced mass m_{ef} equals $m/2$. Due to this simple relation, and the 1D character of particle motion in our current system, it is, however, easier to carry out this transformation again more explicitly, rather that use the results of the more general solution of that problem.

The first of these eigenfunctions is symmetric with respect to the particle permutation, $x_1 \leftrightarrow x_2$, and hence is suitable for their singlet spin state (8.21). Now using the fact that for any of the above eigenfunctions, $\partial^2\psi/\partial x_1^2 = \partial^2\psi/\partial x_2^2$, we may recast the system's Hamiltonian in the single-particle form:

$$\hat{H} = -\frac{\hbar^2}{2m_{\text{ef}}}\frac{\partial^2}{\partial x^2} - \mathscr{W}\,\delta(x), \quad \text{with } m_{\text{ef}} = \frac{m}{2}.$$

The (easy) problem of finding the localized eigenstates of this Hamiltonian was solved in the beginning of section 2.6 of the lecture notes: there is only one such state, with the κ given by Eq. (2.161). With the due replacement $m \to m_{\text{ef}} = m/2$ it is

$$\kappa = \frac{m}{2\hbar^2}\mathscr{W},$$

finally giving the lowest singlet-state energy

$$E_{\text{g}} = -\frac{\hbar^2\kappa^2}{m} = -\frac{m\mathscr{W}^2}{4\hbar^2}.$$

On the other hand, for the triplet spin state, the orbital wavefunction had to be asymmetric, i.e. taken in the second form of Eq. (***). However, this function is discontinuous at $x = 0$, and hence cannot be the solution of the Schrödinger equation for any physically-realistic Hamiltonian. So, the system does not have a localized triplet eigenstate, and all acceptable triplet states are extended, i.e. have energies $E \geqslant 0$, i.e. higher than the negative E_{g}, which is therefore the ground-state energy of the system.

***Problem* 8.15.** Calculate the energy spectrum of the system of two identical spin-½ particles, moving along the x-axis, which is described by the following Hamiltonian:

$$\hat{H} = \frac{\hat{p}_1^2}{2m_0} + \frac{\hat{p}_2^2}{2m_0} + \frac{m_0\omega_0^2}{2}\left(x_1^2 + x_2^2 + \varepsilon x_1 x_2\right),$$

and the degeneracy of each energy level.

Solution: Due the symmetry of the Hamiltonian, and acting just like in problem 8.11, let us introduce two linear combinations of particle coordinates x_1 and x_2: the distance $x \equiv x_1 - x_2$ between the particles, and the coordinate $X \equiv (x_1 + x_2)/2$ of their center of mass. Plugging the reciprocal relations,

$$x_1 = X + \frac{x}{2}, \quad x_1 = X - \frac{x}{2},$$

into the Hamiltonian, we see that it separates into two independent parts:

$$\hat{H} = \left(\frac{\hat{P}^2}{2M} + \frac{M\Omega^2 X^2}{2}\right) + \left(\frac{\hat{p}^2}{2m} + \frac{m\omega^2 x^2}{2}\right), \qquad (*)$$

where

$$\Omega \equiv \omega_0 \left(1 + \frac{\varepsilon}{2}\right)^{1/2}, \qquad \omega \equiv \omega_0 \left(1 - \frac{\varepsilon}{2}\right)^{1/2}, \qquad (**)$$

$M \equiv 2m_0$ is the total mass of the particles, $m \equiv m_0/2$ is the 'reduced' mass of their relative motion, and

$$\hat{P} \equiv \hat{p}_1 + \hat{p}_2, \quad \hat{p} \equiv \frac{\hat{p}_1 - \hat{p}_2}{2}.$$

(As we know from the solution of problem 8.11, the operators so defined are the legitimate generalized momenta corresponding to the generalized coordinates X and x, respectively.) Since the two parts of the right-hand side of Eq. (*) are the usual Hamiltonians of two independent 1D harmonic oscillators with frequencies given by Eq. (**), the total energy of the system is just the sum of their energies:

$$E_{N,n} = \hbar\Omega\left(N + \frac{1}{2}\right) + \hbar\omega\left(n + \frac{1}{2}\right), \quad \text{with } N, n = 0, 1, 2, \ldots \qquad (***)$$

Besides certain exact values of the parameter ε (such as $\varepsilon = 6/5$, when $\Omega = 2\omega$), the frequencies Ω and ω are incommensurate, so that the energy levels (***) are orbitally non-degenerate. In order to analyze their spin degeneracy, let us spell out the orbital wavefunction corresponding to a certain pair of the quantum numbers N and n:

$$\Psi_{N,n}(x_1, x_2) = \psi_N(X)\,\psi_n(x) \equiv \psi_N\left(\frac{x_1 + x_2}{2}\right)\psi_n(x_1 - x_2), \qquad (****)$$

where ψ_N and ψ_n are the single-oscillator's eigenfunctions. According to Eqs. (2.281) and (2.284) of the lecture notes (see also figure 2.35), these functions are symmetric if their index is even, and antisymmetric if it is odd. Since the first operand of the product in Eq. (****) is symmetric with respect to particle permutation ($x_1 \leftrightarrow x_2$) for any N, the total orbital wavefunction's parity depends only on the index n:

$$\Psi_{N,n}(x_2, x_1) = (-1)^n \Psi_{N,n}(x_1, x_2) \equiv \begin{cases} +\,\Psi_{N,n}(x_1, x_2), & \text{for } n \text{ even}, \\ -\,\Psi_{N,n}(x_1, x_2), & \text{for } n \text{ odd}. \end{cases}$$

Since the total (orbital plus spin) state vector of the system of two indistinguishable fermions has to be antisymmetric with respect to particle permutation, each energy level (***) with even n may be only occupied by a spin singlet (8.18), and hence is non-degenerate. On the other hand, each level with odd n may be occupied by any of three triplet states (8.21), and hence is triple-degenerate.

Problem* 8.16. Two indistinguishable spin-½ particles are confined to move around a circle of radius R, and interact only at a very short arc distance $l = R(\varphi_1 - \varphi_2) \equiv R\varphi$ between them, so that the interaction potential U may be well approximated with a delta-function of φ. Find the ground state and its energy, for the following two cases:

(i) the 'orbital' (spin-independent) repulsion: $\hat{U} = \mathscr{w}\,\delta(\varphi)$,
(ii) the spin–spin interaction: $\hat{U} = -\mathscr{w}\,\hat{\mathbf{s}}_1 \cdot \hat{\mathbf{s}}_2\,\delta(\varphi)$,

both with constant $\mathscr{w} > 0$. Analyze the trends of your results in the limits $\mathscr{w} \to 0$ and $\mathscr{w} \to \infty$.

Solutions:

(i) The system's Hamiltonian is

$$\hat{H} = \frac{\hat{L}_1^2}{2mR^2} + \frac{\hat{L}_2^2}{2mR^2} + \hat{U},$$

where $\hat{L}_{1,2}$ are the single-particle operators of angular momentum (or more exactly, its only Cartesian component), in the coordinate representation equal to

$$\hat{L}_{1,2} = -i\hbar\frac{\partial}{\partial\varphi_{1,2}}.$$

Since at $\varphi \neq 0$ the potential energy U vanishes, the two-particle Schrödinger equation corresponding to this Hamiltonian is satisfied by any product of single-particle eigenfunctions given by Eq. (3.129) of the lecture notes:

$$\psi_{\nu_1,\nu_2} = \text{const} \times \exp\{i(\nu_1\varphi_1 + \nu_2\varphi_2)\}, \qquad (*a)$$

with any c-numbers ν_1 and ν_2—in an analogy with Eq. (*) of the model solution of problem 8.14. Similarly to that problem, since U depends only on the difference angle $\varphi \equiv \varphi_1 - \varphi_2$, it is natural to replace the variables, introducing, besides the 'distance angle' $\varphi \equiv \varphi_1 - \varphi_2$, the average angle $\phi \equiv (\varphi_1 + \varphi_2)/2$ (essentially the angular position of the particles' center of mass), so that the individual particle positions are

$$\varphi_1 = \phi + \frac{\varphi}{2}, \quad \varphi_2 = \phi - \frac{\varphi}{2}.$$

Plugging these expressions into the above expression into Eq. (*a), we may recast it as

$$\psi_{\nu_1,\nu_2} = \text{const} \times \exp\{i(\nu\varphi + N\phi)\}, \quad \text{with } \nu \equiv \frac{\nu_1 - \nu_2}{2}, \quad N \equiv \nu_1 + \nu_2, \qquad (*b)$$

where N scales the total angular momentum of the system, $L = L_1 + L_2 = \hbar(\nu_1 + \nu_2) = \hbar N$, in such a state.

For each spin state of the system we 'only' need to form a correct linear superposition of such partial solutions, which would:

- satisfy the fermionic permutation rule—see the second of Eqs. (8.14) of the lecture notes,
- satisfy the proper boundary condition at $\varphi_1 = \varphi_2$, i.e. at the interaction point $\varphi = 0$, and
- be invariant to the physically indistinguishable translations $\varphi_{1,2} \to \varphi_{1,2} + 2\pi$.

Since we do not know *a priori* which spin state of the system has the lowest-energy (ground) state[29], we need to consider both options. In any of the *triplet* spin states, described by Eqs. (8.21) of the lecture notes, the orbital wavefunction $\psi_t(\varphi_1, \varphi_2)$ has to be antisymmetric,

$$\psi_t(\varphi_1, \varphi_2) = -\psi_t(\varphi_2, \varphi_1), \qquad (**)$$

and since it also has to be continuous everywhere (even at potential's singularity point), the wavefunction has to vanish at $\varphi_1 \to \varphi_2$, i.e. at $\varphi \to 0$. As a result, according to Eq. (2.75), applied to ψ_t as a function of any of its arguments, the orbital wavefunction of the triplet state is not affected by the interaction potential at all, and the eigenstates may be constructed as a linear superposition of the states (*) extended to all values of φ_1 and φ_2, from $-\infty$ to $+\infty$. Since such extended states are linearly-independent, the above periodicity-invariance condition has to apply to each of them, giving $\nu_1 = n_1$, $\nu_2 = n_2$, with integer $n_{1,2}$.

For the lowest-energy eigenstates, we have to select the quantum numbers $n_{1,2}$ with the smallest $|n_{1,2}|$ (corresponding to the lowest magnitudes of the corresponding angular momenta $L_{1,2} = \hbar n_{1,2}$), but the choice $n_1 = n_2 = 0$ would give $\psi(\varphi_1, \varphi_2) =$ const and hence violate Eq. (**). The next quantum number sets are $\{n_1 = 0, n_2 = \pm 1\}$ and $\{n_1 = \pm 1, n_2 = 0\}$, whose eigenstates may be readily combined to satisfy Eq. (**):

$$\psi_t = \psi_\pm = \text{const} \times (\exp\{\pm i\varphi_1\} - \exp\{\pm i\varphi_2\}).$$

For each of these two states, each of the component wavefunctions corresponds to the angular orbital momentum $L_{1,2} = \pm\hbar$ of one particle, and $L_{2,1} = 0$ of its counterpart, and hence the total kinetic energies $T_\pm = (L_1^2 + L_2^2)/2mR^2$ of these states may be found even without their formal calculation:

$$E_t = E_\pm = T_\pm = \frac{\hbar^2}{2mR^2} > 0.$$

Conversely, in the *singlet* spin state the orbital wavefunction $\psi_s(\varphi_1, \varphi_2)$ has to be symmetric:

$$\psi_s(\varphi_1, \varphi_2) = \psi_s(\varphi_2, \varphi_1). \qquad (***)$$

Hence, the wavefunction does not necessarily vanish at the interaction point $\varphi_1 = \varphi_2$, i.e. $\varphi = 0$, and is generally affected by the interaction potential U. Here it is more natural to consider it a function of the combinational arguments φ and ϕ introduced above:

[29] On the basis of discussion of the helium atom in section 8.2, and the solution of several prior problems of this chapter, we may *guess* this should be the singlet state, but it is prudent to verify this guess.

$$\Psi_s(\varphi, \phi) \equiv \psi_s(\varphi_1, \varphi_2) = \psi_s\left(\phi + \frac{\varphi}{2}, \phi - \frac{\varphi}{2}\right),$$

and try to compose the correct wavefunction from the partial functions (*b). According to Eq. (***), Ψ_s as a function of φ is symmetric for any ϕ:

$$\Psi_s(-\varphi, \phi) = \Psi_s(\varphi, \phi)$$

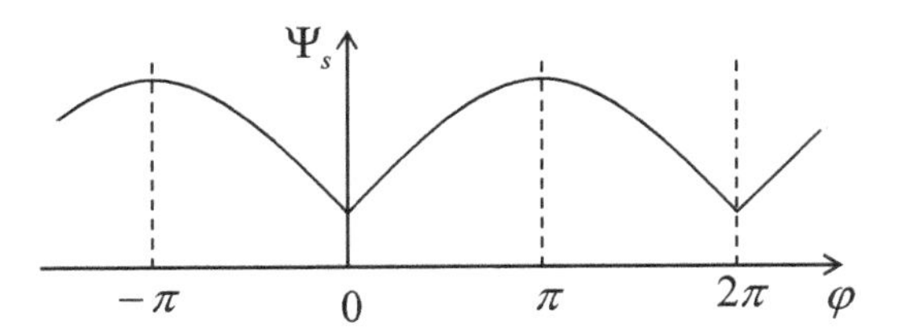

—see the figure above. As a result, its single-side derivatives at the interaction point, $\varphi = 0$, should be equal but sign-opposite:

$$\frac{\partial \Psi_s}{\partial \varphi}\Big|_{\varphi=+0} = -\frac{\partial \Psi_s}{\partial \varphi}\Big|_{\varphi=-0}.$$

Using this equality in Eq. (2.75) of the lecture notes, applied to Ψ_s as a function of φ, with the due replacement $m \to m_{ef} = m/2$, we get the following boundary condition:

$$\frac{\partial \Psi_s}{\partial \varphi}\Big|_{\varphi=+0} = \frac{m}{\hbar^2} w \Psi_s\Big|_{\varphi=\pm 0}. \qquad (****)$$

This relation does not affect the dependence of Ψ_s on ϕ, so that if we want the lowest-energy wavefunction, we should compose it from the exponential functions (*b) with $N = 0$:[30]

$$\Psi_s = \sum_\nu c_\nu \exp\{i\nu\varphi\}.$$

Due to the cusp of function Ψ_s at $\varphi = 0$ (see the figure above), and its 2π-periodicity, it is natural to limit this expansion to a 2π-segment of argument φ, where the function is smooth—say $[0, 2\pi]$, extending it to other similar segments periodically. At this limited segment, the sum of just two terms, with $\nu = \pm\lambda$,

$$\begin{aligned}\Psi_s &= \text{const} \times [\exp\{i\lambda(\varphi - \pi)\} + \exp\{-i\lambda(\varphi - \pi)\}] \\ &\propto \cos\lambda(\varphi - \pi), \quad \text{for } 0 < \varphi < \pi,\end{aligned}$$

satisfies the symmetry condition (***) at any (not necessarily integer!) λ. Similarly to the triplet case, it is unnecessary to calculate the eigenenergy corresponding to this

[30] Physically, this means that we consider the system with zero total angular momentum—apparently, a natural requirement for the lowest-energy state. Note, however, that for the calculated triplet states $\psi_\pm$, the equality $\langle L\rangle = 0$ is valid for each of the states, but *not* for each exponential component of such a state, which correspond to equal and opposite momenta $L = \pm\hbar$.

linear superposition explicitly, because each term has the same angular momentum magnitude $|L_{1,2}| = \hbar\lambda$ of each particle, so that

$$E_s = T_1 + T_2 = 2\frac{L_{1,2}^2}{2mR^2} = \lambda^2\frac{\hbar^2}{mR^2}.$$

The value of λ, and hence that of E_s, may be found by plugging this Ψ_s into the boundary condition (****). This gives the following characteristic equation:

$$\lambda \sin \lambda\pi = \frac{m}{\hbar^2}\mathscr{W} \cos \lambda\pi,$$

which may be rewritten, more conveniently, as

$$\lambda = \alpha \cot \lambda\pi, \quad \text{with } \alpha \equiv \frac{m}{\hbar^2}\mathscr{W}.$$

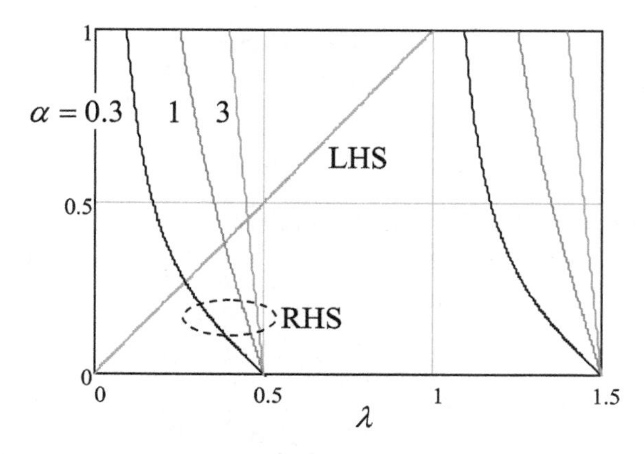

Both sides of the last form of the equation are plotted, in the figure above, as functions of λ, for several values of the normalized interaction strength α. Since the eigenenergy of the state scales as λ^2, the lowest-energy state corresponds to the smallest root of the characteristic equation, i.e. to the curve intersection point that is closest to the origin. As the figure shows, the smallest root is always confined to the segment $0 < \lambda < ½$, so that for any λ,

$$0 \leqslant E_s \leqslant \frac{\hbar^2}{4mR^2} \equiv \frac{1}{2}E_t.$$

Thus, as could be expected, for any parameter α of the system, the spin-singlet state has an energy lower than that of the spin-triplet states, and it is the ground state of the system[31].

[31] A thoughtful reader might ask: how do we know that all eigenstates of the system, in which the particles *do* interact, are limited to the singlet and triplet states, i.e. the states that may be factored according to Eq. (8.12)? An answer to this concern is that each of the four considered states (one singlet and three triplet) are eigenstates of the problem. Since these states may be taken for the full basis in the four-function Hilbert space of two spins-½, this problem cannot have other independent spin eigenstates. However, this fact was *not* clear *a priori*, and may *not* be true for other problems, so that the caution of this (hypothetical :-) reader is justified.

If the interaction is weak, $\alpha \to 0$, then $\lambda \to 0$ as well, and we may find its asymptotic value by approximating $\sin\lambda\pi$ with $\lambda\pi$, and $\cos\lambda\pi$ with 1, so that $\cot\lambda\pi \approx 1/\lambda\pi$. With this approximation, the characteristic equation yields $\lambda \approx (\alpha/\pi)^{1/2}$, so that[32]

$$E_{\rm s} \approx \frac{\alpha}{\pi}\frac{\hbar^2}{mR^2} = \frac{1}{\pi}\frac{\mathcal{W}}{R^2} \to 0.$$

On the other hand, at very strong repulsion ($\alpha \to \infty$), $\lambda \to ½$, so that the singlet state energy tends to

$$(E_{\rm s})_{\max} = \frac{\hbar^2}{4mR^2} = \frac{1}{2}E_{\rm t}.$$

Since $\cos[(\varphi - \pi)/2] = -|\sin(\varphi/2)|$, in this limit the singlet state's eigenfunction may be represented, for any value of φ ($-\infty < \varphi < +\infty$) as

$$\Psi_{\rm s} = \text{const} \times \left|\sin\frac{\varphi}{2}\right|,$$

approaching zero at the interaction point $\varphi = 0$. This fact explains why the state's energy does not increase further as the interaction increases.

(ii) As was argued above, *triplet* spin states of the system are not affected by the delta-functional interaction $U \propto \delta(\varphi)$ of the particles at all. This result does not depend on the particular form of the coefficient before the delta-function, so that even for the local spin–spin interaction, the result for $E_{\rm t}$ remains the same as above:

$$E_{\rm t} = \frac{\hbar^2}{2mR^2}.$$

For the *singlet* state, we may use the result obtained in the solution of problem 8.3,

$$\langle \mathbf{s}_1 \cdot \mathbf{s}_2 \rangle_- = -3\left(\frac{\hbar}{2}\right)^2,$$

for the expectation value of the product $\mathbf{s}_1 \cdot \mathbf{s}_2$, participating in the interaction U. Since, according to this solution, the singlet state is one of the eigenstates of the corresponding operator (the other being the three triplet states), its energy may be obtained from the results of part (i) of the problem by the simple replacement

$$\mathcal{W} \to 3\left(\frac{\hbar}{2}\right)^2 \mathcal{W}, \quad \text{i.e. } \alpha \to 3\left(\frac{\hbar}{2}\right)^2 \alpha,$$

giving, in particular, the following asymptotic values:

$$E_{\rm s} \to \frac{\hbar^2}{mR^2} \times \begin{cases} 3\alpha(\hbar/2)^2/\pi \to 0, & \text{for } \alpha \to 0, \\ 1/4, & \text{for } \alpha \to \infty, \end{cases}$$

so that for any α, $E_{\rm s} < E_{\rm t}$, i.e. the singlet state is the (non-degenerate) ground state of the system.

[32] This result may be also obtained using the perturbation theory—the additional exercise highly recommended to the reader.

Finally, note the both interactions given in the assignment correspond to the effective repulsion of the particles. For the opposite signs of $\mathscr{w}$ (particle attraction), the singlet ground states may be *bound*, i.e. localized at $\varphi = 0$, having imaginary values of $\nu_{1,2}$, and hence negative eigenenergies[33]—cf. problem 8.14.

***Problem* 8.17.** Two particles of mass M, separated by two much lighter particles of mass $m \ll M$, are placed on a ring of radius R—see the figure below. The particles strongly repulse at contact, but otherwise each of them is free to move along the ring. Calculate the lower part of the energy spectrum of the system.

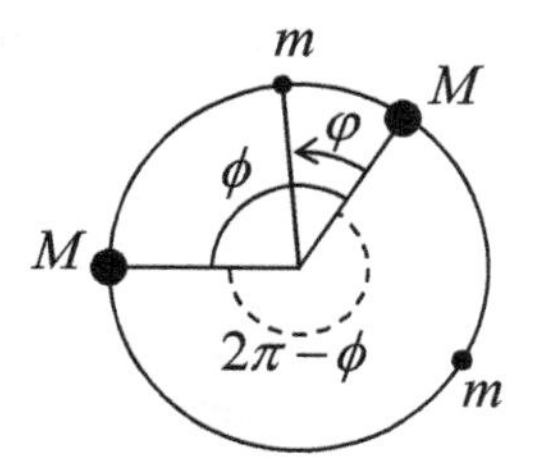

Solution: Due to the strong hierarchy of the masses, we may solve the problem in two stages: first, analyze the 'fast' (high-energy) motion of the lighter particles considering the angular positions ϕ_1 and ϕ_2 of the heavier particles fixed, and then, with the account of the kinetic energy of this motion, analyze the 'slow' motion of the heavier particles[34].

At the first stage, due to the strong particle repulsion at contact, we may consider the motion of each lighter particle as free but confined to a time-independent angular segment of length $\phi \equiv \phi_2 - \phi_1$, with its wavefunction turning to zero at its ends. The Hamiltonian of this problem is the same as for the single-particle plane rotator (see section 3.5 of the lecture notes):

$$\hat{H}_m = \frac{\hat{L}_z^2}{2mR^2}, \quad \text{with } \hat{L}_z = \hat{p}R = -i\hbar\frac{\partial}{\partial\varphi},$$

where φ is the angular position of the light particle, referred, for example, to one of the boundaries of the segment—see the figure above. The solutions of the corresponding Schrödinger eigenproblem,

$$-\frac{\hbar^2}{2mR^2}\frac{d^2\psi_n}{d\varphi^2} = E_n\psi, \quad \text{with } \psi_n\big|_{\varphi=0} = \psi_n\big|_{\varphi=\phi} = 0,$$

[33] Solving this (conceptually, very similar) problem may be one more useful exercise for the reader.

[34] Such *Born–Oppenheimer approximation* is broadly used for the separation of electron and nuclear motion at theoretical analyses of molecular spectra. Due to the large ratio $m_p/m_e \approx 1835$, it works very well, but may need small corrections necessary for the description of the lightest atoms/ions, such as hydrogen—see, e.g. [1].

are also very simple (cf. section 1.7 of the lecture notes):

$$\psi_n = C \sin \frac{\pi n \varphi}{\phi}, \qquad E_n = \frac{\pi^2 \hbar^2}{2mR^2\phi^2} n^2, \qquad n = 0, 1, 2, \dots.$$

At $m \to 0$ the gaps between these energy levels are very large, so that for our purposes we are interested only in the ground state of this motion, with the energy $E_1 = \pi^2\hbar^2/2mR^2\phi^2$. Adding to this energy the similar ground-state energy of motion of the second light particle on its angular segment of length $(2\pi - \phi)$, we get the following total ground state energy of the light particles:

$$E_g = U(\phi) \equiv \frac{\pi^2\hbar^2}{2mR^2}\left[\frac{1}{\phi^2} + \frac{1}{(2\pi - \phi)^2}\right].$$

Since this energy is a function of ϕ only, it plays the role of the potential energy for the slow motion of the heavy particles. The scale of the variable ϕ is of the order of π, so that the quantization of that motion leads to additional energies of the order of $\hbar^2/2MR^2$, which is, at $M \gg m$, much lower than the scale $\hbar^2/2mR^2$ of the function $U(\phi)$. This is why we are only interested in the shape of this function near its minimum, evidently at the point $\phi_0 = \pi$. Near this point we may take $\phi = \pi + \tilde{\phi}$, and Taylor-expand this function,

$$U(\phi) \equiv \frac{\pi^2\hbar^2}{2mR^2}\left[\frac{1}{\phi^2} + \frac{1}{(2\pi - \phi)^2}\right] = \frac{\pi^2\hbar^2}{2mR^2}\left[\frac{1}{(\pi + \tilde{\phi})^2} + \frac{1}{(\pi - \tilde{\phi})^2}\right]$$
$$\equiv \frac{\hbar^2}{mR^2}\frac{1 + (\tilde{\phi}/\pi)^2}{[1 - (\tilde{\phi}/\pi)^2]^2},$$

with respect to small $\tilde{\phi}/\pi$, keeping only two leading terms of the series:

$$U(\phi) \approx \frac{\hbar^2}{mR^2}\left(1 + 3\frac{\tilde{\phi}^2}{\pi^2}\right) = U_0 + \frac{\kappa(R\tilde{\phi})^2}{2}, \quad \text{with}$$
$$U_0 \equiv \frac{\hbar^2}{mR^2}, \quad \text{and} \quad \kappa \equiv \frac{6\hbar^2}{\pi^2 mR^4}. \qquad (*)$$

With that, the Hamiltonian describing the heavier particle motion is

$$\hat{H}_M = \frac{\hat{P}_1}{2M} + \frac{\hat{P}_2}{2M} + U(\phi), \quad \text{with } \hat{P}_{1,2} = -i\hbar\frac{\partial}{\partial(R\phi_{1,2})}.$$

This Hamiltonian may be partitioned, exactly as was done in several previous problems, starting from problem 8.11, into the sum of that of the joint rotation of both particles, with the total mass $2M$, around the circle, and the harmonic oscillator, with the reduced mass $M_r = M/2$ and the effective spring constant κ, and hence the frequency

$$\omega_0 = \left(\frac{\kappa}{M_r}\right)^{1/2} = \frac{\sqrt{12}\,\hbar}{\pi(mM)^{1/2}R^2}.$$

Adding the results of quantization of these independent subsystems, we get the requested energy spectrum[35]:

$$E_{N,m} = U_0 + \frac{\sqrt{12}\,\hbar^2}{\pi R^2(mM)^{1/2}}\left(N + \frac{1}{2}\right) + \frac{\hbar^2}{4MR^2}m^2, \quad \text{with}$$
$$N = 0,\ 1,\ 2,\ \ldots, \quad \text{and} \quad m = 0,\ \pm1,\ \pm2,\ \ldots$$

Due to the mass relation $m \ll M$, the intervals between first rotational values are much narrower than those between the vibrational levels (and those, in turn, are much smaller than U_0), so that the spectrum looks like the one sketched in the model solution of problem 8.12.

Note that if the particles of each pair (of the specific mass) are indistinguishable, the set of quantum numbers m may be decimated, because the period of the rotational wavefunction may change from 2π (as assumed above) to π—as it does, for example, for any permutation-symmetric spin-vibrational state—see the solutions of problems 8.12 and 8.13 for discussion.

***Problem* 8.18.** N indistinguishable spin-½ particles move in a spherically-symmetric quadratic potential $U(\mathbf{r}) = m\omega_0^2 r^2/2$. Neglecting the direct interaction of the particles, find the ground-state energy of the system.

Solution: As was discussed in the beginning of section 8.3 of the lecture notes, solving such problems (for non-interacting particles only!) we may neglect the genuine structure of the spin states, and reduce the spin effects to the Pauli principle, and the spin degeneracy $g = 2s + 1$ of each single-particle orbital state. According to Eq. (3.124) of the lecture notes, the single-particle orbital energy spectrum in this potential is

$$\varepsilon_n = \hbar\omega_0\left(n + \frac{3}{2}\right), \quad \text{with } n \equiv n_x + n_y + n_z, \quad \text{and} \quad n_j = 0,\ 1,\ \ldots$$

As was discussed in the model solution of problem 3.23, the degeneracy of each orbital level may be calculated as the number of different ways to distribute n indistinguishable 'balls' (the partial quantum numbers contributing to integer n) between three distinct 'boxes' (n_x, n_y, and n_z):[36]

$$M_n^{(3)} = \frac{(3 - 1 + n)!}{(3 - 1)!n!} = \frac{1}{2}(n + 1)(n + 2).$$

[35] Let me hope that the difference between the magnetic quantum number m and the lighter particle mass in this expression is absolutely clear from the context.

[36] This is a particular case of a formula for k boxes: $M_n^{(k)} = (n - 1 + k)!/n!(k - 1)!$—see, e.g. Eq. (A.7).

The spin degeneracy $g = 2s + 1 = 2$ doubles this orbital degeneracy. Hence, if the given number N of particles coincides with any of the following special integers,

$$N_m \equiv 2\sum_{n=0}^{m-1} M_n^{(3)},$$

then exactly m single-particle levels (with $n = 0, 1,\ldots, m - 1$) are completely filled in the ground state, and the total energy is

$$E(N_m) = 2\sum_{n=0}^{m-1} \varepsilon_n M_n^{(3)} = 2\hbar\omega_0 \sum_{n=0}^{m-1}\left(n + \frac{3}{2}\right)M_n^{(3)}.$$

Defining a new summation index $n' \equiv n + 1$, we get

$$\begin{aligned}
N_m &\equiv 2\sum_{n=0}^{m-1} M_n^{(3)} = \sum_{n=0}^{m-1}(n + 1)(n + 2) \\
&\equiv \sum_{n'=1}^{m} n'(n' + 1) \equiv \sum_{n'=1}^{m} n'^2 + \sum_{n'=1}^{m} n', \\
E(N_m) &= \hbar\omega_0 \sum_{n=0}^{m-1}\left(n + \frac{3}{2}\right)(n + 1)(n + 2) \\
&\equiv \hbar\omega_0\left(\sum_{n'=1}^{m} n'^3 + \frac{3}{2}\sum_{n'=1}^{m} n'^2 + \frac{1}{2}\sum_{n'=1}^{m} n'\right).
\end{aligned}$$

Now using the well-known formulas for summation of several first natural numbers, their squares and cubes[37], we finally get

$$\begin{aligned}
N_m &= \frac{1}{3}m(m + 1)(m + 2), \\
E(N_m) &= \frac{1}{4}m(m + 1)(m^2 + 3m + 2)\hbar\omega_0.
\end{aligned}$$

As the easiest sanity check, for $m = 1$ (i.e. for only the lowest, ground-state level completely filled), these formulas yield $N_m = 2$, $E(N_m) = 3\hbar\omega_0$, i.e. the evidently correct results.

In the general case, when $N = N_m + N'$, with $0 < N' < 2({M^{(3)}}_{m+1} - {M^{(3)}}_m) \equiv 4(m + 1)(m + 2)$, additional N' particles have to occupy the next level, with $n = m$, and $\varepsilon_n = \varepsilon_m = \hbar\omega_0(m + 3/2)$, so that

$$E(N) = E(N_m) + \varepsilon_m N' = \hbar\omega_0\left[\frac{1}{4}m(m + 1)(m^2 + 3m + 2) + N'\left(m + \frac{3}{2}\right)\right].$$

[37] See, e.g. Eqs. (A.8*b*), (A.9*a*), and (A.9*b*).

In the limit $N \gg 1$, m is large as well, and these results may be approximated as

$$N \approx \frac{m^3}{3}, \quad \text{so that} \quad m \approx (3N)^{1/3}, \quad \text{and} \quad E(N) \approx \frac{\hbar\omega_0}{4} m^4 \approx \frac{\hbar\omega_0}{4}(3N)^{4/3}.$$

***Problem* 8.19.** Use the Hund rules to find the values of the quantum numbers L, S, and J in the ground states of the atoms of carbon and nitrogen. Write down the Russell–Saunders symbols for these states.

Solution: As the table in figure 3.24 of the lecture notes shows, the ground state of the carbon (C) atom has two electrons in the (completely filled) He shell, two electrons in the 2*s* orbital state, and two electrons in the 2*p* sub-shell $\{n = 2, l = 1\}$. As follows from the discussion in section 8.2 of the lecture notes, each pair of the *s*-state electrons have to be in the same orbital state and in the spin-singlet state, with zero net orbital momentum and zero net spin, so they do not contribute to the net quantum numbers L, S, and J of the atom, and we need to discuss only the two 2*p* electrons, with $l = 1$. Due to the triple degeneracy ($m_l = +1, 0, -1$) of their orbital states, and double degeneracy ($m_s = \pm ½$) of the spin states, the Pauli principle allows one to have up to six electrons in this sub-shell, so we need to use the Hund rules to understand which exactly of these states our two electrons would take.

The highest Rule 1 of the Hund's hierarchy says that S has to take the largest value possible. Since, according to Eq. (8.48), $S \equiv \max|M_S| = \max|(m_s)_1 + (m_s)_2|$, such maximum, $S = 1$, is evidently achieved in a linear superposition of two factorable triplet states with $(m_s)_1 = (m_s)_2 = \pm½$—see figure 8.2 of the lecture notes. Both these states are symmetric with respect to the electron permutation; hence in order to satisfy the Pauli principle, the involved orbital states have to be different, i.e. the quantum numbers, $(m_l)_1$ and $(m_l)_2$, have to take different values from the available set $\{+1, 0, -1\}$, so that the quantum number $L \equiv \max|(m_l)_1 + (m_l)_2|$ can only take values 0 or 1.[38] Now using Hund's Rule 2, we have to select the highest of the these values, $L = 1$. Finally, Hund's Rule 3 says that since this sub-shell is filled by less than half, then $J = |L - S| = 0$. As a result, the Russell–Saunders symbol (defined by Eq. (8.59) of the lecture notes) of this ground state is ${}^3P_0$.

The atom of nitrogen (N) is different from that of carbon 'only' by one more, third electron in the same sub-shell $\{n = 2, l = 1\}$, so that again, non-vanishing contributions to the net quantum numbers L, S, and J may be given only by the (now, three) electrons of this sub-shell, with three different orbital states. Indeed, if the electrons are in different orbital states, their magnetic spin numbers m_s may take equal values without violating the Pauli principle. Such a configuration, $(m_s)_1 = (m_s)_2 = (m_s)_3 = \pm½$, is evidently the best one for Hund's Rule 1, providing the highest possible value $S \equiv \max|M_S| = \max|(m_s)_1 + (m_s)_2 + (m_s)_3| = 3/2$. Such a spin state is fully symmetric with respect to the permutation of any two particles, so that the

[38] For this case, as well as for the nitrogen atom discussed below, the physical origin of the first Hund rule is very clear: the Coulomb repulsion of the electrons forces them into different *p*-orbitals, thus maximizing their average distance from each other—see, e.g. the second row of figure 3.20.

orbital state of the system has to be a linear superposition fully asymmetric with respect to such a permutation, similar to the Slater determinant (8.60a), of all possible six states with different $m_l = \{+1, 0, -1\}$. Since in such a superposition, the weight moduli (and hence the spin orientation probabilities), are equal, $M_L = (m_s)_1 + (m_s)_2 + (m_s)_3 = 0$. Translating what we know from section 8.2 about the case of two spins-½, and from the solution of problem 8.10 about two spins-1, to the orbital-momentum language, we may conclude that such an asymmetric state has to have $L = 0$.[39] Hence $J = S = 3/2$ (so that Hund's Rules 2 and 3 are redundant here). Thus, the Russell–Saunders symbol of the ground state of nitrogen is $^4S_{3/2}$.

***Problem* 8.20.** $N \gg 1$ indistinguishable quantum particles, not interacting directly, are placed in a hard-wall, rectangular box with sides a_x, a_y, and a_z. Calculate the ground-state energy of the system, and the average forces it exerts on each face of the box. Can we characterize the forces by certain pressure $\mathcal{P}$?

Hint: Consider separately the cases of bosons and fermions.

Solution: Non-interacting *bosons* may occupy the same single-particle energy level ε, so that the ground-state energy E_g of N of them is just $N\varepsilon_g$. Using Eq. (1.86) of the lecture notes, with $n_x = n_y = n_z = 1$, for ε_g, we get

$$E_g = N\frac{\pi^2\hbar^2}{2m}\left(\frac{1}{a_x^2} + \frac{1}{a_y^2} + \frac{1}{a_z^2}\right).$$

As was discussed in the model solution of problem 1.9, this expression shows that the force-to-area ratios for each wall of the box are generally different. For example, for the faces normal to axis x, of area $A_x = a_y a_z$,

$$\mathcal{P}_x \equiv \frac{\langle F_x\rangle}{A_x} = \frac{1}{a_y a_z}\left(-\frac{\partial E_g}{\partial a_x}\right) = N\frac{\pi^2\hbar^2}{m a_x^3 a_y a_z} = \frac{\pi^2\hbar^2}{m a_x^2}\frac{N}{V},$$

where $V = a_x a_y a_z$ is the box volume. Hence the exerted forces generally *cannot* be characterized by a unique pressure $\mathcal{P}$ (which by definition should be isotropic), and only for a cubic box, with $a_x = a_y = a_z \equiv a$, we may write

$$\mathcal{P}_x = \mathcal{P}_y = \mathcal{P}_z \equiv \mathcal{P} = N\frac{\pi^2\hbar^2}{m a^5} = \frac{\pi^2\hbar^2 N}{m V^{5/3}}. \qquad (*)$$

In contrast, because of the Pauli principle, indistinguishable *fermions* cannot be in the same quantum state. Hence, to form the ground state with the lowest energy, we may place, in each single-particle orbital state, only g fermions, where $g = 2s + 1$ is their spin degeneracy. (For electrons, as spin-½ particles, $g = 2$.) Since, according to Eq. (1.90) of the lecture notes, the density of such states in the wave vector space is

[39] An explicit proof of this result, using the same approach as in problem 8.3, and Eq. (5.164) of the lecture notes (with $l = 1$), is a good additional exercise, recommended to the reader.

constant, the set of $N_o \equiv N/g \gg 1$ orbital states with lowest energies form, in **k**-space, a sphere[40] of certain radius k_F. This radius and the total ground energy of the system may be readily calculated by writing two very similar expressions (using Eqs. (1.90) and, in the second case, Eq. (1.89) as well)[41]:

$$\begin{aligned} N &= \int_{k<k_F} g dN_o = \frac{V}{(2\pi)^3}\int_{k<k_F} g d^3k \\ &= \frac{gV}{(2\pi)^3}\int_{k<k_F} d^3k = \frac{gV}{(2\pi)^3}\int_0^{k_F} 4\pi k^2 dk = \frac{gV}{(2\pi)^3}\frac{4\pi k_F^3}{3}, \\ E_g &= \int_{k<k_F} g\varepsilon_{\mathbf{k}} dN_o = \frac{V}{(2\pi)^3}\int_{k<k_F} g\varepsilon_{\mathbf{k}} d^3k = \frac{gV}{(2\pi)^3}\int_{k<k_F}\frac{\hbar^2k^2}{2m}d^3k \\ &= \frac{gV\hbar^2}{(2\pi)^3 2m}\int_0^{k_F} k^2 4\pi k^2 dk = \frac{gV\hbar^2}{(2\pi)^3 2m}\frac{4\pi k_F^5}{5}, \end{aligned}$$

where in our case $V = a_x a_y a_z$.

Calculating k_F from the first equality, and plugging the result into the second one, we finally get

$$E_g = \frac{3}{5}\varepsilon_F N,$$

where ε_F is the so-called *Fermi energy*—the largest single-particle energy of an occupied level:

$$\varepsilon_F \equiv \frac{\hbar^2 k_F^2}{2m} = \frac{\hbar^2}{2m}\left(\frac{6\pi^2}{g}\frac{N}{V}\right)^{2/3}.$$

This result shows that the energy of the system depends similarly on each linear dimension of the box, and hence the average forces exerted on its walls *may* be characterized by an isotropic pressure $\mathcal{P}$, which may be calculated as

$$\mathcal{P} = -\frac{\partial E_g}{\partial V} = \frac{2}{3}\frac{E_g}{V} = \left(\frac{6\pi^2}{g}\right)^{2/3}\frac{\hbar^2 N^{2/3}}{3mV^{5/2}}.$$

Problem* 8.21. Explore the *Thomas–Fermi model*[42] of a heavy atom, with the nuclear charge $Q = Ze \gg e$, in which the interaction between electrons is limited to their contribution to the common electrostatic potential $\phi(\mathbf{r})$. In particular, derive the ordinary differential equation obeyed by the radial distribution of the potential, and use it to estimate the effective radius of the atom.

---

[40] This is exactly what is called the *Fermi sphere*. (Note that in non-isotropic systems, such as electrons in crystallic lattices, the *Fermi surface* $\varepsilon_{\max}(\mathbf{k})$ = const may have a non-spherical form in the momentum space.)

[41] Due to its importance, this calculation is virtually repeated in *Part SM* section 3.3, where is serves as a background for a discussion of fermion gas properties at non-zero temperatures.

[42] It was suggested in 1927, independently, by L Thomas and E Fermi.

Solution: Due to the condition $Z \gg 1$, we may expect the characteristic radius $r_{TF}(Z)$ of the atom (i.e. of the electron cloud surrounding the point-like nucleus) to be much larger than the characteristic radius r_0 of the single-electron wavefunctions in the Coulomb field of bare nucleus with charge $Q = Ze$:

$$r_0 = \frac{r_B}{Z},$$

where r_B is the Bohr radius[43]. (This assumption, $r_{TF} \gg r_0$, will be confirmed by our solution.) Due to this relation, which means that the electron's electrostatic potential energy $U(\mathbf{r}) = -e\phi(\mathbf{r})$ changes in space slowly on the r_0-scale, we may calculate the electron density $n(\mathbf{r}) \equiv N/V$ in a small local volume V (with $r_0 \ll V^{1/3} \ll r_{TF}$) neglecting the gradient of $U(\mathbf{r})$, i.e. considering the local electrons as free particles with the energy

$$\varepsilon = \frac{p^2}{2m_e} - e\phi(\mathbf{r}), \qquad (*)$$

where the second term is treated, at each point, as a local constant. As a result, we may apply to this small local volume of this gas, at negligible temperature[44], the analysis carried out in the model solution of the previous problem, with the spin degeneracy $g = 2s + 1 = 2$, to write

$$\varepsilon_F(\mathbf{r}) \equiv \frac{\hbar^2 k_F^2(\mathbf{r})}{2m_e} = \frac{\hbar^2}{2m_e}[3\pi^2 n(\mathbf{r})]^{2/3}. \qquad (**)$$

Now comes the most non-trivial point of this solution. If we accept the free electron energy at distance $r \to \infty$ from the nucleus for the reference, the largest value of the full energy (*), for any $\mathbf{r}$, should equal zero, because the atom should be in equilibrium with free electrons in the environment[45]. Hence the maximum value, $\varepsilon_F(\mathrm{r})$, of the local kinetic energy has to be equal to $-q\phi(\mathbf{r}) \equiv e\phi(\mathbf{r})$. Together with Eq. (**), this equality yields

$$n(\mathbf{r}) = \frac{1}{3\pi^2}\left[\frac{2m_e e\phi(\mathbf{r})}{\hbar^2}\right]^{3/2}. \qquad (***)$$

[43] See, e.g. Eqs. (1.10) and (3.192) of the lecture notes.

[44] The scale of temperatures at which this assumption becomes evidently invalid is given by the Hartree energy $E_H \approx 27.2$ eV (see Eq. (1.13) and its discussion), corresponding to $T_K = E_H/k_B \sim 3 \times 10^5$ K—about a thousand times higher than the standard room temperature of 300 K. Actually, the solution of the next problem shows that the actual validity threshold for temperature is even $\sim Z^{7/3} \gg 1$ times higher.

[45] In this sense, the ionization energy of the Thomas–Fermi atom equals zero. Note that in statistical physics, such equilibrium is called *chemical*, and the corresponding value of energy (in our case, accepted for zero), is called the *chemical potential*, commonly denoted as μ. A more general introduction of this notion in statistical physics allows us to streamline such calculations, and also generalize them to the case of non-zero temperatures. This is why I will give this problem and the next one again, and provide their more formal solutions, in *Part SM* of this series.

The second relation between the functions $n(\mathbf{r})$ and $\phi(\mathbf{r})$ is given by the Poisson equation of electrostatics[46],

$$\nabla^2\phi(\mathbf{r}) = -\frac{\rho(\mathbf{r})}{\varepsilon_0} \equiv -\frac{e[Z\delta(\mathbf{r}) - n(\mathbf{r})]}{\varepsilon_0},$$

where the electric charge density $\rho(\mathbf{r})$ consists of the point-like positive charge $Q = Ze$ of the nucleus at the origin, and the space-distributed negative charge of the electron cloud, with the density $-en(\mathbf{r})$. Plugging the $n(\mathbf{r})$ from Eq. (***), and spelling out the Laplace operator for our spherically-symmetric problem[47], we get the following *Thomas–Fermi equation* for the radial distribution of the electrostatic potential ϕ:

$$\frac{1}{r^2}\frac{d}{dr}\left(r^2\frac{d\phi}{dr}\right) = \frac{e}{3\pi^2\varepsilon_0}\left(\frac{2m_e e\phi}{\hbar^2}\right)^{3/2}, \quad \text{for} \quad r > 0.$$

This ordinary differential equation has to be solved with the following boundary conditions. As the above Poisson equation shows, at $r \to 0$, the potential has to approach that of the atomic nucleus:

$$\phi(r) \to \frac{Q}{4\pi\varepsilon_0 r} \equiv \frac{Ze}{4\pi\varepsilon_0 r}, \quad \text{at} \quad r \to 0.$$

On the other hand, due to the atom's neutrality, at large distances its electrostatic potential should not only tend to zero, but also do it faster than that of any non-zero net charge[48]:

$$r\phi(r) \to 0, \quad \text{at} \quad r \to \infty.$$

It is convenient to recast this boundary problem by introducing the dimensionless distance ξ from the origin, defined as

$$\xi \equiv \frac{r}{r_{\rm TF}(Z)}, \quad \text{with } r_{\rm TF}(Z) \equiv b\frac{r_{\rm B}}{Z^{1/3}} = br_0 Z^{2/3} \gg r_0,$$
$$\text{and} \quad b \equiv \frac{1}{2}\left(\frac{3\pi}{4}\right)^{2/3} \approx 0.8853, \qquad (****)$$

and also a dimensionless function $\chi(\xi)$, defined by the following equality:

$$\phi(r) \equiv \frac{Ze}{4\pi\varepsilon_0 r}\chi(\xi).$$

[46] See, e.g. *Part EM* Eq. (1.41). Let me hope that the difference between the single-particle energy ε and the electrostatic constant ε_0 (see, e.g. appendix B) is absolutely clear from the context.

[47] See, e.g. Eq. (A.67) with $\partial/\partial\theta = \partial/\partial\varphi = 0$.

[48] A useful sanity check of the self-consistency of the Thomas–Fermi model is using the above relations to prove that the total number of electrons, calculated as

$$N = \int n(\mathbf{r})d^3r \equiv 4\pi\int_0^\infty n(r)r^2dr,$$

equals exactly Z—a simple exercise, highly recommended to the reader.

With these definitions, the boundary problem becomes 'universal' (i.e. free of parameters, in particular independent of the atomic number Z):

$$\frac{d^2\chi}{d^2\xi} = \frac{\chi^{3/2}}{\xi^{1/2}}, \quad \text{with } \chi(\xi) \to \begin{cases} 1, & \text{at } \xi \to 0, \\ 0, & \text{at } \xi \to \infty. \end{cases}$$

Unfortunately, this nonlinear differential equation may be solved only numerically, but this is not a big loss: the solution shows as its argument ξ is increased, the function $\chi(\xi)$ goes down from unity at $\xi = 0$ to zero at $\xi \to \infty$ monotonically (and very uneventfully), at distances $\xi \sim 1$. (For example, $\chi(1) \approx 0.4$.) This is why, even without the exact solution, we may conclude that Eq. (****) gives a fair scale of the effective atom's size. This relation shows that the effective radius $r_{\rm TF}(Z)$ decreases with the atomic number Z slowly, as $r_{\rm B}/Z^{1/3}$, and hence, at $Z \gg 1$, is much smaller than $r_0 = r_{\rm B}/Z$—confirming, in particular, our initial assumption. This result is in good agreement (for $Z \gg 1$) with those given by more accurate models, describing the quantized energy spectrum of the atoms.

Problem* 8.22. Use the Thomas–Fermi model, explored in the previous problem, to calculate the total binding energy of a heavy atom. Compare the result with that for the simpler model, in which the Coulomb electron–electron interaction is completely ignored.

Solution: The binding energy of an atom may be calculated as

$$E_{\rm b} = \sum_{Z'=Z}^{0} \mathscr{W}(Z'), \tag{*}$$

where $\mathscr{W}(Z')$ is the work necessary to decrease the atomic number from Z' to $(Z' - 1)$. In order to find $\mathscr{W}(Z')$, let us note that the process of decreasing the atomic number by one may be decomposed into two moves: taking one electron to the electron cloud, and one proton, of charge $+e$, out of the nucleus. Making the electron removal from the Fermi surface, i.e. at the energy equal to zero, the first step of the process requires no work, while the second step requires the work $\mathscr{W}(Z') = -e\phi_{\rm e}(0)$, where $\phi_{\rm e}(r)$ is the part of the potential $\phi(r)$ that is due to electrons only[49]. Using the relations derived in the previous problem, with the replacement $Z \to Z'$, we get

$$\phi_{\rm e}(r) = \phi(r) - \frac{Z'e}{4\pi\varepsilon_0 r} \equiv \frac{Z'e}{4\pi\varepsilon_0 r}\chi(\xi) - \frac{Z'e}{4\pi\varepsilon_0 r} \equiv \frac{Z'e}{4\pi\varepsilon_0}\frac{\chi(\xi) - 1}{r}.$$

Since $\chi(0) - 1 = 0$ by construction, at $r \to 0$ the last fraction tends to $(d\chi/dr)_{r=0}$, and we get

[49] Of course, the removed proton also interacts (and very strongly) with the initial Z' protons in the nucleus. However, our goal is to calculate the *binding* energy, i.e. the sum of energies of the 'assembled' nucleus and individual electrons, all far apart from each other, and that of the 'assembled' atom. At the calculation of such energy, the change of the intrinsic energy of the nucleus has to be ignored.

$$\mathscr{W}(Z') = -e\phi_e(0) = -\frac{Z'e^2}{4\pi\varepsilon_0}\left(\frac{d\chi}{dr}\right)_{r=0} \equiv \frac{Z'e^2}{4\pi\varepsilon_0 r_{TF}(Z')}\left(-\frac{d\chi}{d\xi}\right)_{\xi=0}.$$

Due to the properties of the universal function $\chi(\xi)$, discussed at the end of the previous problem's solution, we may expect the derivative $d\chi/d\xi$ to be negative, with a modulus of the order of 1 at $\xi = 0$. Indeed, a numerical solution of the boundary problem for the function $\chi(\xi)$, spelled out in that solution, yields

$$\left(-\frac{d\chi}{d\xi}\right)_{\xi=0} \approx 1.58807,$$

so that $\mathscr{W}(Z') > 0$ for any Z'. As a result, the total binding energy E_b (*) is positive as well. (This means that the atom's components, after they have been brought far apart from each other, have a higher energy than the initial atom—the necessary condition of atom's stability.) Due to the condition $Z \gg 1$, the sum (*) may be calculated as an integral:

$$E_b = \int_0^Z \mathscr{W}(Z')dZ' = \left(-\frac{d\chi}{d\xi}\right)_{\xi=0}\int_0^Z \frac{Z'e^2}{4\pi\varepsilon_0 r_{TF}(Z')}dZ'$$
$$\equiv \frac{1}{b}\left(-\frac{d\chi}{d\xi}\right)_{\xi=0}\frac{e^2}{4\pi\varepsilon_0 r_B}\int_0^Z Z'^{4/3}dZ' = \frac{3}{7b}\left(-\frac{d\chi}{d\xi}\right)_{\xi=0} Z^{7/3}\frac{e^2}{4\pi\varepsilon_0 r_B}.$$

But the last fraction is just the Hartree energy E_H,[50] so that we finally get

$$E_b = \frac{3}{7b}\left(-\frac{d\chi}{d\xi}\right)_{\xi=0} Z^{7/3}E_H \approx 0.7688\ Z^{7/3}E_H \gg E_H, \quad \text{for } Z \gg 1.$$

Note the very nontrivial scaling of the energy with the atomic number Z.

Now let us consider a simpler model[51], in which the electron–electron interaction is completely ignored, so that $\phi(\mathbf{r})$ is the unscreened potential of the nucleus,

$$\phi(\mathbf{r}) = \frac{Ze}{4\pi\varepsilon_0 r},$$

for all r. Here we still may use Eq. (**) of the model solution of the previous problem,

$$\frac{\hbar^2}{2m_e}[3\pi^2 n(\mathbf{r})]^{2/3} = \frac{\hbar^2 k_F^2(\mathbf{r})}{2m_e}, \qquad (**)$$

but since the largest energy μ of the electrons in this case is not known in advance, the local value of $k_F(\mathbf{r})$ should be found from Eq. (*) of the previous problem, with $\varepsilon = \mu$:[52]

[50] See, e.g. Eq. (1.13).

[51] Very unfortunately, this model is sometimes called 'statistical'—why?

[52] Just as in the previous problem, we consider the electron gas to be completely degenerate, $T = 0$.

$$\frac{\hbar^2 k_F^2}{2m_e} = \mu + e\phi(\mathbf{r}) \equiv \mu + \frac{Ze^2}{4\pi\varepsilon_0 r}. \qquad (***)$$

In equilibrium, the energy μ cannot depend on $\mathbf{r}$—otherwise it would be advantageous for some electrons to move to the locations with lower μ. (This constant is exactly what is called the *chemical potential* in statistical physics, and the *local Fermi level* in electronic engineering.) Also, in order to have electrons localized near the nucleus, μ cannot be positive (relative to its value at $r \to \infty$), so that p_F (and hence the electron density) turn to zero at some finite radius r_{ef}, defined as

$$\frac{Ze^2}{4\pi\varepsilon_0 r_{ef}} \equiv -\mu \geqslant 0,$$

and playing the role of atom's radius. With this notation, Eqs. (**) and (***) yield

$$n(\mathbf{r}) = \frac{1}{3\pi^2}\left[\frac{m_e e^2 Z}{2\pi\varepsilon_0\hbar^2}\left(\frac{1}{r} - \frac{1}{r_{ef}}\right)\right]^{3/2}.$$

Now we may calculate r_{ef} (and hence μ) by requiring the atom as a whole to be neutral, i.e. the number of electrons to be equal to Z:[53]

$$\int_{r\leqslant R} n(\mathbf{r})d^3r = Z.$$

Carrying out the integration, we get

$$\int_{r\leqslant R} n(\mathbf{r})d^3r = 4\pi\int_0^R r^2 dr\, n(\mathbf{r}) = \frac{4}{3\pi}\left(\frac{m_e e^2 Z}{2\pi\varepsilon_0\hbar^2}\right)^{3/2}\int_0^R\left(\frac{1}{r} - \frac{1}{r_{ef}}\right)^{3/2} r^2 dr$$
$$= \frac{4}{3\pi}\left(\frac{m_e e^2 Z r_{ef}}{2\pi\varepsilon_0\hbar^2}\right)^{3/2}\int_0^1 (1-\xi)^{3/2}\xi^{1/2}d\xi.$$

This dimensionless integral may be recast into a sum of elementary integrals by the substitution $\xi \equiv \sin^2\alpha$ and then using Eqs. (A.18*d*) and (A.19). The final result is $\pi/16$, so that the electron counting yields

$$Z = \frac{4}{3\pi}\left(\frac{m_e e^2 Z r_{ef}}{2\pi\varepsilon_0\hbar^2}\right)^{3/2}\frac{\pi}{16}, \qquad (****)$$

giving

$$r_{ef} = (18)^{1/3}\frac{4\pi\varepsilon_0\hbar^2}{e^2 m_e}\frac{1}{Z^{1/3}} \equiv (18)^{1/3}\frac{r_B}{Z^{1/3}} \approx 2.621\frac{r_B}{Z^{1/3}}.$$

[53] Evidently, in contrast to the Thomas-Fermi approximation, this rudimentary model is not self-consistent, because it implies $\phi(r_{ef}) = Ze/4\pi\varepsilon_0 r_{ef} \neq 0$, while the potential of a neutral, spherically-symmetrical atom at its effective surface should vanish.

So, for the effective radius of the atom, this simple model gives the same order of magnitude as r_{TF} in the Thomas–Fermi model, though with a significantly larger numerical coefficient. The fact that $r_{ef} \sim r_{TF} \gg r_0 = r_B/Z$ shows that, at least in this system, the implicit interaction of electrons via the Pauli principle, taken into account in this model, is more important that their explicit, Coulomb interaction.

Now let us calculate the binding energy (*) within this simple model. In order to avoid the calculation of the electron potential $\phi_e(0)$ felt by the nuclear charges, the partial work $\mathscr{W}(Z')$ may be calculated differently than for the Thomas–Fermi model. Namely, let us calculate the radius $r_{ef}(Z')$ and $\mu(Z')$ of an ion, with Z' electrons, but with the nuclear charge Q still equal to Z. Reviewing the above calculations, we see that this may be done by replacing Z on the left-hand side of Eq. (****) with Z':

$$\frac{4}{3\pi}\left[\frac{m_e e^2 Z r_{ef}(Z')}{2\pi\varepsilon_0\hbar^2}\right]^{3/2}\frac{\pi}{16} = Z', \quad \text{giving } r_{ef}(Z') = (18)^{1/3}\frac{Z'^{2/3}}{Z}r_B,$$

and then calculating μ as

$$-\mu(Z') = \frac{Ze^2}{4\pi\varepsilon_0 r_{ef}(Z')} = \frac{1}{(18)^{1/3}}\frac{Ze^2}{4\pi\varepsilon_0 r_B}\frac{Z}{Z'^{2/3}} \equiv \frac{E_H}{(18)^{1/3}}\frac{Z^2}{Z'^{2/3}}.$$

The work $\mathscr{W}(Z')$ necessary for the removal of an additional electron from the ion to infinity is $-\mu(Z')$, so that, replacing the sum (*) with the corresponding integral, we get

$$E_b = -\int_0^Z \mu(Z')dZ' = \frac{E_H}{(18)^{1/3}}Z^2\int_0^Z \frac{dZ'}{Z'^{2/3}} = \frac{3}{(18)^{1/3}}Z^{7/3}E_H \approx 1.145\, Z^{7/3}E_H.$$

Very naturally, this value is higher than that calculated in the Thomas–Fermi model, because in the simple model each electron is attracted to the nucleus by its Coulomb field unscreened by other electrons, making their interaction stronger. Note, however, that the difference is not too large—just about 50%.

***Problem* 8.23.** A system of three similar but distinguishable spin-½ particles is described by the Heisenberg Hamiltonian (cf. problems 8.6 and 8.7):

$$\hat{H} = -J(\hat{\mathbf{s}}_1 \cdot \hat{\mathbf{s}}_2 + \hat{\mathbf{s}}_2 \cdot \hat{\mathbf{s}}_3 + \hat{\mathbf{s}}_3 \cdot \hat{\mathbf{s}}_1),$$

where J is the spin interaction constant. Find the stationary states and eigenenergies of this system, and give an interpretation of your results.

Solution: The uncoupled-representation z-basis of the system has $2^3 = 8$ states corresponding to ↑ and ↓ orientations of each spin. Let us see what is the result of the action of the system's Hamiltonian operator on each of these states. Thanks to the intermediate results of the model solution of problem 8.3 for two spins-½,[54]

[54] I am again using the standard shorthand notation, in which the spin's number is coded with its position in the ket-vector.

$$\hat{\mathbf{s}}_1 \cdot \hat{\mathbf{s}}_2 \left|\uparrow\uparrow\right\rangle = \left(\frac{\hbar}{2}\right)^2 \left|\uparrow\uparrow\right\rangle,$$
$$\hat{\mathbf{s}}_1 \cdot \hat{\mathbf{s}}_2 \left|\downarrow\downarrow\right\rangle = \left(\frac{\hbar}{2}\right)^2 \left|\downarrow\downarrow\right\rangle, \qquad (*)$$

$$\hat{\mathbf{s}}_1 \cdot \hat{\mathbf{s}}_2 \left|\uparrow\downarrow\right\rangle = \left(\frac{\hbar}{2}\right)^2 (2\left|\downarrow\uparrow\right\rangle - \left|\uparrow\downarrow\right\rangle),$$
$$\hat{\mathbf{s}}_1 \cdot \hat{\mathbf{s}}_2 \left|\downarrow\uparrow\right\rangle = \left(\frac{\hbar}{2}\right)^2 (2\left|\uparrow\downarrow\right\rangle - \left|\downarrow\uparrow\right\rangle), \qquad (**)$$

this is easy to do. For example, since the operator $\hat{\mathbf{s}}_1 \cdot \hat{\mathbf{s}}_2$ does not affect the state of the third spin, Eqs. (*) yield

$$\hat{\mathbf{s}}_1 \cdot \hat{\mathbf{s}}_2 \left|\uparrow\uparrow\uparrow\right\rangle = \left(\frac{\hbar}{2}\right)^2 \left|\uparrow\uparrow\uparrow\right\rangle, \quad \hat{\mathbf{s}}_1 \cdot \hat{\mathbf{s}}_2 \left|\downarrow\downarrow\downarrow\right\rangle = \left(\frac{\hbar}{2}\right)^2 \left|\downarrow\downarrow\downarrow\right\rangle.$$

These results evidently do not depend on the indices of the operator product components, so that their summation for all three index combinations yields

$$(\hat{\mathbf{s}}_1 \cdot \hat{\mathbf{s}}_2 + \hat{\mathbf{s}}_2 \cdot \hat{\mathbf{s}}_3 + \hat{\mathbf{s}}_3 \cdot \hat{\mathbf{s}}_1) \left|\uparrow\uparrow\uparrow\right\rangle = 3\left(\frac{\hbar}{2}\right)^2 \left|\uparrow\uparrow\uparrow\right\rangle,$$
$$(\hat{\mathbf{s}}_1 \cdot \hat{\mathbf{s}}_2 + \hat{\mathbf{s}}_2 \cdot \hat{\mathbf{s}}_3 + \hat{\mathbf{s}}_3 \cdot \hat{\mathbf{s}}_1) \left|\downarrow\downarrow\downarrow\right\rangle = 3\left(\frac{\hbar}{2}\right)^2 \left|\downarrow\downarrow\downarrow\right\rangle.$$

This means that we have already found two stationary states of the system, with the same energy

$$E_{\uparrow\uparrow\uparrow} = E_{\downarrow\downarrow\downarrow} = E_0 \equiv -3J\left(\frac{\hbar}{2}\right)^2. \qquad (***)$$

Thanks to Eqs. (**), the calculations for the 'mixed' spin states are only slightly more complicated. For example, let us calculate

$$(\hat{\mathbf{s}}_1 \cdot \hat{\mathbf{s}}_2 + \hat{\mathbf{s}}_2 \cdot \hat{\mathbf{s}}_3 + \hat{\mathbf{s}}_3 \cdot \hat{\mathbf{s}}_1) \left|\downarrow\uparrow\uparrow\right\rangle \equiv \hat{\mathbf{s}}_1 \cdot \hat{\mathbf{s}}_2 \left|\downarrow\uparrow\uparrow\right\rangle + \hat{\mathbf{s}}_2 \cdot \hat{\mathbf{s}}_3 \left|\downarrow\uparrow\uparrow\right\rangle$$
$$+ \hat{\mathbf{s}}_3 \cdot \hat{\mathbf{s}}_1 \left|\downarrow\uparrow\uparrow\right\rangle.$$

Applying the second of Eqs. (**) to the first term on the right-hand side, its analog for indices 1 and 3 to the third term, and the analog of the first of Eqs. (*) for indices 2 and 3 to the second term, we get

$$\begin{aligned}
&(\hat{\mathbf{s}}_1 \cdot \hat{\mathbf{s}}_2 + \hat{\mathbf{s}}_2 \cdot \hat{\mathbf{s}}_3 + \hat{\mathbf{s}}_3 \cdot \hat{\mathbf{s}}_1) \left|\downarrow\uparrow\uparrow\right\rangle \\
&= \left(\frac{\hbar}{2}\right)^2 [(2\left|\uparrow\downarrow\uparrow\right\rangle - \left|\downarrow\uparrow\uparrow\right\rangle) + \left|\downarrow\uparrow\uparrow\right\rangle + (2\left|\uparrow\uparrow\downarrow\right\rangle - \left|\downarrow\uparrow\uparrow\right\rangle)] \\
&\equiv \left(\frac{\hbar}{2}\right)^2 (2\left|\uparrow\downarrow\uparrow\right\rangle + 2\left|\uparrow\uparrow\downarrow\right\rangle - \left|\downarrow\uparrow\uparrow\right\rangle).
\end{aligned} \qquad (****)$$

It is instrumental to formulate this result verbally: the action of the sum of all spin operator products on a ket with just one spin down yields, besides the multiplication by $(\hbar/2)^2$, the ket of same state with the minus sign, plus two other possible kets with one spin down, each multiplied by a factor of 2. This rule, which may be readily checked to be valid for any state with one spin down, means that the three states of this group are only coupled to each other, so that their interaction, in the units of $-J(\hbar/2)^2$, may be described by the following 3×3 matrix:

$$\begin{pmatrix} -1 & 2 & 2 \\ 2 & -1 & 2 \\ 2 & 2 & -1 \end{pmatrix}. \qquad (****)$$

(Due to the symmetry of the matrix, it is not even important what exactly states correspond to in its rows/columns.) The eigenvalues of this matrix may be found from the characteristic equation

$$\begin{vmatrix} -1-\lambda & 2 & 2 \\ 2 & -1-\lambda & 2 \\ 2 & 2 & -1-\lambda \end{vmatrix} = 0,$$

$$\text{giving } f(\mu) \equiv \mu^3 - 12\mu - 16 = 0, \quad \text{where } \mu \equiv \lambda + 1.$$

One root $\mu_+ = -2$, i.e. $\lambda_+ \equiv \mu_+ - 1 = -3$, of this cubic equation may be readily guessed[55]; after that, dividing the polynomial $f(\mu)$ by $(\mu - \mu_+) \equiv (\mu + 2)$, we get the quadratic polynomial $g(\mu) = \mu^2 - 2\mu - 8$, whose roots are easy to calculate: $\mu_- = \mu_+ = -2$, $\mu_0 = +4$, so that $\lambda_+ = \lambda_- = -3$, $\lambda_0 = +3$. As the result, the eigenenergies corresponding to the one-spin-down states are $E_+ = E_- = -E_0$ and E_0, where E_0 is given by Eq. (***).

Due to the spin up-down symmetry of the Hamiltonian, it is obvious (and if you like may be checked by the absolutely similar calculation) that the three states with one spin up also are coupled to each other, and have similar eigenenergies: two eigenstates with energy $-E_0$, and one state with energy $+E_0$. To summarize, the system has just two energy levels, each of them four-degenerate: one level with energy $+E_0$, corresponding to the two states with all spins aligned (in any direction), plus two linear superposition of three states with one spin misaligned, while the level with energy $-E_0$ corresponds to four other linear superpositions of the states with one spin misaligned, each superposition consisting of the same number of spins-up and spins-down.

The misaligned-spin superpositions may be readily found by plugging the found values of λ into the linear system of equations with matrix (****), and then the normalization—just as was repeatedly done in examples and exercises of chapter 4. The result for the one-spin-down states may be represented in the form

[55] If you do not like guessing, it is useful to have a look at the numerical plot of function $f(\mu)$, with touches the horizontal axis (i.e. has a double root) at point $\mu = -2$. Another (much harder) option is to use the cumbersome formulas for the roots of an arbitrary cubic equation.

$$|0\rangle = \frac{1}{\sqrt{3}}(|\downarrow\uparrow\uparrow\rangle + |\uparrow\downarrow\uparrow\rangle + |\uparrow\uparrow\downarrow\rangle),$$

$$|\pm\rangle = \frac{1}{\sqrt{3}}\left(|\downarrow\uparrow\uparrow\rangle + \frac{-1 \pm i\sqrt{3}}{2}|\uparrow\downarrow\uparrow\rangle + \frac{-1 \mp i\sqrt{3}}{2}|\uparrow\uparrow\downarrow\rangle\right);$$

the one-spin-up eigenstates are similar. The physical meaning of these three eigenstates becomes more transparent from rewriting them all in the similar wave form:

$$|\alpha\rangle = \frac{1}{\sqrt{3}}\left(\exp\left\{i\alpha\frac{0\cdot 2\pi}{3}\right\}|\downarrow\uparrow\uparrow\rangle + \exp\left\{+i\alpha\frac{2\pi}{3}\right\}|\uparrow\downarrow\uparrow\rangle + \exp\left\{-i\alpha\frac{2\pi}{3}\right\}|\uparrow\uparrow\downarrow\rangle\right), \quad \text{with } \alpha = 0, \pm 1.$$

If we interpret the system's Hamiltonian as a model of interaction of spins located at three equidistant positions on a circle, for example at angles $\varphi = 0$, and $\pm\Delta\varphi = \pm 2\pi/3$, each state α is just a traveling wave of propagation of the spin-down 'excitation', with the equal phase shifts $\alpha\Delta\varphi$ at each excitation's jump to the adjacent site, absolutely similar to that of single-particle Bloch waves—see section 2.7.[56] In this picture, the above eigenvalues of the normalized wave number α (to whom any multiple of 3 may be added without changing the solution) result from the natural cycling-boundary condition $\exp\{i\alpha 2\pi\} = 1$. Such Bloch waves of spin orientations are called either *spin waves* or *magnon waves*[57].

The spin-wave interpretation explains why the state with $\alpha = 0$ has the energy $(+E_0)$,[58] different from that $(-E_0)$ of the two states with $\alpha = \pm 1$: in usual wave systems, with positive kinetic energy, the wave energy grows with the square of the wave number (modulo its period, in our current case $\Delta\alpha = 3$). Note, however, that in the Heisenberg model (and at real spin interactions in solids) the sign of J may be both positive and negative. Only the case $J > 0$, with the ground-state energy $E_0 < 0$, corresponds to classical wave systems. (It also describes the effect of spontaneous spin alignment, the *ferromagnetism*.) For the case $J < 0$,[59] the lowest energy of the

[56] Such constant phase shift between the adjacent sites is the main feature of any traveling waves in periodic structures, including classical systems—see, e.g. *Part CM* section 6.3.

[57] The fact that such waves, possibly with numerous flipped spins, are described by the Heisenberg model is the essence of the so-called *Bethe Anzatz*—named after H Bethe who first suggested this idea in 1931. This fact enables exact analysis of some models for systems of $N \gg 1$ spins, for which the direct approach used above is not practicable. For an introduction to the Bethe Anzatz, I can recommend, for example, either a popular article by M Batchelor [2], or a more detailed three-part review by M Karbach *et al*, available online at http://www.phys.uri.edu/gerhard/introbethe.html.

[58] If the fact that this energy is equal to that of the all-spin-aligned states looks surprising to you, please revisit the discussion of the ^{4}He atom in section 8.2 of the lecture notes (and also the solution of problem 8.3), where (in the absence of the external field) all triplet states (either factorable or entangled) of the two-spin system also have the same energy.

[59] A negative J may result from the exchange interaction of electrons of adjacent atoms.

system is $-E_0 < 0$, corresponding to the alternating-spin ground states. This is the Heisenberg-model description of the effect of *antiferromagnetism*.

One more (parenthetic) remark: since the two spin-misaligned stationary states with $\alpha = \pm 1$, and the same number of spins-up, for example

$$|\pm\rangle = \frac{1}{\sqrt{3}}(|\downarrow\uparrow\uparrow\rangle + e^{\pm i2\pi/3}|\uparrow\downarrow\uparrow\rangle + e^{\mp i2\pi/3}|\uparrow\uparrow\downarrow\rangle),$$

correspond to the same energy $-E_0$, their linear superposition is also a legitimate stationary state. Of such superpositions, notable are the combinations that exclude one of three elementary (factorable) states, for example,

$$|1\rangle \equiv \frac{-i}{\sqrt{2}}(|+\rangle - |-\rangle) = \frac{-i}{\sqrt{6}}\left(2i\sin\frac{2\pi}{3}|\uparrow\downarrow\uparrow\rangle - 2i\sin\frac{2\pi}{3}|\uparrow\uparrow\downarrow\rangle\right)$$
$$\equiv \frac{1}{\sqrt{2}}(|\uparrow\downarrow\uparrow\rangle - |\uparrow\uparrow\downarrow\rangle).$$

While the initial states, $\pm$, may be interpreted as *traveling* spin waves, state 1 represents a *standing* spin wave. On the other hand, we may write

$$|1\rangle = |\uparrow\rangle \otimes \frac{1}{\sqrt{2}}(|\downarrow\uparrow\rangle - |\uparrow\downarrow\rangle),$$

so that this state is based on the singlet of the last two spins. Evidently, there are three such states (with the same energy), which differ only by the number of the particle in the non-entangled spin state, but since they all are linear superpositions of the traveling-wave states $\pm$, only two of them are linearly independent.

***Problem* 8.24.** For a system of three distinguishable spins-½, find the common eigenstates and of the operators $\hat{S}_z$ and $\hat{S}^2$, where

$$\hat{\mathbf{S}} \equiv \hat{\mathbf{s}}_1 + \hat{\mathbf{s}}_2 + \hat{\mathbf{s}}_3$$

is the vector operator of the total spin of the system. Do the corresponding quantum numbers S and M_S obey Eqs. (8.48) of the lecture notes?

Solution: Since the partial operators $\hat{\mathbf{s}}_{1,2,3}$ are defined in different Hilbert spaces, and hence commute, we may transform the operator $\hat{S}^2$ just as in the model solution of problem 8.3:

$$\hat{S}^2 \equiv (\hat{\mathbf{s}}_1 + \hat{\mathbf{s}}_2 + \hat{\mathbf{s}}_3)^2 \equiv \left(\hat{s}_1^2 + \hat{s}_2^2 + \hat{s}_3^2\right) + 2(\hat{\mathbf{s}}_1\cdot\hat{\mathbf{s}}_2 + \hat{\mathbf{s}}_2\cdot\hat{\mathbf{s}}_3 + \hat{\mathbf{s}}_3\cdot\hat{\mathbf{s}}_1). \qquad (*)$$

The eigenstates and the corresponding eigenvalues of the operator in the second parentheses of the last expression have been found in the solution of the previous problem. They all (as well as an arbitrary spin state of this system) may be represented[60] as linear superpositions of $2^3 = 8$ simple (factorable) states, with ket-vectors (in the z-basis for each spin),

$$|\uparrow\uparrow\uparrow\rangle, \quad |\downarrow\uparrow\uparrow\rangle, \quad |\uparrow\downarrow\uparrow\rangle, \quad \text{etc.} \tag{**}$$

forming a full uncoupled-representation basis of the system. Since each of the first three operators, $\hat{s}^2_{1,2,3}$, in the first parentheses of Eq. (*) acts only on the corresponding component of these elementary kets, they all obey the same equality as in the case of a single spin-½ (see chapter 4):

$$\hat{s}^2_{1,2,3} = \left(\hat{s}^2_x + \hat{s}^2_y + \hat{s}^2_z\right)_{1,2,3} = 3\left(\frac{\hbar}{2}\right)^2 \hat{I}.$$

This means that all eigenstates found in the solution of the previous problem are also eigenstates of the operator in the first parentheses of Eq. (*), with the same eigenvalue, $9(\hbar/2)^2$. As a result, these states are eigenstates of the total operator $\hat{S}^2$ as well, with the following eigenvalues[61]:

$$\langle S^2\rangle = \left(\frac{\hbar}{2}\right)^2 \times \begin{cases} (9 + 2\cdot 3) \equiv 15, & \text{for two spin - aligned states,} \\ & \text{and two misaligned states with } \alpha = 0, \\ (9 - 2\cdot 3) \equiv 3, & \text{for four spin - misaligned states with } \alpha = \pm 2\pi/3. \end{cases}$$

Now proceeding to the operator

$$\hat{S}_z \equiv \hat{s}_{1z} + \hat{s}_{2z} + \hat{s}_{3z},$$

we may note that each of the component operators in the right-hand side of this equality acts only on the corresponding spin. As a result, according to Eq. (4.128) of the lecture notes, acting on any of the factorable basis states (**), the net operator results in the same state, multiplied by factor $\hbar M_S$, where[62]

$$M_S \equiv m_1 + m_2 + m_3.$$

For example, for the state with $m_1 = +½$, $m_2 = -½$, and $m_3 = +½$ (and hence $M_S = +½$):

$$\begin{aligned}\hat{S}_z|\uparrow\downarrow\uparrow\rangle &\equiv \hat{s}_{1z}|\uparrow\downarrow\uparrow\rangle + \hat{s}_{2z}|\uparrow\downarrow\uparrow\rangle + \hat{s}_{3z}|\uparrow\downarrow\uparrow\rangle \\ &= \frac{\hbar}{2}|\uparrow\downarrow\uparrow\rangle - \frac{\hbar}{2}|\uparrow\downarrow\uparrow\rangle + \frac{\hbar}{2}|\uparrow\downarrow\uparrow\rangle \equiv +\frac{\hbar}{2}|\uparrow\downarrow\uparrow\rangle.\end{aligned}$$

---

[60] Just as in the case of two spins, discussed at the end of section 8.2 of the lecture notes, such representation is called *uncoupled*.

[61] The (formally, correct) averaging sign is used in the left-hand side of this relation to distinguish it from the square of the quantum number S—see below.

[62] Here, as in the solution of problem 8.4, index s in the magnetic quantum numbers m_s of the component spins has been dropped.

Since each eigenstate of the operator $\hat{S}^2$ is a linear superposition of such elementary states with the same value of M_S (the same number of up- and down-spins), they are also the eigenstates of the operator $\hat{S}_z$, with the following eigenvalues:

$$S_z = \hbar M_S, \quad \text{with } M_S = \begin{cases} +3/2, & \text{for the aligned state with all spins up,} \\ +1/2, & \text{for all misaligned states with two spins up,} \\ -1/2, & \text{for all misaligned states with two spins down,} \\ -3/2, & \text{for the aligned state with all spins down.} \end{cases}$$

Comparing these results with the general Eqs. (8.48) of the lecture notes, we see that they all fit, in particular satisfying the relations

$$\langle S^2 \rangle = \hbar^2 S(S+1), \quad \text{and} \quad -S \leqslant M_S \leqslant +S,$$

provided that we prescribe to the eigenstates (besides the values of M_S specified above) the following values of the quantum number S:

$$S = \begin{cases} 3/2, & \text{for two spin - aligned states, and two misaligned states with } \alpha = 0, \\ 1/2, & \text{for four spin - misaligned states, with } \alpha = \pm 2\pi/3. \end{cases}$$

I believe that these results, together with those for the two-spin system (discussed in section 8.2 of the lecture notes and the solution of problem 8.3) illuminate the physical sense of the net quantum numbers S and M_S, and in particular clarify why these numbers may take not only integer, but also half-integer values.

***Problem* 8.25** Explore basic properties of the Heisenberg model (which was the subject of problems 8.6, 8.7 and 8.23), for a 1D chain of N spins-½:

$$\hat{H} = -J \sum_{\{j,j'\}} \hat{\mathbf{s}}_j \cdot \hat{\mathbf{s}}_{j'} - \gamma \mathscr{B} \cdot \sum_j \hat{\mathbf{s}}_j, \quad \text{with } J > 0,$$

where the summation is over all N spins, with the symbol $\{j, j'\}$ meaning that the first sum is only over the adjacent spin pairs. In particular, find the ground state of the system and its lowest excited states in the absence of external magnetic field $\mathscr{B}$, and also the dependence of their energies on the field.

Hint: For the sake of simplicity, you may assume that the first sum includes the term $\hat{\mathbf{s}}_N \cdot \hat{\mathbf{s}}_1$ as well. (Physically, this means that the chain is bent into a closed loop[63].)

Solution: This problem is a natural generalization of problems 8.6, 8.7, and 8.23, so it may be solved fast using intermediate results of their model solutions. First of all, it is evident from those solutions that at least at $\mathscr{B} = 0$, the ground state of the

[63] Note that for dissipative spin systems, differences between low-energy excitations of open-end and closed-end 1D chains may be substantial even in the limit $N \to \infty$—see, e.g. *Part SM* section 4.5. However, for our Hamiltonian (and hence dissipation-free) system, the differences are relatively small.

system with $J > 0$ (i.e. the Heisenberg model of *ferromagnetism*[64]) is one of two simple (factorable) products of N aligned spin states:

$$|g_+\rangle = |\uparrow \uparrow \dots \uparrow \uparrow\rangle, \quad |g_-\rangle = |\downarrow \downarrow \dots \downarrow \downarrow\rangle. \qquad (*)$$

Indeed, acting on any of these ket-vectors by the first term of our Hamiltonian,

$$\hat{H}\,|_{\mathscr{B}=0} = -J\sum_{\{j,j'\}} \hat{\mathbf{s}}_j \cdot \hat{\mathbf{s}}_{j'} \equiv -J(\hat{\mathbf{s}}_1 \cdot \hat{\mathbf{s}}_2 + \hat{\mathbf{s}}_1 \cdot \hat{\mathbf{s}}_2 + \dots + \hat{\mathbf{s}}_{N-1} \cdot \hat{\mathbf{s}}_N + \hat{\mathbf{s}}_N \cdot \hat{\mathbf{s}}_1),$$

and applying Eq. (*) of the model solution of problem 8.23 to each of N pairs of spins, we get the same ket multiplied by the factor (which is, by definition, the state's energy)

$$E_+ = E_- = E_g \equiv -JN\left(\frac{\hbar}{2}\right)^2.$$

Moreover, as is physically clear from the expressions for kets $|0\rangle$ and $|\pm\rangle$ in the same solution, the lowest excitations of the system should be the Bloch-wave-like linear superpositions,

$$|\alpha\rangle_\pm = C\sum_j e^{i\alpha j}\,|k + j\rangle_\pm, \qquad (**)$$

of the kets $|k\rangle_\pm$ corresponding to the reversal of just one spin, in the kth position:

$$|k\rangle_+ \equiv \left|\underbrace{\uparrow \uparrow \dots \uparrow}_{k-1} \downarrow \underbrace{\uparrow \uparrow \dots \uparrow}_{N-k}\right\rangle, \quad |k\rangle_- \equiv \left|\underbrace{\downarrow \downarrow \dots \downarrow}_{k-1} \uparrow \underbrace{\downarrow \downarrow \dots \downarrow}_{N-k}\right\rangle,$$

where the real parameter α (physically, the normalized wave number[65] of the spin wave) satisfies the cyclic condition

$$e^{iN\alpha} = 1, \qquad (***)$$

and the summation in Eq. (**) is over all N spins of the ring—sequential, but starting from any position (counting the positions modulo N, i.e. identifying $j = j_0 + N$ with j_0). Indeed, applying Eqs. (*) and (**) of the model solution of problem 8.23 to each pair of spins, we get similar results for any of these misaligned spin orientations:

$$\hat{H}|_{\mathscr{B}=0}|k\rangle_\pm = -J\left(\frac{\hbar}{2}\right)^2[(N-4)\,|k\rangle_\pm + 2\,|k-1\rangle_\pm + 2\,|k+1\rangle_\pm].$$

As a result, acting by the same Hamiltonian on any of the kets (**), and then changing the summation indices in the last two partial sums from j to $j' = j \pm 1$, we get

[64] Solutions of the similar Heisenberg model of *antiferromagnetism*, with $J < 0$, are more complex, with the ground state representing alternating spin configurations.
[65] Cf. Eq. (2.193) and/or *Part CM* Eq. (6.26), and the accompanying discussions.

$$\hat{H}\,|_{\mathscr{B}=0}|\,\alpha\rangle_{\pm} = -J\left(\frac{\hbar}{2}\right)^2 C\Bigg[(N-4)\sum_j e^{i\alpha j}\,|k+j\rangle_{\pm}$$

$$+2\sum_j e^{i\alpha j}\,|k-1+j\rangle_{\pm} + 2\sum_j e^{i\alpha j}\,|k+1+j\rangle_{\pm}\Bigg]$$

$$\equiv -J\left(\frac{\hbar}{2}\right)^2 C\Bigg[(N-4)\sum_{j'} e^{i\alpha j'}\,|k+j'\rangle_{\pm}$$

$$+2e^{i\alpha}\sum_{j'} e^{i\alpha j'}\,|k+j'\rangle_{\pm} + 2e^{-i\alpha}\sum_{j'} e^{i\alpha j'}\,|k+j'\rangle_{\pm}\Bigg]$$

$$= -J\left(\frac{\hbar}{2}\right)^2 [(N-4) + 2e^{-i\alpha} + 2e^{+i\alpha}]|\alpha\rangle_{\pm}.$$

Hence the states (**) are indeed the stationary states of our system, with their energies independent (at $\mathscr{B} = 0$ only!) of the misaligned spin's orientation:

$$E_{\pm}(\alpha) = -J\left(\frac{\hbar}{2}\right)^2 [(N-4) + 2e^{+i\alpha} + 2e^{-i\alpha}] \equiv E_{\mathrm{g}} + 4J\left(\frac{\hbar}{2}\right)^2 (1 - \cos\alpha).$$

(As a sanity check, for the particular case $N = 3$, when Eq. (***) gives, modulo 2π, $\alpha = 0, \pm\sqrt{3}/2$, i.e. $\cos\alpha = +1, -\frac{1}{2}$, we recover the results of problem 8.23: $E_{\mathrm{g}} = -3J(\hbar/2)^2$, $E_{\pm}(0) = -3J(\hbar/2)^2 + 4J(\hbar/2)^2 \cdot 0 \equiv -3J(\hbar/2)^2$, $E_{\pm}(\pm\sqrt{3}/2) = -3J(\hbar/2)^2 + 4J(\hbar/2)^2 \cdot (1 + \frac{1}{2}) \equiv +3J(\hbar/2)^2$.)

The last term of this result evidently describes the spin wave excitation energy. For our (ferromagnetic) case $J > 0$, this energy is positive for any α.[66] The excitation energy tends to zero at $\alpha \to 0$, so that the triplet-like solution with $\alpha = 0$ has the same energy as the ground state with all spins aligned. (If this fact looks counter-intuitive, please revisit the discussion of the two-spin triplet in section 8.2 of the lecture notes, and in the solution of problem 8.3.) So, the first actual excitation (with $E(\alpha) > E_{\mathrm{g}}$) corresponds to the two lowest nonvanishing values of α (modulo 2π) of the allowed set (***):

$$E_{\pm}\left(\alpha = \frac{2\pi}{N}\right) - E_{\mathrm{g}} = E_{\pm}\left(\alpha = -\frac{2\pi}{N}\right) - E_{\mathrm{g}} = 4J\left(\frac{\hbar}{2}\right)^2\left(1 - \cos\frac{2\pi}{N}\right).$$

Note that if $N \gg 1$, the energy, and hence frequency, of such low-energy excitations is proportional to the square of α, i.e. of the normalized wave number $k = \alpha/d$ (where d is the spatial period of the natural geometric model of the chain),

$$\hbar\omega \equiv E(\alpha) - E_{\mathrm{g}} \approx 2J\left(\frac{\hbar}{2}\right)^2 \alpha^2 \equiv 2J\left(\frac{\hbar}{2}\right)^2 d^2k^2,$$

[66] For the antiferromagnetic chain ($J < 0$) the term is negative, showing that the spin-aligned states (*) are unstable with respect to a single spin reversal, and hence they are *not* ground states of the system.

meaning that the low-energy spin waves[67] have a quadratic dispersion law $\omega(k)$, similar to that of free non-relativistic particles (see Eq. (1.30) of the lecture notes), but very much different from that of the acoustic waves[68] and the electromagnetic waves in free space[69].

Proceeding to the effects of non-zero magnetic field $\mathscr{B}$, as we already know from the solutions of problems 8.6 and 8.25, the states described by Eqs. (*) and (**) are also eigenstates of the sum $\Sigma_j \hat{s}_{jz}$ (where the z-axis is directed along the field), with the eigenvalues, respectively, $\pm N(\hbar/2)$ and $\pm(N - 2)(\hbar/2)$. As a result, we may immediately write their energies in the field[70]:

$$E_{\pm} = -JN\left(\frac{\hbar}{2}\right)^2 \mp \gamma\mathscr{B}N\frac{\hbar}{2},$$

$$E_{\pm}(\alpha) = -JN\left(\frac{\hbar}{2}\right)^2 + 4J\left(\frac{\hbar}{2}\right)^2(1 - \cos\alpha) \mp \gamma\mathscr{B}(N - 2)\frac{\hbar}{2}.$$

However, at arbitrary $\mathscr{B}$ these (exact!) results do not allow us to say what is the ground state of the system, and what is the lowest excitation energy—not only because the relations between these energies depend on the system's parameters, but also because at $\mathscr{B} \neq 0$, more complex spin waves (with more than one spin inverted) may have comparable, in particular lower, energies.

***Problem* 8.26.** Compose the simplest model Hamiltonians, in terms of the second quantization formalism, for systems of indistinguishable particles moving in the following systems:

(i) two weakly coupled potential wells, with on-site particle-pair interactions (giving additional energy J per each pair of particles in the same potential well), and
(ii) a periodic 1D potential, with the same particle-pair interactions, in the tight-binding limit.

Solutions:

(i) Using Eq. (8.99) of the lecture notes as the baseline, we need to add a term describing on-site interactions. For bosons, this may be done, for example, by keeping the diagonal elements in Eq. (8.117) with all u_{jjjj} equal to J. As a result, the simplest model that satisfies the assignment requirements is

$$\hat{H} = \varepsilon_1\hat{a}_1^\dagger\hat{a}_1 + \frac{J}{2}\hat{a}_1^\dagger\hat{a}_1^\dagger\hat{a}_1\hat{a}_1 + \varepsilon_2\hat{a}_2^\dagger\hat{a}_2 + \frac{J}{2}\hat{a}_2^\dagger\hat{a}_2^\dagger\hat{a}_2\hat{a}_2 + t\left(\hat{a}_1^\dagger\hat{a}_2 + \text{h.c.}\right).$$

---

[67] This means that the particle-like narrow-k wave packets of the spin waves in systems with $N \gg 1$, called *magnons*, may be assigned a finite mass, proportional to J.

[68] See, e.g. *Part CM* sections 6.3 and 6.4, in particular, Eq. (6.31).

[69] See, e.g. *Part EM* section 7.1, in particular, Eq. (7.13).

[70] Note that at $\mathscr{B} \neq 0$, $E_{\pm}(0) \neq E_{\pm}$.

with $\varepsilon_{1,2} = \pm c_z$, $t \equiv |c_-| = |c_+|$. Applying the bosonic operator commutation rule to the middle pair of operators in 4-products, just as was done in Eq. (8.120), we may rewrite the result as

$$\hat{H} = \varepsilon_1 \hat{N}_1 + \frac{J}{2}\hat{N}_1(\hat{N}_1 - \hat{I}) + \varepsilon_2 \hat{N}_2 + \frac{J}{2}\hat{N}_2(\hat{N}_2 - \hat{I})$$
$$+ t\left(\hat{a}_1^\dagger \hat{a}_2 + \text{h.c.}\right), \quad \text{with } \hat{N}_j \equiv \hat{a}_j^\dagger \hat{a}_j.$$

For fermions, such a model would not make much sense, because the eigenvalues N_j may only take values 0 or 1, so that the term proportional to J would equal zero for any state. This is why the simplest model for fermions may be based on the assumption of two quantum states for each site (say with different spin orientations), so that their occupancies $N_\uparrow$ and $N_\downarrow$ may *each* be equal to either 0 or 1. Then, in analogy with the expression for bosons, we may write

$$\hat{H} = \varepsilon_1(\hat{N}_{1\uparrow} + \hat{N}_{1\downarrow}) + J\hat{N}_{1\uparrow}\hat{N}_{1\downarrow} + \varepsilon_2(\hat{N}_{2\uparrow} + \hat{N}_{2\downarrow})$$
$$+ J\hat{N}_{2\uparrow}\hat{N}_{2\downarrow} + t\left(\hat{a}_{1\uparrow}^\dagger \hat{a}_{2\uparrow} + \hat{a}_{1\downarrow}^\dagger \hat{a}_{2\downarrow} + \text{h.c.}\right).$$

This expression takes into account the fact that the orbital motion, whether this is the localized state or interwell tunneling, is typically independent of a particle's spin, so that the terms proportional to the coefficients ε_j and t include creation–annihilation operator products with the same spin orientation, and their coefficients are spin-independent.

(ii) A straightforward generalization of the above result for bosons to an infinite 1D chain of similar localized cites (see also Eq. (8.112) of the notes) is

$$\hat{H} = \sum_{j=-\infty}^{+\infty}\left[\frac{J}{2}\hat{N}_j\left(\hat{N}_j - \hat{I}\right) + t\left(\hat{a}_j^\dagger \hat{a}_{j+1} + \text{h.c.}\right)\right].$$

Here I have used the fact that if the potential is indeed periodic, the core energies ε_j are equal to each other, so that if the total number N of particles in the system is fixed, their sum is equal to εN, and may be taken for the energy reference. This Hamiltonian is known as the *Bose–Hubbard model.*

A close term, *Hubbard model*, is typically reserved for fermions (typically, electrons in condensed matter systems), and may be also obtained by the generalization of the above two-site model. Just as for bosons, the difference between the parameters ε_j and t_j at different sites is typically neglected, giving

$$\hat{H} = \sum_{j=-\infty}^{+\infty}\left[J\hat{N}_{j\uparrow}\hat{N}_{j\downarrow} + t\left(\hat{a}_{j,\uparrow}^\dagger \hat{a}_{j+1,\uparrow} + \hat{a}_{j,\downarrow}^\dagger \hat{a}_{j+1,\downarrow} + \text{h.c.}\right)\right].$$

***Problem* 8.27.** For each of the Hamiltonians composed in the previous problem, derive the Heisenberg equations of motion for particle creation operators, for

(i) bosons, and
(ii) fermions.

Solutions:

(i) The Heisenberg equation contributions due to the terms quadratic in the creation–annihilation operators have been calculated in section 8.3 (see Eqs. (8.101)–(8.106) of the lecture notes), so that the only new terms we should handle for our Bose models are those proportional to $\hat{a}_j^\dagger \hat{a}_j^\dagger \hat{a}_j \hat{a}_j$, all with the same site index. Dropping the index for the sake of notation simplicity, we may write the typical commutator we would need as

$$[\hat{a},\ \hat{a}^\dagger \hat{a}^\dagger \hat{a}\hat{a}] = \hat{a}\hat{a}^\dagger \hat{a}^\dagger \hat{a}\hat{a} - \hat{a}^\dagger \hat{a}^\dagger \hat{a}\hat{a}\hat{a}. \qquad (*)$$

Applying the commutation rule (8.75) twice to move the first annihilation operator in the first term to the back (rightmost) position, we get

$$\begin{aligned}\hat{a}\hat{a}^\dagger \hat{a}^\dagger \hat{a}\hat{a} &= (\hat{I} + \hat{a}^\dagger \hat{a})\hat{a}^\dagger \hat{a}\hat{a} = \hat{a}^\dagger \hat{a}\hat{a} + \hat{a}^\dagger \hat{a}\hat{a}^\dagger \hat{a}\hat{a}\\ &= \hat{a}^\dagger \hat{a}\hat{a} + \hat{a}^\dagger(\hat{I} + \hat{a}^\dagger \hat{a})\hat{a}\hat{a} = 2\hat{a}^\dagger \hat{a}\hat{a} + \hat{a}^\dagger \hat{a}^\dagger \hat{a}\hat{a}\hat{a},\end{aligned}$$

The last term in this expression cancels with that in Eq. (*), so that we finally get

$$[\hat{a},\ \hat{a}^\dagger \hat{a}^\dagger \hat{a}\hat{a}] = 2\hat{a}^\dagger \hat{a}\hat{a}.$$

As a result, the Heisenberg equations of motion take the following forms:

– for two coupled wells:

$$i\hbar\dot{\hat{a}}_1 = \varepsilon_1 \hat{a}_1 + J\hat{a}_1^\dagger \hat{a}_1 \hat{a}_1 + t\hat{a}_2, \qquad i\hbar\dot{\hat{a}}_2 = \varepsilon_2 \hat{a}_2 + J\hat{a}_2^\dagger \hat{a}_2 \hat{a}_2 + t\hat{a}_1,$$

– for the Bose–Hubbard model of a 1D chain of similar sites:

$$i\hbar\dot{\hat{a}}_j = J\hat{a}_j^\dagger \hat{a}_j \hat{a}_j + t\left(\hat{a}_{j-1} + \hat{a}_{j+1}\right).$$

Despite the innocent look of these equations, they are nonlinear, and hence do not allow exact analytical solutions.

(ii) It may look that for the fermions the situation would be easier, because only the first power of each population operator participates in the Hamiltonians. However, the celebration would be premature, because the creation–annihilation operators belonging to each state are not independent now. Indeed, let us crank open a typical commutator we would need (with the well index omitted, as before):

$$[\hat{a}_\uparrow,\ \hat{N}_\uparrow \hat{N}_\downarrow] = \left[\hat{a}_\uparrow,\ \hat{a}_\uparrow^\dagger \hat{a}_\uparrow \hat{a}_\downarrow^\dagger \hat{a}_\downarrow\right] = \hat{a}_\uparrow \hat{a}_\uparrow^\dagger \hat{a}_\uparrow \hat{a}_\downarrow^\dagger \hat{a}_\downarrow - \hat{a}_\uparrow^\dagger \hat{a}_\uparrow \hat{a}_\downarrow^\dagger \hat{a}_\downarrow \hat{a}_\uparrow,$$

and start swapping operators in the first term, using the fermionic commutation rules (8.95)–(8.96):

$$\hat{a}_\uparrow \hat{a}_\uparrow^\dagger \hat{a}_\uparrow \hat{a}_\downarrow^\dagger \hat{a}_\downarrow = \left(\hat{I} - \hat{a}_\uparrow^\dagger \hat{a}_\uparrow\right)\hat{a}_\uparrow \hat{a}_\downarrow^\dagger \hat{a}_\downarrow = \hat{a}_\uparrow \hat{a}_\downarrow^\dagger \hat{a}_\downarrow - \hat{a}_\uparrow^\dagger \hat{a}_\uparrow \hat{a}_\uparrow \hat{a}_\downarrow^\dagger \hat{a}_\downarrow.$$

At this stage, we should notice that the second term gives zero action on any fermion state, and may be ignored. Similarly, twice commuting the operators in the second term of the commutator, we get

$$\hat{a}_\uparrow^\dagger \hat{a}_\uparrow \hat{a}_\downarrow^\dagger \hat{a}_\downarrow \hat{a}_\uparrow = -\hat{a}_\uparrow^\dagger \hat{a}_\downarrow^\dagger \hat{a}_\uparrow \hat{a}_\downarrow \hat{a}_\uparrow = \hat{a}_\uparrow^\dagger \hat{a}_\downarrow^\dagger \hat{a}_\downarrow \hat{a}_\uparrow \hat{a}_\uparrow,$$

so it always gives zero as well. Thus, finally,

$$[\hat{a}_\uparrow, \hat{N}_\uparrow \hat{N}_\downarrow] = \hat{a}_\uparrow \hat{a}_\downarrow^\dagger \hat{a}_\downarrow.$$

This is not more complex than for the bosonic operators, but not simpler either. We may also notice that all commutators like

$$\left[\hat{a}_j, \hat{N}_{j'}\right]$$

vanish if j and j' are different (orbital and/or spin) states. As a result, the Heisenberg equations of motion are:

- for two coupled wells:

$$i\hbar\dot{\hat{a}}_{1\uparrow} = \varepsilon_1 \hat{a}_{1\uparrow} + J\hat{a}_{1\uparrow}\hat{a}_{1\downarrow}^\dagger \hat{a}_{1\downarrow} + t\hat{a}_{2\uparrow},$$

$$i\hbar\dot{\hat{a}}_{2\uparrow} = \varepsilon_2 \hat{a}_{2\uparrow} + J\hat{a}_{2\uparrow}\hat{a}_{2\downarrow}^\dagger \hat{a}_{2\downarrow} + t\hat{a}_{1\uparrow},$$

- for the Hubbard model:

$$i\hbar\dot{\hat{a}}_{j\uparrow} = J\hat{a}_{j\uparrow}\hat{a}_{j\downarrow}^\dagger \hat{a}_{j\downarrow} + t\left(\hat{a}_{j-1,\uparrow} + \hat{a}_{j+1,\uparrow}\right),$$

and similarly for the opposite spin states. We see that equations for the states with opposite spins are coupled via the on-site interaction term, not making their solution any easier.

***Problem* 8.28.** Express the ket-vectors of all possible Dirac states for the system of three indistinguishable

(i) bosons, and
(ii) fermions,

via those of the single-particle states β, β', and β'' they occupy.

Solutions:

(i) *Bosons*: If all three single-particle states occupied in the considered Dirac states are different, we may use Eq. (8.80) of the lecture notes with $N_1 = N_2 = N_3 = 1$, $N_4 = N_5 = \cdots = 0$, so that the left-hand side of this relation may be truncated just to three positions, while its right-hand side is a sum of $N! = 3! = 6$ different kets formed by particle permutations[71]:

$$|1, 1, 1\rangle = \frac{1}{\sqrt{6}}(|\beta\beta'\beta''\rangle + |\beta\beta''\beta'\rangle + |\beta'\beta\beta''\rangle + |\beta'\beta''\beta\rangle + |\beta''\beta\beta'\rangle + |\beta''\beta'\beta\rangle), \qquad (*)$$

where the possible zeros in the left-hand ket (expressing unoccupied states) are suppressed.

If two of the single-particle states (say, β' and β'') are identical, i.e. there are just two different single-particle states, β and β', then the Dirac ket may be truncated even more, to show just the two first positions, with $N_1 = 1$, $N_2 = 2$, and the right-hand side of Eq. (8.80) has only three permutations:

$$|1, 2\rangle = \left(\frac{2!}{3!}\right)^{1/2} (|\beta\beta'\beta'\rangle + |\beta'\beta\beta'\rangle + |\beta'\beta'\beta\rangle) \equiv \frac{1}{\sqrt{3}}(|\beta\beta'\beta'\rangle + |\beta'\beta\beta'\rangle + |\beta'\beta'\beta\rangle).$$

Finally, if all three particles are in the same single-particle state (say, β), then we may take $N_1 = 3$, $N_2 = N_3 = .. = 0$, and Eq. (8.80) yields simply

$$|3\rangle = \left(\frac{3!}{3!}\right)^{1/2} |\beta\beta\beta\rangle \equiv |\beta\beta\beta\rangle.$$

Note that the general state of the three-particle system is a linear superposition of Dirac states of these three types, with various β, β', and β'', i.e. not necessarily one of the Dirac states.

(ii) *Fermions*: According to the Pauli principle, all occupied states β, β', and β'' have to be different, so that suppressing all zeros in the Dirac ket-vector again, we may spell out the Slater determinant (8.60) as

$$|1, 1, 1\rangle = \frac{1}{\sqrt{3!}} \begin{vmatrix} |\beta\rangle & |\beta'\rangle & |\beta''\rangle \\ |\beta\rangle & |\beta'\rangle & |\beta''\rangle \\ |\beta\rangle & |\beta'\rangle & |\beta''\rangle \end{vmatrix} \equiv \frac{1}{\sqrt{6}}(|\beta\beta'\beta''\rangle - |\beta\beta''\beta'\rangle - |\beta'\beta\beta''\rangle + |\beta'\beta''\beta\rangle + |\beta''\beta\beta'\rangle - |\beta''\beta'\beta\rangle),$$

where the last expression on the right-hand side is in the same shorthand notation (with position coding of the particle numbers) as used in part (i) of the solution.

[71] Note again that the symbol positions within the kets on two sides of this relation have very different meaning: in the second quantization language (the left-hand side), they code the single-particle *state* numbers, while those in the 'usual' notation (the right-hand side), code the *particle* numbers.

Evidently, it differs from Eq. (*) only by sign alternation, which ensures the state asymmetry (rather than symmetry) with respect to permutation of any two particles.

***Problem* 8.29.** Explain why the general perturbative result (8.126), when applied to the ^{4}He atom, gives the correct[72] expression (8.29) for the ground singlet state, and correct Eqs. (8.39)–(8.42) (with the minus sign in the first of these relations) for the excited triplet states, but cannot describe these results, with the plus sign in Eq. (8.39), for the excited singlet state.

Solution: In the numbered-particle language, the unperturbed ground (singlet) state, used for the calculation of Eq. (8.29), is represented by Eq. (8.24), and may be described by the ket-vector

$$|gg\rangle\frac{1}{\sqrt{2}}(|\uparrow\downarrow\rangle - |\downarrow\uparrow\rangle) \equiv \frac{1}{\sqrt{2}}(|g_\uparrow\rangle \otimes |g_\downarrow\rangle - |g_\downarrow\rangle \otimes |g_\uparrow\rangle), \qquad (*)$$

where g denotes the orbital factor of the single-particle ground state: $\langle \mathbf{r}|g\rangle = \psi_{100}(\mathbf{r})$, and each ket on the right-hand side of this expression is a spin-orbital—see Eq. (8.125). On the other hand, with the states $g_\uparrow$ and $g_\downarrow$ taken for the single-particle basis (β and β') for the Dirac representation, the two-electron Dirac state ket, $|N_1, N_2\rangle = |1, 1\rangle$, i.e. the Slater determinant (8.60a), is represented by the ket

$$\begin{aligned}\frac{1}{\sqrt{2}}(|\beta\beta'\rangle - |\beta'\beta\rangle) &\equiv \frac{1}{\sqrt{2}}(|\beta\rangle \otimes |\beta'\rangle - |\beta'\rangle \otimes |\beta\rangle) \\ &\equiv \frac{1}{\sqrt{2}}(|g_\uparrow\rangle \otimes |g_\downarrow\rangle - |g_\downarrow\rangle \otimes |g_\uparrow\rangle),\end{aligned}$$

i.e. exactly the same state as Eq. (*).

However, for the analysis of the ^{4}He atom in section 8.2 of the lecture notes, we have described the excited singlet state by its orbital wavefunction (8.35), with the plus sign, i.e. by the following total (orbital + spin) ket-vector:

$$\begin{aligned}&\frac{1}{\sqrt{2}}(|ge\rangle + |eg\rangle)\frac{1}{\sqrt{2}}(|\uparrow\downarrow\rangle - |\downarrow\uparrow\rangle) \\ &\equiv \frac{1}{2}(|g\rangle \otimes |e\rangle + |e\rangle \otimes |g\rangle)(|\uparrow\downarrow\rangle - |\downarrow\uparrow\rangle),\end{aligned}$$

where e means the orbital part of a single-particle excited state, with $\langle \mathbf{r}|e\rangle \equiv \psi_{nlm}(\mathbf{r})$, $n > 1$. Multiplying the parentheses, we may represent this expression as a sum of four terms:

$$\frac{1}{2}(|g_\uparrow\rangle \otimes |e_\downarrow\rangle - |g_\downarrow\rangle \otimes |e_\uparrow\rangle + |e_\uparrow\rangle \otimes |g_\downarrow\rangle - |e_\downarrow\rangle \otimes |g_\uparrow\rangle),$$

which may be re-grouped as

[72] Correct in the sense of the first order of the perturbation theory.

$$\frac{1}{\sqrt{2}}\left[\frac{1}{\sqrt{2}}(|g_\uparrow\rangle \otimes |e_\downarrow\rangle - |e_\downarrow\rangle \otimes |g_\uparrow\rangle) + \frac{1}{\sqrt{2}}(|e_\uparrow\rangle \otimes |g_\downarrow\rangle - |g_\downarrow\rangle \otimes |e_\uparrow\rangle)\right].$$

Each of the two components of this linear superposition is a 2 × 2 Slater determinant in the particle-number representation, i.e. a single Dirac ket $|N_1, N_2\rangle = |1, 1\rangle$, but since they have different spin–orbital state bases ($g_\uparrow$ and $e_\downarrow$ vs. $e_\uparrow$ and $g_\downarrow$), their coherent sum is *not* a single Dirac ket, such as those used at the derivation of Eq. (8.126).

The same is also true for the excited entangled triplet state,

$$\frac{1}{\sqrt{2}}(|ge\rangle - |eg\rangle)\frac{1}{\sqrt{2}}(|\uparrow\downarrow\rangle + |\downarrow\uparrow\rangle)$$
$$\equiv \frac{1}{2}(|g\rangle \otimes |e\rangle - |e\rangle \otimes |g\rangle)(|\uparrow\downarrow\rangle + |\downarrow\uparrow\rangle),$$

which is just another, linearly-independent superposition of the same Dirac kets:

$$\frac{1}{\sqrt{2}}\left[\frac{1}{\sqrt{2}}(|g_\uparrow\rangle \otimes |e_\downarrow\rangle - |e_\downarrow\rangle \otimes |g_\uparrow\rangle) - \frac{1}{\sqrt{2}}(|e_\uparrow\rangle \otimes |g_\downarrow\rangle - |g_\downarrow\rangle \otimes |e_\uparrow\rangle)\right].$$

On the other hand, the two 'simple' (factorable) triplet states, for example

$$\frac{1}{\sqrt{2}}(|ge\rangle - |eg\rangle)|\uparrow\uparrow\rangle \equiv \frac{1}{\sqrt{2}}(|g_\uparrow\rangle \otimes |e_\uparrow\rangle - |e_\uparrow\rangle \otimes |g_\uparrow\rangle),$$

are single Slater determinants, i.e. single Dirac kets.

Hence, the fact that Eq. (8.126), with $N = 2$, does describe a negative contribution from the exchange interaction is only due to the existence of the corresponding factorable triplet states.

***Problem* 8.30.** For a system of two distinct qubits (i.e. two-level systems), introduce a reasonable uncoupled-representation z-basis, and find in this basis the 4 × 4 matrix of the operator that swaps their states.

Solution: The requested basis should obviously include all four possible uncoupled-representation states:

$$\uparrow\uparrow, \quad \uparrow\downarrow, \quad \downarrow\uparrow, \quad \downarrow\downarrow, \qquad (*)$$

but in order to prescribe definite four-component vectors to the states, and definite 4 × 4 matrices to the linear operators acting in this joint Hilbert space, a certain order of the basis states should be selected. (For one spin-½, the traditional order, used in particular in this course, is natural: ↑ first, then ↓.) Generally, the order may be selected at will, but it makes sense to establish it in a way making possible generalization to more than two spins natural. In this sense, the order given by Eq. (*), in which the rightmost spins are altered first, is very reasonable. This may be confirmed by rewiring it in the notation accepted in qubit applications (with ↑ denoted by 0, and ↓ by 1, see the beginning of section 8.5 of the lecture notes):

$$00, \quad 01, \quad 10, \quad 11.$$

If these zeros and ones are understood in the sense of classical bits, this order corresponds to the naturally ordered sequence of binary numbers, in the decimal system equal to 0, 1, 2 and 3.

With this order accepted, the ket-vector of an arbitrary pure state of this composite system,

$$|\alpha\rangle = a\,|\uparrow\uparrow\rangle + b\,|\uparrow\downarrow\rangle + c\,|\downarrow\uparrow\rangle + d\,|\downarrow\downarrow\rangle,$$

is represented with a column

$$\begin{pmatrix} a \\ b \\ c \\ d \end{pmatrix},$$

and the operator $\hat{X}$ swapping the spin states,

$$\begin{aligned}\hat{X}\,|\alpha\rangle &\equiv \hat{X}(a\,|\uparrow\uparrow\rangle + b\,|\uparrow\downarrow\rangle + c\,|\downarrow\uparrow\rangle + d\,|\downarrow\downarrow\rangle) \\ &= a\,|\uparrow\uparrow\rangle + b\,|\downarrow\uparrow\rangle + c\,|\uparrow\downarrow\rangle + d\,|\downarrow\downarrow\rangle,\end{aligned}$$

should just swap the coefficients b and c, so that its matrix X, in the accepted basis, should act as

$$\mathrm{X}\begin{pmatrix} a \\ b \\ c \\ d \end{pmatrix} = \begin{pmatrix} a \\ c \\ b \\ d \end{pmatrix}.$$

Evidently, this operation is achieved by the following 4×4 matrix:

$$\mathrm{X} = \begin{pmatrix} 1 & 0 & 0 & 0 \\ 0 & 0 & 1 & 0 \\ 0 & 1 & 0 & 0 \\ 0 & 0 & 0 & 1 \end{pmatrix}.$$

***Problem* 8.31.** Find a time-independent Hamiltonian that may cause the qubit evolution described by Eqs. (8.155) of the lecture notes. Discuss the relation between your result and the time-dependent Hamiltonian (6.86).

Solution: Eqs. (8.155) has been obtained as solutions of Eqs. (6.94), for the particular case of exact frequency tuning, $\Delta = 0$. In the notation of section 8.5, these equations are

$$i\hbar\dot{a}_0 = a_1 A, \qquad i\hbar\dot{a}_1 = a_0 A^*. \qquad (*)$$

They may be interpreted as the set of Schrödinger equations,

$$i\hbar\dot{a}_0 = a_0H_{00} + a_1H_{01}, \quad i\hbar\dot{a}_1 = a_0H_{10} + a_1H_{11},$$

with the following matrix coefficients of an effective time-independent Hamiltonian (in the basis of the states 0 and 1):

$$\begin{aligned} &H_{00} = H_{11} = 0, \\ &H_{01} = A = |A|\, e^{i\varphi} \equiv |A|\,(\cos\varphi + i\sin\varphi), \\ &H_{10} = A^* \equiv |A|\, e^{-i\varphi} = |A|\,(\cos\varphi - i\sin\varphi). \end{aligned}$$

Recalling the definition (4.105) of the Pauli matrices, we may represent this matrix as

$$\begin{aligned} \mathrm{H} \equiv \begin{pmatrix} H_{00} & H_{01} \\ H_{10} & H_{11} \end{pmatrix} &= |A|\ \cos\varphi \begin{pmatrix} 0 & 1 \\ 1 & 0 \end{pmatrix} - |A|\ \sin\varphi \begin{pmatrix} 0 & -i \\ +i & 0 \end{pmatrix} \\ &\equiv |A|\left(\cos\varphi\sigma_x - \sin\varphi\sigma_y\right). \end{aligned}$$

Hence, for the requested Hamiltonian operator itself we may write the following expression (valid in any basis):

$$\hat{H} = |A|\left(\cos\varphi\,\hat{\sigma}_x - \sin\varphi\,\hat{\sigma}_y\right) \equiv \hat{\sigma}_x\ \mathrm{Re}\,A - \hat{\sigma}_y\ \mathrm{Im}\,A. \qquad (**)$$

This expression result may be interpreted as the result of time averaging of the product of the time-dependent Hamiltonian (6.86), whose perturbative treatment in section 6.5 has led us to Eqs. (*), by the factor $\exp\{i\omega_{10}t\}$ which compensates the re-definition (6.82) of the probability amplitudes. Let me hope that the reader remembers that such an alternative method of performing the rotating wave approximation (RWA) was repeatedly used in chapter 7.

Note also that according to Eq. (4.163*a*) of the lecture notes, for a spin-½ particle with a non-zero gyromagnetic factor γ, the Hamiltonian (**) may be created by applying a time-independent magnetic field directed within the $[x, y]$ plane, at the angle $(-\varphi)$ with the x-axis.

References

[1] Cattaneo L *et al* 2018 *Nature Phys.* **14** 733
[2] Batchelor M 2007 *Phys. Today* **60** 36

IOP Publishing

Quantum Mechanics
Problems with solutions
Konstantin K Likharev

Chapter 9

Elements of relativistic quantum mechanics

***Problem* 9.1.** Prove the Casimir formula, given by Eq. (9.23) of the lecture notes, by calculating the net force $F = \mathcal{P}A$ exerted by the electromagnetic field, in its ground state, on two perfectly conducting parallel plates of area A, separated by a vacuum gap of width $t \ll A^{1/2}$.

Hint: Calculate the field energy in the gap volume with and without the account of the plate effect, and then apply the Euler–Maclaurin formula[1] to the difference of these two results.

Solution: Let us calculate the energy in the volume At of the gap as a sum over all possible sinusoidal standing waves[2] with wave vectors $\mathbf{k} = \{\mathbf{k}_\rho, k_z\}$, where the z-axis is normal to the plates, and $\boldsymbol{\rho} \equiv \{x, y\}$ is the 2D radius-vector in the plane of their surfaces:

$$E = 2\sum_{\omega}\frac{\hbar\omega}{2} \equiv \hbar c\sum_{\mathbf{k}}|\mathbf{k}| = \hbar c\sum_{\mathbf{k}_\rho, k_z}\left[k_\rho^2 + k_z^2\right]^{1/2} \qquad (*)$$

(where the front factor 2 is due to two different electromagnetic field modes, with orthogonal polarizations, for each wave vector $\mathbf{k}$), with and without the effect of the plates. In order to avoid the high-k divergence, we may cut both sums at the same, sufficiently large ($k_{\max} \gg 1/t \gg 1/A^{1/2}$) value of $k \equiv |\mathbf{k}|$, because the higher modes are not affected by the plates.

Thanks to the condition $A^{1/2} \gg t$, the in-plane wave vector quantization may be replaced with an integral, using the electromagnetic-wave analog of Eq. (1.99) of the lecture notes:

[1] See, e.g. Eq. (A.15*a*).

[2] As implied by the discussion in section 9.1 (but perhaps was not sufficiently emphasized there), the analogy between an electromagnetic field mode and that of the simple ('lumped') 1D harmonic oscillator is direct only for a standing (rather than traveling) wave.

$$\sum_{\mathbf{k}_\rho}(\ldots) = \int_{n_x \geq 0} dn_x \int_{n_y \geq 0} dn_y(\ldots) = \frac{a_x}{\pi}\frac{a_y}{\pi}\int_{k_x \geq 0} dk_x \int_{k_y \geq 0} dk_y(\ldots)$$

$$\equiv \frac{A}{\pi^2}\int_{k_x \geq 0} dk_x \int_{k_y \geq 0} dk_y(\ldots). \qquad (**)$$

On the other hand, the values of k_z are significantly limited to a discrete set,

$$k_z = \frac{\pi n}{t}, \quad \text{with } n = 0, 1, 2, \ldots,$$

by the zero-field boundary conditions on the plate surfaces. As a result, Eq. (*) becomes

$$E = \frac{A\hbar c}{\pi^2}\sum_{n \geq 0}{}' \int_{k_x \geq 0} dk_x \int_{k_y \geq 0} dk_y \left[k_\rho^2 + \left(\frac{\pi n}{t}\right)^2\right]^{1/2},$$

where the symbol $\sum'$ means that the term with $n = 0$ is taken with the additional factor ½, because for this special value of n, the boundary conditions allow only one wave polarization, with the electric field normal to the plate surfaces.

Now it is convenient to use the fact that the function under the integral is symmetric with respect to all Cartesian components of the wave vector, and extend the summation and integration to all, positive and negative values of n, k_x, and k_y (compensating this extension by an extra multiplier 1/8), and then make, in the double integral, the transfer from the Cartesian to the polar coordinates:

$$E = \frac{A\hbar c}{\pi^2}\frac{1}{8}\sum_n \int_{\mathbf{k}_\rho}\left[k_\rho^2 + \left(\frac{\pi n}{t}\right)^2\right]^{1/2} d^2k_\rho = \frac{A\hbar c}{8\pi^2}\sum_n \int_{k_\rho \geq 0}\left[k_\rho^2 + \left(\frac{\pi n}{t}\right)^2\right]^{1/2} 2\pi k_\rho dk_\rho.$$

(Note that the exception for $n = 0$ in the sum $\sum'$ is automatically absorbed into this regular sum.) Since $2k_\rho dk_\rho \equiv d(k_\rho^2) \equiv d[k_\rho^2 + (\pi n/t)^2]$, the integration is elementary, giving

$$E = \frac{A\hbar c}{12\pi}\sum_n \left[k_\rho^2 + \left(\frac{\pi n}{t}\right)^2\right]^{3/2} \Bigg|_{k_\rho = 0}^{k = k_{\max}} \equiv \frac{A\hbar c}{12\pi}\sum_{n=-n_{\max}}^{+n_{\max}}\left[k_{\max}^3 - \left|\frac{\pi n}{t}\right|^3\right],$$

where $n_{\max} = k_{\max}t/\pi \gg 1$.

Now let us repeat the calculation, for the same volume, but disregarding the effect of the plates, so that the summation over k_z may be replaced with the integration, similar to that over k_x and k_y:

$$\sum_{k_z > 0}(\ldots) = \frac{t}{\pi}\int_{k_z \geq 0}(\ldots)\, dk_z = \frac{t}{2\pi}\int_{k_z}(\ldots)\, dk_z,$$

where the last integration is for both signs of k_z. The result is

$$E_0 = \hbar c \frac{A}{4\pi^2} \frac{t}{2\pi} \int_{k_z} dk_z \left(k_\rho^2 + k_z^2\right)^{3/2} \Bigg|_{k_\rho = 0}^{k = k_{\max}} \equiv \frac{A\hbar c}{12\pi} \frac{t}{\pi} \int_{-k_{\max}}^{+k_{\max}} \left(k_{\max}^3 - |k_z|^3\right) dk_z.$$

The difference of these two energy values, representing the effect of the conducting plates,

$$\Delta E \equiv E - E_0 = \frac{A\hbar c}{12\pi} \left\{ \sum_{n=-n_{\max}}^{+n_{\max}} \left[k_{\max}^3 - \left| \frac{\pi n}{t} \right|^3 \right] - \frac{t}{\pi} \int_{-k_{\max}}^{+k_{\max}} \left(k_{\max}^3 - |k_z|^3\right) dk_z \right\},$$

may be recast to a simpler form by introducing a continuous dimensionless variable $\xi \equiv k_z t / \pi$:

$$\Delta E \equiv E - E_0 = \frac{A\hbar c}{12\pi} \left[\sum_{n=-n_{\max}}^{+n_{\max}} f(n) - \int_{-n_{\max}}^{+n_{\max}} f(\xi) d\xi \right], \quad \text{where } f(\xi) \equiv k_{\max}^3 - \left| \frac{\pi \xi}{t} \right|^3.$$

The expression in the square brackets may be evaluated using the Euler–Maclaurin formula. However, since the function $f(\xi)$ is not infinitely differentiable (it has a 'soft cusp' at $\xi = 0$), we may only apply the formula separately to two 'wings' of this function, with $-n_{\max} \leqslant \xi \leqslant 0$ and $0 \leqslant \xi \leqslant n_{\max}$. (At this separation, the term $f(0)$ under the total sum may be either delegated to one of the partial sums, or, more logically, equally divided between them.) Summing up the two results, we get

$$\sum_{n=-n_{\max}}^{+n_{\max}} f(n) - \int_{-n_{\max}}^{+n_{\max}} f(\xi) d\xi = \frac{1}{2}[f(+n_{\max}) + f(-n_{\max})]$$
$$+ \frac{1}{6} \cdot \frac{1}{2!} \left[\frac{df}{d\xi}(+n_{\max}) - \frac{df}{d\xi}(+0) + \frac{df}{d\xi}(-0) - \frac{df}{d\xi}(-n_{\max}) \right]$$
$$- \frac{1}{30} \cdot \frac{1}{4!} \left[\frac{d^3 f}{d\xi^3}(+n_{\max}) - \frac{d^3 f}{d\xi^3}(+0) + \frac{d^3 f}{d\xi^3}(-0) - \frac{d^3 f}{d\xi^3}(-n_{\max}) \right] + \ldots$$

At $\xi = \pm n_{\max}$, our function $f(\xi)$ vanishes by the definition of the parameter $n_{\max}$, and its derivatives at these points may be ignored, because the mode sum cutoff at $k_{\max}$ may be made smoothly spread over an interval $\Delta k \ll k_{\max}$, without a substantial effect on the above calculations. As a result, using the derivatives,

$$\frac{df}{d\xi} = -3\left(\frac{\pi}{t}\right)^3 \xi^2 \operatorname{sgn}(\xi), \quad \frac{d^3 f}{d\xi^3} = -6\left(\frac{\pi}{t}\right)^3 \operatorname{sgn}(\xi), \quad \text{and } \frac{d^n f}{d\xi^n} = 0 \text{ for } n > 3 \text{ and } \xi \neq 0,$$

we obtain a finite result (as could be expected, independent of the artificial cutoff $k_{\max}$):

$$\sum_{n=-n_{\max}}^{+n_{\max}} f(n) - \int_{-n_{\max}}^{+n_{\max}} f(\xi) d\xi = -\frac{1}{30} \cdot \frac{1}{4!} \cdot 12\left(\frac{\pi}{t}\right)^3, \quad \text{giving } \Delta E = -\frac{\pi^2 A\hbar c}{720 t^3}. \quad (***)$$

The result shows that the energy difference is negative, and its magnitude grows as t is decreased, i.e. the plates attract each other. The differentiation over t, $F = -\partial(\Delta E)/\partial t$, immediately yields the Casimir formula (9.23):

$$F = -\frac{\pi^2 A\hbar c}{240t^4}.$$

Note also that besides the numerical coefficient, Eq. (***) is a great candidate for its derivation from dimensionality arguments, without solving the problem exactly. Indeed, since we are dealing with quantum mechanics of electromagnetic field in free space, the final expression for the ratio E/A may include the Planck constant $\hbar$, the free-space speed of light c, but hardly anything else—besides the only significant geometric parameter of the problem, t. Now writing the dimensionalities of these constants (in any system of units):

$$[\hbar] = \text{energy} \times \text{time}, \quad [c] = \frac{\text{length}}{\text{time}}, \quad [t] = \text{length},$$

we see that the only way they may be combined to give the dimensionality of the result,

$$\left[\frac{E}{A}\right] = \frac{\text{energy}}{(\text{length})^2},$$

is in the correct fraction $\hbar c/t^3$.

However, the reader should not think that all his/her hard work on the problem was 'only' the calculation of the constant factor before this combination. A much more important reward is that the solution yields a very clear physical picture of the Casimir effect, while the dimensionality-based 'derivation' does not.

***Problem* 9.2.** Electromagnetic radiation of some single-mode quantum sources may have such a high degree of coherence that it is possible to observe the interference of waves from two independent sources with virtually the same frequency, incident on one detector.

(i) Generalize Eq. (9.29) of the lecture notes to this case.
(ii) Use this generalized expression to show that incident waves in different Fock states do not create an interference pattern.

Solutions:

(i) Let us rewrite the second form of Eq. (9.29),

$$\Gamma \propto \langle \text{ini}|\, \hat{a}^\dagger \hat{a} |\text{ini}\rangle e^*(\mathbf{r}) e(\mathbf{r}),$$

as follows:

$$\Gamma \propto \langle \text{ini}^*(\mathbf{r})|\, \hat{a}^\dagger \hat{a} |\text{ini}(\mathbf{r})\rangle, \quad \text{with } |\text{ini}\,(\mathbf{r})\rangle \equiv \mathbf{e}(\mathbf{r})|\text{ini}\rangle, \text{ and } \langle \text{ini}^*(\mathbf{r})| \equiv \langle \text{ini}|\mathbf{e}^*(\mathbf{r}). \qquad (*)$$

A sum of two waves of the same frequency may be described by the ket-vector

$$|\text{ini}\,(\mathbf{r})\rangle = |\text{ini}_1(\mathbf{r})\rangle + |\text{ini}_2(\mathbf{r})\rangle = \mathbf{e}_1(\mathbf{r})|\text{ini}_1\rangle + \mathbf{e}_2(\mathbf{r})|\text{ini}_2\rangle,$$

where each component wave is proportional to its own spatial distribution factor $\mathbf{e}_j(\mathbf{r})$ and may be in its own quantum state. Plugging this expression, and the corresponding expression for the bra-vector, into the first of Eqs. (*), and opening the parentheses, we may represent Γ as a sum of four terms, including the following two terms describing the waves' interference:

$$\Gamma_{\text{int}} \propto \langle \text{ini}_1|\, \hat{a}^\dagger \hat{a} |\text{ini}_2\rangle\ \mathbf{e}_1^*(\mathbf{r}) \cdot \mathbf{e}_2(\mathbf{r}) + \text{c.c.} \qquad (**)$$

For plane waves, $\mathbf{e}(\mathbf{r}) \propto \exp\{i\mathbf{k} \cdot \mathbf{r}\}$, the product $\mathbf{e}_1^*(\mathbf{r}) \cdot \mathbf{e}_2(\mathbf{r})$ is proportional to $\exp\{i\varphi_{12}\}$, with $\varphi_{12} = kl$, where $l = l_1 - l_2 + \text{const}$ is the difference of wave path lengths; the same is approximately true for quasi-plane-wave situations—see section 3.1.[3]

(ii) Let the component waves be initially in Fock states:

$$|\text{ini}_{1,2}\rangle = |n_{1,2}\rangle;$$

then the interference terms in Eq. (**) are proportional to the following bracket:

$$\langle n_1|\hat{a}^\dagger \hat{a}\, |n_2\rangle = \langle n_1|\hat{N}\, |n_2\rangle.$$

But the number operator $\hat{N} \equiv \hat{a}^\dagger \hat{a}$ commutes with the (field) oscillator Hamiltonian (see, e.g. Eq. (5.72) of the lecture notes) and hence is diagonal in the stationary (Fock) state basis. As a result, if the component waves are in different Fock states, $n_1 \neq n_2$, then $\Gamma_{\text{int}} = 0$, i.e. the interference pattern disappears.

***Problem* 9.3.** Calculate the zero-delay value $g^{(2)}(0)$ of the second-order correlation function of a single-mode electromagnetic field in the so-called *Schrödinger-cat state*[4]: a coherent superposition of two Glauber states, with equal amplitudes, equal but sign-opposite parameters α, and a certain phase shift between them.

[3] If the wave frequencies are slightly different, all above formulas remain valid, with the phase shift φ_{12} slowly drifting in time, so that the interference should be measured out relatively fast in order not to be averaged out.
[4] Its name stems from the well-known *Schrödinger cat paradox*, which is (very briefly) discussed in section 10.1 of the lecture notes.

Solution: Such an initial state of the field may be represented by the following ket-vector:

$$|\text{cat}\rangle = C\,(|+\alpha\rangle + |-\alpha\rangle e^{i\varphi}),$$

where each ket $|\alpha\rangle$ represents the Glauber state with the complex 'shift parameter' α, which participates in the basic Eqs. (5.102) and (5.124) of the lecture notes. The normalization constant C may be found by requiring the superposition state to be normalized, $\langle\text{cat}|\text{cat}\rangle = 1$ (just as the component states $\pm\alpha$ are), giving the following equation

$$|C|^2(\langle+\alpha| + \langle-\alpha|e^{-i\varphi})\,(|+\alpha\rangle + |-\alpha\rangle e^{i\varphi}) = 1.$$

Opening the parentheses, we get

$$|C|^2 = \frac{1}{1 + 1 + \langle+\alpha|-\alpha\rangle e^{i\varphi} + \langle-\alpha|+\alpha\rangle e^{-i\varphi}}.$$

The inner products participating is this expression may be calculated, for example, by using the Fock-state expansions (5.134) of the Glauber states[5]:

$$|+\alpha\rangle = \exp\left\{-\frac{|\alpha|^2}{2}\right\}\sum_{n=0}^{\infty}\frac{(+\alpha)^n}{(n!)^{1/2}}|n\rangle, \quad |-\alpha\rangle = \exp\left\{-\frac{|-\alpha|^2}{2}\right\}\sum_{n=0}^{\infty}\frac{(-\alpha)^n}{(n!)^{1/2}}|n\rangle,$$

giving

$$\langle+\alpha|-\alpha\rangle = \langle-\alpha|+\alpha\rangle = \exp\{-|\alpha|^2\}\sum_{n=0}^{\infty}\frac{(-\alpha\alpha^*)^n}{n!} = \exp\{-2\,|\alpha|^2\}, \qquad (*)$$

so that

$$|C|^2 = \frac{1}{2 + 2\exp\{-2\,|\alpha|^2\}\cos\varphi}. \qquad (**)$$

Now we can calculate both the numerator of the ratio (9.35), at the final steps of the calculation using Eqs. (*) and (**) again:

$$\langle\hat{a}^\dagger\hat{a}^\dagger\hat{a}\hat{a}\rangle_{\text{cat}} \equiv \langle\text{cat}|\hat{a}^\dagger\hat{a}^\dagger\hat{a}\hat{a}|\text{cat}\rangle = |C|^2(\langle+\alpha|+\langle-\alpha|e^{-i\varphi})\,\hat{a}^\dagger\hat{a}^\dagger\hat{a}\hat{a}(|+\alpha\rangle + |-\alpha\rangle e^{i\varphi})$$

$$\equiv |C|^2\begin{pmatrix}\langle+\alpha|\hat{a}^\dagger\hat{a}^\dagger\hat{a}\hat{a}|+\alpha\rangle + \langle-\alpha|\hat{a}^\dagger\hat{a}^\dagger\hat{a}\hat{a}|-\alpha\rangle \\ +\langle+\alpha|\hat{a}^\dagger\hat{a}^\dagger\hat{a}\hat{a}|-\alpha\rangle e^{i\varphi} + \langle-\alpha|\hat{a}^\dagger\hat{a}^\dagger\hat{a}\hat{a}|+\alpha\rangle e^{-i\varphi}\end{pmatrix}$$

$$= |C|^2\begin{pmatrix}\alpha^*\alpha^*\alpha\alpha + (-\alpha)^*(-\alpha)^*(-\alpha)(-\alpha) \\ +\alpha^*\alpha^*(-\alpha)(-\alpha)e^{i\varphi}\langle+\alpha|-\alpha\rangle + (-\alpha)^*(-\alpha)^*\alpha\alpha e^{-i\varphi}\langle-\alpha|+\alpha\rangle\end{pmatrix}$$

$$= |C|^2|\alpha|^4(2 + 2\exp\{-2\,|\alpha|^2\}\cos\varphi) = |\alpha|^4,$$

[5] Another way to get the same final result for C is to use the coordinate representation (5.107) of the Glauber state.

and the inner product in its denominator:

$$\langle \hat{a}^\dagger \hat{a} \rangle_{\text{cat}} = \langle \text{cat}|\hat{a}^\dagger \hat{a}|\text{cat}\rangle = |C|^2(\langle +\alpha| + \langle -\alpha| e^{-i\varphi})\, \hat{a}^\dagger \hat{a}\, (|+\alpha\rangle + |-\alpha\rangle e^{i\varphi})$$

$$\equiv |C|^2 \begin{pmatrix} \langle +\alpha|\hat{a}^\dagger \hat{a}|+\alpha\rangle + \langle -\alpha|\hat{a}^\dagger \hat{a}|-\alpha\rangle + \\ \langle +\alpha|\hat{a}^\dagger \hat{a}|-\alpha\rangle e^{i\varphi} + \langle -\alpha|\hat{a}^\dagger \hat{a}|+\alpha\rangle e^{-i\varphi} \langle +\alpha|-\alpha\rangle \end{pmatrix}$$

$$= |C|^2 \begin{pmatrix} \alpha^*\alpha + (-\alpha)^*(-\alpha) + \alpha^*(-\alpha) e^{i\varphi}\langle +\alpha|-\alpha\rangle \\ +(-\alpha)^*\alpha\, e^{-i\varphi}\langle -\alpha|+\alpha\rangle \end{pmatrix}$$

$$= |C|^2 |\alpha|^2 (2 - 2\exp\{-2\,|\alpha|^2\}\cos\varphi) = |\alpha|^2 \frac{1 - \exp\{-2\,|\alpha|^2\}\cos\varphi}{1 + \exp\{-2\,|\alpha|^2\}\cos\varphi},$$

so that, finally,

$$g^{(2)}(0) = \frac{\langle \hat{a}^\dagger \hat{a}^\dagger \hat{a} \hat{a} \rangle}{\langle \hat{a}^\dagger \hat{a} \rangle^2} = \left(\frac{1 + \exp\{-2\,|\alpha|^2\}\cos\varphi}{1 - \exp\{-2\,|\alpha|^2\}\cos\varphi} \right)^2.$$

The result shows that the second-order correlation function is largest at $\varphi = 0$,

$$\max_\varphi g^{(2)}(0) = \left(\frac{1 + \exp\{-2\,|\alpha|^2\}}{1 - \exp\{-2\,|\alpha|^2\}} \right)^2 \equiv \coth^2(|\alpha|^2), \qquad (***)$$

and may be larger than 2.[6] Such *super-bunching*, with $g^{(2)}(0) > 2$, may be obtained for other field states as well—see the next problem.

***Problem* 9.4.** Calculate the zero-delay value $g^{(2)}(0)$ of the second-order correlation function of a single-mode electromagnetic field in the squeezed ground state ζ defined by Eq. (5.142) of the lecture notes.

Solution: Let us reuse the important intermediate result,

$$\hat{b}\hat{b}^\dagger = \hat{b}^\dagger \hat{b} + \hat{I}, \qquad (*)$$

obtained in the model solution of Problem 5.17 for the squeezing creation-annihilation operators $\hat{b}^\dagger$ and $\hat{b}$, defined by Eq. (5.144). According to the same solution, that definition yields

$$\hat{a} = \mu\hat{b} - \nu\hat{b}^\dagger, \quad \hat{a}^\dagger = \mu\hat{b}^\dagger - \nu^*\hat{b},$$

[6] The result (***) even tends to infinity at $\alpha \to 0$, but we should remember that if $\alpha = 0$ exactly, the field is in its ground state, and according to Eq. (9.29) with $n = 0$, there are no photon counts at all to speak about.

where μ and ν are the c-number parameters of the squeezed ground state:

$$\mu \equiv \cosh r, \quad \nu \equiv e^{i\theta}\sinh r,$$

which is an eigenstate of the operator $\hat{b}$, with zero eigenvalue:

$$\hat{b}|\zeta\rangle = 0, \quad \langle\zeta|\hat{b}^\dagger = 0, \qquad (**)$$

and is normalized:

$$\langle\zeta|\zeta\rangle = 1.$$

Using these relations, we see that the average participating in the denominator of the expression for the required correlation function (9.35),

$$g^{(2)}(0) = \frac{\langle\hat{a}^\dagger\hat{a}^\dagger\hat{a}\hat{a}\rangle}{\langle\hat{a}^\dagger\hat{a}\rangle^2},$$

was already calculated in the model solution of problem 5.18:

$$\langle\hat{a}^\dagger\hat{a}\rangle \equiv \langle\zeta|\hat{a}^\dagger\hat{a}|\zeta\rangle = \nu\nu^* \equiv \sinh^2 r,$$

so that we only need to calculate the expectation value in the numerator of this fraction:

$$\begin{aligned}\langle\hat{a}^\dagger\hat{a}^\dagger\hat{a}\hat{a}\rangle &\equiv \langle\zeta|\,\hat{a}^\dagger\hat{a}^\dagger\hat{a}\hat{a}\,|\zeta\rangle = \langle\zeta|\,(\mu\hat{b}^\dagger - \nu^*\hat{b})^2(\mu\hat{b} - \nu\hat{b}^\dagger)^2|\zeta\rangle \\ &= \langle\zeta|[\mu^2\hat{b}^\dagger\hat{b}^\dagger + (\nu^*)^2\hat{b}\hat{b} - \mu\nu^*(\hat{b}^\dagger\hat{b} + \hat{b}\hat{b}^\dagger)] \\ &\quad\times[\mu^2\hat{b}\hat{b} + \nu^2\hat{b}^\dagger\hat{b}^\dagger - \mu\nu\,(\hat{b}^\dagger\hat{b} + \hat{b}\hat{b}^\dagger)]|\zeta\rangle.\end{aligned}$$

The multiplication of the square brackets gives 16 operator-product terms, but due to Eqs. (**) we may immediately discard all operator products either starting with $\hat{b}^\dagger$, or ending with $\hat{b}$, or both, because they vanish. The remaining four terms are

$$\begin{aligned}\langle\hat{a}^\dagger\hat{a}^\dagger\hat{a}\hat{a}\rangle &= \mu^2\nu\nu^*\langle\zeta|\hat{b}\hat{b}^\dagger\hat{b}\hat{b}^\dagger|\zeta\rangle + \nu^2(\nu^*)^2\langle\zeta|\hat{b}\hat{b}\hat{b}^\dagger\hat{b}^\dagger|\zeta\rangle \\ &\quad - \mu\nu^2\nu^*\langle\zeta|\hat{b}\hat{b}^\dagger\hat{b}^\dagger\hat{b}^\dagger|\zeta\rangle - \mu\nu(\nu^*)^2\langle\zeta|\hat{b}\hat{b}\hat{b}\hat{b}^\dagger|\zeta\rangle.\end{aligned}$$

Let us calculate each of the averages, applying the commutation relation (*). It makes sense to start with an auxiliary two-operator product average

$$\langle\zeta|\hat{b}\hat{b}^\dagger|\zeta\rangle = \langle\zeta|(\hat{b}^\dagger\hat{b} + \hat{I})|\zeta\rangle \equiv \langle\zeta|\hat{b}^\dagger\hat{b}|\zeta\rangle + \langle\zeta|\hat{I}|\zeta\rangle = 0 + 1 \equiv 1,$$

and then use this result, together with Eqs. (**), to calculate the four-operator product averages one-by-one:

$$\langle\zeta|\hat{b}\hat{b}^\dagger\hat{b}\hat{b}^\dagger|\zeta\rangle \equiv \langle\zeta|\hat{b}\hat{b}^\dagger(\hat{b}\hat{b}^\dagger)|\zeta\rangle = \langle\zeta|\hat{b}\hat{b}^\dagger(\hat{b}^\dagger\hat{b} + \hat{I})|\zeta\rangle$$
$$\equiv \langle\zeta|\hat{b}\hat{b}^\dagger\hat{b}^\dagger\hat{b}|\zeta\rangle + \langle\zeta|\hat{b}\hat{b}^\dagger|\zeta\rangle = 0 + 1 \equiv 1,$$
$$\langle\zeta|\hat{b}\hat{b}\hat{b}^\dagger\hat{b}^\dagger|\zeta\rangle \equiv \langle\zeta|\hat{b}(\hat{b}\hat{b}^\dagger)\hat{b}^\dagger|\zeta\rangle = \langle\zeta|\hat{b}(\hat{b}^\dagger\hat{b} + \hat{I})\hat{b}^\dagger|\zeta\rangle$$
$$\equiv \langle\zeta|\hat{b}\hat{b}^\dagger\hat{b}\hat{b}^\dagger|\zeta\rangle + \langle\zeta|\hat{b}\hat{b}^\dagger|\zeta\rangle = 1 + 1 \equiv 2,$$
$$\langle\zeta|\hat{b}\hat{b}^\dagger\hat{b}^\dagger\hat{b}^\dagger|\zeta\rangle \equiv \langle\zeta|(\hat{b}\hat{b}^\dagger)\hat{b}^\dagger\hat{b}^\dagger|\zeta\rangle = \langle\zeta|(\hat{b}^\dagger\hat{b} + \hat{I})\hat{b}^\dagger\hat{b}^\dagger|\zeta\rangle$$
$$\equiv \langle\zeta|\hat{b}^\dagger\hat{b}\hat{b}^\dagger\hat{b}^\dagger|\zeta\rangle + \langle\zeta|\hat{b}^\dagger\hat{b}^\dagger|\zeta\rangle = 0 + 0 \equiv 0,$$
$$\langle\zeta|\hat{b}\hat{b}\hat{b}\hat{b}^\dagger|\zeta\rangle \equiv \langle\zeta|\hat{b}\hat{b}(\hat{b}\hat{b}^\dagger)|\zeta\rangle = \langle\zeta|\hat{b}\hat{b}(\hat{b}^\dagger\hat{b} + \hat{I})|\zeta\rangle$$
$$\equiv \langle\zeta|\hat{b}\hat{b}\hat{b}^\dagger\hat{b}|\zeta\rangle + \langle\zeta|\hat{b}\hat{b}|\zeta\rangle = 0 + 0 \equiv 0.$$

Thus, we finally have

$$\langle\hat{a}^\dagger\hat{a}^\dagger\hat{a}\hat{a}\rangle = \mu^2\nu\nu^* + 2\nu^2(\nu^*)^2 \equiv \cosh^2 r \sinh^2 r + 2\sinh^4 r,$$

so that

$$g^{(2)}(0) = \frac{\langle\hat{a}^\dagger\hat{a}^\dagger\hat{a}\hat{a}\rangle}{\langle\hat{a}^\dagger\hat{a}\rangle^2} = \frac{\cosh^2 r \sinh^2 r + 2\sinh^4 r}{(\sinh^2 r)^2} \equiv \coth^2 r + 2. \qquad (***)$$

Since $\coth^2 r$ of a real argument $r \equiv |\zeta|$ is always larger than 1, our result shows that $g^{(2)}(0) \geqslant 3$, indicating the super-bunching effect, mentioned in the end of section 9.2 of the lecture notes, and in the previous problem[7].

Note that, paradoxically, at $r \to 0$ Eq. (***) does not tend to Eq. (9.37), $g^{(2)}(0) = 1$, for the Glauber state, i.e. in the absence of squeezing. However, there is no contradiction here, because at $r = 0$ and $\alpha = 0$, both results yield an uncertainty due to vanishing denominator, physically corresponding to the absence of photon counts from the ground Fock/Glauber state.

***Problem* 9.5.** Calculate the rate of spontaneous photon emission (into unrestricted free space) by a hydrogen atom, initially in the $2p$ state ($n = 2$, $l = 1$) with $m = 0$. Would the result be different for $m = \pm1$? for the $2s$ state ($n = 2$, $l = 0$, $m = 0$)? Discuss the relation between these quantum-mechanical results and those given by the classical theory of radiation, using the simplest classical model of the atom.

Solution: For this single-electron system, we may use the general formula (9.53) of the lecture notes, for the electric-dipole radiation into free space, with $\mathbf{d} = -e\mathbf{r}$:

$$\Gamma_s = \frac{1}{4\pi\varepsilon_0}\frac{4\omega^3}{3\hbar c^3}\langle\text{fin}|\,\hat{\mathbf{d}}|\text{ini}\rangle \cdot \langle\text{ini}|\hat{\mathbf{d}}\,|\text{fin}\rangle^*$$
$$= \frac{e^2}{4\pi\varepsilon_0}\frac{4\omega^3}{3\hbar c^3}\langle\text{fin}|\hat{\mathbf{r}}|\text{ini}\rangle \cdot \langle\text{ini}|\hat{\mathbf{r}}|\text{fin}\rangle^*,$$

[7] For recent experimental observations of super-bunching see, e.g. [1] and references therein.

in the first case with $\langle \text{fin}| \equiv \langle n, l, m|_{\text{fin}} = \langle 1, 0, 0|$, and $|\text{ini}\rangle \equiv |n, l, m\rangle_{\text{ini}} = |2, 1, 0\rangle$. Due to the axial symmetry of these two states, the only nonvanishing contribution to the matrix element of the radius-vector's operator is due to its z-component

$$\begin{aligned}\langle 2, 1, 0|\hat{z}|1, 0, 0\rangle &= \int \psi^*_{2,1,0}\, z\, \psi_{1,0,0} d^3r. \\ &= \int_0^{2\pi} d\varphi \int_0^{\pi} \sin\theta d\theta \int_0^{\infty} r^2 dr \times \mathcal{R}_{1,0}(r) Y_0^0(\theta, \varphi) \\ &\quad \times r\cos\theta \times \mathcal{R}_{2,1}(r) Y_1^0(\theta, \varphi).\end{aligned}$$

Using the formulas of section 3.5, with $r_0 = r_{\text{B}}$, this expression becomes[8]

$$\begin{aligned}&2\pi \int_0^{\pi} \sin\theta d\theta \int_0^{\infty} r^2 dr \times \frac{2}{r_{\text{B}}^{3/2}} e^{-r/r_{\text{B}}} \left(\frac{1}{4\pi}\right)^{1/2} \\ &\qquad \times r\cos\theta \times \frac{1}{(2r_B)^{3/2}} \frac{r}{\sqrt{3}\, r_{\text{B}}} e^{-r/2r_{\text{B}}} \left(\frac{3}{4\pi}\right)^{1/2} \cos\theta \\ &= \frac{r_{\text{B}}}{2\sqrt{2}} \int_{-1}^{+1} \xi^2 d\xi \int_0^{\infty} \zeta^4 \exp\left\{-\frac{3}{2}\zeta\right\} d\zeta = \frac{2^7\sqrt{2}}{3^5} r_{\text{B}},\end{aligned}$$

so that, finally,

$$\Gamma_{\text{s}} = \frac{e^2}{4\pi\varepsilon_0} \frac{2^{15}}{3^{10}} \frac{4\omega^3 r_{\text{B}}^2}{3\hbar c^3}. \qquad (*)$$

Now we may plug in expressions for the Bohr radius r_{B} (see Eq. (1.10) of the lecture notes) and the radiation frequency following from Eqs. (1.12) and (1.13):

$$\omega = \frac{E_{\text{ini}} - E_{\text{fin}}}{\hbar} = \frac{E_{\text{H}}}{2\hbar}\left(-\frac{1}{2^2} + \frac{1}{1^2}\right) \equiv \frac{3E_{\text{H}}}{8\hbar} = \frac{3}{8}\frac{m_{\text{e}}}{\hbar^3}\left(\frac{e^2}{4\pi\varepsilon_0}\right)^2,$$

and represent Eq. (*) in a form more convenient for estimates:

$$\frac{\Gamma_{\text{s}}}{\omega} = \frac{2^{11}}{3^9}\alpha^3 \approx 0.104\ \alpha^3,$$

thus confirming the crude estimate given by Eq. (9.54). Numerically, $\omega \approx 1.6 \times 10^{16}\ \text{s}^{-1}$, $\Gamma_{\text{s}} \approx 6.3 \times 10^8\ \text{s}^{-1}$, so that the relative half-width of the spectral line $\Gamma_{\text{s}}/\omega \approx 4 \times 10^{-8} \ll 1$.

Superficially it may look like for other $2p$ initial states, with $m = \pm 1$, the rate vanishes, because the initial function under the matrix element integral is now proportional to $\exp\{\pm im\varphi\}$, and the integral over the azimuthal angle vanishes. However, this is only true for the z-component of the dipole moment operator; for two other components, the solid-angle integral becomes

[8] Note that the same matrix element was calculated in the model solution of problem 6.26.

$$\int_0^{2\pi} d\varphi \int_0^{\pi} \sin\theta d\theta \;\; Y_0^0(\theta, \varphi)\,(\mathbf{n}_x \cos\varphi + \mathbf{n}_y \sin\varphi)\; Y_1^{\pm 1}(\theta, \varphi)$$

$$= \mp \left(\frac{3}{2}\right)^{1/2} \frac{1}{4\pi} \int_0^{2\pi} (\mathbf{n}_x \cos\varphi + \mathbf{n}_y \sin\varphi)\, e^{\pm i\varphi} d\varphi \int_0^{\pi} \sin\theta d\theta$$

$$= \mp \left(\frac{3}{2}\right)^{1/2} \frac{1}{2}(\mathbf{n}_x \pm i\mathbf{n}_y),$$

so that

$$|\langle 2, 1, \pm 1|\, \mathbf{r}|1, 0, 0\rangle|^2 = |\langle 2, 1, \pm 1|\mathbf{n}_x x + \mathbf{n}_y y|1, 0, 0\rangle|^2$$

$$= \left(\frac{2^7\sqrt{2}}{3^5} r_B\right)^2 \frac{1}{2}\, |\mathbf{n}_x \pm i\mathbf{n}_y|^2 = \left(\frac{2^7\sqrt{2}}{3^5} r_B\right)^2,$$

i.e. exactly the same result as for $m = 0$. In hindsight, this is very natural, because wavefunctions of all three $2p$ states with different m are similar, besides their spatial orientation—see, e.g. the second row of figure 3.20.

On the other hand, for the initial $2s$ state, both the initial and final wavefunctions are independent of θ and φ; and hence all Cartesian components of the solid-angle integral of the product $\psi_f^* \mathbf{r} \psi_i$ are equal to zero, so that the electric dipole transition rate vanishes. This zero result is a particular manifestation of the orbital selection rules (mentioned in 5.6 of the lecture notes, and proved in the model solution of problem 5.35): in the absence of spin effects, the electric dipole transitions are possible only if $\Delta l \equiv l_{\text{ini}} - l_{\text{fin}} = \pm 1$.

In order to compare Eq. (*) with the classical theory of radiation, we may calculate the average[9] electromagnetic radiation power emitted by an excited atom as

$$\mathscr{P}_Q = \hbar\omega\Gamma_s.$$

Using Eq. (*), the average power at the $2p \to 1s$ transition may be represented as

$$\mathscr{P}_Q = \frac{2^{14}}{3^9} \frac{e^2}{4\pi\varepsilon_0} \frac{\omega^4 r_B^2}{3c^3} \approx 0.83 \frac{e^2}{4\pi\varepsilon_0} \frac{\omega^4 r_B^2}{3c^3}.$$

This quantum-mechanical result may be compared with the following classical result for the average power radiated by a charge $q = -e$ moving around a circle of radius R:[10]

$$\mathscr{P}_C = 2\frac{e^2}{4\pi\varepsilon_0} \frac{\omega^4 R^2}{3c^3}.$$

[9] This average has the following sense: if we keep placing $N \gg 1/\Gamma_s$ atoms each unit time into the initial (excited) state, letting them decay spontaneously into the ground state after that, the total radiation power will be almost constant in time, and equal to $N\mathscr{P}$.

[10] This expression follows from the well-known Larmor formula for radiation at 1D motion (see, e.g. *Part EM* Eq. (8.29) for the free space, in which $Z = Z_0 = 1/\varepsilon_0 c$ and $v = c$), with the extra factor of 2, due to the radiation by two nonvanishing oscillating Cartesian components of the rotating electric dipole moment—see the model solution of *Part EM* problem 8.1.

The comparison of the two last formulas shows that the quantum and classical expressions are close if we identify R with r_B.[11] However not only the numerical coefficients, but more importantly, the very spirit of these results are very much different. The classical theory depends only on the initial, but not the final state (and hence there are no selection rules), does not have a good analog of the most important s-states (with $l = 0$), and most importantly, there is no fundamental notion of the ground state (with a finite R)—which produces no spontaneous radiation at all.

Still, it may be shown that the quantum-mechanical results approach the classical ones for transitions between the so-called *Rydberg states*, with $n \gg 1$ and hence $\langle r \rangle \gg r_B$.

***Problem* 9.6.** An electron has been placed on the lowest excited level of a spherically-symmetric, quadratic potential well $U(\mathbf{r}) = m_e\omega^2 r^2/2$. Calculate the rate of its relaxation to the ground state, with the emission of a photon (into unrestricted free space). Compare the rate with that for a similar transition of the hydrogen atom, for the case when the radiation frequencies of these two systems are equal.

Solution: Just like in the previous problem, the spontaneous photon emission rate of this single-particle system, due to its electric dipole moment $\mathbf{d} = q\mathbf{r}$, may be calculated as

$$\Gamma_s = \frac{e^2}{4\pi\varepsilon_0}\frac{4\omega^3}{3\hbar c^3}\langle \text{fin}|\, \hat{\mathbf{r}}|\text{ini}\rangle \cdot \langle \text{ini}|\, \hat{\mathbf{r}}\, |\text{fin}\rangle^*, \qquad (*)$$

so that the problem is reduced to the calculation of this matrix element for the 3D harmonic oscillator. As was discussed in section 3.5 of the lecture notes, and in the model solution of problem 3.23, its eigenfunctions may be represented either in the spherical-coordinate form or the Cartesian form. In the latter representation (which is a bit easier for calculations) the final, ground-state wavefunction is the product of similar 1D functions:

$$\psi_{\text{fin}}(\mathbf{r}) = \psi_0(x)\, \psi_0(y)\, \psi_0(z),$$

while the excited energy level is triple-degenerate, with wavefunctions of the type

$$\psi_{\text{ini}}(\mathbf{r}) = \psi_1(x)\, \psi_0(y)\, \psi_0(z),$$

and two other eigenfunctions different only by the argument swap—and of course giving the same radiation rate. (Here ψ_0 and ψ_1 are, respectively, the ground-state and the first excited state of the 1D oscillator.) Hence the matrix element participating in Eq. (*) may be calculated as

[11] In the view of the original Bohr's theory, and the exact relations (3.210)–(3.211), this is a reasonable (though an approximate) estimate.

$$\langle \text{ini}|\hat{\mathbf{r}}|\text{fin}\rangle \equiv \langle \text{ini}|\mathbf{n}_x\hat{x} + \mathbf{n}_y\hat{y} + \mathbf{n}_z\hat{z}|\text{fin}\rangle$$
$$= \int_{-\infty}^{+\infty} dx \int_{-\infty}^{+\infty} dy \int_{-\infty}^{+\infty} dz \ \psi_1^*(x)\,\psi_0^*(y)\,\psi_0^*(z)(\mathbf{n}_x x + \mathbf{n}_y y + \mathbf{n}_z z)\,\psi_0(x)\,\psi_0(y)\,\psi_0(z).$$

Since ψ_0 is an even function, while ψ_1 is an odd function of its argument (see, e.g. Eq. (2.284) and/or figure 2.35 of the lecture notes), only the first of the resulting three integrals does not vanish, giving

$$\langle \text{ini}|\,\hat{\mathbf{r}}|\text{fin}\rangle = \mathbf{n}_x \int_{-\infty}^{+\infty} \psi_1^*(x) x \psi_0(x) dx \int_{-\infty}^{+\infty} \psi_0^*(y)\psi_0(y) dy \int_{-\infty}^{+\infty} \psi_0^*(z)\,\psi_0(z) dz.$$

Taking all 1D wavefunctions in the normalized form, each the two last integrals equals 1, while the first one, by definition, is the matrix element x_{10}. According to Eq. (5.92), with $n' = 1$ and $n = 0$, for our system (with $\omega_0 = \omega$ and $m = m_e$) it is equal to $(\hbar/2m_e\omega)^{1/2}$. As a result, we get

$$\Gamma_s = \frac{e^2}{4\pi\varepsilon_0}\frac{4\omega^3}{3\hbar c^3}\frac{\hbar}{2m_e\omega}.$$

According to Eq. (5.95), the last fraction is just the ground-state expectation value of x^2, so that comparing this result with the solution of the previous problem, we get a physically very transparent result:

$$\frac{(\Gamma_s)_{\text{oscillator}}}{(\Gamma_s)_{\text{Hatom}}} = \frac{3^{10}}{2^{15}}\frac{\langle x^2\rangle}{r_B^2}.$$

Since the difference between the adjacent energy levels of a harmonic oscillator is $\hbar\omega$, its radiation frequency is just ω, so that for its value matching that of the hydrogen atom's radiation (at the $n = 2$ to $n = 1$ transition),

$$\omega = \frac{E_H}{\hbar}\left[\left(-\frac{1}{2\cdot 2^2}\right) - \left(-\frac{1}{2}\right)\right] = \frac{3}{8}\frac{E_H}{\hbar} \equiv \frac{3}{8}\frac{\hbar}{m_e r_B^2},$$

we get

$$\langle x^2\rangle = \frac{\hbar}{2m_e\omega} = \frac{\hbar}{2m_e}\Big/\frac{3}{8}\frac{\hbar}{m_e r_B^2} = \frac{4}{3}r_B^2,$$

$$\text{so that} \quad \frac{(\Gamma_s)_{\text{oscillator}}}{(\Gamma_s)_{\text{Hatom}}} = \frac{3^{10}}{2^{15}}\frac{4}{3} = \frac{3^9}{2^{13}} \approx 2.403 > 1.$$

This result is natural, because in the hydrogen atom, the wavefunctions are more pressed to the center by the rapidly diverging Coulomb potential $U(r) \propto -1/r$, and hence provide a smaller dipole moment d.

***Problem* 9.7.** Derive an analog of Eq. (9.53) of the lecture notes, for the spontaneous photon emission into the free space, due to a change of the *magnetic* dipole moment $\mathbf{m}$ of a small-size system.

Solution: The derivation is straightforward, due to the similarity between the Hamiltonians describing the interaction of electromagnetic field components with these dipole moments,

$$\hat{H}_{\text{electric}} = -\hat{\mathscr{E}} \cdot \hat{\mathbf{d}}, \quad \hat{H}_{\text{magnetic}} = -\hat{\mathscr{B}} \cdot \hat{\mathbf{m}},$$

and of the expressions of the field operators via the photon creation and annihilation operators—see Eqs. (9.16) of the lecture notes:

$$\hat{\mathscr{E}}(\mathbf{r}, t) = i\sum_j \left(\frac{\hbar\omega_j}{2}\right)^{1/2} \mathbf{e}_j(\mathbf{r}) \left(\hat{a}_j^\dagger - \hat{a}_j\right), \quad \hat{\mathscr{B}}(\mathbf{r}, t) = \sum_j \left(\frac{\hbar\omega_j}{2}\right)^{1/2} \mathbf{b}_j(\mathbf{r}) \left(\hat{a}_j^\dagger + \hat{a}_j\right).$$

(As a reminder, the c-number vector functions $\mathbf{e}_j(\mathbf{r})$ and $\mathbf{b}_j(\mathbf{r})$, describing the spatial distribution of the electric and magnetic fields in the jth mode, are similarly normalized,

$$\int e_j^2(\mathbf{r})d^3r = \frac{1}{\varepsilon_0}, \quad \int b_j^2(\mathbf{r})d^3r = \mu_0,$$

so that they similarly drop out at the averaging over free-space modes[12], besides the replacement of $1/\varepsilon_0$ in the electric field case for μ_0 in the magnetic field case.) As a result, reproducing the calculations carried out in sections 9.2–9.3 for the electric dipole case, we get

$$\Gamma_{\text{magnetic}} = \frac{\mu_0}{4\pi}\frac{4\omega^3}{3\hbar c^3} |\langle \text{fin}|\hat{\mathbf{m}}|\text{ini}\rangle|^2.$$

Due to the identity $\mu_0 \equiv 1/\varepsilon_0 c^2$, this relation may be rewritten as

$$\Gamma_{\text{magnetic}} = \frac{1}{4\pi\varepsilon_0}\frac{4\omega^3}{3\hbar c^3}\frac{1}{c^2} |\langle \text{fin}|\hat{\mathbf{m}}|\text{ini}\rangle|^2. \qquad (*)$$

Its comparison with Eq. (9.53) of the lecture notes,

$$\Gamma_{\text{electric}} = \frac{1}{4\pi\varepsilon_0}\frac{4\omega^3}{3\hbar c^3} |\langle \text{fin}|\,\hat{\mathbf{d}}|\text{ini}\rangle|^2, \qquad (**)$$

shows that their relation parallels that between the classical expressions for the intensity of the electric-dipole and magnetic-dipole radiation, with an extra factor c^2 in the denominator in the latter case[13]. Due to this factor, for an orbital motion of a charged particle with velocity $v \ll c$, the magnetic-dipole radiation power of such a particle is of the order of $\sim (v/c)^2$ of the electric-dipole one. This is the justification for the focus on the latter radiation in section 9.3; however, if a system does not

[12] This is not true for the electromagnetic cavities, where the exact position of the emitting particle in the standing wave field may play an important role, in particular affecting the coupling constant κ—see section 9.4 of the lecture notes.

[13] See, e.g. *Part EM* Eqs. (8.26) and (8.139).

change its electric dipole moment at a quantum transition, its magnetic-dipole radiation may be important—see, e.g. the next problem.

***Problem* 9.8.** A spin-½ particle, with gyromagnetic ratio γ, is in its orbital ground state in dc magnetic field $\mathscr{B}_0$. Calculate the rate of its spontaneous transition from the higher to the lower spin energy level, with the emission of a photon into the free space. Evaluate the rate for an electron in a field of 10 T, and discuss the implications of this result for laboratory experiments with electron spins.

Solution: Since the particle remains in the same (ground) orbital state, its electromagnetic radiation at such quantum transition may be only to the flip of its magnetic dipole moment $\mathbf{m} = \gamma\mathbf{S}$, with the emission of a photon with frequency[14]

$$\omega = \frac{|E_\uparrow - E_\downarrow|}{\hbar} = \frac{2|\mathbf{m}\cdot\mathscr{B}_0|}{\hbar} = \frac{2|\gamma\mathbf{S}\cdot\mathscr{B}_0|}{\hbar} = |\gamma\boldsymbol{\sigma}\cdot\mathscr{B}_0| = |\gamma|\mathscr{B}_0.$$

Hence we may use the result of the previous problem for the rate of such a spontaneous transition:

$$\Gamma_{\text{magnetic}} = \frac{1}{4\pi\varepsilon_0}\frac{4\omega^3}{3\hbar c^3}\frac{1}{c^2}\,|\langle \text{fin}|\hat{\mathbf{m}}|\text{ini}\rangle|^2.$$

In order to evaluate the matrix element participating in this formula, we may use the standard spin z-basis, with the axis z directed along the magnetic field, so that, for $\gamma > 0$, the initial spin state (with a higher energy) is $\downarrow$, and the final one is $\uparrow$, i.e.

$$\langle \text{fin}|\hat{\mathbf{m}}|\text{ini}\rangle = \langle\uparrow|\hat{\mathbf{m}}|\downarrow\rangle = \gamma(\mathbf{n}_x\langle\uparrow|\hat{S}_x|\downarrow\rangle + \mathbf{n}_y\langle\uparrow|\hat{S}_y|\downarrow\rangle + \mathbf{n}_z\langle\uparrow|\hat{S}_z|\downarrow\rangle).$$

By now, the reader probably knows these matrix elements by heart, but just in case, let us calculate them again—for example, using Eqs. (4.128) of the lecture notes:

$$\begin{aligned}\langle \text{fin}|\hat{\mathbf{m}}|\text{ini}\rangle &= \gamma\frac{\hbar}{2}\big[\mathbf{n}_x\langle\uparrow|\,(|\uparrow\rangle\langle\downarrow|+|\downarrow\rangle\langle\uparrow|)\,|\downarrow\rangle - i\,\mathbf{n}_y\langle\uparrow|\,(|\uparrow\rangle\langle\downarrow|-|\downarrow\rangle\langle\uparrow|)\,|\downarrow\rangle \\ &\quad + \mathbf{n}_z\langle\uparrow|\,(|\uparrow\rangle\langle\uparrow|-|\downarrow\rangle\langle\downarrow|)\,|\downarrow\rangle\big] \\ &= \gamma\frac{\hbar}{2}(\mathbf{n}_x - i\mathbf{n}_y),\end{aligned}$$

$$\text{so that}\quad |\langle \text{fin}|\hat{\mathbf{m}}|\text{ini}\rangle|^2 = \left(\gamma\frac{\hbar}{2}\right)^2 |\mathbf{n}_x - i\mathbf{n}_y|^2 = 2\left(\gamma\frac{\hbar}{2}\right)^2.$$

As a result, we get

$$\Gamma_{\text{magnetic}} = \frac{1}{4\pi\varepsilon_0}\frac{4\omega^3}{3\hbar c^3}\frac{\gamma^2\hbar^2}{2c^2}.$$

[14] The conservation of the angular momentum of the system is achieved by the emission of a circularly-polarized photon, carrying away a momentum of magnitude $\hbar$.

For an electron, with $\gamma = \gamma_e = -e/m_e$, this expression may be rewritten as

$$\Gamma_{\text{magnetic}} = \frac{e^2}{4\pi\varepsilon_0}\frac{4\omega^3}{3\hbar c^3}\frac{1}{2}\left(\frac{\hbar}{m_e c}\right)^2 \equiv \frac{e^2}{4\pi\varepsilon_0}\frac{4\omega^3}{3\hbar c^3}\frac{(\alpha r_B)^2}{2},$$

where $\alpha \equiv e^2/4\pi\varepsilon_0\hbar c \approx 1/137$ is the fine structure constant. Comparing this expression with the solution of problem 9.5 for the electric-dipole emission of a hydrogen atom,

$$\Gamma_{\text{electric}} \sim \frac{e^2}{4\pi\varepsilon_0}\frac{4\omega^3}{3\hbar c^3}r_B^2,$$

we see that the decay rate due to the spin flip is of the same order as that due to the electric-dipole radiation of an extremely small system, with a linear size of the order of $\alpha r_B \approx r_B/137$, even if the radiation frequency ω is the same.

Even more importantly for applications, in realistic laboratory magnetic fields the spin-flip radiation frequency $\omega = |\gamma|\mathscr{B}_0 = e\mathscr{B}_0/m_e$ is much lower than those ($\omega \sim 10^{16}\,\text{s}^{-1}$) of the optical transitions in atoms; for example, even for $\mathscr{B}_0 = 10$ T (a pretty high field), $\omega \approx 1.76 \times 10^{12}\,\text{s}^{-1}$. As a result, the rate of the spontaneous spin flips is extremely low: for the above case, $\Gamma \approx 4.4 \times 10^{-8}$ s. This means that the lifetime $\tau = 1/\Gamma$ of the spin states is limited by such flips to $\sim 2.3 \times 10^7\,\text{s} \approx 265$ days—the time more than sufficient to carry out virtually any imaginable experiment (say, of the Stern–Gerlach type) with electron spins.

***Problem* 9.9.** Calculate the rate of spontaneous transitions between the two sublevels of the ground state of a hydrogen atom, formed as a result of its hyperfine splitting. Discuss the implications of the result for the width of the 21-cm spectral line of hydrogen.

Solution: As was discussed in the model solution of problem 8.5, the hyperfine splitting of the ground state energy of the hydrogen atom, due to the interaction between the spins of its electron and nucleus (proton), separates the three spin-triplet states, with the (z-basis) ket-vectors

$$\begin{aligned}|s_+\rangle &= \frac{1}{\sqrt{2}}(|\uparrow\rangle_e|\downarrow\rangle_p + |\downarrow\rangle_e|\uparrow\rangle_p),\\ |s_{\uparrow\uparrow}\rangle &= |\uparrow\rangle_e|\uparrow\rangle_p, \text{ and } \quad |s_{\downarrow\downarrow}\rangle = |\downarrow\rangle_e|\downarrow\rangle_p,\end{aligned} \qquad (*)$$

from the spin singlet state, with the ket-vector

$$|s_-\rangle = \frac{1}{\sqrt{2}}(|\uparrow\rangle_e|\downarrow\rangle_p - |\downarrow\rangle_e|\uparrow\rangle_p), \qquad (**)$$

giving the states (*) a slightly higher energy.

Since the magnitude of the gyromagnetic ratio of the proton, $\gamma_p = g_p e/2m_p$, is much smaller than that of the electron, $\gamma_e = -g_e e/2m_e \approx -e/m_e$ (due to the proton's larger mass, $m_p \gg m_e$), its interaction with the electromagnetic field is proportionally smaller, and may be neglected. As a result, the radiation is due to the flip of electron's spin, and its rate may be calculated using the formula derived in the solution of problem 9.7:

$$\Gamma_s = \frac{1}{4\pi\varepsilon_0}\frac{4\omega^3}{3\hbar c^3}\frac{1}{c^2}\,|\langle \text{fin}|\hat{\mathbf{m}}|\text{ini}\rangle|^2,$$

with the magnetic dipole moment operator $\hat{\mathbf{m}} = \gamma_e \hat{\mathbf{S}}_e$, just as the previous problem. However, this element now has to be calculated for each of the three initial states (*), with the same final state (**).

Let us start from the first, entangled-triplet initial state s_+:

$$\begin{aligned}\langle \text{fin}|\hat{\mathbf{m}}|\text{ini}\rangle &= \langle s_-|\hat{\mathbf{m}}|s_+\rangle = \gamma_e\langle s_-|\hat{S}_e|s_+\rangle \\ &= \gamma_e(\mathbf{n}_x\langle s_-|\hat{S}_{ex}|s_+\rangle + \mathbf{n}_y\langle s_-|\hat{S}_{ey}|s_+\rangle + \mathbf{n}_z\langle s_-|\hat{S}_{ez}|s_+\rangle).\end{aligned}$$

These partial matrix elements may be calculated, for example, just as in the previous problem, using Eqs. (4.128) of the lecture notes, while remembering that the electron spin operators act only on the electron spin kets:

$$\begin{aligned}\langle s_-|\hat{S}_{ex}|s_+\rangle &= \frac{1}{\sqrt{2}}(\langle\uparrow|_e\langle\downarrow|_p - \langle\downarrow|_e\langle\uparrow|_p) \\ &\times\frac{\hbar}{2}(|\uparrow\rangle_e\langle\downarrow|_e + |\downarrow\rangle_e\langle\uparrow|_e)\frac{1}{\sqrt{2}}\,(|\uparrow\rangle_e|\downarrow\rangle_p + |\downarrow\rangle_e|\uparrow\rangle_p).\end{aligned}$$

Due to the orthonormality of the proton spin states, this expression is reduced to

$$\begin{aligned}\langle s_-|\hat{S}_{ex}|s_+\rangle &= \frac{\hbar}{4}\langle\uparrow|_e(|\uparrow\rangle_e\langle\downarrow|_e + |\downarrow\rangle_e\langle\uparrow|_e)\,|\uparrow\rangle_e \\ &- \frac{\hbar}{4}\langle\downarrow|_e(|\uparrow\rangle_e\langle\downarrow|_e + |\downarrow\rangle_e\langle\uparrow|_e)\,|\downarrow\rangle_e = 0.\end{aligned}$$

The result for the y-component of the spin operator is similar, but the z-component bracket is different from zero:

$$\begin{aligned}\langle s_-|\hat{S}_{ez}|s_+\rangle &= \frac{1}{2}(\langle\uparrow|_e\langle\downarrow|_p - \langle\downarrow|_e\langle\uparrow|_p) \\ &\times\frac{\hbar}{2}(|\uparrow\rangle_e\langle\uparrow|_e - |\downarrow\rangle_e\langle\downarrow|_e)\;(|\uparrow\rangle_e|\downarrow\rangle_p + |\downarrow\rangle_e|\uparrow\rangle_p) \\ &= \frac{\hbar}{4}\langle\uparrow|_e(|\uparrow\rangle_e\langle\uparrow|_e - |\downarrow\rangle_e\langle\downarrow|_e)|\uparrow\rangle_e \\ &- \frac{\hbar}{4}\langle\downarrow|_e(|\uparrow\rangle_e\langle\uparrow|_e - |\downarrow\rangle_e\langle\downarrow|_e)|\downarrow\rangle_e = \frac{\hbar}{2}.\end{aligned}$$

As a result, we get

$$\langle \text{fin}|\hat{\mathbf{m}}|\text{ini}\rangle = \gamma_e\mathbf{n}_z\frac{\hbar}{2}, \quad |\langle \text{fin}|\hat{\mathbf{m}}|\text{ini}\rangle|^2 = \left(\gamma_e\frac{\hbar}{2}\right)^2\mathbf{n}_z^2 = \left(\gamma_e\frac{\hbar}{2}\right)^2 = \left(\frac{e\hbar}{2m_e}\right)^2. \qquad (***)$$

(Note that this value is twice smaller than that in the previous problem, due to a different, two-spin character of the initial and final states.)

Now let us perform the similar calculation for the first simple (factorable) triplet state, $s_{\uparrow\uparrow}$. For this state, the z-component's matrix element vanishes:

$$\begin{aligned}\langle s_-|\hat{S}_{ez}|s_{\uparrow\uparrow}\rangle &= \frac{1}{\sqrt{2}}(\langle\uparrow|_e\langle\downarrow|_p - \langle\downarrow|_e\langle\uparrow|_p)\\ &\times\frac{\hbar}{2}(|\uparrow\rangle_e\langle\uparrow|_e - |\downarrow\rangle_e\langle\downarrow|_e)(|\uparrow\rangle_e|\uparrow\rangle_p)\\ &= -\frac{\hbar}{2\sqrt{2}}\langle\downarrow|_e(|\uparrow\rangle_e\langle\uparrow|_e - |\downarrow\rangle_e\langle\downarrow|_e)|\uparrow\rangle_e = 0,\end{aligned}$$

but the x- and y-components do not. Due to the similarity of their operators (see Eq. (4.128) again), it is convenient to calculate them in one shot:

$$\begin{aligned}\langle s_-|\mathbf{n}_x\hat{S}_{ex} + \mathbf{n}_y\hat{S}_{ey}|s_{\uparrow\uparrow}\rangle &= \frac{1}{\sqrt{2}}(\langle\uparrow|_e\langle\downarrow|_p - \langle\downarrow|_e\langle\uparrow|_p)\\ &\times\frac{\hbar}{2}[(\mathbf{n}_x - i\mathbf{n}_y)\,|\uparrow\rangle_e\langle\downarrow|_e + (\mathbf{n}_x + i\mathbf{n}_y)\,|\downarrow\rangle_e\langle\uparrow|_e]\,(|\uparrow\rangle_e|\uparrow\rangle_p)\\ &= -\frac{\hbar}{2\sqrt{2}}\langle\downarrow|_e[(\mathbf{n}_x - i\mathbf{n}_y)|\uparrow\rangle_e\langle\downarrow|_e + (\mathbf{n}_x + i\mathbf{n}_y)\,|\downarrow\rangle_e\langle\uparrow|_e]|\uparrow\rangle_e\\ &= -\frac{\hbar}{2\sqrt{2}}(\mathbf{n}_x + i\mathbf{n}_y),\end{aligned}$$

so that

$$|\langle\text{fin}|\hat{\mathbf{m}}|\text{ini}\rangle|^2 = \gamma_e^2\left(\frac{\hbar}{2\sqrt{2}}\right)^2|\mathbf{n}_x + i\mathbf{n}_y|^2 = \gamma_e^2\left(\frac{\hbar}{2\sqrt{2}}\right)^2\cdot 2 \equiv \left(\gamma_e\frac{\hbar}{2}\right)^2,$$

similarly to Eq. (***). An absolutely similar calculation for the second factorable initial state, $s_{\downarrow\downarrow}$, gives the same final result. So, the spontaneous photon emission rate,

$$\Gamma_s = \frac{e^2}{4\pi\varepsilon_0}\frac{4\omega^3}{3\hbar c^3}\left(\frac{\hbar}{2m_e c}\right)^2, \quad \text{so that} \quad \frac{\Gamma_{ss}}{\omega} = \frac{\alpha}{3}\left(\frac{\hbar\omega}{m_e c^2}\right)^2,$$

where α is the fine structure constant, does not depend on which exactly of the states (*), or their linear superposition, the system initially is.

Plugging in the fundamental constants, and the radiation frequency $\omega = \omega_{ss} \approx 0.892 \times 10^{10}\,\text{s}^{-1}$ (the famous 21-cm line, see the model solution of problem 8.5), we get $\Gamma_s \approx 2.86 \times 10^{-15}\,\text{s}^{-1}$.[15] Thus, the 'natural' (fundamental) broadening of the 21-cm line, due to the spontaneous character of the radiation, is extremely small:

[15] Reformulated into the very long decay time $\tau_s = 1/\Gamma_s \sim 3 \times 10^{14}\,\text{s} \sim 10^7$ years, this result means that such spontaneous radiation events are extremely rare. Indeed, most of the observed 21 line radiation coming from space is due to the transitions both up and down across the hyperfine energy gap of hydrogen atoms, induced by electromagnetic radiation from other sources, including the cosmic microwave background.

$$\frac{\Gamma_s}{\omega_{ss}} \approx 3.2 \times 10^{-25}.$$

The much larger width ($\Delta\omega \sim 10^{-3}\omega_{ss}$) of the 21-cm radiation, observed from various space regions, is mostly due to the Doppler effect, caused by the random magnitudes and directions of thermal velocities of the emitting hydrogen atoms.

***Problem* 9.10.** Find the eigenstates and eigenvalues of the Janes–Cummings Hamiltonian (9.78), and discuss their behavior near the resonance point $\omega = \Omega$.

Solution: The functional form of both eigenstates with eigenenergies close to E_n (9.80) is given by Eq. (9.82) of the lecture notes:

$$|\alpha\rangle = c_+|+\rangle + c_-|-\rangle, \qquad (*)$$

where

$$|+\rangle \equiv |\uparrow\rangle \otimes |n-1\rangle, \quad |-\rangle \equiv |\downarrow\rangle \otimes |n\rangle.$$

Plugging this solution into the stationary Schrödinger equation corresponding to the Janes–Cummings Hamiltonian (9.78),

$$\left(\frac{\hbar\Omega}{2}\hat{\sigma}_z + \hbar\omega\hat{a}^\dagger\hat{a} + \frac{\hbar\kappa}{2}\hat{\sigma}_+\hat{a} + \frac{\hbar\kappa}{2}\hat{\sigma}_-\hat{a}^\dagger\right)|\alpha\rangle = E|\alpha\rangle,$$

and taking into account the following results of operator action on the component states $|\pm\rangle$ (which readily follow from what the reader already knows about two-level systems and harmonic oscillators),

$$\hat{\sigma}_z|\pm\rangle = \pm|\pm\rangle, \qquad \hat{a}^\dagger\hat{a}|+\rangle \equiv (n-1)|+\rangle, \qquad \hat{a}^\dagger\hat{a}|-\rangle \equiv n|-\rangle,$$

$$\hat{\sigma}_+\hat{a}\,|+\rangle = 0, \qquad \hat{\sigma}_+\hat{a}\,|-\rangle = 2n^{1/2}|+\rangle, \qquad \hat{\sigma}_-\hat{a}^\dagger|+\rangle = 2n^{1/2}|-\rangle, \qquad \hat{\sigma}_-\hat{a}^\dagger|-\rangle = 0,$$

we get the following system of two linear algebraic equations for the coefficients $c_\pm$:

$$\frac{\hbar\Omega}{2}c_+ + \hbar\omega(n-1)c_+ + \frac{\hbar\kappa}{2}2n^{1/2}c_- = Ec_+,$$

$$-\frac{\hbar\Omega}{2}c_- + \hbar\omega n c_- + \frac{\hbar\kappa}{2}2n^{1/2}c_+ = Ec_-.$$

Using the definitions (9.70) and (9.80) of the parameters E_n and ξ, and a natural notation for the eigenenergy deviation from the central value E_n,

$$\tilde{E} \equiv E - E_n,$$

the system takes a very simple form,

$$\begin{aligned}(\hbar\xi/2 - \tilde{E})\,c_+ + \hbar\kappa n^{1/2}c_- &= 0,\\ \hbar\kappa n^{1/2}c_+ + (-\hbar\xi/2 - \tilde{E})\,c_- &= 0,\end{aligned} \qquad (**)$$

so that the condition of its consistency,

$$\begin{vmatrix} \hbar\xi/2 - \tilde{E} & \hbar\kappa n^{1/2} \\ \hbar\kappa n^{1/2} & -\hbar\xi/2 - \tilde{E} \end{vmatrix} = 0,$$

gives a simple result for the Janes–Cummings eigenenergies:

$$\tilde{E}_\pm \equiv E - E_n = \pm\left[\left(\frac{\hbar\xi}{2}\right)^2 + (\hbar\kappa)^2 n\right]^{1/2}. \qquad (***)$$

At $\kappa \to 0$, this result is reduced to Eq. (9.79) of the lecture notes, while at $\kappa \neq 0$ it describes the anticrossing diagram that was encountered in the course so many times (see, e.g. figure 5.1 of the lecture notes), with the minimum sublevel splitting (at the exact resonance, $\Omega = \omega$),

$$\Delta E_n \equiv \tilde{E}_+ - \tilde{E}_- = 2\hbar\kappa n^{1/2}, \quad \text{at } \xi = 0.$$

Now we need to find the eigenstate vectors $|\alpha\rangle_\pm$ of the system, corresponding to the calculated eigenenergies. Since we already know their functional form (*), it is sufficient to calculate the pair of the coefficients $c_\pm$ for each eigenstate[16]. For that, we need to plug each of two solutions (***) back into any of Eqs. (**), use it to calculate the corresponding coefficient ratios,

$$\left(\frac{c_+}{c_-}\right)_\pm = \frac{\hbar\kappa n^{1/2}}{\tilde{E}_\pm - \hbar\xi/2},$$

and then use the normalization condition $|c_+|^2 + |c_-|^2 = 1$ to complete the calculation. The result may be best represented in the following form,

$$(c_+)_+ = (c_-)_- = \cos\theta, \quad (c_-)_+ = \sin\theta, \quad (c_+)_- = -\sin\theta,$$

$$\text{where} \quad \tan 2\theta \equiv \frac{\kappa n^{1/2}}{\xi}.$$

From here, it is easy to analyze the most important properties of the distribution coefficients. In particular:

(i) If the detuning's magnitude $|\xi|$ is much larger than the coupling factor $\kappa n^{1/2}$, the angle θ is close to 0, and the magnitude of one of the coefficients is much larger than that of the other one, i.e. the system essentially resides in one of the component states $|\pm\rangle$, meaning that the interaction of the spin and oscillator subsystems has virtually no effect on the system's properties.

(ii) In the opposite, most interesting limit $|\xi| \ll \kappa n^{1/2}$, $\theta \to \pm\pi/4$, i.e. the distribution coefficients are equal by magnitude, $|c_\pm| = 1/\sqrt{2}$, with opposite signs

[16] Note that they are frequently called *distribution coefficients*, because they specify how exactly each eigenstate is distributed between the simple 'partial' states of the system, in our particular case, between $|\pm\rangle$. Such coefficients are also an important notion of the classical theory of oscillations and waves—see, e.g. *Part CM* sections 5.5–5.6 and 6.1–6.3.

of the ratio c_+/c_- at the upper and lower sublevels, so that (besides an inconsequential common phase multiplier), we may write

$$|\alpha\rangle = \frac{1}{\sqrt{2}} \times \begin{cases} (|\uparrow\rangle \otimes |n-1\rangle + |\downarrow\rangle \otimes |n\rangle), & \text{with } E = E_n + \hbar\kappa n^{1/2}, \\ (|\uparrow\rangle \otimes |n-1\rangle - |\downarrow\rangle \otimes |n\rangle), & \text{with } E = E_n - \hbar\kappa n^{1/2}, \end{cases}$$

in exact analogy with Eqs. (2.169) and (2.175), and also Eqs. (4.114) of the lecture notes. This means that near the exact resonance, the partial states $|\pm\rangle$ interact most. As was discussed in section 9.4 of the lecture notes, the physics of this fact is that at $\omega \approx \Omega$, the partial energies of the two components of the coherent superposition (*) are very close, enabling their strong coherent mixing ('hybridization').

***Problem* 9.11.** Analyze the Purcell effect, mentioned in sections 9.3 and 9.4 of the lecture notes, qualitatively; in particular, calculate the so-called *Purcell factor* F_P, defined as the ratio of the spontaneous emission rates Γ_s of an atom to a resonant cavity (tuned exactly to the quantum transition frequency) and that to the free space.

Solution: If a cavity's coupling to environment is sufficiently large to suppress the phase coherence of components of the entangled states (9.82),[17] the photon emission by the atom is an incoherent effect, which may be described by the Golden Rule, just as has been done for free space in the beginning of section 9.3. The main necessary modification is that of the density of states, because Eq. (9.50) is valid only in the free space.

As we know from classical electrodynamics, in the absence of dissipation, i.e. of the coupling to environment, the frequency spectrum of a resonance cavity is discrete, so that its density of states, participating in Eq. (9.49) as ρ_{fin}, is a sum of delta-functions $\delta(E - \hbar\omega_j)$ at energies corresponding to resonance frequencies ω_j. Each of these delta-functions is essentially an infinitely narrow resonance curve, well known from the classical theory of oscillations—see, e.g. *Part CM* section 5.1. The theory says that small but nonvanishing dissipation broadens and gives the resonance a small but nonvanishing width—see, e.g. *Part CM* Eq. (5.22):

$$\rho_{\text{fin}}(E) = \delta(E - \hbar\omega_j) \to \frac{\delta}{\pi\hbar}\frac{1}{\delta^2 + \xi^2}, \quad \text{with } \xi \equiv \frac{E}{\hbar} - \omega_j, \tag{*}$$

where δ is the damping constant, related as $\delta = \omega_j/2Q$ to the Q-factor of the resonance[18], and the constant factor before the fraction is selected to satisfy the normalization condition

[17] The corresponding condition is $\delta \gg \kappa$, $|\xi|$, where the constant κ is defined in section 9.4, and δ is the damping constant—see below.

[18] Generally, the factors δ and Q depend on the resonance number j; I do not use the indices only to avoid formula cluttering. Note that Eq. (*) is strictly valid only if both δ and $|\xi|$ are much less than ω_j, meaning in particular that $Q \gg 1$.

$$\int_{E \approx \hbar\omega_j} \rho_{\text{fin}}(E)\, dE = 1.$$

In particular, at the exact tuning ($E = \hbar\omega_j$), Eq. (*) yields

$$\rho_{\text{fin}}(\hbar\omega_j) = \frac{1}{\pi\hbar\delta} = \frac{2Q}{\pi\hbar\omega}.$$

One more change to be made in the formulas of section 9.3 for the case of cavity with a non-degenerate resonance, Eq. (9.52) has to be replaced with some $e_d \leqslant e_{\max}$, where

$$e_{\max}^2 = \frac{1}{\varepsilon_0 V_{\text{cavity}}},$$

reflecting the facts that, first, the atom may not be necessarily placed into the point where the electric field of the cavity mode is largest, and, second, that only one mode is involved. With these replacements, the Purcell factor is

$$F_{\text{P}} \equiv \frac{(\Gamma_{\text{s}})_{\text{cavity}}}{(\Gamma_{\text{s}})_{\text{free space}}} = \frac{\left(\rho_{\text{fin}} e_d^2\right)_{\text{cavity}}}{\left(\rho_{\text{fin}} e_d^2\right)_{\text{free space}}} = 6\pi \frac{Q}{V_{\text{cavity}}} \left(\frac{\omega}{c}\right)^3 \left(\frac{e_d}{e_{\max}}\right)^2.$$

At the 'best' location of the atom (in the field's maximum), the last factor equals 1, and we get

$$F_{\text{P}} = 6\pi \frac{Q}{V_{\text{cavity}}} \left(\frac{\omega}{c}\right)^3 = \frac{3}{4\pi^2} Q \frac{\lambda^3}{V_{\text{cavity}}}.$$

In the lowest-frequency modes of a typical resonant cavity, the last ratio is of the order of 1 (see, e.g. *Part EM* section 7.9), so that the Purcell factor of a cavity scales as its Q-factor. For transitions between the adjacent Rydberg states, with very high values of the principal quantum number n, the electric dipole transition frequencies may be lowered to the microwave band ($\omega \sim 10^{11}\ \text{s}^{-1}$), in which superconducting cavities may have very high values of Q (up to $\sim 10^{12}$), so that the Purcell factor may be made very large.

***Problem* 9.12.** Prove that the Klein–Gordon equation (9.84) may be rewritten in the form similar to the non-relativistic Schrödinger equation (1.25), but for a two-component wavefunction, with the Hamiltonian represented (in the usual z-basis) by the following 2×2-matrix:

$$\text{H} = -(\sigma_z + i\sigma_y)\frac{\hbar^2}{2m}\nabla^2 + mc^2\sigma_z.$$

Use your solution to discuss the physical meaning of the wavefunction's components.

Solution: Plugging the general form of the two-component wavefunction[19],

$$\psi = \begin{pmatrix} \psi_+ \\ \psi_- \end{pmatrix},$$

where (in contrast with the apparently similar Eq. (9.123) of the lecture notes) $\psi_\pm$ are some *scalar* wavefunctions, into the Schrödinger equation to be proved,

$$i\hbar\frac{\partial\psi}{\partial t} = -(\sigma_z + i\sigma_y)\frac{\hbar^2}{2m}\nabla^2\psi + mc^2\sigma_z\psi,$$

and spelling out the Pauli matrices, we get

$$i\hbar\frac{\partial}{\partial t}\begin{pmatrix} \psi_+ \\ \psi_- \end{pmatrix} = -\begin{pmatrix} 1 & 1 \\ -1 & -1 \end{pmatrix}\frac{\hbar^2}{2m}\nabla^2\begin{pmatrix} \psi_+ \\ \psi_- \end{pmatrix} + mc^2\begin{pmatrix} 1 & 0 \\ 0 & -1 \end{pmatrix}\begin{pmatrix} \psi_+ \\ \psi_- \end{pmatrix}.$$

Performing the matrix-by-vector multiplications, and requiring both components of the resulting columns on the left-hand and right-hand sides to be equal, we get the following two scalar equations:

$$\begin{aligned} i\hbar\frac{\partial\psi_+}{\partial t} &= -\frac{\hbar^2}{2m}\nabla^2(\psi_+ + \psi_-) + mc^2\psi_+, \\ i\hbar\frac{\partial\psi_-}{\partial t} &= +\frac{\hbar^2}{2m}\nabla^2(\psi_+ + \psi_-) - mc^2\psi_-. \end{aligned} \qquad (*)$$

Comparing these equations with Eq. (9.84), rewritten in the form

$$\frac{1}{c^2}\frac{\partial^2\Psi}{\partial t^2} = \nabla^2\Psi - \frac{m^2c^2}{\hbar^2}\Psi, \qquad (**)$$

we see that their major difference is the absence of the second derivative over time in Eq. (*). This is why it is natural to look for the relation between the functions $\psi_\pm$ and Ψ in the following simple form:

$$\psi_\pm = C_\pm\Psi + D_\pm\frac{\partial\Psi}{\partial t}, \qquad (***)$$

where $C_\pm$ and $D_\pm$ are some *c*-number constants, so that the left-hand sides of Eqs. (*) would have terms proportional to $\partial^2\Psi/\partial t^2$. However, the comparison of Eqs. (*) and (**) shows that the sum $(\psi_+ + \psi_-)$ should not contain the derivative $\partial\Psi/\partial t$; this is only possible if the coefficients $D_\pm$ in Eqs. (***) are equal and opposite. Also, in order for two Eqs. (*) to give the same single Eq. (**), the coefficients $C_\pm$ have to be equal. With these observations, Eq. (***) is reduced to

[19] Here $\psi_\pm$ are functions of both **r** and t, and the lower-case letter is used only to distinguish this two-component function from the scalar (single-component) function $\Psi(\mathbf{r}, t)$ obeying the same Klein–Gordon equation, but written in the scalar form (9.84).

$$\psi_\pm = C\Psi \pm D\frac{\partial\Psi}{\partial t} \equiv C\left(\Psi \pm \lambda\frac{\partial\Psi}{\partial t}\right), \quad \text{with } \lambda \equiv \frac{D}{C}.$$

Plugging this relation into Eqs. (*), we see that they, after the cancellation of the normalization constant C, give two equivalent equations:

$$i\hbar\frac{\partial}{\partial t}\left(\Psi + \lambda\frac{\partial\Psi}{\partial t}\right) = -\frac{\hbar^2}{2m}\nabla^2(2\Psi) + mc^2\left(\Psi + \lambda\frac{\partial\Psi}{\partial t}\right),$$
$$i\hbar\frac{\partial}{\partial t}\left(\Psi - \lambda\frac{\partial\Psi}{\partial t}\right) = +\frac{\hbar^2}{2m}\nabla^2(2\Psi) - mc^2\left(\Psi - \lambda\frac{\partial\Psi}{\partial t}\right),$$

which differ from Eq. (**) only by the presence of the terms proportional to $\partial\Psi/\partial t$. However, these terms cancel if the constant λ is selected as

$$\lambda = \frac{i\hbar}{mc^2}, \quad \text{so that } \psi_\pm \equiv C\left(\Psi \pm \frac{i\hbar}{mc^2}\frac{\partial\Psi}{\partial t}\right),$$

giving two equivalent equations,

$$\mp\frac{\hbar^2}{mc^2}\frac{\partial^2\Psi}{\partial t^2} = \mp\frac{\hbar^2}{m}\nabla^2\Psi \pm mc^2\Psi,$$

which, after the division of all terms by $\mp\hbar^2/m$, coincide with the Klein–Gordon equation (**).

In order to interpret the spinor ψ, note a substantial (though incomplete) similarity between Eqs. (*) for the stationary states, i.e. with $\partial/\partial t \to E/i\hbar$:

$$(E - mc^2)\,\psi_+ = -\frac{\hbar^2}{2m}\nabla^2(\psi_+ + \psi_-),$$
$$(E + mc^2)\,\psi_- = \frac{\hbar^2}{2m}\nabla^2(\psi_+ + \psi_-),$$

and Eqs. (9.126) of the lecture notes, with $U(\mathbf{r}) = 0$, for a free spin-½ particle:

$$(E - mc^2)\,\psi_+ = c\hat{\boldsymbol{\sigma}} \cdot \hat{\mathbf{p}}\psi_-,$$
$$(E + mc^2)\,\psi_- = c\hat{\boldsymbol{\sigma}} \cdot \hat{\mathbf{p}}\psi_+.$$

From this analogy, and the discussion of the Dirac equation in sections 9.6 and 9.7 of the lecture notes, we may conclude that in both cases, $\psi_\pm$ describe, respectively, the particle and antiparticle components of the composite wavefunction. Not surprisingly, for spinless particles, which obey the Klein–Gordon equation, these components are just scalar functions of $\mathbf{r}$ and t, while for spin-½ particles, described by the Dirac equation, each of the $\psi_\pm$ is itself a two-component column (spinor), with each component describing an orbital wavefunction of a state with $S_z = \pm\hbar/2$.

***Problem* 9.13.** Calculate and discuss the energy spectrum of a relativistic, spinless, charged particle placed into an external uniform, time-independent magnetic field $\mathscr{B}$. Use the result to formulate the condition of validity of the non-relativistic theory in this situation.

Solution: In the absence of spin and particle creation/annihilation, we may use for analysis the relativistic Schrödinger equation that may be obtained by making the general replacements (9.90), with the momentum and Hamiltonian operators given by Eqs. (9.83), in the Klein–Gordon equation—see, e.g. the bottom-right cell of table 9.1. The replacement yields

$$\left(i\hbar\frac{\partial}{\partial t} - q\phi\right)^2\Psi = c^2(-i\hbar\nabla - q\mathbf{A})^2\Psi + (mc^2)^2\Psi. \qquad (*)$$

In a time-independent magnetic field, with $\mathbf{A} = \mathbf{A}(\mathbf{r})$ and $\phi = 0$, this equation has a set of stationary solutions, similar to those in the non-relativistic case,

$$\Psi(\mathbf{r}, t) = \psi(\mathbf{r})\exp\left\{-i\frac{E}{\hbar}t\right\}.$$

Plugging such a solution into Eq. (*), we get the following stationary relativistic Schrödinger equation

$$E^2\psi = c^2(-i\hbar\nabla - q\mathbf{A})^2\psi + (mc^2)^2\psi,$$

$$\text{i.e.}\quad \frac{1}{2m}(-i\hbar\nabla - q\mathbf{A})^2\psi = \frac{E^2 - (mc^2)^2}{2mc^2}\psi.$$

Let us compare the last equation with the non-relativistic Schrödinger equation for the same problem (see, e.g. Eq. (3.27) with $\phi = 0$), with some effective energy, referred (as usual in the non-relativistic theory) to the rest energy mc^2 of the particle:

$$\frac{1}{2m}(-i\hbar\nabla - q\mathbf{A})^2\psi = E_{\text{ef}}\psi,$$

We see that these two equations are identical, if

$$E_{\text{ef}} = \frac{E^2 - (mc^2)^2}{2mc^2}, \qquad \text{i.e. } E = \pm[(mc^2)^2 + 2mc^2E_{\text{ef}}]^{1/2},$$

where the two signs before the square root describe, respectively, the particle and the antiparticle[20].

Hence we may use the solution of the non-relativistic problem (see, e.g. section 3.2 of the lecture notes and the solution of problem 3.4 with $\mathscr{E} = 0$), in particular, its energy spectrum,

[20] Comparing the last expression with Eq. (9.1) of the lecture notes, we can see that E_{ef} is just $p^2/2m$, where $\mathbf{p} = \mathbf{P}—q\mathbf{A}$ is the relativistic kinetic ('Mv-') momentum of the particle.

$$\varepsilon_n(p_z) = \hbar\omega_c\left(n + \frac{1}{2}\right) + \frac{p_z^2}{2m},$$

$$\text{with} \quad \omega_c \equiv \frac{|q|\mathscr{B}}{m}, \; n = 0, 1, 2, \ldots, \text{and} \; -\infty < p_z < +\infty,$$

where the z-axis is directed along the magnetic field, to get the (exact!) energy spectrum of the relativistic problem:

$$E_n(p_z) = \pm[(mc^2)^2 + 2mc^2\varepsilon_n]^{1/2} \equiv \pm\left[(mc^2)^2 + c^2\hbar q\mathscr{B}(2n + 1) + c^2p_z^2\right]^{1/2}.$$

This means that as in the non-relativistic case, due to the free motion of the particle along the magnetic field, the spectrum consists of a series of parabolic continuous bands (each corresponding to a specific quantum number n), with shifted bottom values, corresponding to the purely transverse motion with $p_z = 0$:

$$|E_n|_{\min} = \pm[(mc^2)^2 + c^2\hbar q\mathscr{B}(2n + 1)]^{1/2} \equiv \pm mc^2\left[1 + \frac{\hbar\omega_c}{mc^2}(2n + 1)\right]^{1/2}.$$

From here we may return to the non-relativistic result for the Landau levels by expanding the right-hand side in the small dimensionless parameter $\hbar\omega_c/mc^2 \ll 1$, and keeping only two leading terms. Hence, the non-relativistic theory is valid only if, first, $cp_z \ll mc^2$ (i.e. $v_z \ll c$), and second, if

$$\hbar\omega_c \ll mc^2, \quad \text{i.e.} \; \mathscr{B} \ll \frac{m^2c^2}{\hbar|q|}.$$

Even for a particle as light as the electron, the high-hand part of the last relation is close to 4×10^9 T—the value much higher than the strongest static fields ($\sim 10^2$ T) created in laboratory. However, it is comparable to the fields conjectured in the so-called *magnetars*—a specific type of neutron stars[21]. (In a 'usual' neuron star, $\mathscr{B} \sim 10^6$T.)

***Problem* 9.14.** Prove Eq. (9.91) of the lecture notes for the energy spectrum of a hydrogen-line atom, starting from the relativistic Schrödinger equation.

Hint: A mathematical analysis of Eq. (3.193) shows that its eigenvalues are still given by Eq. (3.201), $\varepsilon_n = -1/2n^2$, with $n = l + 1 + n_r$, where $n_r = 0, 1, 2, \ldots$, even if the parameter l is not integer.

Solution: For the Coulomb field of the nucleus with charge $Q = +Ze$,

$$\varphi = \frac{Ze}{4\pi\varepsilon_0 r}, \qquad \mathbf{A} = 0,$$

the relativistic Schrödinger equation (see, e.g. Eq. (*) of the model solution of the previous problem), for an electron with the electric charge $q = -e$, takes the form

[21] For a recent review, see, e.g. [2].

$$\left(i\hbar\frac{\partial}{\partial t}+\frac{Ze^2}{4\pi\varepsilon_0 r}\right)^2\Psi=-\hbar^2c^2\nabla^2\Psi+(m_ec^2)^2\Psi,$$

so that the stationary Schrödinger equation is

$$\left(E+\frac{Ze^2}{4\pi\varepsilon_0 r}\right)^2\psi=-\hbar^2c^2\nabla^2\psi+(m_ec^2)^2\psi.$$

Looking for its solution in the same form (3.200) that was used in the non-relativistic case,

$$\psi(\mathbf{r})=\mathcal{R}_l(r)Y_l^m(\theta,\varphi),$$

we get the following equation for the radial function $\mathcal{R}_l$:

$$\left(E+\frac{Ze^2}{4\pi\varepsilon_0 r}\right)^2\mathcal{R}_l=-\hbar^2c^2\left[\frac{1}{r^2}\frac{d}{dr}\left(r^2\frac{d}{dr}\right)-\frac{l(l+1)}{r^2}\right]\mathcal{R}_l+(m_ec^2)^2\mathcal{R}_l.$$

Grouping the terms with $1/r$ in the same power, and dividing all terms by $2m_ec^2$, this equation may be rewritten in the form

$$-\frac{\hbar^2}{2m_er^2}\left\{\frac{d}{dr}\left(r^2\frac{d}{dr}\right)-\left[l(l+1)-\frac{1}{\hbar^2c^2}\left(\frac{Ze^2}{4\pi\varepsilon_0}\right)^2\right]\right\}\mathcal{R}_l-\frac{E}{m_ec^2}\frac{Ze^2}{4\pi\varepsilon_0 r}\mathcal{R}_l$$
$$=\frac{E^2-(m_ec^2)^2}{2m_ec^2}\mathcal{R}_l,$$

which is similar to the non-relativistic equation (see the Eq. (3.181) of the lecture notes, with $U(r)$ given by Eq. (3.190), and $C=Z_{\text{ef}}e^2/4\pi\varepsilon_0$ and $m=m_e$),

$$-\frac{\hbar^2}{2m_er^2}\left[\frac{d}{dr}\left(r^2\frac{d}{dr}\right)-l_{\text{ef}}(l_{\text{ef}}+1)\right]\mathcal{R}_l-\frac{Z_{\text{ef}}e^2}{4\pi\varepsilon_0 r}\mathcal{R}_l=E_{\text{ef}}\mathcal{R}_l,$$

with the following effective parameters:

$$l_{\text{ef}}(l_{\text{ef}}+1)\equiv l(l+1)-Z^2\alpha^2,\quad Z_{\text{ef}}\equiv\frac{E}{m_ec^2}Z,\quad E_{\text{ef}}\equiv\frac{E^2-(m_ec^2)^2}{2m_ec^2},\qquad (*)$$

where $\alpha\equiv(e^2/4\pi\varepsilon_0)/\hbar c\approx 1/137$ is the fine structure constant. Hence, we may use the energy spectrum of the non-relativistic problem, given by Eq. (3.201) with $C=Ze^2/4\pi\varepsilon_0$, and the fact provided in the Hint, to write

$$\begin{aligned}E_{\text{ef}}&=-\frac{E_{\text{H}}}{2n_{\text{ef}}^2}Z_{\text{ef}}^2\\&\equiv-\frac{\alpha^2m_ec^2}{2[(n_r+½)+(l_{\text{ef}}+½)]^2}Z_{\text{ef}}^2\qquad(**)\\&\equiv-\frac{\alpha^2m_ec^2}{2\{(n_r+½)+[l_{\text{ef}}(l_{\text{ef}}+1)+¼]^{1/2}\}^2}Z_{\text{ef}}^2.\end{aligned}$$

Now using the similar identity, $l(l+1) + ½ \equiv (l + ½)^2$, we get from Eqs. (*) and (**) the following simple equation for E:

$$\frac{E^2 - (m_e c^2)^2}{2m_e c^2} = -\frac{\alpha^2 m_e c^2}{2\{(n_r + ½) + [(l + ½)^2 - Z^2\alpha^2]^{1/2}\}^2}\left(\frac{E}{m_e c^2}Z\right)^2.$$

Solving it, we get

$$E = m_e c^2\left(1 + \frac{Z^2\alpha^2}{\{(n_r + ½) + [(l + ½)^2 - Z^2\alpha^2]^{1/2}\}^2}\right)^{-1/2};$$

with the notation

$$\lambda \equiv (n_r + ½) + [(l + ½)^2 - Z^2\alpha^2]^{1/2} \equiv n + [(l + ½)^2 - Z^2\alpha^2]^{1/2} - (l + ½),$$

this is just Eq. (9.91) of the lecture notes.

***Problem* 9.15.** Derive the general expression for the differential cross-section of the elastic scattering of a spinless relativistic particle by a static potential $U(\mathbf{r})$, in the Born approximation, and formulate the conditions of its validity. Use these results to calculate the differential cross-section of scattering of a particle with the electric charge $-e$ by the Coulomb electrostatic potential $\phi(\mathbf{r}) = Ze/4\pi\varepsilon_0 r$.

Solution: In the absence of spin, and of particle creation/annihilation, we may use the relativistic Schrödinger equation, which may be obtained by using the replacements (9.90), and the momentum and Hamiltonian operators given by Eqs. (9.83), in the Klein–Gordon equation—see, e.g. the bottom-right cell of table 9.1 of the lecture notes[22]. In our case, with $q\phi = U(\mathbf{r})$ and $\mathbf{A} = 0$, the replacement yields

$$\left[i\hbar\frac{\partial}{\partial t} - U(\mathbf{r})\right]^2\Psi = -\hbar^2c^2\nabla^2\Psi + (mc^2)^2\Psi. \qquad (*)$$

As was discussed in section 3.3 of the lecture notes, the elastic scattering may be analyzed using the definite-energy wavefunctions—the wave packets of a formally infinite spatial extension. As is evident from Eq. (*), it is satisfied by the usual variable-separated wavefunctions

$$\Psi(\mathbf{r}, t) = \psi(\mathbf{r})\exp\left\{-i\frac{E}{\hbar}t\right\},$$

with the spacial factor $\psi(\mathbf{r})$ satisfying the following stationary equation:

$$[E - U(\mathbf{r})]^2\psi = -\hbar^2c^2\nabla^2\psi + (mc^2)^2\psi.$$

Dividing both sides of this equation by $2mc^2$, we may rewrite it as

$$-\frac{\hbar^2}{2m}\nabla^2\psi + U_{\text{ef}}(\mathbf{r})\psi = E_{\text{ef}}\psi,$$

[22] See also the model solutions of the two previous problems.

with

$$U_{\text{ef}}(\mathbf{r}) \equiv \frac{U(\mathbf{r})[2E - U(\mathbf{r})]}{2mc^2}, \quad \text{and} \quad E_{\text{ef}} \equiv \frac{E^2 - (mc^2)^2}{2mc^2}.$$

But this equation exactly coincides with the non-relativistic Schrödinger equation (with the effective potential energy profile $U_{\text{ef}}(\mathbf{r})$ and effective particle energy E_{ef}) that was used, in particular, in section 3.3 of the lecture notes to derive Eq. (3.88) for the differential cross-section of scattering in the Born approximation. Hence we may use this equation, in the form

$$\frac{d\sigma}{d\Omega} = \left(\frac{m}{2\pi\hbar^2}\right)^2 \left|\int U_{\text{ef}}(\mathbf{r})\, e^{-i\mathbf{q}\cdot\mathbf{r}} d^3r\right|^2, \quad \text{with } \mathbf{q} \equiv \mathbf{k} - \mathbf{k}_0, \tag{**}$$

keeping in mind that the $\mathbf{k}$ and $\mathbf{k}_0$ are now effective rather than the actual wave vectors, whose magnitude is related, by the very familiar Eq. (1.89), to not the actual, but the effective energy of the particle:

$$k = k_0 = \left(\frac{2mE_{\text{ef}}}{\hbar^2}\right)^{1/2} = \frac{[E^2 - (mc^2)^2]^{1/2}}{\hbar c}.$$

The last expression has a simple physical sense: $k = k_0 = p/\hbar$, where the free-particle momentum p is given by the basic Eq. (9.1):

$$E^2 = (pc)^2 + (mc^2)^2.$$

The condition of validity of Eq. (**) may be obtained by the corresponding replacements in the non-relativistic condition (3.77):

$$(U_{\text{ef}})_0 \ll \frac{\hbar^2}{ma^2} \max[1, \ ka], \tag{***}$$

where $(U_{\text{ef}})_0$ is the proper scale of the effective potential's magnitude, and a denotes its characteristic spatial extension. Let us specify these conditions for our particular case of the Coulomb potential, when

$$U(\mathbf{r}) = -\frac{Ze^2}{4\pi\varepsilon_0 r}, \quad \text{so that } U_{\text{ef}}(\mathbf{r}) = -\frac{Ze^2}{4\pi\varepsilon_0 r}\frac{E}{mc^2} + \left(\frac{Ze^2}{4\pi\varepsilon_0 r}\right)^2 \frac{1}{2mc^2}.$$

As was discussed in the model solution of problem 3.8, the effective radius a of an exponentially screened Coulomb potential tends to infinity as the screening is gradually removed ($\lambda \to 0$); as a result the expression on the right-hand side of Eq. (***) is reduced to $(\hbar^2/ma^2)(ka) \equiv \hbar^2 k/ma$. Repeating the arguments given at the end of that solution, with the additional factor $E/mc^2 = M/m$ (where $M \equiv m/(1 - v^2/c^2)^{1/2}$ is the relativistic mass), and v is the particle's velocity, and replacing the non-relativistic relation $p = mv$ used there with the relativistic one, $p = Mv$,[23] the relation takes exactly the same form as in the non-relativistic case:

$$Z\alpha \ll \frac{v}{c} \leqslant 1,$$

where $\alpha \approx 1/137$ is the fine structure constant.

Now let us notice that if this condition is satisfied, the second term of $U_{ef}(\mathbf{r})$ is much smaller than the first one at all distances of interest, $r > 1/k$. Indeed,

$$\left|\frac{2^{\text{nd}}\text{ term}}{1^{\text{st}}\text{ term}}\right| = \frac{Ze^2}{4\pi\varepsilon_0 r}\frac{1}{2E} < \frac{Ze^2 k}{4\pi\varepsilon_0}\frac{1}{2E} \equiv \frac{Ze^2}{4\pi\varepsilon_0 c\hbar}\frac{cp}{2E} \equiv Z\alpha\frac{cp}{2E} \leqslant \frac{Z\alpha}{2} \ll 1.$$

Hence we may not only ignore the effect of this term on the Born approximation's validity condition (***), but also, in this approximation, drop it at the calculation of the differential cross-section (**). Since the remaining (first) term of $U_{ef}(\mathbf{r})$ differs from $U(\mathbf{r})$ only by the additional factor E/mc^2, we may reuse the solution of the same problem 3.8, getting

$$\frac{d\sigma}{d\Omega} = \left(\frac{Ze^2}{4\pi\varepsilon_0}\frac{E}{mc^2}\frac{2m}{\hbar^2 q^2}\right)^2 \equiv \left(\frac{Ze^2}{4\pi\varepsilon_0}\frac{2E}{\hbar^2 q^2 c^2}\right)^2.$$

Since the relativistic replacements keep intact the geometric relation between the magnitude k of vectors $\mathbf{k}$ and $\mathbf{k}_0$, and that of the vector $\mathbf{q}$,

$$q = 2k\sin\frac{\theta}{2},$$

where θ is the scattering angle (see the figure below), we may rewrite our result for the differential cross-section as

$$\frac{d\sigma}{d\Omega} = \left(\frac{Ze^2}{4\pi\varepsilon_0}\frac{E}{2\hbar^2 k^2 c^2}\right)^2 \frac{1}{\sin^4(\theta/2)} \equiv \left(\frac{Ze^2}{4\pi\varepsilon_0}\frac{E}{2p^2 c^2}\right)^2 \frac{1}{\sin^4(\theta/2)}.$$

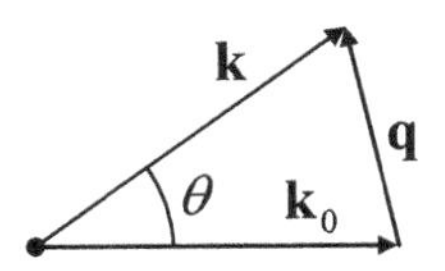

Hence the angular dependence of $d\sigma/d\Omega$ remains the same as in the non-relativistic quantum mechanics (and as in the classical mechanics), with a strong (non-integrable) divergence at $\theta \to 0$, and hence a formally infinite full cross-section.

---

[23] See, e.g. *Part EM* section 9.3, in particular Eqs. (9.70), (9.71), and (9.73). Note again that since the relativistic Schrödinger equation is the quantum-mechanical generalization of the classical relations of special relativity, they are *exactly* valid for the definite values $\mathbf{p}$ and E characterizing its plane-wave solutions $\psi \propto \exp\{i(\mathbf{p}\cdot\mathbf{r} - Et)/\hbar\}$ for a free particle.

***Problem* 9.16.** Starting from Eqs. (9.95)–(9.98) of the lecture notes, prove that the probability density w given by Eq. (9.101) and the probability current density $\mathbf{j}$ defined by Eq. (9.102) do indeed satisfy the continuity equation (1.52): $\partial w/\partial t + \nabla \cdot \mathbf{j} = 0$.

Solution: Very similarly to what was done in section 1.4 of the lecture notes for the non-relativistic Schrödinger equation, let us write Eq. (9.95), with the Dirac Hamiltonian (9.97), in the coordinate representation:

$$i\hbar\frac{\partial \Psi}{\partial t} = c\hat{\boldsymbol{\alpha}} \cdot \hat{\mathbf{p}}\, \Psi + mc^2\hat{\beta}\Psi, \quad \text{i.e.} \quad i\hbar\frac{\partial \Psi}{\partial t} = -i\hbar(c\hat{\boldsymbol{\alpha}} \cdot \nabla\, \Psi) + mc^2(\hat{\beta}\Psi), \qquad (*)$$

and left-multiply all its terms by the Hermitian conjugate wavefunction $\Psi^\dagger$, defined by the second of Eqs. (9.96):

$$i\hbar\Psi^\dagger\frac{\partial \Psi}{\partial t} = -i\hbar c\Psi^\dagger\hat{\boldsymbol{\alpha}} \cdot \nabla\, \Psi + mc^2\Psi^\dagger\hat{\beta}\Psi. \qquad (**)$$

(Since according to Eq. (9.96), both Ψ and $\Psi^\dagger$ are matrix rows/columns, while according to Eqs. (9.98), $\hat{\boldsymbol{\alpha}}$ and $\hat{\beta}$ are square matrices, all multiplications here and below should be understood in the matrix sense.)

Now let us write the Hermitian conjugate of Eq. (*):[24]

$$-i\hbar\frac{\partial \Psi^\dagger}{\partial t} = i\hbar c\, (\nabla\, \Psi^\dagger \cdot \hat{\boldsymbol{\alpha}}) + mc^2(\Psi^\dagger\hat{\beta}),$$

right-multiply all its terms by Ψ:

$$-i\hbar\frac{\partial \Psi^\dagger}{\partial t}\Psi = i\hbar c\nabla\, \Psi^\dagger \cdot \hat{\boldsymbol{\alpha}}\Psi + mc^2\Psi^\dagger\hat{\beta}\Psi,$$

and subtract this relation from Eq. (**). After the cancellation of the last terms, and of the common factor $i\hbar$ in the remaining terms, the result is

$$\left(\Psi^\dagger\frac{\partial \Psi}{\partial t} + \frac{\partial \Psi^\dagger}{\partial t}\Psi\right) + c(\Psi^\dagger\hat{\boldsymbol{\alpha}} \cdot \nabla\, \Psi + \nabla\, \Psi^\dagger \cdot \hat{\boldsymbol{\alpha}}\Psi) = 0. \qquad (***)$$

Now taking into account the differentiation identities[25]

$$\frac{\partial}{\partial t}(\Psi^\dagger\Psi) = \Psi^\dagger\frac{\partial \Psi}{\partial t} + \frac{\partial \Psi^\dagger}{\partial t}\Psi, \quad \text{and} \quad \nabla \cdot (\Psi^\dagger\hat{\boldsymbol{\alpha}}\Psi) = \Psi^\dagger\hat{\boldsymbol{\alpha}} \cdot \nabla\, \Psi + \nabla\, \Psi^\dagger \cdot \hat{\boldsymbol{\alpha}}\Psi,$$

we may spell out Eq. (***) as

$$\frac{\partial}{\partial t}(\Psi^\dagger\Psi) + c\nabla \cdot (\Psi^\dagger\hat{\boldsymbol{\alpha}}\Psi) = 0,$$

i.e. as the continuity equation for the probability density (9.101) and the probability current (9.102).

[24] Note the change of operators' order, following from the basic rule of the Hermitian conjugation of operator products: $(\hat{A}\hat{B})^\dagger = \hat{B}^\dagger\hat{A}^\dagger$—see, e.g. problem 4.1(iii).

[25] See, e.g. Eqs. (A.22) and (A.74*a*). (It is straightforward to check that these relations remain valid for multi-component functions like Ψ and vector-matrices like $\boldsymbol{\alpha}$.)

***Problem* 9.17.** Calculate the commutator of the operator $\hat{L}^2$ and the Dirac's Hamiltonian of a free particle. Compare the result with that for the non-relativistic Hamiltonian, and interpret the difference.

Solution: As was discussed in section 9.6 of the lecture notes, since the vector operator $\hat{\mathbf{L}} \equiv \hat{\mathbf{r}} \times \hat{\mathbf{p}}$ is defined in the Hilbert space of orbital states of a particle, it commutes with the spin operators (9.98), and hence with the second term of the Dirac Hamiltonian (9.97). Let us calculate the commutator of its jth Cartesian component L_j with the first term of that Hamiltonian[26]:

$$[\hat{L}_j, \hat{H}] = [\hat{L}_j, c\hat{\boldsymbol{\alpha}} \cdot \hat{\mathbf{p}}] = c\sum_{j'=1}^{3}[\hat{L}_j, \hat{\alpha}_{j'}\hat{p}_{j'}] = c\sum_{j'=1}^{3}\hat{\alpha}_{j'}[\hat{L}_j, \hat{p}_{j'}].$$

According to the second of Eqs. (5.149), the last commutator equals $i\hbar p_{j''}\varepsilon_{jj'j''}$, where $\varepsilon_{jj'j''}$ is the Levi-Civita symbol[27], so that

$$[\hat{L}_j, \hat{H}] = i\hbar c\sum_{j'=1}^{3}\hat{\alpha}_{j'}\hat{p}_{j''}\varepsilon_{jj'j''}. \qquad (*)$$

Rewriting this commutation relation as

$$\hat{L}_j\hat{H} = \hat{H}\hat{L}_j + i\hbar c\sum_{j'=1}^{3}\hat{\alpha}_{j'}\hat{p}_{j''}\varepsilon_{jj'j''},$$

we may use it to calculate

$$\begin{aligned}
\left[\hat{L}_j^2, \hat{H}\right] &= \hat{L}_j\hat{L}_j\hat{H} - \hat{H}\hat{L}_j\hat{L}_j \equiv \hat{L}_j\,(\hat{L}_j\hat{H}) - \hat{H}\hat{L}_j\hat{L}_j \\
&= \hat{L}_j\left(\hat{H}\hat{L}_j + i\hbar c\sum_{j'=1}^{3}\hat{\alpha}_{j'}\hat{p}_{j''}\varepsilon_{jj'j''}\right) - \hat{H}\hat{L}_j\hat{L}_j \\
&= (\hat{L}_j\hat{H})\hat{L}_j - \hat{H}\hat{L}_j\hat{L}_j + i\hbar c\sum_{j'=1}^{3}\hat{\alpha}_{j'}\hat{L}_j\hat{p}_{j''}\varepsilon_{jj'j''} \\
&= \left(\hat{H}\hat{L}_j + i\hbar c\sum_{j'=1}^{3}\hat{\alpha}_{j'}\hat{p}_{j''}\varepsilon_{jj'j''}\right)\hat{L}_j - \hat{H}\hat{L}_j\hat{L}_j + i\hbar c\sum_{j'=1}^{3}\hat{\alpha}_{j'}\hat{L}_j\hat{p}_{j''}\varepsilon_{jj'j''} \\
&= i\hbar c\sum_{j'=1}^{3}\hat{\alpha}_{j'}\left\{\hat{L}_j, \hat{p}_{j''}\right\}\varepsilon_{jj'j''}.
\end{aligned}$$

[26] Actually, this is just a straightforward generalization of the calculation made for L_x in the lecture notes—see Eq. (9.106).

[27] If still necessary, see Eq. (A.83) again.

Now the sum of these relations over all $j = 1, 2, 3$ yields the required commutator:

$$[\hat{L}^2, \hat{H}] = \left[\sum_{j=1}^{3} \hat{L}_j^2, \hat{H}\right] = \sum_{j=1}^{3}\left[\hat{L}_j^2, \hat{H}\right] = i\hbar c \sum_{j,j'=1}^{3} \hat{\alpha}_{j'}\left\{\hat{L}_j, \hat{p}_{j''}\right\} \varepsilon_{jj'j''}. \qquad (**)$$

Since this commutator is not equal to zero, the expectation value of L^2 of a spin-½ particle is *not* conserved at its free motion—and neither are the Cartesian components L_j. This result differs dramatically from the situation in non-relativistic quantum mechanics, where the free-particle's Hamiltonian,

$$\hat{H} = \frac{\hat{p}^2}{2m},$$

commutes with the operators of L^2 and all L_j, so that the orbital angular momentum is conserved—see, for example, the solution of problem 5.20. This difference shows that in the Dirac theory (and in the physical reality) the orbital and spin degrees of the particle are generally related, and only in the non-relativistic limit does this relation weaken, manifesting itself only in small artifacts such as the spin–orbit coupling (9.122), and its result—the fine structure of atomic levels (see section 6.3 of the lecture notes).

***Problem* 9.18.** Calculate the commutators of the operators $\hat{S}^2$ and $\hat{J}^2$ with the Dirac's Hamiltonian (9.97), and give an interpretation of the results.

Solution: Let us first generalize the calculation of Eqs. (9.109) of the lecture notes, for arbitrary Cartesian components of the vector operators $\hat{\mathbf{S}}$, $\hat{\boldsymbol{\alpha}}$, and $\hat{\boldsymbol{\beta}}$. This may be readily done not only in the explicit 4×4 matrix form, using Eqs. (9.98*b*) and (9.107*b*), but even in the shorthand 2×2 matrix form, using Eqs. (9.98*a*) and (9.107*a*):

$$\mathrm{S}_j\alpha_{j'} = \frac{\hbar}{2}\begin{pmatrix} \sigma_j & 0 \\ 0 & \sigma_j \end{pmatrix}\begin{pmatrix} 0 & \sigma_{j'} \\ \sigma_{j'} & 0 \end{pmatrix} = \frac{\hbar}{2}\begin{pmatrix} 0 & \sigma_j\sigma_{j'} \\ \sigma_j\sigma_{j'} & 0 \end{pmatrix},$$

$$\alpha_{j'}\mathrm{S}_j = \frac{\hbar}{2}\begin{pmatrix} 0 & \sigma_{j'} \\ \sigma_{j'} & 0 \end{pmatrix}\begin{pmatrix} \sigma_j & 0 \\ 0 & \sigma_j \end{pmatrix} = \frac{\hbar}{2}\begin{pmatrix} 0 & \sigma_{j'}\sigma_j \\ \sigma_{j'}\sigma_j & 0 \end{pmatrix}. \qquad (*)$$

From here, we may find the commutator of these matrices:

$$[\mathrm{S}_j, \alpha_{j'}] \equiv \mathrm{S}_j\alpha_{j'} - \alpha_{j'}\mathrm{S}_j = \frac{\hbar}{2}\begin{pmatrix} 0 & [\sigma_j, \sigma_{j'}] \\ [\sigma_j, \sigma_{j'}] & 0 \end{pmatrix}.$$

With the Pauli matrix commutation rule $[\sigma_j, \sigma_{j'}] = 2i\sigma_{j''}\varepsilon_{jj'j''}$ (whose calculation was one of the tasks of problem 4.3), this relation becomes

$$[\mathrm{S}_j, \alpha_{j'}] = \frac{\hbar}{2}\begin{pmatrix} 0 & 2i\sigma_{j''}\varepsilon_{jj'j''} \\ 2i\sigma_{j''}\varepsilon_{jj'j''} & 0 \end{pmatrix} \equiv i\hbar\begin{pmatrix} 0 & \sigma_{j''} \\ \sigma_{j''} & 0 \end{pmatrix}\varepsilon_{jj'j''}.$$

Using the definition (9.98a) of the vector matrix $\boldsymbol{\alpha}$ again, we may rewrite this result just as[28]

$$[\mathrm{S}_j, \alpha_{j'}] = i\hbar\alpha_{j''}\varepsilon_{jj'j''}, \quad \text{i.e. } [\hat{S}_j, \hat{\alpha}_{j'}] = i\hbar\hat{\alpha}_{j''}\varepsilon_{jj'j''}.$$

An absolutely similar calculation for $\hat{\boldsymbol{\beta}}$ yields quite a different result:

$$[\mathrm{S}_j, \beta_{j'}] \equiv \mathrm{S}_j\beta_{j'} - \beta_{j'}\mathrm{S}_j = \frac{\hbar}{2}\left[\begin{pmatrix}\sigma_j & 0\\ 0 & \sigma_j\end{pmatrix}\begin{pmatrix}\mathrm{I} & 0\\ 0 & -\mathrm{I}\end{pmatrix} - \begin{pmatrix}\mathrm{I} & 0\\ 0 & -\mathrm{I}\end{pmatrix}\begin{pmatrix}\sigma_j & 0\\ 0 & \sigma_j\end{pmatrix}\right]$$
$$= \frac{\hbar}{2}\left[\begin{pmatrix}\sigma_j & 0\\ 0 & -\sigma_j\end{pmatrix} - \begin{pmatrix}\sigma_j & 0\\ 0 & -\sigma_j\end{pmatrix}\right] = 0, \quad \text{i.e. } [\hat{S}_j, \hat{\beta}_{j'}] = \hat{0}.$$

Now using Eq. (9.97), we get

$$\left[\hat{S}_j, \hat{H}\right] = \left[\hat{S}_j, c\sum_{j'=1}^{3}\hat{\alpha}_{j'}\hat{p}_{j'} + mc^2\hat{\beta}\right] = c\sum_{j'=1}^{3}[\hat{S}_j, \hat{\alpha}_{j'}]\hat{p}_{j'} = i\hbar c\sum_{j'=1}^{3}\hat{\alpha}_{j''}\hat{p}_{j'}\varepsilon_{jj'j''}.$$

For the purposes of comparison with the commutator $[\hat{L}_j, \hat{H}]$, which was calculated as a by-product in the model solution of the previous problem, let us swap the indices j' and j', and then use the property $\varepsilon_{jj''j'} = -\varepsilon_{jj'j''}$ of the Levi-Civita symbol, to write

$$[\hat{S}_j, \hat{H}] = i\hbar c\sum_{j''=1}^{3}\hat{\alpha}_{j'}\hat{p}_{j''}\varepsilon_{jj''j'} = -i\hbar c\sum_{j''=1}^{3}\hat{\alpha}_{j'}\hat{p}_{j''}\varepsilon_{jj'j''} = -i\hbar c\sum_{j'=1}^{3}\hat{\alpha}_{j'}\hat{p}_{j''}\varepsilon_{jj'j''}. \qquad (**)$$

(The last change of the summation index is legitimate, because due to the convention $j'' \neq j, j'$, both sums cover the same three terms, one of which equals 0.) The comparison shows that the commutator $[\hat{S}_j, \hat{H}]$ differs from $[\hat{L}_j, \hat{H}]$ only by the sign.

So, one might expect that the commutators $[\hat{S}^2, \hat{H}]$ should also be similar to operator $[\hat{L}^2, \hat{H}]$, whose calculation was another task of the previous problem. However, this is not so. Indeed, rewriting Eq. (**) as

$$\hat{S}_j\hat{H} = \hat{H}\hat{S}_j - i\hbar c\sum_{j'=1}^{3}\hat{\alpha}_{j'}\hat{p}_{j''}\varepsilon_{jj'j''},$$

we may use this relation twice to calculate

[28] Note that in the 'Dirac basis', used in these lecture notes (and most texts), the notions of an operator and the corresponding 2 × 2 matrix are identical, so that we may switch between the operator and matrix notations at will.

$$\begin{aligned}
\left[\hat{S}_j^2, \hat{H}\right] &\equiv \hat{S}_j\hat{S}_j\hat{H} - \hat{H}\hat{S}_j\hat{S}_j \equiv \hat{S}_j(\hat{S}_j\hat{H}) - \hat{H}\hat{S}_j\hat{S}_j \\
&= \hat{S}_j\left(\hat{H}\hat{S}_j - i\hbar c\sum_{j'=1}^{3}\hat{\alpha}_{j'}\hat{p}_{j''}\varepsilon_{jj'j''}\right) - \hat{H}\hat{S}_j\hat{S}_j \\
&= (\hat{S}_j\hat{H})\hat{S}_j - \hat{H}\hat{S}_j\hat{S}_j - i\hbar c\sum_{j'=1}^{3}\hat{\alpha}_{j'}\hat{S}_j\hat{p}_{j''}\varepsilon_{jj'j''} \\
&= \left(\hat{H}\hat{S}_j - i\hbar c\sum_{j'=1}^{3}\hat{\alpha}_{j'}\hat{p}_{j''}\varepsilon_{jj'j''}\right)\hat{S}_j - \hat{H}\hat{S}_j\hat{S}_j - i\hbar c\sum_{j'=1}^{3}\hat{\alpha}_{j'}\hat{S}_j\hat{p}_{j''}\varepsilon_{jj'j''} \\
&\equiv - i\hbar c\sum_{j'=1}^{3}\left\{\hat{S}_j, \hat{\alpha}_{j'}\right\}\hat{p}_{j''}\,\varepsilon_{jj'j''}.
\end{aligned}$$

The last anticommutator may be readily calculated using Eq. (*) and the well-known property $\{\sigma_j, \sigma_{j'}\} = 2\mathrm{I}\delta_{jj'}$ of the Pauli matrices[29]:

$$\begin{aligned}
\left\{\mathrm{S}_j, \alpha_{j'}\right\} &\equiv \mathrm{S}_j\alpha_{j'} + \alpha_{j'}\mathrm{S}_j = \frac{\hbar}{2}\begin{pmatrix} 0 & \sigma_j\sigma_{j'} \\ \sigma_j\sigma_{j'} & 0 \end{pmatrix} + \frac{\hbar}{2}\begin{pmatrix} 0 & \sigma_{j'}\sigma_j \\ \sigma_{j'}\sigma_j & 0 \end{pmatrix} \\
&= \frac{\hbar}{2}\begin{pmatrix} 0 & \{\sigma_j, \sigma_{j'}\} \\ \{\sigma_j, \sigma_{j'}\} & 0 \end{pmatrix} = \hbar\begin{pmatrix} 0 & \mathrm{I} \\ \mathrm{I} & 0 \end{pmatrix}\delta_{jj'}.
\end{aligned}$$

But by the very definition of the Kronecker and Levi-Civita symbols, the product $\delta_{jj'}\varepsilon_{jj'j''}$ equals zero for any combination of indices, so that

$$\left[\hat{S}_j^2, \hat{H}\right] = \hat{0}, \quad \text{and hence } [\hat{S}^2, \hat{H}] \equiv \left[\sum_{j=1}^{3}\hat{S}_j^2, \hat{H}\right] \equiv \sum_{j=1}^{3}\left[\hat{S}_j^2, \hat{H}\right] = \hat{0}, \qquad (***)$$

i.e. the observable S^2 is conserved (at least) at the free motion of the particle. This result might be expected, because in the non-relativistic theory this observable is firmly fixed by the particle's spin s, $S^2 = \hbar^2 s(s+1)$, so that for a spin-½ particle $S^2 = (3/4)\hbar^2$ in any of its quantum states. As an easy direct calculation using Eq. (9.107) shows[30], the last result is also valid in the relativistic Dirac theory (which covers only spin-½ particles).

Now the second task of the problem's assignment is very easy. With the definition (9.111) of the total angular momentum, which may be represented in the Cartesian components as

$$\hat{J}_j = \hat{L}_j + \hat{S}_j,$$

[29] See, e.g. the model solution of the same problem 4.3.

[30] This additional exercise is highly recommended to the reader.

we may use the relation (**) derived above, together with Eq. (**) of the model solution of the previous problem, to get[31]

$$[\hat{J}_j, \hat{H}] \equiv [\hat{L}_j + \hat{S}_j, \hat{H}] = [\hat{L}_j, \hat{H}] + [\hat{S}_j, \hat{H}] = i\hbar c \sum_{j'=1}^{3} \hat{\alpha}_{j'}\hat{p}_{j''}\varepsilon_{jj'j''} - i\hbar c \sum_{j'=1}^{3} \hat{\alpha}_{j'}\hat{p}_{j''}\varepsilon_{jj'j''} = \hat{0}.$$

Since all operators of J_j commute with the Hamiltonian, so do the operators of J_j^2 and of $J^2 = J_1^2 + J_2^2 + J_3^2$. So, in contrast to the orbital momentum **L**, the expectation values of any Cartesian component, and of the square of the *total* angular momentum $\mathbf{J} = \mathbf{L} + \mathbf{S}$ of a free particle are conserved in the Dirac theory.

***Problem* 9.19.** In the Heisenberg picture of quantum dynamics, derive an equation describing the time evolution of free electron's velocity in the Dirac theory. Solve the equation for the simplest state, with definite energy and momentum, and discuss the solution.

Solution: Let us start with finding what operator corresponds to the particle's velocity in the Dirac theory. With the Dirac's Hamiltonian (9.97),

$$\hat{H} = c\hat{\boldsymbol{\alpha}} \cdot \hat{\mathbf{p}} + \hat{\beta}\, mc^2, \qquad (*)$$

the Heisenberg equation (4.199) for the jth Cartesian component of electron's radius-vector becomes

$$i\hbar \frac{d\hat{r}_j}{dt} = [\hat{r}_j, \hat{H}] \equiv [\hat{r}_j, c\hat{\boldsymbol{\alpha}} \cdot \hat{\mathbf{p}}] + [\hat{r}_j, \hat{\beta}]\, mc^2.$$

The operators $\hat{\boldsymbol{\alpha}}$ and $\hat{\beta}$, defined in the Hilbert space of the spin states, commute with the operators of coordinates, defined in the Hilbert space of the orbital states, while the momentum operator in the Dirac theory is defined just in the non-relativistic quantum mechanics, and hence their Cartesian components obey the usual commutation relations (2.14). As a result, the second commutator in the last expression vanishes, while the first one is

$$[\hat{r}_j, c\hat{\boldsymbol{\alpha}} \cdot \hat{\mathbf{p}}] = \left[\hat{r}_j, \sum_{j'=1}^{3} c\hat{\alpha}_{j'}\hat{p}_{j'}\right] = \sum_{j'=1}^{3} c\hat{\alpha}_{j'}[\hat{r}_j, \hat{p}_{j'}] = \sum_{j'=1}^{3} c\hat{\alpha}_{j'} i\hbar\delta_{jj'} = c\hat{\alpha}_j i\hbar.$$

So, for a particle's velocity we get a very simple result:

$$\hat{v}_j \equiv \frac{d\hat{r}_j}{dt} = c\hat{\alpha}_j, \quad \text{i.e. } \hat{\mathbf{v}} \equiv \frac{d\hat{\mathbf{r}}}{dt} = c\hat{\boldsymbol{\alpha}}.$$

[31] This result confirms the fact already stated in section 9.6 of the lecture notes: at the free motion of a particle, any component of its full angular moment $\mathbf{J} \equiv \mathbf{L} + \mathbf{S}$ is conserved.

This result is a formal confirmation of the conclusion, made in the lecture notes on the basis of comparison of Eqs. (9.101) and (9.102), that the operator $c\hat{\boldsymbol{\alpha}}$ corresponds to the particle's velocity[32].

Now let us use the same Eqs. (4.199) and (9.97) to find the equation of motion of this operator (or rather its jth Cartesian component):

$$i\hbar\frac{d\hat{v}_j}{dt} = [\hat{v}_j, \hat{H}] = [c\hat{\alpha}_j, c\hat{\boldsymbol{\alpha}}\cdot\hat{\mathbf{p}}] + [c\hat{\alpha}_j, \hat{\beta}]\,mc^2 \equiv c^2\sum_{j'=1}^{3}\left[\hat{\alpha}_j, \hat{\alpha}_{j'}\right]\hat{p}_{j'} + c\left[\hat{\alpha}_j, \hat{\beta}\right]mc^2$$

$$\equiv c^2\sum_{\substack{j'=1\\ j'\neq j}}^{3}(\hat{\alpha}_j\hat{\alpha}_{j'} - \hat{\alpha}_{j'}\hat{\alpha}_j)\hat{p}_{j'} + c\,(\hat{\alpha}_j\hat{\beta} - \hat{\alpha}_j\hat{\beta})\,mc^2.$$

Using the anticommutation relations (9.99)–(9.100), we get

$$i\hbar\frac{d\hat{v}_j}{dt} = c^2\sum_{\substack{j'=1\\ j'\neq j}}^{3}(2\hat{\alpha}_j\hat{\alpha}_{j'})\hat{p}_{j'} + c\,(2\hat{\alpha}_j\hat{\beta})\,mc^2 \equiv 2\hat{v}_j\left(\sum_{\substack{j'=1\\ j'\neq j}}^{3}c\hat{\alpha}_{j'}\hat{p}_{j'} + \hat{\beta}mc^2\right). \qquad (**)$$

Now noticing that for an arbitrary choice of the index j, Eq. (*) may be rewritten as

$$\hat{H} = \left(\sum_{\substack{j'=1\\ j'\neq j}}^{3}c\hat{\alpha}_{j'}\hat{p}_{j'} + \hat{\beta}\,mc^2\right) + c\hat{\alpha}_j\hat{p}_j,$$

we may simplify Eq. (**) as

$$i\hbar\frac{d\hat{v}_j}{dt} = 2\hat{v}_j(\hat{H} - c\hat{\alpha}_j\hat{p}_j) = 2\hat{v}_j\hat{H} - 2c^2\alpha_j\alpha_j\hat{p}_j = 2\hat{v}_j\hat{H} - 2c^2\hat{p}_j,$$

where for the last transition, Eq. (9.99) was used again. Thus, we have got a simple equation, which may be rewritten in the vector form:

$$i\hbar\frac{d\hat{\mathbf{v}}}{dt} = 2\hat{\mathbf{v}}\hat{H} - 2c^2\hat{\mathbf{p}}. \qquad (***)$$

For a free particle, the Hamiltonian and the momentum operators do not depend on time. (The first fact follows from the Heisenberg equation of motion for *any* system with time-independent Hamiltonian, while the second one is obvious from Eq. (*), because the momentum operator commutes with itself and both spin operators.) Hence a free Dirac particle may be placed into a simple state with

[32] Still please remember that in the spin Hilbert space, each Cartesian component of this operator is represented by a 4×4 matrix (see Eqs. (9.98) of the lecture notes), and acts upon the four-component bispinor (9.96), rather than on a scalar wavefunction as in the wave mechanics.

definite, constant values of the energy E and the momentum $\mathbf{p}$. For such a state, Eq. (***) becomes

$$i\hbar\frac{d\hat{\mathbf{v}}}{dt} = 2\hat{\mathbf{v}}E - 2c^2\mathbf{p}.$$

This linear differential equation may be readily integrated, giving

$$\hat{\mathbf{v}}(t) = \frac{\mathbf{p}}{M} + \left(\hat{\mathbf{v}}(0) - \frac{\mathbf{p}}{M}\right)e^{-i\omega t}, \quad \text{with } M \equiv \frac{E}{c^2} \quad \text{and} \quad \omega \equiv \frac{2E}{\hbar}.$$

So, on the top of the possible constant velocity $\mathbf{p}/M$ (where M is its relativistic, i.e. velocity-dependent mass), which might be expected from a free particle, the electron performs sinusoidal oscillations. Averaging this solution over the ensemble of initial states of the particle,

$$\langle\mathbf{v}\rangle(t) = \frac{\mathbf{p}}{M} + \left(\langle\mathbf{v}\rangle(0) - \frac{\mathbf{p}}{M}\right)e^{-i\omega t},$$

we see that this conclusion is valid for the expectation value of velocity of the particle (and as a result, of its average spatial position) as well, unless the initial state, i.e. the initial set of the spin bispinor components, is selected so that $\langle\mathbf{v}\rangle(0)$ is exactly equal to $\mathbf{p}/M$.

This curious effect of *Zitterbewegung* (German for 'trembling motion')[33] is similar to the 'usual' quantum oscillations, with the frequency $\omega = (E_1 - E_2)/\hbar$ between two coupled states of the particle, which were repeatedly discussed in this course, starting from section 2.6. In this particular case, one of the energies is that of the particle, $E_1 = E$, while another one is that of the antiparticle, $E_2 = -E$ (see figure 9.6 in the lecture notes), so that $E_1 - E_2 = 2E$. So, the oscillations may be interpreted as a result of the periodic conversion of the electron from the particle to the antiparticle (positron) form.

Unfortunately, an experimental observation of this effect would require using either a very specific measurement tool, or an intermediate agent (a photon or another particle/field) with its energy quantum $\hbar\omega$ at least high as $2E$, i.e. larger than $2m_ec^2 \approx 1$ MeV. However, such an agent, interacting with the original electron, might create multiple other electrons—the situation not described by the Dirac equation. As a result, to the best of my knowledge, the Zitterbewegung has not been directly observed yet—though a few of its close analogs have been[34].

***Problem* 9.20.** Calculate the eigenstates and eigenenergies of a relativistic spin-½ particle with charge q, placed into a uniform, time-independent external magnetic field $\mathscr{B}$. Compare the calculated energy spectrum with those following from the non-relativistic theory and the relativistic Schrödinger equation.

[33] This feature of the Dirac theory was first revealed in 1930 by E Schrödinger.

[34] See, e.g. [3].

Solution: Just as was done in section 9.7 of the lecture notes, let us look for the solution of the Dirac equation (9.112) for a particle in a time-independent magnetic field (i.e. for $\phi = 0$, but $\mathbf{A} \neq 0$),

$$[\hat{\boldsymbol{\alpha}} \cdot c(\hat{\mathbf{P}} - q\mathbf{A}) + \hat{\beta}mc^2 - \hat{H}]\Psi = 0, \quad \text{where} \quad \hat{\mathbf{P}} = -i\hbar\nabla,$$

in the form (9.125):

$$\Psi(\mathbf{r}, t) = \begin{pmatrix} \psi_+(\mathbf{r}) \\ \psi_-(\mathbf{r}) \end{pmatrix} \exp\left(-i\frac{E}{\hbar}t\right)$$

where each of $\psi_\pm$ is a two-component column (spinor) of the type (9.123), representing two spin states of the particle (index +) and the antiparticle (index −). Plugging this solution into the Dirac equation, instead of Eqs. (9.126) of the lecture notes (which are valid in the opposite case when $\phi \neq 0$, but $\mathbf{A} = 0$), we get the following set of two equations:

$$(E - mc^2)\,\psi_+ - c\hat{\boldsymbol{\sigma}} \cdot (-i\hbar\nabla - q\mathbf{A})\,\psi_- = 0,$$
$$(E + mc^2)\,\psi_- - c\hat{\boldsymbol{\sigma}} \cdot (-i\hbar\nabla - q\mathbf{A})\,\psi_+ = 0.$$

Eliminating the antiparticle's ψ_- from this system, we get the following equation for the particle's wavefunction:

$$\{[E^2 - (mc^2)^2] - c^2[\hat{\boldsymbol{\sigma}} \cdot (-i\hbar\nabla - q\mathbf{A})]^2\}\psi_+ = 0. \qquad (*)$$

Let us spell out the second term, using the fact that the Pauli operators (9.98), defined in the spin Hilbert space, and the momentum operator, defined in the orbital Hilbert space, commute:

$$[\hat{\boldsymbol{\sigma}} \cdot (-i\hbar\nabla - q\mathbf{A})]^2 = \left[\sum_{j=1}^{3}\hat{\sigma}_j(-i\hbar\nabla_j - qA_j)\right]^2 = \sum_{j,j'=1}^{3}\hat{\sigma}_j\hat{\sigma}_{j'}(-i\hbar\nabla_j - qA_j)(-i\hbar\nabla_{j'} - qA_{j'}),$$

where $\nabla_j \equiv \partial/\partial r_j$. Now using the facts that $\hat{\sigma}_j\hat{\sigma}_{j'} = \hat{I}\delta_{jj'} + i\hat{\sigma}_{j''}\varepsilon_{jj'j''}$, where $j'' \neq j, j'$,[35] we get

$$\begin{aligned}
[\hat{\boldsymbol{\sigma}} \cdot (\hat{\mathbf{P}} - q\mathbf{A})]^2 &= \sum_{j,j'=1}^{3}(\hat{I}\delta_{jj'} + i\hat{\sigma}_{j''}\varepsilon_{jj'j''})(-i\hbar\nabla_j - qA_j)(-i\hbar\nabla_{j'} - qA_{j'}) \\
&\equiv \sum_{j=1}^{3}(-i\hbar\nabla_j - qA_j)^2 + i\sum_{\substack{j,j'=1 \\ j\neq j'}}^{3}\hat{\sigma}_{j''}(i\hbar\nabla_j + qA_j)(i\hbar\nabla_{j'} + qA_{j'})\varepsilon_{jj'j''} \\
&\equiv (-i\hbar\nabla - q\mathbf{A})^2 - \hbar q\sum_{j''=1}^{3}\hat{\sigma}_{j''}(\nabla_j A_{j'} - \nabla_{j'}A_j)\,\varepsilon_{jj'j''}.
\end{aligned}$$

[35] See, e.g. the solution of problem 4.3.

But according to the definition of the curl of a vector, $(\nabla_j A_{j'} - \nabla_{j'} A_j)\varepsilon_{jj'j''}$ is just the (j'')th component of the vector $\nabla \times \mathbf{A} \equiv \mathscr{B}$, so that

$$[\hat{\boldsymbol{\sigma}} \cdot (-i\hbar\nabla - q\mathbf{A})]^2 = (-i\hbar\nabla - q\mathbf{A})^2 - \hbar q \sum_{j''=1}^{3} \hat{\sigma}_{j''}\mathscr{B}_{j''}$$
$$\equiv (-i\hbar\nabla - q\mathbf{A})^2 - \hbar q \hat{\boldsymbol{\sigma}} \cdot \mathscr{B} = (-i\hbar\nabla - q\mathbf{A})^2 - 2q\hat{\mathbf{S}} \cdot \mathscr{B},$$

where the last step used the basic Eq. (4.116) for the spin-½ operator. Plugging this expression into Eq. (*), and dividing all its terms by $2mc^2$, we may rewrite that equation as

$$-\frac{\hbar^2}{2m}\left(\nabla + i\frac{q}{\hbar}\mathbf{A}\right)^2 \psi_+ - \frac{q}{m}\hat{\mathbf{S}} \cdot \mathscr{B}\,\psi_+ = E_{\text{ef}}\psi_+, \qquad (**)$$

where $E_{\text{ef}} = p^2/2m$ is the same effective energy that appears in solutions of the relativistic Schrödinger equation[36]:

$$E_{\text{ef}} \equiv \frac{E^2 - (mc^2)^2}{2mc^2}. \qquad (***)$$

But, besides the replacement $E \to E_{\text{ef}}$, Eq. (**) is exactly the non-relativistic Schrödinger equation of a particle with the gyromagnetic ratio $\gamma \equiv gq/2m$ (i.e. with the g-factor equal to exactly 2) in the magnetic field. As we know from the discussion in chapters 3 and 5,[37] its energy spectrum consists of continuous bands, each corresponding to various values of the momentum p_z (where the z-axis is directed along the magnetic field), with discrete interband offsets due to the Landau-level quantization within the $[x, y]$ plane, plus the spin energy quantization with the same energy spacing[38]:

$$E_{\text{ef}} = \frac{p_z^2}{2m} + \hbar\omega_c\left(n + \frac{1}{2} - m_s \operatorname{sgn} q\right),$$

with

$$-\infty < p_z < +\infty, \qquad \omega_c \equiv \frac{|q|\mathscr{B}}{m}, \qquad n = 0, 1, 2, \ldots, \qquad \text{and} \qquad m_s = \pm 1/2.$$

So, in order to get the spectrum of the relativistic energy, we need just to plug this result into the relation following from Eq. (***):

$$E = \pm mc^2\left[1 + \left(\frac{2E_{\text{ef}}}{mc^2}\right)\right]^{1/2};$$

[36] See, e.g. the model solutions of problems 9.13–9.15.

[37] See the model solutions of problem 5.48.

[38] As Eq. (**) shows, the eigenstates of the relativistic problem are also similar to those of the corresponding non-relativistic problem.

in the non-relativistic limit, this recalculation becomes trivial:

$$E \approx \pm(mc^2 + E_{\text{ef}}), \quad \text{at } E_{\text{ef}} \ll mc^2.$$

On the other hand, as has been shown in the solution of problem 9.13, the analysis of the same situation using the relativistic Schrödinger equation gives for E_{ef} (and hence for E) a similar spectrum, but without the m_s-term, describing the spin.

Problem* 9.21. Following the discussion at the very end of section 9.7 of the lecture notes, introduce quantum field operators $\hat{\psi}$ that would be related to the usual wavefunctions ψ just as the electromagnetic field operators (9.16) are related to the classical electromagnetic fields, and explore the basic properties of these operators. (For this preliminary study, consider the fixed-time situation.)

Solution: In an analogy with Eqs. (9.16) of the lecture notes, but taking into account the scalar nature of the 'matter field' (wavefunction) ψ, we may define the field operator and its Hermitian conjugate as

$$\hat{\psi}(\mathbf{r}) \equiv \sum_j \psi_j(\mathbf{r})\hat{a}_j, \quad \hat{\psi}^\dagger(\mathbf{r}) \equiv \sum_j \psi_j^*(\mathbf{r})\, \hat{a}^\dagger.$$

Here $\psi_j(\mathbf{r})$ is a full, orthonormal set of single-particle wavefunctions of a multi-particle system (where the index j numbers both the orbital and spin degrees of freedom)[39], while $\hat{a}_j^\dagger$ and $\hat{a}_j$ are the particle creation and annihilation operators discussed in sections 8.3–8.4 of the lecture notes.

Let us explore properties of the state (called, say, α) that is created by the action of the operator $\hat{\psi}^\dagger(\mathbf{r})$ upon the vacuum Dirac state:

$$|\alpha\rangle = \hat{\psi}^\dagger(\mathbf{r})|0, 0, \ldots, 0\rangle \equiv \sum_j \psi_j^*(\mathbf{r})\, \hat{a}_j^\dagger|0, 0, \ldots, 0\rangle$$
$$= \sum_j \psi_j^*(\mathbf{r})|0, 0, \ldots, 1_j, \ldots, 0\rangle.$$

The last ket describes a single particle in the state number j, so in the single-particle language we may represent it just as $|j\rangle$, and rewrite the above relation just as

$$|\alpha\rangle = \sum_j \psi_j^*(\mathbf{r})\, |j\rangle. \qquad (*)$$

Let us calculate the wavefunction $\psi_\alpha(\mathbf{r}')$ of this state at some arbitrary point $\mathbf{r}'$, at this point not necessarily equal to the argument $\mathbf{r}$ of the field operator. The evident

[39] When dealing with free (or nearly free) particles, the natural, and hence very popular choice of the base functions $\psi_j(\mathbf{r})$ are the plane waves $\psi_\mathbf{k}(\mathbf{r}) \propto \exp\{i\mathbf{k}\cdot\mathbf{r}\}$. (Since their spectrum is continuous, the summation over j in the above formulas is then replaced with the integration over the 3D space of the wave vectors $\mathbf{k}$, plus the summation over spin states.)

3D generalization of the definition (4.233) of the wavefunction, in the single-particle representation, is[40]

$$\psi_\alpha(\mathbf{r}') = \langle \mathbf{r}'|\alpha\rangle, \quad \text{while } \psi_j^*(\mathbf{r}) = \langle \mathbf{r}|j\rangle^* = \langle j|\mathbf{r}\rangle. \qquad (**)$$

Now combining Eqs. (*) and (**), we get

$$\psi_\alpha(\mathbf{r}') = \langle \mathbf{r}'|\sum_j \langle j|\mathbf{r}\rangle|j\rangle \equiv \langle \mathbf{r}'|\sum_j |j\rangle\langle j|\mathbf{r}\rangle.$$

Using the closure relation (4.44), and the evident 3D generalization of Eq. (4.231), $\langle x|x'\rangle = \delta(x - x')$, we get

$$\psi_\alpha(\mathbf{r}') = \langle \mathbf{r}'|\hat{I}|\mathbf{r}\rangle \equiv \langle \mathbf{r}'|\mathbf{r}\rangle = \delta(\mathbf{r} - \mathbf{r}').$$

This means that the operator $\hat{\psi}^\dagger(\mathbf{r})$, acting upon the vacuum, creates a particle localized definitely at point $\mathbf{r}$. An absolutely similar calculation shows that its Hermitian conjugate, the operator $\hat{\psi}(\mathbf{r})$, annihilates a particle in this position. This is very natural, because the field operators are just the sums of the creation/annihilation operators, weighed by the wavefunctions of the corresponding states.

In contrast with these general properties, the commutation relations of the field operators depend on whether we are dealing with bosons or fermions. In the former case, we may use Eqs. (8.75) and (8.76) of the lecture notes to get, for example,

$$[\hat{\psi}(\mathbf{r}), \hat{\psi}^\dagger(\mathbf{r}')] = \sum_{j,j'} \psi_j(\mathbf{r})\, \psi_{j'}^*(\mathbf{r}') \left[\hat{a}_j, \hat{a}_{j'}^\dagger\right] = \sum_{j,j'} \psi_j(\mathbf{r})\, \psi_{j'}^*(\mathbf{r}')\, \hat{I}\delta_{jj'} = \sum_j \psi_j(\mathbf{r})\, \psi_j^*(\mathbf{r}')\, \hat{I}.$$

Now using Eqs. (**), with the argument replacements $\mathbf{r} \leftrightarrow \mathbf{r}'$, we may complete the calculation as

$$[\hat{\psi}(\mathbf{r}), \hat{\psi}^\dagger(\mathbf{r}')] = \sum_j \langle \mathbf{r}|j\rangle\langle j|\mathbf{r}'\rangle\, \hat{I} = \hat{I}\delta(\mathbf{r} - \mathbf{r}').$$

Acting absolutely similarly, for the bosonic operators we may also get

$$[\hat{\psi}(\mathbf{r}), \hat{\psi}(\mathbf{r}')] = [\hat{\psi}^\dagger(\mathbf{r}), \hat{\psi}^\dagger(\mathbf{r}')] = \hat{0},$$

while for the fermionic operators, Eqs. (8.95) and (8.96) of the lecture notes yield similar relations for the anticommutators:

$$\{\hat{\psi}(\mathbf{r}), \hat{\psi}^\dagger(\mathbf{r}')\} = \hat{I}\delta(\mathbf{r} - \mathbf{r}'), \quad \{\hat{\psi}(\mathbf{r}), \hat{\psi}(\mathbf{r}')\} = \{\hat{\psi}^\dagger(\mathbf{r}), \hat{\psi}^\dagger(\mathbf{r}')\} = \hat{0}.$$

Next, let us consider the following operator integral,

$$\hat{F} \equiv \int \hat{\psi}^\dagger(\mathbf{r}) \hat{f}(\mathbf{r})\, \hat{\psi}(\mathbf{r})\, d^3r, \qquad (***)$$

[40] Since all calculations in this solution are for a fixed moment of time, the wavefunction may be denoted as ψ, rather than Ψ in Eq. (4.233).

where $\hat{f}(\mathbf{r})$ is some single-particle operator. Plugging into this expression the field operators definitions, we obtain

$$\hat{F} = \sum_{j,j'} \hat{a}_j^\dagger \hat{a}_{j'} \int \psi_{j'}^*(\mathbf{r}) \hat{f}(\mathbf{r}) \psi_j(\mathbf{r})\, d^3r \equiv \sum_{j,j'} f_{jj'} \hat{a}_j^\dagger \hat{a}_{j'}.$$

where $f_{jj'}$ are the usual matrix elements of the operator $\hat{f}$. But the last expression exactly coincides with the right-hand side of Eq. (8.87) of the lecture notes; hence the integral (***) is an equivalent representation of the similar single-particle components:

$$\hat{F} = \sum_{k=1}^{N} \hat{f}(\mathbf{r}_k).$$

The most important particular cases of such operators are those of the full momentum and the full kinetic energy of the system, which are equal to, respectively,

$$\hat{P} \equiv \sum_{k=1}^{N} \hat{p}_k = -i\hbar \sum_{k=1}^{N} \nabla_k, \quad \text{and} \quad \hat{T} \equiv -\frac{\hbar^2}{2m} \sum_{k=1}^{N} \nabla_k^2$$

in the 'usual' (particle-number) representation[41]. According to Eq. (***), in the second-quantization language these operators may be represented as

$$\hat{P} = -i\hbar \int \hat{\psi}^\dagger(\mathbf{r}) \nabla \hat{\psi}(\mathbf{r})\, d^3r \quad \text{and} \quad \hat{T} = -\frac{\hbar^2}{2m} \int \hat{\psi}^\dagger(\mathbf{r}) \nabla^2 \hat{\psi}(\mathbf{r})\, d^3r$$

(where m is the mass of a *single* particle). Very similarly, the pair-interaction operators of the type (8.113),

$$\hat{U}_{\text{int}} = \frac{1}{2} \sum_{\substack{k,k'=1 \\ k \neq k'}}^{N} \hat{u}_{\text{int}}(\mathbf{r}_k, \mathbf{r}_{k'}),$$

may be expressed via the field operators as

$$\hat{U}_{\text{int}} = \frac{1}{2} \int d^3r \int d^3r' \; \hat{\psi}^\dagger(\mathbf{r}) \hat{\psi}^\dagger(\mathbf{r}') \hat{u}_{\text{int}}(\mathbf{r}, \mathbf{r}') \hat{\psi}(\mathbf{r}')\, \hat{\psi}(\mathbf{r}). \qquad (****)$$

Using the relations (***) and (****), one may express the Hamiltonians of many important models of interacting particle systems via the field operators. After that, the equations of motion of these operators (for example, in the Heisenberg picture) may be obtained and analyzed, forming the fabric of the quantum field theory. Evidently, the most important new feature of such a theory, in comparison with the plain-vanilla quantum mechanics discussed in this course, is its ability to describe, in

[41] A useful (and simple) optional exercise for the reader is to explore how these operators commute with the field operators, separately for bosons and fermions.

a natural way, the creation and annihilation of particles—just as the creation and annihilation of photons was studied in sections 9.2–9.4. These phenomena are most significant at particle energies higher than $2mc^2$, but also show up as minor corrections to the results of the 'usual' (particle-number-conserving) quantum mechanics at lower energies.

References

[1] Bromberg Y *et al* 2010 *Nature Photonics* **4** 721
[2] Turolla R *et al* 2015 *Rep. Prog. Phys.* **78** 116901
[3] Gerritsma R *et al* 2010 *Nature* **463** 68

IOP Publishing

Quantum Mechanics
Problems with solutions
Konstantin K Likharev

Appendix A

Selected mathematical formulas

This appendix lists selected mathematical formulas that are used in this lecture course series, but not always remembered by students (and some instructors :-).

A.1 Constants

- Euclidean circle's *length-to-diameter ratio*:

$$\pi = 3.141\,592\,653\ldots; \qquad \pi^{1/2} \approx 1.77. \tag{A.1}$$

- *Natural logarithm base*:

$$e \equiv \lim_{n\to\infty}\left(1 + \frac{1}{n}\right)^n = 2.718\,281\,828\ldots; \tag{A.2a}$$

from that value, the logarithm base conversion factors are as follows ($\xi > 0$):

$$\frac{\ln \xi}{\log_{10}\xi} = \ln 10 \approx 2.303, \qquad \frac{\log_{10}\xi}{\ln \xi} = \frac{1}{\ln 10} \approx 0.434. \tag{A.2b}$$

- The *Euler* (or 'Euler–Mascheroni') *constant*:

$$\gamma \equiv \lim_{n\to\infty}\left(1 + \frac{1}{2} + \frac{1}{3} + \ldots \frac{1}{n} - \ln n\right) = 0.577\,156\,649\,0\ldots; \quad e^\gamma \approx 1.781. \tag{A.3}$$

A.2 Combinatorics, sums, and series

(i) *Combinatorics*

- The number of different *permutations*, i.e. *ordered* sequences of k elements selected from a set of n distinct elements ($n \geqslant k$), is

$$^nP_k \equiv n \cdot (n-1) \cdots (n-k+1) = \frac{n!}{(n-k)!}; \tag{A.4a}$$

doi:10.1088/2053-2563/aaf3a6ch10

in particular, the number of different permutations of *all* elements of the set ($n = k$) is

$$^{k}P_k = k \cdot (k-1)\cdots 2 \cdot 1 = k!. \qquad \text{(A.4}b\text{)}$$

- The number of different *combinations*, i.e. *unordered* sequences of k elements from a set of $n \geqslant k$ distinct elements, is equal to the binomial coefficient

$$^{n}C_k \equiv \binom{n}{k} \equiv \frac{^{n}P_k}{^{k}P_k} = \frac{n!}{k!(n-k)!}. \qquad \text{(A.5)}$$

In an alternative, very popular 'ball/box language', $^{n}C_k$ is the number of different ways to put in a box, in an arbitrary order, k balls selected from n distinct balls.

- A generalization of the binomial coefficient notion is the multinomial coefficient,

$$^{n}C_{k_1,k_2,\ldots k_l} \equiv \frac{n!}{k_1!k_2!\ldots k_l!}, \qquad \text{with } n = \sum_{j=1}^{l} k_j, \qquad \text{(A.6)}$$

which, in the standard mathematical language, is a number of different permutations in a multiset of l distinct element types from an n-element set which contains k_j ($j = 1, 2,\ldots l$) elements of each type. In the 'ball/box language', the coefficient (A.6) is the number of different ways to distribute n distinct balls between l distinct boxes, each time keeping the number (k_j) of balls in the jth box fixed, but ignoring their order inside the box. The binomial coefficient $^{n}C_k$ (A.5) is a particular case of the multinomial coefficient (A.6) for $l = 2$ - counting the explicit box for the first one, and the remaining space for the second box, so that if $k_1 \equiv k$, then $k_2 = n - k$.

- One more important combinatorial quantity is the number $M_n^{(k)}$ of ways to place n *indistinguishable* balls into k distinct boxes. It may be readily calculated from Eq. (A.5) as the number of different ways to select $(k-1)$ partitions between the boxes in an imagined linear row of $(k - 1 + n)$ 'objects' (balls in the boxes *and* partitions between them):

$$M_n^{(k)} = {}^{n-1+k}C_{k-1} \equiv \frac{(k-1+n)!}{(k-1)!n!}. \qquad \text{(A.7)}$$

(ii) *Sums and series*

- *Arithmetic progression*:

$$r + 2r + \cdots + nr \equiv \sum_{k=1}^{n} kr = \frac{n(r+nr)}{2}; \qquad \text{(A.8}a\text{)}$$

in particular, at $r = 1$ it is reduced to the sum of n first natural numbers:

$$1 + 2 + \cdots + n \equiv \sum_{k=1}^{n} k = \frac{n(n+1)}{2}. \qquad \text{(A.8}b\text{)}$$

- Sums of squares and cubes of n first natural numbers:

$$1^2 + 2^2 + \cdots + n^2 \equiv \sum_{k=1}^{n} k^2 = \frac{n(n+1)(2n+1)}{6}; \qquad \text{(A.9}a\text{)}$$

$$1^3 + 2^3 + \cdots + n^3 \equiv \sum_{k=1}^{n} k^3 = \frac{n^2(n+1)^2}{4}. \qquad \text{(A.9}b\text{)}$$

- The *Riemann zeta function*:

$$\zeta(s) \equiv 1 + \frac{1}{2^s} + \frac{1}{3^s} + \cdots \equiv \sum_{k=1}^{\infty} \frac{1}{k^s}; \qquad \text{(A.10}a\text{)}$$

the particular values frequently met in applications are

$$\zeta\left(\frac{3}{2}\right) \approx 2.612, \quad \zeta(2) = \frac{\pi^2}{6}, \quad \zeta\left(\frac{5}{2}\right) \approx 1.341,$$
$$\zeta(3) \approx 1.202, \quad \zeta(4) = \frac{\pi^4}{90}, \quad \zeta(5) \approx 1.037. \qquad \text{(A.10}b\text{)}$$

- Finite geometric progression (for real $\lambda \neq 1$):

$$1 + \lambda + \lambda^2 + \cdots + \lambda^{n-1} \equiv \sum_{k=0}^{n-1} \lambda^k = \frac{1-\lambda^n}{1-\lambda}; \qquad \text{(A.11}a\text{)}$$

in particular, if $\lambda^2 < 1$, the progression has a finite limit at $n \to \infty$ (called the *geometric series*):

$$\lim_{n\to\infty} \sum_{k=0}^{n-1} \lambda^k = \sum_{k=0}^{\infty} \lambda^k = \frac{1}{1-\lambda}. \qquad \text{(A.11}b\text{)}$$

- *Binomial sum* (or the 'binomial theorem'):

$$(1+a)^n = \sum_{k=0}^{n} {}^nC_k a^k, \qquad \text{(A.12)}$$

where ${}^nC_k$ are the binomial coefficients defined by Eq. (A.5).

- The *Stirling formula*:

$$\lim_{n\to\infty} \ln(n!) = n(\ln n - 1) + \frac{1}{2}\ln(2\pi n) + \frac{1}{12n} - \frac{1}{360n^3} + \ldots; \qquad \text{(A.13)}$$

for most applications in physics, the first term[1] is sufficient.

- The *Taylor* (or 'Taylor–Maclaurin') *series*: for any infinitely differentiable function $f(\xi)$:

$$\begin{aligned}\lim_{\tilde{\xi}\to 0} f(\xi+\tilde{\xi}) &= f(\xi) + \frac{df}{d\xi}(\xi)\,\tilde{\xi} + \frac{1}{2!}\frac{d^2f}{d\xi^2}(\xi)\,\tilde{\xi}^2 + \cdots \\ &= \sum_{k=0}^{\infty}\frac{1}{k!}\frac{d^kf}{d\xi^k}(\xi)\,\tilde{\xi}^k;\end{aligned} \tag{A.14a}$$

note that for many functions this series converges only within a limited, sometimes small range of deviations $\tilde{\xi}$. For a function of several arguments, $f(\xi_1,\xi_2,\ldots,\xi_N)$, the first terms of the Taylor series are

$$\begin{aligned}\lim_{\tilde{\xi}_k\to 0} f(\xi_1+\tilde{\xi}_1,\ \xi_2+\tilde{\xi}_2,\ \cdots) &= f(\xi_1,\ \xi_2,\ \cdots) \\ &+ \sum_{k=1}^{N}\frac{\partial f}{\partial\xi_k}(\xi_1,\ \xi_2,\ \cdots)\,\tilde{\xi}_k \\ &+ \frac{1}{2!}\sum_{k,k'=1}^{N}\frac{\partial^2 f}{\partial_k\xi\,\partial\xi_{k'}}\tilde{\xi}_k\tilde{\xi}_{k'} + \cdots\end{aligned} \tag{A.14b}$$

- The *Euler–Maclaurin formula*, valid for any infinitely differentiable function $f(\xi)$:

$$\begin{aligned}\sum_{k=1}^{n} f(k) &= \int_0^n f(\xi)d\xi + \frac{1}{2}[f(n)-f(0)] + \frac{1}{6}\cdot\frac{1}{2!}\left[\frac{df}{d\xi}(n)-\frac{df}{d\xi}(0)\right] \\ &- \frac{1}{30}\cdot\frac{1}{4!}\left[\frac{d^3f}{d\xi^3}(n)-\frac{d^3f}{d\xi^3}(0)\right] \\ &+ \frac{1}{42}\cdot\frac{1}{6!}\left[\frac{d^5f}{d\xi^5}(n)-\frac{d^5f}{d\xi^5}(0)\right] + \cdots;\end{aligned} \tag{A.15a}$$

the coefficients participating in this formula are the so-called *Bernoulli numbers*[2]:

$$\begin{aligned}&B_1=\frac{1}{2},\quad B_2=\frac{1}{6},\quad B_3=0,\quad B_4=\frac{1}{30},\quad B_5=0,\\ &B_6=\frac{1}{42},\quad B_7=0,\quad B_8=\frac{1}{30},\quad \cdots\end{aligned} \tag{A.15b}$$

[1] Actually, this leading term was derived by A de Moivre in 1733, before J Stirling's work.

[2] Note that definitions of B_k (or rather their signs and indices) vary even among the most popular handbooks.

A.3 Basic trigonometric functions

- Trigonometric functions of the sum and the difference of two arguments[3]:

$$\cos(a \pm b) = \cos a \cos b \mp \sin a \sin b, \tag{A.16a}$$

$$\sin(a \pm b) = \sin a \cos b \pm \cos a \sin b. \tag{A.16b}$$

- Sums of two functions of arbitrary arguments:

$$\cos a + \cos b = 2\cos\frac{a+b}{2}\cos\frac{b-a}{2}, \tag{A.17a}$$

$$\cos a - \cos b = 2\sin\frac{a+b}{2}\sin\frac{b-a}{2}, \tag{A.17b}$$

$$\sin a \pm \sin b = 2\sin\frac{a \pm b}{2}\cos\frac{\pm b-a}{2}. \tag{A.17c}$$

- Trigonometric function products:

$$2\cos a\ \cos b = \cos(a+b) + \cos(a-b), \tag{A.18a}$$

$$2\sin a\ \cos b = \sin(a+b) + \sin(a-b), \tag{A.18b}$$

$$2\sin a\ \sin b = \cos(a-b) - \cos(a+b); \tag{A.18c}$$

For the particular case of equal arguments, $b = a$, these three formulas yield the following expressions for the squares of trigonometric functions, and their product:

$$\cos^2 a = \frac{1}{2}(1 + \cos 2a), \qquad \sin a \cos a = \frac{1}{2}\sin 2a, \qquad \sin^2 a = \frac{1}{2}(1 - \cos 2a). \tag{A.18d}$$

- Cubes of trigonometric functions:

$$\cos^3 a = \frac{3}{4}\cos a + \frac{1}{4}\cos 3a, \qquad \sin^3 a = \frac{3}{4}\sin a - \frac{1}{4}\sin 3a. \tag{A.19}$$

- Trigonometric functions of a complex argument:

$$\sin(a + ib) = \sin a \cosh b + i\cos a \sinh b, \qquad \cos(a + ib) = \cos a \cosh b - i\sin a \sinh b. \tag{A.20}$$

[3] I am confident that the reader is quite capable of deriving the relations (A.16) by representing the exponent in the elementary relation $e^{i(a \pm b)} = e^{ia}e^{\pm ib}$ as a sum of its real and imaginary parts, Eqs. (A.18) directly from Eqs. (A.16), and Eqs. (A.17) from Eqs. (A.18) by variable replacement; however, I am still providing these formulas to save his or her time. (Quite a few formulas below are included because of the same reason.)

- Sums of trigonometric functions of n equidistant arguments:

$$\sum_{k=1}^{n}\begin{Bmatrix}\sin\\ \cos\end{Bmatrix}k\xi = \begin{Bmatrix}\sin\\ \cos\end{Bmatrix}\left(\frac{n+1}{2}\xi\right)\sin\left(\frac{n}{2}\xi\right)\Big/\sin\left(\frac{\xi}{2}\right). \tag{A.21}$$

A.4 General differentiation

- Full differential of a product of two functions:

$$d(fg) = (df)g + f(dg). \tag{A.22}$$

- Full differential of a function of several independent arguments, $f(\xi_1, \xi_2, \ldots, \xi_n)$:

$$df = \sum_{k=1}^{n}\frac{\partial f}{\partial \xi_k}d\xi_k. \tag{A.23}$$

- Curvature of the Cartesian plot of a 1D function $f(\xi)$:

$$\kappa \equiv \frac{1}{R} = \frac{|\, d^2f/d\xi^2 \,|}{[1 + (df/d\xi)^2]^{3/2}}. \tag{A.24}$$

A.5 General integration

- Integration *by parts* - immediately follows from Eq. (A.22):

$$\int_{g(A)}^{g(B)} f\ dg = fg\,\Big|_A^B - \int_{f(A)}^{f(B)} g\ df. \tag{A.25}$$

- Numerical (approximate) integration of 1D functions: the simplest *trapezoidal rule*,

$$\begin{aligned}\int_a^b f(\xi)d\xi &\approx h\left[f\left(a+\frac{h}{2}\right)+f\left(a+\frac{3h}{2}\right)+\cdots+f\left(b-\frac{h}{2}\right)\right]\\ &= h\sum_{n=1}^{N}f\left(a-\frac{h}{2}+nh\right), \qquad h \equiv \frac{b-a}{N}.\end{aligned} \tag{A.26}$$

has relatively low accuracy (error of the order of $(h^3/12)d^2f/d\xi^2$ per step), so that the following *Simpson formula*,

$$\begin{aligned}\int_a^b f(\xi)d\xi &\approx \frac{h}{3}[f(a) + 4f(a+h) + 2f(a+2h) + \cdots + 4f(b-h) + f(b)],\\ h &\equiv \frac{b-a}{2N},\end{aligned} \tag{A.27}$$

whose error per step scales as $(h^5/180)d^4f/d\xi^4$, is used much more frequently[4].

A.6 A few 1D integrals[5]

(i) *Indefinite integrals*:

- Integrals with $(1 + \xi^2)^{1/2}$:

$$\int (1 + \xi^2)^{1/2} d\xi = \frac{\xi}{2}(1 + \xi^2)^{1/2} + \frac{1}{2}\ln|\xi + (1 + \xi^2)^{1/2}|, \tag{A.28}$$

$$\int \frac{d\xi}{(1 + \xi^2)^{1/2}} = \ln|\xi + (1 + \xi^2)^{1/2}|, \tag{A.29a}$$

$$\int \frac{d\xi}{(1 + \xi^2)^{3/2}} = \frac{\xi}{(1 + \xi^2)^{1/2}}. \tag{A.29b}$$

- Miscellaneous indefinite integrals:

$$\int \frac{d\xi}{\xi(\xi^2 + 2a\xi - 1)^{1/2}} = \arccos \frac{a\xi - 1}{|\xi|(a^2 + 1)^{1/2}}, \tag{A.30a}$$

$$\int \frac{(\sin\xi - \xi\cos\xi)^2}{\xi^5} d\xi = \frac{2\xi\sin 2\xi + \cos 2\xi - 2\xi^2 - 1}{8\xi^4}, \tag{A.30b}$$

$$\int \frac{d\xi}{a + b\cos\xi} = \frac{2}{(a^2 - b^2)^{1/2}} \tan^{-1}\left[\frac{(a - b)}{(a^2 - b^2)^{1/2}} \tan\frac{\xi}{2}\right], \tag{A.30c}$$

for $a^2 > b^2$.

$$\int \frac{d\xi}{1 + \xi^2} = \tan^{-1}\xi. \tag{A.30d}$$

(ii) *Semi-definite integrals*:

- Integrals with $1/(e^\xi \pm 1)$:

$$\int_a^\infty \frac{d\xi}{e^\xi + 1} = \ln(1 + e^{-a}), \tag{A.31a}$$

[4] Higher-order formulas (e.g. the *Bode rule*), and other guidance including ready-for-use codes for computer calculations may be found, for example, in the popular reference texts by W H Press *et al* [1]. In addition, some advanced codes are used as subroutines in the software packages listed in the same section. In some cases, the Euler–Maclaurin formula (A.15) may also be useful for numerical integration.

[5] A powerful (and free) interactive online tool for working out indefinite 1D integrals is available at http://integrals.wolfram.com/index.jsp.

$$\int_{a>0}^{\infty} \frac{d\xi}{e^{\xi} - 1} = \ln \frac{1}{1 - e^{-a}}. \quad \text{(A.31}b\text{)}$$

(iii) *Definite integrals*:

- Integrals with $1/(1 + \xi^2)$:[6]

$$\int_0^{\infty} \frac{d\xi}{1 + \xi^2} = \frac{\pi}{2}, \quad \text{(A.32}a\text{)}$$

$$\int_0^{\infty} \frac{d\xi}{(1 + \xi^2)^{3/2}} = 1; \quad \text{(A.32}b\text{)}$$

more generally,

$$\int_0^{\infty} \frac{d\xi}{(1 + \xi^2)^n} = \frac{\pi}{2}\frac{(2n-3)!!}{(2n-2)!!} \equiv \frac{\pi}{2}\frac{1 \cdot 3 \cdot 5 \ldots (2n-3)}{2 \cdot 4 \cdot 6 \ldots (2n-2)}, \quad \text{(A.32}c\text{)}$$
$$\text{for } n = 2, 3, \ldots$$

- Integrals with $(1 - \xi^{2n})^{1/2}$:

$$\int_0^1 \frac{d\xi}{(1 - \xi^{2n})^{1/2}} = \frac{\pi^{1/2}}{2n}\Gamma\left(\frac{1}{2n}\right)\Big/\Gamma\left(\frac{n+1}{2n}\right), \quad \text{(A.33}a\text{)}$$

$$\int_0^1 (1 - \xi^{2n})^{1/2} d\xi = \frac{\pi^{1/2}}{4n}\Gamma\left(\frac{1}{2n}\right)\Big/\Gamma\left(\frac{3n+1}{2n}\right), \quad \text{(A.33}b\text{)}$$

where $\Gamma(s)$ is the *gamma-function,* which is most often defined (for $\text{Re}\, s > 0$) by the following integral:

$$\int_0^{\infty} \xi^{s-1} e^{-\xi}\, d\xi = \Gamma(s). \quad \text{(A.34}a\text{)}$$

The key property of this function is the recurrence relation, valid for any $s \neq 0, -1, -2, \ldots$:

$$\Gamma(s + 1) = s\Gamma(s). \quad \text{(A.34}b\text{)}$$

Since, according to Eq. (A.34*a*), $\Gamma(1) = 1$, Eq. (A.34*b*) for non-negative integers takes the form

$$\Gamma(n + 1) = n!, \quad \text{for } n = 0, 1, 2, \cdots \quad \text{(A.34}c\text{)}$$

[6] Eq. (A.32*a*) follows immediately from Eq. (A.30*d*), and Eq. (A.32*b*) from Eq. (A.29*b*)—a couple more examples of the (intentional) redundancy in this list.

(where $0! \equiv 1$). Because of this, for integer $s = n + 1 \geqslant 1$, Eq. (A.34*a*) is reduced to

$$\int_0^\infty \xi^n e^{-\xi} d\xi = n!. \tag{A.34d}$$

Other frequently met values of the gamma-function are those for positive semi-integer arguments:

$$\Gamma\left(\frac{1}{2}\right) = \pi^{1/2}, \quad \Gamma\left(\frac{3}{2}\right) = \frac{1}{2}\pi^{1/2}, \quad \Gamma\left(\frac{5}{2}\right) = \frac{1}{2}\cdot\frac{3}{2}\pi^{1/2},$$
$$\Gamma\left(\frac{7}{2}\right) = \frac{1}{2}\cdot\frac{3}{2}\cdot\frac{5}{2}\pi^{1/2}, \quad \ldots. \tag{A.34e}$$

- Integrals with $1/(e^\xi \pm 1)$:

$$\int_0^\infty \frac{\xi^{s-1} d\xi}{e^\xi + 1} = (1 - 2^{1-s})\,\Gamma(s)\zeta(s), \quad \text{for } s > 0, \tag{A.35a}$$

$$\int_0^\infty \frac{\xi^{s-1} d\xi}{e^\xi - 1} = \Gamma(s)\zeta(s), \quad \text{for } s > 1, \tag{A.35b}$$

where $\zeta(s)$ is the Riemann zeta-function—see Eq. (A.10). Particular cases: for $s = 2n$,

$$\int_0^\infty \frac{\xi^{2n-1} d\xi}{e^\xi + 1} = \frac{2^{2n-1} - 1}{2n}\pi^{2n} B_{2n}, \tag{A.35c}$$

$$\int_0^\infty \frac{\xi^{2n-1} d\xi}{e^\xi - 1} = \frac{(2\pi)^{2n}}{4n} B_{2n}. \tag{A.35d}$$

where B_n are the Bernoulli numbers—see Eq. (A.15). For the particular case $s = 1$ (when Eq. (A.35*a*) yields uncertainty),

$$\int_0^\infty \frac{d\xi}{e^\xi + 1} = \ln 2. \tag{A.35e}$$

- Integrals with $\exp\{-\xi^2\}$:

$$\int_0^\infty \xi^s e^{-\xi^2} d\xi = \frac{1}{2}\Gamma\left(\frac{s+1}{2}\right), \quad \text{for } s > -1; \tag{A.36a}$$

for applications the most important particular values of s are 0 and 2:

$$\int_0^\infty e^{-\xi^2} d\xi = \frac{1}{2}\Gamma\left(\frac{1}{2}\right) = \frac{\pi^{1/2}}{2}, \tag{A.36b}$$

$$\int_0^\infty \xi^2 e^{-\xi^2} d\xi = \frac{1}{2}\Gamma\left(\frac{3}{2}\right) = \frac{\pi^{1/2}}{4}, \tag{A.36c}$$

although we will also run into the cases $s = 4$ and $s = 6$:

$$\int_0^\infty \xi^4 e^{-\xi^2} d\xi = \frac{1}{2}\Gamma\left(\frac{5}{2}\right) = \frac{3\pi^{1/2}}{8},$$
$$\int_0^\infty \xi^6 e^{-\xi^2} d\xi = \frac{1}{2}\Gamma\left(\frac{7}{2}\right) = \frac{15\pi^{1/2}}{16}; \tag{A.36d}$$

for odd integer values $s = 2n + 1$ (with $n = 0, 1, 2,\ldots$), Eq. (A.36*a*) takes a simpler form:

$$\int_0^\infty \xi^{2n+1} e^{-\xi^2} d\xi = \frac{1}{2}\Gamma(n + 1) = \frac{n!}{2}. \tag{A.36e}$$

- Integrals with cosine and sine functions:

$$\int_0^\infty \cos(\xi^2)\, d\xi = \int_0^\infty \sin(\xi^2)\, d\xi = \left(\frac{\pi}{8}\right)^{1/2}. \tag{A.37}$$

$$\int_0^\infty \frac{\cos \xi}{a^2 + \xi^2} d\xi = \frac{\pi}{2a} e^{-a}. \tag{A.38}$$

$$\int_0^\infty \left(\frac{\sin \xi}{\xi}\right)^2 d\xi = \frac{\pi}{2}. \tag{A.39}$$

- Integrals with logarithms:

$$\int_0^1 \ln \frac{a + (1 - \xi^2)^{1/2}}{a - (1 - \xi^2)^{1/2}} d\xi = \pi[a - (a^2 - 1)^{1/2}], \quad \text{for } a \geqslant 1. \tag{A.40}$$

$$\int_0^1 \ln \frac{1 + (1 - \xi)^{1/2}}{\xi^{1/2}} d\xi = 1. \tag{A.41}$$

- Integral representations of the Bessel functions of integer order:

$$J_n(\alpha) = \frac{1}{2\pi}\int_{-\pi}^{+\pi} e^{i(\alpha \sin \xi - n\xi)} d\xi,$$
$$\text{so that } e^{i\alpha \sin \xi} = \sum_{k=-\infty}^{\infty} J_k(\alpha) e^{ik\xi}; \tag{A.42a}$$

$$I_n(\alpha) = \frac{1}{\pi}\int_0^\pi e^{\alpha \cos \xi} \cos n\xi \; d\xi. \tag{A.42b}$$

A.7 3D vector products

(i) *Definitions*:

- *Scalar* ('dot-') *product*:

$$\mathbf{a} \cdot \mathbf{b} = \sum_{j=1}^{3} a_j b_j, \tag{A.43}$$

where a_j and b_j are vector components in any orthogonal coordinate system. In particular, the vector squared (the same as the norm squared):

$$a^2 \equiv \mathbf{a} \cdot \mathbf{a} = \sum_{j=1}^{3} a_j^2 \equiv \| \mathbf{a} \|^2. \tag{A.44}$$

- *Vector* ('cross-') *product*:

$$\mathbf{a} \times \mathbf{b} \equiv \mathbf{n}_1(a_2b_3 - a_3b_2) + \mathbf{n}_2(a_3b_1 - a_1b_3) + \mathbf{n}_3(a_1b_2 - a_2b_1) = \begin{vmatrix} \mathbf{n}_1 & \mathbf{n}_2 & \mathbf{n}_3 \\ a_1 & a_2 & a_3 \\ b_1 & b_2 & b_3 \end{vmatrix}, \tag{A.45}$$

where $\{\mathbf{n}_j\}$ is the set of mutually perpendicular unit vectors[7] along the corresponding coordinate system axes[8]. In particular, Eq. (A.45) yields

$$\mathbf{a} \times \mathbf{a} = 0. \tag{A.46}$$

(ii) *Corollaries* (readily verified by Cartesian components):

- Double vector product (the so-called *bac minus cab* rule):

$$\mathbf{a} \times (\mathbf{b} \times \mathbf{c}) = \mathbf{b}(\mathbf{a} \cdot \mathbf{c}) - \mathbf{c}(\mathbf{a} \cdot \mathbf{b}). \tag{A.47}$$

- Mixed scalar–vector product (the *operand rotation rule*):

$$\mathbf{a} \cdot (\mathbf{b} \times \mathbf{c}) = \mathbf{b} \cdot (\mathbf{c} \times \mathbf{a}) = \mathbf{c} \cdot (\mathbf{a} \times \mathbf{b}). \tag{A.48}$$

- Scalar product of vector products:

$$(\mathbf{a} \times \mathbf{b}) \cdot (\mathbf{c} \times \mathbf{d}) = (\mathbf{a} \cdot \mathbf{c})(\mathbf{b} \cdot \mathbf{d}) - (\mathbf{a} \cdot \mathbf{d})(\mathbf{b} \cdot \mathbf{c}); \tag{A.49a}$$

[7] Other popular notations for this vector set are $\{\mathbf{e}_j\}$ and $\{\hat{\mathbf{r}}_j\}$.

[8] It is easy to use Eq. (A.45) to check that the direction of the product vector corresponds to the well-known 'right-hand rule' and to the even more convenient *corkscrew rule*: if we rotate a corkscrew's handle from the first operand toward the second one, its axis moves in the direction of the product.

in the particular case of two similar operands (say, $\mathbf{a} = \mathbf{c}$ and $\mathbf{b} = \mathbf{d}$), the last formula is reduced to

$$(\mathbf{a} \times \mathbf{b})^2 = (ab)^2 - (\mathbf{a} \cdot \mathbf{b})^2. \quad \text{(A.49}b\text{)}$$

A.8 Differentiation in 3D Cartesian coordinates

- Definition of the *del* (or 'nabla') vector-operator ∇:[9]

$$\nabla \equiv \sum_{j=1}^{3} \mathbf{n}_j \frac{\partial}{\partial r_j}, \quad \text{(A.50)}$$

where r_j is a set of linear and orthogonal (*Cartesian*) coordinates along directions $\mathbf{n}_j$. In accordance with this definition, the operator ∇ acting on a *scalar* function of coordinates, $f(\mathbf{r})$,[10] gives its gradient, i.e. a new *vector*:

$$\nabla f \equiv \sum_{j=1}^{3} \mathbf{n}_j \frac{\partial f}{\partial r_j} \equiv \mathbf{grad}\, f. \quad \text{(A.51)}$$

- The *scalar product* of del by a *vector* function of coordinates (a *vector field*),

$$\mathbf{f}(\mathbf{r}) \equiv \sum_{j=1}^{3} \mathbf{n}_j f_j(\mathbf{r}), \quad \text{(A.52)}$$

compiled formally following Eq. (A.43), is a *scalar* function—the *divergence* of the initial function:

$$\nabla \cdot \mathbf{f} \equiv \sum_{j=1}^{3} \frac{\partial f_j}{\partial r_j} \equiv \text{div}\, \mathbf{f}, \quad \text{(A.53)}$$

while the *vector product* of ∇ and $\mathbf{f}$, formed in a formal accordance with Eq. (A.45), is a new vector - the *curl* (in European tradition, called rotor and denoted **rot**) of **f**:

$$\nabla \times \mathbf{f} \equiv \begin{vmatrix} \mathbf{n}_1 & \mathbf{n}_2 & \mathbf{n}_3 \\ \frac{\partial}{\partial r_1} & \frac{\partial}{\partial r_2} & \frac{\partial}{\partial r_3} \\ f_1 & f_2 & f_3 \end{vmatrix} = \mathbf{n}_1\left(\frac{\partial f_3}{\partial r_2} - \frac{\partial f_2}{\partial r_3}\right) + \mathbf{n}_2\left(\frac{\partial f_1}{\partial r_3} - \frac{\partial f_3}{\partial r_1}\right) + \mathbf{n}_3\left(\frac{\partial f_2}{\partial r_1} - \frac{\partial f_1}{\partial r_2}\right) \equiv \mathbf{curl}\, \mathbf{f}. \quad \text{(A.54)}$$

[9] One can run into the following notation: $\nabla \equiv \partial/\partial\mathbf{r}$, which is convenient is some cases, but may be misleading in quite a few others, so it will be not used in these notes.

[10] In this, and four next sections, all scalar and vector functions are assumed to be differentiable.

- One more frequently met 'product' is $(\mathbf{f}\cdot\nabla)\mathbf{g}$, where $\mathbf{f}$ and $\mathbf{g}$ are two arbitrary vector functions of $\mathbf{r}$. This product should be also understood in the sense implied by Eq. (A.43), i.e. as a vector whose jth Cartesian component is

$$[(\mathbf{f}\cdot\nabla)\,\mathbf{g}]_j = \sum_{j'=1}^{3} f_{j'}\frac{\partial g_j}{\partial r_{j'}}. \qquad \text{(A.55)}$$

A.9 The Laplace operator $\nabla^2 \equiv \nabla\cdot\nabla$

- Expression in Cartesian coordinates—in the formal accordance with Eq. (A.44):

$$\nabla^2 = \sum_{j=1}^{3}\frac{\partial^2}{\partial r_j^2}. \qquad \text{(A.56)}$$

- According to its definition, the Laplace operator acting on a *scalar* function of coordinates gives a new scalar function:

$$\nabla^2 f \equiv \nabla\cdot(\nabla f) = \operatorname{div}(\mathbf{grad}\, f) = \sum_{j=1}^{3}\frac{\partial^2 f}{\partial r_j^2}. \qquad \text{(A.57)}$$

- On the other hand, acting on a *vector* function (A.52), the operator ∇^2 returns another *vector*:

$$\nabla^2\mathbf{f} = \sum_{j=1}^{3}\mathbf{n}_j\nabla^2 f_j. \qquad \text{(A.58)}$$

Note that Eqs. (A.56)–(A.58) are only valid in Cartesian (i.e. orthogonal and linear) coordinates, but generally not in other (even orthogonal) coordinates—see, e.g. Eqs. (A.61), (A.64), (A.67) and (A.70) below.

A.10 Operators ∇ and ∇^2 in the most important systems of orthogonal coordinates[11]

(i) *Cylindrical*[12] *coordinates* $\{\rho, \varphi, z\}$ (see figure below) may be defined by their relations with the Cartesian coordinates:

$$r_1 = \rho\cos\varphi,\quad r_2 = \rho\sin\varphi,\quad r_3 = z. \qquad \text{(A.59)}$$

[11] Some other orthogonal curvilinear coordinate systems are discussed in *Part EM*, section 2.3.

[12] In the 2D geometry with fixed coordinate z, these coordinates are called *polar*.

- Gradient of a scalar function:

$$\nabla f = \mathbf{n}_\rho \frac{\partial f}{\partial \rho} + \mathbf{n}_\varphi \frac{1}{\rho} \frac{\partial f}{\partial \varphi} + \mathbf{n}_z \frac{\partial f}{\partial z}. \tag{A.60}$$

- The Laplace operator of a scalar function:

$$\nabla^2 f = \frac{1}{\rho} \frac{\partial}{\partial \rho}\left(\rho \frac{\partial f}{\partial \rho}\right) + \frac{1}{\rho^2} \frac{\partial^2 f}{\partial \varphi^2} + \frac{\partial^2 f}{\partial z^2}, \tag{A.61}$$

- Divergence of a vector function of coordinates ($\mathbf{f} = \mathbf{n}_\rho f_\rho + \mathbf{n}_\varphi f_\varphi + \mathbf{n}_z f_z$):

$$\nabla \cdot \mathbf{f} = \frac{1}{\rho} \frac{\partial(\rho f_\rho)}{\partial \rho} + \frac{1}{\rho} \frac{\partial f_\varphi}{\partial \varphi} + \frac{\partial f_z}{\partial z}. \tag{A.62}$$

- Curl of a vector function:

$$\nabla \times \mathbf{f} = \mathbf{n}_\rho \left(\frac{1}{\rho} \frac{\partial f_z}{\partial \varphi} - \frac{\partial f_\varphi}{\partial z}\right) + \mathbf{n}_\varphi \left(\frac{\partial f_\rho}{\partial z} - \frac{\partial f_z}{\partial \rho}\right) + \mathbf{n}_z \frac{1}{\rho}\left(\frac{\partial(\rho f_\varphi)}{\partial \rho} - \frac{\partial f_\rho}{\partial \varphi}\right). \tag{A.63}$$

- The Laplace operator of a vector function:

$$\nabla^2 \mathbf{f} = \mathbf{n}_\rho \left(\nabla^2 f_\rho - \frac{1}{\rho^2} f_\rho - \frac{2}{\rho^2} \frac{\partial f_\varphi}{\partial \varphi}\right) + \mathbf{n}_\varphi \left(\nabla^2 f_\varphi - \frac{1}{\rho^2} f_\varphi + \frac{2}{\rho^2} \frac{\partial f_\rho}{\partial \varphi}\right) + \mathbf{n}_z \nabla^2 f_z. \tag{A.64}$$

(ii) *Spherical coordinates* $\{r, \theta, \varphi\}$ (see figure below) may be defined as:

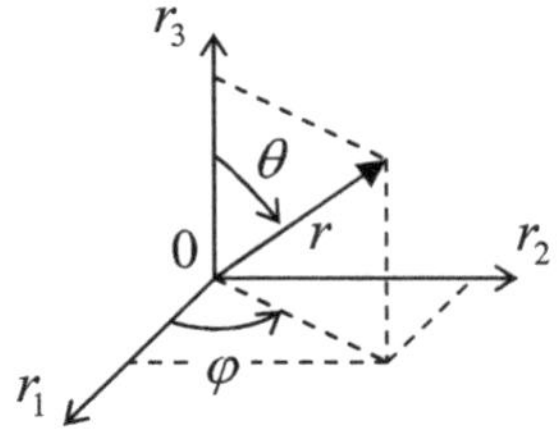

$$r_1 = r \sin\theta \cos\varphi,$$
$$r_2 = r \sin\theta \sin\varphi, \tag{A.65}$$
$$r_3 = r \cos\theta.$$

- Gradient of a scalar function:

$$\nabla f = \mathbf{n}_r \frac{\partial f}{\partial r} + \mathbf{n}_\theta \frac{1}{r} \frac{\partial f}{\partial \theta} + \mathbf{n}_\varphi \frac{1}{r \sin\theta} \frac{\partial f}{\partial \varphi}. \tag{A.66}$$

- The Laplace operator of a scalar function:

$$\nabla^2 f = \frac{1}{r^2} \frac{\partial}{\partial r}\left(r^2 \frac{\partial f}{\partial r}\right) + \frac{1}{r^2 \sin\theta} \frac{\partial}{\partial \theta}\left(\sin\theta \frac{\partial f}{\partial \theta}\right) + \frac{1}{(r \sin\theta)^2} \frac{\partial^2 f}{\partial \varphi^2}. \tag{A.67}$$

- Divergence of a vector function $\mathbf{f} = \mathbf{n}_r f_r + \mathbf{n}_\theta f_\theta + \mathbf{n}_\varphi f_\varphi$:

$$\nabla \cdot \mathbf{f} = \frac{1}{r^2}\frac{\partial(r^2 f_r)}{\partial r} + \frac{1}{r\sin\theta}\frac{\partial(f_\theta \sin\theta)}{\partial\theta} + \frac{1}{r\sin\theta}\frac{\partial f_\varphi}{\partial\varphi}. \tag{A.68}$$

- Curl of a similar vector function:

$$\begin{aligned}\nabla \times \mathbf{f} = \mathbf{n}_r \frac{1}{r\sin\theta}\left(\frac{\partial(f_\varphi \sin\theta)}{\partial\theta} - \frac{\partial f_\theta}{\partial\varphi}\right) + \mathbf{n}_\theta \frac{1}{r}\left(\frac{1}{\sin\theta}\frac{\partial f_r}{\partial\varphi} - \frac{\partial(r f_\varphi)}{\partial r}\right) \\ + \mathbf{n}_\varphi \frac{1}{r}\left(\frac{\partial(r f_\theta)}{\partial r} - \frac{\partial f_r}{\partial\theta}\right).\end{aligned} \tag{A.69}$$

- The Laplace operator of a vector function:

$$\begin{aligned}\nabla^2 \mathbf{f} = \mathbf{n}_r &\left(\nabla^2 f_r - \frac{2}{r^2} f_r - \frac{2}{r^2\sin\theta}\frac{\partial}{\partial\theta}(f_\theta \sin\theta) - \frac{2}{r^2\sin\theta}\frac{\partial f_\varphi}{\partial\varphi}\right) \\ + \mathbf{n}_\theta &\left(\nabla^2 f_\theta - \frac{1}{r^2\sin^2\theta} f_\theta + \frac{2}{r^2}\frac{\partial f_r}{\partial\theta} - \frac{2\cos\theta}{r^2\sin^2\theta}\frac{\partial f_\varphi}{\partial\varphi}\right) \\ + \mathbf{n}_\varphi &\left(\nabla^2 f_\varphi - \frac{1}{r^2\sin^2\theta} f_\varphi + \frac{2}{r^2\sin\theta}\frac{\partial f_r}{\partial\varphi} + \frac{2\cos\theta}{r^2\sin^2\theta}\frac{\partial f_\theta}{\partial\varphi}\right).\end{aligned} \tag{A.70}$$

A.11 Products involving ∇

(i) *Useful zeros*:

- For any scalar function $f(\mathbf{r})$,

$$\nabla \times (\nabla f) \equiv \mathbf{curl}(\mathbf{grad} f) = 0. \tag{A.71}$$

- For any vector function $\mathbf{f}(\mathbf{r})$,

$$\nabla \cdot (\nabla \times \mathbf{f}) \equiv \mathrm{div}(\mathbf{curl} f) = 0. \tag{A.72}$$

(ii) The *Laplace operator* expressed via the curl of a curl:

$$\nabla^2 \mathbf{f} = \nabla(\nabla \cdot \mathbf{f}) - \nabla \times (\nabla \times \mathbf{f}). \tag{A.73}$$

(iii) Spatial differentiation of a product of a *scalar* function by a *vector* function:x

- The scalar 3D generalization of Eq. (A.22) is

$$\nabla \cdot (f\,\mathbf{g}) = (\nabla f) \cdot \mathbf{g} + f(\nabla \cdot \mathbf{g}). \tag{A.74a}$$

- Its vector generalization is similar:

$$\nabla \times (f\,\mathbf{g}) = (\nabla f) \times \mathbf{g} + f(\nabla \times \mathbf{g}). \tag{A.74b}$$

(iv) Spatial differentiation of products of *two vector* functions:

$$\nabla \times (\mathbf{f} \times \mathbf{g}) = \mathbf{f}(\nabla \cdot \mathbf{g}) - (\mathbf{f} \cdot \nabla)\mathbf{g} - (\nabla \cdot \mathbf{f})\mathbf{g} + (\mathbf{g} \cdot \nabla)\,\mathbf{f}, \tag{A.75}$$

$$\nabla(\mathbf{f} \cdot \mathbf{g}) = (\mathbf{f} \cdot \nabla)\mathbf{g} + (\mathbf{g} \cdot \nabla)\,\mathbf{f} + \mathbf{f} \times (\nabla \times \mathbf{g}) + \mathbf{g} \times (\nabla \times \mathbf{f}), \tag{A.76}$$

$$\nabla \cdot (\mathbf{f} \times \mathbf{g}) = \mathbf{g} \cdot (\nabla \times \mathbf{f}) - \mathbf{f} \cdot (\nabla \times \mathbf{g}). \tag{A.77}$$

A.12 Integro-differential relations

(i) For an *arbitrary surface S* limited by closed contour C:

- The *Stokes theorem*, valid for any differentiable vector field $\mathbf{f}(\mathbf{r})$:

$$\int_S (\nabla \times \mathbf{f}) \cdot d^2\mathbf{r} \equiv \int_S (\nabla \times \mathbf{f})_n d^2r = \oint_C \mathbf{f} \cdot d\mathbf{r} \equiv \oint_C f_\tau dr, \tag{A.78}$$

where $d^2\mathbf{r} \equiv \mathbf{n}d^2r$ is the elementary area vector (normal to the surface), and $d\mathbf{r}$ is the elementary contour length vector (tangential to the contour line).

(ii) For an *arbitrary volume V* limited by closed surface S:

- *Divergence* (or 'Gauss') *theorem*, valid for any differentiable vector field $\mathbf{f}(\mathbf{r})$:

$$\int_V (\nabla \cdot \mathbf{f})\, d^3r = \oint_S \mathbf{f} \cdot d^2\mathbf{r} \equiv \oint_S f_n d^2r. \tag{A.79}$$

- *Green's theorem*, valid for two differentiable scalar functions $f(\mathbf{r})$ and $g(\mathbf{r})$:

$$\int_V (f\,\nabla^2 g - g\nabla^2 f)\, d^3r = \oint_S (f\,\nabla g - g\nabla f)_n d^2r. \tag{A.80}$$

- An identity valid for any two scalar functions f and g, and a vector field $\mathbf{j}$ with $\nabla\cdot\mathbf{j} = 0$ (all differentiable):

$$\int_V [f(\mathbf{j} \cdot \nabla g) + g(\mathbf{j} \cdot \nabla f)]\; d^3r = \oint_S fgj_n d^2r. \tag{A.81}$$

A.13 The Kronecker delta and Levi-Civita permutation symbols

- The *Kronecker delta symbol* (defined for integer indices):

$$\delta_{jj'} \equiv \begin{cases} 1, & \text{if } j' = j, \\ 0, & \text{otherwise.} \end{cases} \tag{A.82}$$

- The *Levi-Civita permutation symbol* (most frequently used for 3 integer indices, each taking one of values 1, 2, or 3):

$$\varepsilon_{jj'j''} \equiv \begin{cases} +1, & \text{if the indices follow in the 'correct' ('even')} \\ & \text{order: } 1 \to 2 \to 3 \to 1 \to 2 \ldots, \\ -1, & \text{if the indices follow in the 'incorrect' ('odd')} \\ & \text{order: } 1 \to 3 \to 2 \to 1 \to 3 \ldots, \\ 0, & \text{if any two indices coincide.} \end{cases} \quad \text{(A.83)}$$

- Relation between the Levi-Civita and the Kronecker delta products:

$$\varepsilon_{jj'j''}\varepsilon_{kk'k''} = \sum_{l,l',l''=1}^{3} \begin{vmatrix} \delta_{jl} & \delta_{jl'} & \delta_{jl''} \\ \delta_{j'l} & \delta_{j'l'} & \delta_{j'l''} \\ \delta_{j''l} & \delta_{j''l'} & \delta_{j''l''} \end{vmatrix}; \quad \text{(A.84}a\text{)}$$

summation of this relation, written for 3 different values of $j = k$, over these values yields the so-called *contracted epsilon identity*:

$$\sum_{j=1}^{3} \varepsilon_{jj'j''}\varepsilon_{jk'k''} = \delta_{j'k'}\delta_{j''k''} - \delta_{j'k''}\delta_{j''k'}. \quad \text{(A.84}b\text{)}$$

A.14 Dirac's delta-function, sign function, and theta-function

- Definition of 1D *delta-function* (for real $a < b$):

$$\int_a^b f(\xi)\delta(\xi)d\xi = \begin{cases} f(0), & \text{if } a < 0 < b, \\ 0, & \text{otherwise,} \end{cases} \quad \text{(A.85)}$$

where $f(\xi)$ is any function continuous near $\xi = 0$. In particular (if $f(\xi) = 1$ near $\xi = 0$), the definition yields

$$\int_a^b \delta(\xi)d\xi = \begin{cases} 1, & \text{if } a < 0 < b, \\ 0, & \text{otherwise.} \end{cases} \quad \text{(A.86)}$$

- Relation to the *theta-function* $\theta(\xi)$ and *sign function* $\text{sgn}(\xi)$

$$\delta(\xi) = \frac{d}{d\xi}\theta(\zeta) = \frac{1}{2}\frac{d}{d\xi}\text{sgn}(\xi), \quad \text{(A.87}a\text{)}$$

where

$$\theta(\xi) \equiv \frac{\text{sgn}(\xi) + 1}{2} = \begin{cases} 0, & \text{if } \xi < 0, \\ 1, & \text{if } \xi > 1, \end{cases}$$
$$\text{sgn}(\xi) \equiv \frac{\xi}{|\xi|} = \begin{cases} -1, & \text{if } \xi < 0, \\ +1, & \text{if } \xi > 1. \end{cases} \quad \text{(A.87}b\text{)}$$

- An important integral[13]:

$$\int_{-\infty}^{+\infty} e^{is\xi} ds = 2\pi\delta(\xi). \tag{A.88}$$

- 3D generalization of the delta-function of the radius-vector (the 2D generalization is similar):

$$\int_V f(\mathbf{r})\delta(\mathbf{r})d^3r = \begin{cases} f(0), & \text{if } 0 \in V, \\ 0, & \text{otherwise;} \end{cases} \tag{A.89}$$

it may be represented as a product of 1D delta-functions of Cartesian coordinates:

$$\delta(\mathbf{r}) = \delta(r_1)\delta(r_2)\delta(r_3). \tag{A.90}$$

A.15 The Cauchy theorem and integral

Let a complex function $f(z)$ be analytic within a part of the complex plane z, that is limited by a closed contour C and includes point z'. Then

$$\oint_C f(z)dz = 0, \tag{A.91}$$

$$\oint_C f(z)\frac{dz}{z - z'} = 2\pi i f(z') \tag{A.92}$$

The first of these relations is usually called the *Cauchy integral theorem* (or the 'Cauchy–Goursat theorem'), and the second one—the *Cauchy integral* (or the 'Cauchy integral formula').

A.16 Literature

(i) Properties of some *special functions* are briefly discussed at the relevant points of the lecture notes; in the alphabetical order:

- Airy functions: *Part QM* section 2.4;
- Bessel functions: *Part EM* section 2.7;
- Fresnel integrals: *Part EM* section 8.6;
- Hermite polynomials: *Part QM* section 2.9;
- Laguerre polynomials (both simple and associated): *Part QM* section 3.7;

[13] The coefficient in this relation may be readily recalled by considering its left-hand part as the Fourier-integral representation of function $f(s) \equiv 1$, and applying Eq. (A.85) to the reciprocal Fourier transform

$$f(s) \equiv 1 = \frac{1}{2\pi}\int_{-\infty}^{+\infty} e^{-is\xi}[2\pi\delta(\xi)]d\xi.$$

- Legendre polynomials, associated Legendre functions: *Part EM* section 2.8, and *Part QM* section 3.6;
- Spherical harmonics: *Part QM* section 3.6;
- Spherical Bessel functions: *Part QM* sections 3.6 and 3.8.

(ii) For *more formulas*, and their discussion, I can recommend the following handbooks[14]:

- *Handbook of Mathematical Formulas* [2];
- *Tables of Integrals, Series, and Products* [3];
- *Mathematical Handbook for Scientists and Engineers* [4];
- *Integrals and Series* volumes 1 and 2 [5];
- A popular textbook *Mathematical Methods for Physicists* [6] may be also used as a formula manual.

Many formulas are also available from the symbolic calculation modules of the commercially available software packages listed in section (iv) below.

(iii) Probably the most popular collection of *numerical calculation codes* are the twin manuals by W Press *et al* [1]:

- *Numerical Recipes in Fortran 77*;
- *Numerical Recipes* [in C++—KKL].

My lecture notes include very brief introductions to numerical methods of differential equation solution:

- ordinary differential equations: *Part CM*, section 5.7;
- partial differential equations: *Part CM* section 8.5 and *Part EM* section 2.11, which include references to literature for further reading.

(iv) The following are the most popular *software packages* for numerical and symbolic calculations, all with plotting capabilities (in the alphabetical order):

- Maple (www.maplesoft.com/products/maple/);
- MathCAD (www.ptc.com/engineering-math-software/mathcad/);
- Mathematica (www.wolfram.com/mathematica/);
- MATLAB (www.mathworks.com/products/matlab.html).

References

[1] Press W *et al* 1992 *Numerical Recipes in Fortran 77* 2nd edn (Cambridge: Cambridge University Press)
Press W *et al* 2007 *Numerical Recipes* 3rd edn (Cambridge: Cambridge University Press)

[2] Abramowitz M and Stegun I (eds) 1965 *Handbook of Mathematical Formulas* (New York: Dover), and numerous later printings. An updated version of this collection is now available online at http://dlmf.nist.gov/.

[14] On a personal note, perhaps 90% of all formula needs throughout my research career were satisfied by a tiny, wonderfully compiled old book [7], used copies of which, rather amazingly, are still available on the Web.

[3] Gradshteyn I and Ryzhik I 1980 *Tables of Integrals, Series, and Products* 5th edn (New York: Academic)
[4] Korn G and Korn T 2000 *Mathematical Handbook for Scientists and Engineers* 2nd edn (New York: Academic)
[5] Prudnikov A *et al* 1986 *Integrals and Series* vol 1 (Boca Raton, FL: CRC Press)
Prudnikov A *et al* 1986 *Integrals and Series* vol 2 (Boca Raton, FL: CRC Press)
[6] Arfken G *et al* 2012 *Mathematical Methods for Physicists* 7th edn (New York: Academic)
[7] Dwight H 1961 *Tables of Integrals and Other Mathematical Formulas* 4th edn (London: Macmillan)

IOP Publishing

Quantum Mechanics
Problems with solutions
Konstantin K Likharev

Appendix B

Selected physical constants

The listed numerical values of the constants are from the most recent (2014) International CODATA recommendation (see, e.g. http://physics.nist.gov/cuu/Constants/index.html), besides a newer result for k_B*—see* [1]. *Please note the recently announced (but, by this volume's press time, not yet official) adjustment of the SI values - see, e.g. https://www.nist.gov/si-redefinition/meet-constants. In particular, the Planck constant will also get a definite value (within the interval specified in table B.1), enabling a new, fundamental standard of the kilogram.*

Table B.1.

Symbol	Quantity	SI value and unit	Gaussian value and unit	Relative rms uncertainty
c	speed of light in free space	$2.99\,792\,458 \times 10^{8}$ m s^{-1}	$2.99\,792\,458 \times 10^{10}$ cm s^{-1}	0 (defined value)
G	gravitation constant	6.6741×10^{-11} m^3 kg^{-1} s^{-2}	6.6741×10^{-8} cm^3 g^{-1} s^{-2}	$\sim 5 \times 10^{-5}$
$\hbar$	Planck constant	$1.05\,457\,180 \times 10^{-34}$ J s	$1.05\,457\,180 \times 10^{-27}$ erg s	$\sim 2 \times 10^{-8}$
e	elementary electric charge	$1.6\,021\,762 \times 10^{-19}$ C	$4.803\,203 \times 10^{-10}$ statcoulomb	$\sim 6 \times 10^{-9}$
m_e	electron's rest mass	$0.91\,093\,835 \times 10^{-30}$ kg	$0.91\,093\,835 \times 10^{-27}$ g	$\sim 1 \times 10^{-8}$
m_p	proton's rest mass	$1.67\,262\,190 \times 10^{-27}$ kg	$1.67\,262\,190 \times 10^{-24}$ g	$\sim 1 \times 10^{-8}$
μ_0	magnetic constant	$4\pi \times 10^{-7}$ N A^{-2}	–	0 (defined value)
ε_0	electric constant	$8.854\,187\,817 \times 10^{-12}$ F m^{-1}	–	0 (defined value)
k_B	Boltzmann constant	$1.380\,649 \times 10^{-23}$ J K^{-1}	$1.3\,806\,490 \times 10^{-16}$ erg K^{-1}	$\sim 2 \times 10^{-6}$

Comments:

1. The fixed value of c was defined by an international convention in 1983, in order to extend the official definition of the second (as 'the duration of 9 192 631 770 periods of the radiation corresponding to the transition between the two hyperfine levels of the ground state of the cesium-133 atom') to that of the meter. The values are back-compatible with the legacy definitions of the meter (initially, as 1/40 000 000th of the Earth's meridian length) and the second (for a long time, as $1/(24 \times 60 \times 60) = 1/86\ 400$th of the Earth's rotation period), within the experimental errors of those measures.
2. ε_0 and μ_0 are not really the fundamental constants; in the SI system of units one of them (say, μ_0) is selected arbitrarily[1], while the other one is defined via the relation $\varepsilon_0\mu_0 = 1/c^2$.
3. The Boltzmann constant k_B is also not quite fundamental, because its only role is to comply with the independent definition of the kelvin (K), as the temperature unit in which the triple point of water is exactly 273.16 K. If temperature is expressed in energy units k_BT (as is done, for example, in *Part SM* of this series), this constant disappears altogether.
4. The dimensionless *fine structure* ('Sommerfeld's') *constant* α is numerically the same in any system of units:

$$\alpha \equiv \left\{ \begin{array}{ll} e^2/4\pi\varepsilon_0\hbar c & \text{in SI units} \\ e^2/\hbar c & \text{in Gaussian units} \end{array} \right\} \approx 7.297\ 352\ 566 \times 10^{-3}$$
$$\approx \frac{1}{137.035\ 999\ 14},$$

and is known with a much smaller relative rms uncertainty (currently, $\sim 3 \times 10^{-10}$) than those of the component constants.

References

[1] Gaiser C *et al* 2017 *Metrologia* **54** 280
[2] Newell D 2014 *Phys. Today* **67** 35–41

[1] Note that the selected value of μ_0 may be changed (a bit) in a few years—see, e.g., [2].

Bibliography

This section presents a partial list of textbooks and monographs used in the work on the EAP series[1,2].

Part CM: Classical Mechanics

Fetter A L and Walecka J D 2003 *Theoretical Mechanics of Particles and Continua* (New York: Dover)

Goldstein H, Poole C and Safko J 2002 *Classical Mechanics* 3rd edn (Reading, MA: Addison Wesley)

Granger R A 1995 *Fluid Mechanics* (New York: Dover)

José J V and Saletan E J 1998 *Classical Dynamics* (Cambridge: Cambridge University Press)

Landau L D and Lifshitz E M 1976 *Mechanics* 3rd edn (Oxford: Butterworth-Heinemann)

Landau L D and Lifshitz E M 1986 *Theory of Elasticity* (Oxford: Butterworth-Heinemann)

Landau L D and Lifshitz E M 1987 *Fluid Mechanics* 2nd edn (Oxford: Butterworth-Heinemann)

Schuster H G 1995 *Deterministic Chaos* 3rd edn (New York: Wiley)

Sommerfeld A 1964 *Mechanics* (New York: Academic)

Sommerfeld A 1964 *Mechanics of Deformable Bodies* (New York: Academic)

Part EM: Classical Electrodynamics

Batygin V V and Toptygin I N 1978 *Problems in Electrodynamics* 2nd edn (New York: Academic)

Griffiths D J 2007 *Introduction to Electrodynamics* 3rd edn (Englewood Cliffs, NJ: Prentice-Hall)

Jackson J D 1999 *Classical Electrodynamics* 3rd edn (New York: Wiley)

Landau L D and Lifshitz E M 1984 *Electrodynamics of Continuous Media* 2nd edn (Auckland: Reed)

Landau L D and Lifshitz E M 1975 *The Classical Theory of Fields* 4th edn (Oxford: Pergamon)

[1] The list does not include the sources (mostly, recent original publications) cited in the lecture notes and problem solutions, and the mathematics textbooks and handbooks listed in section A.16.

[2] Recently several high-quality teaching materials on advanced physics became available online, including R. Fitzpatrick's text on *Classical Electromagnetism* (farside.ph.utexas.edu/teaching/jk1/Electromagnetism.pdf), B Simons' 'lecture shrunks' on *Advanced Quantum Mechanics* (www.tcm.phy.cam.ac.uk/~bds10/aqp.html), and D Tong's lecture notes on several advanced topics (www.damtp.cam.ac.uk/user/tong/teaching.html).

doi:10.1088/2053-2563/aaf3a6ch12

Panofsky W K H and Phillips M 1990 *Classical Electricity and Magnetism* 2nd edn (New York: Dover)
Stratton J A 2007 *Electromagnetic Theory* (New York: Wiley)
Tamm I E 1979 *Fundamentals of the Theory of Electricity* (Paris: Mir)
Zangwill A 2013 *Modern Electrodynamics* (Cambridge: Cambridge University Press)

Part QM: Quantum Mechanics

Abers E S 2004 *Quantum Mechanics* (London: Pearson)
Auletta G, Fortunato M and Parisi G 2009 *Quantum Mechanics* (Cambridge: Cambridge University Press)
Capri A Z 2002 *Nonrelativistic Quantum Mechanics* 3rd edn (Singapore: World Scientific)
Cohen-Tannoudji C, Diu B and Laloë F 2005 *Quantum Mechanics* (New York: Wiley)
Constantinescu F, Magyari E and Spiers J A 1971 *Problems in Quantum Mechanics* (Amsterdam: Elsevier)
Galitski V *et al* 2013 *Exploring Quantum Mechanics* (Oxford: Oxford University Press)
Gottfried K and Yan T-M 2004 *Quantum Mechanics: Fundamentals* 2nd edn (Berlin: Springer)
Griffith D 2005 *Quantum Mechanics* 2nd edn (Englewood Cliffs, NJ: Prentice Hall)
Landau L D and Lifshitz E M 1977 *Quantum Mechanics (Nonrelativistic Theory)* 3rd edn (Oxford: Pergamon)
Messiah A 1999 *Quantum Mechanics* (New York: Dover)
Merzbacher E 1998 *Quantum Mechanics* 3rd edn (New York: Wiley)
Miller D A B 2008 *Quantum Mechanics for Scientists and Engineers* (Cambridge: Cambridge University Press)
Sakurai J J 1994 *Modern Quantum Mechanics* (Reading, MA: Addison-Wesley)
Schiff L I 1968 *Quantum Mechanics* 3rd edn (New York: McGraw-Hill)
Shankar R 1980 *Principles of Quantum Mechanics* 2nd edn (Berlin: Springer)
Schwabl F 2002 *Quantum Mechanics* 3rd edn (Berlin: Springer)

Part SM: Statistical Mechanics

Feynman R P 1998 *Statistical Mechanics* 2nd edn (Boulder, CO: Westview)
Huang K 1987 *Statistical Mechanics* 2nd edn (New York: Wiley)
Kubo R 1965 *Statistical Mechanics* (Amsterdam: Elsevier)
Landau L D and Lifshitz E M 1980 *Statistical Physics, Part 1* 3rd edn (Oxford: Pergamon)
Lifshitz E M and Pitaevskii L P 1981 *Physical Kinetics* (Oxford: Pergamon)
Pathria R K and Beale P D 2011 *Statistical Mechanics* 3rd edn (Amsterdam: Elsevier)
Pierce J R 1980 *An Introduction to Information Theory* 2nd edn (New York: Dover)
Plishke M and Bergersen B 2006 *Equilibrium Statistical Physics* 3rd edn (Singapore: World Scientific)
Schwabl F 2000 *Statistical Mechanics* (Berlin: Springer)
Yeomans J M 1992 *Statistical Mechanics of Phase Transitions* (Oxford: Oxford University Press)

Multidisciplinary/specialty

Ashcroft W N and Mermin N D 1976 *Solid State Physics* (Philadelphia, PA: Saunders)
Blum K 1981 *Density Matrix and Applications* (New York: Plenum)

Breuer H-P and Petruccione E 2002 *The Theory of Open Quantum Systems* (Oxford: Oxford University Press)

Cahn S B and Nadgorny B E 1994 *A Guide to Physics Problems, Part 1* (New York: Plenum)

Cahn S B, Mahan G D and Nadgorny B E 1997 *A Guide to Physics Problems, Part 2* (New York: Plenum)

Cronin J A, Greenberg D F and Telegdi V L 1967 *University of Chicago Graduate Problems in Physics* (Reading, MA: Addison Wesley)

Hook J R and Hall H E 1991 *Solid State Physics* 2nd edn (New York: Wiley)

Joos G 1986 *Theoretical Physics* (New York: Dover)

Kaye G W C and Laby T H 1986 *Tables of Physical and Chemical Constants* 15th edn (London: Longmans Green)

Kompaneyets A S 2012 *Theoretical Physics* 2nd edn (New York: Dover)

Lax M 1968 *Fluctuations and Coherent Phenomena* (London: Gordon and Breach)

Lifshitz E M and Pitaevskii L P 1980 *Statistical Physics, Part 2* (Oxford: Pergamon)

Newbury N *et al* 1991 *Princeton Problems in Physics with Solutions* (Princeton, NJ: Princeton University Press)

Pauling L 1988 *General Chemistry* 3rd edn (New York: Dover)

Tinkham M 1996 *Introduction to Superconductivity* 2nd edn (New York: McGraw-Hill)

Walecka J D 2008 *Introduction to Modern Physics* (Singapore: World Scientific)

Ziman J M 1979 *Principles of the Theory of Solids* 2nd edn (Cambridge: Cambridge University Press)

CPSIA information can be obtained
at www.ICGtesting.com
Printed in the USA
BVHW062047080819
555457BV00005B/46/P

9 780750 314145